GENSTAT 5 RELEASE 3 REFERENCE MANUAL

Genstat™ 5 Release 3
Reference Manual

Planned and written by the

Genstat 5 Committee

of the

Statistics Department, Rothamsted Experimental Station,
AFRC Institute of Arable Crops Research, Harpenden, Hertfordshire AL5 2JQ.

R.W. Payne (*Chairman*)
P.W. Lane (*Secretary*)
P.G.N. Digby
S.A. Harding
P.K. Leech
G.W. Morgan

A.D. Todd
R. Thompson
G. Tunnicliffe Wilson
S.J. Welham
R.P. White

Other contributors: A.E. Ainsley, K.E. Bicknell, M.F. Franklin, J.C. Gower, T.J. Hastie, S.K. Haywood, J.H. Maindonald, J.A. Nelder, H.D. Patterson, D.L. Robinson, G.J.S. Ross, H.R. Simpson, R.J. Tibshirani, L.G. Underhill, P.J. Verrier.

CLARENDON PRESS · OXFORD

Oxford University Press, Walton Street, Oxford OX2 6DP

Oxford New York Toronto
Delhi Bombay Calcutta Madras Karachi
Kuala Lumpur Singapore Hong Kong Tokyo
Nairobi Dar es Salaam Cape Town
Melbourne Auckland Madrid
and associated companies in
Berlin Ibadan

Oxford is a trade mark of Oxford University Press

Published in the United States by
Oxford University Press Inc., New York

A catalogue record for this book is available from the British Library

Library of Congress Cataloging in Publication Data
(data available)
ISBN 0–19–852312–2

Printed in Great Britain from text supplied by the authors
by the Alden Press, Oxford

Preface

Genstat is a very general computer program for statistical analysis, with all the facilities of a general-purpose statistics package. All the usual analyses are readily available using the standard Genstat commands, or *directives*. However, Genstat is not just a collection of pre-programmed commands for selecting from fixed recipes of available analyses. It has a very flexible command language, which you can use to write your own programs to cover the occasions when the standard analyses do not give exactly what you want, or when you want to develop a new technique. Most users will need to do this only occasionally, since the standard facilities in Genstat are extremely comprehensive. However, the ability to extend Genstat removes the temptation, that occurs with some other packages, to use an inappropriate or approximate technique when an unusual set of data has to be analysed. Programs can be formed into procedures, to simplify their future use or to make them easily available to other users. You can use Genstat either interactively, or in batch, or via a conversational menu system. Menus are available for many simple analyses, and the system is written using the Genstat language itself, so that it is easy to customize and extend.

Chapter 1 explains the basic rules of the Genstat syntax. These apply in an identical way to every Genstat directive and procedure. Once you have learned these rules, you can learn the directives and procedures as you find that you need them. Chapter 1 also describes the on-line help facilities which allow you to obtain information about syntax, directives, and procedures from within Genstat.

The data structures that are used to store information within Genstat are described in Chapter 2. There is a wide variety of structures, ranging from simple vectors of numbers or of text, to matrices and multi-way tables. There are also structures that point to other structures, allowing compound and hierarchical forms of data to be represented.

Chapter 3 describes input and output. Genstat can take input from more than one file, and it can have several output files too. There are also special files: for example, backing-store files allow you to store the contents of data structures, together with all their defining information, thus simplifying future analyses that examine the same set of data. You can also dump all the information about the current state of a Genstat run into a file, so that you can resume the run later on.

Genstat has many facilities for doing calculations and for manipulating data; these are covered in Chapter 4. You can do arithmetic calculations on any numerical data structure; you can also make logical tests on data values. There is a wide range of mathematical functions including density, cumulative, and inverse functions for many probability distributions. Matrix operations include the calculation of inverses, eigenvalues, and singular value decompositions. You can form tables to summarize observations, for example, from surveys; these tables can be saved within Genstat for further calculations and manipulation.

Chapter 5 describes the facilities for programming in Genstat: for example, there are directives to loop over a sequence of commands, or to select between alternative sets of commands. It also explains how to write *procedures*; these are analogous to subroutines in other programming languages. Genstat has an accompanying Library of standard procedures, contributed by ordinary users as well as the Genstat developers. The Library is controlled by an Editorial Board who check that the procedures are useful and reliable, and maintain standards for the documentation. It is updated at regular intervals, thus allowing new facilities to be made available independently of new releases of Genstat itself.

Chapter 6 covers graphics. For simple investigatory work, Genstat enables you to produce graphs, histograms, and contour plots on a terminal or a line-printer. They can also be plotted on graphics monitors or plotters, if available, to give greater resolution or the higher quality required for publication. Pie charts, three-dimensional histograms, and perspective views of surfaces can also be plotted. Information can be "read" from interactive graphics devices, allowing interactive graphical procedures to be developed.

These first six chapters, then, describe all the non-statistical facilities in Genstat. They may be useful also to those who have no wish to do any statistics. For example you may want to use Genstat just as a graphics package, or to use it as a powerful language for calculations on vectors and matrices.

For statistical work, you will not need to know all the information in Chapters 1-6 immediately, although you will probably find most of it useful at some time or other. To get started, you need to read at least Sections 1.1 to 1.3 in Chapter 1. You should probably also read the sections of Chapter 2 that describe the data structures required for the particular type of analysis that you want: for example, you need know only about variates for simple regression, or variates and factors for analysis of variance. Section 3.1 in Chapter 3 describes how to read data into Genstat. Later on you may want to learn how to print the contents of data structures, so that you can check their values; this is described in Section 3.2 in Chapter 3. The most useful part of Chapter 4 is Section 4.1 which describes how to do numerical calculations, for example to transform data, and Section 4.2 which lists the available functions. You will find loops useful (Subsection 5.2.1 in Chapter 5), once you start to analyse several sets of similar data in the same program. Graphics (Chapter 6) can give you useful insight into any patterns in the data, or allow you to present the results more clearly.

The statistical facilities in Genstat are divided into six areas: basic statistics (Chapter 7), regression (Chapter 8), analysis of designed experiments (Chapter 9), fitting of mixed models by residual maximum likelihood (Chapter 10), multivariate and cluster analysis (Chapter 11), and time series (Chapter 12). Examples are given to illustrate how to use the techniques, but there is no attempt to teach statistics. References are given to suitable statistical text books.

The discussion of basic statistics covers simple techniques for displaying data such as histograms and boxplots, sample statistics, facilities for studying probability distributions and generating random numbers, and a range of parametric and non-parametric tests for single and multiple samples.

The regression facilities in Genstat allow you to do simple and multiple linear regression, and to fit separate or related regression lines where your data are partitioned into groups. As well as the usual estimated regression coefficients, tables of residuals, fitted values, and so on, you can also obtain tables of predicted values; these fulfil the same purpose as tables of adjusted means in an analysis of variance. The output also includes regression diagnostics to warn about potential problems with the model. The directives are designed to make it easy to explore a series of models, by adding and deleting explanatory variables. Genstat keeps a record of all the modifications that you make, and you can print a table summarizing the changes at any point in the series. The facilities are all available not just for ordinary linear regression, but also for the wider class of *generalized linear models*; these allow the fitting, for example, of probit and logit models for bioassay, or log-linear models for contingency tables. Non-parametric relationships can be included using smoothing techniques; the models are then described as *additive* or *generalized additive models*. Nonlinear models can be fitted: rational functions, exponential and logistic growth curves are available as standard options; you can also specify and fit more general curves.

Genstat can do analysis of variance for virtually all the standard experimental designs: for example, completely randomized orthogonal designs, randomized block designs, split plots, Latin and Graeco-Latin squares, repeated-measures designs, balanced incomplete blocks and other designs with balanced confounding, and lattices. The output includes the analysis-of-variance table, tables of means with standard errors, and estimates of polynomial and other contrasts. You can do analysis of covariance with any of these designs. Genstat can also handle missing values. There are facilities to assist with the generation and selection of effective experimental designs, and to provide randomization.

The *residual maximum likelihood* (REML) algorithm estimates treatment effects and variance components in a linear mixed model (that is, a linear model containing both fixed and random effects). It can thus be used to analyse unbalanced data sets when the model contains more than one error term. It can also be used to assess the relative sizes and importance of the different sources of variation.

Directives are available for all the standard multivariate techniques: principal component analysis, canonical variate analysis, factor rotation, principal coordinate analysis, multidimensional scaling, and Procrustes rotation. Other techniques, including biplots, canonical correlation analysis, correspondence analysis, multivariate analysis of variance, and the analysis of skew symmetry, are available as procedures. Hierarchical cluster analysis can be done using one of several methods: single linkage, average linkage, median, centroid, or furthest neighbour. Minimum spanning trees can be produced, and nearest neighbours can be listed. For non-hierarchical clustering Genstat uses an exchange algorithm, which aims to maximize one of four available criteria.

Time series can be analysed using the range of ARIMA and seasonal ARIMA models defined by Box and Jenkins. The relationship between series can be investigated by transfer-function models, relating one output series to several input series. Directives are provided to help with model selection and checking, as well as the estimation of parameters and forecasts from a model. There is a directive for calculating Fourier transforms, allowing you to do spectral analyses.

The final Chapter, 13, gives background information about the environment in which Genstat executes your program, and explains how it can be modified and customized. Chapter 13 also describes how you can extend Genstat by adding your own source code, and how you can transfer results and data between Genstat and other programs.

There are two Appendices. The first gives brief details of all the Genstat directives, and indicates the page number of the full definition. The second similarly lists the procedures in the current release of the Procedure Library.

The initials of the original authors are listed at the end of each Chapter; generally this also indicates responsibility for the corresponding source code of Genstat. R.W. Payne acted as the overall Editor.

This Manual provides the definitive description of Genstat 5 syntax and directives. It also covers many of the procedures in the standard library. The Procedure Library is updated at regular intervals, independently of Genstat itself. Full details of the current release are given in the *Genstat 5 Procedure Library Manual Release 3[1]* (ed. Payne, Arnold, and Morgan 1993) published by the Numerical Algorithms Group (address below). For learning Genstat, there are two other books (also published by Oxford University Press) which may be useful: *Genstat 5: an Introduction* (Alvey, Galwey, and Lane 1987) and *Genstat 5: a Second Course* (Digby, Galwey, and Lane 1989). A Newsletter is published twice a year by the Numerical Algorithms Group describing new features of Genstat and some of the many ways in which it is being used. A *Users' Note* is distributed with Genstat giving the details of each particular implementation. There may also be local documentation giving information about how Genstat has been installed at your site: for example how to run Genstat on your computer, which graphics devices are available, information about the local procedure library, and so on.

Genstat 5 is currently developed (at Rothamsted) on DEC VAX and Sun SPARC computers, but implementations are available for many other ranges of computer including mainframes, mini-computers, Unix-based Workstations and IBM-compatible PC's. Full details can be obtained from the Numerical Algorithms Group Ltd, who market Genstat and collaborate in its development, via any of the following addresses.

The Numerical Algorithms Group Ltd, Wilkinson House, Jordan Hill Road, Oxford OX2 8DR, UK. Tel: +44 (0)865 511245. Fax: +44 (0)865 310139.

or The Numerical Algorithms Group Inc, 1400 Opus Place, Suite 200, Downers Grove, IL 60515-5702, USA. Tel: +1 708 971 2337. Fax: +1 708 971 2706.

or The Numerical Algorithms Group (Deutschland) GmbH, Schleißheimerstraße 5, D-85748 Garching, Deutschland. Tel: +49 (0)89 3207395. Fax: +49 (0)89 3207396.

There are also local distributors in many countries; please contact NAG for details.

Genstat users form a community world-wide, and there is a regular programme of conferences and workshops, and an EMAIL discussion list for exchanging news, hints, and ideas. Details are obtainable from the Genstat Secretary, at Rothamsted.

Although Genstat 5 has a completely redesigned syntax from Genstat 4 and the earlier versions, it nevertheless builds on many of their concepts. We would like to acknowledge those who contributed to them but who are no longer closely involved with Genstat, in particular, J.A. Nelder, N.G. Alvey, C.F. Banfield, R.I. Baxter, W.J. Krzanowski, H.R. Simpson, the late R.W.M. Wedderburn, and G.N. Wilkinson.

R.W.P.
Rothamsted
May 1993

Contents

9 Design and analysis of experiments 461

1 Introduction, terminology, and syntax

A brief guide to the statistical and mathematical facilities in Genstat is given in the Preface. This chapter describes the basic rules, terminology, and conventions of the Genstat syntax; you can think of these as the *grammatical rules* of the Genstat language. Later chapters will then describe the vocabulary required for all the operations that Genstat can perform. Section 1.1 illustrates some of the concepts using a simple example, while the later sections contain the rigorous definitions.

1.1 Interactive, batch, or conversational?

Genstat is equally convenient to use interactively or in batch. Whichever of these two modes you prefer, you can use the same command language described in this manual. In addition, in interactive mode there is also a conversational interface which covers many standard types of analysis; it is based on menus and so requires no knowledge of the command language. The menus can be modified or extended by anyone familiar with the command language, and customized to suit the special requirements of a group of users (Chapter 13).

To illustrate these different ways of using Genstat, we shall show how you could do a simple task, using each mode in turn. Suppose that you have taken nine samples of polluted soil and measured the level of Zinc in each of them. Initially you wish just to calculate some summary statistics, the mean, median and variance, of the amounts of Zinc in the samples.

1.1.1 Interactive mode

First, we shall show how to work interactively using commands. To start Genstat running on your computer, you should need only to give the instruction

    GENSTAT

If this does not work, you will need to refer to the local documentation to find out how to get started. For example, PC implementations of Genstat come with a file called USER.DOC, while on mainframes, information may be available using the Help system on the computer; in any case, a printed copy of the documentation is supplied to everyone who buys Genstat.

On some computers, the instruction will automatically set up some windows on the screen to separate input and output. The exact style of the windows differs according to the computer so, to avoid the description being too specific to any one type, we shall assume here that input and output are recorded consecutively in the same window. Example 1.1.1 reproduces the interactive session, with the information that has been typed from the keyboard printed in bold type to distinguish it from that displayed by Genstat.

Example 1.1.1

```
Genstat 5  Release 3.1  (IBM-PC 80386/DOS)     Wednesday, 14 October  1992
Copyright 1992, Lawes Agricultural Trust (Rothamsted Experimental Station)

**********************************************************************
```

```
* You can use Genstat interactively in command-mode or in menu-mode.    *
* You are now in command-mode:                                          *
*    type HELP for on-line help about the command language of Genstat;  *
*    type LIBHELP for on-line help about standard procedures;           *
*    type MENU to enter menu-mode - Genstat will prompt for information; *
*    type NOTICE to get news and information about Genstat Release 3;    *
*    type STOP to finish.                                               *
* The standard menu system covers some of the standard analyses that can *
* be done in command-mode, and is designed so that you can extend it.    *
*************************************************************************
> READ Zinc
Zinc/1> 164.2 160.6 163 166 159.8
Zinc/6> 163.9 161 161.3 165.8 :

    Identifier   Minimum      Mean    Maximum     Values    Missing
         Zinc     159.8      162.8      166.0          9          0

> CALCULATE Zmed = MEDIAN(Zinc)
> & Zvar = VARIANCE(Zinc)
> PRINT Zmed,Zvar; DECIMALS=1,2

        Zmed        Zvar
       163.0        5.22

> STOP

******** End of job.  Maximum of 5090 data units used at line 7 (45220 left)
```

Genstat starts by printing some initial information. This shows what version of Genstat you are using and notes that the name "Genstat" is the copyright of the Lawes Agricultural Trust (the governing body of Rothamsted). It also mentions the different modes of use of Genstat, and how to change mode, get information, and stop. Then it prints the prompt > and waits for you to type the first command:

> **> READ Zinc**

At the end of the line you should press the carriage-return key (referred to as <RETURN> in this manual) to tell Genstat to process this line of input. The prompt now changes to Zinc/1> requesting you to type in the data values for Zinc. The colon at the end of the second line of numbers indicates the end of the data, and the READ command finishes by printing a brief summary. The prompt now reverts to > allowing you to type more commands. The READ command has already provided the mean, so it remains only to calculate the median and variance (in interactive mode, you have the chance to see the results of one command before deciding on what to do next!). The PRINT command then displays the required summary statistics. Finally, the command STOP ends this run of Genstat.

Genstat keeps a log of the interactive session in a file on the computer. The name of this file depends on the computer, but is given in the local documentation. The file contains all the commands and data values that you have typed, up to and including STOP. This allows you to check what you have done, or to keep a record for future reference, though on some computers it may be necessary to copy the file to avoid it being overwritten by the log of the next session. It also allows you to modify the commands using an editor and then execute them all again using batch mode.

1.1.2 Batch mode

To run Genstat in batch mode, you need to set up a file on the computer containing all the commands for Genstat to process. You can do this using any of the facilities for creating files on your computer, such as a text editor. Then you give an instruction to the computer to run Genstat with that file attached as input and another file allocated to receive the output; details of the necessary instruction will again be in your local documentation, but it will probably be very like this:

```
GENSTAT input-filename,output-filename
```

Here is the input file to produce a similar analysis of the contamination data using Genstat in batch.

```
VARIATE [NVALUES=9] IDENTIFIER=Zinc
READ STRUCTURE=Zinc
164.2 160.6 163 166 159.8 163.9 161 161.3 165.8 :
SCALAR IDENTIFIER=Zmed,Zvar
CALCULATE Zmed = MEDIAN(Zinc)
CALCULATE Zvar = VARIANCE(Zinc)
PRINT STRUCTURE=Zmed,Zvar; DECIMALS=1,2
STOP
```

In the interactive example, we have exploited the ability of Genstat to determine some aspects of the input implicitly in order to save typing, but here the details are all given in full. The details of these ways of simplifying the input are described in Section 1.8 where the general syntactic rules are given. Here is the output produced from the batch run.

Example 1.1.2

```
1   VARIATE [NVALUES=9] IDENTIFIER=Zinc
2   READ STRUCTURE=Zinc

  Identifier   Minimum      Mean   Maximum    Values   Missing
       Zinc     159.8     162.8     166.0         9         0

4   SCALAR IDENTIFIER=Zmed,Zvar
5   CALCULATE Zmed = MEDIAN(Zinc)
6   CALCULATE Zvar = VARIANCE(Zinc)
7   PRINT STRUCTURE=Zmed,Zvar; DECIMALS=1,2

     Zmed        Zvar
    163.0        5.22
```

Notice that in batch mode Genstat echoes the input lines, each one prefixed by the number of that line in the input file. A sequence of commands to Genstat, like those used in this example, is called a Genstat *program*. Each command is known as a Genstat *statement*, and requests Genstat to perform some sort of action. Generally statements will use a Genstat *directive* (our term for a standard command). You can also use procedures. These are self-contained sets of statements, like a sub-program in the Genstat language. Genstat has a library of standard procedures, and you can also write your own. Details are given in 5.3, but all you need to know for now is that the syntax for using procedures is the same as that for using directives.

The program in Example 1.1.2 first declares a data structure with the identifier Zinc to store the amounts of zinc in the samples. Several different types of data structure are available in Genstat. This one is known as a *variate*, and can be defined using the VARIATE directive. Variates are used to store a list of numbers, in this case of length nine. Example 1.1.2 also has two *scalar* data structures, Zmed and Zvar, each of which stores a single number.

The example shows that values can be assigned to data structures in various ways. The values for Zinc are input using the READ directive. In the interactive program, these can be seen following the prompts Zinc/1> and Zinc/6>, which indicate that values are being read into Zinc starting firstly at unit 1 and then (after the first five values have been read) at unit 6. The data values in line 3 of the batch program have not been echoed to the output file and so just the summary is shown; however READ allows all of this to be controlled as is explained in more detail in 3.1. The values for the scalars are assigned from the results of two calculations, using the CALCULATE directive. In the interactive program, we have exploited the fact that CALCULATE is also able to make implicit declarations of data structures, and Zmed and Zvar are declared automatically as scalars so that they can store the single value generated by each calculation. Finally, the STOP directive is used to indicate the end of the program in both the interactive and the batch run.

1.1.3   Conversational mode

We now show how you can use the conversational interface to do the same task. Again, we shall suppress the use of windows, but on many types of computer the questions will appear in a separate pop-up window.

You start running Genstat in the same way as for interactive command mode, but then give the single command

        MENU

to switch into conversational mode. Genstat will respond with some information about files that will be used to record the session, and then display the Base Menu of the conversational system.

Example 1.1.3a

```
Welcome to Release 3[1] of the Genstat menu system.
The system will use three files to record the work you do:
        G31RESLT:  results produced by the operations you choose;
        G31COMND:  copy of Genstat commands that do the operations;
        G31STORE:  quick-access binary storage of your data.

BASE menu
What would you like to do next ?
                (null) chooses default response, if any;
?               help;  ?? lists current structures of allowed type;   Special
                       ?code gives specific help, if any;             responses
&               repeat the question;                                  to any
<               return to previous layer of menus;                    question
>>              exit to command mode.
```

```
i          input data
c          calculate new data, edit data, or define groups
t          display or summarize data in tables
p          display data in pictures
a          analyse data by standard statistical methods
q          quit using menu system

Code (i,c,t,p,a,q; Default:  i) > i
```

The conversational system consists of a simple tree of menus like this. After going down one of the branches shown, such as by choosing "i" for Input, and then inputting the values, the system will bring you back to the Base Menu. On some computers you will be able to choose from the menu using a mouse to control a pointer on the screen: on others you might have to type the letter identifying the choice. In addition, you can respond to any menu with the special symbols listed in the Base Menu, instead of with the information requested, in order to get help or move back up the tree, for example.

So to start, you should choose "i", or just press <RETURN> since this is the default choice for the Base Menu. As in Example 1.1.1, the typed responses are printed in bold type.

Example 1.1.3b

```
INPUT menu

Where are the data values ?

b          in a binary file previously set up using Genstat
s          in a character-type file, with values separated by spaces
t          to be typed at the terminal

Code (b,s,t; Default:  b) > t
```

To enter data at the terminal, you should then choose "t".

Example 1.1.3c

```
IDENTIFIER menu (from INPUT menu)

What identifiers will you use for the variables to be input ?
   If more than one, separate the identifiers with commas;
      end a line with backslash (\) to continue on the next line.

Identifiers > ?
```

This menu asks for you to give names to refer to the data about to be entered. You can choose "?" to find out the rules for such names.

Example 1.1.3d

An identifier is a name that you give to a data structure. It must start
with a letter and contain only letters and digits; use no more than eight
letters and digits in all. Upper and lower case letters are considered to
be different. Make sure that you do not use the same identifier for more
than one structure.

Identifiers > **Zinc**

On answering with the name Zinc, you are asked what type of data are to be input.

Example 1.1.3e

```
STRUCTURE-TYPE menu (from INPUT menu)

What type of variable is  Zinc

v         Variate: for quantitative measurements (integers or real numbers)
t         Text: for textual data (simple or quoted strings of characters)
f         Factor: for group codes (integers or real numbers)

Code (v,t,f; Default  v) >
```

The default is to input a series of numbers, for a variate data structure, so you can simply
press <RETURN>. The system will then prompt for values to be input, in exactly the same way
that the READ directive prompts for values in interactive command mode. After the colon has
been entered, the system prints the summary, including the mean, as before and then asks if
you want to store the data permanently, for possible re-use in later Genstat runs.

Example 1.1.3f

```
Following the READ> prompt, type values for the following structures:
       Zinc
Leave at least one space or press RETURN between values.
   When you have typed all the values, type a colon (:) and then RETURN.

   Zinc/1> 164.2 160.6 163 166 159.8
   Zinc/6> 163.9 161 161.3 165.8 :

   Identifier   Minimum      Mean   Maximum     Values   Missing
        Zinc     159.8      162.8     166.0          9         0

STORE menu (from INPUT menu)

Do you want to store the data in the binary file:
G31STORE
for later access ?

Code (y,n; Default:  y) > y

***** Catalogue *****
```

```
catalogue of structures in the subfile SUBFILE

    entry  identifier      type
    1      Zinc            variate
```

On choosing "y", it will display the current contents of the storage file (see 3.5.5), and then return to the Base Menu; we do not print this in full in Example 1.1.3g, but simply show the choice "t" (display or summarize data in tables).

Example 1.1.3g

```
Code (i,c,t,p,a,q; Default:  i) > t

TABULATION menu

How do you want the data to be displayed ?

1          List the values of one or more variates, texts and factors
s          Summary statistics of the values in a variate
g          Grouped summary of a variate, for groups defined by factors

Code (l,s,g); Default  l) > s
```

You can now type "s" to request summary statistics.

Example 1.1.3h

```
DATA menu (from TABULATION menu)

What variate do you want to summarize ?

Identifier > Zinc
```

On giving the name Zinc, the system will respond with a summary.

Example 1.1.3i

```
  Summary statistics for Zinc

      Number of observations = 9
    Number of missing values = 0
                        Mean = 162.844
                      Median = 163.000
                     Minimum = 159.800
                     Maximum = 166.000
              Lower quartile = 160.800
              Upper quartile = 165.000
                    Variance = 5.220
          Standard deviation = 2.285
```

There is no control over the contents of the summary, but it does include all the information that we want. This illustrates though that the conversational system offers only a limited range of techniques, so if for example you also wanted the 5 and 95 percentiles of the set of measurements it would be unable to help. However, anyone familiar with the Genstat language can easily break into command mode and type whatever statements are necessary to obtain any extra information that is required (in this case a TABULATE statement, see 4.11.1). Moreover, the system can be modified so that it does offer an option to display these percentiles, or indeed any results that can be produced in command mode (Chapter 13).

After displaying the results, the system returns to the Base Menu (which again is not reprinted), and you can finish by choosing "q" (quit using menu system) followed by the command STOP.

Example 1.1.3j

```
Code (i,c,t,p,a,q; Default:  i) > q

Quit from menu system.
   Type STOP to leave Genstat;
   type MENU to re-enter menu system;
   or type Genstat commands.

> STOP
```

As with interactive command mode, Genstat keeps a log of what you have done. This file has the same name as with command mode, and it contains Genstat commands to perform the same tasks as were done by the menus; however, the commands may be a little more complicated than necessary because of the automatic way in which they are constructed.

In addition, another file has a record of the output produced during the session, and a third file contains the data that was stored during the session. The names of all these files were given at the start of the session, and are also given in the local documentation.

As the examples above show, Genstat can conveniently be used interactively, in batch, or through the conversational interface. In later chapters, we shall use a mixture of batch and interactive use as appropriate for the application.

The conversational interface is likely to be most useful to those that use Genstat only occasionally. Extensions will be added during Release 3 and, in fact, Chapter 13 describes the ways in which you yourself can modify it. So we will not discuss the precise details any further here, but suggest that you explore it for yourself.

Otherwise, your preferred mode of use is likely to depend on whether you wish to be able to leave Genstat to do your analyses while you do something else, or whether you prefer the greater involvement and the immediacy of interactive use. You can even mix batch and interactive use. Within an interactive run, you can include statements that you have prepared in advance, and have stored in a computer file; this is achieved using the INPUT directive, described in 3.4.1. Likewise, data can be read from a computer file instead of being typed in during the interactive run.

1.2   Genstat programs

As you have seen, a *program* consists of a series of instructions to Genstat, or *statements* in our terminology. Before describing the general syntax of statements (1.4 to 1.9), we shall summarize some of the basic things that they can do.

1.2.1   Declarations

A statement specifying the type and identifier of a data structure is called a *declaration*. Declarations can be explicit or implicit. An example of an explicit declaration is the VARIATE statement in Example 1.1.2. Examples of implicit declarations are shown in the CALCULATE statements: the particular calculations done here produce a single-valued result, and so implicitly define scalar structures.

Other kinds of calculation produce other kinds of results, thus implicitly defining other kinds of structures. Implicit declarations are called *default declarations*: the rules for these are described in this manual at the same point as the directives that can make them.

1.2.2   Assigning values

You can define data values in the declaration itself: for example, the Zinc values could be defined by

```
VARIATE [VALUES=164.2,160.6,163,166,159.8,163.9,161,161.3,165.8] Zinc
```

Alternatively you can read the values, as was done for Zinc in Example 1.1.2a, or you can derive values as the results of calculations, as for the scalars Zmed and Zvar. On some computers, there is a command SPREADSHEET that will allow you to input and edit data interactively using a spreadsheet-like interface.

Later you will see that statistical analyses too can derive values to be assigned to data structures. You will see also that the data can be read from files other than the main input file (that is, instead of listing the data immediately after the READ statement).

1.2.3   Calculations

Calculations can be done with many kinds of data structure. Genstat contains many flexible tools for analysing data, ranging from the simplest (taking means, for example) to the advanced (such as nonlinear regression). But the essential point is that they do calculations with data held in data structures and referred to by their identifiers.

1.2.4   Printing

Many of the statistical directives in Genstat produce their own output. For example, the ANOVA directive will produce an analysis-of-variance table, tables of means, standard errors, and so on. But you may often want to produce output of your own. PRINT is merely one way of doing that. Genstat can also for example produce tables, graphs, and histograms.

1.2.5   Statements

The one thing that all these features of Genstat have in common is that you get access to them by means of the statements that make up a program. All statements have the same rules of syntax: first you give the name of the directive (or procedure, see 5.3) that you wish to use, then perhaps some options, and then usually some parameters.

The names are intended to be natural, or to refer naturally to common statistical techniques. But since there may be many natural words for the same operation, you should become familiar with the particular ones used by Genstat. For example, there is no directive called PLOT but there is one called GRAPH. Appendix 1 lists the directives, and Appendix 2 the procedures that were in the standard library, when Release 3 first became available.

Options are enclosed in square brackets, as in the declaration of the variate Zinc in Example 1.1.2. Each option has a name (for example NVALUES), and you can give it an appropriate *setting* (for example 9 here): the general form for setting an option is "name=setting". Another example is the INDENTATION option in the following statement:

        PRINT [INDENTATION=7] STRUCTURE=Zmed,Zvar

This would indent the printed values seven spaces from the left-hand margin of the page. Some options have default settings, that are the settings assumed by Genstat if you do not specify any explicitly. For example, the default for indentation is zero (the values are printed from the left margin). Conversely, the VARIATE directive has an option called VALUES, which can be used to define the values in a variate when it is declared; this has no default, and if it is omitted no values are defined.

Parameters are set in a similar way to the options, coming after the close of the square brackets (if any). One parameter that is nearly always part of a statement is a list of identifiers or an expression on which the statement is to operate. Some directives allow no more than this – for example CALCULATE; in such cases, there is no name for the parameter. Other directives have several parameters. For example, in the PRINT statement in Example 1.1.2 two parameters are set: STRUCTURE and DECIMALS. Sometimes the names of the parameters can be left out. The full rules that Genstat then uses to determine which parameter is being set are described in 1.8.1; but the simplest rule is that if no name is included for the first parameter given in a statement, Genstat takes this as the setting for the main parameter of the statement. You can see in the interactive example that the main parameter of the VARIATE directive specifies the identifiers of the variates that are being defined, while for PRINT it lists the structures whose values are to be printed.

If more than one parameter is set, each one must be separated from the next by a semicolon. The same rule applies if several options are to be set. The lists specified for each of the parameters are taken by Genstat in parallel, so the statement

        PRINT [INDENTATION=7] Zmed,Zvar; DECIMALS=1,2

will print Zmed with one decimal place and Zvar with two, and all this information will be indented by seven spaces. The parameters after the main parameter are thus used to supply further information for each item in the main list. Conversely, options supply information that applies to all the parameters in the list.

In this manual we generally give names of directives, options, and parameters in capitals and in full. But small letters and abbreviations can be used if you prefer. In particular, it is always enough to give four characters, but option and parameter names can often be abbreviated beyond that (1.8.1).

1.2.6   Punctuation

Items in lists are separated by commas: see for example the list of identifiers in the SCALAR declaration in the first program, or the list of values in the same declaration.

Option settings are separated by semicolons, as are parameter settings.

The usual way of ending a statement is with the carriage-return key, usually labelled (<RETURN>) on the keyboard. But you can also end with a colon, and thus get several statements on one line. The continuation symbol \ allows you to continue the statement onto the next line (1.4.6).

1.2.7   Comments

You can put comments into your programs to help other people to understand them, or to help you yourself understand them if you need to return to them later on. The series of comments can then give a running description of what the program is doing. You tell Genstat that you are making a comment by using the double-quote character ("); notice that this is not the same as two single quotes (' '). In Example 1.1.2, we could add the comment

```
"This program calculates some simple statistics
to summarize the amount of zinc in the samples."
```

You can type anything you like between the double quotes; Genstat simply ignores it. In longer programs you might want to put comments at several places in the program, to describe what different sets of statements are doing. In an interactive run Genstat will add the double-quote character to the prompt to remind you when you are in the middle of a comment "> while in a batch run the line number is prefixed by a minus sign.

1.3   On-line help

Help is available on-line while you are running Genstat. The HELP directive allows you to obtain general information about Genstat, as well as the specific details of any Genstat directive.

There are also several procedures that give information. NOTICE supplies news about Genstat activities, such as future conferences or recent changes to the standard procedure library, and allows you to obtain advice about any errors in the current release of Genstat. There is a PRINT option with settings news, library, and errors to control what is produced: for example

```
NOTICE [PRINT=news,library]
```

to produce the general news and news about the procedure library. LIBINFORM provides information about the contents of the procedure library: for example

```
LIBINFORM [PRINT=index] MODULE=repeatedmeasures
```

to list the index lines of the procedures relevant to the analysis of repeated measurements. LIBHELP gives details of the syntax and facilities of individual procedures in the library, and LIBEXAMPLE allows you to obtain an example of how to use any library procedure, or a copy of the Genstat statements that it contains. Further details are given in 5.3.1.

Another useful facility, available in most implementations of Genstat, is the ability of the SUSPEND directive to suspend the use of Genstat so that you can give commands instead to the computer's operating system (13.3.1). This can be useful for example if you have forgotten the precise name and location of a computer-file file from which you want to read some data.

1.3.1   The **HELP** directive

HELP   Prints details about the Genstat language and environment. The information is hierarchical; the initial menu, given by default if no parameters are specified, lists the choices at the top level: information about general areas of the Genstat facilities, or about specific directives. Within each menu, HELP allows you a choice of keywords to select information from the next level. Responding with a star (*) causes all the information at the next level to be displayed, the percent character (%) repeats the information from the current level, and less-than (<) moves up a level, as does <RETURN>. When HELP is used interactively, you remain within the HELP system until you type colon (:); alternatively, when at the top level, you can exit by typing <RETURN>.

Option

CHANNEL = *identifier*   Channel number of file, or identifier of a text to store output; default current output file

Parameter

strings   Directive names or keywords indexing the desired details

---

The HELP information in Genstat is arranged as a hierarchical system, where the information becomes more specific as you move down the hierarchy. At every stage Genstat lists the information available at the next level. It is thus very easy to browse through the system when using Genstat interactively, and indeed we shall not give any detailed examples here but suggest that you try it for yourself. To enter the system at the top level in an interactive run, you merely type the directive name, HELP, on its own.

Example 1.3

```
> HELP

    You are now in the Genstat Help System.

    The top level of HELP contains information on all individual directives
    and on the following subjects:
```

```
STATISTICS: Calculating statistics          LIBRARY: Procedure library
AOV: Analysis of experiments                TSA: Time-series analysis
REGRESSION: Linear and nonlinear regression  MVA: Multivariate analysis

DATAHANDLING: Calculations and manipulations PICTURES: Graphics
PROGRAMCONTROL: Control of program execution SYNTAX: Genstat syntax
COMMUNICATION: Input, output, backing-store  STRUCTURE: Data structures
ENVIRONMENT: Current environment             DESCRIBE: How to use HELP

If you do not know how to use help type DESCRIBE for further details.
```

```
Further information available on any of the top-level keywords
 or on individual directive names
```

```
HELP>
```

Wherever you are in the system, Genstat will print the information that you have requested, followed by the list of words (strings of characters) that you can use to move to the next level and select further information; you then get the prompt:

```
    HELP>
```

In response to the prompt you can do one of four things.

(a) Type one of the suggested words and obtain further information. When doing this you need include only enough characters to distinguish the word from those earlier in the list. Genstat will then print the requested information, followed by a further list of words and a prompt ready for your next choice. If you are already familiar with HELP, you can always skip levels in the prompting hierarchy by giving a list of words, each separated from the next by a comma. For example,

```
    HELP> read,options
```

skips the information that you would get if you specified read only; it thus takes you straight to the information on the options of the READ directive. You could, indeed, have done this at the outset, when you first typed HELP, by putting

```
    HELP read,options
```

The words can be typed in either lower or upper case, or in any mixture.

(b) Type an asterisk (*) to see the information relevant to all the possible words.

(c) Type carriage-return (<RETURN>) to move back to the previous level of the hierarchy. If you are already at the top level this takes you out of the HELP system.

(d) Type a colon (:) to exit from HELP from any level.

Whenever there is more than a screenful of information, HELP pauses and gives a question mark (?) as a prompt. You can respond either by pressing carriage-return (<RETURN>) to continue with the current information, or by selecting any of the possibilities (a) to (d). The words allowed under (a) are now those that would be permitted if the current information had been completed: these are the words available at the next level down if one exists, or from the current level if there is none below this.

In an interactive run, a HELP statement continues either until you type a colon, or until you type (<RETURN>) at the top level; otherwise newlines merely generate further prompts of HELP>. In batch, however, newline will be interpreted just as in any other statement. Thus each time that you use HELP in batch, you can specify just a single list, although you can still put asterisk to obtain all the information from a particular level. For example,

        HELP regression

to obtain general information about the facilities in Genstat for regression, or

        HELP fit,*

to learn about the options and parameters of the FIT directive, or

        HELP print,parameters

for details of the parameters of the PRINT directive.

1.4   Characters

Sections 1.4 to 1.9 contain a rigorous definition of the Genstat language, starting with the simple aspects and moving gradually to the more complicated. You may prefer not to read these sections immediately, but to return to them when you need to know more details of the rules; however, it will probably be useful to read about the conventions used in this manual, given in Section 1.10. There is much cross-referencing among Sections 1.4 to 1.9, and there are also references forward to the rest of the manual.

The characters in Genstat statements are a subset of the ASCII set available on most computers. Inside Genstat they are classified as in Subsections 1.4.1 to 1.4.6.

1.4.1   Letters

A *letter* is any of the alphabetic characters A, B, up to Z, a, b, up to z, the underline character (_), and the percent character (%).

1.4.2   Digits

A *digit* is one of the numerical characters 0, 1, 2, up to 9.

1.4.3   Simple operators

These occur in arithmetic *expressions* or in the *formulae* that define statistical models. The *simple operators* are:

        +   -   *   /   .   =   <   >

Equals (=), less than (<), and greater than (>) occur only in expressions (1.7.2). Dot (.) occurs only in formulae (1.7.3).

The meanings of the simple operators, and of the *compound operators* made up of more than one character, are given in 1.5.6.

1.4.4 Brackets

There are two kinds of *bracket*.

Round brackets (or) are used in lists (1.6) and expressions (1.7.2), and to enclose the arguments of functions (1.7.1).

Square brackets [or] enclose option settings, and are also used for suffixed identifiers (1.5.3). Left curly bracket { is synonymous with left square bracket [, and right curly bracket } with right square bracket]; these provide alternatives if square brackets are unavailable on your keyboard.

1.4.5 Punctuation symbols

Punctuation is used to separate different components of statements.

The *space* character can be used to improve the layout and readability of your programs. Statements use *free format*: that is, there may be any number of spaces between items; items are described in 1.5. Spaces can be left out altogether if the items are already separated by another punctuation symbol, by a bracket, by an operator, or by a special symbol (1.4.6). Most keyboards have a tab key (<TAB>), which has the effect of inserting spaces before subsequent characters on the terminal screen. Genstat treats the tab character as a synonym of space everywhere except within strings (1.5.2) and comments, or if reading in fixed format when it is treated as a fault (3.1.5).

Comma (,) is used to separate items in lists; lists are described in 1.6.

Equals (=) separates an option name or a parameter name from the list of settings. This character can thus have two meanings – separator or operator – but it will always be clear which is intended from the context.

Semicolon (;) is used to separate one list from another.

Colon (:) marks the end of a statement.

Newline is obtained by pressing the carriage-return key (<RETURN>). It is another way of marking the end of a statement. (But the SET directive can be used to request that newlines be ignored; see 13.1.1).

Single quote (') marks the start and finish of a string (1.5.2). On many computer terminals, there are two kinds of quote (' and '); these are synonymous.

Double quote (") marks the beginning and end of a comment (1.2.7).

1.4.6 Special symbols

Some characters have more specialized meanings; details of the ways in which they are used are given later in this chapter.

Ampersand (&) indicates that the directive name or procedure name from the previous statement is to be repeated, together with any option settings that are not explicitly changed (1.8.4).

Asterisk (*) is used to denote a missing value (1.5.5); this is another character with two meanings – missing value or operator – which again are easily distinguished by context.

Backslash (\) indicates that a statement is continued on the next line (1.8).

Dollar ($) is used to define subsets of the data in some structure. The dollar is followed by a list enclosed in square brackets, which specifies the subsets (1.7.2).

Exclamation mark (!) introduces an *unnamed* data structure (1.6.3). The vertical bar (|), available on some keyboards, is synonymous with exclamation mark.

Hash (#) is followed by the identifier of a data structure whose values are to be inserted at the current point of the program (1.6.4). It can also be used to indicate the default setting of an option (1.8.3). On some terminals and printers, # is replaced by £.

1.5   Items

A *Genstat* statement can contain various pieces of information; we shall call these *items*. There are six kinds of item, illustrated in these statements that calculate and print the area of a circle:

```
CALCULATE Area = 3.142 * Radius**2
PRINT [IPRINT=*] 'The area is',Area
```

The option setting IPRINT=* stops the name of the data structure being printed.

The words CALCULATE, PRINT, and IPRINT are *system words*, whereas Radius and Area are *identifiers* and the quoted characters make up a *string*. There are two *numbers* (3.142 and 2) and three *operators* (=, *, and **). The asterisk inside the square brackets is a *missing value*. The statements contain some other characters, which separate the items: these are the square brackets, the equals sign in the option, and the comma.

1.5.1   Numbers

A *number* conveys numerical information, and in its simplest form consists of digits only. For example,

      0   245609

A number can also have a *sign* (+ or –) and a decimal point (.).

      –2   4.5   +33.   –.2

However, a number must not contain any commas. Thus you must write one thousand as 1000 not 1,000.

To avoid lots of zeroes in large or small numbers, you can use an *exponent*. For example 2E–20 means 2×10^{-20}. Another example is 2D–5 which means 0.00002. D and E have the same meaning, and can also be replaced by d or e: these four are all called *exponent codes*. In general, a number can have the form

      xEy

which means that the number x is to be multiplied by 10 to the power y. The number x can have a sign, as can the exponent y. There must not be any spaces between x and the exponent code, nor between a sign and the exponent. But there can be spaces between the exponent code and the exponent: for example   2d   –5   again means 0.00002.

1.5.2 Strings

A *string* is a piece of textual information. Some examples are

```
apple;
five apples;
5 apples.
```

The spaces and punctuation here are part of the string. Important uses of strings are to form the values of text data structures (2.3.2) and to annotate output from a program (3.2.1).

More formally, we define a string to be a series of characters conveying textual information. In most places *quoted strings* are required: there, the characters are placed between single quotes ('); for example

```
'apple'
```

There is also the *unquoted string*, which must have its first character as a letter and all its characters as letters or digits.

Upper-case and lower-case letters are distinct within strings; so the strings Apple and apple are not the same.

If you want to put a single quote itself into a quoted string, you must put it in twice; otherwise Genstat thinks the string is ending. For example

```
'don''t do that'
```

will be interpreted as

```
don't do that
```

Similarly, a quoted string cannot contain a double-quote character on its own, because this is interpreted inside a string as the start of a comment (1.2.7): a comment inside a string is not interpreted as part of the string but is ignored. So to include a double quote in the string, you need to put two double quotes.

A continuation symbol (\) on its own in a quoted string continues the string onto the next line. However, a pair of backslash characters is interpreted as a single appearance of the character. For example

```
'C:\\EXAMPLES\\REGRESS.GEN'
```

is interpreted as

```
C:\EXAMPLES\REGRESS.GEN
```

If a quoted string contains a newline (<RETURN>) that does not follow an unduplicated continuation symbol, then it becomes a *string list* (1.6.2), unless you have used the SET directive to specify that newlines are to be ignored (13.1.1).

1.5.3 Identifiers

An *identifier* is the name used to refer to a data structure. An *unsuffixed* identifier is made up of letters and digits, starting with a letter. For example,

```
Cost    Yield1985    Yield1986
```

Any characters beyond the first eight are ignored; thus the second and third identifiers here refer to the same structure. By default, Genstat will treat capital letters as distinct from small letters. However, the SET directive (13.1.1) can been used to request that Genstat regards them as equivalent.

Identifiers can have suffixes; for example

```
Yield[1985]
```

The suffix is enclosed in square brackets, and can be a number, a quoted string, or another identifier. You can put spaces on either side of either of the square brackets.

A *suffixed* identifier is a value, or a set of values, of a pointer data structure (2.6). Thus Yield[1985] and Yield[1986] are two structures which are pointed to by the pointer structure Yield. When you use a suffixed identifier Genstat will automatically define the necessary pointer (2.6).

1.5.4   System words

A *system word* is the name of a directive, or an option, or a parameter, or a function (1.7.1). The first character is a letter; subsequent characters are letters or digits. For example,

```
PRINT     print     Log     Log10
```

You can use capital and small letters interchangeably: thus the first two system words here are equivalent. System words can always be abbreviated to four letters; option and parameter names can often be abbreviated more than that (see 1.8.1).

1.5.5   Missing values

A *missing value* indicates unknown information, and is represented by a single asterisk (*). When reading or printing data, missing values are represented by asterisks by default, but other representations can be chosen if preferred (3.1.1 and 3.2.1).

1.5.6   Operators

An *operator* stands for an arithmetic or logical operation, or some relationship between other kinds of item. Some operators have different meanings according to whether they appear in expressions or in formulae (1.7). Here is a list of all the operators and their names: more details are given in 4.1.1, 8.3.1, and 9.1.1.

> *Arithmetic operators*
> addition   +
> subtraction   −
> multiplication   *
> division   /
> exponentiation   **
> matrix product   *+

Assignment operator
assignment   =

Relational operators
equality   `.EQ.` or `==`
string equality   `.EQS.`
non-equality   `.NE.` or `/=` or `<>`
string non-equality   `.NES.`
less than   `.LT.` or `<`
less than or equals   `.LE.` or `<=`
greater than   `.GT.` or `>`
greater than or equals   `.GE.` or `>=`
identifier equivalence   `.IS.`
identifier non-equivalence   `.ISNT.`
inclusion   `.IN.`
non-inclusion   `.NI.`

Logical operators
negation   `.NOT.`
conjunction   `.AND.`
disjunction   `.OR.`
exclusive disjunction   `.EOR.`

Formula operators
summation   `+`
dot product   `.`
cross product   `*`
nested product   `/`
deletion   `−`
crossed deletion   `−*`
nested deletion   `−/`
linkage of pseudo terms   `//`

Upper-case and lower-case letters can be used interchangeably for relational and logical operators. However the characters making up any of these operators must be contiguous; thus, for example, there must be no spaces between the dots and the letters of a relational operator.

1.6 Lists

A *list* is a set of items that are to be treated in the same way in a statement. The items are usually separated by commas (but not always: see 1.6.4).

Here are some examples of the three kinds of list:

```
VARIATE [VALUES=18,28,27,19,21] IDENTIFIER=Temp
TEXT [VALUES='London','Madrid','New York','Ottawa','Paris'] \
    IDENTIFIER=City
PRINT STRUCTURE=City,Temp
```

The first set of values constitutes a *number list*, the second set a *string list*, and the STRUCTURE list for PRINT is an *identifier list*.

Missing values can occur in any of these lists. Their meanings and the ways of indicating them are described in 1.6.1, 1.6.2, and 1.6.3.

1.6.1   Number lists

Number lists appear in statements when values are put into a numerical data structure. Each item in a number list must be a number (1.5.1), a missing value, or the identifier of a data structure storing only one number; if an identifier, it stands for the value currently stored there. Missing values are interpreted as unknown observations in all directives that deal with numbers.

When numbers are to be listed in a repetitive or patterned series, you can save space and effort by compacting the lists as described in 1.6.4. Moreover, a set of numbers that form an arithmetic progression within a list can be written compactly using an *ellipsis*: this is three contiguous dots (. . .). For example,

```
1,2...10                    means                    1,2,3,4,5,6,7,8,9,10
```

In general, if k, m, and n are numbers and d $(=m-k)$ is the difference between m and k, then $k,m...n$ stands for k, $k+d$, $k+2d$, $k+3d$, up to n. If n is not in the progression defined by k and m, then the progression ends at the value beyond which n would be passed (and this can sometimes be k itself). Here are two more examples:

```
-2, -1.5 ... 0.4            means                    -2, -1.5, -1, -0.5, 0
-2, -1.5 ... -1.6           means                    -2
```

If the last value in the progression is close to n, but not quite equal, this may be due to rounding error on the computer. For this reason, the last value of the progression will then be set to n itself; the precise criterion is to check if the last value is within $d/100$ of n.

When the step length d is plus or minus 1, you can compact the list even further. For example,

```
10...1                      is the same as            10,9...1
```

In general, the construction $(k...n)$ is the same as $k,m...n$, where m is $k+1$ or $k-1$ depending on whether n is greater or less than k. If k equals n, the construction gives the single number k. You can leave out the brackets so long as there is no number preceding k that is not itself preceded by an ellipsis. For example,

`1...3,5...8`	means	`1,2,3,5,6,7,8`
`1,2,3,5...8`	means	`1,2,3,5,7`
`1,2,3,(5...8)`	means	`1,2,3,5,6,7,8`

1.6.2   String lists

String lists appear in two places. They occur when values are assigned to a structure that is to store text (as opposed to numbers), and they occur when the setting of an option or parameter is one or more words chosen from the restricted set that is valid for the directive. The second of these uses is described in 1.8.3.

For the first, each item in the string list must be a string (1.5.2), or a missing value. The latter is equivalent to the empty string ' '. An example of a string list with six items is

```
purple,'black and white','blue-green',so_dark,'2bright',F16
```

You can compact repetitive strings similarly to numbers (1.6.4).

Provided you have not used the SET directive (13.3.1) to request that newlines be ignored, the intermediate quotes and commas in a list of quoted strings can be replaced by newlines. For example

```
'Jack and Jill\
 went up the hill
To fetch a pail of water.'
```

is the same as

```
'Jack and Jill went up the hill','To fetch a pail of water.'
```

If the continuation symbol (\) were omitted, you would obtain

```
Jack and Jill',' went up the hill','To fetch a pail of water.'
```

while if you had previously specified SET [NEWLINE=ignored] (13.3.1), this would give the single string:

```
'Jack and Jill went up the hillTo fetch a pail of water.'
```

1.6.3   Identifier lists

Identifier lists are needed by many options and parameters of statements; they name the structures that are to be operated on. Each item in an identifier list must be an identifier, a missing value (representing an unset item), or an *unnamed structure*. Examples are Zinc and Chromium in

```
VARIATE IDENTIFIER=Zinc,Chromium
```

and in

```
PRINT STRUCTURE=Zinc,Chromium
```

The following types of structure can be unnamed: scalar, variate, text, pointer, expression, and formula (see Chapter 2).

An *unnamed scalar* may simply be a number. The other forms have a common style: they start with an exclamation mark, then a *type code*, and then a list enclosed in round brackets. The type codes are:

(a) V (or v) for an *unnamed variate*. For example,

    !V(1...10)

is an unnamed variate containing the numbers 1 to 10. If you do not specify any type code, V is assumed by default. So this example is the same as

    !(1...10)

(b) S (or s) provides another way of specifying an unnamed scalar; this is likely to be useful mainly when defining procedures (5.3).

(c) T (or t) for an *unnamed text*. (Each value of a text is a string: see 2.3.2). For example,

    !T(apples,pears)

is an unnamed text containing two strings: apples and pears. For a text containing a single string, an alternative is to give just the string within quotes. For example:

    'apples'

(d) P (or p) for an *unnamed pointer* (2.6), when the list is of identifiers. For example,

    !P(N,M,Q)

is a pointer containing the identifiers N, M, and Q.

(e) E (or e) for an *unnamed expression*.

(f) F (or f) for an *unnamed formula*.

These last two are explained in 1.7.

You can compact unnamed scalars in progression with the ellipsis (1.6.1); other repetitive or patterned identifier lists can be compacted by the methods described in 1.6.4.

You can compact a list of identifiers with suffixes by using a *suffix list*. For example,

    A[1,2]

is the same as

    A[1],A[2]

Identifier lists and suffix lists can be combined. For example,

    (A,B)[1,2]

is the same as

    A[1], A[2], B[1], B[2]

The lists are matched in lexicographic order: the items in the second list are matched in turn with the first item of the first list, then they are matched with the second item of the first list, and so on.

The empty suffix list [] stands for all suffixes of the identifiers preceding it. For example, if P[1], P[2], and P[3] are the only current suffixed identifiers involving P, then P[] is the same as P[1,2,3]. Further examples are described in 2.6.

1.6.4 Ways of compacting lists

All three types of lists can be compacted by any of the methods described below, as well as those methods described individually for each type of list earlier in this section.

The values of a structure can be substituted into a list using the substitution symbol (#). If I is an identifier, then #I is a list whose items are the values of the structure identified by I. If I is a pointer, then any item in #I that is itself a pointer is replaced by the values of that pointer (2.6).

For example, suppose I is a variate holding values 3,4. Then

        1,2,#I

is the same as

        1,2,3,4

If J is another variate with values 1,2 then the same list could be written as

        #J,#I

Notice that this list is quite different from the list of two identifiers

        J,I

You can do the same with lists of strings. For example, if Letters is a text containing the strings 'b','c','d', then

        'a',#Letters

is the same as

        'a','b','c','d'

If you put a dummy (2.2.2) in a list, it is automatically replaced by the identifier that it is currently storing. So if the identifier of the dummy is preceded by #, then the values put into the list are those of the structure that the dummy is storing.

When a list is to contain a set of items repeated several times, you can use a *multiplier*. A multiplier is a number without a sign; it can also be #identifier, where the identifier is of a structure storing one non-negative number. A pre-multiplier repeats each item in the list in turn. For example,

        2(A,B,C)

is the same as

        A,A,B,B,C,C

A post-multiplier repeats the whole list:

        ('a','b')2

is the same as

        'a','b','a','b'

These can be combined. Thus,

        2(1...3)3

is the same as

```
1,1,2,2,3,3,1,1,2,2,3,3,1,1,2,2,3,3
```

If the multiplier has the value 0, the construction contributes no items to the list. A multiplier with value 1 can be left out to give the form

```
(list)
```

(You might want to use such a matched pair of brackets to indicate some grouping of items to anyone reading the program.)

1.7   Expressions and formulae

Expressions contain arithmetic and logical operations, and are allowed only in directives that allow explicit calculations. For example

```
CALCULATE Boundary = 2*(Width+Height)
```

Formulae define the structure of a model in directives for some kinds of statistical analysis. For example

```
TREATMENTSTRUCTURE Drugs*Rates
```

Both these constructions may contain *functions*. Details of functions are given in 4.2 for expressions and in 8.4 and 9.5 for formulae.

1.7.1   Functions

Functions have the form

function name (sequence of arguments)

A function name is a system word of one of the standard functions (4.2, 8.4, and 9.5). For example:

```
SQRT(X)     SUM(Y + 4*Z)
```

The function name can be abbreviated to four characters. If you give further characters up to the eighth they must match the full form; characters beyond the eighth are ignored. The *arguments* of a function are either lists or expressions; if there are several arguments they are separated by semicolons. For example:

```
CHISQ(2.5; 6)
```

1.7.2   Expressions

An *expression* consists of identifier lists, operators (1.5.6) and functions. Identifier lists must not include any missing identifiers, and the operators

```
.    -*    -/    //
```

cannot be used in expressions.

The simplest form of expression is an identifier list by itself, or a function by itself. You build up expressions from identifiers and functions by mixtures of three rules. Let E and F be expressions. Then these are also expressions:

    (E)
    monadic operator E
    E *dyadic operator* F

The first means that putting brackets round an expression makes another expression. For the second, a *monadic operator* is an operator that works on just a single item: an example is minus in -1. In Genstat there are two monadic operators:

.NOT.   which negates a logical expression, and

\-   which changes the sign of a numerical expression.

All the other operators work on pair of items: that this, they are *dyadic*. These operators (including the use of minus to mean subtract) can be used with the third rule. Other examples of expressions, illustrating the rules, are

    5,6
    A,B = -(C)
    SUM(X) .EQ. 4
    A = (B = C + 1) + 1

The precedence rules of operators are very similar to those of computer languages like Fortran, but are not exactly the same. The following list shows the precedence of all operators in expressions when brackets are not used to make the order of evaluation explicit:

(1) .NOT. Monadic -
(2) .IS.   .ISNT.   .IN.   .NI.   *+
(3) **
(4) *   /
(5) +   Dyadic -
(6) < > == <= >= /= <> .LT. .GT. .EQ. .LE. .GE. .NE. .NES.
(7) .AND. .OR. .EOR.
(8) =

Within each class, operations are done from left to right within an expression. For example,

    A > B+C/D*E

is the same as

    A > (B + ((C/D) * E)

An identifier list in an expression can contain *qualified identifiers*; these select subsets of values of the structures (4.1.6). For example

    V$[3,4]

means take the third and fourth items of data from the variate V. This is the same as

    V$[3],V$[4]

which is a list of two scalars. If the structure has textual labels (2.3), these can be used in the qualification. For example

    V$['c','d']

In general, a qualified identifier is

> identifier $ [sequence of identifier lists]

If there are two lists, their elements are taken in parallel: for example,

> M$[1,2; 4,5]

refers to a matrix M, and selects two elements: column 4 of row 1 and column 5 of row 2. So this is the same as

> M$[1;4],M$[2;5]

You can compact a list of qualified identifiers in an expression by specifying a *qualified identifier* list:

> (A,B)$[1,2]

is the same as

> A$[1],A$[2],B$[1],B$[2]

The general form is

> (identifier list) $ [sequence of identifier lists]

If any list in the sequence is shorter than the others, it is repeated as often as is necessary; so for example

> M$[1,2; 4,5,6]

is the same as

> M$[1;4],M$[2;5],M$[1;6]

1.7.3   Formulae

A *formula* defines a statistical model; it consists of identifier lists, operators, and functions. The identifier lists must not include any missing identifiers, and only the operators

>   +   -   *   /   .   -/   */   //

can be used. The simplest form of a formula is an identifier list; also a formula can be a function by itself.

You build up other formulae by mixtures of two rules: if M and N are formulae, then so are

> (M)
> M operator N

For example

```
Sex * Diet
(Group / Variety) * Fertilizer
Drug * POL(Dose; 2)
```

The operators in a formula have the following precedence:

(1) .
(2) //
(3) /
(4) *
(5) +   –   –/   –*

Within each class, operations are done from left to right within a formula.

A formula is expanded into a series of *model terms*, linked by the summation operator (+). A model term contains one or more elements, separated from each other by the operator dot (.), each element being either an identifier or a function whose arguments are single identifiers. For example, the expanded form of the first formula above is

```
Sex + Diet + Sex.Diet
```

The interpretation of the terms is described in 8.3.1 and 9.1.1.

Identifiers in a list within a formula are treated as if they were separated by the summation operator and enclosed within brackets. For example

```
A,B * C
```

is the same as

```
(A + B) * C
```

The following table shows how operators combine terms, using L and M to represent two sums of terms.

Construction	Expansion
L.M	Sum of all pairwise combinations of terms in L with terms in M using the dot operator, with the terms ordered as explained below. For example: `(A+B).(C+D.E)` is the same as `A.C+B.C+A.D.E+B.D.E`
L*M	`L+M+L.M` ordered as explained below. For example `(A+B)*C` is the same as `A+B+C+A.C+B.C`
L/M	`L + L.M` where L is a term formed by combining all terms in L with the dot operator, ordered as explained below. For example `(A+B)/(C+D.E)` is the same as `A+B+A.B.C+A.B.D.E`
L–M	L without any terms that appear in M. For example `(A+B)–(A+C)` is the same as B
L–/M	L without any terms that consist of a term appearing in M combined with any other identifiers. For example `(A+B+B.C)–/B` is the same as A+B
L–*M	`L–M–/M` For example `(A+B+B.C)–*B` is the same as A

After expansions for the dot, slash, and star operators, the terms are rearranged in order of increasing numbers of identifiers. Terms with the same number of identifiers are arranged in lexicographical order with respect to the order in which the identifiers first occurred in the formula itself.

1.8   Statements

A *statement* is an instruction to Genstat, and has the general form:

statement-name [option-sequence] parameter-sequence terminator

For example,

```
READ [CHANNEL=2] STRUCTURE=Zinc,Chromium
```

If there are no options, the square brackets can be left out; but there must then be at least one space between the statement name and the first parameter setting: for example

```
PRINT STRUCTURE=Zinc; DECIMALS=2
```

Some directives have options but no parameters: for example,

```
SET [CASE=ignored]
```

makes upper-case and lower-case letters equivalent in identifiers (13.1.1). Others have neither options nor parameters. For example:

```
STOP
```

The statement name is one of three things: the name of one of the standard Genstat directives, or the name of a *procedure* (1.9.1 and 5.3), or the repetition symbol (&) described in 1.8.4.

The name of a directive is a *system word* (1.5.4) and can be abbreviated to four characters. Names of procedures can be abbreviated only to eight characters. If you give more than four characters of a directive name, the fifth to the eighth must match the full form, but characters beyond the eighth are ignored. This allows you to define procedures whose names differ from those of standard Genstat directives only within the fifth to eighth characters. For example, the fact that there are directives called CALCULATE and STOP does not prevent you from defining procedures called CALCULUS and STOP_IT.

The terminator of a statement is colon (:). Thus the line

```
VARIATE [NVALUES=12] Sales : READ Sales
```

contains two statements.

Alternatively, you can usually end a statement by pressing the carriage-return key (<RETURN>). In other words, newline is normally synonymous with colon. You can change this with the SET directive (13.1.1):

```
SET [NEWLINE=ignored]
```

indicates that newlines are to be ignored in the rest of the program.

Even if newlines are not ignored, there are still three situations when a newline will not end a statement.

(a) When newline occurs within a string, it terminates that string and begins another (1.6.2).

(b) A newline within a comment is ignored (along with the rest of the comment): for example

```
PRINT STRUCTURE=Zmin,Zmean,Zmax; FIELDWIDTH=8,9,8; "
"DECIMALS=1,2,1
```

is a single statement.

(c) You can indicate that a statement is to continue onto the next line by putting a continuation symbol (\) before pressing <RETURN> for example,

```
PRINT STRUCTURE=Zmin,Zmean,Zmax; FIELDWIDTH=8,9,8;\
DECIMALS=1,2,1
```

is again a single statement. Any characters between the continuation symbol and the end of the line are ignored. Genstat does however have the limitation that a statement must not exceed 2048 characters, after deletion of extraneous spaces.

1.8.1   Syntax of options and parameters

The sequences of options and parameters specify the items upon which the statement is to operate: these items are called the *arguments* of the statement. A sequence consists of one or more settings, each separated from the next by a semicolon (;). You can see an example of a sequence of parameter settings in the PRINT statement above. Each setting, whether of an option or a parameter, has one of the general forms:

> name = list
> name = expression
> name = formula

The list, expression, or formula can be null (length zero). Rules by which the "name=" can be left out are defined below; the types of setting are discussed further in 1.8.3.

An *option name* is a system word, which can be abbreviated to the minimum number of letters needed to distinguish it from the options that precede it in the prescribed order for the directive or procedure concerned. Characters up to the eighth must match the appropriate part of the full form; those after the eighth are ignored. For example, here are the options of the TABULATE directive (4.11.1), with the minimum form of each name printed in bold:

> **PR**INT,   **C**LASSIFICATION,   **CO**UNTS,   **S**EQUENTIAL,   **M**ARGINS,   **I**PRINT,
> **W**EIGHTS,   **P**ERCENTQUANTILES,   **O**WN,   **OWNF**ACTORS,   **OWNV**ARIATES,
> **INC**HANNEL,   **INF**ILETYPE

Notice for example that the minimum for COUNTS is CO, since C on its own would not distinguish it from CLASSIFICATION which precedes it in this prescribed order.

A *parameter name* is also a system word, and has the same abbreviation rule as the option names. For example, the parameters for TABULATE are (with minimum forms again in bold):

> **D**ATA,   **T**OTALS,   **N**OBSERVATIONS,   **ME**ANS,   **MI**NIMA,   **MA**XIMA,   **V**ARIANCES,
> **Q**UANTILES

You usually need type no more than one or two characters for any option or parameter name; there are no directives that require more than four characters for their option and parameter names. However, if you are likely to need to refer to a statement again in future (as for example if it is part of a procedure), remember that it may be difficult to understand if it is abbreviated too heavily.

You can omit the name and the equals character altogether by taking account of the prescribed order of options, or of parameters, within the directive or procedure. The rules for parameters are the same as those for options, and are as follows:

(a) If the first option setting in a statement is for the first option defined for that directive or procedure, then "name=" can be omitted.

(b) The "name=" can also be omitted for later option settings if the preceding setting is for the option immediately before that option in the prescribed order. For example,

```
TABULATE [PRINT=totals,means; COUNTS=Rep; SEQUENTIAL=Sval] \
    DATA=Spending; MEANS=Meansp
```

can be abbreviated to

```
TABULATE [totals,means; COUNTS=Rep; Sval] Spending; MEANS=Meansp
```

You can omit "PRINT=" here by rule (a) as it is the first option in the order prescribed for TABULATE. Similarly, "DATA=" can be omitted as DATA is the first parameter. You can omit "SEQUENTIAL=" by rule (b), because COUNTS which precedes it here is also the option that precedes SEQUENTIAL in the definition of TABULATE. However, you cannot omit "COUNTS=", because in the prescribed order there is another option between COUNTS and PRINT. The same is true for "MEANS=".

An option or parameter setting can be *null*: that is, it can have a list of length zero, or a null expression, or a null formula. Thus, by putting a null setting for the CLASSIFICATION option and the TOTALS and NOBSERVATIONS parameters, all the names can be omitted:

```
TABULATE [totals,means; ; Repl; Sval] Spending; ; ; Meansp
```

If a directive has a single parameter, no name is defined. For example, there is no name for the expression that is the only parameter for the CALCULATE directive (4.1.1).

1.8.2   Roles of options and parameters

Parameters specify parallel series of arguments that are operated on in turn when the statement is carried out. For example, in

```
TABLE [CLASSIFICATION=Age,Sex] IDENTIFIER=Income,Cars,Spending; \
    DECIMALS=2,0,2
```

there are two parameters: IDENTIFIER and DECIMALS. The statement declares three tables: when they are printed later in the program Income will have two decimal places, Cars will have none, and Spending will have two.

The main information in a directive or procedure is usually given by the first parameter, and so this is said to define the series of *primary arguments*. In the example, they are Income, Cars, and Spending. Usually the parameter setting is an identifier list, and so the series is a list of identifiers of data structures.

Alternatively, the setting of the first parameter can be an expression, in which case the first identifier list in the expression is the series of primary arguments. For example, if A, B, M, N, P, and Q are variates, then in

```
CALCULATE A,B = M,N + P,Q
```

the primary arguments are A and B.

Another possibility is that the setting may be a formula, in which case the expanded list of model terms (1.7.3) is the series of primary arguments. For example, in the regression statement

    FIT A * B

the primary arguments are the terms A, B, and A.B (representing the main effects of A and B and their interaction, 8.3.1),

    Later parameters, or other lists within an expression, specify *secondary arguments* which run in parallel with the primary arguments, and provide ancillary information. Examples of secondary arguments are in the TABLE statement above, and on the right-hand side of the assignment operator in the CALCULATE statement.

    The series of primary arguments should always be the longest; if a series of secondary arguments is longer, you are given a warning and elements beyond the length of the primary series are ignored. Any series that is shorter is recycled: that is, the series is traversed again, as many times as is necessary to match the length of the primary series. Thus the TABLE declaration above means exactly the same if it is written

    TABLE [CLASSIFICATION=Age,Sex] IDENTIFIER=Income,Cars,Spending; \
       DECIMALS=2,0

Options, on the other hand, specify information that applies to all the primary arguments (with their corresponding secondary arguments). Thus in the TABLE example above, all three tables are classified by Age and Sex.

    Many options have *default* values, namely values that are assumed if the option is not set explicitly in a statement. But some options have to be set, for example the OLDSTRUCTURE and NEWSTRUCTURE options of the COMBINE directive (4.11.4). Some parameters also have defaults: for example, the METHOD option of GRAPH (6.1.2) assumes point plots. In a very few directives, the primary parameter also has a default; for example, the Y parameter of RKEEP (8.1.4) assumes the list of current response variates.

1.8.3 Types of option and parameter settings

An option or parameter setting may need a formula, or an expression, or a list (1.8.1).

    When the setting is an expression, and the parameter or option has a defined name, the name and its accompanying equals character cannot be left out if the expression begins with "unsuffixed identifier=". This is because there would then be confusion between the name of the option or parameter and the unsuffixed identifier. For example, you could not leave out the name CONDITION in the statement

    RESTRICT STRUCTURE=Income; CONDITION=Agecond=Age>30

That is, if you wrote

    RESTRICT Income; Agecond=Age>30

Genstat would try to interpret Agecond as a parameter name, and a message would be printed alerting you to this syntax error. You could, however, put the expression in brackets:

    RESTRICT Income; (Agecond=Age>30)

No such problem arises with directives like CALCULATE (4.1.1), CASE, IF, and ELSIF (5.2), because in these the expression is the only parameter, and thus has no defined name. For example, you can write

        CALCULATE Agecond = Age>30

Many options and parameters need lists of identifiers. Any restrictions on the types of identifiers for particular lists are mentioned along with the descriptions of the syntax in later chapters. For example, the specification of the CLASSIFICATION option of the TABLE directive (2.5) states

        CLASSIFICATION = *factors*   Factors classifying the tables; default *

No structure other than a factor can be used here.

   Apart from the VALUES option of the TEXT directive (2.3.2), all options or parameters that require string lists use them to choose one or more textual values from a set defined by Genstat for that option or parameter. For example, the PRINT options of ADISPLAY and ANOVA have possible values:

   aovtable, **i**nformation, **c**ovariates, **e**ffects, **r**esiduals, **con**trasts, **m**eans, **cb**effects, **cbm**eans, **s**tratumvariances, **%**cv, **mi**ssingvalues

These let you choose which components of output are to be printed from an analysis of variance (9.1.3). The rules for such sets of values defined by Genstat are exactly the same as those for option and parameter names (1.8.1): they may be typed in capital or small letters (or mixtures), and each one can be abbreviated to the minimum number of characters necessary to distinguish it from earlier values in the list. If more than that number is given, the extra characters must match the full form up to the eighth.

   The minimum forms of the values for the PRINT options, above, are marked in bold. Thus

        PRINT=Aovtable,Effects,MissingValues

is the same as

        PRINT=aovtable,effects,missingvalues

and both of these can be abbreviated, for example, to

        PRINT=a,e,mi

To prevent any printing at all, with the PRINT option of any directive, you specify a missing string:

        PRINT=*

or

        PRINT=''

The special symbol # provides a succinct way of specifying the default setting of any option. This is most useful in options like PRINT above, where you might want to ask for the default plus some extra output. The default of the PRINT option for ANOVA is aovtable, information,covariates,means,missing so, to print the residuals as well, you can simply put

        PRINT=#,residuals

which will have the same effect as

```
PRINT=aovtable,information,covariates,means,missing,residuals
```

Number lists are needed by the VALUES options of the directives that define numerical data structures (2.1.1), but not by the options or parameters of any other directive.

1.8.4 Repetition of a statement and its options

You can repeat a directive name by typing the ampersand character (&). At the same time you can reset as many options as you want; those that you do not mention remain as in the previous statement. For example, after

```
READ [PRINT=data; CHANNEL=2] Costs
```

the statement

```
& Profits
```

is equivalent to

```
READ [PRINT=data; CHANNEL=2] Profits
```

while the statement

```
& [CHANNEL=3; REWIND=yes] Profits
```

is equivalent to

```
READ [PRINT=data; CHANNEL=3; REWIND=yes] Profits
```

You need not type a colon or newline before an ampersand, as it automatically terminates the previous statement.

1.9 Ways of compacting programs

You can store Genstat statements in two ways: in a procedure, or in a macro.

1.9.1 Procedures

A *procedure* is a series of complete Genstat statements. It is like a subroutine in Fortran, or a procedure in Basic or Pascal. These statements are self-contained, in that all the data structures that they use are accessible only within the procedure, apart from those explicitly defined as options or parameters of the procedure. Rules for writing and defining procedures are described in 5.3. Apart from the fact that procedure names must not be abbreviated beyond eight letters, the rules of syntax for using a procedure are identical to those for the standard Genstat directives (1.8); indeed, since you can get access to procedures automatically from libraries, you do not have to know whether a particular statement uses a directive or a procedure.

1.9.2   Macros

A *macro* is a Genstat text into which you have placed a section of Genstat program. The text must have an unsuffixed identifier. You can substitute the contents of the macro into the program by a contiguous pair of hash characters ##; the substitution takes place immediately after Genstat reads the statement that contains the hash characters.

A simple kind of macro would be a part of a Genstat statement. For example,

```
TEXT [VALUES='[PRINT=data,summary; CHANNEL=2]'] Optset
```

assigns to a text with identifier Optset the string between the single quotes. If you later type

```
READ ##Optset Patient,Sex,Weight
READ ##Optset Calories,Wtgain
```

then Optset is treated as a macro and its contents are inserted into each of the two statements; so the named structures are read using the options for PRINT and CHANNEL defined in the string that has been put in Optset. Defining Optset in this way saves effort in typing the two READ statements; it would also allow you to change the options of both statements simultaneously.

More complicated macros may contain complete statements. For example, suppose that the computer file ALG.DAT contains three lines, each a quoted string (1.5.2):

```
'CALCULATE Previous = Root'
'& Root = (X/Previous + Previous)/2'
'PRINT STRUCTURE=Root,Previous; DECIMALS=4'  :
```

These three statements can be read into a text for use as a macro. A simple program for calculating the square root of 48 (without using the standard function SQRT) can then conveniently be written as follows:

```
SET [INPRINT=statements,macros]
SCALAR IDENTIFIER=X,Root; VALUE=48
TEXT [NVALUES=3] Estsqrt
OPEN NAME='ALG.DAT'; CHANNEL=2
READ [CHANNEL=2] STRUCTURE=Estsqrt
##Estsqrt
##Estsqrt
##Estsqrt
PRINT [IPRINT=*] '3 iterations estimate square root of 48 as',Root
STOP
```

Output from running this program in batch is shown below.

Example 1.9.2

```
Genstat 5   Release 3.1   (IBM-PC 80386/DOS)          Tuesday, 4 May 1993
Copyright 1993, Lawes Agricultural Trust (Rothamsted Experimental Station)

    1   SET [INPRINT=statements,macros]
    2   SCALAR IDENTIFIER=X,Root; VALUE=48
    3   TEXT [NVALUES=3] Estsqrt
    4   OPEN NAME='ALG.DAT'; CHANNEL=2
    5   READ [CHANNEL=2] STRUCTURE=Estsqrt
    6   ##Estsqrt
```

```
   1   CALCULATE Previous = Root
   2   & Root = (X/Previous + Previous)/2
   3   PRINT STRUCTURE=Root,Previous; DECIMALS=4

          Root    Previous
      24.5000     48.0000

   7  ##Estsqrt
      1   CALCULATE Previous = Root
      2   & Root = (X/Previous + Previous)/2
      3   PRINT STRUCTURE=Root,Previous; DECIMALS=4

             Root    Previous
         13.2296     24.5000

   8  ##Estsqrt
      1   CALCULATE Previous = Root
      2   & Root = (X/Previous + Previous)/2
      3   PRINT STRUCTURE=Root,Previous; DECIMALS=4

             Root    Previous
          8.4289     13.2296

   9   PRINT [IPRINT=*] '3 iterations estimate square root of 48 as',Root

 3 iterations estimate square root of 48 as       8.429

  10   STOP

******** End of job.   Maximum of 9034 data units used at line 1 (41276 left)
```

The first statement arranges to print statements and contents of macros. Then X and Root are defined as scalars, and both are given the value 48. Estsqrt is defined as a text with three values (or lines), and read from ALG.DAT in lines 4 to 5. The macro is substituted into the program three times: because of SET, its contents are printed each time, with line numbers indented by two characters. The IPRINT option of PRINT in line 9 prevents printing of the identifier Root: all that appears is the number stored in Root.

As you can see from the output, the value is still some way from convergence. Methods of testing for convergence in iterative algorithms like this are described in 5.2.4.

Substitution using ## takes effect immediately after Genstat has read the relevant input line. For macros that contain complete statements, like Estsqrt, an alternative is to use the EXECUTE directive (5.3.5). The substitution will then take place only when the EXECUTE statement is executed. This makes no difference in ordinary programs, but is very useful inside procedures or loops, where the statements are defined before they are executed (5.3 and 5.2.1).

1.10   Conventions for examples in later chapters

You have now seen that you can lay out your programs in many ways. You can include spaces to make them more readable, or you can leave spaces out to make them compact (1.4.5). You can type statements spread over several lines, or you can have more than one to a line (1.8). You can write system words in capital letters, or in small letters, or in a mixture, and you can use the full or the abbreviated forms (1.8.1 and 1.9). You can do the same with strings in options and parameters (1.8.3). You can write identifiers with capital letters or with small letters, or in a mixture, and you can control whether or not these are equivalent (1.5.3).

In this manual, however, we have imposed some conventions. The use of spaces is standardized. System words are given in full and in capitals; the only exception is that the name, and corresponding equals character, of the main parameter of a directive will usually be left out in later chapters. Option strings are given in full and in small letters. Identifiers will begin with a capital; any other letters are small. There is usually only one statement per line, unless this is very wasteful of space; continuation lines are indented.

We hope these conventions will help you to recognize the items, both in the descriptions of syntax and in the examples. However, in your own programs, you should develop your own style according to what you find most convenient.

P.W.L.
R.W.P.

2 Data structures

Data structures store the information on which a Genstat program operates. Examples include data for statistical analyses, coordinates for graphs, text for annotation, and so on. You can also store almost anything that can be printed in an analysis. This enables you to extend the range of facilities that Genstat offers, by taking information from one directive and using it as input for another. To allow you to do this, Genstat has a comprehensive set of different structures. However, there are many similarities across the directives that are used to define them, and the more complicated structures are required only for the more advanced uses of Genstat.

The simplest structures store a single piece of information or *value*. The most important of these is the *scalar*, which stores a single number (2.2.1).

Many types of analysis require structures that store several values. For example a *variate* contains a list of numbers (2.3.1), and may be used for response and explanatory variables in a linear regression (Chapter 8), or for the y-variables and covariates in an analysis of variance (Chapter 9).

There are three different kinds of matrix: rectangular, symmetric, and diagonal (2.4). These are used most often in multivariate analysis (Chapter 11).

There are also multi-way tables (2.5), which can be filled with various sorts of summaries; for example, means, totals, minima, and maxima (4.11).

Not all structures store numbers. Other *modes* for their values are strings (2.3.2), identifiers (2.2.2 and 2.6), expressions (2.2.3), and formulae (2.2.4). The combination of the shape of a data structure and the mode of its values is determined by its *type*. There are 15 types available. Your program can contain as many data structures of each type as you like, limited only by the total amount of workspace that they occupy within Genstat.

2.1 Declarations

Most data structures have a name; the exceptions are called *unnamed structures* and are described in 1.6.3. The name is called an *identifier*, and this is used to refer to the structure within your program. As explained in (1.5.3), an ordinary identifier starts with a letter (1.4.1) and then contains digits (1.4.2) or letters (or both). Genstat stores only the first eight characters; subsequent characters are ignored. Identifiers can also have suffixes, enclosed in square brackets; further details are given in 1.5.3 and 2.6.

You can define the identifier of a structure, together with its type, using a directive known as a *declaration*. There is a directive available to declare each type of structure. For example the declaration

```
SCALAR Length
```

uses the SCALAR directive to define a scalar with identifier Length.

You can declare several structures in a single statement: for example

```
SCALAR Length,Width,Height
```

declares Length, Width, and Height all to be scalars. A declaration need define only the identifier and the type. However, you can also specify values for the data structures (2.1.1), as well as various attributes that carry ancillary information about the structures.

Some attributes must be specified before the structure can be given values, for example the number of rows and columns of a matrix (2.4.1). Others need be set only if you choose to use them; for example, the number of decimal places (2.1.2) to be used by default when printing the values.

Options and parameters that apply generally to several different directives are described in this section; the others are described with the directive concerned, later in the chapter.

2.1.1   The VALUES option and parameter

In any declaration, you can assign values either by an option or by a parameter. The same name is used for both purposes: it is VALUE if the structure is of a type that stores a single value, and VALUES if it stores several. The option defines a common value (or set of values) for all the structures in the declaration, while the parameter allows the structures each to be given different values.

With the option you must supply a list of values. With the parameter, however, you must give a list of identifiers of data structures of the appropriate mode; the unnamed structures described in 1.6.3 are particularly useful for this. Thus, to declare variates X and Xsq each with its own set of values, you can put:

```
VARIATE X,Xsq; VALUES=!(1,2,3,4),!(1,4,9,16)
```

X then contains the values 1 up to 4, and Xsq contains 1, 4, 9, and 16.

If both the option and the parameter are specified, the parameter takes precedence. Thus

```
SCALAR [VALUE=12.5] Length,Width,Height; VALUE=*,*,200
```

gives Length and Width the value 12.5 and Height the value 200. (The asterisk in the identifier list for the VALUE parameter means an omitted entry: see 1.6.3.)

2.1.2   The DECIMALS parameter

You can use the DECIMALS parameter to define the number of decimal places that Genstat will use by default whenever the values of the structure are printed. This applies to output either by PRINT or from an analysis (but it does not affect the accuracy with which the numbers are stored). For example,

```
SCALAR Length,Width,Height; VALUE=12.5,6.25,120; DECIMALS=1,2,0
```

specifies that Length, Width, and Height should in future be printed with one, two, and zero decimal places respectively, although you can of course override this within the PRINT directive itself (3.2). This parameter occurs only in the declarations of structures that contain numbers.

Procedure DECIMALS can be used to set the DECIMALS parameter automatically to the minimum number of decimal places to print the structure exactly. For example

```
DECIMALS Length,Width,Height
```

2.1.3   The **EXTRA** parameter

You can associate a text with each data structure by means of the parameter EXTRA. This text is then used by many Genstat directives to give a fuller annotation of output. For example:

```
SCALAR Length,Weight; EXTRA=' in centimetres',' in grams'
```

2.1.4   The **MINIMUM** and **MAXIMUM** parameters

These two parameters allow you to define lower and upper limits on the values expected for any structure that stores numbers. Genstat then prints warnings if any values outside that range are assigned to the structure.

2.1.5   The **MODIFY** option

Normally if you declare a data structure for a second time, you will lose all its existing attributes and values. If you want to retain them you should set option MODIFY=yes. Thus, to redeclare the scalar Length, changing only its number of decimals to two, you would need to put

```
SCALAR [MODIFY=yes] Length; DECIMALS=2
```

The one attribute that you cannot readily redefine is the type. Before you can redeclare an identifier to refer to a structure of a different type, you must delete all its attributes. (See 2.9, in which there is an example redeclaring a variate as a text.)

2.2   Single-valued data structures

2.2.1   Scalars

A scalar data structure stores a single number (1.5.1). The SCALAR directive which declares scalars has only the general options and parameters already described in 2.1.

SCALAR declares one or more scalar data structures.

Options

VALUE = *scalar*	Value for all the scalars; default is a missing value
MODIFY = *string*	Whether to modify (instead of redefining) existing structures (yes, no); default no

Parameters

IDENTIFIER = *identifiers*	Identifiers of the scalars
VALUE = *scalars*	Value for each scalar
DECIMALS = *scalars*	Number of decimal places for printing
EXTRA = *texts*	Extra text associated with each identifier
MINIMUM = *scalars*	Minimum value for the contents of each structure
MAXIMUM = *scalars*	Maximum value for the contents of each structure

---

SCALAR is the one type of declaration where values are defined by default: if you do not define a value explicitly for a scalar, Genstat gives it a missing value.

Examples are given in 2.1. Unnamed scalars are described in 1.6.3.

2.2.2   Dummies

A *dummy* is a data structure that itself stores the identifier of some other structure. You will find this useful in identifier lists, where in nearly all cases Genstat replaces a dummy by the identifier that it stores. The only exceptions are the IDENTIFIER parameter of the DUMMY directive itself (see below), the STRUCTURE parameter of ASSIGN (4.9.1), the parameters of FOR, and in the UNSET function in expressions.

Dummies are particularly useful when you want the same series of statements to be used with several different data structures. By using a dummy structure within the statements, you can make them apply to whichever structure you require. The dummy structure is like a plug which can be connected to the structure that you need; the important point is that you can then connect another structure without changing the statements themselves.

The most obvious occasions where this is useful are in loops and procedures, and there the dummies are declared automatically as explained in 5.2.1 and 6.3.2.

To declare a dummy explicitly, you use the DUMMY directive. This has only the general options and parameters already described in 2.1.

DUMMY declares one or more dummy data structures.

Options

VALUE = *identifier*	Value for all the dummies; default *
MODIFY = *string*	Whether to modify (instead of redefining) existing structures (yes, no); default no

Parameters

IDENTIFIER = *identifiers*	Identifiers of the dummies
VALUE = *identifiers*	Value for each dummy
EXTRA = *texts*	Extra text associated with each identifier

For example:

```
DUMMY Xdum,Ydum; VALUE=Day,Growth
```

2.2.3   Expression data structures

The expression data structure stores a Genstat expression (1.7.2), for example

```
Hours = Minutes/60
```

Usually you will find it easiest to type out an expression like this explicitly whenever you need it. The main use, then, for this rather specialized data structure is to supply an expression as the argument of a procedure.

Options and parameters of the EXPRESSION directive, which declares expressions, are already described in 2.1.

EXPRESSION declares one or more expression data structures.

Options

VALUE = *expression*	Value for all the expressions; default *
MODIFY = *string*	Whether to modify (instead of redefining) existing structures (yes, no); default no

Parameters

IDENTIFIER = *identifiers*	Identifiers of the expressions
VALUE = *expressions*	Value for each expression
EXTRA = *texts*	Extra text associated with each identifier

Here are two examples using the VALUE option:

```
EXPRESSION [VALUE=Length*Width*Height] Vcalc
EXPRESSION [VALUE=Dose=LOG10(Dose)] Dtrans
```

These put the expression Length*Width*Height into the identifier Vcalc, and the expression Dose=LOG10(Dose) into Dtrans. Both expressions could be declared simultaneously, using the VALUE parameter, by putting

```
EXPRESSION Vcalc,Dtrans; VALUE=!E(Length*Width*Height), \
   !E(Dose=LOG10(Dose))
```

Rules for omitting "VALUE=" when the expression contains an assignment are described in 1.8.3. Unnamed expressions like !E(Length*Width*Height) are described in 1.6.3.

2.2.4 Formula data structures

The formula data structure stores a Genstat formula. As explained in 1.7.3, these can be used to define the model to be fitted in a statistical analysis. Like the expression data structure (2.2.3), its main use is to give a formula as the argument of a procedure (5.3). The FORMULA directive which declares formulae has only the general options and parameters described in 2.1.

FORMULA declares one or more formula data structures.

Options

VALUE = *formula*	Value for all the formulae; default *
MODIFY = *string*	Whether to modify (instead of redefining) existing structures (yes, no); default no

Parameters

IDENTIFIER = *identifiers*	Identifiers of the formulae
VALUE = *formulae*	Value for each formula

EXTRA = *texts*	Extra text associated with each identifier

For example:

```
FORMULA [VALUE=Drug*Logdose] Model
FORMULA BModel,Tmodel; VALUE=!F(Litter/Rat),!F(Vitamin*Protein)
```

The construction !F(Litter/Rat) is an example of an unnamed formula, as described in 1.6.3.

2.3   Vectors

Most Genstat directives operate on structures that store several values. The most important of these contain a list of values, which you can imagine as being arranged as a *vector* in a column. Genstat has three different types of vector: variates (2.3.1), texts (2.3.2), and factors (2.3.3). Also, the pointer structure (2.6), which stores a list of identifiers, is treated like a vector in some directives.

The directives that declare vectors all have an option called NVALUES, with which you can specify a scalar to define the number of values to be stored in the vector or pointer. Alternatively, you can set NVALUES to another text or variate; this then defines both the length of the new vectors and provides labels for use in output (4.1.6, 8.1.2, and 8.1.4). Finally, if you set NVALUES to a factor, the number of levels defines the length and its labels if available, or otherwise its levels, provide labelling.

If NVALUES is omitted in the declaration of a vector, Genstat takes the value or vector specified by the preceding UNITS statement if you have given one (2.3.4). In Genstat we call the elements of a vector its *units*. If you define values in the declaration and omit the NVALUES option, Genstat will deduce the appropriate setting from the number of values specified. However, it is safest to define both, since Genstat can then check that you have specified as many values as you intended. Thus, for example, if you were to type

```
VARIATE [NVALUES=5; VALUES=1,2,3.4,5] X
```

Genstat would be able to tell you that x has been given only four values instead of the five that were required. Further examples are given in the subsections below.

2.3.1   Variates

The variate is probably the structure that you will use most often in Genstat. You can think of this as being just a list of numbers – a vector, in mathematical language. Variates occur for example as the response and explanatory variables in regression (Chapter 8), as covariates and y-variables in analysis of variance (Chapter 9), and can be used to form the matrices of correlations, similarities, or sums of squares and products required for multivariate analyses (Chapter 11). Unnamed variates, for example !(1,2,3,4,5), are described in 1.6.3. To declare a variate you use the VARIATE directive.

VARIATE declares one or more variate data structures.

Options

NVALUES = *scalar* or *vector*	Number of units, or vector of labels; default * takes the setting from the preceding UNITS statement, if any
VALUES = *numbers*	Values for all the variates; default *
MODIFY = *string*	Whether to modify (instead of redefining) existing structures (yes, no); default no

Parameters

IDENTIFIER = *identifiers*	Identifiers of the variates
VALUES = *identifiers*	Values for each variate
DECIMALS = *scalars*	Number of decimal places for output
EXTRA = *texts*	Extra text associated with each identifier
MINIMUM = *scalars*	Minimum value for the contents of each structure
MAXIMUM = *scalars*	Maximum value for the contents of each structure

For example:

```
VARIATE Weight; EXTRA='in grams'
VARIATE Volume,Price; VALUES=!(60,75,88),!(5,2,1.75); DECIMALS=0,2
```

2.3.2   Texts

Each unit of a Genstat text structure is a string (1.5.2) which you can regard as a line of textual description. Texts can be used to label vectors and pointers (2.3 and 2.6), for captions or pieces of explanation within output (3.2.1), to store Genstat statements (1.9.2 and 5.3.5), and to store output (3.2.1). The various operations which you can perform with texts are described in 4.7. You declare texts with the TEXT directive.

TEXT   declares one or more text data structures.

Options

NVALUES = *scalar* or *vector*	Number of strings, or vector of labels; default * takes the setting from the preceding UNITS statement, if any
VALUES = *strings*	Values for all the texts; default *
MODIFY = *string*	Whether to modify (instead of redefining) existing structures (yes, no); default no

Parameters

IDENTIFIER = *identifiers*	Identifiers of the texts
VALUES = *texts*	Values for each text
CHARACTERS = *scalars*	Numbers of characters of the lines of each text to be printed by default
EXTRA = *texts*	Extra text associated with each identifier

For example:

```
TEXT [NVALUES=5] Name; VALUES=!T(Ferrari,Lotus,'Aston Martin',MG)
```

Unnamed texts, like that in the VALUES parameter in this example, are described in 1.6.3. Notice that the third value has to be enclosed in single quotes as it contains a space. The rules governing when strings need to be quoted and when the quotes can be omitted are described in 1.5.2.

You may be unable to define all the values of a long text in its declaration, because of the restriction on the total length of a statement (1.8). One possibility then is to read the values (3.1.3). Alternatively, you could define several texts each containing a section of the full text and then use EQUATE (4.3) to join them together. Or you could form the values from within the editor (4.7.2).

2.3.3 Factors

Factors are used to indicate groupings of units. The commonest occurrence is in designed experiments (Chapter 9). For example, suppose you had 12 observations in an experiment, the first four on one treatment, the next four on a second treatment, and the last four on a third. Then you could record which treatment went with which observation by declaring a factor with the values

```
1,1,1,1,2,2,2,2,3,3,3,3
```

Thus a factor is a vector that has only a limited set of possible values, one for each group; this limitation distinguishes factors from variates and texts. In Genstat, the groups are referred to by numbers known as *levels*. Unless otherwise specified these are the integers 1 up to the number of groups, as in our example; however, you can specify any other numbers by the LEVELS option of the FACTOR directive (see below). You can also give textual labels to the groups, using the LABELS option of FACTOR: these might, for example, be mnemonics for the biochemical names of treatments in an experiment. The full syntax of FACTOR is:

FACTOR   declares one or more factor data structures.

Options

NVALUES = *scalar* or *vector*	Number of units, or vector of labels; default * takes the setting from the preceding UNITS statement, if any
LEVELS = *scalar* or *vector*	Number of levels, or series of numbers which will be used to refer to levels in the program; default *
VALUES = *numbers*	Values for all the factors, given as levels; default *
LABELS = *text*	Labels for levels, for input and output; default *
MODIFY = *string*	Whether to modify (instead of redefining) existing structures (yes, no); default no

Parameters

IDENTIFIER = *identifiers*	Identifiers of the factors

VALUES = *identifiers*	Values for each factor, specified as levels or labels
DECIMALS = *scalars*	Number of decimals for printing levels
CHARACTERS = *scalars*	Number of characters for printing labels
EXTRA = *texts*	Extra text associated with each identifier

Use of the VALUES parameter to assign values has the advantage that you can refer either to labels or to levels; the VALUES option lets you refer only to levels. So, to summarize, the LEVELS and LABELS options list the groups that can occur, while the VALUES option or parameter specifies which groups actually do occur, and in what pattern over the units.

Our simple explanatory example would therefore be:

```
FACTOR [LEVELS=3; VALUES=4(1...3)] Treatment
```

Other examples are:

```
FACTOR [LEVELS=!(2,4,8,16); VALUES=8,4,2,16,4,2,16,8,2] Dose
FACTOR [LABELS=!T(male,female)] Sex; VALUES=!T(4(male,female))
FACTOR [LEVELS=!(0,2.5,5); LABELS=!T(none,standard,double)] Rate \
    VALUES=!(0,5,2.5,5,0,2.5)
```

Notice that if we had assigned the values using the VALUES option in the second of these, we would have needed to use the (numerical) levels:

```
FACTOR [LABELS=!T(male,female); VALUES=4(1,2)] Sex
```

Conversely, in the VALUES parameter in the declaration of Rate, we can use either the labels or the levels; so the following statement gives Rate exactly the same values:

```
FACTOR [LEVELS=!(0,2.5,5); LABELS=!T(none,standard,double)] Rate \
    VALUES=!T(none,double,standard,double,none,standard)
```

When reading or printing the values of factors, you can use either the levels or labels (see the FREPRESENTATION parameter of the READ and PRINT directives: 3.1 and 3.2).

Factors can be defined automatically from variates and texts by the GROUPS directive (4.6.1). You can also use factors for example to specify groups for tabulation (4.11.1), to fit parallel regression lines (8.3 and 8.6.3), and to store groupings from cluster analysis (11.5 and 11.6).

2.3.4 The **UNITS** directive

UNITS defines an auxiliary vector of labels and/or the length of any vector whose length is not defined when a statement needing it is executed.

Option

| NVALUES = *scalar* | Default length for vectors |

Parameter

| *variate* or *text* | Vector of labels |

The UNITS directive can be used to define a default length which will then be used, if necessary, for any new vectors encountered later in the job. For example, in the statements

```
UNITS [NVALUES=20]
TEXT Subject
VARIATE [VALUES=0,1,2,4,8] Dlev
FACTOR [LEVELS=Dlev] Drug
VARIATE Age,Response; DECIMALS=0,2
```

the text Subject, the factor Drug, and the variates Age and Response are all defined to be of length 20. However, the length of the variate Dlev does not need to be set by default, but is deduced to be five from the number of values that have been specified by the VALUES option.

The READ directive (3.1) will use UNITS if values are to be read into a previously undeclared vector, as will the RESTRICT directive (4.4.1) if you use it to restrict a structure that has not yet been declared. The UNITS setting is also used by the CALCULATE directive with the EXPAND and URAND functions if their secondary argument is not specified (4.2.8 and 4.2.9).

The parameter of the UNITS directive allows you to specify the units structure, which is a variate or a text whose values will then be used as labels for output from regression or time-series directives, provided the vectors in the analysis have the same length as the units structure and provided also that these vectors do not have labels associated with them already.

The length of the units structure must match the value set by the NVALUES option if both are set. However, either one can be used to define the other. Thus, either

```
TEXT [VALUES=Sun,Mon,Tue,Wed,Thur,Fri,Sat] Day
UNITS Day
```

or

```
TEXT Day
UNITS [NVALUES=7] Day
```

would specify the default length for vectors to be seven. In the second example this default would be applied to Day too but, of course, its (seven) values would need to be read or defined in some other way before it could be used for labelling. If the type of the units structure has not been declared, UNITS will define it as a variate.

You can cancel the effect of a UNITS statement by

```
UNITS [NVALUES=*]
```

This means that statements that require a units structure will fail, which is the situation at the start of each job in a program. Similarly, the statement

```
UNITS *
```

cancels any reference to a units structure, but retains the default length if that has already been defined.

2.4 Matrices

A matrix stores a set of numbers as a two-dimensional array indexed by rows and columns. For example, the array

```
1    2    3    4
5    6    7    8
9   10   11   12
```

is called a three-by-four matrix.

You specify the size of the matrix by saying how many rows and columns it is to have; the total number of values is obtained by multiplying the number of rows by the number of columns. In the example there are 12 values. If the numbers of rows and columns are equal the matrix is said to be square.

Any matrix can be stored as an ordinary rectangular matrix. Genstat also has special structures to store diagonal matrices (2.4.2) and symmetric matrices (2.4.3): these are needed in many statistical contexts.

You can assign values to matrices when they are declared, just as with vectors. But you must also set the size of the matrix, and it must correspond exactly to the number of values that you assign. Genstat stores the values of the matrix in row order: that is, all of the first row, followed by all of the second row, and so on. So you must assign the values in this order.

You are most likely to use matrices in multivariate analysis (Chapter 11), where you may need them either to input data, or to save the results of an analysis. Genstat also provides many facilities for matrix calculations: for example, you can add and multiply matrices, find their inverses, and decompose them to diagonal form (4.1.3 and 4.10).

2.4.1 Rectangular matrices

The Genstat matrix structure is a rectangular array. It can be declared using the MATRIX directive.

MATRIX declares one or more matrix data structures.

Options

ROWS = *scalar, vector,* or *pointer*	Number of rows, or labels for rows; default *
COLUMNS = *scalar, vector,* or *pointer*	
	Number of columns, or labels for columns; default *
VALUES = *numbers*	Values for all the matrices; default *
MODIFY = *string*	Whether to modify (instead of redefining) existing structures (yes, no); default no

Parameters

IDENTIFIER = *identifiers*	Identifiers of the matrices
VALUES = *identifiers*	Values for each matrix

DECIMALS = *scalars*	Number of decimal places for printing
EXTRA = *texts*	Extra text associated with each identifier
MINIMUM = *scalars*	Minimum value for the contents of each structure
MAXIMUM = *scalars*	Maximum value for the contents of each structure

You use the ROWS and COLUMNS options to specify the size of the matrix. The simplest way of doing this is to use scalars to define the numbers of rows and columns explicitly. Alternatively, you can set ROWS (or COLUMNS) to a variate, text, or pointer, whose length then defines the number of rows (or columns) and whose values will then be used as labels, for example when the matrix is printed. Finally, if you specify a factor, the number of levels defines the number of rows or columns and the labels if available, or otherwise the levels, are used for labelling. Here is an example:

Example 2.4.1

```
1   TEXT [VALUES=Beer,Lager,Orange] Drink
2   VARIATE [VALUES=0.5,1.0] Quantity
3   MATRIX [ROWS=Drink; COLUMNS=Quantity; \
4      VALUES=1.1,0.6,1.2,0.65,0.8,0.45] Cost
5   PRINT Cost; DECIMALS=2
```

	Cost	
Quantity	0.50	1.00
Drink		
Beer	1.10	0.60
Lager	1.20	0.65
Orange	0.80	0.45

In some contexts Genstat will interpret a variate as being equivalent to a matrix with a single column; this is described with each directive, such as CALCULATE (4.1.3).

2.4.2   Diagonal Matrices

A square matrix that has zero entries except on its leading diagonal is called a diagonal matrix: for example,

```
2   0   0
0   1   0
0   0   3
```

Another example is the identity matrix, which has a diagonal of values equal to 1. To save space, Genstat has a special structure for diagonal matrices. You will probably use them most often to store latent roots in multivariate analysis (4.10.2, 11.2.1, 11.2.2, and 11.3.1). You can declare diagonal matrices using the DIAGONALMATRIX directive.

DIAGONALMATRIX declares one or more diagonal matrix data structures.

Options

ROWS = *scalar, vector, or pointer*	Number of rows, or labels for rows (and columns); default *
VALUES = *numbers*	Values for all the diagonal matrices; default *
MODIFY = *string*	Whether to modify (instead of redefining) existing structures (yes, no); default no

Parameters

IDENTIFIER = *identifiers*	Identifiers of the diagonal matrices
VALUES = *identifiers*	Values for each diagonal matrix
DECIMALS = *scalars*	Number of decimal places for printing
EXTRA = *texts*	Extra text associated with each identifier
MINIMUM = *scalars*	Minimum value for the contents of each structure
MAXIMUM = *scalars*	Maximum value for the contents of each structure

Because a diagonal matrix is square, Genstat requires you to specify only the number of rows. The ROWS option can be set to either a scalar or a labels vector or a pointer, as in the MATRIX directive (2.4.1).

When you give the values of a diagonal matrix, either in a declaration or when its values are read, you should specify only the diagonal elements. (Genstat does not store the off-diagonal elements, but assumes them to be zero.) Similarly, when a diagonal matrix is printed it appears as a column of numbers; Genstat omits the off-diagonal zeros. For example:

Example 2.4.2

```
   1   DIAGONALMATRIX [ROWS=3; VALUES=2,1,3] D
   2   PRINT D

                         D

          1        2.000
          2        1.000
          3        3.000
```

2.4.3   Symmetric Matrices

A symmetric square matrix is symmetric about its leading diagonal: that is, the value in column i of row j is the same as that in column j of row i. For example:

```
   1   2   3
   2   1   4
   3   4   1
```

Symmetric matrices often occur in statistics. Suppose, for example, that we have n random variables $X_1 \dots X_n$. Then the covariance of X_i with X_j is the same as the covariance of X_j with X_i. The covariance matrix of the random variables is therefore symmetric: the off-diagonal elements of the matrix are the covariances (and the diagonal elements are the variances).

Because of this symmetry, Genstat stores only the diagonal elements and those below it; this is called the *lower triangle*. So you must specify only these values, whether in the declaration or in a READ statement (3.1). (As always, you give them in row order: so if there are *n* rows, then for the first you supply one value, for the second two, and so on.) Likewise, Genstat prints only the lower triangle in output, for example with PRINT (3.2).

The syntax for the declaration of symmetric matrices is as follows:

SYMMETRICMATRIX declares one or more symmetric matrix data structures.

Options

ROWS = *scalar, vector,* or *pointer*	Number of rows, or labels for rows (and columns); default *
VALUES = *numbers*	Values for all the symmetric matrices; default *
MODIFY = *string*	Whether to modify (instead of redefining) existing structures (yes, no); default no

Parameters

IDENTIFIER = *identifiers*	Identifiers of the symmetric matrices
VALUES = *identifiers*	Values for each symmetric matrix
DECIMALS = *scalars*	Number of decimal places for printing
EXTRA = *texts*	Extra text associated with each identifier
MINIMUM = *scalars*	Minimum value for the contents of each structure
MAXIMUM = *scalars*	Maximum value for the contents of each structure

---

The ROWS option defines both the number of rows and the number of columns. You can use a vector or pointer to specify row and column labels, as with MATRIX (2.4.1). For example:

Example 2.4.3

```
  1   VARIATE Weight,Height,Reach
  2   POINTER [VALUES=Weight,Height,Reach] Vars
  3   SYMMETRICMATRIX [ROWS=Vars; VALUES=1.0,0.68,1.0,0.43,0.72,1.0] Correl
  4   PRINT Correl
```

```
                  Correl

       Weight     1.0000
       Height     0.6800     1.0000
        Reach     0.4300     0.7200     1.0000

                  Weight     Height      Reach
```

2.5 Tables

Tables are used to store numerical summaries of data that are classified into groups. With Genstat, the classification into groups is specified by a set of factors (2.3.3). The table contains an element, called a *cell*, for each combination of the levels of the factors that classify it.

You can specify the values of a table when you declare it. More often, you may wish to calculate the values within Genstat. The TABULATE directive (4.11.1) allows you to summarize observations, for example from surveys. The observed values are supplied in a variate, and the levels of the factors classifying the table indicate the group to which each observed unit belongs. The table can contain, in each of its cells, either the total of the observations with the corresponding levels of the classifying factors, or perhaps the mean, or the minimum value, or the maximum value, or the variance. In an analysis of variance, you can save tables of means and tables of replications by the AKEEP directive (9.6.1). Calculations with tables are described in 4.1.4. The full list of facilities available for tables is given in 4.11. Tables are declared using the TABLE directive.

TABLE declares one or more table data structures.

Options

CLASSIFICATION = *factors*	Factors classifying the tables; default *
MARGINS = *string*	Whether to add margins (yes, no); default no
VALUES = *numbers*	Values for all the tables; default *
MODIFY = *string*	Whether to modify (instead of redefining) existing structures (yes, no); default no

Parameters

IDENTIFIER = *identifiers*	Identifiers of the tables
VALUES = *identifiers*	Values for each table
DECIMALS = *scalars*	Number of decimal places for printing
EXTRA = *texts*	Extra text associated with each identifier
UNKNOWN = *identifiers*	Identifier for scalar to hold summary of unclassified data associated with each table
MINIMUM = *scalars*	Minimum value for the contents of each structure
MAXIMUM = *scalars*	Maximum value for the contents of each structure

---

The example below shows a table called Classnum which stores numbers of children of each sex in the classes of a school. Here there are two factors defined in lines 1 and 2: Class with levels 1 to 5 and Sex with levels labelled 'boy' and 'girl'. The CLASSIFICATION option of the TABLE declaration (line 3) defines them to be the factors classifying the table, and the VALUES option defines a value for each of the 10 cells (two sexes × five classes) of the table. As you can see from the printed form of the table, the cells are arranged with both levels of Sex for Class 1, then both levels of Sex for Class 2, and so on. If there were three classifying factors, the table would have cells for all the levels of the third factor at level 1 of the first and second factors, then cells for all the levels of the third factor at level 1 of the first factor and level 2 of the second factor, and so on. In other words, the right-most factor in the classification rotates fastest, followed by the second from the right, and so on. This is illustrated by the second table, Schoolnm, which has a further factor School before Class and Sex in the list of classifying factors. Tables can be classified by up to nine factors.

Example 2.5a

```
 1   FACTOR [LABELS=!T(boy,girl)] Sex
 2   FACTOR [LEVELS=5] Class
 3   TABLE [CLASSIFICATION=Class,Sex; VALUES=15,17,29,31,34,30,33,35,28,27] \
 4      Classnum
 5   PRINT Classnum; DECIMALS=0
```

```
                      Classnum
           Sex          boy           girl
         Class
             1           15            17
             2           29            31
             3           34            30
             4           33            35
             5           28            27
```

```
 6   FACTOR [LEVELS=2] School
 7   TABLE [CLASSIFICATION=School,Class,Sex; \
 8      VALUES=15,17,29,31,34,30,33,35,28,27,18,16,33,31,35,36,34,33,31,32] \
 9      Schoolnm
10   PRINT Schoolnm; DECIMALS=0
```

```
                             Schoolnm
                     Sex        boy           girl
         School    Class
             1        1           15            17
                      2           29            31
                      3           34            30
                      4           33            35
                      5           28            27
             2        1           18            16
                      2           33            31
                      3           35            36
                      4           34            33
                      5           31            32
```

A table can also have *margins*. There is then a margin for each classifying factor; this contains some sort of summary over the levels of that factor. For example, if you have a table in which the cells contain totals of the observations, you would want the marginal cells to contain totals across the levels of the factor: see the next section of the example. You can define a table to have margins when you declare it, using the MARGINS option of the TABLE directive. Or you can add margins later by the MARGIN directive (4.11.2), as shown in Example 2.5b.

Example 2.5b

```
11   MARGIN Classnum,Schoolnm
```

```
12   PRINT Classnum; DECIMALS=0
```

	Classnum		
Sex	boy	girl	Margin
Class			
1	15	17	32
2	29	31	60
3	34	30	64
4	33	35	68
5	28	27	55
Margin	139	140	279

The margin row of Classnum contains the total numbers of boys and girls in the school (totalled over classes), and the margin column contains the total numbers (boys plus girls) in each class. The cell where this column and row coincide contains the total number in the school. With Schoolnm, there are marginal summaries over each classifying factor individually, over each pair of factors, and over all three factors. Thus the margin over a single factor is itself a two-dimensional array, classified by the other two factors, as shown in Example 2.5c.

Example 2.5c

```
13   PRINT Schoolnm; DECIMALS=0
```

		Schoolnm		
	Sex	boy	girl	Margin
School	Class			
1	1	15	17	32
	2	29	31	60
	3	34	30	64
	4	33	35	68
	5	28	27	55
	Margin	139	140	279
2	1	18	16	34
	2	33	31	64
	3	35	36	71
	4	34	33	67
	5	31	32	63
	Margin	151	148	299
Margin	1	33	33	66
	2	62	62	124
	3	69	66	135
	4	67	68	135
	5	59	59	118
	Margin	290	288	578

Tables also have an associated scalar which collects a summary of all the observations for which any of the classifying factors has a missing value; these observations cannot be assigned to any cell of the table itself. This scalar can be given an identifier, so that you can refer to it, using the UNKNOWN parameter of the TABLE directive.

2.6   Pointers

A pointer is a data structure that points to other structures: that is, each of its elements is the identifier of some other Genstat data structure. You use pointers in Genstat wherever you have to specify a collection of structures; for example in EQUATE (4.3), COMBINE (4.11.4), in some functions (4.2.3), and for a data matrix specified via the variates forming its columns (11.2). You can use them as a convenient means of compacting lists (1.6.4), and they are also involved in the use of suffixed identifiers (see below). You can declare pointers using the POINTER directive.

POINTER   declares one or more pointer data structures.

Options

NVALUES = *scalar* or *text*	Number of values, or labels for values; default *
VALUES = *identifiers*	Values for all the pointers; default *
SUFFIXES = *variate*	Defines an integer number for each of the suffixes; default * indicates that the numbers 1,2,... are to be used
MODIFY = *string*	Whether to modify (instead of redefining) existing structures (yes, no); default no

Parameters

IDENTIFIER = *identifiers*	Identifiers of the pointers
VALUES = *pointers*	Values for each pointer
EXTRA = *texts*	Extra text associated with each identifier

Thus, for example,

        POINTER [VALUES=Yield,Costs,Profit] Info

sets up a pointer Info with values Yield, Costs, and Profit. These three are themselves data structures, which can be assigned values, operated on, and so forth. You can refer to individual elements of pointers by suffixes, enclosed in square brackets (1.5.3): so Info[3] is Profit, and Info[1,2] is the list of structures Yield, Costs. Thus if Yield held the values 5.6 and 6.1, then

        PRINT Info[1]

would print the values of Yield, as shown below:

Example 2.6

```
1   VARIATE [NVALUES=2] Yield,Costs,Profit
2   READ Yield,Costs,Profit
```

```
      Identifier    Minimum      Mean   Maximum       Values    Missing
           Yield      5.600     5.850     6.100            2          0
           Costs       1200      1365      1530            2          0
          Profit      455.0     537.5     620.0            2          0
    4  POINTER [VALUES=Yield,Costs,Profit] Info
    5  PRINT Info[1]

        Yield
        5.600
        6.100
```

In fact, when Genstat meets a suffixed identifier, it sets up a pointer automatically if necessary. For example if your program contains a suffixed identifier `Data[4]`, Genstat first checks whether or not a pointer called `Data` already exists and, if not, creates it; then if there is no element for suffix 4 it creates one. If the pointer `Data` already exists but does not have a fourth element, then an appearance of `Data[4]` automatically extends `Data`. So you can add elements to pointers without redeclaring them.

The suffixes need not run from 1, nor be a complete list, although they must be integers; if you give a decimal number it will be rounded to the nearest integer (for example, −27.2 becomes −27). You specify the list of suffixes that you require by the SUFFIXES option; if you omit this, they are assumed to run from 1 up to the number of values. You can also label the elements of pointers by supplying a text in the NVALUES option (2.3): for example

```
    POINTER [NVALUES=!T(workstations,PCs,laptops)] Sales
```

allows you to refer to `Sales['PCs']`, `Sales['laptops','workstations']`, and even to `Sales[1,2,'laptops']`. The suffix list within the square brackets is a list of identifiers, so the strings must be quoted: they are then treated as unnamed texts each with a single value (1.6.3).

The identifiers in a suffix list can be of scalars, variates, or texts; this of course includes numbers and strings as unnamed scalars and texts respectively. If one of these structures contains several values, it defines a sub-pointer: for example `Info[!(3,2)]` is a pointer with two elements, `Profit` and `Costs`. You can also give a null list to mean all the elements of the pointer: for example `Info[]` is `Yield,Costs,Profit`. You must be careful not to confuse a sub-pointer with a list of some of the elements of a pointer: for example `Info[!(3,2)]` is a single pointer with two elements, whereas `Info[3,2]` is a list of the two structures `Profit` and `Costs`.

Elements of pointers can themselves be pointers, allowing you to construct trees of structures. For example

```
    VARIATE A,B,C,D,E
    POINTER R; VALUES=!P(D,E)
    &  S; VALUES=!P(B,C)
    &  Q; VALUES=!P(A,S)
    &  P; VALUES=!P(Q,R)
```

defines the tree

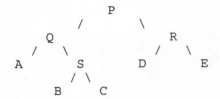

You can refer to elements within the tree by giving several levels of suffixes: for example P[2][1] is R[1] which is D; P[2,1][1,2] is (R,Q)[1,2] or D,E,A,S. The special symbol # (1.4.6 and 1.6.4) allows you to list all the structures at the ends of the branches of the tree: #P replaces P by the identifiers of the structures to which it points (Q and R); then, if any of these is a pointer, it replaces it by its own values, and so on. Thus #P is the list A,B,C,D,E.

2.7   Compound structures

You can use the pointer structure (2.6) to group together related data structures, so that you can refer to them as a single structure. Some Genstat directives expect standard combinations of data structures for their input or output; in these cases you use special pointers called *compound structures*. These differ from ordinary pointers in that they have a fixed number of elements which must be of the correct types, and must form a consistent set (in terms of their sizes and so on).

You can refer to elements of these structures in exactly the same way as the elements of pointers: for example if L is an LRV (2.7.1) then L refers to the set of structures L[1], L[2], L[3]. The suffixes run from 1 upwards, and Genstat does not allow you to change that. Neither can you change the labels that Genstat gives to the structures; details of these labels come later in this section. However, unlike pointers, the labels are not case sensitive; Genstat will recognize the label in either uppercase or lowercase letters or in any mixture of the two.

You can give the individual elements of a compound structure identifiers in their own right, just as with pointers. Indeed, you can use all the features of pointer syntax: for example, you can use the null list, or the substitution symbol #, to list all the elements of the structure (2.6).

When you declare a compound structure, you conveniently declare, simultaneously and automatically, a whole collection of structures. At the same time you ensure that they match the requirements of whatever form of analysis you want to use.

2.7.1   The LRV structure

The LRV structure is used to store latent roots and vectors resulting from the decomposition of a matrix (4.10.2), or produced in multivariate analysis (Chapter 11). You need not store all the latent roots; usually Genstat will select the largest ones. The LRV structure points to three structures (identified by their suffixes):

[1] or ['VECTORS'] is a matrix whose columns are the latent vectors: the word "VECTOR" is used here in its mathematical sense rather than in the more specific Genstat sense; in fact, latent vectors are most conveniently stored in matrices rather than in Genstat vectors;

[2] or ['ROOTS'] is a diagonal matrix whose elements are the latent roots;

[3] or ['TRACE'] is a scalar holding the trace of the matrix, which is the sum of all its latent roots.

To declare an LRV you use the LRV directive.

LRV declares one or more LRV data structures.

Options

ROWS = *scalar*, *vector*, or *pointer* Number of rows, or row labels, for the matrix; default *

COLUMNS = *scalar*, *vector*, or *pointer*

Number of columns, or column labels, for matrix and diagonal matrix; default *

Parameters

IDENTIFIER = *identifiers*   Identifiers of the LRVs

VECTORS = *matrices*   Matrix to contain the latent vectors for each LRV

ROOTS = *diagonal matrices*   Diagonal matrix to contain the latent roots for each LRV

TRACE = *scalars*   Trace of the matrix

The length of each latent vector is specified by the ROWS option; this then defines the number of rows in the 'VECTORS' matrix. The COLUMNS option defines the number of latent roots to be stored; this is also the number of latent vectors, and so indicates the number of columns in the 'VECTORS' matrix and the number of elements in the 'ROOTS' matrix. If you do not specify the number of columns Genstat will set it to be the same as the number of rows. The value of COLUMNS can be less than the value of ROWS; however, it must not exceed than that of COLUMNS, otherwise Genstat gives an error diagnostic. Row and column labels can be defined, as in the declaration of matrices (2.4).

You can specify identifiers for the three individual elements of the LRV by using the VECTORS, ROOTS, and TRACE parameters. If you have declared them already they must be of the correct type (and you can also have given them values). If you have given these identifiers row or column settings, then these will be used for the LRV declaration and must match any of the corresponding options of LRV that you choose to set.

Example 2.7.1 declares an LRV, and then forms its values (see 4.10.2).

Example 2.7.1

```
1   POINTER [VALUES=stem,leaf,root,petal,pollen] Vars
2   SYMMETRICMATRIX [ROWS=Vars] Symm
3   READ Symm

   Identifier   Minimum      Mean   Maximum   Values   Missing
         Symm   -0.9820    0.1974    1.0000       15         0

9   PRINT Symm
```

```
              Symm
   stem      1.0000
   leaf     -0.6550    1.0000
   root     -0.9450    0.8660    1.0000
  petal     -0.7560    0.0000    0.5000    1.0000
 pollen      0.5000   -0.9820   -0.7560    0.1890    1.0000

              stem      leaf      root     petal    pollen
```

```
10   LRV [ROWS=Vars;COLUMNS=2] Latent; VECTORS=Lvecs
11   FLRV Symm; Latent
12   PRINT Latent['Vectors','Roots']
```

```
              Lvecs
                 1                  2
  Vars
   stem     -0.4875             0.3372
   leaf      0.4875             0.3372
   root      0.5335            -0.0770
  petal      0.2227            -0.7383
 pollen     -0.4366            -0.4707
```

```
          Latent['Roots']

      1        3.482
      2        1.518
```

2.7.2   The SSPM structure

The SSPM structure stores a matrix of corrected sums of squares and products, and associated information, as used for regression (Chapter 8) and some multivariate analyses (Chapter 11). You can form values for SSPM structures by the FSSPM directive (4.10.3). However, most multivariate and regression analyses can be done without declaring and forming an SSPM explicitly.

An SSPM comprises four structures (identified by their suffixes).

[1] or ['SUMS'] is a symmetric matrix containing the sums of squares and products. The number of rows and columns of this matrix will equal the number of parameters defined by the expanded terms list: that is, the number of variates plus the number of dummy variates generated by the model formula. (See the TERMS directive: 8.2.2.)

[2] or ['MEANS'] is a variate containing the mean for each variate or dummy variate.

[3] or ['NUNITS'] is a scalar holding the total number of units used in constructing the sums of squares and products matrix. If the SSPM is weighted, this scalar will hold the sum of the weights.

A within-group SSPM has one additional element:

[4] or ['WMEANS'] is a pointer, pointing to variates holding within-group means. There is one variate for each row of the 'SUMS' matrix plus one extra. They are all of the same length, namely the number of levels of the GROUPS factor. The extra variate holds counts of the number of units in each group.

The syntax for the declaration of SSPM structures is as follows:

SSPM declares one or more SSPM data structures.

Options

TERMS = *formula*	Terms for which sums of squares and products are to be calculated; default *
FACTORIAL = *scalar*	Maximum number of vectors in a term; default 3
FULL = *string*	Full factor parameterization (yes, no); default no
GROUPS = *factor*	Groups for within-group SSPMs; default *
DF = *scalar*	Number of degrees of freedom for sums of squares; default *

Parameters

IDENTIFIER = *identifiers*	Identifiers of the SSPMs
SSP = *symmetric matrices*	Symmetric matrix to contain the sums of squares and products for each SSPM
MEANS = *variates*	Variate to contain the means for each SSPM
NUNITS = *scalars*	Number of units or sum of weights for each SSPM
WMEANS = *pointers*	Pointers to variates of group means for each SSPM

The TERMS option defines the model for whose components the sums of squares and products are to be calculated. In the simplest case the model is just a list of variates, but you can use more complex model formulae, involving variates and factors; this is done in conjunction with the FACTORIAL and FULL options. Details of how formulae are interpreted in regression are given in 8.3.1.

You can form a within-group matrix of sums of squares and products by specifying the relevant factor with the GROUPS option.

Sometimes you may already have calculated values for the matrix of sums of squares and products. You can then assign them to the component structures of the SSPM for example by READ (3.1). You would still, however, need to set the number of degrees of freedom associated with the matrix, and for that you use the DF option.

The parameter lists let you specify identifiers for the four components of an SSPM. You can have declared them previously (and you can have given them values), but if so they must be of the correct type.

Example 2.7.2 shows the declaration and formation (4.10.3) of an SSPM.

Example 2.7.2

```
 1  READ [SETNVALUES=yes] V[1...5]

  Identifier   Minimum      Mean   Maximum    Values    Missing
        V[1]     1.000     2.667     4.000         3          0
        V[2]     0.000     2.000     4.000         3          0
        V[3]     1.000     3.000     7.000         3          0
        V[4]    0.0000    0.6667    1.0000         3          0
```

```
        V[5]      0.000      1.333      3.000          3          0
    5  SSPM [TERMS=V[1...5]] Ssp
    6  FSSPM [PRINT=sspm] Ssp
```

*** Degrees of freedom ***

Sums of squares: 2
Sums of products: 1

*** Sums of squares and products ***

```
V[1]      1        4.6667
V[2]      2       -4.0000       8.0000
V[3]      3      -10.0000      12.0000      24.0000
V[4]      4       -1.3333       0.0000       2.0000      0.6667
V[5]      5        2.3333      -6.0000      -8.0000      0.3333      4.6667

                    1            2            3            4            5
```

*** Means ***

```
V[1]      1        2.667
V[2]      2        2.000
V[3]      3        3.000
V[4]      4        0.6667
V[5]      5        1.333
```

*** Number of units used ***

```
          3
```

2.7.3   The TSM structure

The TSM structure stores a time-series model which you can use in Box-Jenkins modelling of time series (see Chapter 12). The information that you give to specify the model is stored in two variates, called the *orders* and the *parameters*; an optional third variate contains *lags*. A complete description of how these structures are defined and assigned values is given in Chapter 12.

The elements of a TSM are:

[1] or ['ORDERS'];

[2] or ['PARAMETERS'];

[3] or ['LAGS'].

To declare a TSM you use the TSM directive.

TSM   declares one or more TSM data structures.

Option

MODEL = *string*   Type of model (arima, transfer); default arim

Parameters

IDENTIFIER = *identifiers*   Identifiers of the TSMs

ORDERS = *variates*	Orders of the autoregressive, integrated, and moving-average parts of each TSM
PARAMETERS = *variates*	Parameters of each TSM
LAGS = *variates*	Lags, if not default

The TSM directive sets up a compound structure pointing to the variates that will later be used to define the model. You set the type of model by the MODEL option. You can use the parameters of TSM to supply previously declared identifiers as the elements of the TSM, just as with the LRV and SSPM. In this way you can specify a variate of lags, to give the TSM three elements rather than the default of two.

Here are some examples:

```
TSM [MODEL=arima] T1
TSM [MODEL=transfer] T2; ORDERS=!(1,0,1)
TSM T3; ORDERS=O; PARAMETERS=P; LAGS=L
```

2.8 Save structures

Genstat has several special-purpose structures for saving the information from an analysis. These cannot be declared explicitly, but are defined automatically by the directives that perform the analysis. For example, the ASAVE structure (9.6) can be defined by ANOVA and then used by ADISPLAY or AKEEP.

```
ANOVA Gain; SAVE=Gsave
ADISPLAY [PRINT=residuals] SAVE=Gsave
AKEEP [SAVE=Gsave] Source.Amount; MEANS=Meangain
```

In many cases the structure need not be mentioned explicitly. For example, Genstat automatically stores the ASAVE structure from the last y-variate analysed by ANOVA, and ADISPLAY and AKEEP will use this by default if no other ASAVE structure is specified. Save structures are also available from regression and generalized linear models (RSAVE, 8.1.1), REML (VSAVE, 10.3.1), time series (TSAVE, 12.3.2), and to store the environment for high-resolution graphics (DSAVE) They can all be accessed using the GET 13.1.2 and reset using the SET directive (13.1.2).

2.9 Deleting or redefining data structures

Genstat stores the values and attributes of data structures in internal arrays. Usually there will be ample space but, if not, it may be possible to request extra space when you start to run Genstat (see the *Users' Note*).

Another possibility is to check whether there are data structures in your program whose values are no longer required. The DELETE directive allows these values to be deleted so that Genstat can recover the space that they occupy. It may also make the program execute more efficiently as Genstat will then need to keep track of less information.

Additionally, DELETE allows the attributes of data structures to be deleted. This may be worthwhile merely to save further space. However, the main advantage is that the structures can then be redefined to be of different types.

Each time that DELETE is used, Genstat will also remove any unnamed structures that are no longer required and recover any space that has been used for temporary storage. This sort of tidying of workspace will happen automatically if Genstat sees in time that the space is becoming short. However, to avoid unnecessary computation, this does not occur after every statement. Thus, if the space appears to be exhausted, it may be worth using DELETE, even if you have no named structures to delete.

The amount of space recovered by DELETE thus depends both on the size of the data structures deleted and on the number and complication of operations that have been performed with them. Its use should solve most space problems but, if you are still having difficulty, the *Installer's Note* may describe how to expand the workspace of your version of Genstat.

2.9.1   The DELETE directive

DELETE   deletes the attributes and values of structures.

Options

REDEFINE = *string*	Whether or not to delete the attributes of the structures so that the type etc can be redefined (yes, no); default no
LIST = *string*	How to interpret the list of structures (inclusive, exclusive, all); default incl
PROCEDURE = *string*	Whether the list of identifiers is of procedures instead of data structures (yes, no); default no

Parameter

identifiers	Structures whose values (and attributes, if requested) are to be deleted

---

The REDEFINE option controls whether the attributes of the structures are deleted as well as their values. If REDEFINE is set to yes, the only information that is still stored is the identifier and the internal reference number of the structure. The default, REDEFINE=no, deletes only the values of the structures. For example, suppose we have defined a variate Dose by

        VARIATE [VALUES=0,0,2,2,4,4] IDENTIFIER=Dose

This gives Dose the values 0, 0, 2, 2, 4, and 4. If we then put

        DELETE Dose

only the values of Dose are deleted; so we could now assign a new set: for example

        READ Dose
        2 4 0 4 2 0 :

Dose remains a variate but now has the values 2, 4, 0, 4, 2, and 0.

Alternatively, if we set REDEFINE=yes in the above example, we could then redefine Dose as (for example) a text with seven values.

```
DELETE [REDEFINE=yes] Dose
TEXT [VALUES=none,double,standard,double,none,standard,none] Dose
```

Once you have defined the type of a structure in a job (as variate, factor, or whatever), you cannot redeclare it as a structure of any other type unless you have first used DELETE to delete its values and attributes. The only exception to this rule is that the GROUPS directive also has a REDEFINE option, which allows a variate or text to be redefined as a factor.

The LIST option defines how the parameter list is to be interpreted. With the default setting, LIST=inclusive, attributes or values are deleted only for the structures in the list (as well those of any unnecessary unnamed structures). If there is no parameter list, then only unnamed structures are deleted. LIST=exclusive means that the parameter list is the complement of the set of structures that are deleted: that is, all named or unnamed structures that are not in the list are deleted. LIST=all causes the attributes or values of all structures to be deleted. Thus, if LIST=all, any parameter list is ignored; and LIST=exclusive with no parameter is equivalent to LIST=all.

2.10 Accessing or copying details of data structures

It can sometimes be difficult to remember all the details of your data structures. For example, in a long interactive session you might forget the identifiers of certain structures, or some of the attributes that you have given them. The DUMP directive (2.9.1) allows you to display lists of structures, their attributes and their values. It can also display internal information about Genstat but this is useful mainly for those extending Genstat; see Chapter 13.

On other occasions you may need to store and not just display the attributes of structures. This can be done using the GETATTRIBUTE directive (2.10.2). Alternatively, the DUPLICATE directive allows you to define new structures with attributes like those of existing structures. These two directives are particularly useful to writers of procedures (5.3).

2.10.1 The DUMP directive

DUMP prints information about data structures, and internal system information.

Options

PRINT = *strings*	What information to print about structures (attributes, values, identifiers); default attr
CHANNEL = *identifier*	Channel number of file, or identifier of a text to store output; default current output file
INFORMATION = *string*	What information to print for each structure (brief, full, extended); default brie
TYPE = *strings*	Which types of structure to include in addition to those in the parameter list (all, diagonalmatrix, dummy, expression, factor, formula, LRV,

matrix, pointer, scalar, SSPM,
symmetricmatrix, table, text, TSM,
variate); default * i.e. none

COMMON = *strings*   Which internal Fortran commons to display (all,
banks, fdg, ich, iin, iot, jdd, jix, jrt,
lcp, lfn, opr, out, ucs, usy, uws); default *
i.e. none

SYSTEM = *string*   Whether to display Genstat system structures (yes,
no); default no

UNNAMED = *string*   Whether to display unnamed structures (yes, no);
default no

Parameter

   identifiers or *numbers*   Identifier or reference number of a structure whose
information is to be printed

The structures for which the information is to be displayed are specified by the parameter of
DUMP. The PRINT option indicates what is to be presented: you can ask for just the identifiers,
or values and identifiers, or attributes (the identifier is itself an attribute), or for all three. For
example, to get all three for the structures A and B you would put:

    DUMP [PRINT=attributes,values] A,B

Example 2.10.1a

```
  1   VARIATE [VALUES=1...8,*] A
  2   FACTOR [NVALUES=9; LEVELS=!(0,1.2,2.4)] B
  3   DUMP [PRINT=attributes,values] A,B

***** DUMP *****

Identifier      Type  Length   Values Missing  Ref.No.
       A    Variate       9  Present       1     -496
       1.0000      2.0000           3.0000      4.0000       5.0000
       6.0000      7.0000           8.0000           *

       B     Factor       9  Absent        *     -499
No values
```

If the CHANNEL option is set to a scalar, this specifies the output channel to which the
information is sent. Alternatively, if you specify the identifier of a text structure, the lines of
information will be stored in the text instead of being printed; likewise if you specify the
identifier of a structure that has not yet been declared, it will be defined automatically as a text
to store the information. If CHANNEL is not specified, the information is displayed on the
current output channel.

   The INFORMATION option selects which attributes are presented. The default setting brief selects only the most important ones. The setting full causes all the attributes to be presented, and the setting extended also gives details of the structures associated with listed structures.

Example 2.10.1b

```
  4   DUMP [INFORMATION=extended; PRINT=attributes,values] B

***** DUMP *****

       IDENT VECNO  ATTOR    VALOR TYPE  NVAL NVALUE MODE MVPTR OWNER
           B  -499  12224        *    2     9      9    3     *     *
   LEVELS =  -498  NLEV   = 3
No values

              -498  12190    12186    4     3      3    2     0     *
         0.0000         1.2000          2.4000
```

Some of the attributes may be set to unnamed structures. You can obtain further information about any of these by giving its (negative) reference number (as displayed by DUMP when indicating its association with another structure) in the parameter list. This is likely to be useful mainly to advanced users.

The TYPE option lets you display, in addition, lists of all structures of a particular type, or of several types. For example, if you had forgotten the identifier of a factor, you could give the statement

```
        DUMP [TYPE=factor; PRINT=identifiers]
```

Example 2.10.1c

```
  5   FACTOR [NVALUES=9; LEVELS=3; VALUES=3(1...3)] F1
  6   & [LEVELS=2; VALUES=(1,2)4,1] F2
  7   DUMP [TYPE=factor; PRINT=identifiers]

***** DUMP *****

 List of structure names
B           F1          F2
```

This lists all the current factors. When PRINT=attributes or values (or both), the setting TYPE=all provides a list of all named and unnamed structures, except system structures. "PRINT=identifiers; TYPE=all" lists only named structures.

The COMMON option is provided to allow those developing or extending Genstat to display useful internal information. Similarly, the SYSTEM option allows all the system structures to be dumped: there are many of these, so it is not a good idea to set this option frivolously.

2.10.2 The GETATTRIBUTE directive

GETATTRIBUTE accesses attributes of structures.

Option

ATTRIBUTE = *strings*   Which attributes to access (nvalues, nlevels, nrows, ncolumns, type, levels, labels {of factors or pointers}, nmv, present, identifier, refnumber {structure number}, extra, decimals, characters, minimum, maximum, restriction, mode {integer code 1 - 5 denoting type of values: double real, real, integer, character, and word}, maxline {of a text or factor}, rows, columns, classification, margins {of tables}, associatedidentifier {of a table}, unknown {cell of table}, suffixes {of pointers}, owner, terms {of an SSPM}, groups {of an SSPM}, weights {of an SSPM}, SSPMauxiliary, SSPrst, tsmmodel, rstat {of an RSAVE}); default * i.e. none

Parameters

STRUCTURE = *identifiers*   Structures whose attributes are to be accessed

SAVE = *pointers*   Pointer to store copies of the attributes of each structure; these are labelled by the ATTRIBUTE strings

The GETATTRIBUTE directive allows you to access attributes of each of the structures that are listed with its STRUCTURE parameter. It refers to the list of structures by pointers, which are set up by the SAVE parameter. You must always set the option and both parameters. Thus, in Example 2.10.2a, P is defined to be a pointer with an element for each of the two attributes requested by the ATTRIBUTE option. The first is P['nvalues'], alternatively referred to as P[1], storing the value 4; and the second is P['nmv'], or P[2], storing the value 1.

Example 2.10.2a

```
  1   VARIATE [VALUES=1,2,*,4] X
  2   GETATTRIBUTE [ATTRIBUTE=nvalues,nmv] X; P
  3   PRINT P[]

P['nvalues']     P['nmv']
         4             1
```

If you request an attribute that is not relevant to a structure, it is omitted from the pointer. Thus for example the nlevels, levels, and labels settings are relevant only for factors, and nrows and ncolumns only for matrices. The references to those attributes that you do specify are always stored in the order shown in the definition of the ATTRIBUTE option at the beginning of this subsection.

For attributes that are single numbers, the information is copied into an unnamed scalar which is pointed to by the appropriate element of the pointer; if the attribute has not been set, then the corresponding scalar will contains a missing value. With non-scalar attributes the corresponding element of the pointer will store a reference to the attribute itself. One example is the labels vector of a factor (2.3.3). However, if the factor has no labels vector the corresponding entry of the pointer will be set to the missing value; the same will be true for the levels attribute if the factor has not been declared with levels other than the default integers. Thus, Example 2.10.2b sets up P as a pointer with two values, the first being Lev and the second missing.

Example 2.10.2b

```
  4   VARIATE [VALUES=4,8,12] Lev
  5   FACTOR [LEVELS=Lev] F
  6   GETATTRIBUTE [ATTRIBUTE=levels,labels] F; P
  7   DUMP [PRINT=attributes,values] P,Lev

***** DUMP *****

Identifier      Type  Length   Values Missing  Ref.No.
          P   Pointer       2  Present       1    -491
     -498         *

        Lev   Variate       3  Present       0    -498
       4.0000       8.0000      12.0000
```

The setting type produces a scalar value denoting the type of structure, according to the code:

1	scalar	11	expression
2	factor	12	formula
3	text	13	dummy
4	variate	14	pointer
5	matrix	15	LRV
6	diagonal matrix	16	SSPM
7	symmetric matrix	17	TSM
8	table		

2.10.3 The DUPLICATE directive

DUPLICATE forms new data structures with attributes taken from an existing structure.

Options

ATTRIBUTES = *strings*   Which attributes to duplicate (all, nvalues, values, nlevels, levels, labels (of factors or pointers), extra, decimals, characters, rows,

columns, classification, margins,
suffixes, minimum, maximum, restriction);
default all

Parameters

OLDSTRUCTURE = *identifiers*	Data structures to provide attributes for the new structures
NEWSTRUCTURE = *identifiers*	Identifiers of the new structures
VALUES = *identifiers*	Values for each new structure
DECIMALS = *scalars*	Number of decimals for printing numerical structures
CHARACTERS = *scalars*	Number of characters for printing texts or labels of a factor
EXTRA = *texts*	Extra text associated with each identifier
MINIMUM = *scalars*	Minimum value for numerical structures
MAXIMUM = *scalars*	Maximum value for numerical structures

The DUPLICATE directive allows you to define new data structures with attributes like those of existing structures. The attributes to be duplicated are defined by the ATTRIBUTES option. The structures from which the attributes are to be taken are specified by the OLDSTRUCTURES parameter, while the structures that are to be defined are specified by the NEWSTRUCTURES parameter. The other parameters allow some of the more important attributes to be reset at the same time. This is illustrated in Example 2.10.3, where the factor Species2 takes its levels (and thus its number of levels) from the factor Species1. However, the labels are not transferred, and other values are defined using the VALUES parameter.

Example 2.10.3

```
1   FACTOR [LEVELS=!(0,1); LABELS=!T(absent,present); \
2     VALUES=0,1,1,0,0,0,1] Species1
3   DUPLICATE [ATTRIBUTES=levels] Species1; \
4     NEWSTRUCTURE=Species2; VALUES=!(1,0,1,1,0,1,0)
5   PRINT Species1,Species2
```

```
Species1    Species2
  absent           1
 present           0
 present           1
  absent           1
  absent           0
  absent           1
 present           0
```

S.A.H.
R.W.P.

3   Input and output

This chapter describes how to read data values into Genstat and how to print them out. It also looks at some of the more general aspects of input and output, such as the use of files for storing different kinds of information.

As already mentioned (1.1), Genstat statements may be typed in at the keyboard or stored in files and executed as a complete program. Similarly, data can be typed in directly or read, from files that have been prepared in advance. Many directives produce output that can either be displayed on the screen or stored in output files. Usually Genstat programs, data, and output are stored in *character files*. These are easily manipulated by other programs, such as text editors or word processors; so they can easily be viewed or modified on the screen of the PC or terminal, or printed out if hard copy is required. When data are to be used many times by Genstat, it can be more efficient to store them in *binary files*, in which the values are represented by a coded form that is faster to access from disk; however, these may not be suitable for use by other programs or transferable to other implementations of Genstat running on different types of computer.

Section 3.1 describes how to read values into data structures. It also explains how data structures can be defined automatically from the values that are read. Genstat can handle a wide variety of formats. You can also rescale and sort the data values as they are read. Section 3.2 describes how to print the contents of data structures, and the ways in which you can control the format in which they appear. Section 3.3 gives further information about how to access the different types of file, extending the simple methods illustrated in Section 3.1, and Section 3.4 shows how you can take statements from other input files, and send output to other output files. Section 3.5 describes how to store data structures and their contents in structured binary files (*backing store*), and retrieve them later. Finally, Section 3.6 explains how you can preserve all your data structures and the current environment when you exit from Genstat, so that you can restart your activities from that point when you next run Genstat; it also describes an alternative and quicker method of storing just the values in a binary file.

3.1   Reading data

Although you can define values for data structures when you declare them, using the VALUES option or parameter (2.1.1), it is usually more convenient to read the values – especially with large sets of data. Using the READ directive, data can be typed at the keyboard, read from the file containing your Genstat program, or read from a separate data file. The following simple example shows how to read the values for a variate called Weights:

```
VARIATE [NVALUES=10] Weights
READ Weights
24.3 25.6 57.3 43.8 45.3
46.5 47.9 97.0 77.5 64.3 :
```

There are many options and parameters to allow control over most aspects of data input, so data can be read in almost any form. We first describe the more straightforward uses of READ, with most of the options retaining their default settings. Then, in 3.1.1, we give the full details of the syntax of READ showing how you can use the options and parameters to read data that may be arranged in many other ways.

Unless specified otherwise, Genstat assumes that the data values will be found immediately after the READ statement. The values are usually specified in *free format*: that is, they are separated by one or more spaces (or tabs) and can be arranged any way you like, on one or more lines, so long as the correct order is maintained. Genstat reads the data one line at a time, so the first element of Weights is 24.3, the second element is 25.6, and so on. There is no need to use the continuation character \ when data for READ is spread over several lines; in fact \ should occur only when it is part of a string that is being read into a text. To show that the end of the data has been reached a *terminator* is needed, which by default is a colon (:). This may be at the end of the last line of data or on a line of its own. Once the terminator has been read a simple summary of the data is printed and a quick examination can indicate if READ was successful, or if there were any problems such as incorrectly typed values.

Example 3.1a

```
 1   VARIATE [NVALUES=10] Weights
 2   READ Weights

    Identifier    Minimum       Mean    Maximum    Values    Missing
       Weights      24.30      52.95      97.00        10          0
```

If the minimum value of Weights was less than zero, you might assume there was some kind of problem with the data!

When you are working interactively, Genstat produces a prompt indicating the name of the data structure and the unit number of the next value it expects to read:

Example 3.1b

```
> VARIATE [NVALUES=10] Weights
> READ Weights
Weights/1> 24.3 25.6 57.3
Weights/4> 43.8 45.3 46.5
Weights/7> 47.9 97.0 77.5 64.3
***** sufficient data input, READ terminated *****

    Identifier    Minimum       Mean    Maximum    Values    Missing
       Weights      24.30      52.95      97.00        10          0
>
```

READ prompts for the first data value, Weights/1>, and the first three values are typed in. The next prompt is Weights/4>, requesting values for the fourth and subsequent units of Weights. Because Weights was declared to have 10 values, Genstat will know to stop reading data once 10 values have been typed in. Once the 10th value (64.3) has been typed and the <RETURN> key pressed, READ automatically terminates input, without asking for the terminating colon, although it is quite correct to include it at the end of the last line of data. If you type too many values by mistake you will get a warning message telling you that the extra data has been ignored.

When running Genstat in batch, unless you set the END option (3.1.1), you must mark the end of the data with a colon; READ then checks that you have given the correct number of values. If there are too few values a warning is printed and the data structure is completed by using missing values, whereas a fault will be produced if there are too many values.

Genstat will also perform range checks when reading data if you have set the MINIMUM or MAXIMUM parameters when declaring data structures.

Note that, whether you are running Genstat interactively or in batch, READ will immediately take a fresh line of input, so the data cannot be on the same line as the READ statement; also any characters after the terminating colon will be ignored.

Any numerical structure can be read in this way: scalars, variates, matrices, symmetric and diagonal matrices, and tables. The values can be entered in any of the forms described in 1.5.1, that is,

```
1.20   -.2   3e1   -1.25E-2   27   *
```

are all valid, with * indicating a missing value.

The values for rectangular and symmetric matrices and multi-way tables must be given in the order described in 2.4 and 2.5. The rules for free format allow you to arrange them in any way you like, as long as you maintain the correct order, but you will probably find data files easier to manage if the layout corresponds to the dimensions of the data structure: for example

```
SYMMETRIC [NROWS=10] Galaxy
READ Galaxy
0
1.87 0
2.24 0.91 0
4.03 2.05 1.51 0
4.09 1.74 1.59 0.68 0
5.38 3.41 3.15 1.86 1.27 0
7.03 3.85 3.24 2.25 1.89 2.02 0
6.02 4.85 4.11 3.00 2.11 1.71 1.45 0
6.88 5.70 5.12 3.72 3.01 2.97 1.75 1.13 0
4.12 3.77 3.86 3.93 3.27 3.77 3.52 2.79 3.29 0   :
```

Note, however, that the shortcuts for compacting number lists described in 1.6.1 are not allowed within READ. All the values must be given; that is, pre- and post- multipliers and progressions are not recognized. Thus in some cases it will be easier to assign values when declaring your data structures. For example,

```
VARIATE [VALUES=1...365] Day
```

is simpler to type than

```
READ Day
```

followed by 365 individual values.

Textual values (strings) must be enclosed within single quotes if they contain any characters that have special meaning to READ (space, tab, comma, colon, asterisk, backslash, single or double quote). The quotes can be omitted for other strings. For example:

```
TEXT [NVALUES=5] Country
READ Country
Australia   Canada   'Great Britain'   U.S.A.   'New Zealand'  :
```

The rules for strings in READ are thus slightly different to those for lists of strings (1.6.2), where quotes are required for any string that does not start with a letter or contains any character other than letters or digits. Thus Newcastle-on-Tyne and 500Km are both valid when read in as data, but not in a TEXT declaration.

Factors can be read using either their numeric levels or the associated textual labels (but you cannot use both methods for the same factor within a single READ statement). You can also let READ set up the factor levels or labels according to the values that it finds when reading the data (Example 3.1d).

If you want to read the values of a pointer (that is, a list of identifier names) the rules are rather stricter than for other types of data, as explained in 3.1.3.

You cannot read formulae or expressions directly. The easiest way to do this is to read the required value into a text which can then be used in an appropriate declaration using either the macro-substitution symbols ## (1.9.2) or the EXECUTE directive (5.3.5). You cannot read values into the compound data structures described in 2.7 (SSPMs, LRVs and TSMs); these should be formed using the appropriate directives (FSSPM, FLRV, FTSM), or by reading the individual components of these structures.

You can read values for more than one structure in a single READ statement. The values can be taken either *serially* or in *parallel*. The default is to take the values in parallel: the first element of each structure is read, then the second element of each, until all the data are read. For example:

```
a₁ b₁ c₁                                a₁ b₁ c₁ a₂
a₂ b₂ c₂            or                  b₂ c₂
a₃ b₃ c₃                                a₃ b₃ c₃ a₄ b₄ c₄ :
a₄ b₄ c₄ :
```

Here A, B, and C are in parallel, each with four values. The complete set of values for all three structures is given, followed by one terminating colon. The term *parallel* merely indicates the order in which READ is to read the values: that is, the first element of each structure, then the second element of each, and so on. It is not necessary for the data to be laid out in neat columns, although this may make a data file easier to work with.

Different types of structures can be read in parallel and they may have different kinds of values (numerical or text), as shown in Example 3.1c.

Example 3.1c

```
> VARIATE [NVALUES=5] Area
> TEXT [NVALUES=5] Country
> READ Country,Area
Country/1> Australia 2975.0 Bolivia 424.18 Canada
   Area/3> 3851.9 Denmark 16.618 Ethiopia 457.28 :
```

Notice how the name of the first country is followed by its area, then the name of the second country and so on. Working interactively, the prompt helps you to keep track of which values you need to type next. If you want to read data in parallel, all the data structures must be the same length.

When reading in serial mode, all the values of the first structure are read, then all the values for the second structure, until all the data structures have been read. For example

```
X₁ X₂ X₃ :
Y₁ Y₂ :
Z₁ Z₂ Z₃ Z₄ Z₅ Z₆ :
```

Here all the values of X are given first, followed by all the values for Y, and then all the values for Z. Unlike the parallel layout, each set of values must end with the terminating colon, so that READ can tell when to move on to the next structure; this means that the structures can be of different lengths.

In all the examples so far we have defined the type and size of the data structures in advance of the READ statement. However, READ can make some declarations and definitions by default. Any identifier that has not been declared previously will be set up as a variate. Vectors (variates, texts, and factors) of previously unspecified size will be set up to the current units length, if set by UNITS (2.3.4); otherwise READ sets their length to match the number of values read. Also, factors can be generated automatically from the values found, with LEVELS or LABELS set up as appropriate. The exact rules are described below (3.1.2) but a simple illustration, in Example 3.1d, shows how to use READ to set labels (for factor Location) and levels (for factor Year).

Example 3.1d

```
1    FACTOR Location,Year
2    READ [PRINT=data,errors,summary] Location,Year; \
3      FREPRESENTATION=labels,levels

4    England 1979  Australia 1979  Netherlands 1981  France 1983
5    England 1985  Italy 1987  Australia 1988  Scotland 1989
6    Netherlands 1991 'New Zealand' 1992  Canada 1993  England 1993 :

  Identifier   Values   Missing   Levels
    Location      12        0         8
        Year      12        0        10
```

You can also use option SETNVALUES=yes to ensure that any previous setting of length of a vector is reset according to the numbers of values in the new data. This option can be used only when reading variates, texts, or factors; more complex structures such as matrices and tables must be declared in advance.

With small amounts of data it may be convenient to type it in directly, or to include it within your Genstat program. However, when you analyse larger data sets it may be more convenient to read the data from a separate file. The use of different files and input channels is explained in full in 3.3, but to use data files with READ only the simpler features are required. All you need do is open the data file on another input channel and then tell Genstat to read from that channel. Suppose, for example, the data are stored in a file called WEIGHTS.DAT:

```
24.3 25.6 57.3 43.8 45.3
46.5 47.9 97.0 77.5 64.3 :
```

You need to decide which input channel to use (here channel 4), and then set CHANNEL appropriately in OPEN and READ:

```
OPEN 'WEIGHTS.DAT'; CHANNEL=4; FILETYPE=input
READ [CHANNEL=4] Weights
```

The data file is just an ordinary text file, which may have been created within an editor or data-entry system, or perhaps as output from another program. You can still use the other options of READ, to read multiple data structures in serial or parallel format and so on. You may need to edit the file, for example to insert a colon after each set of data, or you can use the facilities described in 3.1.1 for reading data sets that do not meet the default rules. If the data are laid out in a particularly simple way, in neat columns for example, you can use the FILEREAD procedure to set up appropriate data structures and read the data into them.

We now explain the various options and parameters of READ in more detail and introduce some other ways in which data may be read: in *fixed format*, or from *unformatted* (binary) files, or from Genstat text structures. There are options available to make it easier to read very large amounts of data and to skip over unwanted sections of data. Also you can specify your own characters or strings to separate data values, indicate missing values and mark the end of data.

3.1.1   Main features of the **READ** directive

READ   reads data from an input file, an unformatted file, or a text.

Options

PRINT = *strings*	What to print (data, errors, summary); default erro,summ
CHANNEL = *identifier*	Channel number of file, or text structure from which to read data; default current file
SERIAL = *string*	Whether structures are in serial order, i.e. all values of the first structure, then all of the second, and so on (yes, no); default no, i.e. values in parallel
SETNVALUES = *string*	Whether to set number of values of vectors from the number of values read (yes, no); default no causes the number of values to be set only for structures whose lengths are not defined already (e.g. by declaration or by UNITS)
LAYOUT = *string*	How values are presented (separated, fixedfield); default sepa
END = *text*	What string terminates data (* means there is no

	terminator); default ':'
SEQUENTIAL = *scalar*	To store the number of units read (negative if terminator is met); default *
ADD = *string*	Whether to add values to existing values (yes, no); default no (available only in serial read)
MISSING = *text*	What character represents missing values; default '*'
SKIP = *scalar*	Number of characters (LAYOUT=full) or values (LAYOUT=sepa) to be skipped between units (* means skip to next record); default 0 (available only in parallel read)
BLANK = *string*	Interpretation of blank fields with LAYOUT=full (missing, zero, error); default miss
JUSTIFIED = *string*	How values are to be assumed justified with LAYOUT=full (left, right); default righ
ERRORS = *scalar*	How many errors to allow in the data before reporting a fault rather than a warning, a negative setting, $-n$, causes reading of data to stop after the nth error; default 0
FORMAT = *variate*	Allows a format to be specified for situations where the layout varies for different units, option SKIP and parameters FIELDWIDTH and SKIP are then ignored (in the variate: 0 switches to fixed format; 0.1, 0.2, 0.3, or 0.4 to free format with space, comma, colon, or semi-colon respectively as separators; * skips to the beginning of the next line; in fixed format, a positive integer n indicates an item in a field width of n, $-n$ skips n characters; in free format, n indicates n items, $-n$ skips n items); default *
QUIT = *scalar*	Channel number of file to return to after a fatal error; default * i.e. current input file
UNFORMATTED = *string*	Whether file is unformatted (yes, no); default no
REWIND = *string*	Whether to rewind the file before reading (yes, no); default no
SEPARATOR = *text*	Text containing the (single) character to be used in free format; default ' '
SETLEVELS = *string*	Whether to define factor levels or labels (according to the setting of FREPRESENTATION) automatically from those that occur in the data (yes, no); default no causes them to be set only when they are not defined already

Parameters

STRUCTURE = *identifiers*	Structures into which to read the data

FIELDWIDTH = *scalars*	Field width from which to read values of each structure (LAYOUT=fixe only)
DECIMALS = *scalars*	Number of decimal places for numerical data containing no decimal points
SKIP = *scalars*	Number of values (LAYOUT=sepa) or characters (LAYOUT=fixe) to skip before reading a value
FREPRESENTATION = *string*	How factor values are represented (labels, levels, ordinals); default levels

The PRINT option has three settings, data, errors, and summary, which control printed output from the READ directive. The default is PRINT=errors,summary. This produces a printed summary of the data that has been read, and asks for warning messages to be printed about any errors in the data (such as an incorrect number of values); 3.1.11 explains what happens after errors have occurred. The setting data will print a copy of each line of input as it is read; this may be useful if data are being read from a file, especially if there are errors.

For numerical structures the printed summary includes the message Skew if the values have a markedly skew distribution; that is, if the difference between mean and minimum is more than three times, or less than a third of, the difference between maximum and mean. The summaries can be useful as a quick check that the data have been read successfully, and do not contain any gross errors such as a mistyped number with the decimal point in the wrong position. A separate summary is produced for factors which indicates how many levels are defined for each; you can use this to check that READ has defined the factors correctly when the option SETLEVELS=yes has been set. The summary also indicates the number of missing values read into each structure; these may affect the results of subsequent analyses.

If you set PRINT=* no output is produced; however, you should do this only if you are sure there are no errors in the data.

By default, READ will expect to find the data on the current input channel. Working interactively this is the terminal, so a prompt is produced indicating that data is required. When Genstat is being run in batch, the data should start on the line following the READ statement. If you want to read data from another file it should first be opened on another input channel (3.3.1), then the CHANNEL option should be set to that channel number. You can also use CHANNEL to read from a text structure (3.1.7), and by setting UNFORMATTED=yes you can read from an unformatted binary file. In the last case, CHANNEL will refer to a file opened specifically for unformatted access; this is discussed separately in 3.6.

If you specify more than one structure to be read, it is assumed that you want to read the data in parallel. If you want to read in the structures one at a time (for example when they are of different lengths) you should set the option SERIAL=yes.

The default terminator for marking the end of data is the colon (:) but you can use the END option to change this to any string of up to eight characters, for example ENDDATA. If you have defined the size of data structures in advance you can set END=* to indicate that there is no terminator; Genstat then reads the required number of values. You can omit the terminator from the data if it is stored at the end of a file as the read will be terminated by the end-of-file marker; end-of-file will always terminate the data, whatever the setting of END.

By default, a missing value should be indicated by an asterisk (*); this means that any data item that begins with * is treated as missing. For example, any of the three strings

```
*     ***     *789
```

will be treated as missing. You can use the MISSING option to change this to any other single character; for example, if you set MISSING='-' then any negative numbers will be read as missing values.

In free format, values are usually separated by spaces or tabs. The SEPARATOR option can be used to specify another character to use as a separator. For example you can use a comma:

```
READ [SEPARATOR=','] Weights
24.3, 25.6, 57.3, 43.8, 45.3,
46.5, 47.9, 97.0, 77.5, 64.3 :
```

You can use spaces and tabs in addition to the specified separator, so long as the separator is present between each pair of values (except at the end of line, when it may be omitted).

The SEPARATOR, END, and MISSING strings are all case-sensitive; for example, END=enddata is different from END=EndData. The missing-value and separator characters must be distinct and neither may be part of the END string. This is so that READ can make sense of the input data.

A file can contain several sets of data: for example, it might contain 50 measurements on heights of plants, followed by 50 values of weights. You could read the first 50 by one statement, and the next 50 by another. Genstat maintains a pointer to the current position in each input channel, and so returns to the correct place for the second READ (note that if the first READ finished part-way through a line of data the next READ will start at the next line of the data file). Occasionally you may want to go right back to the beginning of the file; you can do this by setting the REWIND option to yes. For example, if you are working interactively and make a mistake in READ so that the data in a file is read incorrectly, it may be easiest to start all over again with a new READ statement rewinding to the beginning of the file.

Although READ is probably easiest to use when the data are in free format, you may sometimes need to read data using a fixed format. This is selected by the option setting LAYOUT=fixed, described in 3.1.5. You can use the options BLANK and JUSTIFIED and parameters FIELDWIDTH and DECIMALS to control reading in fixed format. Alternatively, the FORMAT option caters for more complex examples of free-format or fixed-format data and also allows you to switch between the methods whilst reading; this is discussed in greater detail in 3.1.6.

3.1.2   Implicit declaration of structures

READ can define some of the properties of vectors automatically from the values that are read. More complicated data structures, such as matrices or tables, must be fully defined in advance; for the remainder of this section it is assumed that you are reading vectors.

If the structures to be read have not previously been declared, they will be set up to be variates. If you have already used the UNITS directive (2.3.4) to define a default length then this will apply to any vectors of unknown length in READ. When the structures are being read in parallel (that is, according to the default setting SERIAL=no), they must all be the same length; any vectors of unknown size will be set to the same length as the other vectors being read. If none of the structures has a previously defined length, then READ will act as if SETNVALUES had been set to yes, so that vectors will have their lengths defined from the number of data values found. When reading serially (SERIAL=yes), the structures are treated individually, and any structure of unknown length will be defined from the number of values read in, as if you had set SETNVALUES=yes. You can of course also set SETNVALUES=yes explicitly, to ensure that vector lengths are set from the data, even when they had previously been set to a different size. If you use SETNVALUES when reading structures in parallel with the units vector, slightly different rules apply (3.1.10).

The following examples illustrate some of these rules. X and Y are assumed to be undeclared previously, unless otherwise shown:

```
VARIATE [NVALUES=5] X
READ X,Y
```

declares Y to be a variate of length 5 (like X);

```
VARIATE [NVALUES=5] X
READ [SERIAL=yes] X,Y
```

expects five values for X, and defines Y as a variate with its length defined from the number of values found in the second set of data;

```
READ X,Y
```

defines X and Y as variates of the same length, calculated from the number of values found (which must be a multiple of 2);

```
READ [SERIAL=yes] X,Y
```

defines X and Y from the number of values found, which may be different for each variate.

You can also let READ define the levels and labels of factors automatically. If you just define an identifier to be a factor, and do not mention either levels or labels, READ can set these from the values that are read. The FREPRESENTATION parameter (3.1.3) controls how this is done. If FREPRESENTATION is set to ordinals, the values should all be positive integers, and the number of levels is set equal to the largest number that is read. With the default setting, levels, the values can be any real numbers; the levels of the factor are formed from all the distinct values in the data. Similarly, with FREPRESENTATION=labels, the factor values are supplied as strings, and Genstat forms factor labels from the different strings that are found. (With levels or labels, the method is the same as that used by the GROUPS directive (4.6.1) when neither the NGROUPS option nor the LIMITS parameter are set: you could obtain the same factors by reading a variate or text and then using GROUPS to form the factor yourself.) You can use the option SETLEVELS=yes to force the definition of factors in READ so that any previous labels or levels are overwritten. The lengths of factors can also be set by READ, according to the rules already defined. For example,

```
FACTOR [NVALUES=5] Age
READ [SETLEVELS=yes] Age; FREPRESENTATION=ordinals
21 22 21 24 29 :
```

sets up the factor AGE with 29 levels, that is, as if the FACTOR statement had the option setting LEVELS=29. In contrast, the setting FREPRESENTATION=levels would form Age as a factor with the four levels (21,22,24,29).

If you have defined your data structures in advance, READ implicitly includes a check on the validity of your data: that it has the correct number of values and, when reading factors, that the correct values are given. Although it may be more convenient to let READ set up your data structures, you need to be careful as there is then no longer any check on the input values. It is unwise to suppress the printed summary (3.1.1); this will tell you how many values have been read, how many levels have been set up for factors, and so on.

One point to watch when defining factor labels automatically is that the labels are case-sensitive, so for example Male and male are regarded as different. You would probably want to allow only one form in your data, and treat the other as an error. To do this you could define the desired labels in a FACTOR statement. Alternatively, you could use the NEWLEVELS function (4.2.1) and FACTOR to redefine your factor after letting READ define the labels.

Another danger is that a misspelt value, for example Mael, would also generate another unwanted level for your factor.

There is a library procedure called FILEREAD that provides an easy way of defining and reading data from a file. The file must obey some simple conventions regarding layout, so that FILEREAD can work out how many data structures to read, and set them up as variates, texts, or factors as appropriate.

3.1.3 Reading non-numerical data: texts, factors, and pointers

The rules for the interpretation of strings in READ are different from those when string lists occur in a statement (1.6.2). Double quotes and backslashes are accepted as ordinary characters, and the strings cannot be continued over a line.

In free format, quotes are required around any string that contains a space, quote, tab, colon, asterisk, backslash, double quote, or character specified in the SEPARATOR or MISSING options. If you have set END, the end string would need to be quoted if you also wanted to read it as a data value. In a quoted string, any of the aforementioned characters are treated literally, except for the single quote which must be repeated. A textual missing value can be represented by either a quoted empty string (' '), or the missing value character (*, unless set otherwise by the MISSING option). An asterisk (or any other character representing the missing value) can still be read, provided it is put within quotes: '*'.

```
TEXT Heading
READ Heading
'*** Latent Roots of X''X ***':
```

The value stored in Heading is *** Latent Roots of X'X ***.

The values of factors are usually represented by their levels. You can change this by setting the FREPRESENTATION parameter. If you set it to labels, READ will accept as values the labels of the factor, using the rules for reading text described above. The strings given as data values must match exactly the labels of the factor if they have been declared. The setting FREPRESENTATION=ordinals causes READ to expect an integer in the range 1 up to n, the number of levels declared for the factor. As FREPRESENTATION is a parameter it can be set to a list of values which are cycled in parallel with the structures to be read. Thus, you are allowed to read several factors in one READ statement, possibly using a different method for reading each one. The setting of this parameter is ignored for any structures that are not factors, but remember that the list will still be cycled in parallel with these other structures.

The values of pointers are identifiers, that is, names of other data structures. When reading a pointer only simple identifiers are allowed: suffixes cannot be used. For example, Winston is allowed but Orwell[1984] is not.

The rules for reading text and factor labels are slightly different if you are using fixed format (LAYOUT=fixed). These is explained at the end of 3.1.5.

3.1.4   Skipping unwanted data (in free format)

You may sometimes find that a data file contains more data than you want to read in to Genstat. For example, there may be several lines at the beginning of a file to describe the data set. You can use the SKIP directive (3.3.5) to skip over these lines before using READ to read in the actual data. Alternatively, you can embed the description in double-quotes (") and make it into a comment that READ will ignore. You can also use comments to annotate your data or to remove some values temporarily from the data.

If you want to skip over some of the data systematically, as for example when there are several columns and only some are required for your analysis, there is an option and a parameter that you can use, both of which are called SKIP.

The SKIP option indicates how many values to skip between complete units of data. For example, with a file in channel 2 containing five columns of data, the statement

```
READ [CHANNEL=2; SKIP=3] X,Y
```

would read X and Y from the first two columns, and then skip the final three columns: Genstat reads the first value for X and Y, the next three values are skipped before reading the second value of X; so READ moves onto the next line of the file, and so on. You can also set SKIP=* to skip directly to the next line of data; you could use this if there were varying numbers of additional columns in the file. By default, SKIP is zero, so no values are skipped.

The SKIP parameter is interpreted in parallel with the structures whose values are to be read. It indicates how many values should be skipped before reading the value for the corresponding structure. This is easiest to explain in terms of parallel columns (although the rules for free format do allow other actual layouts of the data).

```
31 91 11 81 21
32 92 12 82 22
33 93 13 83 23
34 94 14 84 24
35 95 15 85 25:
```

To read only the first, third, and fifth columns, we could type

```
READ A,C,E; SKIP=0,1,1
```

The SKIP parameter tells Genstat to skip no values before reading A and one value before reading C and reading E. Thus Genstat reads the values shown in bold. This statement would work in exactly the same way if the data had been laid out differently: for example

31 91 **11** 81 **21 32** 92 **12** 82 **22 33** 93 **13** 83 **23**
34 94 **14** 84 **24 35** 95 **15** 85 **25**:

The SKIP option can be used in conjunction with the parameter when additional values need to be skipped between units of data. In the example above, to skip over the values shown in bold and read the intervening columns instead, the statement

    READ [SKIP=1] B,D; SKIP=1

could be used with either layout of values. With the parallel layout of data, setting option SKIP=* would work equally well, but this would not work with the data in the more compressed layout.

The FORMAT option (3.1.6) also allows you to skip unwanted values or lines of data, but is most useful when the data file contains more complex arrangements of data. If you set FORMAT, the SKIP option and parameter will be ignored. In fixed format data is skipped one character at a time, rather than one value at a time; this is described in the next section.

3.1.5 Reading fixed-format data

In fixed format, data values are arranged in specific *fields* on each line of the file. Each field consists of a fixed number of characters. There is no need for separating spaces; the tab character is not permitted, nor are comments. So, depending on how the fields are defined, the sequence of digits 123456 could be interpreted for example as the single number 123456, or two numbers 123 and 456, or three numbers 123, 4 and 56. Data like this are usually produced by special-purpose programs or equipment; for example, automatic data recorders.

To read data in fixed format you set the LAYOUT option to fixed, and then specify the format to be used. If the values for a structure always occupy the same number of character positions, you can do this with the FIELDWIDTH parameter. For example,

    READ [CHANNEL=2; LAYOUT=fixed] Weight,Height; FIELDWIDTH=3,5

takes data from channel 2 in fixed format. The data are in parallel: that is, reading across lines of the file, values for Weight and Height appear alternately. The FIELDWIDTH parameter is processed in parallel with the structures to be read, so each item of Weight data takes up three characters, and each item of Height data takes up five. If the fieldwidth for a structure is not constant, that is if different layouts are used for different units of the data, then you need to use the FORMAT option, described in the next section (3.1.6).

Suppose there are 80 characters per line in the file; each pair of Weight and Height values takes up 8, and so you have 10 pairs per line. The first line looks like:

    Weight_1Height_1Weight_2Height_2 ... Weight_{10}Height_{10}

Suppose that the first two values for Weight were 1 and 200, and that the first two for Height were 10 and 1200. Then, using ⎵ to represent a space, the first four items on this line would be:

    ⎵⎵1⎵⎵⎵10200⎵1200

Genstat is able to identify the separate values 10 and 200 because it is reading a fixed number of characters for each structure.

Genstat input files have a nominal width, set by default to 80. This can be altered by an OPEN statement (3.3.1) to a different value if necessary. When reading in fixed format, each line of input is taken to be exactly this width; shorter lines are extended with spaces (blanks). It is important to make sure that you account for this when setting the options for READ, otherwise you may read some values from these blank fields (the BLANK option, described below, explains how the blank fields would be interpreted). In the example above, if the values for Height occupied four characters instead of five there would be 11 pairs of values per line of 77 characters. Using the default settings, the final three characters on the first line would be read as the 12th value of Weight, and READ would then be out of step as the 12th value of Height would be read in from the beginning of the next line. The simplest solution is to set the file width to 77 in the OPEN statement, but you can also use the SKIP option and parameter (see below) or the FORMAT option (3.1.6) to avoid this sort of problem.

When you are using fixed format, the data terminator must begin within the first field to be read after the final data value: so you must ensure that you set the field widths and position the terminator appropriately. If you are using either the SKIP option or parameter, you must take care not to skip accidentally over the terminator, as READ will continue to take input - and probably generate many error messages.

Normally Genstat treats a blank field in fixed-format data as a missing value, and the only indication will be in the count of missing values in the printed summary. You can request warning messages for blank fields by setting the option BLANK=error. Alternatively, you can cause blanks to be interpreted as zeroes, by setting BLANK=zero.

Data in fixed format are normally taken to be right-justified: that is, their right-hand ends are flush with the right-hand end of the field; you can have either blanks or leading zeroes (for numbers) in the redundant spaces at the left of the field. You can change this default by setting the JUSTIFIED option. For example the value 123 can appear in a field of width 5 as:

␣␣123	JUSTIFIED=right	there may be leading blanks (the default)
123␣␣	JUSTIFIED=left	there may be trailing blanks
00123	JUSTIFIED=left,right	there must be no blanks
␣123␣	JUSTIFIED=*	there may be leading or trailing blanks

In this way, JUSTIFIED allows you to check the blanks in each field. If a data field contains any blanks that are not allowed by the current setting, an error will be reported. Note that when reading numerical data embedded blanks are never permitted. So a field containing, for example 1␣2␣3, will always produce an error message.

As an example, we can read the values of five scalars using a fixed format with values left-justified in their fields by the following:

```
SCALAR V,W,X,Y,Z
READ [LAYOUT=fixed;JUSTIFIED=left] V,W,X,Y,Z; \
                      FIELDWIDTH=4,5,7,4,5
1.235.62␣678.9␣␣3.7810.31:
```

This reads the values 1.23, 5.62, 678.9, 3.78, and 10.31 into V, W, X, Y, and Z respectively.

The general principles of the SKIP option and parameter are discussed in the context of a free format read in the previous section. When reading in fixed format the same ideas apply, but the SKIP settings now specify numbers of characters to be ignored, instead of numbers of values. Thus, you can obtain exactly the same effect as in the example above by putting

```
READ [LAYOUT=fixed] V,W,X,Y,Z; FIELDWIDTH=4,4,5,4,5; \
                    SKIP=0,0,1,2,0
```

Sometimes fixed format data can be further compressed by omitting the decimal point. The DECIMALS parameter allows you to re-scale data automatically when it is read; details are given in 3.1.9.

When reading textual data in fixed format, the contents of each field are taken exactly as they appear in the input file. There is no need to enclose values in quotes; in fact if you do so, the quotes are treated as part of the data. For example,

```
TEXT [NVALUES=1] T1,T2,T3,T4
READ [LAYOUT=fixed; SKIP=*] T1,T2,T3,T4; FIELDWIDTH=6,3,4,7
'What's␣it␣all␣about?':
```

gives text T1 the value 'What's, text T2 the value ␣it, text T3 the value ␣all, and text T4 the value ␣about?'.

Consequently, the only way to represent a missing string in fixed format is by a blank field, as ' ' or * would both be treated literally and stored as data values.

The rules for reading textual data in fixed format also affect the reading of factors. If you set FREPRESENTATION=labels, the width of the field must equal the number of characters in the label, as for example no␣ is not the same as no. This means that fixed format can be used to read the labels of a factor only if they all contain the same number of characters.

3.1.6 Reading data with variable formats

When you are responsible for producing your own data files you can ensure that they are arranged so that they can be read using simple combinations of the options and parameters of READ. Usually the default settings will be sufficient. However, when you obtain data from other sources this may not be the case. For example, you might find it necessary to read in fixed format as described in 3.1.5. Sometimes even this may not provide sufficient flexibility, so you can set the FORMAT option and use a *variable format*. By this we mean that the layout of the values may vary from unit to unit of the data, and may also vary within each unit. For example, suppose you have some meteorological data which was measured daily and that the file also contains some additional summary values at the end of each week. The first eleven lines are reproduced to illustrate the structure of the file:

```
Monday            5.5     -0.4      0.0      1.9      10.0
Tuesday          -1.1     -2.1      0.0      0.0      34.0
Wednesday         0.6     -8.3      1.3      5.4     142.0
Thursday          6.8     -5.7      1.1      0.0     158.0
Friday           10.6      0.5      8.1      0.0     141.0
Saturday         10.7      6.4      8.3      0.0     152.0
Sunday           10.0      1.9      1.0      0.1     237.0
Summary week 1>  10.7  -8.3   4   19.8   7.4   10.0  124.8  237.0
Monday            9.9      2.5      0.0      4.4     229.0
Tuesday          11.4      2.1      8.5      0.3     237.0
```

Wednesday	11.9	6.3	18.7	0.0	520.0

Suppose the file contains data for 28 days. If you try to read a text and five variates of length 28 then the summaries found after the 7th, 14th, 21st and 28th days would cause an error in READ. You need to read seven lines, skip one, read seven more, and so on. This can be done by setting the option FORMAT=!((6)7,*,*). This means "read six values, do this seven times, skip to the next line, skip again, then return to the beginning of the format and repeat, until enough data has been read". The format is made clear by using (6)7 which corresponds to the physical layout of the data, but 42 could have been specified instead, meaning read the next 42 values.

You can use FORMAT when reading in either free format or fixed format, and can also switch between the two during the READ. When you have set FORMAT, Genstat ignores the SKIP option and the FIELDWIDTH and SKIP parameters, and READ is controlled entirely by the values of the FORMAT. These values are not in parallel with the list of structures: they apply to data values in turn, recycling from the beginning when necessary.

You set FORMAT to a variate, which may be declared in advance or can be an unnamed structure as shown above. Each value of this variate is interpreted as follows (where *n* is a positive integer):

+*n* read *n* values (in free format) or one value from a field of *n* characters (in fixed format);
-*n* skip the next *n* values (in free format) or *n* characters (in fixed format)
*   skip to the beginning of the next line
0.0 switch to fixed format
0.1 switch to free format using space as a separator
0.2 switch to free format using comma as a separator
0.3 switch to free format using colon as a separator
0.4 switch to free format using semicolon as a separator
0.5 switch to free format using the setting of the SEPARATOR option

Using the FORMAT variate READ will start in either free format or fixed format, according to the setting of LAYOUT (by default, LAYOUT=separated; that is, free format). You can switch between these at any time by specifying a value in the range 0-0.5. Remember that if you use free format, spaces and tabs can also be used in addition to the specified separator, and you must use a separator that is distinct from the END and MISSING indicators (see 3.1.1).

3.1.7   Reading from a text structure

You can use READ to read data that has been stored in a text structure, by giving the identifier of the text as the setting of the CHANNEL option. Each string of the text is treated as a line of input, as if it had been read from a file. The length of each string defines the length of line that is read; this may vary from line to line, so you will find that reading in fixed format is rather difficult to specify correctly, and is perhaps better avoided here.
For example:

```
TEXT [VALUES=\
  '35 ''J. Smith'' 24000',\
```

```
       '24 ''G. Brown'' 11500:',\
       '22 33 44 55',\
       '66 55 77 88 :'] Data
TEXT Name
READ [CHANNEL=Data; SETNVALUES=yes] Age,Name,Income
 & X
```

This gives `Age`, `Name`, and `Income` each two values, and `X` eight.

Care is needed if you define the values of the text in the declaration as, in a string list, any sequences of the single-quote, double-quote, or backslash characters will be halved in length when they are assigned to the text structure (1.5.2). In the example above, the first line that is stored in `Data` and then read is actually

```
     35 'J. Smith' 24000
```

Just as when reading from a file, READ keeps a records of its current position when reading from a text, so that a subsequent READ from the same text will continue at the next line. This means that you can read more than one set of data from a text, but you too need to remember the position particularly when writing general programs or procedures. If you need to start again from the beginning you can set REWIND=yes, or you can use the CLOSE directive (3.3.2) to close the text. If the text is redefined, for example by a TEXT, READ, or CONCATENATE statement, an implicit CLOSE is carried out, so that the input buffers are not inconsistent with the new values of the text.

3.1.8   Reading large data sets

You may sometimes have more data to read than can be stored in the space available within Genstat. You may be able to increase the storage space (see the *Users' Note* for your implementation of Genstat), or you can use the SEQUENTIAL option of READ to process the data in smaller batches. This works by reading in some of the data, partially processing it to form an intermediate result, and then overwriting the original data with a new batch that is used to update the intermediate results. This can be repeated until all the data has been read and the final summary is obtained. There are two directives that include facilities specifically designed to work with sequential data input: TABULATE which forms tabular summaries (4.11.1), and FSSPM which forms SSPM data structures for use in linear regression (4.10.3). You can also use other directives, such as CALCULATE, to process data sequentially, but you will have to program the sequential aspects yourself.

You should first declare the structures to be of some convenient size, such that you will not use up all the work space. You then use READ as normal, but with the SEQUENTIAL option set to the identifier of a scalar, which will be used to keep track of how the input is progressing. For example, to read in 10 variates of length 272500:

```
VARIATE [NVALUES=10000] X[1...10]
READ [CHANNEL=2; SEQUENTIAL=N] [1...10]
```

The number of values declared for `X[1...10]` defines the size of batch to read (10000 in this example). So, READ will read the first 10000 units of data (100,000 values), and set N to 10000 to indicate that is the number of units read. This should be followed by the statements to process the first batch of data, then the READ can be repeated. Once again N is set to 10000,

indicating that another 10000 units have been read. This can be continued until READ finds
the data terminator, when it sets the sequential indicator to minus the number of values found
in the last batch. If this is less than the declared size of the data structures they will be filled
out with missing values. In the example given above, after the 28th READ the variates will
each contain 2500 values followed by 7500 missing values, and N will be set to −2500,
indicating that all the data has been read and that the final batch contains only 2500 values.
Usually you will use the SEQUENTIAL facility in conjunction with FSSPM or TABULATE which
are designed to recognize the different settings of the scalar N.

The SEQUENTIAL option is best used within a FOR loop (5.2.1). You should set the NTIMES
option to a value large enough to ensure that sufficient batches of data are read. The loop
should contain the READ statement and any other statements required to process the data. For
example

```
VARIATE [NVALUES=10000] X[1...10]
SSPM [TERMS=X[]] S
FOR [NTIMES=9999]
   READ [PRINT=*;CHANNEL=2;SEQUENTIAL=N] X[]
   FSSPM [SEQUENTIAL=N] S
   EXIT N.LE.0
ENDFOR
```

The EXIT directive is used to jump out of the loop once all the data has been read and
processed; this is safer than trying to program an exact number of iterations for the loop. The
exit condition includes the case when N is equal to zero, as this will arise when the batch size
exactly divides the total number of units. In the above example, if there were 280000 units
of data altogether, the 28th READ would terminate with N set to 10000. This is because READ
is unable to look ahead for the terminator, as there may be other statements in the loop, such
as SKIP, which affect how the file is read. The next READ would immediately find the data
terminator, so would exit with N set to zero. This special case is treated appropriately by
FSSPM and TABULATE, but you should remember to allow for it if you are programming the
sequential processing explicitly.

You can use the SEQUENTIAL option to read data from more than one input channel,
perhaps when a large data set is split into two or more files, but you are not allowed to read
data from the current input channel (that is, the channel containing the READ statement). If you
want to process several structures sequentially from the same file, you must read them in
parallel. You must also be careful not to modify the value of the scalar, N, within the loop
when using sequential data input with FSSPM or TABULATE, as that could interfere with the
sequential processing.

Another means of handling large amounts of data is provided by the ADD option. This
allows you to add values to those already stored in a structure, thus forming cumulative totals
without having to store all the individual data values. You must set SERIAL=yes with
ADD=yes; and it is allowed only for variates. For example:

```
VARIATE [NVALUES=6] A
READ [ADD=yes; SERIAL=yes] 3(A)
5 12 9 * * 9 :
8  1 3 * 2 10 :
```

```
3  4  0  *  11  *  :
```

This starts by assigning the values 5, 12, 9, *, *, and 9 to A. Then A is read again, and its values become 13, 13, 12, *, 2, 19: with ADD=yes (and only then) missing values are interpreted as zeroes when being added to non-missing values. Finally A contains the values 16, 17, 12, *, 13, 19.

When you read large quantities of data it may be worth using the ERRORS and QUIT options, described in 3.1.11, to control error recovery from READ.

3.1.9   Automatic re-scaling of data

You can scale values with the DECIMALS parameter. For example, suppose you put

```
READ [SETNVALUES=yes] A; DECIMALS=3
2523   2.1 376 0.78  :
```

The values of A would then be 2.523, 2.1, 0.376, 0.78. DECIMALS specifies a power of 10 by which any value that does not contain a decimal point is scaled down. Negative powers are not allowed.

3.1.10   Automatic sorting of data (using the **UNITS** structure)

If you have used the UNITS directive (2.3.4) to specify a variate or text containing unit labels, READ will respect the order of these values when reading other structures in parallel with the units structure; in other words the data is re-ordered to match the order of the unit labels. In Example 3.1.10 the unit structure Item is read in parallel with variate Stock. This does not alter the values of Item, but its values are used to indicate which unit of the data is being read, and thus the order in which to store the values of Stock.

Example 3.1.10

```
1   TEXT   [VALUES=Beans,Carrots,Peas,Sardines,Tuna] Cans
2   UNITS Cans
3   READ   [PRINT=data,errors] Cans,Stock

4   Tuna 2  Peas 3  Beans 4  Carrots 0  Sardines 6 :

5   PRINT Cans,Stock; DECIMALS=0

        Cans     Stock
       Beans       4
     Carrots       0
        Peas       3
    Sardines       6
        Tuna       2
```

If the units structure does not already have values, READ will define order of the units as the order in which it finds them in the data. This means that if you are reading several sets of data, each having a column for the unit number (or label), the first use of READ will define the unit order and subsequent READ statements will ensure that this order is maintained consistently in the remaining data.

If a value is specified more than once when defining the units structure, READ will only ever locate the first occurrence of that unit label. If a unit label is repeated in the data then only the final set of values corresponding to that unit will be stored; earlier occurrences are overwritten by subsequent ones. If you try to read a value that is not present in the units structure this is regarded as a fault. Also, if the units structure contains missing values it cannot be used to re-order the data and will instead be overwritten by the new values: a warning message is printed out to tell you if this occurs. If you use the option SETNVALUES=yes when reading structures in parallel with the units vector, the other structures will all be set to the current unit length.

3.1.11 Errors while reading

There are various kinds of error that may arise during execution of a READ statement. There are those that immediately inhibit the read, such as an attempt to read in a structure that is not sufficiently defined. For example, if you declare a matrix M, without specifying its dimensions, READ will not know how many values are required. Other examples include trying to read incompatible structures in parallel (for example variates of different lengths), or specifying a channel that has not been opened. If you make an error of this kind, READ will generate an appropriate diagnostic just like any other directive.

There are some checks that READ will make after it has read all the data. For example, it checks whether you have supplied the correct number of values, generating a fault if there are too many. If there are too few, the structures are completed with missing values and a warning is printed. If you are reading in parallel, this check is extended to ensure that the number of values supplied is a multiple of the number of structures. For example, suppose that values for five structures of length 10 are being read in parallel. If 45 values are found, then the structures will be completed with missing values; but if only 43 values are read in READ assumes that something more serious must be wrong with the data and generates a fault.

The rest of this section looks at errors that can arise while reading the data, and assumes that the READ statement has been specified correctly.

When you are working interactively and typing data at the terminal, READ will halt immediately it finds an invalid value. You should type the correct value and then continue with the rest of the data. If you had typed several items of data then all those before the erroneous value will have been read and stored, but any remaining values will have been discarded, and so will need to be retyped. For example, suppose you misspell a factor label:

Example 3.1.11a

```
> FACTOR [LABELS=!T(Avon,Bedford,Cornwall,Devon)] County
> READ County; FREPRESENTATION=labels
County/1> Avon Avon Cornwall
County/4> Bedford devon Cornwall :
******** Warning (Code IO 11). Unit 5 of County is incorrect.
Input:     devon      Code IO 44: Factor value not found in LABELS

Please input the correct value and subsequent data (the remainder of the last
line will be ignored).
County/5> Devon Cornwall :
```

```
Identifier    Values   Missing    Levels
    County        6         0         4
```

\>

The message indicates which unit is incorrect and also gives an explanation of the error (in this case devon was invalid because it should have started with a capital letter). The prompt indicates where READ is restarting its input; note that the value for the sixth unit has to be given again even though it was correctly specified in the original input.

   When you are reading data in batch, it is not possible to recover from errors in this way. Instead, READ will continue processing the data, substituting missing values for any data that it cannot read, and printing out a message for every error that is found.

Example 3.1.11b

```
1   VARIATE Speed
2   FACTOR [LEVELS=!(30,40,50,70)] Limit
3   READ Speed,Limit

******** Warning (Code IO 11). Statement 1 on Line 3
Command: READ Speed,Limit
Errors in data values

Unit Identifier   Input:
  1      Speed    I          Code SX 39: Invalid character in number
  4      Limit    60         Code IO 43: Factor value not found in LEVELS
  5      Speed    1.0e99     Code IO 3:  Real number too large

    Identifier   Minimum     Mean   Maximum   Values   Missing
        Speed      35.00    47.33     55.00        5         2

    Identifier    Values   Missing    Levels
        Limit         5         1         4


******** Fault (Code IO 8). Statement 1 on Line 3
Command: READ Speed,Limit
Too many errors in data
A fatal fault has occurred - the rest of this job will be ignored
  10   STOP

******** End of job.  Maximum of 3460 data units used at line 47 (48062 left)
```

The first value of Speed was incorrect as a letter I had been typed, rather than the number 1. Subsequent messages illustrate some of the other errors that may occur when reading data. Notice that the data summaries indicate the presence of missing values, which were inserted by READ. Of course, if you get errors when reading data it may be due to incorrectly specified options or parameters in the READ statement, rather than actual errors in the data file. This is especially likely if you are reading in fixed format or using the FORMAT option.

   If errors occur when running in batch, a fault will be generated when READ terminates, thus terminating the job. This is to avoid spurious output being produced from analyses based on incorrect data. You can override this by using the options ERRORS and QUIT.

If you set ERRORS=*n*, where *n* is a positive integer, then up to *n* errors are allowed in the data before READ generates a fault. You might want to do this if you knew certain items of data were going to generate errors, but were prepared to accept them as missing values so that you could analyse the rest of the data. Obviously, you need to be very careful when doing this, as there may be other unexpected errors in the data. Usually you would have to try reading the data once without setting ERRORS, so you could check all the messages, and find what value of *n* is appropriate. Then the READ statement would have to be repeated, setting ERRORS and REWIND (3.1.1) in order to read the data. For example, if missing values of a factor had been typed in as the letter X, you would not want to define X as an extra level of the factor, but if you set MISSING='X' any numerical data that used * for missing value could not be read either.

As already explained, READ produces a message for every data value that contains an error. This can be very useful, as you then have the opportunity to correct all the errors at once, before trying to read the data again. However, the error messages may not be due to errors in the data, but may be caused by an incorrectly specified READ statement. For example, if you are reading many structures in parallel and specify texts and variates in the wrong order in the list of structures to be read, you will get an error message every time Genstat finds a piece of text rather than a number in the position specified for a variate. This is not likely to be a problem, unless you are reading large amounts of data, when you might end up with thousands of lines of needless error messages. A sensible precaution then is to request Genstat to abort the READ if more than a specified number of errors occur. You can do this by setting ERRORS to a negative integer, −*n*. This means that up to *n* errors are allowed in the data, but READ will abort if any more occur, switching control to the channel specified by QUIT (that is, starting or continuing to read Genstat statements from that channel). If you are working in batch a fault will be generated that inhibits execution of further statements, but interactively you have the opportunity to examine the data that have been read in so far, which may help identify any problems in the original READ statement or declarations of your data. For example:

Example 3.1.11c

```
> OPEN 'DATA.DAT';CHANNEL=2; FILETYPE=input
> FACTOR [LABELS=!T(Die,Sand)] Casting
> VARIATE Breakage
> READ [CHANNEL=2;ERRORS=-3] Breakage,Casting

******** Warning (Code IO 11). Statement 1 on Line 4
Command: READ [CHANNEL=2;ERRORS=-3] Breakage,Casting
Errors in data values

Unit Identifier   Input:
   1    Casting    Die        Code SX 39: Invalid character in number
   2    Casting    Die        Code SX 39: Invalid character in number
   3    Casting    Die        Code SX 39: Invalid character in number
   4    Casting    Sand       Code SX 39: Invalid character in number

******** Fault (Code IO 8). Statement 1 on Line 4
Command: READ [CHANNEL=2;ERRORS=-3] Breakage,Casting
Too many errors in data
```

```
3 allowed
> PRINT Breakage,Casting

    Breakage    Casting
      147.2         *
      119.1         *
      127.8         *
       97.3         *

> READ [CHANNEL=2;ERRORS=-3;REWIND=yes;SETNVALUES=yes] Breakage,Casting; \
                                              FREPRESENTATION=labels

   Identifier   Minimum     Mean   Maximum    Values   Missing
    Breakage     61.20     131.1     164.6      1247         0
   Identifier    Values   Missing    Levels
     Casting      1247         0         2
```

The READ terminated after the fourth error in the data. Control returned to channel 1, the terminal (using the default setting of QUIT). Printing out the two structures showed that they had been set up with four values, the number of units that had been completely read before quitting. All the errors had occurred in the factor values: in this case the mistake was easily identified, the FREPRESENTATION parameter had been omitted so that the default levels were expected rather than the labels which were in the data file. The READ statement was then repeated, specifying FREPRESENTATION=labels, and using REWIND to start again from the beginning of the file and SETNVALUES to reset their lengths.

3.2 Printing data

The contents of Genstat data structures can be displayed, with appropriate labelling, using the PRINT directive. Output can be printed in the current output channel, or sent to other channels, or put into a text structure. PRINT has many options and parameters to allow you to control the style and format of the output but, in most cases, these can be left with their default settings. We start by describing these simple uses of PRINT before moving to the more sophisticated features.

3.2.1 Main features of the **PRINT** directive

PRINT prints data in tabular format in an output file, unformatted file, or text.

Options

CHANNEL = *identifier*	Channel number of file, or identifier of a text to store output; default current output file
SERIAL = *string*	Whether structures are to be printed in serial order, i.e. all values of the first structure, then all of the second, and so on (yes, no); default no, i.e. values in parallel
IPRINT = *string*	What identifier (if any) to print for the structure (identifier, extra, associatedidentifier),

	for a table `associatedidentifier` prints the identifier of the variate from which the table was formed (e.g. by TABULATE), IPRINT=* suppresses the identifier altogether; default `iden`
RLPRINT = *strings*	What row labels to print (`labels, integers`), RLPRINT=* suppresses row labels altogether; default `labe`
CLPRINT = *strings*	What column labels to print (`labels, integers`), CLPRINT=* suppresses column labels altogether; default `labe`
RLWIDTH = *scalar*	Field width for row labels; default 13
INDENTATION = *scalar*	Number of spaces to leave before the first character in the line; default 0
WIDTH = *scalar*	Last allowed position for characters in the line; default width of current output file
SQUASH = *string*	Whether to omit blank lines in the layout of values (`yes, no`); default `no`
MISSING = *text*	What to print for missing value; default `'*'`
ORIENTATION = *string*	How to print vectors or pointers (`down, across`); default `down`, i.e. down the page
ACROSS = *scalar* or *factors*	Number of factors or list of factors to be printed across the page when printing tables; default for a table with two or more classifying factors prints the final factor in the classifying set and the notional factor indexing a parallel list of tables across the page, for a one-way table only the notional factor is printed across the page
DOWN = *scalar* or *factors*	Number of factors or list of factors to be printed down the page when printing tables; default is to print all other factors down the page
WAFER = *scalar* or *factors*	Number of factors or list of factors to classify the separate "wafers" (or slices) used to print the tables; default 0
PUNKNOWN = *string*	When to print unknown cells of tables (`present, always, zero, missing, never`); default `pres`
UNFORMATTED = *string*	Whether file is unformatted (`yes, no`); default `no`
REWIND = *string*	Whether to rewind unformatted file before printing (`no,yes`); default `no`
WRAP = *string*	Whether to wrap output that is too long for one line onto subsequent lines, rather than putting it into a subsequent "block" (`yes, no`); default `no`

Parameters

| STRUCTURE = *identifiers* | Structures to be printed |

FIELDWIDTH = *scalars*	Field width in which to print the values of each structure (a negative value −*n* prints numbers in E-format in width *n*); if omitted, a default is determined (for numbers, this is usually 12; for text, the width is one more character than the longest line)
DECIMALS = *scalars*	Number of decimal places for numbers; if omitted, a default is determined which prints the mean absolute value to 4 significant figures
CHARACTERS = *scalars*	Number of characters to print in strings
SKIP = *scalars* or *variates*	Number of spaces to leave before each value of a structure (* means newline before structure)
FREPRESENTATION = *strings*	How to represent factor values (labels, levels, ordinals); default is to use labels if available, otherwise levels
JUSTIFICATION = *strings*	How to position values within the field (right, left); if omitted, right is assumed
MNAME = *strings*	Name to print for table margins (margin, total, nobservd, mean, minimum, maximum, variance, count, median, quantile); default marg

For a quick display of the contents of a list of data structures, you need only give the name of the directive, PRINT, and then list their identifiers. For example

```
PRINT Source,Amount,Gain
```

The output is fully annotated with the identifiers, and with row and column labels or numbers, where appropriate. Factors are represented by their labels if available, and otherwise by their levels. The layout of the values is determined automatically by the size and shape of the structures to be printed, and by the space needed to print individual values. The output is arranged in columns; the structures are split if the page is not wide enough, so that one set of columns is completed before the next is printed. Example 3.2.1a prints the values of two factors, Source and Amount, and a variate Gain.

Example 3.2.1a

```
1  UNITS [NVALUES=12]
2  FACTOR [LABELS=!T(beef,cereal,pork); \
3    VALUES=1,3,2,3,1,2,2,1,3,1,2,3] Source
4  & [LEVELS=!(25,50); LABELS=!T(low,high); \
5    VALUES=50,25,50,50,25,25,50,25,50,50,25,25] Amount
6  VARIATE [VALUES=73,49,98,94,90,107,74,76,79,102,95,82] Gain
7  PRINT Source,Amount,Gain

    Source     Amount       Gain
      beef       high      73.00
      pork        low      49.00
    cereal       high      98.00
      pork       high      94.00
      beef        low      90.00
```

```
        cereal          low         107.00
        cereal          high         74.00
          beef          low          76.00
          pork          high         79.00
          beef          high        102.00
        cereal          low          95.00
          pork          low          82.00
```

As the three vectors all contain the same number of values, the default is to print their values in parallel. Alternatively, you can request that structures are printed in series, one below another, by setting option SERIAL=yes. Of course, if the structures to be printed have different shapes or sizes, their values can be printed only in series. The setting SERIAL=no is then ignored except that, to save space, any vectors or pointers are then printed across the page (that is as though you had set ORIENTATION=across: see Example 3.2.1d).

You can use the RESTRICT directive (4.4.1) to specify that only a subset of the units of a vector should be printed. When printing in series the vectors can be restricted to different subsets; but with parallel printing any restriction is applied to all the vectors (and any pointers) so, if more than one vector is restricted, they must all have been restricted in the same way.

Genstat annotates each set of values by the identifier of the structure (but this can be controlled by option IPRINT described below) and automatically chooses a suitable format. For a numerical structure, the default is to use a field of 12 characters. If the DECIMALS parameter was set when the structure was declared (2.1.2), this will define the number of decimal places in the output; otherwise, the number of decimal places is determined by calculating the number that would be required to print its mean absolute value to at least four significant figures. Texts (and labels of factors) are usually printed in a field of 12 characters but this is extended if any of the strings in the text requires a wider field. You can define your own formats using the parameters FIELDWIDTH, DECIMALS, CHARACTERS, SKIP, and JUSTIFICATION.

Example 3.2.1b

```
    8   PRINT Source,Amount,Gain; FIELDWIDTH=7,7,6; DECIMALS=0

Source Amount   Gain
  beef   high     73
  pork    low     49
cereal   high     98
  pork   high     94
  beef    low     90
cereal    low    107
cereal   high     74
  beef    low     76
  pork   high     79
  beef   high    102
cereal    low     95
  pork    low     82
```

Example 3.2.1b illustrates the use of the parameters FIELDWIDTH and DECIMALS. These both

operate in a straightforward way. The only potential complication is that a negative
FIELDWIDTH can be used obtain to print numbers in scientific format (for example 7.3 E1 for
the first unit of Gain) with DECIMALS significant places. The DECIMALS parameter is ignored
for strings, like the labels of the factors Source and Amount.

In the same way, the CHARACTERS parameter is ignored for numbers; for strings, it allows
you to control the number of characters that are printed. So, we could put CHARACTERS=1 in
Example 3.2.1b to print only the first letter of each factor label. By default, Genstat prints all
the characters in each string of a text or factor label, unless the CHARACTERS parameter was
set to a lesser number when the text or factor was declared (2.3.2).

The SKIP parameter allows you to place extra spaces between the values of each structure.
By default, no extra spaces are inserted unless a value fills the field completely, when a single
space will be inserted; there is also a blank line before the first printed line. SKIP can be set
to either a scalar or a variate in which a positive integer *n* requests that *n* spaces are left and
a missing value can be used to request a blank line. So, for example, we could put
SKIP=0,2,2 to move the columns in Example 3.2.1b two further spaces apart. The zero value
for Source would mean that there were no extra spaces to the left of the block of output.
There would also be no blank line before the output. This can be reinstated by specifying a
scalar (or variate) containing a missing value in the SKIP setting for Source. However, there
is the limitation that these missing values are ignored for the second and subsequent structures
when printing in parallel.

Example 3.2.1c

```
   9   PRINT Source,Amount,Gain; FIELDWIDTH=7,7,6; DECIMALS=0
  10     FREPRESENTATION=labels,levels; \
  11     JUSTIFICATION=left,right,right        &   '(measurements in grammes)'

Source Amount   Gain
beef        50     73
pork        25     49
cereal      50     98
pork        50     94
beef        25     90
cereal      25    107
cereal      50     74
beef        25     76
pork        50     79
beef        50    102
cereal      25     95
pork        25     82


(measurements in grammes)
```

The values can be left-justified by setting the JUSTIFICATION parameter to left, as has
been done for the factor Source in Example 3.2.1c. This example also shows how to use the
FREPRESENTATION parameter to control the printing of the factor values. By default Genstat
will print labels if there are any; if there are none, it prints the levels. In the example, labels
are printed for Source, levels are printed for Amount, and FREPRESENTATION is ignored for

the variate Gain. The other available setting, ordinals, would represent the values by the integers 1 upwards; so for example beef, cereal, and pork, would be represented by the numbers 1, 2, and 3, respectively. Line 11 shows how you can insert a caption into your output, by printing a string.

Example 3.2.1d illustrates the ORIENTATION option, which is relevant only when you are printing vectors or pointers. By setting ORIENTATION=across, the values are printed in alternate lines, across the page. To ensure that these line up correctly, the fieldwidth is taken as the maximum of those specified for the printed structures, while the field used to print their identifiers is given by the RLWIDTH option (by default 13).

Example 3.2.1d

```
12   PRINT [ORIENTATION=across; RLWIDTH=8] Source,Amount,Gain; \
13      FIELDWIDTH=7,7,6; DECIMALS=0


 Source   beef   pork cereal   pork   beef cereal cereal   beef   pork   beef
 Amount   high    low   high   high    low    low   high    low   high   high
   Gain     73     49     98     94     90    107     74     76     79    102

 Source cereal   pork
 Amount    low    low
   Gain     95     82
```

Notice that Genstat now has to print the output in more than one block. This will happen whenever there is too much output to fit across the page, unless option WRAP is set to yes. Then Genstat simply wraps each line onto subsequent lines. This is likely to be useful mainly if you are printing the contents of the structures to be read by another program. You might then also wish to suppress the identifiers by setting option IPRINT=* and remove blank lines by setting option SQUASH=yes.

By default, IPRINT=identifier will label the output with the identifier of the structure. Putting IPRINT=identifier,extra will also include any text that has been associated with the structure by the EXTRA parameter when it was declared, while the setting associatedidentifier can be used when a table has been produced by the TABULATE (4.11.1) and AKEEP (9.6.1) directives, to request that the output be labelled with the identifier of the variate from which the table was formed.

The width of each line can be controlled by the WIDTH option; the default is to take the full available width. The INDENTATION option specifies the number of spaces to leave before each line; by default there are none.

There are two other options that apply to any type of structure. The CHANNEL option determines where the output appears. By default, the output is placed in the current output channel, but CHANNEL can be set to a scalar to send it to another output channel; the correspondence between channels and files on the computer is described in 3.3. Alternatively, you can set CHANNEL to the identifier of a text to store the output. The text need not be declared in advance; any undeclared structure that is specified by CHANNEL will be defined automatically as a text. Each line of output becomes one value of the text and if the text already has values they will be replaced. You are most likely to want to do this in order to manipulate the text further. Remember, however, that if you print the text later on, its strings will be right-justified by default, so you will need to set JUSTIFICATION=left in the later PRINT statement to achieve the normal appearance of your output. The maximum (and default) line length of this text is the length of what is called the *output buffer*. This is likely to be 200 on most computers. If you intend to print it to an output file, you should set the WIDTH option as appropriate.

The MISSING option allows you to specify a string to be used instead of the default asterisk symbol to represent missing values. For example, you could set MISSING='unknown' or MISSING=' '.

PRINT can similarly be used for the straightforward printing of tables and the various types of matrix, as well as formulae and expressions. The options and parameters that control the layout of multi-way structures are described in 3.2.2, while 3.6.3 explains the UNFORMATTED and REWIND options which are used to send output to unformatted files.

3.2.2 Printing of multi-way structures

PRINT can easily be used to print matrices and tables, by taking the default layout and labelling. Examples of a two-way table and of a three-way table are shown in 2.5. For tables with more than one dimension, the usual layout has one factor across the page and the others down the page (see Example 2.5a); tables with only one dimension are printed down the page. Several tables can be printed in parallel, provided they all have the same classifying factors. As shown in Example 3.2.2a, the tables are then printed in alternate columns, as though they formed a larger table with an extra factor (called the table-factor) representing the list of tables. This extra factor thus becomes another (in fact, the final) factor to be printed across the page.

Example 3.2.2a

```
1   FACTOR [LEVELS=2] Lab
2   & [LEVELS=3; LABELS=!T(beef,cereal,pork)] Source
3   & [LEVELS=!(25,50); LABELS=!T(low,high)] Amount
4   TABLE [CLASSIFICATION=Lab,Source,Amount; \
5       VALUES=162.4,171.2,173.6,160.8,148.4,157.6, \
6             154.4,168.8,149.8,159.0,170.4,160.4] Startwt
7   & [VALUES=243.6,286.8,260.4,286.2,222.6,281.4, \
8             231.6,313.2,217.2,255.0,249.6,315.6] Finalwt
9   PRINT Startwt,Finalwt
```

	Amount	low Startwt	Finalwt	high Startwt	Finalwt
Lab	Source				
1	beef	162.4	243.6	171.2	286.8
	cereal	173.6	260.4	160.8	286.2
	pork	148.4	222.6	157.6	281.4
2	beef	154.4	231.6	168.8	313.2
	cereal	149.8	217.2	159.0	255.0
	pork	170.4	249.6	160.4	315.6

This default layout can be changed using the ACROSS, DOWN, and WAFER options. You may wish to do this simply by changing the factors which appear down and across the page. The ACROSS option can be set to a scalar to specify how many factors should be printed across the page, or to a list of factors to say which ones they should be. DOWN similarly specifies the factors to be printed down the page. However, you cannot specify a list of factors for one of these options and a scalar for any of the others. The table-factor can be represented in these lists by inserting a * in the required position; if you do not mention the table-factor in either list it remains as the last factor in the ACROSS list. In Example 3.2.2b the table-factor, Lab, and Amount are printed across the page (in that order), and Source is printed down the page.

Example 3.2.2b

```
  10   PRINT [ACROSS=*,Lab,Amount; DOWN=Source] Startwt,Finalwt;F=

                  Startwt                                Finalwt
           Lab       1                 2                 1                 2
        Amount    low      high     low      high     low      high     low      high
        Source
          beef   162.4    171.2    154.4    168.8    243.6    286.8    231.6    313.2
        cereal   173.6    160.8    149.8    159.0    260.4    286.2    217.2    255.0
          pork   148.4    157.6    170.4    160.4    222.6    281.4    249.6    315.6
```

The WAFER option allows you to split the output up into subtables or "wafers". This is particularly useful if the tables have many classifying factors, or if the factors have very long labels. The setting can again be either a scalar or a list of factors (possibly including the table-factor). As shown in Example 3.2.2c, each subtable has a heading its position in the full table. If the table-factor is included in the wafer, the identifier of the appropriate table will be printed at the beginning of the label for that wafer; this does not mean that the table-factor itself has been moved, simply that the labelling has been rearranged to make it easier to read.

Example 3.2.2c

```
  11   PRINT [ACROSS=Amount; DOWN=Lab; WAFER=Source] Startwt,Final

Source beef.
        Amount           low                      high
                   Startwt    Finalwt       Startwt    Finalwt
          Lab
            1        162.4      243.6         171.2      286.8
            2        154.4      231.6         168.8      313.2

Source cereal.
        Amount           low                      high
                   Startwt    Finalwt       Startwt    Finalwt
          Lab
            1        173.6      260.4         160.8      286.2
            2        149.8      217.2         159.0      255.0
```

```
Source pork.
        Amount          low                  high
                    Startwt    Finalwt    Startwt    Finalwt
           Lab
             1      148.4      222.6      157.6      281.4
             2      170.4      249.6      160.4      315.6
```

You need not specify all the options DOWN, ACROSS and WAFER. If you leave any of them out PRINT will deduce the missing information.

You can control the space allowed for labels of the DOWN factors by using the RLWIDTH option. By default this is set to 13, but you might want something else if the labels are very small. If the width provided (by you, or implicitly) is inadequate, PRINT automatically resets it to accommodate the longest row label. You can suppress the labelling by the down factors by setting option RLPRINT=*, and the labelling of the across factors by setting CLPRINT=*.

When tables are produced by TABULATE (4.11.1) Genstat sets an internal indicator for use by PRINT to indicate the appropriate label for any margins. When a single table is printed this name will be used by default. When printing tables in parallel, if they all have the same setting of the margin name indicator, the appropriate name is used. If they have different settings, or none at all (tables from sources other than TABULATE) the margins will be labelled Margin by default. You can change the label by setting the MNAME parameter. Tables printed in parallel must have the same label throughout, and Genstat will take the one specified for the first table in the list. But in serial printing, you can use a different margin name for each table.

The TABULATE (4.11.1) and AKEEP (9.6.1) directives also record the identifier of the variate from which the table was formed, and you can request that this be used to label the output, instead of the identifier of the table itself, by setting the IPRINT option to associatedidentifier.

The PUNKNOWN option controls the printing of the "unknown" cell of a table (see 3.5). The default action is to print this cell, labelled with the table identifier, but only if it contains a value other than missing value or zero. You can select one of five settings:

present (default)	print value if not missing or zero
always	print the unknown cell regardless of value
zero	print unless the value is zero
missing	print unless the value is missing
never	do not print the unknown cell whatever its value

Options ACROSS, DOWN, WAFER, RLPRINT, and CLPRINT also apply to matrices. By default, though, if you have several matrices they will be printed one after another on the page.

With symmetric matrices the only options of these that are relevant are RLPRINT and CLPRINT; a further setting integer is available for these to request that the rows or columns be labelled by the integers 1 onwards, as well as, or instead of the labels provided with the symmetric matrix: for example setting RLPRINT=integers and CLPRINT=integers, labels would identify the rows by integers and the columns with integers and labels.

3.3   Accessing external files

Genstat makes use of various types of file. These are classified according to the information that they store. Some files are in the standard text format recognized by many other programs such as editors, which you can use to prepare your Genstat program and data files. Other files (binary files) are produced by Genstat in formats specific to Genstat. Graphics output files use standard formats and may be written as ordinary text or as unformatted binary files; usually a number of formats are provided by Genstat, these are described further in 6.5.1.

Genstat accesses the files via *channels*. For each type there is a set of numbered channels that can be used to reference different files in the relevant directives. For example, there are five input channels, numbered 1 up to 5. Likewise, there are five output channels. Genstat distinguishes between the different types of channel, so you can have one file attached to output channel 3 and a different file simultaneously attached to backing store channel 3. Then, setting the option CHANNEL=3 in PRINT and STORE statements will send the different kinds of output to the appropriate files. The table below gives details of the channel numbers that are generally available in most versions of Genstat. It is possible that a particular version may allow additional channels to be used for some types of file; your local documentation will give information about any differences for your version, or you can obtain details from the Genstat on-line help by typing

```
HELP environment,channel
```

Graphics channels use a slightly different numbering system, explained in Chapter 5.

Type of file	Channels	Purpose
input	1...5	text files containing Genstat instructions
output	1...5	text files to contain Genstat output
backingstore	0...5	structured binary files for storage of data
procedure library	1...3	backing store files containing procedures to be accessed automatically (5.3.3)
graphics	see 6.2	text or binary files for storing graphical output that can subsequently be printed on a plotter or laser printer
unformatted	0...5	binary files for rapid storage and retrieval of data

When you run Genstat it starts taking input from input channel 1 and produces output on output channel 1. In an interactive run, these will be keyboard and screen, while in a batch run they will be files on the computer (1.1). Another file that is attached automatically is the start-up file of instructions that are executed at the outset of each job (13.1.4); this is attached to input channel 5. The start-up file may attach other files, for example to hold a transcript of input or output; your local documentation will contain details. Transcript files can also be set up, automatically, by the menu system (1.1.3).

The command that you use to run Genstat may allow you to arrange for other files to be attached when Genstat starts running. Alternatively, within Genstat, you can use the OPEN directive. OPEN also lets you define additional characteristics of the file, such as the maximum length of each line. When you have finished using a file you can tell Genstat to CLOSE it (but note that all files are automatically closed by the STOP directive).

Transcripts of input or output can be saved in output files using the COPY directive (3.3.3). The SKIP directive can be used to skip over part of an input file or to print extra blank lines in an output file (3.3.5), while PAGE allows you to advance to a new page before starting the next section of output (3.3.4). The ENQUIRE directive can be used to find out about the files that are currently connected to any of Genstat's channels; this is likely to be useful particularly within general programs and procedures.

One special type of file is the Genstat text structure, which may be thought of as an "internal" file. This can be used for input or output by certain directives. Those directives that can create texts to contain their output (for example TEXT or PRINT) leave them in a "closed" state. There is no need to open a text explicitly if you want to use it for input (for example in READ), but you may need to CLOSE it afterwards; see 3.1.7 for further details. The contents of the texts are lost at the end of the job unless you save them, for example by using backing store (3.5). The use of texts as internal files is a more advanced facility that is likely to be required only for more complicated programs and procedures.

3.3.1   The OPEN directive

OPEN   opens files.

No options

Parameters

NAME = *texts*	External names of the files
CHANNEL = *scalars*	Channel number to be used to refer to each file in other statements (numbers for each type of file are independent); if this is set to a scalar containing a missing value, the first available channel of the specified type is opened and the scalar is set to the channel number
FILETYPE = *strings*	Type of each file (input, output, unformatted, backingstore, procedurelibrary, graphics); default inpu
WIDTH = *scalars*	Maximum width of a record in each file; if omitted, 80

	is assumed for input files, the full line-printer width (usually 132) for output files
INDENTATION = *scalar*	Number of spaces to leave at the start of each line; default 0
PAGE = *scalars*	Number of lines per page (relevant only for output files)
ACCESS = *string*	Allowed type of access (readonly, writeonly, both); default both

The OPEN directive enables you to connect files to the various available channels within Genstat. Usually you need specify only the name of the file, the channel number and type of file, and leave the other parameters to take their default settings. For example, the following statements attach a file called WEATHER.DAT to the second input channel, and then read data from it, as explained in 3.1.1.

```
OPEN 'WEATHER.DAT'; CHANNEL=2; FILETYPE=input
READ [CHANNEL=2] Rain,Temperature,Sunshine
```

The filename can be anything that is acceptable to your computer system. You should, however, check for any constraints: for example, plotting software may require HPGL graphics files to have the extension .HPGL. You should check in your local documentation for information regarding any features that are specific to your computer or version of Genstat. For example, logical or symbolic names may be automatically translated by Genstat before files are accessed; upper and lower case characters may be significant, as on Unix systems. The filename may involve characters that have special meaning within Genstat. For example, the character \ may be required to specify directories and sub-directories on a PC. As explained in 1.5.2, this character needs to be duplicated in a string to avoid Genstat interpreting it as the continuation symbol. For example

```
OPEN 'C:\\RES\\WEATHER.DAT'; CHANNEL=2; FILETYPE=input
```

to open the file 'C:\RES\WEATHER.DAT'. As a more convenient alternative, the PC version of Genstat allows you to use / instead. Again, this should all be explained in the local documentation.

You are free to choose which channels you want to use (within the range available for the specified type of file), apart from input and output channel 1 which are "reserved" for use by the files specified on the command line. Also, input channel 5 is used for the start-up file (13.1.4) and, if you are working interactively, the standard start-up file arranges for output channel 5 to store a transcript of your output. However, you can use the CLOSE directive (3.3.2) to disconnect these files if you want to use the channels for some other purpose. The backing-store and unformatted work files are attached to channel 0, and this channel cannot be used in OPEN or CLOSE. Graphics files must be opened on the channel corresponding to the device number (see 6.5.1).

Obviously you cannot open more than one file on a channel, so if you wish to open a file on a channel that is currently in use you must first close that channel (3.3.2). Sometimes, in general programs or procedures, you may not know which channels are available. You can then let OPEN find a free channel: if CHANNEL is set to a scalar containing a missing value, the file is opened on the next available channel of the appropriate type, and the scalar is set to the number of the channel. The scalar need not be declared in advance; if CHANNEL is set to an undeclared structure, this will be defined as a scalar automatically.

```
SCALAR FreeChan
OPEN 'WEATHER.DAT'; CHANNEL=FreeChan; FILETYPE=input
READ [CHANNEL=FreeChan] Rain,Temperature,Sunshine
```

Another constraint is that you cannot open the same file on more than one channel at once.

Input files must already exist when they are opened, whereas output files will be created by Genstat. If an output file with the specified name exists already, Genstat may create an extra "version" of the file, or report a fault, or cause the file to be *overwritten*, depending on the usual conventions on your type of computer. Your local documentation will describe what rules apply in this situation, and should also explain if there are any system variables you can set to control this action.

When you open a file for use by backing store (3.5) or unformatted input and output (3.6), you can both read from it and send output to it, unless you set the ACCESS parameter (see below). Procedure libraries are a special type of backing-store file, described in 5.3.4.

The WIDTH parameter sets the maximum number of characters per line for input and output files. It is ignored for other types of file. The default values for WIDTH are designed to be appropriate for each implementation of Genstat and may differ between input and output; details will be found in your local documentation. For input and output with screen displays that use windows WIDTH may be set automatically from the size of the appropriate window.

For input files the default is normally 80, reflecting the size of most screen displays. You can change this if necessary, to read either fewer characters from each line, or longer lines. If the WIDTH is set to be too small any extra characters will be lost, which may cause unexpected action or syntax errors. Remember that if you use READ with LAYOUT=fixed to read fixed-format data, short lines are extended with spaces up to the WIDTH setting. If you want to read data from a file with, say, 64 characters per line, setting WIDTH=64 when you open the file may make the format specification easier (rather than taking the default width of 80 and having to remember to skip 16 characters at the end of each line).

For output files, the default is the largest number of characters that can usually be displayed in a single line. This number is typically 80 for terminals but for files it is likely to be either 80, 120 or 132, depending on the type of computer. You can use the WIDTH parameter to restrict the number of output characters to a smaller number, or to a larger number up to 200. The statement

```
HELP environment,channel
```

can be used to obtain information about current width settings for input and output channels.

The PAGE parameter specifies the size of page in an output file. This can be used to ensure that graphs and statistical analyses each start on a new page; see the OUTPRINT option of JOB (5.1.1) and SET (13.1.1). It also affects output from GRAPH, QUESTION, and HELP. For output to files, the default value of PAGE is designed to be suitable for printers. For windowed displays Genstat will, if possible, detect the size of the window and set the page size appropriately.

The INDENTATION parameter can be used to leave a specified number of blank characters to the left of each line of an output file, so that printed output can be bound for example. The indentation is subtracted from the WIDTH setting, so if you set WIDTH=80 and INDENTATION=10 then only 70 characters will be printed on each line of output.

The ACCESS parameter is used to control the way in which unformatted and backing-store files can be accessed, on computers that allow this; for details see your local documentation.

3.3.2   The CLOSE directive

CLOSE   closes files.

No options

Parameters

CHANNEL = *scalars* or *texts*	Numbers of the channels to which the files are attached, or identifiers of texts used for input (which, after "closing", can then be re-read)
FILETYPE = *strings*	Type of each file (input, output, unformatted, backingstore, procedurelibrary, graphics); default inpu
DELETE = *strings*	Whether to delete the file on closure (yes, no); default no

---

When you have finished with a file you can use CLOSE to release the channel to which it was attached, so that the channel is available for use with some other file. However, you do not need to close every file before you stop running Genstat; files are automatically closed at the end of every Genstat programme.

Parameters CHANNEL and FILETYPE are similar to those of the OPEN directive. The DELETE parameter is useful if you are using files to store data temporarily, perhaps to release workspace within Genstat. When you have finished with the file you can set DELETE=yes to request that it be deleted on closure so that disk space is not wasted. For example,

```
OPEN 'temp.bin'; CHANNEL=3; FILETYPE=unformatted
PRINT [CHANNEL=3;UNFORMATTED=yes] Surveys[1900,1910...1990]
DELETE Surveys[1900,1910...1990]

" ... and later on when you wish to retrieve the data ... "
READ [CHANNEL=3;UNFORMATTED=yes] Surveys[1900,1910...1990]
CLOSE 3; FILETYPE=unformatted; DELETE=yes
```

You cannot close a channel to which the terminal is attached, nor the current input or output channels. Also you cannot use CLOSE to delete files that have been opened with ACCESS=readonly or that are protected by the computer's file system.

3.3.3   Saving a transcript of input or output (COPY)

COPY   forms a transcript of a job.

Option

PRINT = *strings*	What to transcribe (statements, output); default stat

Parameter

scalar	Channel number of output file

The COPY directive can be used to save a copy of either input statements, or output, or both, in an output file. For example

```
OPEN 'GEN.REC','GEN.OUT'; CHANNEL=2,3; FILETYPE=output
COPY [PRINT=statements] 2
COPY [PRINT=output] 3
```

will keep a record of all the statements in the file GEN.REC and of all the output in the file GEN.OUT. A later statement

```
COPY [PRINT=statements,output] 2
```

will stop output from being directed to GEN.OUT (because information can be copied to only one file at a time), and send it instead to GEN.REC together with the statements. Setting PRINT=* stops any copying to the specified channel. For example

```
COPY [PRINT=*] 2
```

3.3.4   The PAGE directive

PAGE   moves to the top of the next page of an output file.

Option

CHANNEL = *scalar*	Channel number of file; default * i.e. current output file

No parameters

When output is to a file, graphs and output from statistical analyses will automatically start on a new page, unless you have requested otherwise using the OUTPRINT option of JOB (5.1.1) or SET (13.1.1). With other directives, such as PRINT or TABULATE, you can request a new page using the PAGE directive. By default, PAGE works on the current output channel, but you can use the CHANNEL option if you are sending output to another file.

PAGE has no effect unless output is to a file, and it achieves its effect by printing a line consisting of just the control code for a form feed (ASCII character 12). The effect of PAGE is therefore independent of the page size set by the OPEN directive (3.3.1).

3.3.5   The SKIP directive

SKIP   skips lines in input or output files.

Options

CHANNEL = *scalar*   Channel number of file; default current channel of the
                                  specified type

FILETYPE = *string*   Type of the file concerned (input, output); default
                                  inpu

Parameter

   identifiers   How many lines to skip; for input files, a text means
                                  skip until the contents of the text have been found,
                                  further input is then taken from the following line

---

This directive can be used for both input and output files. The FILETYPE and CHANNEL options indicate which file is to be skipped. By default this is the current input channel.

For input files you can skip over unwanted lines, which might be comments describing the data that is to follow, or might be some statements that you do not want to use in your current job. You can skip a specified number of lines, *n* say, by setting the parameter to a scalar containing the value *n*. Alternatively, you can skip everything up to and including a particular string of characters by setting the parameter to a text containing that string. For example,

       SKIP [CHANNEL=2] 'Section 2'

will skip the contents of the input file on channel 2 from the current position until the string Section 2 is found. The next line to be read from channel 2 will then be the one immediately after the line containing Section 2.

For output files you can use SKIP to print blank lines to separate one section of output from another. You might want to do this if you had set the PRINT option SQUASH=yes (3.2.1) to suppress the automatic blank lines within a section of output. For example,

       PRINT [CHANNEL=2; IPRINT=*; SQUASH=yes] Heading
       SKIP [CHANNEL=2; FILETYPE=output] 2
       PRINT [CHANNEL=2; IPRINT=*; SQUASH=yes] Table

places two blank lines between Heading and Table when printing their values to channel 2.

3.3.6   The ENQUIRE directive

ENQUIRE   provides details about files opened by Genstat.

No options

Parameters

CHANNEL = *scalars*   Channel numbers to enquire about; for

	FILETYPE=input or output, a scalar containing a missing value will be set to the number of the current channel of that type
FILETYPE = *strings*	Type of each file (input, output, unformatted, backingstore, procedurelibrary, graphics); default inpu
OPEN = *scalars*	To indicate whether or not the corresponding channels are currently open (0=closed, 1=open)
NAME = *texts*	External name of the file, if channel is open
EXIST = *scalars*	To indicate whether files on corresponding channels currently exist (0=not yet created, 1=exist)
WIDTH = *scalars*	Maximum width of records in each file (relevant only for input and output files, set to * for other types)
PAGE = *scalars*	Number of lines per page (relevant only for output files)
ACCESS = *texts*	Allowed type of access: set to 'readonly', 'writeonly' or 'both'
LINE = *scalars*	Number of the current line (input files only)

ENQUIRE allows you to ascertain whether a particular channel is already in use and, if so, what properties are defined for aspects like the width of each line or the number of lines per page. This is likely to be of most use within general programs and procedures.

You specify the channel using the parameters CHANNEL and FILETYPE, in the usual way (3.3.1); the other parameters allow you to save the required information in data structures of the appropriate type. This is illustrated in Example 3.3.6.

Example 3.3.6

```
1   OPEN 'WEATHER.DAT','SUMMARY.OUT'; CHANNEL=2; FILETYPE=input,output
2   ENQUIRE 2,2; FILETYPE=input,output; \
3     NAME=In2,Out2; WIDTH=InW2,OutW2; ACCESS=InAcc2,OutAcc2
4   PRINT In2,InW2,InAcc2,Out2,OutW2,OutAcc2; \
5     FIELDWIDTH=20,7,10; DECIMALS=0; JUSTIFICATION=left
```

In2	InW2	InAcc2	Out2	OutW2	OutAcc2
C:\DAT\WEATHER.DAT	80	readonly	C:\DAT\SUMMARY.OUT	80	writeonly

---

3.4   Transferring input and output control

Genstat always starts by processing statements from the keyboard or from the file attached to input channel 1. However, you can use the INPUT directive to change this within a job, to take statements from another file. Subsequently, you can use INPUT to switch to yet another channel, or RETURN to go back to the original file of statements. You can also use the EXECUTE directive (5.3.5) or macro substitution (1.9.2) to take input from a text structure containing Genstat statements.

Similarly, any output from Genstat will be directed initially to output channel 1. When you start Genstat this will be connected either to the screen, in an interactive run, or to a file. You can change the current output channel at any time by using the OUTPUT directive. Once you have given an OUTPUT statement all output will appear in the file on this channel, until another OUTPUT statement is executed. Many directives, like PRINT, have a CHANNEL option that lets you specify where the output from the directive is to go. This provides an alternative method of selectively diverting some output to a secondary file.

3.4.1   Taking input statements from other files (**INPUT**)

INPUT specifies the input file from which to take further statements.

Options

PRINT = *strings*	What output to generate from the statements in the file (statements, macros, procedures, unchanged); default stat
REWIND = *string*	Whether to rewind the file (yes, no); default no

Parameter

scalar	Channel number of input file

Having opened a file of Genstat statements on another input channel you can switch control to that channel at any time using an INPUT statement. You specify the channel as a number or as a scalar containing that number. For example,

```
OPEN 'MYPROCS.GEN'; CHANNEL=4; FILETYPE=input
INPUT 4
```

The file can contain any valid Genstat statements: they will be executed just as if they had been on the original input channel. In this file you could use an INPUT statement to switch back to channel 1 after a while. Alternatively, you may have set up several input files and jump from one to another, again using INPUT. You can use RETURN to go back to the previous channel or STOP to end this run of Genstat. If the end of the file is reached without finding any of these statements, control will be passed back to the previous input channel as described below in 3.4.2. Note that if you use INPUT to go back to an earlier channels you may affect the way in which RETURN works; details are given in 3.4.2.

The PRINT option can be used to specify whether the statements read from the file should be echoed to the current output channel. This is used in the same way as INPRINT in JOB (5.1.1) and SET (13.1.1).

The REWIND option allows you to return to the beginning of the file. You might need to do this, for example, if you had made an error, so that the statements on the secondary input file were executed wrongly. After correcting your error you could set REWIND=yes to start again from the beginning of the file.

3.4.2 The **RETURN** directive

RETURN returns to a previous input stream (text vector or input channel).

Option
NTIMES = *scalar* — Number of streams to ascend; default 1
Parameter
   expression — Logical expression controlling whether or not to return to the previous input stream; default 1 (i.e. *true*)

In its simplest form, you type

    RETURN

to make Genstat stop taking statements from the current input channel and to go back to the channel that was previously active, and contained the INPUT statement that switched to the secondary file. Input then continues from the line following the original INPUT statement, but a marker is left in the channel that contains the RETURN statement, so that you can use INPUT to continue from the next line after RETURN later in your programme.

Sometimes you may want to return only if a particular condition is satisfied, for example if you have discovered that the data are unsatisfactory for whatever operations occur later in the file. To do this, you set the parameter to an appropriate logical expression; this must return a scalar result, which is interpreted as *true* if it is equal to 1, and *false* otherwise. For example

    RETURN MIN(Height)<0

If you have use INPUT several times, you may wish to return through several channels. The NTIMES option can be set to a number, or a scalar, to control how many returns take place. For example, with input starting on channel 1, supposing you had used INPUT 2 to switch to a file on channel 2, and then INPUT 3 to switch to a further file (on channel 3). If this file then contained the statement RETURN [NTIMES=2] you would return to channel 1. You can never return from input channel 1, so if you set NTIMES to a number greater than the number of currently active input channels, Genstat simply returns to channel 1.

If Genstat meets the end of the file on the current input channel, it will try to return control to the channel from which it was called. This is called an *implicit return*. The channel is closed automatically when this happens, and a warning message will be printed.

In order to maintain control over the different input channels, and know where to go after a RETURN, Genstat keeps an internal stack of input channels. Suppose you specify channel k, by typing INPUT k. There are three possible actions:

(a) if k is the current input channel, the statement is ignored;

(b) if k is not in the stack, it is added to it;

(c) if k is already in the stack (that is, the current state is: $1 \rightarrow ... \rightarrow k \rightarrow k_1 \rightarrow k_2 \rightarrow ... \rightarrow k_n$) then the intermediate channels $k_1 ... k_n$ are suspended at their current positions and removed from the stack.

Input then switches to channel k, taking statements from the beginning of the file if it has never been used before, or from the point at which it was last suspended. Subsequent INPUT statements will re-start the other channels from where they were suspended. When a RETURN statement is used, Genstat steps back NTIMES through the stack, removing any intermediate channels from the stack. This means that, using the above representation of the input stack, if channel k_n contained the statement INPUT k_2 and channel k_2 then had a RETURN, this would return to channel k_1.

If you use ## to execute macros (1.9.2), these are treated in the same way as input channels and added to the input stack. You can use INPUT to temporarily halt a macro and switch to a file, and RETURN to get back to the macro.

3.4.3   Sending output to another file (OUTPUT)

OUTPUT   defines where output is to be stored or displayed.

Options

PRINT = *strings*	Additions to output (dots, page, unchanged); default dots,page
DIAGNOSTIC = *strings*	What diagnostic printing is required (messages, warnings, faults, extra, unchanged); default faul,warn
WIDTH = *scalar*	Limit on number of characters per record; default width of output file
INDENTATION = *scalar*	Number of spaces to leave at the start of each line; default 0
PAGE = *scalar*	Number of lines per page

Parameter

scalar	Channel number of output file

An OUTPUT statement changes the current output channel and thus re-defines where the output will be sent by the subsequent statements in a program, until another OUTPUT statement is given (excluding any statements that use a CHANNEL option to redirect their output). Thus

```
OUTPUT 2
PRINT X
PRINT [CHANNEL=3] Y
ANOVA X
```

sends the values of X, and the analysis of X by the ANOVA statement, to the file on the second output channel, and the values of Y to the file on the third.

The PRINT option controls two aspects of the output produced for example from statistical analyses: whether a line of dots is printed at the start, and whether the output begins on a new page; this can also be controlled by the OUTPRINT option of SET (13.1.1). Similarly, the DIAGNOSTIC option has exactly the same effect as the DIAGNOSTIC option of SET (13.1.1).

The WIDTH option specifies the maximum width to be used when producing output. The default value is the width specified when the file was opened (3.3.1), but you can subsequently decrease it; you cannot use OUTPUT to set the width to a greater value than that specified when the file was opened. The PAGE option allows you to reset the number of lines per page.

3.5 Storing and retrieving structures

You will frequently want to save information that you have put into a data structure. This section explains how to transfer information to various other storage media on the computer, so that you can access the information easily later on.

There is an important difference between *storing* and merely printing. When you give the statement

```
PRINT [CHANNEL=2] X
```

you put only the identifier and the values of X into the character file attached to input channel 2. But if you give the statement

```
STORE [CHANNEL=2] X
```

you put all the details about X into the binary file attached to backing-store channel 2. So all the attributes of X are stored there too: for example, what type of structure it is, how long it is, and so on.

Subsection 3.5.1 describes the simplest use of storage and retrieval, which may be enough for most of your needs. Subsection 3.5.2 describes how backing-store files are arranged, with details of subfiles, userfiles, and workfiles. Subsections 3.5.3 to 3.5.6 describe the four directives that are relevant: STORE, RETRIEVE, CATALOGUE, and MERGE.

3.5.1 Simple use of backing store

Here is an example to illustrate the simplest way of storing data, and then retrieving it. First, to store the scalar A and the variate B:

```
OPEN 'EXAMPLE'; CHANNEL=1; FILETYPE=backingstore
SCALAR A; VALUE=2
VARIATE [VALUES=1...4] B
"Store structures A and B"
STORE [CHANNEL=1] A,B
```

The information about A and B is stored in the file named EXAMPLE which is opened on backing-store channel 1 (3.3.1). There is actually an invisible intermediate stage here: A and B are first stored in a *subfile* by the STORE statement; this subfile is then stored in the file EXAMPLE. The default name for the subfile is SUBFILE.

Example 3.5.1a shows how A can be retrieved in a subsequent job.

Example 3.5.1a

```
1   OPEN 'EXAMPLE'; CHANNEL=1; FILETYPE=backingstore
2   "Retrieve structure A only"
3   RETRIEVE [CHANNEL=1] A
4   PRINT A
```

```
               A
         2.000
```

So far, the file consists of only one subfile, but you can add others if you want. To do this, you must give a subfile name:

```
     OPEN 'EXAMPLE'; CHANNEL=1; FILETYPE=backingstore
     TEXT [VALUES='Storing more data','on backing store'] T
     "Add new subfile called Newset to file"
     STORE [CHANNEL=1; SUBFILE=Newset] T
```

There are now two subfiles in the file, called SUBFILE and Newset. Example 3.5.1b shows how to retrieve the text structure T.

Example 3.5.1b

```
     1   OPEN 'EXAMPLE'; CHANNEL=1; FILETYPE=backingstore
     2   "Retrieve T and print it"
     3   RETRIEVE [CHANNEL=1; SUBFILE=Newset] T
     4   PRINT T
```

```
                   T
     Storing more data
       on backing store
```

You can add as many new subfiles as you want, exactly as shown above, but you must keep the subfile names distinct within each file.

3.5.2 Subfiles, userfiles, and workfiles

Before going any further, you need to know how structures are stored.

A subfile is itself merely a portion of the backing-store file. Each subfile starts with a *catalogue*, recording which structures it stores. Then come the attributes (see 2.1) and the values of each structure. Note that a subfile cannot be changed by subsequent statements, for example to include further structures, although it can be completely replaced. Thus to add a set of structures, you would need to recover all the existing structures in the subfile and then recreate it to include also the extra structures, as shown in 3.5.3. Alternatively, library procedure BSUPDATE can be used to do this automatically.

There are two types of subfiles. *Ordinary subfiles* can hold any type of structures except procedures; *procedure subfiles* hold only procedures (and their dependent structures).

Whenever you store a structure in a subfile, Genstat automatically stores also all the associated structures to which it points. If these latter also point to further structures, then they are stored too, and so on. Some of the structures may be unnamed (1.6.3) and some structures may be system structures (2.10.1). For example

```
     TEXT [VALUES=A,B,C] T
     FACTOR [LABELS=T; VALUES=1...3] F
     STORE F
```

creates a subfile containing factor F. The complete definition of factor F depends on text T to supply level names. So T is stored too. The text T depends on a system structure (indicating the length of each line), which is therefore also stored. Hence to save factor F, Genstat has actually had to save three structures. However, this is all automatic, so you do not need to worry about any of the details of the system structures, and so on.

When you store a structure with a suffixed identifier, Genstat may have to set up a series of pointer structures if they are not already present (1.5.3 and 2.6). An example is:

```
VARIATE [VALUES=1,2] V[1,2]
STORE [PRINT=catalogue] V[1]
```

The first line sets up a pointer structure V, pointing to V[1] and V[2]. To store variate V[1], a pointer structure V has to be set up in the subfile, pointing to V[1] only. Thus two structures are saved on backing store, namely V and V[1]. The original pointer V in the program is left unchanged. (If the example had stored the whole of V, no such complications would have arisen.)

You can retrieve any pointer structure that you have set up in this way, and use it subsequently in the same way as any other pointer. But when a smaller pointer has to be set up only so that a suffixed identifier can be stored, no textual suffixes will be defined. So, if you want to store textual suffixes, you must define and store the pointer explicitly.

A backing-store file then consists of several subfiles; in fact the file can exist even if it is empty. However, if a file containing anything that has not been stored by a Genstat backing-store statement is attached as a backing-store file, it will be rejected.

A file that can be read by another job is called a *userfile*; it is permanent, in the sense that it will continue to exist after you have finished the job that created it. The userfile contains a catalogue of subfiles followed by the contents of each subfile. The number of userfiles that can be open at any one time will depend on your local computer; you can find this out by putting:

```
HELP environment,channel
```

Each job can have one temporary file called the *backing-store workfile* which also consists of a set of subfiles. The workfile's catalogue is deleted at the end of each job. The workfile itself may be overwritten in a later job in the same Genstat program, and on most computers it will be deleted automatically by STOP. However, if you abandon a run of Genstat before it has ended (for example by rebooting a PC), the workfile may survive.

A subfile name can be either an unsuffixed identifier or a suffixed identifier (1.5.3) with a numerical suffix. The identifiers of subfiles are kept in a separate catalogue to the identifiers of data structures, so you do not to worry about keeping the identifiers data structures and subfile distinct. However, if you use a suffixed identifier for a subfile, Sub[1] say, you cannot also use the identifier Sub.

3.5.3 The **STORE** directive

STORE stores structures in a subfile of a backing-store file.

Options

PRINT = *string*	What to print (catalogue); default *
CHANNEL = *scalar*	Channel number of the backing-store file where the subfile is to be stored; default 0, i.e. the workfile
SUBFILE = *identifier*	Identifier of the subfile; default SUBFILE
LIST = *string*	How to interpret the list of structures (inclusive, exclusive, all); default incl
METHOD = *string*	How to append the subfile to the file (add, overwrite, replace); default add, i.e. clashes in subfile identifiers cause a fault (note: replace overwrites the complete file)
PASSWORD = *text*	Password to be stored with the file; default *
PROCEDURE = *string*	Whether subfile contains procedures only (yes, no); default no
UNNAMED = *string*	Whether to list unnamed structures (yes, no); default no

Parameters

IDENTIFIER = *identifiers*	Identifiers of the structures to be stored
STOREDIDENTIFIER = *identifiers*	Identifier to be used for each structure when it is stored

The structures to be stored are specified by the IDENTIFIER parameter. The CHANNEL option indicates the backing-store file to use, and the SUBFILE option specifies the subfile that is created. Both these options can be omitted; by default the file will be the workfile, and the subfile will be called SUBFILE. The structures that are stored in the subfile are merely copies of the structures in the job, so the original structures remain available for further use within the job.

The STOREDIDENTIFIER parameter allows you to give a structure a different name within the subfile: For example,

```
VARIATE [VALUES=10.2,15.3,21.4,16.8,22.3] Weight
STORE Weight; STOREDIDENTIFIER=WtWeek2
```

stores a structure with identifier Weight within Genstat as a structure with identifier WtWeek2 in the backing-store file. If you want to rename only some of the structures, you can either respecify the existing identifier, or insert * at the appropriate point in the list. For example, you could store X and Y, renaming only Y as Yy, by

```
STORE X,Y; STOREDIDENTIFIER=X,Yy
```

or by

```
STORE X,Y; STOREDIDENTIFIER=*,Yy
```

You can give an unnamed structure in the list of either parameter. For example

```
STORE !(10.2,15.3,21.4,16.8,22.3); STOREDIDENTIFIER=WtWeek2
```

But of course you will not be able to retrieve any structure that has been stored as an unnamed structure (except perhaps as a dependent structure of another structure, see 3.5.2).

All the structures in a subfile must have distinct identifiers, and Genstat will report a fault if you try to give two the same name. You thus need to be careful if you are storing structures inside a procedure, as the same identifier can be used for one structure within the procedure, and for another one outside; you cannot store both in the same subfile.

Procedures that have been retrieved automatically from libraries (5.3.3) cannot be stored by STORE.

You can set option PRINT=catalogue to obtain a catalogues of the subfiles in the backing-store file, and of the structures in the subfile just created. If you also set option UNNAMED=yes Genstat will also list any unnamed structures, with details of how they depend on each other.

The LIST option controls how the IDENTIFIER list is interpreted. The default setting inclusive simply stores the structures that have been listed.

Alternatively, if you set LIST=all Genstat will store all the structures in the current job that have identifiers and whose types have been defined. If the statement is inside a procedure, then only the structures defined within the procedure are stored (5.3). If you are storing procedures, then this setting will store all procedures that you have created explicitly in this job, by PROCEDURE or RETRIEVE statements.

Finally, you can see LIST=exclusive to store everything that you have not included in the IDENTIFIER parameter: that is, all the other named structures that are currently accessible, or all the other procedures that have been created in this job. Note, though, that some of the structures in the IDENTIFIER list may be stored if they are needed to complete the set of structures to be stored. If you use this setting, the STOREDIDENTIFIER parameter is ignored. For example

```
TEXT [VALUES=a,b] T
FACTOR [LABELS=T] F
TEXT [VALUES='variate  text'] Vt
VARIATE V; EXTRA=Vt
```

creates four named structures, T, F, V and Vt. The statement

```
STORE [LIST=inclusive] T
```

stores the text T;

```
STORE [LIST=all]
```

stores all the four structures that have identifiers;

```
STORE [LIST=exclusive] F,T
```

stores Vt and V; and

```
STORE [LIST=exclusive] Vt,T
```

results in all four structures being saved, because V points to Vt, and F points to T.

If a subfile of the specified name already exists on the backing-store file, the storing operation will usually fail. You can then set option METHOD=overwrite to overwrite the old subfile, that is, to replace the old subfile with a new subfile; alternatively, you can put METHOD=replace to form a new backing-store file containing only the new subfile.

To make your files secure, you can specify a password using the PASSWORD option. Once you have done this, you must include the same password in any future use of STORE or MERGE with this same userfile; spaces, case, and newlines are significant in the password. You cannot change the password in a userfile once you have set it, but you can use the MERGE directive to create a new userfile with no password or with a new password. If you set the password to be a text whose values have been have restricted (4.4.1), the restriction is ignored.

The PROCEDURE option indicates whether the subfile is to store procedures (PROCEDURE=yes), or ordinary data structures.

3.5.4   The **RETRIEVE** directive

RETRIEVE retrieves structures from a subfile.

Options

CHANNEL = *scalar*	Specifies the channel number of the backing-store or procedure-library file containing the subfile (FILETYPE settings 'back' or 'proc'); default 0 (i.e. the workfile) for FILETYPE=back, no default for FILETYPE=proc, not relevant with other FILETYPE settings
SUBFILE = *identifier*	Identifier of the subfile; default SUBFILE
LIST = *string*	How to interpret the list of structures (inclusive, exclusive, all); default incl
MERGE = *string*	Whether to merge structures with those already in the job (yes, no); default no, i.e. a structure whose identifier is already in the job overwrites the existing one, unless it has a different type
FILETYPE = *string*	Indicates the type of file from which the information is to be retrieved (backingstore, procedurelibrary, siteprocedurelibrary, Genstatprocedurelibrary); default back

Parameters

IDENTIFIER = *identifiers*	Identifiers to be used for the structures after they have been retrieved
STOREDIDENTIFIER = *identifiers*	Identifier under which each structure was stored

---

You recover information from a subfile of a backing-store file using the RETRIEVE directive. The CHANNEL option specifies the backing-store file, and the SUBFILE option indicates the subfile. Both these options can be omitted; by default the file will be the workfile, and the subfile will be called SUBFILE.

When you retrieve a structure Genstat may also retrieve a chain of associated structures: that is, all the structures to which it points, and the structures to which they point, and so on. For example, suppose you store the three structures with identifiers T, V, and F, along with an unnamed structure storing information about T, in a subfile called SUBFILE in backing-store file FILE1:

```
OPEN 'FILE1'; CHANNEL=1; FILETYPE=backingstore
TEXT [VALUES=a,b,c] T
VARIATE V; EXTRA=T
FACTOR [LABELS=T] F
STORE [CHANNEL=1] T,V,F
```

Then the statement

```
RETRIEVE [CHANNEL=1] V
```

will retrieve not only V but also T (which was associated with T by the EXTRA parameter of the VARIATE statement), and the unnamed structure that is associated with T. The structures V, T, and the unnamed structure, are said to be a *complete set* from the subfile.

The IDENTIFIER parameter specifies the structures to be retrieved. You can use the STOREDIDENTIFIER parameter to give a structure a different name from the one within the subfile. For example

```
RETRIEVE IDENTIFIER=Weeks; STOREDIDENTIFIER=Time
```

You are not allowed to give identical identifiers to two retrieved structures, nor are you allowed to have the same identifier referring to a structure of one type in a subfile, and to a structure of a different type in your job.

As with STORE, if you want to rename only some of the structures, you can either respecify the existing identifier, or insert * at the appropriate point in the STOREDIDENTIFIER list.

Genstat knows whether you are retrieving a procedure by the type of SUBFILE that you have are accessing. You are not allowed to rename a procedure as a suffixed identifier or as the name of a directive.

You can even rename a structure so that it is unnamed in the job. Suppose, for example, that a structure T already exists within Genstat, and that you want to retrieve the variate V stored in the file FILE1 above. Then, as we have seen, the structure T will also be retrieved. However, you can avoid the existing structure T job being overwritten by making the retrieved version of T unnamed:

```
OPEN 'FILE1'; CHANNEL=1; FILETYPE=backingstore
RETRIEVE [CHANNEL=1] V,!T(a); STOREDIDENTIFIER=V,T
```

The value, a, of the unnamed text !T(a) will be replaced by the values stored for T, and this unnamed text will become the EXTRA text for V. Alternatively you could rename T to be Tnew by

```
RETRIEVE [CHANNEL=1] V,TNew; STOREDIDENTIFIER=V,T
```

When you are retrieving a suffixed identifier, Genstat matches the numerical suffix only, and not the whole structure that is denoted by the identifier. For example, suppose pointer P stored in a subfile points to structures with identifiers A, B, C, and D, and that P has numerical suffixes 1 to 4 respectively. Also suppose that in your current job, you have never mentioned pointer P either directly or indirectly. Then the statement

```
RETRIEVE [CHANNEL=1] P[2]
```

will retrieve the structure B from backing store but, as it has not been referenced only as P[2] in the RETRIEVE statement, the identifier B will not be recovered and it will be known only as P[2] within Genstat.

A structure that you are retrieving from a subfile may sometimes overwrite the values of an existing structure in your program. If this structure is a pointer or a compound structure, the existing suffixes will be overwritten by those of the stored structure, so some existing structures with suffixed identifiers may in effect be lost. For example, suppose that userfile FILE2 contains a pointer P, with suffixes 1 and 2 pointing to structures A and B. If we set up a variate P[3], and then retrieve the pointer P

```
OPEN 'FILE2'; CHANNEL=1; FILETYPE=backingstore
VARIATE [VALUES=1...6] P[5,6,7]
RETRIEVE [CHANNEL=1] P
```

P will now have suffixes 1 and 2 pointing to A and B, but the variate P[3] will have been lost. For more details about pointers, see 2.6.

The LIST option is similar to its namesake in the STORE directive (3.5.3), but it now refers to the named structures in the subfile.

The FILETYPE option specifies whether you wish to retrieve information from backing store files that have been attached as normal backing store files or as procedure libraries by the OPEN directive (3.3.1), or from Genstat Procedure library or from the site procedure library. The CHANNEL setting is ignored if the siteprocedurelibrary or Genstatprocedurelibrary settings are used. The source code of the procedures in the Genstat Procedure library can be accessed using the LIBEXAMPLE procedure.

Normally when you retrieve a complete subset of structures, Genstat overwrites all structures in the job that have the same identifier (after any renaming). As a result, some other structures already in the job may become inconsistent and will be destroyed. You can avoid this happening by setting the MERGE option to yes. Genstat then does not overwrite any structures with the same name and type. However, a consequence is that some of the retrieved structures may now be inconsistent and thus need to be destroyed in the program (although they will of course remain in the subfile).

3.5.5   The CATALOGUE directive

CATALOGUE   displays the contents of a backing-store file.

Options

PRINT = *strings*   What to print (subfiles, structures); default
                           subf, stru

CHANNEL = *scalar*   Channel number of the backing-store file; default 0, i.e.
                           the workfile

LIST = *string*   How to interpret the list of subfiles (inclusive,
                           exclusive, all); default incl

SAVESUBFILE = *text*	To save the subfile identifiers; default *
UNNAMED = *string*	Whether to list unnamed structures (yes, no); default no

Parameters

SUBFILE = *identifiers*	Identifiers of subfiles in the file to be catalogued
SAVESTRUCTURE = *texts*	To save the identifiers of the structures in each subfile

You can use CATALOGUE to obtain details of the subfiles contained in a backing-store file, or the structures within an ordinary subfile, or the procedures within a procedure subfile. The file is indicated by the CHANNEL option, and the SUBFILE parameter specifies the subfiles (of ordinary structures or of procedures) that are to be catalogued.

The PRINT option specifies which catalogues are to be printed. The subfiles setting prints the catalogue of subfiles in the backing-store file attached to the channel specified by the CHANNEL option, while the structures setting prints the catalogue of structures or procedures that are in the subfiles specified by the SUBFILE parameter.

If you set option UNNAMED=yes the unnamed structures in each subfile will also be listed, together with details of how the structures depend on each other.

The LIST option is similar to its namesake in the STORE directive, but it now refers to the identifiers in the SUBFILE list.

The SAVESTRUCTURE parameter allows you to set up texts, one for each subfile in the SUBFILE parameter. Each text contains the identifiers of all structures with an unsuffixed identifier in the subfile. Each identifier is put on a separate line, and the characters , \ are appended to all but the last line. You would normally use these texts as a macro; the , \ makes them useable as lists of identifiers. If the text is used as a macro, it is subject to the restriction on the length of statements (1.8). The SAVESUBFILE option allows you to save a similar text containing the identifiers of all the subfiles in a backing-store file.

Suppose we have a userfile called FILE3 which already containing three subfiles: two ordinary subfiles (called Sub[1] and Sub[2]), and a procedure subfile called Sub3 containing a procedure called TRANSFOR. Example 3.5.5a adds a fourth subfile called Sub4.

Example 3.5.5a

```
  1   TEXT [VALUES='vector_1','vector_3'] L
  2   & [VALUES='heading'] T
  3   POINTER [SUFFIXES=!(1,3); NVALUES=L] P
  4   VARIATE P[],V; EXTRA=T
  5   OPEN 'FILE3'; CHANNEL=1; FILETYPE=backingstore
  6   STORE [CHANNEL=1; SUBFILE=Sub4] P,V
  7   CATALOGUE [PRINT=Structures; CHANNEL=1] Sub4

***** Catalogue *****

catalogue of structures in the subfile Sub4

      entry   identifier      type
      1       P               pointer
      2       V               variate
```

```
3          P[1]              variate
4          P[3]              variate
5          T                 text
```

The subfile contains five named structures with identifiers P, V, P[1], P[2] and T. There are also two unnamed structures associated with P and T, as can be seen when we set option UNNAMED=yes in Example 3.5.5b, and obtain a more detailed catalogue. This also gives details of any dependencies among structures, referenced by their index in the entry column. The identifier column gives the numerical suffix, and the labels column gives textual suffixes. This information is particularly helpful with complicated trees of pointers (2.6) or with compound structures (2.7).

Example 3.5.5b

```
   8   CATALOGUE [PRINT=Structures; CHANNEL=1; UNNAMED=Yes] \
   9      Sub4; SAVESTRUCTURE=Tsub4

***** Catalogue *****

catalogue of structures in the subfile Sub4

      entry   identifier      type        points to
      1       P               pointer     3
                                          pointer values
                                          unit          entry       labels
                                          1             4           vector_1
                                          3             5           vector_3
      2       V               variate     6
      3                       text        7
      4       P[1]            variate     6
      5       P[3]            variate     6
      6       T               text        8
      7                       system
      8                       system

  10    PRINT Tsub4; JUSTIFICATION=left

Tsub4
P,\
V,\
T
```

The text Tsub4, saved by the SAVESTRUCTURE parameter of CATALOGUE, gives the names of all the structures in the subfile that can be accessed directly. These texts can be used as macros. Example 3.5.5c shows how you can effectively add a variate Vnew to the subfile Sub4.

Example 3.5.5c

```
  11   VARIATE [VALUES=1,2] Vnew
  12   RETRIEVE [CHANNEL=1; SUBFILE=Sub4] ##Tsub4
  13   STORE [CHANNEL=1; SUBFILE=Sub4; METHOD=overwrite] ##Tsub4,Vnew
```

```
 14   CATALOGUE [PRINT=structures; CHANNEL=1] Sub4

***** Catalogue *****

catalogue of structures in the subfile Sub4

      entry   identifier      type
      1       P               pointer
      2       V               variate
      3       T               text
      4       Vnew            variate
      5       P[1]            variate
      6       P[3]            variate
```

Example 3.5.5d firstly produces a catalogue of the procedure subfile Tsub3 (line 15), and then produces a catalogue of the subfiles (line 16), at the same time using the SAVESUBFILE option to place the subfile names into the text Tsubf. Notice that Tsubf excludes the system-type subfile Sub, which exists because two of the subfiles (Sub[1] and Sub[2]) have suffixed identifiers, and cannot be used as a subfile in its own right.

Example 3.5.5d

```
 15   CATALOGUE [PRINT=structures; CHANNEL=1] Sub3

***** Catalogue *****

catalogue of procedures in the procedure subfile Sub3

      entry   identifier
      1       TRANSFOR

 16   CATALOGUE [PRINT=subfiles; CHANNEL=1; SAVESUBFILE=Tsubf

***** Catalogue *****

catalogue of subfiles in the userfile  1

      entry   identifier      type
      1       Sub4            ordinary
      2       Sub3            procedure
      3       Sub             system
      4       Sub[2]          ordinary
      5       Sub[1]          ordinary

 17   PRINT Tsubf; JUSTIFICATION=left

Tsubf
Sub4,\
Sub3,\
Sub[2],\
Sub[1]
```

3.5.6   The MERGE directive

MERGE   copies subfiles from backing-store files into a single file.

Options

PRINT = *string*	What to print (catalogue); default *
OUTCHANNEL = *scalar*	Channel number of the backing-store file where the subfiles are to be stored; default 0, i.e. the workfile
METHOD = *string*	How to append subfiles to the OUT file (add, overwrite, replace); default add, i.e. clashes in subfile identifiers cause a fault (note: replace overwrites the complete file)
PASSWORD = *text*	Password to be checked against that stored with the file; default *

Parameters

SUBFILE = *identifiers*	Identifiers of the subfiles
INCHANNEL = *scalars*	Channel number of the backing-store file containing each subfile
NEWSUBFILE = *identifiers*	Identifier to be used for each subfile in the new file

---

The MERGE directive is used to copy subfiles into another backing-store file. You can either add the subfiles to an existing backing-store file, or form a new backing-store file.

The OUTCHANNEL option specifies the backing-store channel of the file to which the subfiles are to be copied; by default this is the workfile (channel 0).

The SUBFILE parameter specifies the list of subfiles that are to be copied, and the INCHANNEL parameter indicates the channel of the backing-store file where each one is currently stored. If you do not specify the INCHANNEL parameter, Genstat assumes that the subfiles are coming from the workfile. You are not allowed to include the OUTCHANNEL among the channels in the INCHANNEL list. Also, you cannot store two subfiles with the same names, and should use the NEWSUBFILE parameter to rename any that clash. For example

```
MERGE [OUTCHANNEL=3] JanData,JulyData,JanData; INCHANNEL=1,1,2; \
      NEWSUBFILE=Jan92dat,Jul92dat,Jan93dat
```

To rename only some of the subfiles, you can either respecify the existing identifier, or insert * at the appropriate point in the NEWSUBFILE list.

If you specify a missing identifier * in the SUBFILE list, Genstat will include all the subfiles from the relevant INCHANNEL. If you want to rename any of these subfiles, you can also mention it explicitly. For example, this statement will take all the subfiles from channel 1 and rename subfile Sub as Subf.

```
MERGE *,Sub; INCHANNEL=1; NEWSUBFILE=*,Subf
```

You can set option PRINT=catalogue to produce a catalogue of the subfiles in the new backing-store file (3.5.5).

If a subfile of the specified name already exists on the backing-store file, the storing operation will usually fail. However, you can set option METHOD=overwrite to overwrite the old subfile, that is, to replace the old subfile with a new subfile. Alternatively, you can put METHOD=replace to form a new backing-store file containing only the new subfiles.

Subfiles are merged in a fixed order. Genstat first takes the subfiles from the backing-store file with the lowest channel number, in the order in which they occur there, then it takes the subfiles the next lowest channel number, and so on. If OUTCHANNEL=0 (that is, the new file is the workfile), the original subfiles that are to be retained from that file will be followed by the new subfiles; otherwise, if OUTCHANNEL is non-zero, the original subfiles are placed after the new subfiles. If you want to put the subfiles into a particular order, you should merge them into the workfile in that order, and then merge the workfile into a new userfile.

To keep the new file secure, you can use the PASSWORD option to incorporate a password, as explained in 3.5.3.

3.6   Storing and retrieving data and programs in unformatted files

The RECORD directive (3.6.1) allows you to produce an unformatted file containing all the details required to recreate the current state of a Genstat job. You can then use the RESUME directive (3.6.2), either later in your program, or during a completely different Genstat run, to recover all this information and continue your use of Genstat from that point. This can be useful if you need to abandon an analysis and resume it at some later date, or if you want to save the current state of a program in case your next operations turn out to be unsuccessful.

You can also use unformatted files for temporary data storage (3.6.3).

3.6.1   The **RECORD** directive

RECORD   dumps a job so that it can later be restarted by a RESUME statement.

Option

CHANNEL = *scalar*   Channel number of the unformatted file where information is to be dumped; default 1

No parameters

RECORD sends all the relevant information about the current state of your Genstat job to the unformatted file specified by the CHANNEL option. You can then use the RESUME directive to re-establish that situation either in a future Genstat run, or later in the same run. The information includes the attributes and values of all your data structures, procedures, and the current graphics settings (6.5), but no details are kept of the files that are open on any of the channels. If you use RECORD with the same channel number again, the earlier information is overwritten.

3.6.2 The RESUME directive

RESUME restarts a recorded job.

Option

CHANNEL = *scalar*   Channel number of the unformatted file where the information was dumped; default 1

No parameters

RESUME recovers the information stored by a previous RECORD statement so that you can continue your use of Genstat as though nothing had happened in between. Thus, for example, Genstat deletes all the data structures that were created in the current job prior to RESUME, and reinstates the data structures that were available in Genstat at the time the RECORD statement took place. Similarly, the current graphics settings are replaced by those that were in force when RECORD was used, but any external files that are attached to Genstat remain unaffected.

If the RECORD directive was used within a procedure or a FOR loop, the job is not resumed at that point. Instead, it restarts at the statement after the procedure call, or after the outermost ENDFOR statement.

The amount of space available for data in the current job need not be the same as that in the recorded job. However, you will get a fault if the available space is too small, that is, if the space needed by the recorded job is greater than space available in the current job.

Example 3.6.2 illustrates how RECORD and RESUME are used.

Example 3.6.2

```
 1   OPEN 'Dump'; CHANNEL=1; FILETYPE=unformatted
 2   VARIATE [VALUES=1,2] A
 3   RECORD
 4   PRINT A

          A
      1.000
      2.000

 5   STOP

 1   OPEN 'Dump'; CHANNEL=1; FILETYPE=unformatted
 2   RESUME
 3   CALCULATE A = A+1
 4   PRINT A

          A
      2.000
      3.000
```

3.6.3   Storing and reading data with unformatted files

Unformatted files can also be used to store values of data structures using PRINT (3.2.1), so that they can later be input again using READ (3.1.1). This provides a convenient of way to free some space temporarily. It can also save computing time if you have a large data set that may need to be read several times. Input from character files is slow. So after vetting a large data set, it will be read more efficiently on future occasions if you transfer its contents to an unformatted file. As an alternative you could use backing store, but this stores the attributes of the structures as well as their values, and so access will take longer. Finally, you can use these facilities to transfer data between Genstat and other programs; see 13.3.2.

Unformatted files are selected in READ and PRINT by setting option UNFORMATTED=yes. The only options that are then relevant are CHANNEL, REWIND, and SERIAL.

Genstat automatically creates an *unformatted workfile*, on channel 0, to which unformatted output is sent by default (by PRINT), and from which unformatted input is taken by default (by READ). This file is deleted automatically by the STOP directive.

It is usually quicker to read and write structures in series. Also the values of the structures transferred in parallel must all be of the same *mode*. Neither texts nor factors can be stored in parallel with values of the other, numerical, structures: scalars, variates, matrices or tables. As an example, we first open a file, and declare some variates, matrices, and factors.

```
OPEN 'BDAT'; CHANNEL=3; FILETYPE=unformatted
VARIATE X,Y,Z; VALUES=!(11...19),!(21...29),!(31...39)
MATRIX [ROWS=2; COLUMNS=3; VALUES=11,12,13,21,22,23] M
FACTOR [LEVELS=3; VALUES=1,3,2,3,1,2,2,2,1,3] F
```

The next three statements store data for M and F on the file named BDAT and data for X, Y and Z (in parallel) on the workfile.

```
PRINT [CHANNEL=3; SERIAL=yes; UNFORMATTED=yes] M,F
PRINT [UNFORMATTED=yes] X,Y,Z
```

You can now free the space for numerical data for other purposes, by putting

```
DELETE X,Y,Z,F,M
```

By rewinding the files we can read the data back into Genstat.

```
READ [UNFORMATTED=yes; REWIND=yes] X,Y,Z
READ [CHANNEL=3; SERIAL=yes; UNFORMATTED=yes; REWIND=yes] M,F
```

You can also re-use the external file BDAT in a later job.

If you change the lengths of structures, you must remember to reset them to their original values before you use unformatted READ to recover the data values from the file. Only the data values are stored in unformatted files, and not the attributes (such as lengths) as in backing-store files.

S.A.H.
P.K.L.
A.D.T.

4 Data handling

Genstat has many directives for doing calculations or for manipulating data. There are also many procedures, mainly in the `Manipulation` module of the procedure library. You may wish to use these facilities as part of a statistical analysis; for example, you may want to transform your data before fitting a regression or doing an analysis of variance. However, they can be useful even if you have no intention of doing a statistical analysis but merely wish to use Genstat as a package for data-handling and arithmetic.

The `CALCULATE` directive (4.1) allows you to perform straightforward arithmetic operations on any numerical data structure. It also enables you to make logical tests on data: for example, you may want to check whether two variates contain the same values; similar checks can be done with factors, texts, and pointers. You can use `CALCULATE` for matrix operations: for example, matrix multiplication, inversion, and Choleski decompositions (4.1.3 and 4.2.4). Other directives are available to form eigenvalues (latent roots) and singular values (4.10). `CALCULATE` can do calculations with tables, and these need not have identical sets of classifying factors (4.1.4). You can also form marginal summaries over any of the dimensions of a table (4.2.5 and 4.11.2). Other directives allow you to combine "slices" in any dimension (4.11.4) and to tabulate values, for example, to summarize data from surveys (4.11.1). Genstat also provides a full range of mathematical and statistical functions (4.2).

There are several directives and procedures for transferring and sorting values. You can transfer data from one set of structures into another: for example you may want to copy the columns of a matrix into a list of variates (4.3). You can re-order the units in vectors according to numerical or textual "keys" (4.4.3). Facilities for putting them into random order are described in Chapter 9, together with the ways in which values can be generated for factors in experimental designs (9.8).

Genstat has several directives for manipulating textual data. You can omit complete lines, or append one text onto the end of another (4.3). You can form a text each of whose lines is made up from sections of lines from several texts concatenated together (4.7.1). For more complicated operations you can use the general text editor (4.7.2).

4.1 Numerical calculations

The main directive for calculations in Genstat is called `CALCULATE`, and this is described in the first part of this section. The calculation to be done is defined by a Genstat expression. The formal rules for these are given in 1.7.2, but below we give examples that explain in more detail exactly how they are used. Expressions also occur in `RESTRICT` (4.4.1), the directives for program control (5.2), and `FITNONLINEAR` (8.7); so this section is relevant also to several other areas of Genstat.

Subsection 4.1.1 contains general information about `CALCULATE`, describing its options and illustrating the operators that can occur in expressions. Information about calculations with particular data structures is given in 4.1.2 (scalars, factors, variates, and texts), 4.1.3 (matrices) and 4.1.4 (tables). The rules for implicit declarations in `CALCULATE` are given in 4.1.5.

Subsection 4.1.6 describes how to define subsets of vectors or matrices using qualified identifiers. The functions that can be used in expressions are described in 4.2.

4.1.1   The **CALCULATE** directive

CALCULATE   calculates numerical values for data structures.

Options

PRINT = *string*	Printed output required (summary); default * i.e. no printing
ZDZ = *string*	Value to be given to zero divided by zero (missing, zero); default miss
TOLERANCE = *scalar*	If the scalar is non missing, this defines the smallest non-zero number; otherwise it accesses the default value, which is defined automatically for the computer concerned

Parameter

expression	Expression defining the calculations to be performed

---

The parameter of CALCULATE is unnamed, and is an expression. An expression (1.7.2) consists of identifier lists, operators and functions. However, for an expression to be valid in CALCULATE, it must include the assignment operator (=). For example

```
CALCULATE 5,6
```

will fail, with an error message, even though the list 5,6 is an expression.

The simplest form of expression in CALCULATE merely assigns values from one structure to another of the same type. For example:

```
VARIATE [VALUES=1...4] V1
CALCULATE V2 = V1
```

The values of variate V1 are copied into the structure V2; since V2 has not been declared previously Genstat defines it implicitly, here as a variate. The rules for implicit declarations in CALCULATE are described in 4.1.5, but you may prefer to declare everything explicitly until you are confident in the use of CALCULATE.

A complete list of the operators available for expressions is given in 1.5.6. Most of the operators in this list act element-by-element on the values of data structures of the same type, the exceptions being the compound operator *+ (matrix multiplication) and the four relational operators: .IS., .ISNT., .IN. and .NI. The assignment operator (=) has been demonstrated above; the next example shows the arithmetic operators +, −, *, /, and ** operating element-by-element on variates, X and Y:

Example 4.1.1a

```
 1   VARIATE [VALUES=10,12,14,16,*,20] X
 2   VARIATE [VALUES=4,3,2,1,0,-1] Y
 3   CALCULATE Vadd = X + Y
 4   & Vsub = X - Y
 5   & Vmult = X * Y
 6   & Vdiv = X / Y
 7   & Vexp = X ** Y
 8   PRINT X,Y,Vadd,Vsub,Vmult,Vdiv,Vexp; FIELDWIDTH=9; DECIMALS=2
```

X	Y	Vadd	Vsub	Vmult	Vdiv	Vexp
10.00	4.00	14.00	6.00	40.00	2.50	10000.00
12.00	3.00	15.00	9.00	36.00	4.00	1728.00
14.00	2.00	16.00	12.00	28.00	7.00	196.00
16.00	1.00	17.00	15.00	16.00	16.00	16.00
*	0.00	*	*	*	*	*
20.00	-1.00	19.00	21.00	-20.00	-20.00	0.05

A missing value in either or both of the variates produces a missing value in the resulting variate.

You can use the operator minus (–) in two ways: either as a *dyadic* minus, to subtract one operand from another, as shown above; or as a *monadic* minus, to change the sign of a single operand. Genstat gives the monadic minus high precedence, which means that when it appears in an expression, it is one of the first operations to be done. Thus you need to be careful when using monadic minus to change the sign of the result of an expression. In particular, these two CALCULATE statements will give the same values to both Vb and Vc:

```
CALCULATE Vb = Va**2
CALCULATE Vc = -Va**2
```

This is because the operator – appears as a monadic minus, and so the signs of the values of Va are changed *before* being squared; to obtain the negative of the square of Va you need

```
CALCULATE Vc = -(Va**2)
```

In logical and relational expressions, Genstat uses the value 0 to represent false, and the value 1 to represent true. In fact any non-zero non-missing value is taken to represent a true value.

For numerical structures, Genstat has the relational operators .EQ., .NE., .LT., .LE., .GT., and .GE. with their symbolic equivalents. Note that the symbolic equivalent for .EQ. is the compound operator ==. Genstat requires the two (adjacent) equals signs to distinguish this from the assignment operator, which would generate rather different results! For text structures, the appropriate relational operators are .EQS. and .NES. The two texts must have the same number of units (or lines).

Example 4.1.1b

```
 1   VARIATE [VALUES=1,2,3,4,5,*,*,1] X
 2   & [VALUES=5,4,3,2,1,1,*,*] Y
 3   TEXT [VALUES=a,b,c,d,e,'','',a] Tx
 4   & [VALUES=a,x,c,d,y,a,'',''] Ty
```

```
 5   CALCULATE Veq = X.EQ.Y
 6   & Vne = X.NE.Y
 7   & Vlt = X.LT.Y
 8   & Vle = X.LE.Y
 9   & Vgt = X.GT.Y
10   & Vge = X.GE.Y
11   & Veqs = Tx.EQS.Ty
12   & Vnes = Tx.NES.Ty
13   PRINT X,Y,Veq,Vne,Vlt,Vle,Vgt,Vge,Tx,Ty,Veqs,Vnes; \
14      FIELDWIDTH=5; DECIMALS=0
```

X	Y	Veq	Vne	Vlt	Vle	Vgt	Vge	Tx	Ty	Veqs	Vnes
1	5	0	1	1	1	0	0	a	a	1	0
2	4	0	1	1	1	0	0	b	x	0	1
3	3	1	0	0	1	0	1	c	c	1	0
4	2	0	1	0	0	1	1	d	d	1	0
5	1	0	1	0	0	1	1	e	y	0	1
*	1	0	1	*	*	*	*		a	0	1
*	*	1	0	*	*	*	*			1	0
1	*	0	1	*	*	*	*	a		0	1

With most of the relational operators, a missing value in either operand, or in both, gives a missing result. The exceptions are .EQ. and .NE., .EQS., and .NES. When both operands are missing, .EQ. gives a true result and .NE. gives a false result. The same is true with .EQS. and .NES. when they encounter missing values (or null strings) in texts.

The relational operators .IS. and .ISNT. test whether or not a dummy points to a particular identifier. For example, to store in Sca the result of a test to check whether dummy D points to Va, you would put

    CALCULATE Sca = D.IS.Va

while to test that D does not point to Vb, you would put

    CALCULATE Sca = D.ISNT.Vb

The final pair of relational operators, .IN. and .NI., represent *inclusion* and *non-inclusion*. These two operators differ from the other relational operators in that each value in the structure on the left-hand side is compared in turn with every value in the structure on the right-hand side.

For .IN., the result is true if the value on the left-hand side is included in the set of values in the right-hand structure; otherwise the result is false. The .NI. operator is the opposite of .IN.

The length of the result is taken from the length of the left-hand structure, since it is the values of the left-hand structure that are being tested. For a very simple example, suppose that the variate X contains the values 1,2,1,1,3,5,1,2,1,4, and that the variate Evens contains 0,2,4,6,8. The statement

    CALCULATE S = X.IN.Evens

will store in S an indication of whether each element of X is odd or even: that is, S will be given the values 0,1,0,0,0,0,0,1,0,1.

When there is a factor on the left-hand side of .IN. or .NI. and a variate on the right-hand side, Genstat checks the levels of the factor against the values in the variate.

Alternatively, if the factor has a labels vector, you can specify a text against which Genstat will then compare the labels. In the next example, the variate Large records which elements of the factor Size have values that lie in the set {4.8, 6}, and the variate NotAB records which elements of the factor Type have values that lie outside the set {A, B}:

Example 4.1.1c

```
1   FACTOR [LEVELS=!(1.2,2.4,3.6,4.8,6)] Size; \
2      VALUES=!(1.2,4.8,6,2.4,3.6,2.4,1.2,6)
3   FACTOR [LABELS=!T(A,B,C,D)] Type; VALUES=!T(2(A,B,C,D))
4   CALCULATE Large = Size .IN. !(4.8,6)
5   & NotAB = Type .NI. !T(A,B)
6   PRINT Size,Large,Type,NotAB; FIELDWIDTH=6; DECIMALS=1,0,0,0
```

Size	Large	Type	NotAB
1.2	0	A	0
4.8	1	A	0
6.0	1	B	0
2.4	0	B	0
3.6	0	C	1
2.4	0	C	1
1.2	0	D	1
6.0	1	D	1

You can use the logical operators .AND., .NOT., .OR., and .EOR. to combine the results of the relational operators, and form a single logical result: .NOT. reverses true and false results; .OR. gives a true result only if one or both operands are true; .AND. gives a true result if both operands are true; and .EOR. gives a true result if one of the operands is true, but a false result if both are true or both false. Example 4.1.1d shows the results of the four logical operators. Notice that a missing value in either operand gives a missing value in the result.

Example 4.1.1d

```
1   VARIATE [VALUES=3(0,1,2),1,*] X
2   & [VALUES=(0,1,2)3,*,*] Y
3   CALCULATE Vnot = .NOT. Y
4   & Vor = X .OR. Y
5   & Vand = X .AND. Y
6   & Veor = X .EOR. Y
7   PRINT X,Y,Vnot,Vor,Vand,Veor; FIELDWIDTH=5; DECIMALS=0
```

X	Y	Vnot	Vor	Vand	Veor
0	0	1	0	0	0
0	1	0	1	0	1
0	2	0	1	0	1
1	0	1	1	0	1
1	1	0	1	1	0
1	2	0	1	1	0
2	0	1	1	0	1
2	1	0	1	1	0
2	2	0	1	1	0
1	*	*	*	*	*
*	*	*	*	*	*

If the expression contains lists, Genstat does several calculations. For example,

```
CALCULATE A,B,C = X,Y,Z + 1,2,3
```

is equivalent to the three CALCULATE statements:

```
CALCULATE A = X + 1
CALCULATE B = Y + 2
CALCULATE C = Z + 3
```

Genstat takes the items in the lists in parallel, and recycles any lists that are shorter than the list of primary arguments. In CALCULATE, the primary arguments are the identifiers on the left-hand side of the assignment operator (=). In the above example, each list had three identifiers, and so no recycling was done; but in the statement

```
CALCULATE A,B,C = X,Y + 1,2,3
```

the second list is of length only two, and so is recycled to give the calculations:

```
CALCULATE A = X + 1
CALCULATE B = Y + 2
CALCULATE C = X + 3
```

If the longest list is not on the left-hand side of the assignment operator, CALCULATE gives a fault diagnostic.

We must stress that Genstat operates on lists of *data structures* in its calculations and not on lists of *expressions*; if you want to specify several expressions, you must separate them by semicolons and not commas. As an example, suppose that you have two variates X and Y, where X is to be multiplied by 10, and Y is to be divided by 180. Naively, you might write the statement:

```
CALCULATE X,Y = X*10,Y/180
```

This statement is syntactically correct, but it does not do what you want: in fact it corresponds to the pair of statements

```
CALCULATE X = X*10/180
CALCULATE Y = X*Y/180
```

which is quite different from the intention. Genstat interprets the elements of the list on the right-hand side as 10 and Y, and not as "X*10" and "Y/180". If you really want to combine these two operations together in a single expression, you need to put

```
CALCULATE X,Y = X,Y * 10,1 / 1,180
```

Then the three lists on the right-hand side are taken in parallel: firstly X, 10, and 1, and then Y, 1, and 180. If you want to execute more than one expression in a CALCULATE statement, you must separate each one from the next by a semicolon: for example

```
CALCULATE X,Y = X*10; Y/180
```

CALCULATE has three options: PRINT, ZDZ, and TOLERANCE. If you set the PRINT option to summary, Genstat will print some summary information every time that values are assigned to a structure. The information has the same form as in the READ directive (3.1.1): identifier, minimum value, mean value, maximum value, number of values, number of missing values, and whether or not the set of values is skew. In Example 4.1.1e two assignments are made, and summaries are printed for the variates B and C.

Example 4.1.1e

```
1   VARIATE [VALUES=1,4,*,7,10] A
2   CALCULATE [PRINT=summary] C = (B = 2*A) + 1

  Identifier   Minimum    Mean    Maximum    Values    Missing
           B      2.00    11.00      20.00         5          1
           C      3.00    12.00      21.00         5          1
```

If you try to use CALCULATE to do something invalid, such as the logarithm or the square root of a negative number, Genstat generates a warning diagnostic and inserts a missing value in the offending unit. The one exception is the division of zero by zero, which is regarded as deliberate. Genstat thus does not print a diagnostic, but uses option ZDZ to determine whether the result should be a missing value (ZDZ=missing) or zero (ZDZ=zero); the default is missing. In this example, the variate %dm is formed with zeroes in the positions where Fresh_wt and Dry_wt both have zeroes.

Example 4.1.1f

```
1   VARIATE [VALUES=15.74,88.61,48.70,0,49.37] Fresh_wt
2   & [VALUES=3.21,11.3,7.83,0,7.23] Dry_wt
3   CALCULATE [ZDZ=zero] %dm = 100*Dry_wt / Fresh_wt
4   PRINT Dry_wt,Fresh_wt,%dm; FIELDWIDTH=9; DECIMALS=2

Dry_wt Fresh_wt      %dm
  3.21    15.74    20.39
 11.30    88.61    12.75
  7.83    48.70    16.08
  0.00     0.00     0.00
  7.23    49.37    14.64
```

Arithmetic operations with real numbers can suffer from rounding errors. Genstat uses real arithmetic for all its operations in CALCULATE, and so makes allowance for cases where rounding error may cause problems: in other words, very small numbers are taken to be zero. Sometimes, however, you may want to do calculations with numbers that are genuinely very small, and so the TOLERANCE option allows you to change the value that Genstat uses to assess the rounding-off.

4.1.2 Expressions with scalars and vectors

Example 4.1.2a shows a calculation involving scalars and variates. Several scalars are used to transform the variate Mpg (miles per gallon) into its metric counterpart Lp100k (litres per 100 kilometres). The scalar values are applied to every unit of the variate. You will see that the zero value in Mpg causes Genstat to print the warning "Attempt to divide by zero"; a missing value is placed in the corresponding position in Lp100k. This warning is printed only once per operation; so subsequent zero values in Mpg do not trigger it again.

Example 4.1.2a

```
1   VARIATE [VALUE=0,10,20,30,32...40,0,50] Mpg
2   VARIATE Lp100km
3   SCALAR Lpt,Cmin,Ydm,Inyd,Mkm; VALUE=0.568,2.54,1760,36,1000
4   CALCULATE Lp100km = 8 * Lpt * 100 * Mkm * 100 / \
5      ( Mpg * Ydm * Inyd * Cmin )
```

```
******** Warning (Code CA 18). Statement 1 on Line 5
Command: CALCULATE Lp100km = 8 * Lpt * 100 * Mkm * 100 /   ( Mpg * Ydm * Inyd
* Cmin)
Attempt to divide by zero
Attempt to divide by zero occurs at unit 1
```

```
6   PRINT Lp100km,Mpg; FIELDWIDTH=8; DECIMALS=2
```

```
Lp100km      Mpg
      *      0.00
  28.24     10.00
  14.12     20.00
   9.41     30.00
   8.82     32.00
   8.30     34.00
   7.84     36.00
   7.43     38.00
   7.06     40.00
      *      0.00
   5.65     50.00
```

If you use CALCULATE with variates that are restricted, Genstat applies the same restriction to all the variates involved in each calculation. Thus if more than one of these variates is restricted, they must all be restricted in the same way. In Example 4.1.2b, the variate Fresh_wt is restricted to those of its values that correspond to the first level of the factor Block. Only these units are involved in the calculation; the other units are left unchanged. Here the variate %dm had no values, and so the units in block 2 are given missing values.

Example 4.1.2b

```
1  VARIATE [VALUES=15.74,88.61,48.70,49.37,18.96,12.13,23.38,48.16] Fresh_wt
2  & [VALUES=3.21,11.3,7.83,7.23,3.55,2.6,4.0,6.43] Dry_wt
3  VARIATE [NVALUES=8] %dm
4  FACTOR [LEVELS=2; VALUES=4(1,2)] Block
5  RESTRICT Fresh_wt; CONDITION=Block.EQ.1
6  CALCULATE %dm = Dry_wt*100/Fresh_wt
7  PRINT %dm
```

```
     %dm
   20.39
   12.75
   16.08
   14.64
       *
       *
```

```
        *
        *
```

When Genstat implicitly declares a structure during a CALCULATE operation, it also by default sets its attributes to match those of the structures in the calculation: for details, see 4.1.5. We now do the same calculation, but leave Genstat to declare the structure %dry_m as a variate. In particular, the length of %dry_m will be the same as that of Fresh_wt and %dry_m will become restricted in the same way as Fresh_wt. Thus, the PRINT statement shows only the values of %dry_m corresponding to block 1; in fact, all the other values of %dry_m are missing. Genstat would also have carried the restriction across if we had declared %dry_m as a variate but had left the CALCULATE statement to set its number of values.

Example 4.1.2c

```
8   CALCULATE %dry_m = Dry_wt*100/Fresh_wt
9   PRINT %dry_m

    %dry_m
    20.39
    12.75
    16.08
    14.64
```

If you put a factor in a calculation, Genstat will use its levels, as shown when variate V takes its values from the factor F in line 2 of Example 4.1.2d. The function NEWLEVELS (5.2.1) allows you to specify an alternative levels variate to be used instead in the calculation. Line 3 of Example 4.1.3d uses the values 3.5 and 6.4, instead of the values 2 and 4 in the levels variate of the factor F, when forming the values of the variate Vn.

Example 4.1.2d

```
1   FACTOR [LEVELS=!(2,4)] F; VALUES=!((2,4)4)
2   CALCULATE V = F
3   & Vn = NEWLEVELS(F; !(3.5,6.4))
4   PRINT V,Vn

         V          Vn
     2.000       3.500
     4.000       6.400
     2.000       3.500
     4.000       6.400
     2.000       3.500
     4.000       6.400
     2.000       3.500
     4.000       6.400
```

If the factor is on the left–hand side of the equals sign, Genstat checks that each of the results of the calculation is an acceptable level. This allows you to define the values of a factor from a variate, or from another factor. However you must already have declared the factor, with its levels and labels vectors; factors cannot be declared implicitly. Example 4.1.2e first sets the values of the factor Rate from the variate Setting; it then uses the NEWLEVELS function to form the values of the factor Amount, whose first level corresponds to levels 1 and 2 of the factor Rate and whose second level corresponds to levels 3 and 4.

Example 4.1.2e

```
1   VARIATE [VALUES=1,3,2,1,4,3,1,2] Setting
2   FACTOR [LEVELS=!(1.25,2.5,3.75,5)] Rate
3   CALCULATE Rate = Setting*1.25
4   FACTOR [LABELS=!T(lower,higher)] Amount
5   CALCULATE Amount = NEWLEVEL(Rate; !(1,1,2,2))
6   PRINT Setting,Rate,Amount; FIELDWIDTH=8; DECIMALS=2
```

Setting	Rate	Amount
1.00	1.25	lower
3.00	3.75	higher
2.00	2.50	lower
1.00	1.25	lower
4.00	5.00	higher
3.00	3.75	higher
1.00	1.25	lower
2.00	2.50	lower

Text structures are allowed only with the relational operators .EQS., .NES., .IN., and .NI. described in 4.1.1, or in the functions CHARACTERS, GETFIRST, GETLAST, GETPOSITION, NOBSERVATIONS, NMV, NVALUES, and POSITION. The result of any expression is a number, so you cannot create a text with CALCULATE, even if the structures on which the operations are being done are texts.

4.1.3   Expressions with matrices

All the arithmetic, relational, and logical operators that we have now seen in use with variates can also be used with rectangular matrices, symmetric matrices, and diagonal matrices. The basic rule when using these with different types of matrix is that their dimensions must conform. This means that, for each pair of matrices, row dimension must match row dimension, and column dimension must match column dimension. Consider the matrices Mx, My, and Mz, and the symmetric matrix Smz declared here:

```
MATRIX [ROWS=3; COLUMNS=4] Mz,My
MATRIX [ROWS=3; COLUMNS=3] Mx
SYMMETRICMATRIX [ROWS=3] Smz
```

The dimensions of Mz and My conform; but the dimensions of Mx and Mz do not, since Mx and Mz have different numbers of columns, three and four respectively. Similarly the dimensions of the symmetric matrix Smz and the matrix Mx conform; but the dimensions of Smz and Mz do not.

For simplicity, our examples mostly involve addition; but remember that you can replace the operator + with any of the other arithmetic, logical, or relational operators. Matrix multiplication is described towards the end of this subsection.

In Example 4.1.3a, two rectangular matrices, Ma and Mb (each with four rows and three columns) are added together to form Mc. Note that Genstat operates in turn on each element of these two matrices, and that the new structure Mc is a matrix also with four rows and three columns.

Example 4.1.3a

```
 1   MATRIX [ROWS=4; COLUMNS=3] Ma,Mb; \
 2      VALUES=!(1...12),!(5,4,6,12,10,11,7,9,8,3,1,2)
 3   & Mc
 4   CALCULATE Mc = Ma + Mb
 5   PRINT Mc
```

	Mc 1	2	3
1	6.00	6.00	9.00
2	16.00	15.00	17.00
3	14.00	17.00	17.00
4	13.00	12.00	14.00

When you do calculations with two diagonal matrices, each one must have the same number of rows. Similarly, with symmetric matrices, the row dimensions must match. When you add, subtract, multiply, divide, or exponentiate a symmetric matrix, only those elements that are stored by Genstat are operated on. Here the two symmetric matrices Sma and Smb are added together to form another symmetric matrix Smc; this is done element by element.

Example 4.1.3b

```
 6   SYMMETRICMATRIX [ROWS=4] Sma,Smb; \
 7      VALUES=!(1...10),!(7,8,4,9,5,2,10,6,3,1)
 8   & Smc
 9   CALCULATE Smc = Sma + Smb
10   PRINT Smc
```

	Smc			
1	8.00			
2	10.00	7.00		
3	13.00	10.00	8.00	
4	17.00	14.00	12.00	11.00
	1	2	3	4

If you use a symmetric matrix in a calculation together with a matrix, it will be extended to include the values above the diagonal, before the calculation is done. Similarly, diagonal matrices are extended for calculations with matrices or symmetric matrices. Example 4.1.3c adds the diagonal matrix Da to the symmetric matrix Sma and puts the results in the matrix Md.

Example 4.1.3c

```
11   DIAGONALMATRIX [ROWS=4; VALUES=3,2,4,1] Da
12   MATRIX [ROWS=4; COLUMNS=4] Md
13   CALCULATE Md = Sma + Da
14   PRINT Md
```

	Md 1	2	3	4
1	4.000	2.000	4.000	7.000
2	2.000	5.000	5.000	8.000
3	4.000	5.000	10.000	9.000
4	7.000	8.000	9.000	11.000

You can also use variates together with matrices, provided their dimensions conform. Genstat treats variates as column matrices: that is, with *n* rows and one column. Example 4.1.3d adds the variate Va to the four-by-one matrix Me.

Example 4.1.3d

```
15   VARIATE [NVALUE=4; VALUES=4,2,1,3] Va
16   MATRIX [ROWS=4; COLUMNS=1; VALUES=10,4,7,2] Me
17   CALCULATE Me = Me + Va
18   PRINT Me
```

	Me 1
1	14.000
2	6.000
3	8.000
4	5.000

You can use a scalar with any of the matrix structures; the scalar is applied to every element of the matrix, in exactly the same way as when scalars and variates occur together in a calculation (4.1.2). Here the scalar Sca is added to every element of the symmetric matrix Sma.

Example 4.1.3e

```
19   SCALAR Sca; VALUE=3
20   CALCULATE Sma = Sma + Sca
21   PRINT Sma
```

	Sma			
1	4.000			
2	5.000	6.000		
3	7.000	8.000	9.000	
4	10.000	11.000	12.000	13.000

1	2	3	4

The multiplication operator (*) means element-by-element multiplication for the two matrices, not matrix multiplication.

Example 4.1.3f

```
22  MATRIX [ROWS=4; COLUMNS=3] Mf
23  CALCULATE Mf = Ma * Mb
24  PRINT Mf
```

```
        Mf
         1          2          3

1      5.00       8.00      18.00
2     48.00      50.00      66.00
3     49.00      72.00      72.00
4     30.00      11.00      24.00
```

For matrix multiplication you can use the compound operator *+ or the function PRODUCT (4.2.4). The column dimension of the first matrix must then match the row dimension of the second. In Example 4.1.3g, the four-by-four matrix Mh is formed from the matrix product of Ma with Mg, a matrix with three rows and four columns.

Example 4.1.3g

```
25  MATRIX [ROWS=3; COLUMNS=4; VALUES=1,4,7,10,2,5,8,11,3,6,9,12]
26  MATRIX [ROWS=4; COLUMNS=4] Mh
27  CALCULATE Mh = Ma *+ Mg
28  PRINT Mh
```

```
        Mh
         1          2          3          4

1      14.0       32.0       50.0       68.0
2      32.0       77.0      122.0      167.0
3      50.0      122.0      194.0      266.0
4      68.0      167.0      266.0      365.0
```

To summarize then, *+ is used in Genstat for matrix multiplication while * allows the corresponding elements of two matrices to be multiplied together.

The rules for implicit declarations when combining matrices are in 4.1.5. The rules for qualified identifiers of matrices are in 4.1.6. Genstat also provides several special matrix functions, including the INVERSE function, which can be included in CALCULATE statements: details are given in 4.2.4.

4.1.4   Expressions with tables

You can use tables in expressions in much the same way as you would any other numerical structure. Arithmetic, relational, and logical operators act element-by-element, as do the general functions (4.2.1).

Tables in expressions must be either all without margins or all with margins. If you try to mix tables with and without margins, Genstat will report an error.

Calculations with tables are very straightforward when they have the same factors in their classifying sets. In Example 4.1.4a two tables are added together:

Example 4.1.4a

```
 1   FACTOR [LEVELS=2; LABELS=!T(Woburn,Rothamsted)] Soil
 2   & [LEVELS=2; LABELS=!T(low,medium)] Acidity
 3   TABLE [CLASSIFICATION=Soil,Acidity] Ta,Tb; \
 4     VALUES=!(6.91,4.98,4.86,*),!(6.38,4.68,6.49,*)
 5   & Tc
 6   CALCULATE [PRINT=summary] Tc = Ta + Tb

    Identifier    Minimum      Mean   Maximum     Values     Missing
           Tc        9.66     11.43     13.29          4           1

 7   PRINT Tc

                        Tc
       Acidity         low      medium
          Soil
        Woburn       13.29        9.66
    Rothamsted       11.35           *
```

When tables have different classifying sets, there are two cases to consider. We illustrate them with the assignment operator, but the rules apply to any operation. The first case is when the table on the left-hand side has a factor in its classifying set that is not in the classifying set of the table on the right-hand side. In this case, the left-hand table is expanded to include that factor, by duplicating its values across the levels of the factor and any margin. Thus, in Example 4.1.4b, the values of the table Tb are repeated over the levels of the factor Block, which is the factor additional in the table Td. In other words the table Tb has been extended to include the factor Block: perhaps the easiest way of thinking about what happens is that each level of the extra factor contains a whole copy of the table on the right-hand side.

Example 4.1.4b

```
 8   FACTOR [LEVELS=2] Block
 9   TABLE [CLASSIFICATION=Soil,Acidity,Block] Td
10   CALCULATE Td = Tb
11   PRINT Td

                                 Td
                     Block         1            2
       Soil        Acidity
     Woburn            low       6.380        6.380
                   medium        4.680        4.680
```

```
Rothamsted              low      6.490      6.490
                       medium       *          *
```

The second case is when the table on the right-hand side has a factor in its classifying set that is not in the classifying set of the table on the left-hand side. Now the values in the margin over that factor are taken for the left-hand table. If the table has no margins, they must be calculated first. By default Genstat forms marginal totals, but you can use the special table functions (4.2.5) to form other types of margin. In Example 4.1.4c, marginal totals are calculated for table Td over the factor Block, and the results are placed in the previously declared table Tc.

Example 4.1.4c

```
   12   CALCULATE Tc = Td
   13   PRINT Tc

                        Tc
       Acidity         low      medium
          Soil
         Woburn       12.76       9.36
      Rothamsted      12.98         *
```

The classifying set of a table has two forms – one taken from the sequence in which the factors were listed in the CLASSIFICATION option of the TABLE declaration, the other determined by the order in which the identifiers of the factors are stored within Genstat. The second of these is called the *ordered classifying* set, and is the one used by CALCULATE for all operations on tables. CALCULATE permutes the values of tables so that they correspond to the ordered classifying set.

There are two consequences. The first is that if a fault occurs while an operation on a table is being done, its values may have been permuted, and so may no longer be in the order corresponding to the classifying set specified in the CLASSIFICATION option of the TABLE declaration. However, this occurs only if there has been a fault, since CALCULATE does not permute the values permanently.

The second consequence concerns implicit declarations. When a table is declared implicitly there is no obvious order for the factors other than the order in which their identifiers are stored (which generally reflects the order in which they were defined within the job). Thus Genstat will define the classifying set to be the same as the ordered classifying. So, if the resulting table Tc had not already been declared in Example 4.1.4a, its classifying set would in this case have been Acidity, Soil. So, if you want your table printed with the factors in a particular order, you must declare the table before its values are assigned in CALCULATE, or else use the DOWN, ACROSS and WAFER options of PRINT (3.2.2).

4.1.5   Rules for implicit declarations

Undeclared structures on the left-hand side of an assignment (=) in an expression are declared automatically: this is known as an *implicit declaration*. The type of structure is chosen to be the one most appropriate to the results that have been produced. This can be described according to a few straightforward rules. However, you do not need to know about the rules unless you intend to let Genstat perform these declarations for you.

The assignment operator (=) can appear anywhere in an expression, and so you need to be aware of the order of evaluation. For example, in the CALCULATE statement

```
CALCULATE Vc = Va*Vb
```

the result of Va*Vb is not placed directly in Vc: CALCULATE forms an intermediate structure whose values in this case are the results of Va*Vb; then the values of the intermediate structure are assigned to Vc. On assignment, the type and other relevant attributes of the resultant structure are also taken from the intermediate structure if these have not been defined previously (either implicitly or explicitly).

When structures of the same type are combined, the rule is that the intermediate structure will be of the same type; the same rule applies to tables with identical classifying sets. When structures of different types are combined, you need to know what form the intermediate structure takes.

In list below, .OP. refers to any arithmetic, logical, or relational operator, except .IS., .ISNT., .IN., .NI., .EQS., and .NES. which have their own rules described earlier (4.1.1). The dimensions of operands must conform in any operation involving matrices and variates. The second column indicates the type of structure resulting from the operation and the third column lists the types of structures to which it can be assigned.

Combination	Type of intermediate structure	Assignment
Scalar .OP. Scalar	Scalar	*any*
Variate .OP. Scalar	Variate	Variate,Factor
Variate .OP. Variate	Variate	Variate,Factor
Factor .OP. Scalar	Variate	Variate,Factor
Factor .OP. Variate	Variate	Variate,Factor
Factor .OP. Factor	Variate	Variate,Factor
Diagonal .OP. Scalar	Diagonal	Diagonal,Symmetric
Diagonal .OP. Variate	*invalid*	–
Diagonal .OP. Factor	*invalid*	–
Diagonal .OP. Diagonal	Diagonal	Diagonal,Symmetric,Matrix
Symmetric .OP. Scalar	Symmetric	Diagonal,Symmetric,Matrix
Symmetric .OP. Variate	*invalid*	–
Symmetric .OP. Factor	*invalid*	–
Symmetric .OP. Diagonal	Symmetric	Diagonal,Symmetric,Matrix
Symmetric .OP. Symmetric	Symmetric	Diagonal,Symmetric,Matrix
Matrix .OP. Scalar	Matrix	Matrix
Matrix .OP. Variate	Matrix	Matrix,Variate

Matrix .OP. Factor	Matrix	Matrix,Variate
Matrix .OP. Diagonal	Matrix	Diagonal,Symmetric,Matrix
Matrix .OP. Symmetric	Matrix	Diagonal,Symmetric,Matrix
Matrix .OP. Matrix	Matrix	Matrix
Table .OP. Scalar	Table	Table
Table .OP. Variate	*invalid*	–
Table .OP. Factor	*invalid*	–
Table .OP. Diagonal	*invalid*	–
Table .OP. Symmetric	*invalid*	–
Table .OP. Matrix	*invalid*	–
Table .OP. Table	Table	Table

In the last rule, Table .OP. Table, the classifying set of the intermediate table is the union of the two classifying sets. For example, in

```
FACTOR [LEVELS=2] Fa,Fb,Fc
TABLE [CLASSIFICATION=Fa,Fb] Ta
TABLE [CLASSIFICATION=Fa,Fc] Tb
CALCULATE Tc = Ta+Tb
```

The resulting table, Tc, will have the classifying set Fa, Fb, and Fc. As explained at the end of 4.1.4, the classifying set of a table has two forms. All tables in CALCULATE have their values permuted according to the ordered classifying set. On assignment, the ordered classifying set is transferred to the new table, which Genstat declares implicitly. So the classifying set and ordered classifying set are the same for tables declared implicitly.

The third column, headed "Assignment", lists the types of structure to which the values in the intermediate structure can be assigned. Genstat allows a fair amount of flexibility in this. All the intermediate structures contain numbers, and so you cannot declare factors implicitly in CALCULATE. However, you can assign a variate to a factor, so long as the values of the variate all occur as valid levels of the factor (4.1.2).

Most functions produce a result with the same type as their first argument, but there are some exceptions; see 4.2.

4.1.6 Rules for qualified identifiers

Qualified identifiers were introduced in 1.7.2, together with the rules for expanding them into lists. The rules for their use are similar to the rules for the arguments of the ELEMENTS function (4.2.8). the number of qualifiers that a structure can have is determined by its dimensionality. The dimensionality of scalars is defined to be zero, and so they cannot be qualified. Tables have varying numbers of dimensions, up to nine, but in Release 3.1 of Genstat 5 cannot be qualified. The dimensionalities of the stuctures that can be qualified are as follows.

1: variate, text, factor, diagonal matrix, and symmetric matrix.
2: matrix and symmetric matrix.

Notice that a symmetric matrix can have a dimensionality of either one or two, and so can be qualified in two ways; these are described below.

When an expression contains several qualified vector structures, you define a different subset for each vector; but for the calculation to work, the number of values contributed from each vector must be the same: see lines 5 to 6 of Example 4.1.6a. Genstat then ignores any restrictions on the vectors; in fact qualified identifiers provide an alternative way of specifying subsets of vectors. Example 4.1.6a illustrates the use of qualifications with variates, texts, and a factor. In each case the qualified vector is a vector with fewer values, but of the same type as the original structure: for example, Ta$[!(1,3,5)] is a text with three values instead of six.

Example 4.1.6a

```
1    VARIATE [NVALUES=5] Va; VALUES=!(1...5)
2    TEXT [NVALUES=6] Ta,Tb; VALUES=!T(a,b,c,d,e,f),!T(a,a,c,c,f,f)
3    FACTOR [NVALUES=8; LEVELS=3] Fa; VALUES=!(1,3,2,3,1,2,3,1)
4    VARIATE [VALUES=12(0)] Vb
5    CALCULATE Vb$[!(3,6,10)] = Va$[!(1,2,5)] * \
6       (Ta$[!(1,3,5)] .EQS. Tb$[!(2,4,6)]) + Fa$[!(5,7,2)]
7    PRINT Vb; DECIMALS=0

     Vb
      0
      0
      2
      0
      0
      5
      0
      0
      0
      3
      0
      0
```

When you have a qualified diagonal matrix, the subset of values is itself a diagonal matrix. Similarly a symmetric matrix, qualified by a single list, is also a symmetric matrix. The qualifier indicates which rows and columns are to be included; see line 3 of Example 4.1.6b.

Example 4.1.6b

```
1    SYMMETRICMATRIX [ROWS=4] Sma; VALUES=!(1...10)
2    & [ROWS=3] Smb
3    CALCULATE Smb = Sma$[!(1,4,2)]
4    PRINT Sma,Smb; FIELDWIDTH=6; DECIMALS=0

     Sma

1      1
2      2    3
3      4    5    6
4      7    8    9    10

       1    2    3    4

     Smb
```

```
1      1
2      7    10
3      2     8     3

       1     2     3

5    MATRIX [ROWS=4; COLUMNS=5] Ma; VALUES=!(1...20)
6    & [ROWS=2; COLUMNS=2] Mb
7    CALCULATE Mb = Sma$[!(1,4);!(2,3)] + Ma$[!(1,4);!(3,4)]
8    PRINT Ma,Mb; FIELDWIDTH=6; DECIMALS=0

              Ma
               1     2     3     4     5

         1     1     2     3     4     5
         2     6     7     8     9    10
         3    11    12    13    14    15
         4    16    17    18    19    20

              Mb
               1     2

         1     5     8
         2    26    28
```

Symmetric matrices can also have two qualifiers, in which case Genstat treats the result as a rectangular matrix. Rectangular matrices must have two qualifiers. In line 7 of Example 4.1.6b, the values of the rectangular matrix Mb are formed from the addition of the values in rows 1 and 4, and columns 2 and 3, of the symmetric matrix Sma to the values in rows 1 and 4, and columns 3 and 4, of the matrix Ma.

All the examples above show how to form vectors and matrices that have fewer values than the original: that is, the vectors and matrices take their values from subsets of the source structures. You can form also larger vectors and matrices, by using repeated values in the qualifier set. In Example 4.1.6c the matrix Mc, with four rows and three columns is formed from the two-by-two matrix Mb.

Example 4.1.6c

```
1    MATRIX [ROWS=2; COLUMNS=2; VALUES=5,7,6,2] Mb
2    MATRIX [ROWS=4; COLUMNS=3] Mc
3    VARIATE [NVALUES=4; VALUES=1,2,2,1] Va
4    & [NVALUES=3; VALUES=1,1,2] Vb
5    CALCULATE Mc = Mb$[Va; Vb]
6    PRINT Mc; FIELDWIDTH=6; DECIMALS=0

              Mc
               1     2     3

         1     5     5     7
         2     6     6     2
         3     6     6     2
         4     5     5     7
```

Instead of using variates to qualify the structures, you can use any numerical structure, and these structures can be qualified too. Genstat treats any structure used as a qualifier as a one-dimensional list of values. You can build very complicated qualifications in this way. The only limitation is that the set of values of the qualifiers must form a valid address list for the parent structure. In Example 4.1.6d, the complicated qualification reduces to assigning the value 3 to the element in row 3 and column 4 of the matrix Ma.

Example 4.1.6d

```
1    VARIATE [NVALUES=6] Va; VALUES=!(1,4,3,2,4,3)
2    MATRIX [ROWS=4; COLUMNS=6] Ma; VALUES=!(1...24)
3    CALCULATE Ma$[Va$[2]; Ma$[1; 3]] = 3
4    PRINT Ma; FIELDWIDTH=6; DECIMALS=0
```

```
           Ma
            1      2      3      4      5      6

     1      1      2      3      4      5      6
     2      7      8      9     10     11     12
     3     13     14     15     16     17     18
     4     19     20      3     22     23     24
```

You can use text to qualify structures, since it can label the rows and columns of matrices and the units of vectors. In Example 4.1.6e the matrix Mb is formed with numbers of rows and columns equal to the number of values (that is lines) of the texts Tsa and Tsb.

Example 4.1.6e

```
1    TEXT [NVALUES=6] Ta; VALUES=!T(a,b,c,d,e,f)
2    &    [NVALUES=4] Tb; VALUES=!T(g,h,i,j)
3    &    [NVALUES=3] Tsa; VALUES=!T(d,a,f)
4    &    Tsb; VALUES=!T(i,h,j)
5    MATRIX [ROWS=Ta; COLUMNS=Tb] Ma; VALUES=!(1...24)
6    CALCULATE Mb = Ma$[Tsa; Tsb]
7    PRINT Ma,Mb; FIELDWIDTH=6; DECIMALS=0
```

```
              Ma
     Tb        g      h      i      j
     Ta
      a        1      2      3      4
      b        5      6      7      8
      c        9     10     11     12
      d       13     14     15     16
      e       17     18     19     20
      f       21     22     23     24

              Mb
               1      2      3

       1      15     14     16
       2       3      2      4
       3      23     22     24
```

You can put in a missing identifier (*) to mean the complete set of elements from the dimension concerned. Example 4.1.6f shows how to transfer the values from columns 1 and 2 of the matrix Ma into the variates Vc1 and Vc2 respectively. Using qualified identifiers for transferring rows and columns of matrices to and from variates is more straightforward than using the EQUATE directive (4.3). The missing identifier (*) in the first qualifier for Ma indicates that Genstat is to take all the rows.

Example 4.1.6f

```
1   MATRIX [ROWS=5; COLUMNS=4] Ma; VALUES=!(1...20)
2   VARIATE [NVALUES=5] Vc1,Vc2
3   CALCULATE Vc1,Vc2 = Ma$[*; 1,2]
4   PRINT Ma; FIELDWIDTH=6; DECIMALS=0

                Ma
                 1      2      3      4

            1    1      2      3      4
            2    5      6      7      8
            3    9     10     11     12
            4   13     14     15     16
            5   17     18     19     20

5   & Vc1,Vc2; FIELDWIDTH=6; DECIMALS=0

 Vc1    Vc2
   1      2
   5      6
   9     10
  13     14
  17     18
```

Single values from a qualified variate are treated as scalars, but those from the various types of matrices have the same type as their parent. If you want these one-by-one matrices to be used as scalars, you can include an embedded assignment in the expression. For example, to multiply the variate Va by the value in row 2 and column 1 of the matrix Ma, you should put:

```
SCALAR Sca
CALCULATE Vb = Va * (Sca = Ma$[2;1])
```

If you tried to use the expression Va*Ma$[2;1], you would get an error message, since Genstat would object to multiplying the variate Va by the one-by-one matrix Ma$[2;1].

4.2   Functions for use in expressions

This section lists and describes the functions that can be used in expressions. The general form is illustrated by the statement:

```
CALCULATE y = LOG10(x)
```

Here LOG10 is the name of a function, and the identifier enclosed in brackets is its argument. Throughout this section we use lower case for identifiers that are arguments or results of functions, such as x and y above, to contrast with the upper-case conventionally used in this manual for function names, such as LOG10.

The argument of a function can be a list of identifiers, or even an expression. Some functions may need two arguments, in which case the arguments are separated by a semicolon (;). For example:

```
CALCULATE w = SORT(x; y+z)
```

(For an explanation of SORT, see below.) Genstat checks that you have given the correct number of arguments. With some functions, you do not need to set the second and subsequent arguments; in that case, you should omit the semicolons that would follow the last argument that you do use.

The functions in Genstat are divided into classes as follows: general and mathematical functions (4.2.1), scalar functions (4.2.2), variate functions (4.2.3), matrix functions (4.2.4), table functions (4.2.5), dummy functions (4.2.6), character functions (4.2.7), elements of structures (4.2.8), and statistical functions (4.2.9). They are described in alphabetical order within each subsection. At the beginning of each class we set out the valid types of argument for each function, and the type of the result. We give synonyms, and abbreviations for the function names where these have fewer than four letters: for example, the matrix function INVERSE has the two abbreviations INV and I. You can abbreviate any function to four letters (1.7.1): for example, LOG10 could be written as LOG1 – although this particular abbreviation might be a little misleading!

Some operations, such as the formation of generalized inverses of matrices or the calculation of ranks and quantiles of variates, are provided by directives and procedures instead of by functions. These are described in later sections of this chapter.

4.2.1   General and mathematical functions

In this subsection, x and y represent identifiers, or lists of identifiers, of any structures containing numerical data: that is, scalars, variates, factors, tables, matrices, diagonal matrices or symmetric matrices; s represents a scalar, f a factor and v a variate. Where x and y occur together as arguments they must be of the same type. Apart from NEWLEVELS, which produces a variate from a factor, the result of any of these functions has the same type as that of the first argument.

ABS(x)	gives the absolute value of x: $\lvert x \rvert$.
ARCCOS(x)	gives the inverse cosine of x ($-1 \le x \le 1$); the result is in radians.
ARCSIN(x)	give the inverse sine of x ($-1 \le x \le 1$); the result is in radians.
CIRCULATE(x; s)	treats x as a circular list and shifts its values round the list according to the value and sign of s. For example, if x contains 1,2,3,4,5, and s is −2, then the result is 3,4,5,1,2; if s were 2, the result would be 4,5,1,2,3. If you omit the second operand, CIRCULATE moves the values by one place to the right: that is, s=1.

COS(x)	gives the cosine of x, for x in radians.
CUMULATE(x) or CUM(x)	forms the cumulative sum of the values of x: for example, the result from x with values 1,5,4 is 1,6,10. If the operand is a scalar, the result is the value of the scalar.
DIFFERENCE(x; s)	forms the differences between consecutive elements of x: that is, the *i*th element of the result is $x_i - x_{i-s}$. If you omit the second operand, first differences are formed (s=1). If $i-s<1$ or $i-s>n$, where *n* is the number of values of x, the *i*th element is set to missing.
EXP(x)	gives the exponential function of x: e^x.
INTEGER(x) or INT(x)	gives the integer part of x: [x].
LOG(x)	gives the natural logarithm of x (x>0).
LOG10(x)	gives the logarithm to base 10 of x (x>0).
MODULO(x; y)	Form modulus of x to base y.
MVINSERT(x; y)	replaces values in x by missing value wherever the second identifier stores a non-zero value (representing the logical result .TRUE.).
MVREPLACE(x; y)	replaces missing values in x with corresponding values from y. Elements with missing values in both x and y produce a warning message.
NEWLEVELS(f; x)	forms a variate from the factor f; the variate x contains values to correspond to the levels, and should be of the same length as the number of levels of the factor. The result of this function is a variate of the same length as f. For an example see 4.1.2.
REVERSE(x)	reverses the values of x: for example, the result from x with values 1,2,3 is 3,2,1.
ROUND(x)	rounds to nearest integer.
SHIFT(x; s)	shifts the values of x by s places (to the right or left according to the sign of s). This is not a circular shift, and so some positions lose values; these are replaced with missing values. That is, the *i*th element of the result is the value that was in element *i*−s unless *i*−s≤0.
SIN(x)	gives the sine of x, for x in radians.
SORT(x; y)	sorts the elements of x into the order that would put the values of y into ascending order; the values of y are left unchanged. If the second argument is omitted, the values of x are sorted into ascending order. x can be the same structure as y. See below for an example.
SQRT(x)	gives the square root of x (x≥0).

Example 4.2.1 illustrates the functions DIFFERENCE, INTEGER, ROUND, MVREPLACE, SIN, and SORT. In the example of SORT, Genstat sorts the missing values in variate Va to the beginning

of the array; the sorted order within the missing values is completely arbitrary. Tied values, too, are sorted arbitrarily, although in this example the tied values are by chance listed in their order of occurrence in the variate Va.

Example 4.2.1

```
  1   VARIATE [VALUES=-0.4,4.1,8.4,*,-1.6,5.7,-2.3]
  2   CALCULATE Vb = DIFFERENCE(Va; 2)
  3   PRINT Va,Vb; FIELDWIDTH=6; DECIMALS=1

    Va      Vb
  -0.4       *
   4.1       *
   8.4     8.8
     *       *
  -1.6   -10.0
   5.7       *
  -2.3    -0.7

  4   CALCULATE Iva = INTEGER(Va)
  5   & Rva = ROUND(Va)
  6   PRINT Va,Iva,Rva; FIELDWIDTH=6; DECIMALS=1

    Va     Iva     Rva
  -0.4     0.0     0.0
   4.1     4.0     4.0
   8.4     8.0     8.0
     *       *       *
  -1.6    -1.0    -2.0
   5.7     5.0     6.0
  -2.3    -2.0    -2.0

  7   VARIATE [VALUES=1,2,3,27.3,5,6,7] Vb
  8   CALCULATE Vc = MVREPLACE(Va; Vb)
  9   PRINT Vc; DECIMALS=2

         Vc
      -0.40
       4.10
       8.40
      27.30
      -1.60
       5.70
      -2.30

 10   CALCULATE Ve = SIN(Vc)
 11   PRINT Ve; FIELDWIDTH=8; DECIMALS=3

       Ve
  -0.389
  -0.818
   0.855
   0.827
  -1.000
  -0.551
  -0.746

 12   VARIATE [VALUES=3,1,*,*,1,4,7,4,*] Vsa
 13   & [VALUES=1...9] Vsb
 14   CALCULATE Vsc = SORT(Vsb; Vsa)
```

```
15  PRINT Vsc; FIELDWIDTH=6; DECIMALS=0

   Vsc
    9
    4
    3
    2
    5
    1
    6
    8
    7
```

4.2.2 Scalar functions

The scalar functions generate a scalar result from other types of structure. Some of these functions calculate a summary value describing some aspect of the contents of the structure such as the maximum value, the median value, the mean, the variance, or the area under a curve. Other functions allow you to copy attributes of the structure in the argument: for example, NVALUES gives the number of values. Finally, the CONSTANTS function, which has a single-valued text (or a string) as its argument, provides an easy and accurate way of specifying various scalar constants such as π and the value used by Genstat to represent missing values.

In this subsection, x again represents any numerical structure (scalar, variate, factor, rectangular matrix, symmetric matrix, diagonal matrix, or table), f is a factor, and m is either a rectangular matrix, a symmetric matrix, or a diagonal matrix; y is a structure of the same type as x. All the functions produce a scalar result from each structure in the argument list; all except NMV and NOBSERVATIONS ignore missing values in the structure. Thus, the function MEAN is equivalent to SUM divided by NOBSERVATIONS, and the function NOBSERVATIONS is equivalent to NVALUES minus NMV.

Restrictions on a variate within a scalar function do not carry over to the expression outside.

AREA(y; x)	numerically integrates the curve running through the points specified by variates x and y; x must be monotonically increasing or decreasing.
CONSTANTS(t) or C(t)	provides the value of various constants, according to the contents of the string in the single-valued text t: e (for a string of 'e' or 'E'), π ('pi' or 'PI'), or missing value ('*' or 'missingvalue'). The string can be specified in either upper or lower case (or any mixture) and can be abbreviated just like the string settings of options such as PRINT.
CORRELATION(x; y)	if both x and y are specified, returns a scalar giving the correlation between the values of x and y; if y is omitted, CORRELATION is a matrix function which forms a matrix of correlations from a (symmetric) matrix of sums of squares

and products (4.2.4).

COVARIANCE(x; y)	calculates the covariance between the values of x and y.
MAXIMUM(x) or MAX(x)	finds the maximum of the values of x.
MEAN(x)	gives the mean of the values of x.
MEDIAN(x) or MED(x)	finds the median of the values of x.
MINIMUM(x) or MIN(x)	finds the minimum of the values of x.
NCOLUMNS(m)	gives the number of columns of matrix m.
NLEVELS(f)	gives the number of levels of factor f.
NMV(x)	counts the number of missing values in x.
NOBSERVATIONS(x)	counts the number of observations (non-missing values) in x.
NROWS(m)	gives the number of rows of matrix m.
NVALUES(x)	gives the number of values, including missing values, of x (the length of x).
SUM(x) or TOTAL(x)	gives the sum of the values in x.
VARIANCE(x) or VAR(x)	gives the variance of the values in x (the divisor being the number of non-missing values in x, minus 1).

For example:

Example 4.2.2

```
 1   VARIATE [VALUES=8,2,16,4,1,10,*,30] Va
 2   " Med, Mn, Tot, Obs, and Nv are declared implicitly (as scalars). "
 3   CALCULATE Med = MEDIAN(Va)
 4   & Mn = MEAN(Va)
 5   & Tot = SUM(Va)
 6   & Obs = NOBSERVATIONS(Va)
 7   & Nv  = NVALUES(Va)
 8   PRINT Med,Mn,Tot,Obs,Nv; FIELDWIDTH=8; DECIMALS=2

     Med        Mn       Tot       Obs        Nv
    8.00     10.14     71.00      7.00      8.00

 9   FACTOR [LEVELS=!(1,2,4,8)] Ff
10   CALCULATE Nl = NLEVELS(Ff)
11   PRINT Nl; FIELDWIDTH=6; DECIMALS=1

    Nl
   4.0
```

Other summaries, including quartiles and coefficients of skewness, can be produced by the DESCRIBE procedure (7.1.1).

4.2.3   Variate functions

Variate functions produce summaries across a set of variates or a set of scalars. They each have a single argument, which is a pointer to the set of variates or scalars to be summarized. The variates in a set must all be of the same length. If any of them is restricted, that restriction is applied to all of them; if several are restricted, each restriction must be to the same set of units. For a set of variates the result of each function is a variate of the same length as the variates in the set, while for a set of scalars the result is a scalar. For example, if p points to the variates X1, X2, and X3, each of length *n*, VMEANS(p) produces a variate of length *n*, whose *i*th unit contains the mean of the values in the unit *i* of X1, X2, and X3.

All the functions except VNMV and VNOBSERVATIONS ignore missing values. Thus, the function VMEANS is equivalent to VSUMS divided by VNOBSERVATIONS, and the function VNOBSERVATIONS is equivalent to VNVALUES minus VNMV.

VCORRELATION(p1; p2)	gives the correlation, at every unit, between the values of the corresponding structures in pointers p1 and p2.
VCOVARIANCE(p1; p2)	gives the covariance, at every unit, between the values of the corresponding structures in pointers p1 and p2.
VMAXIMA(p)	finds the maximum of the values in each unit over the variates (or scalars) in pointer p.
VMEANS(p)	gives the mean of the non-missing values in each unit over the variates (or scalars) in pointer p.
VMEDIANS(p)	finds the median of the values in each unit over the variates (or scalars) in pointer p.
VMINIMA(p)	finds the minimum of the values in each unit over the variates (or scalars) in pointer p.
VNMV(p)	counts the number of missing values in each unit over the variates (or scalars) in pointer p.
VNOBSERVATIONS(p)	counts the number of observations (non-missing values) in each unit over the variates (or scalars) in pointer p.
VNVALUES(p)	gives the total number of values in each unit over the variates (or scalars) in pointer p: that is the number of variates (or scalars) in p.
VSUMS(p) or VTOTAL(p)	gives the sum of the non-missing values in each unit over the variates (or scalars) in pointer p.
VVARIANCES(p)	gives the variance of the non-missing values in each unit over the variates (or scalars) in pointer p.

Example 4.2.3

```
 1   VARIATE [NVALUES=6] X,Y,Z; \
 2     VALUES=!(28,*,18,26,*,17),!(12,27,*,34,*,15),!(17,25,3(*),20)
 3   & Min,Mean,Max,Obs,Nval,Tot
 4   POINTER [VALUES=X,Y,Z] P
 5   CALCULATE Min = VMINIMA(P)
 6   & Mean = VMEANS(P)
 7   & Max = VMAXIMA(P)
 8   & Obs = VNOBSERVATIONS(P)
 9   & Nval = VNVALUES(P)
10   & Tot = VTOTALS(P)
11   PRINT X,Y,Z,Min,Mean,Max,Obs,Nval,Tot; FIELDWIDTH=8; DECIMALS=1
```

X	Y	Z	Min	Mean	Max	Obs	Nval	Tot
28.0	12.0	17.0	12.0	19.0	28.0	3.0	3.0	57.0
*	27.0	25.0	25.0	26.0	27.0	2.0	3.0	52.0
18.0	*	*	18.0	18.0	18.0	1.0	3.0	18.0
26.0	34.0	*	26.0	30.0	34.0	2.0	3.0	60.0
*	*	*	*	*	*	0.0	3.0	*
17.0	15.0	20.0	15.0	17.3	20.0	3.0	3.0	52.0

4.2.4   Matrix functions

These functions operate on the various types of matrix available in Genstat. The type of the resulting structure depends on the function concerned. For some of the functions you can specify a variate, which is treated as a rectangular matrix with one column. Any restriction on the variate is then ignored. (Remember that matrices cannot be restricted.) A *matrix* is a rectangular, symmetric or diagonal matrix structure; a *square matrix* is a rectangular matrix with the same number of rows as of columns.

CORRELATION(x) or CORRMAT(x)

forms a correlation matrix from a symmetric matrix x that contains sums of squares and products: the values of the resulting symmetric matrix c are formed by $c_{ij} = x_{ij} / \sqrt{(x_{ii} x_{jj})}$. Note, CORRELATION with two arguments, x and y, can also be used to produce the (scalar) correlation between the values in two structures (4.2.2).

CHOLESKI(x)

forms the Choleski decomposition of a symmetric matrix x; this produces a square matrix L such that x = LL′ and such that upper off-diagonal elements are zero. The symmetric matrix x must be positive semi-definite.

DETERMINANT(x) or DET(x) or D(x)

forms the determinant of a symmetric matrix or a square matrix; the result is a scalar. Genstat uses the decomposition x = LU, and the determinant is defined to be $\Pi\{l_{ii} u_{ii}\}$.

INVERSE(x) or INV(x) or I(x)

forms the inverse of a non-singular symmetric matrix, or a square matrix; the result is a square matrix or a symmetric matrix, according to the type of x. For a square matrix, Genstat uses Crout's method by forming the lower and upper triangular decomposition of the matrix, x = LU, and inverting L and U separately. Genstat uses the equivalent decomposition (Choleski) for symmetric matrices, which must be positive semi-definite. To form the generalized inverses of singular rectangular matrices you can use the procedure GINVERSE.

LTPRODUCT(x; y)

forms the left transposed product of x and y: that is, the matrix product of the transpose of x with y, which can also be written $T(x)*+y$. The structures x and y can be matrices or variates. The number of rows of x must equal the number of rows of y. The result is a rectangular matrix with number of rows equal to the number of columns of x and number of columns equal to the number of columns of y, unless both x and y are diagonal matrices when the result is also a diagonal matrix.

PRODUCT(x; y)   forms the matrix product of x and y; this can also be written $x*+y$ using the operator *+. The structures x and y can be matrices or variates. The number of columns of x must equal the number of rows of y. The result is a rectangular matrix with number of rows equal to the number of rows of x and number of columns equal to the number of columns of y, unless both x and y are diagonal matrices when the result is also a diagonal matrix.

QPRODUCT(x; y)   forms the quadratic product of x and y; it can thus be written as $x*+y*+T(x)$, but the use of QPRODUCT is more efficient. x is a rectangular matrix or a variate, and y is a symmetric matrix or a diagonal matrix or a scalar. The number of columns of x must be the same as the number of rows of y. The result is a symmetric matrix with number of rows equal to the number of rows of x.

RTPRODUCT(x; y)   forms the right transposed product of x and y: that is, the matrix product of x with the transpose of y, which can also be written $x*+T(y)$. The structures x and y can be matrices or variates. The number of columns of x must equal the number of columns of y. The result is a rectangular matrix with number of rows equal to the number of rows of x and number of columns equal to the number of rows of y, unless both x and y are diagonal matrices when the result is also a diagonal matrix.

SOLUTION(x; y)   solves a set of simultaneous linear equations $x*+b=y$:

$$x_{11} b_1 + x_{12} b_2 + ... + x_{1n} b_n = y_1$$
$$...$$
$$x_{n1} b_1 + x_{n2} b_2 + ... + x_{nn} b_n = y_n$$

The function thus finds b, as in the alternative expression
CALCULATE b = PRODUCT(INVERSE(x); y)
but the use of SOLUTION is more efficient and numerically stable than using PRODUCT and INVERSE: x is a square matrix and y is a rectangular matrix or a variate. The number of rows of x must be the same as the number of rows of y. The result is a rectangular matrix with numbers of rows and columns the same as y.

SUBMAT(x)   forms sub-triangles or sub-rectangles of a rectangular or
                                     symmetric matrix x, whose dimensions must be labelled by
                                     pointers. The structure to receive the values must have been
                                     declared already, as a rectangular or symmetric matrix
                                     according to the type of x, and have each of its dimensions
                                     also labelled by a pointer whose values are included in the
                                     pointer of the corresponding dimension of x. The
                                     correspondence between the values of the pointers that label
                                     the resulting matrix and those labelling x determines which
                                     rows and columns of x appear in the result. The same effect
                                     can be obtained by using the function ELEMENTS with a
                                     single list or expression for symmetric matrices, and with two
                                     lists for rectangular matrices. Just as with the ELEMENTS
                                     function, the resulting matrix can be made larger than x, by
                                     specifying repeated identifiers in its pointers.

TRACE(x)   forms the trace of matrix x: that is, the sum of its diagonal
                                     elements. x can be a square matrix, a diagonal matrix or a
                                     symmetric matrix. The result is a scalar.

TRANSPOSE(x) or T(x)   forms the transpose of x, where x is a rectangular matrix or
                                     a variate. The result is a rectangular matrix.

Example 4.2.4

```
 1   SYMMETRICMATRIX [ROWS=4] Sma; \
 2      VALUES=!(36,40,64,65,90,144,80,110,175,225)
 3   MATRIX [ROWS=4; COLUMNS=4] Chsma
 4   CALCULATE  Chsma = CHOLESKI(Sma)
 5   PRINT Chsma; FIELDWIDTH=8; DECIMALS=3

              Chsma
                  1        2        3        4

           1   6.000    0.000    0.000    0.000
           2   6.667    4.422    0.000    0.000
           3  10.833    4.020    3.237    0.000
           4  13.333    4.774    3.511    3.479

 6   MATRIX [ROWS=3; COLUMNS=3] Ma; VALUES=!(1,1,2,3,4,5,1,4,2)
 7   & Mainv
 8   CALCULATE Mainv = INVERSE(Ma)
 9   PRINT Ma; FIELDWIDTH=8; DECIMALS=3

                 Ma
                  1        2        3

           1   1.000    1.000    2.000
           2   3.000    4.000    5.000
           3   1.000    4.000    2.000

10   & Mainv; FIELDWIDTH=8; DECIMALS=3
```

```
            Mainv
                1         2         3

        1   -4.000     2.000    -1.000
        2   -0.333     0.000     0.333
        3    2.667    -1.000     0.333

11   MATRIX [ROWS=3; COLUMNS=3] Mx; VALUES=!(1,1,2,3,4,5,1,4,2)
12   & [ROWS=3; COLUMNS=1] My; VALUES=!(4,5,6)
13   & Bxy
14   CALCULATE Bxy = SOLUTION(Mx; My)
15   PRINT Bxy; FIELDWIDTH=8; DECIMALS=3

               Bxy
                 1

        1   -12.000
        2     0.667
        3     7.667

16   VARIATE Va,Vb,Vc,Vd,Ve,Vf,Vg,Vh,Vi,Vj
17   POINTER Pa,Pb,Pc,Pd; VALUES=!P(Va,Vb,Vc,Vd,Ve,Vf),!P(Vg,Vh,Vi,Vj), \
18        !P(Vc,Va,Vf,Ve),!P(Vi,Vh,Vg)
19   MATRIX [ROWS=Pa; COLUMNS=Pb] Ma ; VALUES=!(1...24)
20   & [ROWS=Pc; COLUMNS=Pd] Mb
21   CALCULATE Mb = SUBMAT(Ma)
22   PRINT Ma; FIELDWIDTH=8; DECIMALS=1

               Ma
        Pb     Vg        Vh        Vi        Vj
        Pa
        Va     1.0       2.0       3.0       4.0
        Vb     5.0       6.0       7.0       8.0
        Vc     9.0      10.0      11.0      12.0
        Vd    13.0      14.0      15.0      16.0
        Ve    17.0      18.0      19.0      20.0
        Vf    21.0      22.0      23.0      24.0

23   & Mb; FIELDWIDTH=8; DECIMALS=1

               Mb
        Pd     Vi        Vh        Vg
        Pc
        Vc    11.0      10.0       9.0
        Va     3.0       2.0       1.0
        Vf    23.0      22.0      21.0
        Ve    19.0      18.0      17.0
```

Other matrix operations and decompositions are described in 4.10.

4.2.5   Table functions

The table functions operate on tables to produce new values for extended or summarized tables; for example,

```
CALCULATE tr = TMEANS(ta)
```

takes means of certain of the cells in table `ta` and puts them in the table `tr`. If the resulting table, `tr` above, has already been declared, it must have the same status for margins as the corresponding in the function (ta above). But if tr is left to be declared implicitly, it will be given margins whether or not they occur in ta. Summaries are produced over the levels of the factors that occur in ta but not in tr; the type of summary depends on which function is used. Then, if there are factors that occur in tr but not in ta, these are given duplicate values as described in 4.1.4. Finally, if tr has margins, these are filled in according to the function specified. For example, if tr is classified by factors A and B but ta is classified by A, B and C,

        CALCULATE tr = TMEANS(ta)

will put, in each cell of tr, means over the levels of factor C, as shown in Example 4.2.5.

TMAXIMA(t)	forms margins of maxima for table t.
TMEDIANS(t)	forms margins of medians for table t.
TMEANS(t)	forms margins of means for table t.
TMINIMA(t)	forms margins of minima for table t.
TNOBSERVATIONS(t)	forms margins counting the numbers of observations (non-missing values) in table t.
TNMV(t)	forms margins counting the numbers of missing values in table t.
TNVALUES(t)	forms margins counting the numbers of values, missing or non-missing, in table t.
TSUMS(t) or TTOTALS(t)	forms margins of totals for table t.
TVARIANCES(t)	forms margins of between-cell variances for table t.

Example 4.2.5

```
    1   FACTOR [LEVELS=2] A,B,C
    2   TABLE [CLASSIFICATION=A,B,C] Ta; VALUES=!(1...8)
    3   & [CLASSIFICATION=A,B] Tr
    4   CALCULATE Tr = TMEANS(Ta)
    5   PRINT Ta,Tr; FIELDWIDTH=6; DECIMALS=1
```

```
                                  Ta
                       C     1      2
          A            B
          1            1    1.0    2.0
                       2    3.0    4.0
          2            1    5.0    6.0
                       2    7.0    8.0

                  Tr
          B       1      2
          A
          1      1.5    3.5
          2      5.5    7.5
```

4.2.6   Dummy functions

The function UNSET allows you to check whether a dummy is set; this is useful particularly in procedures (5.3) and FOR loops (5.2.1).

UNSET(d)   gives a scalar logical value (0 or 1) indicating whether or not the dummy d is set: that is, whether or not d points to another structure.

4.2.7   Character functions

This subsection describes the functions in Genstat that allow you to obtain information about text structures. As already mentioned, in 4.2.2, you can ascertain the number of lines in a text using the NVALUES function, the number of missing lines (null strings) by NMV, and the number of non-missing lines by NOBSERVATIONS. The functions described here produce variates from a text, giving details of the contents of each of its lines.

The CHARACTER function indicates the length of each line of the text, while GETFIRST and GETLAST find the position of the first or last non-space character in each line respectively.

GETPOSITION lets you find the position, in each line of the text in its first argument, of the corresponding line from the text in its second argument. This implies that the lines from the second text are shorter than or equal to the lines of the first text. In addition, there is an optional third argument (a logical), which allows you to specify whether or not comparisons of characters/letters are case sensitive. The default is false (that is, 0), which means that comparisons are case sensitive. If the third argument is set to true (a non-zero value), either as a scalar or in a variate with the same number of values as there are lines in the first argument, then lower and upper case letters are treated as the same; that is, comparisons are case insensitive.

CHARACTERS(t)   returns a variate giving the length of each line of text t.
GETFIRST(t)   gives a variate containing the position of the first non-space character in each string of text t.
GETLAST(t)   gives a variate containing the position of the last non-space character in each string of the text t.
GETPOSITION(t1; t2; x)   for each unit, if the string in t2 occurs as a substring of the string in t1, this returns the position at which the substring starts; otherwise it returns the value zero. t2 may contain a single string to be checked against every string of t1. x can be either a scalar or a variate, and supplies a logical value to indicate whether to ignore the case of any letters; if x is omitted the logical is assumed to be false (case not ignored).

4.2.8   Elements of structures

The ELEMENTS function has a similar role to qualified identifiers (4.1.6). Two functions, EXPAND and RESTRICTION, are available to derive sets of values from the results of a RESTRICT statement (4.4.1), while POSITION allows you to determine the position at which the values of one vector occur within another.

ELEMENTS(x; e1; e2)   specifies a set of elements of x; e1 and e2 are expressions. As with qualified identifiers, you cannot specify elements of scalars or tables. You cannot use a text in any of the arguments of ELEMENTS. However the ability to specify expressions in the second and third arguments, instead of merely structures, is one way in which the use of ELEMENTS is more powerful that the use of qualified identifiers.

EXPAND(x; s)   forms a variate of zeroes and ones from the values of x, which Genstat takes to be a list of unit numbers; usually x will have been formed as the save structure from a RESTRICT statement. The second argument, s, is a scalar defining the length of the result; if s is omitted and EXPAND cannot determine the length of the result from its context within the expression, the resulting variate will take its length from the units structure (2.3.4).

POSITION(x; y)   finds the position, within the vector y, of each value of x.

RESTRICTION(x)   forms a variate with ones in the positions of the set of units to which x is currently restricted; the other units of the result are left unchanged (or left as missing values if no values have been set previously). If this variate is declared implicitly here, it will be restricted in the same way and have the same number of values as x. If you use the RESTRICTION function on its own in the CONDITION parameter of the RESTRICT directive (4.4.1), the restriction on x is passed to all the vectors listed with first parameter of RESTRICT.

The rules of dimensionality of the structures to which ELEMENTS is applied, and the specification of the expressions e1 and e2, which identify the elements in each dimension, are similar to those for qualified identifiers (4.1.6). If x is a variate, a factor, or a diagonal matrix, you should not specify the third argument e2; the type of the result is the same as that of x. You can also omit the third argument if x is a symmetric matrix, in which case the result is also a symmetric matrix; or you can specify both expressions, in which case the result is a rectangular matrix. For rectangular matrices, both e1 and e2 must be specified, and the result is a rectangular matrix. Genstat evaluates each expression and treats the result as a one-dimensional list of values. In line 4 of Example 4.2.7a, the values of the symmetric matrix Smb are taken from the rows and columns of the symmetric matrix Sma indicated by the variate Va.

Example 4.2.7a

```
1  SYMMETRICMATRIX [ROWS=5] Sma; VALUES=!(15...1)
2  & [ROWS=3] Smb
3  VARIATE Va; VALUES=!(5,4,2)
4  CALCULATE Smb = ELEMENTS(Sma; Va)
5  PRINT Sma,Smb; FIELDWIDTH=5; DECIMALS=0
```

```
   Sma

1   15
2   14   13
3   12   11   10
4    9    8    7    6
5    5    4    3    2    1

     1    2    3    4    5


   Smb

1    1
2    2    6
3    4    8   13

     1    2    3
```

```
6  VARIATE Vb,Vc; VALUES=!(5,3,1),!(1,4,3)
7  MATRIX [ROWS=3; COLUMNS=3; VALUES=1...9] Ma
8  CALCULATE ELEMENTS(Sma; Vb; Vc) = Ma
9  PRINT Sma; FIELDWIDTH=4; DECIMALS=0
```

```
   Sma

1   7
2  14  13
3   9  11   6
4   8   8   5   6
5   1   4   3   2   1

    1   2   3   4   5
```

ELEMENTS is the only function that you are allowed to put on the left-hand side of an assignment. This is illustrated in line 8 of Example 4.2.7a, where the values of the matrix Ma are assigned to the elements of the symmetric matrix Sma indicated by the variates Va and Vb. Since Sma is symmetric, any values above the main diagonal indicated by Va and Vb are automatically transposed to their corresponding position below the diagonal.

Example 4.2.7b illustrates the use of the functions EXPAND, RESTRICTION, and POSITION.

Example 4.2.7b

```
1  VARIATE [VALUES=35,24,27,26,42,57] Age
2  RESTRICT Age; CONDITION=Age>30; SAVESET=Va
3  CALCULATE Vb = EXPAND(Va; 8)
4  PRINT Va,Vb; FIELDWIDTH=6; DECIMALS=0
```

```
          Va     1     5     6

          Vb     1     0     0     0     1     1     0     0
    5   VARIATE [VALUES=6(-1)] Rest
    6   CALCULATE Rest = RESTRICTION(Age)
    7   " Cancel the restriction on Age. "
    8   RESTRICT Age
    9   PRINT Age,Rest; FIELDWIDTH=6; DECIMALS=0

    Age  Rest
     35     1
     24    -1
     27    -1
     26    -1
     42     1
     57     1
```

4.2.9   Statistical functions

The statistical functions cover various activities relevant to statistical analyses.

There are functions to transform percentage data: LOGIT, CLOGLOG (complementary log-log), NED (equivalent to the probit), and ANGULAR. The inverse transformations are also available: ILOGIT, ICLOGLOG, NORMAL, and IANGULAR.

Cumulative lower and upper probabilities, and equivalent deviates are available for various probability distributions: Normal, F, chi-square, t, binomial, Poisson, hypergeometric, beta, gamma, lognormal and bivariate Normal. In addition, point probabilities are provided for the discrete distributions (binomial, Poisson, and hypergeometric). These functions all have a standard form: first a prefix (for example CL for cumulative lower probabilities) and then the name of the distribution. There are also various natural synonyms, such as NORMAL for CLNORMAL.

Log-likelihoods can be calculated for samples from either binomial, gamma, Normal or Poisson distributions (LLBINOMIAL, LLGAMMA, LLNORMAL, LLPOISSON).

Function URAND allows you to generate sets of uniform pseudo-random numbers. Random numbers from other distributions can be obtained using the procedure GRANDOM.

Unless otherwise stated in the descriptions below, the arguments of the functions can be any compatible numerical data structures. Any constraints on their possible values are given with each description. Except for the log-likelihood functions and the function URAND, the result is a structure of the same type, dimension, and number of values as the structure in the first argument.

The log-likelihood functions produce a scalar result. Their first arguments must be variates. The second and third arguments can be scalars or variates; if they are variates, they must be of the same length as the variate in the first argument. The meaning of the second and third arguments is given with each description, as well as the form of the expression used to calculate the log-likelihood.

ANGULAR(%p) or ANG(%p)   provides the angular transformation: %p is a percentage with

	$0<\%p<100$. The function forms
	$x = (180/\pi) \times \text{arcsine}(\sqrt{(\%p/100)})$
	and so the result x is in degrees $0<x<90$.

CED(p; df) — gives the deviate for probability p ($0<p<1$), for a chi-square distribution with df degrees of freedom. (Synonym of EDCHISQUARE.)

CHISQ(x; df) — gives the probability that a random variable, distributed as chi-square with df degrees of freedom, is less than x. (Synonym of CLCHISQUARE.)

CLBETA(x; a; b) — cumulative lower probability for a beta distribution with parameters a and b.

CLBINOMIAL(x; n; p) — probability of x or fewer successes out of n binomial trials with probability of success p.

CLBVARIATENORMAL(x; y; r) — cumulative lower probability for a bivariate Normal distribution with means 0, variances 1, and correlation r.

CLCHISQUARE(x; df) — cumulative lower probability for a chi-square distribution with df degrees of freedom.

CLF(x; df1; df2) — cumulative lower probability for an F distribution with df1 and df2 degrees of freedom.

CLGAMMA(x; m; d) — cumulative lower probability for a gamma distribution with mean m and index d.

CLHYPERGEOMETRIC(j; l; m; n) — probability of x or fewer positive samples out of a total sample of size m from a population of size n of which l are positive (hypergeometric distribution).

CLLOGNORMAL(x) — cumulative lower probability for a lognormal distribution corresponding to a Normal distribution with mean 0 and variance 1.

CLNORMAL(x) — cumulative lower probability for a Normal distribution with mean 0 and variance 1.

CLOGLOG(p) — takes the complementary log-log transformation of the percentages p ($0<p<100\%$).

CLPOISSON(j; m) — probability of value of x or less for a Poisson distribution with mean m.

CLT(x; df) — cumulative lower probability for a t distribution with df degrees of freedom.

CUBETA(x; a; b) — cumulative upper probability for a beta distribution with parameters a and b.

CUBINOMIAL(j; n; p) — probability of more than x successes out of n binomial trials with probability of success p.

CUBVARIATENORMAL(x; y; r) — cumulative upper probability for a bivariate Normal distribution with means 0, variances 1, and correlation r.

CUCHISQUARE(x; df)   cumulative upper probability for a chi-square distribution with df degrees of freedom.

CUF(x; df1; df2)   cumulative upper probability for an F distribution with df1 and df2 degrees of freedom.

CUGAMMA(x; m; d)   cumulative upper probability for a gamma distribution with mean m and index d.

CUHYPERGEOMETRIC(j; l; m; n)
                           probability of more than x positive samples out of a total sample of size m from a population of size n of which l are positive (hypergeometric distribution).

CULOGNORMAL(x)   cumulative upper probability for a lognormal distribution corresponding to a Normal distribution with mean 0 and variance 1.

CUNORMAL(x)   cumulative upper probability for a Normal distribution with mean 0 and variance 1.

CUPOISSON(j; m)   probability of a value greater than x for a Poisson distribution with mean m.

CUT(x; df)   cumulative upper probability for a t distribution with df degrees of freedom.

EDBETA(p; a; b)   equivalent deviate corresponding to cumulative lower probability p for a beta distribution with parameters a and b.

EDCHISQUARE(p; df)   equivalent deviate corresponding to cumulative lower probability p for a chi-square distribution with df degrees of freedom.

EDF(p; df1; df2)   equivalent deviate corresponding to cumulative lower probability p for an F distribution with df1 and df2 degrees of freedom.

EDGAMMA(p; m; d)   equivalent deviate corresponding to cumulative lower probability p for a gamma distribution with mean m and index d.

EDLOGNORMAL(p)   equivalent deviate corresponding to cumulative lower probability p for a lognormal distribution corresponding to a Normal distribution with mean 0 and variance 1.

EDNORMAL(p)   equivalent deviate corresponding to cumulative lower probability p for a Normal distribution with mean 0 and variance 1.

EDT(p; df)   equivalent deviate corresponding to cumulative lower probability p for a t distribution with df degrees of freedom.

FED(p; df1; df2)   gives the equivalent deviate at probability p ($0<p<1$) for an F distribution with df1 and df2 degrees of freedom, that is, the positive real number x to the left of which the area under the F distribution curve is p. (Synonym of EDF.)

FRATIO(x; df1; df2) or FPROBABILITY(x; df1; df2)

	gives the probability that a random variable with the F distribution, with numbers of degrees of freedom df1 and df2, is less than x. (Synonym of CLF.)
IANGULAR(x)	gives the inverse of the angular transformation (result in percentages).
ICLOGLOG(x)	gives the inverse of the complementary log-log transformation (result in percentages).
ILOGIT(x)	gives the inverse of the logit transformation (result in percentages).
LLBINOMIAL(x; n; p) or LLB(x; n; p)	provides the log-likelihood function for the binomial distribution with sample size n and mean proportion p (n and p are scalars or variates): $\Sigma \{ x \, \mathrm{Log}(n \, p \, / \, x) + (n{-}x) \, \mathrm{Log}(n \, (1{-}p) \, / \, (n{-}x)) \}$
LLGAMMA(x; m; d) or LLG(x; m; d)	provides the log-likelihood function for the gamma distribution with mean m and index d (m and d are scalars or variates): $\Sigma \{ d \, (\mathrm{Log}(d \, x \, / \, m) - x \, / \, m) - \mathrm{Log}(\Gamma (d)) \}$
LLNORMAL(x; m; v) or LLN(x; m; v)	provides the log-likelihood function for the Normal distribution with mean m and variance v (m and v are scalars or variates): $-\tfrac{1}{2} \Sigma \{ \mathrm{Log}(v) + (x{-}m)(x{-}m)/v \}$
LLPOISSON(x; m) or LLP(x; m)	provides the log-likelihood function for the Poisson distribution with sample size m (m is a scalar or a variate): $\Sigma \{ x \, \mathrm{Log}(m/x) + x - m \}$
LOGIT(p)	takes the logit transformation $\log(p/(100{-}p))$ of the percentages p $(0<p<100\%)$.
NED(p)	gives the Normal equivalent deviate for probability p $(0<p<1)$: that is, the real number x to the left of which the area under the standard Normal curve is p. The probit transformation was originally defined as NED(p)+5, but nowadays most applications (including the GLM section of Genstat, 8.5) omit the 5. (Synonym of EDNORMAL.)
NORMAL(x)	provides the Normal probability integral; that is, the probability that a random variable with the standard Normal distribution is less than x. (Synonym of CLNORMAL).
PRBETA(x; a; b)	probability density for a beta distribution with parameters a and b.
PRBINOMIAL(x; n; p)	probability of x successes out of n binomial trials with probability of success p.
PRHYPERGEOMETRIC(j; l; m; n)	probability of x successes out of a sample of m from a

PRPOISSON(j; m)

URAND(s1; s2)

population of size n of which l are positive (hypergeometric distribution).

probability of obtaining the value x for a Poisson distribution with mean m.

provides a uniform pseudo-random number generator, giving values in the range (0, 1). the algorithm is a modified version of that presented by Wichman and Hill (1982). The same algorithm is used by RANDOMIZE (9.8.3). s1 is a scalar which specifies the seed for the random numbers. The seed must have a non-zero value on the first occasion that you use URAND in a job; subsequently you can give a zero value to continue the sequence of random numbers. s2 is also a scalar; if you set this, the result is a variate of length equal to the value of the scalar. If you omit s2, the type of the result of URAND is determined from the context of the expression: that is from the type of the structure that is to receive the values that are generated; if the receiving structure has not been declared already, it will be declared implicitly as a variate with the length of the units structure (2.3.4)

Example 4.2.8 illustrates the functions LLNORMAL, NED, PRPOISSON, and CLPOISSON.

Example 4.2.8

```
  1   " Normal log-likelihood for X with mean 0.6 and variance 1.9 "
  2   VARIATE [VALUES=4.0,-3.5,-1.3,-2.8,1.9,2.5,0.3,-0.8,1.2,0.9] X
  3   CALCULATE Loglik = LLNORMAL(X; 0.6; 1.9)
  4   PRINT Loglik

      Loglik
      -16.72

  5   " Transform Pr to Normal equivalent deviates "
  6   VARIATE Pr; VALUES=!(0.1,0.45,*,0.2,0.83,-0.3,0.95)
  7   " There is an invalid value in unit 6; this is
 -8     given a missing value and a warning is printed. "
  9   CALCULATE Tran = NED(Pr)

******* Warning (Code CA 7). Statement 1 on Line 9
Command: CALCULATE Tran = NED(Pr)
Invalid value for argument of function
The first argument of the NED     function in unit 6 has the value    -0.3000

 10   PRINT Tran,Pr; FIELDWIDTH=8,10; DECIMALS=2,3

      Tran        Pr
     -1.28     0.100
     -0.13     0.450
         *         *
     -0.84     0.200
      0.95     0.830
         *    -0.300
```

```
 1.64     0.950
 11   " Calculate probabilities and cumulative probabilities
-12     for a Poisson distribution with mean 2.5 "
 13   VARIATE [V=1...10] N
 14   CALCULATE Prob = PRPOISSON(N; 2.5)
 15   & Cumprob = CLPOISSON(N; 2.5)
 16   PRINT N,Prob,Cumprob; DECIMALS=0,3,3
```

N	Prob	Cumprob
1	0.205	0.287
2	0.257	0.544
3	0.214	0.758
4	0.134	0.891
5	0.067	0.958
6	0.028	0.986
7	0.010	0.996
8	0.003	0.999
9	0.001	1.000
10	0.000	1.000

4.3 Transferring values between structures of different types (EQUATE)

The EQUATE directive copies values from one set of data structures to another. For example, you may wish to copy the values from a one-way table into a variate, or from a matrix into a set of variates (one variate for each row, or for each column), or the other way round, from variates into a matrix. Alternatively, you may want to append values from several data structures into a single one (see Example 4.3a). The only constraint is that the structures in the respective sets must all contain the same kind of values.

EQUATE   transfers data between structures of different sizes or types (but the same modes i.e. numerical or text) or where transfer is not from single structure to single structure.

Options

OLDFORMAT = *variate*	Format for values of OLDSTRUCTURES; within the variate, a positive value *n* means take *n* values, −*n* means skip *n* values and a missing value means skip to the next structure; default * i.e. take all the values in turn
NEWFORMAT = *variate*	Format for values of NEWSTRUCTURES; within the variate, a positive value *n* means fill the next *n* positions, −*n* means skip *n* positions and a missing value means skip to the next structure; default * i.e. fill all the positions in turn
FREPRESENTATION = *strings*	How to interpret factor values (labels, levels, ordinals); default leve

Parameters

OLDSTRUCTURES = *identifiers*	Structures whose values are to be transferred; if values of several structures are to be transferred to one item in the NEWSTRUCTURES list, they must be placed in a pointer
NEWSTRUCTURES = *identifiers*	Structures to take each set of transferred values; if several structures are to receive values from one item in the OLDSTRUCTURES list, they must be placed in a pointer

The general idea with EQUATE is that the values in the structures in the OLDSTRUCTURES list are copied into the structures in the NEWSTRUCTURES list. Each item in OLDSTRUCTURES list specifies a single data structure, or a single set of data structures, containing the values to be transferred. A single structure can be a factor, or a text, or any one of the structures that contain numbers (scalar, variate, rectangular matrix, diagonal matrix, symmetric matrix, or table). If you want to give a set of structures you must put them into a pointer. As already mentioned, all the structures in the set must contain the same kind of values: that is, they must all be texts, or all factors, or must all contain numbers (but they need not all be the same kinds of numerical structure – they could, for example, be a mixture of variates and matrices).

The corresponding entry in the NEWSTRUCTURE list indicates where the transferred data are to be placed. It is either a single structure or a pointer to a set of structures; the structures must be of a type suitable to store the values to be transferred.

In Example 4.3a, information about the employees of a firm has been typed in series in two separate sections, and the statement in lines 19 and 20 copies them into one; for each employee there are three pieces of information – name, grade, and hours.

Example 4.3a

```
  1   OPEN 'EMPLOYEE.DAT'; CHANNEL=2
  2   " Read values for the first 6 employees,
 -3      in series, into Name1, Grade1 and Hours1."
  4   TEXT [NVALUES=6] Name1
  5   FACTOR [NVALUES=6; LEVELS=3] Grade1
  6   VARIATE [NVALUES=6] Hours1
  7   READ [PRINT=data,errors; CHANNEL=2; SERIAL=yes] Name1,Grade1,Hours1

  1   Clarke Innes Adams Jones Day Grey :
  2   2 1 2 1 1 3 :
  3   45 51 40 46 44 40 :

  8   " Read values for the final 4 employees,
 -9      in series, into Name2, Grade2 and Hours2."
 10   TEXT [NVALUES=4] Name2
 11   FACTOR [NVALUES=4; LEVELS=3] Grade2
 12   VARIATE [NVALUES=4] Hours2
 13   READ [PRINT=data,errors; CHANNEL=2; SERIAL=yes] Name2,Grade2,Hours2

  4   Edwards Baker Hill Foster :
  5   2 2 3 1 :
  6   47 42 40 41 :
```

```
 14   " Use EQUATE to put information about all the employees
-15     into single vectors Name, Grade and Hours."
 16   TEXT [NVALUES=10] Name
 17   FACTOR [NVALUES=10; LEVELS=3] Grade
 18   VARIATE [NVALUES=10] Hours
 19   EQUATE !P(Name1,Name2),!P(Grade1,Grade2),!P(Hours1,Hours2); \
 20     NEWSTRUCTURES=Name,Grade,Hours
 21   PRINT Name,Grade,Hours
```

Name	Grade	Hours
Clarke	2	45.00
Innes	1	51.00
Adams	2	40.00
Jones	1	46.00
Day	1	44.00
Grey	3	40.00
Edwards	2	47.00
Baker	2	42.00
Hill	3	40.00
Foster	1	41.00

Except with a format (see below) Genstat ignores where each structure within a set from the OLDSTRUCTURES list ends and another one begins: that is, it treats the set as being a concatenated list of values. Similarly, it treats the structures in each NEWSTRUCTURES set as an unstructured list of positions that are to receive values. The old values are repeated as often as is necessary to traverse all the new positions. Example 4.3b forms a matrix M with repeated and alternating rows taken from variates R1 and R2.

Example 4.3b

```
1   VARIATE [VALUES=1...6] R1
2      & [VALUES=101...106] R2
3   " Form a matrix M whose rows are R1, R2, R1 and R2."
4   MATRIX [ROWS=4; COLUMNS=6] M
5   EQUATE !P(R1,R2); NEWSTRUCTURES=M
6   PRINT M; FIELDWIDTH=6; DECIMALS=0
```

	M					
	1	2	3	4	5	6
1	1	2	3	4	5	6
2	101	102	103	104	105	106
3	1	2	3	4	5	6
4	101	102	103	104	105	106

You can use the OLDFORMAT and NEWFORMAT options to control how the old values and new positions are traversed. The setting for each of these is a variate whose values are interpreted as follows:

(a)   a positive integer *n* means take the next *n* values (OLDFORMAT) or fill the next *n* positions (NEWFORMAT);

(b)   a negative integer *−n* means skip the next *n* values or positions;

(c)   a missing value * means skip to the end of the structure.

As usual, Genstat recycles when it runs out of values. That is, if the contents of one of the variates is exhausted before all the NEWSTRUCTURES positions have either been filled or skipped, then that variate is repeated. For example:

Example 4.3c

```
 7    " Form variates C[1...6] containing the values in the columns of M."
 8    VARIATE [NVALUES=4] C[1...6]
 9    EQUATE [OLDFORMAT=!((1,-5)4,-1)] M; NEWSTRUCTURES=C
10    PRINT C[1...6]; FIELDWIDTH=6; DECIMALS=0

 C[1]   C[2]   C[3]   C[4]   C[5]   C[6]
    1      2      3      4      5      6
  101    102    103    104    105    106
    1      2      3      4      5      6
  101    102    103    104    105    106
```

This gives the variates C[1...6] the values in the columns of M. It does it by taking one column at a time from M, skipping the values in the other columns. (Remember that the values of M are held row-by-row.) In detail, what happens is this. For C[1], the format !((1,-5)4,-1) in line 30 takes the value in row 1 column 1, then skips the elements in the remaining five columns of row 1 before taking the value from column 1 of row 2. For C[1] this continues for each row of M, until the final element of the format, -1, skips column 1 of row 1, so that C[2] is given the values in column 2, and so on.

Notice that, as pointer C is automatically available to refer to C[1...6] (see 3.3.4), there is no need to put, for example, !P(C[1...6]).

The final part of the example shows how to form a matrix from a set of variates that contain the values for the columns.

Example 4.3d

```
11    " Reform values of M so that its columns are C[1...6] in reverse order."
12    EQUATE [OLDFORMAT=!((1,-3)6,-1)] !P(C[6...1]); NEWSTRUCTURES=M
13    PRINT M; FIELDWIDTH=6; DECIMALS=0

                   M
                   1      2      3      4      5      6

          1        6      5      4      3      2      1
          2      106    105    104    103    102    101
          3        6      5      4      3      2      1
          4      106    105    104    103    102    101
```

If you are transferring values between factors, Genstat will check that each value to be transferred is valid for the factor in the NEWSTRUCTURES list. By default, Genstat will try to match the values using the levels of the factors, but you can set option FREPRESENTATION=labels to match by their labels, or FREPRESENTATION=ordinals to match them merely according to the ordinal position in the levels vector of each factor.

Example 4.3e illustrates the various possibilities.

Example 4.3e

```
 1  FACTOR [LEVELS=!(2,4); LABELS=!T(standard,double); VALUES=2,4,4,2] Dose1
 2  FACTOR [NVALUES=8; LEVELS=3; LABELS=!T(none,standard,double)] Dose2
 3  " Form Dose2 from values of Dose 1, repeated twice, matching by labels."
 4  EQUATE [FREPRESENTATION=labels] Dose1; NEW=Dose2
 5  PRINT [SERIAL=yes; ORIENTATION=across; RLWIDTH=6] Dose1,Dose2; FIELD=9

Dose1 standard    double    double standard

Dose2 standard    double    double standard standard    double    double standard

 6  " Form Dose3 from Dose1, matching by levels (the default)
-7     and then Dose4, matching by ordinal positions of the levels."
 8  FACTOR [NVALUES=4; LEVELS=!(0,1,2,3,4); LABELS=!T(none,d1,d2,d3,d4)] Dose3
 9  FACTOR [NVALUES=4; LEVELS=!(20,40,60)] Dose4
10  EQUATE Dose1; NEW=Dose3
11  EQUATE [FREPRESENTATION=ordinals] Dose1; NEW=Dose4
12  PRINT Dose1,Dose3,Dose4

        Dose1       Dose3       Dose4
     standard          d2       20.00
       double          d4       40.00
       double          d4       40.00
     standard          d2       20.00
```

The values of factors that have labels can be copied into texts. In addition, values of texts can be copied into factors, provided all the strings are valid labels for the factor concerned.

Factor values can also be copied into variates; the FREPRESENTATION option controls whether Genstat uses the levels or the ordinal values.

An alternative type of copying is provided by the VEQUATE procedure. This copies the values in a set of structures into a set of variates, one for each element of the original structures. The original structures (which must be of the same type and length) are input in a pointer, using the OLDSTRUCTURES parameter. The variates to take the values are returned in a pointer whose identifier is specified by the NEWSTRUCTURES parameter.

4.4 Operations on vectors

The directives and procedures described below can be used with any of the vector structures that Genstat supports: variates, factors, or texts. More specific facilities are described in 4.5 (variates), 4.6 (factors), and 4.7 (texts).

The RESTRICT directive (4.4.1) allows you to indicate that future statements should operate only on a subset of the units of the specified vectors. (The precise way in which RESTRICT affects the operation of other directives is described in the chapters that are devoted to these directives.) This is a convenient way of saving space when you wish to examine successive subsets, as there is no need to create a copy of the subset; the vectors themselves are unchanged and merely have some *restriction* associated with them. The restriction can be changed or cancelled at any time by specifying RESTRICT again. This would, for example enable you to analyse a data variate, taking one subset at a time while building up full variates of residuals and fitted values that contain the information from all the subset analyses.

Alternatively, if you wish to look at a single subset, the SUBSET procedure allows you to create a set of vectors which contain only a subset of the original vectors.

The SORT directive allows you to reorder the units of vectors according to the values of one or more index vectors (4.4.3). You can use the RANDOMIZE directive, described in 9.8.3, to put units into random order.

4.4.1   Applying a restriction to the units of a vector (**RESTRICT**)

RESTRICT defines a restricted set of units of vectors for subsequent statements.

No options

Parameters

VECTOR = *vectors*	Vectors to be restricted
CONDITION = *expression*	Logical expression defining the restriction for each vector; a zero (false) value indicates that the unit concerned is not in the set
SAVESET = *variates*	List of the units in each restricted set

---

The RESTRICT directive defines a *restriction* on the units of a vector, so that future operations will involve only a subset of the units. The directives that take account of RESTRICT are listed at the end of this subsection.

The VECTOR parameter specifies the vector or vectors that are to be restricted. These can be variates, factors, or texts, but all the vectors listed must be of the same length.

The CONDITION parameter specifies a logical expression which indicates which units of the vectors are in the defined subset. For example,

```
VARIATE [VALUES=1,2,3,2,3,4,3,4,5] V
RESTRICT V; CONDITION=V.EQ.2
```

restricts the vector V to those units with the value 2. Genstat evaluates the expression to generate internally a variate of zeroes and ones, of the same length as the vectors being restricted. A zero value indicates that the corresponding unit is to be excluded. The logical expression can involve any vector of the same length ar the vector to be restricted. For example, to restrict variate V and text T to the units with levels 1 or 2 or 4 of factor F, you could use the statement

```
RESTRICT V,T; CONDITION=(F.LE.2).OR.(F.EQ.4)
```

When using a text to define a restriction, remember that you cannot use logical operators like .EQ. and .NE. Instead you should use operators .IN., .NI., .EQS., and .NES. (4.1.2):

```
TEXT [VALUES=London,Madrid,Nairobi,Ottawa,Paris,Quito,Rome] City
& [VALUES=London,Madrid,Paris,Rome] Europe
RESTRICT City; CONDITION=City.IN.Europe
```

restricts the text City to lines 1, 2, 5, and 7 only.

Of course, the expression may just contain a single variate of the of the same length as the vectors to be restricted. Again a zero indicates that the corresponding unit in the vector to be restricted is excluded, while any non-zero entry causes inclusion. Thus the restriction above on the text `City` could also be specified by

```
RESTRICT T; CONDITION=!(1,1,0,0,1,0,1)
```

The same effect can be achieved by using the EXPAND function (4.2.8):

```
RESTRICT City; CONDITION=EXPAND(!(1,2,5,7))
```

Another function that may be useful is RESTRICTION; this allows you to generate a variate of ones and zeros indicating the units to which a vector is currently restricted (4.2.8). It thus provides a very convenient way of transferring a restriction from one vector to another. For example,

```
RESTRICT Timezone,Distance; CONDITION=RESTRICTION(City)
```

restricts the vectors `Timezone` and `Distance` to the same units as those to which `City` is currently restricted.

Finally, if you omit the CONDITION parameter, this removes any restrictions on the vectors are removed. For example

```
RESTRICT City,Timezone,Distance
```

removes any restrictions that have been set on `City`, `Timezone`, and `Distance`.

Note that if the vectors used in the CONDITION expression are themselves restricted these restrictions will remain in force during the current calculation of the condition. A danger here, therefore, is that you may accidentally end up restricting out all the elements of a vector by using RESTRICT repeatedly. The safest way to avoid this is to remove the restrictions on any vectors to be used in the CONDITION expression before you use them to restrict vectors in some different way.

The SAVESET parameter can be used to save the numbers of the units that are in the restricted set. These are saved in a variate with one value for each unit retained by the restriction. Thus, if the example above with variate V were to become

```
VARIATE [VALUES=1,2,3,2,3,4,3,4,5] V
RESTRICT V; CONDITION=V.EQ.2; SAVESET=S
```

`S` would be created as a variate of length 2, with values 2 and 4.

Not all directives take account of RESTRICT. For those that do, usually only one vector in the list of parameters has to be restricted for the directive to treat them all as being restricted in the same way. A fault is reported if any vectors in such a list are restricted in different ways.

The table below summarizes which directives obey restrictions and which do not. A general guideline is that RESTRICT is obeyed by all directives that operate on vectors except in those statements where explicit identification of elements is possible: for example EQUATE (4.3) and READ (3.1), and when qualified identifiers or ELEMENTed identifiers are used in CALCULATE (4.1.6 and 4.2.8). However, this guideline does not always operate, so you should check each directive in this manual to confirm precisely what happens.

Restrictions on texts are obeyed only by PRINT (3.2), CALCULATE (4.1.2), CONCATENATE (4.7.1), and EDIT (4.7.2).

RESTRICT obeyed	**RESTRICT not obeyed**
Data structures	
SSPM [TERMS= ; GROUPS=]	FACTOR
	TEXT
	VARIATE
Control structures	
CASE expression	
ELSIF expression	
EXIT expression	
IF expression	
Input and output	
PRINT STRUCTURE=	DUMP
	READ
Graphics	
CONTOUR GRID=	AXES
DCONTOUR GRID=	PEN
DGRAPH Y= ; X=	SKIP
DHISTOGRAM DATA=	
GRAPH Y= ; X=	
HISTOGRAM DATA=	
Backing store	
RETRIEVE	
STORE	
Calculation and manipulation	
CALCULATE expression	ASSIGN
(except using qualified	COMBINE
or ELEMENTed identifiers)	DELETE
CONCATENATE	EQUATE
EDIT	GENERATE
FSSP [WEIGHTS=]	UNITS
INTERPOLATE OLDVALUES= ;	
OLDINTERVALS= ; NEWVALUES= ;	
NEWINTERVALS=	
RANDOMIZE parameter	
RESTRICT VECTOR= ; CONDITION=	
SORT OLDVECTOR=	
TABULATE [CLASSIFICATION=] DATA=	
Regression analysis	
ADD parameter	
DROP parameter	
FIT parameter	

```
FITCURVE parameter
FITNONLINEAR parameter
MODEL [WEIGHTS= ; OFFSET= ; GROUPS= ]
  Y= ; NBINOMIAL=
PREDICT CLASSIFY=
RKEEP RESIDUALS=; FITTEDVALUES=; LEVERAGE=
STEP parameter
SWITCH parameter
TERMS parameter
TRY parameter
```

Analysis of variance
```
ANOVA Y=                          BLOCKSTRUCTURE
                                  COVARIATE
                                  TREATMENTSTRUCTURE
```

Multivariate and cluster analysis
```
FSIMILARITY DATA=                 CLUSTER
HLIST DATA=                       PCO
HSUMMARIZE DATA=                  REDUCE
PCP DATA=
RELATE DATA=
```

Time-series analysis
```
CORRELATE SERIES= ; LAGGEDSERIES= FORECAST
ESTIMATE SERIES=                  FTSM
FILTER OLDSERIES=
FOURIER SERIES= ; ISERIES= ;
  TRANSFORM= ; ITRANSFORM=
TRANSFER SERIES=
```

4.4.2   Forming a subset of the units in a vector (SUBSET)

Procedure SUBSET forms vectors containing subsets of the values in other vectors. Example 4.4.2 uses procedures LIBHELP and LIBEXAMPLE to print the on-line help for its options and parameters, and to run an example.

Example 4.4.2

```
  1   LIBHELP [PRINT=options,parameters] 'SUBSET'

    Procedure       SUBSET

Help['options']
CONDITION  = expression   Logical expression to define which units are to be
                          included; no default - this option must be set
SETLEVELS  = string       Whether to reform the levels (and labels) of factors
                          to exclude those that do not occur in the subset

Help['parameters']
OLDVECTORS = vectors      Vectors from which subsets are to be formed
NEWVECTORS = vectors      Vectors to store the subsets; if none are specified,
                          the OLDVECTORS are redefined to store the subsets
```

```
2   LIBEXAMPLE 'SUBSET'; EXAMPLE=Exsub
3   " Request printing of macros (as well as ordinary statements)."
4   SET [INPRINT=statements,macros
5   " Run Exsub as a macro."
6   ##Exsub
    1   PRINT    'Example of how to use procedure SUBSET.'
```

Example of how to use procedure SUBSET.

```
2   VARIATE [VALUES=101...126] X
3   TEXT    [VALUES=a,b,c,d,e,f,g,h,i,j,k,l,m,n,o,p,q,r,s,t,u,v,w,x,y,z] T
4   FACTOR  [LEVELS=26; VALUES=1...26] F
5   SUBSET  [CONDITION=X<111] OLDVECTOR=X,T,F; NEWVECTOR=Xs,Ts,Fs
6   PRINT   Xs,Ts,Fs
```

Xs	Ts	Fs
101.0	a	1
102.0	b	2
103.0	c	3
104.0	d	4
105.0	e	5
106.0	f	6
107.0	g	7
108.0	h	8
109.0	i	9
110.0	j	10

The subset is defined by a logical condition which must be specified by the CONDITION option; units with a non-zero value (true) for the condition are included in the subset, others are omitted. Subsets can be formed for factors, texts and variates. Relevant attributes will also be transferred across to the new structures but, if the subset excludes some of the levels of a factor, a new reduced set of levels (and labels) can be requested by setting option SETLEVELS=yes. The original vectors are specified by the OLDVECTOR parameter and identifiers for the vectors to contain the subsets are specified by the NEWVECTORS parameter. If NEWVECTORS is not set, the OLDVECTORS are redefined to store the subsets instead of their original values.

4.4.3   Sorting vectors into numerical or alphabetical order (SORT)

SORT sorts units of vectors according to an index vector.

Options

INDEX = *vectors*	Variates, texts or factors whose values are to define the ordering; default is to use the first vector in the OLDVECTOR list
DIRECTION = *string*	Order in which to sort (ascending, descending); default asce
DECIMALS = *scalar*	Number of decimal places to which to round before sorting numbers; default * i.e. no rounding

Parameters

OLDVECTOR = *vectors* or *pointers*	Factors, pointers, texts, or variates whose values are to be sorted
NEWVECTOR = *vectors* or *pointers*	Structure to receive each set of sorted values; if any are omitted, the values are placed in the corresponding OLDVECTOR

The SORT directive allows you to reorder the units of a list of vectors or pointers according to one or more "index" vectors. These can be specified explicitly using the INDEX option. If you omit the INDEX option, Genstat uses the first vector in the OLDVECTOR list. The DECIMALS option allows you to define the number of decimal places that are taken into account for an index variate: for example DECIMALS=0 would round each value to the nearest integer. If you do not set this, there is no rounding. The DIRECTION option controls whether the ordering is into ascending or descending order; by default DIRECTION=ascending.

The vectors or pointers whose values are to be sorted are listed by the OLDVECTOR parameter. The units of each structure are permuted in exactly the same way, into an ordering determined from the index vectors. In Example 4.4.3a, the units of the variates Age and Income, the text Name, and the factor Sex are sorted to put the names into alphabetical order.

Example 4.4.3a

```
 1   VARIATE [VALUES=18,50,24,49,61,29,32,42,36,40] Age
 2   & [VALUES=3000,17500,5000,20000,7000,4500, \
 3      12000,18000,15500,17500] Income
 4   TEXT [VALUES=Clarke,Innes,Adams,Jones,Day, \
 5      Grey,Edwards,Baker,Hill,Foster]    Name
 6   FACTOR [LABELS=!T(male,female); VALUES=2,1,1,1,2,2,1,1,2,1] Sex
 7   PRINT Age,Income,Name,Sex
```

Age	Income	Name	Sex
18.00	3000	Clarke	female
50.00	17500	Innes	male
24.00	5000	Adams	male
49.00	20000	Jones	male
61.00	7000	Day	female
29.00	4500	Grey	female
32.00	12000	Edwards	male
42.00	18000	Baker	male
36.00	15500	Hill	female
40.00	17500	Foster	male

```
 8   SORT [INDEX=Name] Age,Income,Name,Sex
 9   PRINT Age,Income,Name,Sex
```

Age	Income	Name	Sex
24.00	5000	Adams	male
42.00	18000	Baker	male
18.00	3000	Clarke	female
61.00	7000	Day	female
32.00	12000	Edwards	male
40.00	17500	Foster	male
29.00	4500	Grey	female
36.00	15500	Hill	female

```
50.00         17500        Innes        male
49.00         20000        Jones        male
```

Here the index vector Name is also one of the vectors being sorted; indeed if you list it first, then you can omit the INDEX option:

    SORT Name,Age,Income,Sex

However it need not be among the vectors sorted. Moreover, you can specify new vectors to contain the sorted values, and thus keep the unsorted values in the original vectors. For example

    SORT [INDEX=Name] Age,Income,Name,Sex; NEWVECTOR=A,*,N,S

would place the sorted values of Age, Name, and Sex into A, N, and S; as there is a null entry (*) corresponding to Income in the NEWVECTOR list, the sorted incomes would replace the original values of Income. Any undeclared vector in the NEWVECTOR list is declared implicitly to match the corresponding OLDVECTOR.

    We now sort the units into order of descending age.

Example 4.4.3b

```
10   SORT [DIRECTION=descending] Age,Income,Name,Sex
11   PRINT Age,Income,Name,Sex
```

Age	Income	Name	Sex
61.00	7000	Day	female
50.00	17500	Innes	male
49.00	20000	Jones	male
42.00	18000	Baker	male
40.00	17500	Foster	male
36.00	15500	Hill	female
32.00	12000	Edwards	male
29.00	4500	Grey	female
24.00	5000	Adams	male
18.00	3000	Clarke	female

Here there is a variate as index vector. The DIRECTION option can also apply to textual index vectors, when ascending order is interpreted as alphabetical order. The default for DIRECTION is to sort into ascending order.

    Finally, to illustrate the use of more than one index vector, we sort into increasing order of incomes, taking alphabetic ordering of the names where two people earn the same amount.

Example 4.4.3c

```
12   SORT [INDEX=Income,Name] Age,Income,Name,Sex
13   PRINT Age,Income,Name,Sex
```

Age	Income	Name	Sex
18.00	3000	Clarke	female

29.00	4500	Grey	female
24.00	5000	Adams	male
61.00	7000	Day	female
32.00	12000	Edwards	male
36.00	15500	Hill	female
40.00	17500	Foster	male
50.00	17500	Innes	male
42.00	18000	Baker	male
49.00	20000	Jones	male

4.5 Operations on variates

Most of the facilities described earlier in this chapter can be used with variates: CALCULATE to perform calculations on their values, RESTRICT to operate only on a subset of their units, SORT to reorder their units, and EQUATE to transfer values between variates and other numerical structures. This section describes directives that operate only on variates. The INTERPOLATE directive allows you to interpolate values at intermediate points of an observed sequence, and MONOTONIC performs monotonic regressions.

Procedures for operating on variates include the following: DAYCOUNT to convert a date to a daycount, or vice versa; DAYLENGTH to calculate daylengths at a given period of the year HEATUNITS to calculate accumulated heat units of a temperature dependent process; ORTHPOL to calculate orthogonal polynomials from the values in a variate; QUANTILE to calculate quantiles of the values in a variate; RANK to produce ranks, from the values in a variate, allowing for ties; VINTERPOLATE to performs linear & inverse linear interpolation between variates; and VTABLE to form a variate and set of classifying factors from a table.

4.5.1 Interpolation

INTERPOLATE interpolates values at intermediate points.

Options

CURVE = *string*	Type of curve to be fitted to calculate the interpolated value (linear, cubic); default line
METHOD = *string*	Type of interpolation required (interval, value, missing): for METHOD=valu, values are interpolated for each point in the NEWINTERVAL variate and stored in the NEWVALUE variate; for METHOD=inte, points are estimated in the NEWINTERVAL variate for the observations in the NEWVALUE variate; while for METHOD=miss, the NEWVALUE and NEWINTERVAL lists are irrelevant, INTERPOLATE now interpolates for missing values in the OLDVALUE and OLDINTERVAL variates (except those missing in both variates). Default inte

Parameters

OLDVALUES = *variates*	Observations from which interpolation is to be done
NEWVALUES = *variates*	Results of each interpolation
OLDINTERVALS = *variates*	Points at which each set of OLDVALUES was observed
NEWINTERVALS = *variates*	Points for each set of NEWVALUES

If you have a set of pairs of observations (x, y), you can use interpolation to estimate either a value y for a value x that need not be in the set, or a value x for a value y that likewise need not be in the set. The simplest way to interpolate is by joining successive pairs of observations by straight lines and reading off the appropriate values in between: then the two cases are called *linear interpolation* (obtaining y from x) and *inverse linear interpolation* (obtaining x from y). Genstat can alternatively join the points by cubic functions instead of straight lines. Genstat uses the term *values* to describe the set of y-values and *intervals* for the set of x-values, no matter whether you are doing direct or inverse interpolation.

Genstat does the interpolation for each parallel set of variates in the parameter lists. Each variate in the OLDINTERVALS list specifies the x-values of a set of observed points; the corresponding variate in the OLDVALUES list specifies the corresponding y-values. The variates in the NEWINTERVALS and NEWVALUES lists are for the x-values and y-values of the interpolated points.

If you set METHOD=value, Genstat does ordinary interpolation, and you use the NEWINTERVALS variate to specify the x-values for which you require interpolated y-values. Genstat calculates the y-values and stores them in the corresponding NEWVALUES variate; this variate will be declared implicitly if you have not declared it already.

For the interpolation to take place, the x-values must be in either monotonically increasing or decreasing order; thus, if necessary, Genstat takes a copy of the x-values and y-values and sorts these (in parallel) to put the x-values into ascending order.

In Example 4.5.1a, wheat plants have been sampled on five occasions and their growth stage (Zadoks) assessed. The program interpolates values, which it stores in variate Nzad, to estimate the growth stage that the plant has reached after 50, 100, and 150 days.

Example 4.5.1a

```
  1   VARIATE [NVALUES=6] Zadoks,Days; \
  2     VALUES=!(0,15,23,35,65,95),!(0,50,84,119,147,182)
  3   & [NVALUES=3] Nzadoks,Ndays; VALUES=!(25,50,75),!(50,100,150)
  4   INTERPOLATE [METHOD=value] Zadoks; NEWVALUES=Nzad; \
  5     OLDINTERVALS=Days; NEWINTERVALS=Ndays
  6   PRINT Ndays,Nzad; FIELDWIDTH=8; DECIMALS=2

  Ndays     Nzad
  50.00    15.00
 100.00    28.49
 150.00    67.57
```

Similarly, if you set METHOD=interval, Genstat does inverse interpolation. You must then specify the y-values in the NEWVALUES variate. Genstat calculates the x-values and stores them in the corresponding NEWINTERVALS variate, which will be declared implicitly if necessary. Again the *x*-values must be in monotonically increasing or decreasing order, and Genstat will produce a sorted copy if necessary. Inverse interpolation is the default.

Example 4.5.1b uses the same data as above, but does inverse linear interpolation to estimate how long after planting we have to wait for the plant to reach growth stages 25, 50, and 75 Zadoks.

Example 4.5.1b
_____

```
7   INTERPOLATE [METHOD=interval] Zadoks; NEWVALUES=Nzadoks;
8      OLDINTERVALS=Days; NEWINTERVALS=Nd
9   PRINT Nzadoks,Nd; FIELDWIDTH=8; DECIMALS=2

Nzadoks       Nd
  25.00    89.83
  50.00   133.00
  75.00   158.67
```

If you set METHOD=missing, Genstat ignores the NEWVALUES and NEWINTERVALS parameters; it estimates values for *x* or *y* when the other is missing, placing the results in the previously missing position of the OLDVALUES or the OLDINTERVALS variates. Ordinary interpolation is used when the missing value is in *y*, and inverse interpolation when it is in *x*. If both the *x*-value and the *y*-value are missing for a particular unit, no values can be interpolated for it, and it remains missing. To do linear interpolation requires that both the *x*-value and the *y*-value should be non-missing for the point on each side of the unit with the missing value. For cubic interpolation, there must be two non-missing points on each side of the unit. In Example 4.5.1c the missing value in Yval at unit 2 is replaced with the interpolated value 2.85, while the one at unit 4 remains missing because the *x*-value is missing there too. The missing value at unit 9 of Xint is replaced by 5.96, while the one at unit 4 again stays missing. Notice also that Genstat ignores the NEWINTERVALS setting Xnewint.

Example 4.5.1c
_____

```
1   VARIATE [NVALUES=9] Yval,Xint; \
2      VALUES=!(2.5,*,3.2,*,4.3,4.8,7.2,7.3,8.7),!(1,2,3,*,4,5,*,6,7)
3   PRINT Xint,Yval; FIELDWIDTH=8; DECIMALS=2

   Xint     Yval
   1.00     2.50
   2.00        *
   3.00     3.20
      *        *
   4.00     4.30
   5.00     4.80
      *     7.20
   6.00     7.30
   7.00     8.70

4   INTERPOLATE [METHOD=missing] OLDVALUE=Yval; OLDINTERVAL=Xint ;\
5      NEWINTERVAL=Xnewint
6   PRINT Xint,Yval; FIELDWIDTH=8; DECIMALS=2
```

```
Xint     Yval
1.00     2.50
2.00     2.85
3.00     3.20
   *        *
4.00     4.30
5.00     4.80
5.96     7.20
6.00     7.30
7.00     8.70
```

The CURVE option has two settings, linear and cubic. By default, CURVE=linear, and successive pairs of observations are connected by straight-line segments for linear, or inverse-linear, interpolation. For cubic interpolation you set CURVE=cubic; there must then be at least four values in each of the OLDVALUES and OLDINTERVALS variates.

4.5.2   Monotonic regression

MONOTONIC fits an increasing monotonic regression of y on x.

No options

Parameters

Y = *variates*	Y-values of the data points
X = *variates*	X-values of the data points; default is to assume that the x-values are monotonically increasing
RESIDUALS = *variates*	Variate to save the residuals from each fit
FITTEDVALUES = *variates*	Variate to save the fitted values from each fit

Monotonic regression plays a key role in non-metric multidimensional scaling, which is available in Genstat via the MDS directive (11.3.3). However, it can be useful in its own right, so the method has been made accessible by the MONOTONIC directive. A monotonic regression through a set of points is simply the line that best fits the points subject to the constraint that it never decreases: of course the line need not be straight, in fact it rarely will be. If you need a monotonically decreasing line, you can simply subtract all the y-values from their maximum, find the monotonically increasing regression, and then back-transform the data and fitted line, and change the sign of the residuals.

The MONOTONIC directive has no options. It has four parameters: Y to specify the y-values, X for the x-values, RESIDUALS to save the residuals, and FITTEDVALUES to save the fitted values. The x-values need not be supplied, in which case the directive assumes that the y-values are in increasing order of the x-values. In common with the other regression directives, the variates to save the residuals and fitted values need not be declared in advance.

In Example 4.5.2, the MONOTONIC directive is first used with the data in their original order. The fitted values are saved and plotted as a line, with the data. You can see what happens with the coincident x-values of 4; notice also the horizontal fitted lines that occur when the y-values decrease. Once the data are sorted into increasing order of X, at line 6, there is no need to specify the X parameter when the MONOTONIC directive is used at line 7; as shown by the PRINT statement at line 8, the fitted values remain the same.

Example 4.5.2

```
1   VARIATE [VALUES=2,6,4,4, 9,1,12,15,13,18] X
2   &        [VALUES=1,5,3,6,10,0,11,14,16,18] Y
3   MONOTONIC Y=Y; X=X; FITTED=Fvals
4   GRAPH [TITLE='Monotonic regression'; NROWS=25; NCOLUMNS=61] \
5         Fvals,Y; X; METHOD=line,point
```

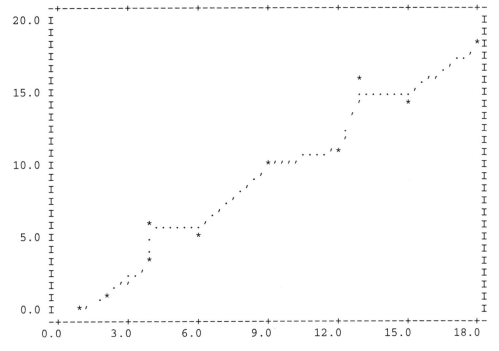

```
6   SORT X,Y
7   MONOTONIC Y; RESIDUALS=Res; FITTED=Fvals
8   PRINT X,Y,Fvals,Res
```

X	Y	Fvals	Res
1.000	0.000	0.000	0.0000
2.000	1.000	1.000	0.0000
4.000	3.000	3.000	0.0000
4.000	6.000	5.500	0.5000
6.000	5.000	5.500	-0.5000
9.000	10.000	10.000	0.0000
12.000	11.000	11.000	0.0000
13.000	16.000	15.000	1.0000
15.000	14.000	15.000	-1.0000
18.000	18.000	18.000	0.0000

4.6   Operations on factors

You use factors in Genstat to indicate groupings of the units of vectors. You would need to do this, for example, in the analysis of designed experiments (Chapter 9), or when forming tabular summaries of group totals, means, maxima, minima, and so on (4.11.1).

This section describes the GROUPS directive, which enables you to construct a factor from a variate or a text. The groups can cover every distinct value of the variate or text, or ranges of values; you can specify these ranges yourself, or have them defined automatically. Genstat can define the levels and labels vectors from either the minimum, the maximum or the median of the units allocated to each group.

Other facilities, for forming factors in experimental designs, are described elsewhere (9.8). The GENERATE directive (9.8.1) allows you to define factor values in a systematic order. You can also use it to form values of treatment factors, using the design-key method, or to define values for the pseudo-factors required to specify partially balanced experimental designs. Other facilities for generating factors in experimental designs are provided by the procedures in the Design module of the procedure library. The RANDOMIZE directive (9.8.3) can put the units of factors and variates into random order; this randomization can take account of the block structure of a designed experiment, if required.

The use of factors within expressions, and in CALCULATE in particular, is described in 4.1.1 and 4.1.2. This allows you to form a variate from a factor, either taking its declared levels or by taking an alternative set of levels using the NEWLEVELS function (4.2.1). CALCULATE also allows you to assign values of a variate to a factor, provided you have already declared the factor with levels including all the values taken by the variate. But a more satisfactory method is to use the GROUPS directive, already mentioned.

Procedure for manipulating factors include FACPRODUCT which forms a factor with a level for every combination of other factors (9.8.2), and VTABLE which forms a variate and set of classifying factors from a table.

4.6.1   Forming factors from variates and texts (GROUPS)

GROUPS forms a factor (or grouping variable) from a variate or text, together with the set of distinct values that occur.

Options

NGROUPS = *scalar*	Number of groups to form when LIMITS is not specified; if NGROUPS is also unspecified, each distinct value (allowing for rounding) defines a group; default *
LMETHOD = *string*	Defines how to form the levels variate if the setting of the VECTOR parameter is a variate, or the labels if it is a text; if LMETHOD=* no levels/labels are formed, and existing labels (for a variate VECTOR) or labels (for a text VECTOR) of an already declared FACTOR will be

	retained if still appropriate (given, minimum, median, maximum); default medi
DECIMALS = *scalar*	Number of decimal places to which to round the VECTOR before forming the groups; default * i.e. no rounding
BOUNDARIES = *string*	Whether to interpret the LIMITS as upper or lower boundaries (upper, lower); default lowe
REDEFINE = *string*	Whether to allow a structure in the FACTOR list that has already been declared (e.g. as a variate or text) to be redefined (yes, no); default no

Parameters

VECTOR = *variates* or *texts*	Vectors whose values are to define the groups
FACTOR = *factors*	Structures to be defined as factors to save details of the groups; default * will, if REDEFINE=yes, cause the corresponding VECTOR itself to be defined as a factor
LIMITS = *variates* or *texts*	Limits to define the groups
LEVELS = *variates*	Variate to define the levels of each FACTOR if LMETHOD=give, or to save them otherwise
LABELS = *texts*	Text to define the labels of each FACTOR if LMETHOD=give, or to save them otherwise

The GROUPS directive is designed to form factors from variates or texts. The variates and texts are specified by the VECTOR parameter, and the factors by the FACTOR parameter. With the simplest use of GROUPS you need specify no more than that, and each factor is defined to have a level for every distinct value of its corresponding variate or text. You need not have declared the factor already; it will be declared automatically if necessary.

Alternatively, you can divide the values of the variate or text into groups to be represented by the factor. You can use the LIMITS option to specify the range of values for each group. The limits vector is a text or a variate, depending whether the factor is being defined from a variate or a text; its values specify boundaries for the ranges. The BOUNDARIES option controls whether these are regarded as upper or lower boundaries; by default BOUNDARIES=lower. In Example 4.6.1 below, to divide the ages into the ranges 0-19, 20-29, 30-39, 40-49, 50-59, and over 60, the limits vector contains the five boundaries 20, 30, 40, 50, and 60. You can also ask GROUPS itself to set limits that will partition the units into groups of nearly equal size. You should then specify the NGROUPS option and leave the LIMITS parameter unset. (If you give both LIMITS and NGROUPS, NGROUPS is ignored.)

If you are defining a factor from a variate VECTOR, the LMETHOD option controls how the levels vector is formed. The default LMETHOD=median forms the levels from the median of the units in each group. There are also settings to allow them to be formed from minima or maxima. With any of these settings (median, minumum, or maximum) you can specify a variate, using the LEVELS parameter, to store the levels that are produced; this can be done even if no factor is being formed, that is if no identifier is supplied for the factor by the

FACTOR list. Alternatively, if you put LMETHOD=given, you can use the LEVELS parameter to supply your own levels. Finally, for LMETHOD=*, no levels are formed and any existing levels of the factor will be retained if they are still appropriate; otherwise the levels will be the integers 1 upwards. With any of these settings, you can use the LABELS parameter to specify labels for the factor.

Similar rules apply if you have a text VECTOR except that LMETHOD then governs how the labels are defined for the factor, and LEVELS can be used to specify its levels.

You can set the DECIMALS option to request that the values of a variate VECTOR be rounded to a particular number of decimal places before the groups are formed: for example DECIMALS=0 would round each value to the nearest integer.

You can redefine a VECTOR structure as a factor by setting option REDEFINE=yes and omitting to specify any corresponding identifier in the FACTOR list. This can be very useful on occasions when you are unable to define in advance which levels will occur in a set of data. In line 14 of Example 4.6.1, the text National (which contains details of the nationality of a list of people) is redefined as a factor so that we can produce a table with the mean ages of the people with each of the nationalities represented in the data.

Example 4.6.1

```
 1   VARIATE Age
 1   TEXT National
 3   READ National,Age
```

Identifier	Minimum	Mean	Maximum	Values	Missing
National				16	0
Age	5.00	32.94	63.00	16	0

```
11   GROUPS Age; FACTOR=Ageclass; LIMITS=!(20,30,40,50,60); \
12     LABELS=!t('under 20','20-9','30-9','40-9','50-9','over 60')
13   PRINT Age,Ageclass,National
```

Age	Ageclass	National
32.00	30-9	British
29.00	20-9	British
7.00	under 20	British
5.00	under 20	British
51.00	50-9	French
49.00	40-9	French
22.00	20-9	British
24.00	20-9	British
35.00	30-9	British
41.00	40-9	British
25.00	20-9	French
24.00	20-9	French
33.00	30-9	Italian
29.00	20-9	Italian
63.00	over 60	British
58.00	50-9	British

```
14   GROUPS [REDEFINE=yes] National
15   TABULATE [CLASSIFICATION=National] Age; MEAN=MeanAge
16   PRINT MeanAge
```

```
          MeanAge
National
  British     31.60
   French     37.25
  Italian     31.00
```

GROUPS takes account of any restrictions (4.4.1) on variates or texts in the VECTOR list, and will give missing values to the excluded units. If more than one vector is restricted, then each such restriction must be the same.

4.7 Operations on text

A text structure (2.3.2) is a vector each line of which contains a string of characters. So you might use it to label the units of other vectors, or to contain a complete piece of description.

The first part of this section describes the CONCATENATE directive (4.7.1) which allows you to concatenate several texts together side by side so that each line of the new text is formed by joining together a series of lines, one from each of the original texts. You can omit characters at the beginning and end of the component lines; so this also gives you a way of truncating the lines of a text. You can also change letters from upper to lower case, and vice versa.

An alternative form of concatenation places whole texts one after another, possibly omitting some of their lines: you can do this with the EQUATE directive (4.3).

The remaining parts of the section describe the EDIT directive (4.7.2). This is a sub-system within Genstat; it has its own command syntax (4.7.3), allowing you to delete and insert series of characters, or to substitute one series for another, or to delete and insert complete lines, and so on.

Some general directives, described elsewhere, are also useful for manipulating text. The SORT directive allows you to sort the units of a text into alphabetical order or to form a factor from a text (4.4.3). You can test for equality and inequality of the lines of texts in the expressions that occur in CALCULATE (4.1), in RESTRICT (4.4.1), and in the directives for program control (5.2). CALCULATE also allows you to determine the number of characters in each line, and to find the positions of strings within lines (4.2.7). READ can take its input from a text (3.1.7), and you can direct output from the PRINT directive (3.2) into a text. PRINT thus allows you to place numerical values into a text. An alternative, for variates, is to use procedure FTEXT, which will also determine an appropriate number of decimal places.

4.7.1 Text concatenation (CONCATENATE)

CONCATENATE concatenates and truncates lines (units) of text structures; allows the case of letters to be changed.

Options

NEWTEXT = *text* Text to hold the concatenated/truncated lines; default is the first OLDTEXT vector

CASE = *string*	Case to use for letters (`given`, `lower`, `upper`, `changed`); default `give` leaves the case of each letter as given in the original string

Parameters

OLDTEXT = *texts*	Texts to be concatenated
WIDTH = *scalars* or *variates*	Number of characters to take from the lines of each text; if * or omitted, all the (unskipped) characters are taken
SKIP = *scalars* or *variates*	Number of characters to skip at the left-hand side of the lines of each text; if * or omitted, none are skipped

The CONCATENATE directive joins lines of several texts together, side by side, to form a new text. You can specify the identifier of this text by the NEWTEXT option, in which case it need not already have been declared as a text. If you do not specify NEWTEXT, Genstat places the new textual values into the first text in the OLDTEXT parameter list (replacing its existing values).

The texts to be concatenated are specified by OLDTEXT; they should all contain the same number of lines, unless you want to insert an identical series of characters into every line of the new text: a series of characters that is to be duplicated within every line can be specified either as a string, or in a single-valued text. In line 6 of Example 4.7.1, the string ', ' inserts a comma and a space into every line of the NEWTEXT Fullname.

Example 4.7.1

```
 1    TEXT [VALUES='1. Adams','2. Baker','3. Clarke','4. Day', \
 2     '5. Edwards','6. Field','7. Good','8. Hall','9. Irving','10. Jones'] Name
 3    TEXT [VALUES='B.J.','J.S.','K.R.','A.T.','R.S.', \
 4     'T.W.','S.I.','D.M.','H.M.','C.C.'] Initials
 5    " Form text Fullname containing the number, name, and initials."
 6    CONCATENATE [NEWTEXT=Fullname] OLDTEXT=Name,', ',Initials
 7    PRINT Fullname; JUSTIFICATION=left

Fullname
1. Adams, B.J.
2. Baker, J.S.
3. Clarke, K.R.
4. Day, A.T.
5. Edwards, R.S.
6. Field, T.W.
7. Good, S.I.
8. Hall, D.M.
9. Irving, H.M.
10. Jones, C.C.

 8    " Now reform Fullname to contain just the first initial and the name."
 9    CONCATENATE [NEWTEXT=Fullname] OLDTEXT=Initials,Name; \
10      WIDTH=2,*; SKIP=*,!(9(2),3)
11    PRINT Fullname; JUSTIFICATION=left

Fullname
```

```
B.  Adams
J.  Baker
K.  Clarke
A.  Day
R.  Edwards
T.  Field
S.  Good
D.  Hall
H.  Irving
C.  Jones
```

If you give a variate in the SKIP list, then it must contain a value for each line of the text in the OLDTEXT list; the value indicates the number of characters to be omitted at the beginning of that line. Alternatively, you can give a scalar if the same number of characters is to be omitted at the start of every line. In line 10 of the example, the null entry for Initials (indicated by *) specifies that no characters are to be omitted.

Similarly the WIDTH parameter specifies how many characters are to be taken, after omitting any initial characters as specified by SKIP. In line 10, WIDTH has a scalar setting of 2 for Initials, so that only the first initial followed by a dot is taken for each name.

The CASE option enables you to change the case of letters. By default, CASE=given to leave the case of each letter as given in the existing text. To change all letters to upper case (or capitals) you can put CASE=upper, or CASE=lower to change all letters to lower case. Alternatively, CASE=changed puts lower-case letters into upper case, and upper-case letters into lower case!

CONCATENATE takes account of restrictions (4.4.1) on any of the vectors that occur in the statement. If more than one vector is restricted, then each such restriction must be the same. The values of the units that are excluded by the restriction are left unchanged.

4.7.2   Editing text (EDIT)

The EDIT directive provides a line editor for modifying text structures.

EDIT edits text vectors.

Options

CHANNEL = *scalar* or *text*	Text structure containing editor commands or a scalar giving the number of a channel from which they are to be read; default is the current input channel
END = *text*	Character(s) to indicate the end of the commands read from an input channel; default is the character colon (:)
WIDTH = *scalar*	Limit on the line width of the text; default *
SAVE = *text*	Text to save the editor commands for future use; default *

Parameters

OLDTEXT = *texts*	Texts to be edited

NEWTEXT = *texts*   Text to store each edited text; if any of these is omitted,
                                     the corresponding OLDTEXT is used

The EDIT directive edits each text in the OLDTEXT list, storing the results in the corresponding structure in the NEWTEXT list. It both edits and stores each text before moving on to the next. If you have not already declared any of the texts in the NEWTEXT list, it will be declared implicitly. If you give a missing identifier (*) in the NEWTEXT list, the edited version simply replaces the values of the original: thus the old text will be overwritten by the new text. You can also omit a text from the OLDTEXT list; you might do this if you wanted to form the values of the new text entirely from within the editor. If any of the old texts are restricted, they must all be restricted to exactly the same set of units. Then only those units will be involved in the edit. When a restriction is in force, you cannot add or delete any units (or lines).

The CHANNEL option tells Genstat where to find the editing commands. A scalar specifies the number of an input channel from which the commands are to be read. Alternatively, you can specify a text structure containing the commands. In either case the commands should be terminated by the string specified by the END option. The end string can be more that one character; the default is the single character colon (:). Genstat gives a warning if you have forgotten to specify the end string in a text of commands. The default for the CHANNEL option is to take input from the current input channel.

The WIDTH option specifies the maximum line length for vectors of commands and of text, the default being 80 and the maximum being 255.

The SAVE option allows you to specify a text structure to store the edit commands, so that you can save them for future EDIT statements.

4.7.3   Commands for the EDIT directive

The commands that you can use to edit text are described in this subsection. You can give commands to the editor in upper or lower case. You can put as many commands as you like on a line, subject only to the width restriction set by the WIDTH option. Commands must be separated by at least one space. You cannot put spaces into the middle of a command, unless they are part of a character string (or part of a sequence of commands).

The character that separates the parts of a command is written here as /, but you can use any character for this other than a space or a digit.

Genstat puts the lines from the old text into an internal *buffer*, where they are modified according to the commands that you specify. While you are editing, Genstat moves a notional *marker* around the buffer. The marker can be moved backwards or forwards along a line or between lines. So you can move around the text and modify the lines in any order. Some commands move the marker automatically, as explained in the definitions below. If the marker is before the first line of text it is at the [start] position; if it is after the last line of text it is at the [end] position. The line that currently contains the marker is called the *current line*. Genstat does not write anything to the new text until the edit has been completed (so if you use the Q command, the new text is left unaltered).

Some commands allow you to specify a number: for example Dn deletes the next *n* lines.

Genstat gives a warning message if this number is zero or is not an integer.

The command definitions are as follows.

A   Insert the next line of text from the buffer, immediately after the marker within the current line.

B   Break the current line at the marker position. Text before the marker is written as a new line to the internal buffer and text after the marker becomes the new current line with the marker at character position 1.

C   Cancel edits performed on the current line by restoring it to the form in which it was most recently read from the buffer. Note that if you have previously edited the line and then moved to some other line, it is the previously edited form that will be given, not the form as originally in the old text; also, if you have given any A or B commands during your modification of the current line, their effects are not negated, so for example any lines that have been inserted into the current line by A will be lost.

D   Delete the current line, and make the next line the current line with the marker at character position 1.

D*n*   Delete the next *n* lines (including the current line), making the next line after that the current line with the marker at position 1.

D+*n* Synonymous with D*n*.

D+   is a synonym for D or D+1.

D+*   Delete from the current line to end of text. The current line is then [end].

D*   Synonymous with D+*.

D-   Delete the current line, making the previous line the current line with the marker at character position 1.

D-*n* Delete the current and previous *n* lines, making the line before that the current line with the marker at character position 1.

D-   is a synonym for D-1.

D-*   Delete the current line and all previous lines, the current line is then [start].

D/*s*/   Delete from the current line to the line with the next occurrence of the character string *s*. The marker is placed immediately before the character string *s* in the located line. If *s* occurs after the marker on the current line, the marker is moved up to *s* and no lines are deleted.

D-/*s*/ The same as D/*s*/, except that it moves backwards through the text, deleting all lines from and including the current one until the first occurrence of a line containing the character string *s*. The marker is placed immediately before the located character string *s*. If *s* occurs before the marker on the current line, the marker placed before *s* and no lines are deleted.

F/*i*/ Inserts the contents of the text structure with identifier *i* immediately before the current line. The marker is not moved.

G+/*s*/*t*/ substitutes string *s* for all occurrences of string *t* found after the marker on the current and subsequent lines, and moves the marker to the end of the text.

G/*s*/*t*/ is a synonym for G/*s*/*t*/.

G-/*s*/*t*/ substitutes string *s* for all occurrences of string *t* found before the marker on the current and previous lines, and moves the marker to the start of the text.

I/*s*/ Inserts the string *s* as a new line immediately before the current line. The marker is not moved.

L   Moves the marker to the start of the next line, which can be [end].

L*n*   Moves the marker to the start of the *n*th line after the current line. So L1 gives the next line.

L+*n*   Is synonymous with L*n*.

L+   Is synonymous with L or L+1.

L+*   Moves the marker to [end].

L*   is a synonym for L+*.

L-*n*   Moves the marker to the start of the *n*th line before the current line, which can be [start]. L-1 gives the line immediately before the current line.

L-   Is synonymous with L-1.

L-*   Moves the marker to [start].

L+/*s*/ Moves the marker to the position immediately before the next occurrence of the character string *s* after the current marker position; this occurrence need not be on the current line. If the string *s* is not found, the marker will be located at [end].

L-/*s*/ Moves the marker to the position immediately before the first occurrence of the string *s* before the current marker position; this occurrence need not be on the current line. If the string *s* is not found, the marker will be located at [start].

P   moves the marker one character to the right along the current line.

P+*n*   Moves the marker *n* characters to the right of the current position within the current line. You cannot move the marker beyond the maximum line length (which will vary between computers, but is normally the same as the width of your local line–printer).

P+   is a synonym for P or P+1.

P+*   Moves the marker to the position immediately after the last non-blank character in the current line. This can be to the left of the current marker position.

P-*n*   Moves the marker *n* characters to the left of the current position within the current line. The marker cannot be moved to the left of character position 1.

P-   is a synonym for P-1.

P-*   Moves the marker to the position immediately before the first non-blank character after character position 1. This can be to the right of the current marker position.

P*n*   Moves the marker to the character position *n* within the current line, counting from the left and starting at 1. The maximum value of *n* varies between computers but is normally the same as the width of your local line-printer.

Q   Abandons the current edit, leaving the original text unaltered.

R+/*s*/*t*/ substitutes character string *t* for the next occurrence of character string *s* after the marker on the current or subsequent lines, and moves the marker to the position immediately after *t*.

R/*s*/*t*/ is a synonym for R+/*s*/*t*/.

R-/*s*/*t*/ substitutes string *t* for the nearest occurrence of string *s* before the marker on the current or previous lines; the marker moves to be immediately before string *t*.

S/*s*/*t*/ Substitutes the string *t* for the next occurrence of string *s* after the marker within the current line. The marker is moved to the character position immediately after the last character in *t*. If *s* is null (when the command is S//*t*/) then *t* is inserted immediately after the marker. If *t* is null (when the command is S/*s*//), then *s* is deleted from the line.

V   Turns on the verification mode. Then, if you are working interactively, the current line will be displayed each time that Genstat prompts you for commands. By default the marker is indicated by the character > but you can change this by the command V*c* or V+*c*.

V*c*   Turns on the verification mode (see V), and changes the marker character to *c*.

V+*c*   Is synonymous with V*c*.

V–   Turns verification mode off (see V).

(*cseq*)*n*   Repeats the command sequence, *cseq*, *n* times. The command sequence *cseq* can be any valid combination of editing commands, each separated by at least one space. The complete sequence, including brackets and repeat count, must all be on a single line. You can nest sequences up to a depth of 10.

(*cseq*)*   Repeats the command sequence cseq until [end] or [start] is encountered. In all other respects (*cseq*)* behaves exactly as (*cseq*)*n*; so it would be equivalent to putting *n* equal to some very large number.

Example 4.7.3

```
> " An interactive run of the editor."
> TEXT Name
> OPEN 'NAMES.DAT'; CHANNEL=2
> READ [CHANNEL=2] Name

    Identifier   Minimum     Mean   Maximum    Values   Missing

         Name                                      10         0

> " Edit Name: within the editor the prompt will be 'EDIT> '."
> EDIT Name
>B.J. Adams
EDIT> S//Mr. /
Mr. >B.J. Adams
EDIT> L
>J.S. Baker
EDIT> (S//Dr. / L)4
>T.W. Field
EDIT> S//Ms. /
Ms. >T.W. Field
EDIT> L
>S.I. Good
EDIT> S//Mr. /
Mr. >S.I. Good
EDIT> L S//Miss. /
Miss. >D.M. Hall
EDIT> (L S//Dr. /)2
Dr. >C.C. Jones
EDIT> :
> PRINT Name; JUSTIFICATION=left

 Name
 Mr. B.J. Adams
 Dr. J.S. Baker
```

```
Dr.  K.R.  Clarke
Dr.  A.T.  Day
Dr.  R.S.  Edwards
Ms.  T.W.  Field
Mr.  S.I.  Good
Miss.  D.M.  Hall
Dr.  H.M.  Irving
Dr.  C.C.  Jones
```

4.8   Operations on formulae

If you are writing procedures, for example for statistical analyses, the model to be fitted will often be specified by a Genstat formula structure (2.2.4). Unless the algorithm within the procedure merely involves straightforward use of one of Genstat's statistical directives, you may wish to know more about the formula: how many model terms does it contain, which factors do they involve, and so on. The FCLASSIFICATION directive is designed to provide the answers to these questions.

FCLASSIFICATION   forms a classification set for each term in a formula, breaks a formula up into separate formulae (one for each term), and applies a limit to the number of factors and variates in the terms of a formula.

Options

FACTORIAL = *scalar*	Limit on the number of factors and variates in each term; default 3
NTERMS = *scalar*	Outputs the number of terms in the formula
CLASSIFICATION = *pointer*	Saves a list of all the factors and variates in the TERMS formula
OUTFORMULA = *formula*	Identifier of a formula to store a new formula, omitting terms with too many factors and variates

Parameters

TERMS = *formula*	Formula from which the classification sets, individual model terms and so on are to be formed
CLASSIFICATION = *pointers*	Identifiers giving a pointer to store the factors and variates composing each model term of the TERMS formula
OUTTERMS = *formulae*	Identifiers giving a formula to store each individual term of the TERMS formula

The FCLASSIFICATION directive enables you to manipulate a formula data structure; the formula is specified using the TERMS parameter.

As explained in 1.7.3, when Genstat uses a formula in a statistical analysis, it is expanded into a series of model terms, linked by the operator +. FCLASSIFICATION allows you to save this expanded form, in another formula, using the OUTFORMULA option.

You can use the FACTORIAL option to apply a limit to the number of factors and variates in the resulting terms, similarly to the FACTORIAL option in the ANOVA, REML, and regression directives (9.1.2, 10.3.1, and 8.3.1). The number of terms in the formula can be saved (in a scalar) using the NTERMS option, and a list of the factors that occur in the formula can be saved (in a pointer) using the CLASSIFICATION option.

The other parameters allow you to save information about the individual model terms in the formula. The identifiers in the lists that they specify are taken in parallel with the model terms in the expanded form of the formula. For each model term, the corresponding identifier in the list for the CLASSIFICATION parameter is defined as a pointer storing the factors that occur in the term; and the identifier in OUTTERMS list is defined as a formula containing just that model term.

The use of FCLASSIFICATION is illustrated in Example 4.8. At line 2, formula ABC2 is formed, to contain the expanded form of the formula A*B*C subject to the limit of FACTORIAL=2. Lines 4 and 6 obtain information about the individual terms in the formula. In line 4, the NTERMS option is used to ascertain how many terms there are. The resulting scalar, NT, can then be used in line 6 to specify the necessary lists of identifiers: Class[1...NT] for the CLASSIFICATION parameter, and Term[1...NT] for the OUTTERMS parameter.

Example 4.8

```
   1   FACTOR [NVALUES=32; LEVELS=2] A,B,C
   2   FCLASSIFICATION [FACTORIAL=2; OUTFORMULA=ABC2] A*B*C
   3   PRINT ABC2

ABC2
((((A + B) + C) + (A . B)) + (A . C)) + (B . C)

   4   FCLASSIFICATION [FACTORIAL=2; NTERMS=NT] A*B*C
   5   PRINT NT

          NT
       6.000

   6   FCLASSIFICATION [FACTORIAL=2] A*B*C; CLASSIFICATION=Class[1...NT]; \
   7      OUTTERMS=Term[1...NT]
   8   FOR Ci=Class[]; Oi=Term[]
   9      PRINT [SERIAL=yes] Ci,Oi
  10   ENDFOR

     Class[1]
             A

Term[1]
A

     Class[2]
             B

Term[2]
B
```

```
     Class[3]
             C

Term[3]
C

     Class[4]
             A
             B

Term[4]
A . B

     Class[5]
             A
             C

Term[5]
A . C

     Class[6]
             B
             C

Term[6]
B . C
```

4.9   Operations on dummies and pointers

You use dummies (2.2.2) when you want the same series of statements to operate on different data structures on different occasions. By referring to a dummy instead of any specific structure, you can make the statements apply to whichever structure you want. The commonest use of dummies is in loops (5.2.1), and in procedures (5.3).

In this section we describe an alternative way of specifying a value for a dummy, by using the ASSIGN directive. ASSIGN also enables you to change the values of elements of pointers, which are used mainly to specify collections of data structures for directives such as EQUATE (4.3), or as a convenient way of specifying lists of structures (1.6.4 and 2.6). A further use of ASSIGN is to control the labelling of structures that exist as subscripted identifiers of two or more pointers but which do not possess identifiers in their own right (see Example 4.9.1c).

You can make tests on the values of dummies and pointers using the .IS. and .ISNT. operators (4.1.1).

4.9.1   Assigning values to dummies and individual elements of pointers (ASSIGN)

ASSIGN sets elements of pointers and dummies.

Option

NSUBSTITUTE = *scalar*	Number of times *n* to substitute a dummy in order to determine which structure to assign (if *n* is negative, the

	assigned structure is the −*n*th from the bottom of the chain of dummies, like the NTIMES option of EXIT); default * implies no substitution
METHOD = *string*	Whether to replace or preserve the existing value in each dummy or pointer element (replace, preserve); default repl (note, pointer elements are never unset so METHOD=preserve with a pointer simply causes the assignment to be ignored)
RENAME = *string*	Whether to reset the default name for the structure if it has only a suffixed identifier (no, yes); default no
SCOPE = *string*	If SCOPE=external, dummies or pointer elements within a procedure will be set to point to structures in the program that called the procedure rather than to structures within the procedure (local, external); default loca

Parameters

STRUCTURE = *identifiers*	Values for the dummies or pointer elements
POINTER = *dummies* or *pointers*	Structure that is to point to each of those in the STRUCTURE list
ELEMENT = *scalars* or *texts*	Unit or unit label indicating which pointer element is to be set; if omitted, the first element is assumed

ASSIGN allows you to set individual elements of pointers, or to assign a value to a dummy. The parameter POINTER lists the pointers or dummies whose values you want to set; the values that you want to give them are listed by the STRUCTURE parameter. You pick out the individual elements of pointers by the ELEMENT parameter; a scalar identifies the element by its suffix number, while a text identifies it by its label. This example sets the dummy Yvar to point to the variate Height, and elements 1 and 2 of the pointer Xvars to Protein and Vitamins, respectively.

```
VARIATE Height,Protein,Vitamins
POINTER [NVALUES=2] Xvars
DUMMY Yvar
ASSIGN Height,Protein,Vitamins;POINTER=Yvar,2(Xvar);ELEMENT=1,1,2
```

Element 1 is assumed unless you specify otherwise; so to set just Yvar we need only put

```
ASSIGN Height; POINTER=Yvar
```

Options NSUBSTITUTE and METHOD are likely to be most useful when setting dummies within a procedure. By setting METHOD=preserve, any dummies that are already set will have their existing settings preserved. Hence this provides a very convenient and effective of making default assignments while leaving any explicit assignments unchanged. Suppose, for example, that a procedure has dummy arguments FITTEDVALUES, RESIDUALS, and RSS available to save various aspects of the analysis, and that we wish to use these as working variables while calculating this information within the procedure. By specifying

```
ASSIGN [METHOD=preserve] LocalF,LocalR,LocalRSS; \
    FITTEDVALUES,RESIDUALS,RSS
```

any of the dummies that is not set when the procedure is called will be assigned to the corresponding local structure, either LocalF, LocalR, or LocalRSS. Note, however, that elements of pointers cannot be unset; they will always point to some identifier, even if it is unnamed. Thus, ASSIGN has no effect on elements of pointers when METHOD=preserve.

The NSUBSTITUTE option is useful when you have dummies pointing to other dummies, in a chain. This can often happen when one procedure calls another, passing one of its own arguments as the argument to the procedure that it calls. The NSUBSTITUTE option allows the dummies in the POINTER list to be substituted a set number of times in order to determine which dummy in a chain is to be assigned a value.

When the procedure SETARG is called from procedure ADDONE at line 15 of Example 4.9.1a, the dummy ARG which is the first parameter of procedure SETARG, points to the dummy RESULT which is the second parameter of procedure ADDONE. (See 5.3 for more details about procedures.) By setting option NSUBSTITUTE=1 in line 5 of the example, the dummy ARG is substituted once before it is assigned, so that the value is assigned to the dummy RESULT. Notice that this option affects only the dummies in the POINTER list, and not any that appear elsewhere; thus the dummy DEFAULT will be substituted to the variate Xplus1 to which DEFAULT is set at line 15.

Example 4.9.1a

```
 1    PROCEDURE 'SETARG'
 2    "Assigns a default to an unset dummy argument."
 3      PARAMETER NAME='ARG','DEFAULT'; MODE=p; TYPE='dummy',*
 4      IF UNSET(ARG)
 5        ASSIGN [NSUBSTITUTE=1] DEFAULT; ARG
 6      ENDIF
 7    ENDPROCEDURE
 8
 9    PROCEDURE 'ADDONE'
10    "Adds one to the values of variate X and prints results
-11     (which can also be saved using the RESULTS parameter)."
12      PARAMETER 'X','RESULT'; MODE=p; SET='yes','no'; DECLARED='yes','no'; \
13        TYPE='variate'; COMPATIBLE=!t(nvalues,type); PRESENT='yes','no'
14      VARIATE    Xplus1
15      SETARG     RESULT; DEFAULT=Xplus1
16      CALCULATE  RESULT=X+1
17      PRINT      RESULT
18    ENDPROCEDURE
19
20    VARIATE [VALUES=1,3,5,7] Y
21    ADDONE  Y

      Xplus1
        2.000
        4.000
        6.000
        8.000
```

Sometimes it may be easier to specify which dummy to assign by counting up from the bottom of the chain of dummies, instead of down from the top. You should then set NSUBSTITUTE to a negative integer. In Example 4.9.1b, dummy A points to dummy B, which in turn points to dummy C, and dummy C then points to dummy D, which points to the scalar X (line 2). Thus, at line 3

```
ASSIGN [NSUBSTITUTE=-1] Y; A
```

will assign Y to the dummy one from the bottom of the chain, that is C, and so

```
PRINT C,D
```

at line 4, prints the values of Y and X.

Example 4.9.1b

```
1   SCALAR X,Y; VALUE=1,2
2   DUMMY  A,B,C,D; VALUE=B,C,D,X
3   ASSIGN [NSUBSTITUTE=-1] Y; A
4   PRINT C,D

        Y              X
    2.000          1.000
```

The RENAME option enables you control what identifier is used for data structures in the rare occasions when your program contains structures that can be referred to by more than one suffixed identifier and which do not have identifiers in their own right. This is illustrated in Example 4.9.1c.

At line 2 of Example 4.9.1c, pointer P is defined to have two elements: suffix 1 refers to the scalar X and suffix 2 to the scalar Y. Line 3 introduces the identifier P[3], so Genstat expands P to have a third suffix; but this structure can be referred to only as P[3] – as is shown when the three structures are printed in line 4. Line 5 defines a new pointer, Q, and sets its values to be the same as those of P; P[3] can now be referred to as Q[3] but, when Genstat prints this structure, it uses the original identifier P[3] (see line 7). Line 8 shows another way of defining the values of Q, using ASSIGN. At the same time, we can change the identifier that Genstat will then use, by setting option RENAME=yes. This is confirmed when Q[3] is printed in line 9.

Example 4.9.1c

```
1   SCALAR X,Y; VALUE=1,2
2   POINTER [VALUES=X,Y] P
3   SCALAR P[3]; VALUE=3
4   PRINT P[]

         X              Y           P[3]
     1.000          2.000          3.000
```

```
5   POINTER [VALUES=P[1,2,3]] Q
6   CALCULATE Q[1,2,3] = Q[1,2,3]*10
7   PRINT Q[]
```

```
            X              Y            P[3]
        10.00          20.00          30.00
```

```
8   ASSIGN [RENAME=yes] P[3]; POINTER=Q; ELEMENT=3
9   PRINT Q[]
```

```
            X              Y            Q[3]
        10.00          20.00          30.00
```

Finally, the SCOPE option enables you to assign a dummy within a procedure to a structure in the program that called the procedure. The dummy will thus operate as though it was a dummy option or parameter, except that the decision about the structure that it references in the outer program has been made within the procedure instead of outside it. This facility allows you to define new data structures in the outer program; however, care needs to be taken to ensure that there is no conflict with any existing structures.

4.10   Operations on matrices and compound structures

The CALCULATE directive (4.1.1 and 4.1.3) allows you to do arithmetic operations on matrices element by element: addition, subtraction, multiplication, division, and exponentiation, as well as logical operations of testing for equality and inequality, and so on; you can also do matrix multiplication. There are several functions for standard operations on matrices, such as taking inverses, described in 4.2.4. You can combine and omit rows or columns of a rectangular matrix using the COMBINE directive (4.11.4). EQUATE allows you to transfer values to matrices from another structure, and vice versa (4.3), or you can select sub-matrices with CALCULATE, using qualified identifiers (4.1.6). Procedures that operate on matrices include GINVERSE to calculate the generalized inverse of a matrix, LINDEPENDENCE to find the linear relations associated with matrix singularities, MPOWER to form integer powers of a square matrix, and ROBSSPM to form robust estimates of sums-of-squares-and-products matrices.

You cannot do calculations directly with a complete compound structures like an LRV or an SSPM, but you can do calculations with the individual elements. For example, to take the diagonal matrix of latent roots from an LRV structure, L, and divide it by the trace, you could put

```
CALCULATE L['Roots'] = L['Roots'] / L['Trace']
```

This section describes the SVD and FLRV directives, which allow you to form singular value and eigenvalue decompositions; it also describes the FSSPM directive, which calculates sums of squares and products and all the associated information stored in an SSPM structure. These operations form the basis of many common statistical methods. The FTSM directive, which forms preliminary values of a time-series model in a TSM structure, is described in 12.7.1.

4.10.1 The singular value decomposition (SVD)

SVD calculates singular value decompositions of matrices i.e. (LEFT *+ SINGULAR *+ TRANSPOSE(RIGHT)).

Option

PRINT = *strings*	Printed output required (left, singular, right); default * i.e. no printing

Parameters

INMATRIX = *matrices*	Matrices to be decomposed
LEFT = *matrices*	Left-hand matrix of each decomposition
SINGULAR = *diagonal matrices*	Singular values (middle) matrix
RIGHT = *matrices*	Right-hand matrix of each decomposition

Suppose that we have a rectangular matrix A with m rows and n columns, and that p is the minimum of m and n. The singular value decomposition can be defined as

$$_m A_n \ = \ _m U_p \ _p S_p \ _p V_n$$

The diagonal matrix S contains the p singular values of A, ordered such that

$$s_1 \geq s_2 \geq \ldots \geq s_p \geq 0$$

The matrices U and V contain the left and right singular vectors of A, and are orthonormal:

$$U'U = V'V = I_p$$

The smaller of U and V will be orthogonal. So, if A has more rows than columns, $m>n$, $p=n$ and $VV'=I_p$.

The least-squares approximation of rank r to A can be formed as

$$A_r = U_r S_r V_r'$$

where U_r and V_r are the first r columns of U and V, and S_r contains the first r singular values of A (Eckart and Young 1936).

The INMATRIX parameter specifies the matrices to be decomposed. The algorithm uses Householder transformations to reduce A to bi-diagonal form, followed by a QR algorithm to find the singular values of the bi-diagonal matrix (Golub and Reinsch 1971). The other parameters allow you to save the component parts of the decomposition: LEFT, SINGULAR, and RIGHT for U, S, and V respectively.

The PRINT option allows you to print any of the components of the decomposition; by default, nothing is printed. If any of the matrices is to be printed, all p columns are shown, even if you are storing only the first r columns. See Example 4.10.1a.

Example 4.10.1a

```
1  MATRIX [ROWS=6; COLUMNS=4] A; VALUES=\
2     !(15,5,9,16,3,20,7,12,22,17,10,11,13,8,1,23,2,4,6,14,18,21,24,19)
3  SVD [PRINT=LEFT,SINGULAR,RIGHT] A
```

3 . . . .

```
*****   Singular value decomposition   *****

***   Singular Values   ***

                  1            2            3            4
               65.30        17.75        14.29        10.82

***   Left Singular Vectors   ***

                  1            2            3            4
        1      0.35066     -0.33717     -0.30338      0.26324
        2      0.32642      0.30654      0.69925     -0.39495
        3      0.45861      0.18847     -0.51086     -0.52922
        4      0.37075     -0.71706      0.15091     -0.29662
        5      0.20711     -0.27069      0.36641      0.39876
        6      0.61629      0.41157     -0.03151      0.49765

***   Right Singular Vectors   ***

                  1            2            3            4
        1      0.50011     -0.13783     -0.80932     -0.27549
        2      0.50254      0.53368      0.40553     -0.54607
        3      0.40479      0.48070     -0.09456      0.77210
        4      0.57749     -0.68199      0.41425      0.17258
```

Genstat will decide how many columns and singular values r to store, and will store that number for any of the components that you specify. If none of the matrices in the LEFT, SINGULAR, and RIGHT lists has been declared in advance, the full number of singular values ($r=p$) is stored; otherwise Genstat sets r to the maximum number of columns contained in any of the matrices. If $r<p$, the first r singular values will be saved, along with the corresponding columns of singular vectors.

Example 4.10.1b

```
  4    DIAGONALMATRIX [ROWS=2] Sa
  5    SVD A; LEFT=Ua; SINGULAR=Sa; RIGHT=Va
  6    CALCULATE A2 = Ua *+ Sa *+ TRANSPOSE(Va)
  7    PRINT [RLWIDTH=6] A; FIELDWIDTH=9; DECIMALS=3

                  A
                  1          2          3          4

        1     15.000      5.000      9.000     16.000
        2      3.000     20.000      7.000     12.000
        3     22.000     17.000     10.000     11.000
        4     13.000      8.000      1.000     23.000
        5      2.000      4.000      6.000     14.000
        6     18.000     21.000     24.000     19.000

  8    & A2; FIELDWIDTH=9; DECIMALS=3

                  A2
                  1          2          3          4

        1     12.276      8.313      6.392     17.304
        2      9.910     13.615     11.243      8.598
        3     14.515     16.834     13.730     15.012
```

```
4    13.861     5.373     3.681    22.660
5     7.426     4.232     3.165    11.087
6    19.119    24.122    19.801    18.257
```

In Example 4.10.1b, the diagonal matrix Sa saves the first two singular values, while the first two left singular vectors are stored in the matrix Ua. A2 is a least-squares approximation to A, based on $r=2$ singular values (known as an Eckart-Young approximation, of rank 2).

One practical application of the singular value decomposition is to form generalized inverses of matrices. If you use the singular value decomposition you obtain the Moore-Penrose generalized inverse, sometimes called the *pseudo-inverse*, and this is the method used by the GINVERSE procedure.

Example 4.10.1c verifies that the necessary properties of the Moore-Penrose inverse are satisfied. You need to set the ZDZ option of CALCULATE to zero when calculating Splus, the generalized inverse of the diagonal matrix of singular values, in case any of the singular values is zero. The default for ZDZ would set the corresponding elements of Splus to be missing (4.1.1).

Example 4.10.1c

```
 9   SVD A; LEFT=Uda; SINGULAR=Sda; RIGHT=Vda
10   CALCULATE [ZDZ=zero] Splus = Sda / Sda / Sda
11   & Aplus = Vda *+ Splus *+ TRANSPOSE(Uda)
12   & Aa,Aap = A,Aplus *+ Aplus,A *+ A,Aplus
13   & CheckAa,CheckAp = MAX(ABS(A,Aplus-Aa,Aap))
14   " If Aplus is the generalized inverse of A, then
-15    Aa and Aap should be identical to A and Aplus."
16   PRINT CheckAa,CheckAp; DECIMALS=3

     CheckAa      CheckAp
      0.000        0.000

17   CALCULATE Asa,Aspa = A,Aplus *+ Aplus,A
18   PRINT Asa; FIELDWIDTH=9; DECIMALS=3
```

```
               Asa
                 1        2        3        4        5        6

          1    0.398   -0.305    0.113    0.248    0.158    0.218
          2   -0.305    0.845    0.059    0.124    0.083    0.109
          3    0.113    0.059    0.787    0.115   -0.354    0.113
          4    0.248    0.124    0.115    0.762    0.208   -0.219
          5    0.158    0.083   -0.354    0.208    0.409    0.203
          6    0.218    0.109    0.113   -0.219    0.203    0.798
```

```
19   & Aspa; FIELDWIDTH=9; DECIMALS=3
```

```
              Aspa
                 1        2        3        4

          1    1.000    0.000    0.000    0.000
          2    0.000    1.000    0.000    0.000
          3    0.000    0.000    1.000    0.000
          4    0.000    0.000    0.000    1.000
```

4.10.2 Eigenvalue decompositions (FLRV)

FLRV forms the values of LRV structures.

Options

PRINT = *strings*	Printed output required (roots, vectors); default * i.e. no printing
NROOTS = *scalar*	Number of roots or vectors to print; default * i.e. print them all
SMALLEST = *string*	Whether to print the smallest roots instead of the largest (yes, no); default no

Parameters

INMATRIX = *symmetric matrices*	Matrices whose latent roots and vectors are to be calculated
LRV = *LRVs*	LRV to store the latent roots and vectors from each INMATRIX
WMATRIX = *symmetric matrices*	(Generalized) within-group sums of squares and products matrix used in forming the two-matrix decomposition, defined in the Reference Manual (Section 4.10.2); if any of these is omitted, it is taken to be the identity matrix, giving the usual spectral decomposition

---

This directive solves the two related eigenvalue problems
1) $AX = XL$
2) $AX = WXL$

of which the first gives the spectral decomposition of the n-by-n matrix A. Here A and W are both symmetric matrices, L is a diagonal matrix, and X is a square matrix. The structures X and L will usually be the first two elements of an LRV structure.

In problem 1, the *one-matrix problem*, XLX' is the spectral decomposition of the symmetric matrix A. Here L is a diagonal matrix containing the n latent roots, or eigenvalues, of A ordered such that

$$l_1 \geq l_2 \geq ... \geq l_n$$

The columns of the n-by-n matrix X are the corresponding latent vectors, or eigenvectors. The matrix X is orthogonal:

$$X'X = XX' = I_n$$

The spectral decomposition is the basis for several multivariate methods (Chapter 11).

In problem 2, the *two-matrix problem*, both A and W must have the same number of rows, n, and W must be positive semi-definite. Now the latent roots are the n elements of the diagonal matrix L and are the successive maxima of

$$l = (x'Ax) / (x'Wx)$$

where x is the corresponding column of the n-by-n matrix X, normalized so that $X'WX=I$. The two-matrix decomposition is particularly relevant for canonical variate analysis (11.2.2).

For either problem, the sum of the latent roots is stored in the element of the LRV labelled 'Trace'. In the one-matrix problem, this is also the trace of the original matrix A; but for the two-matrix problem, it is the trace of $W^{-1}A$. Latent roots are often expressed as percentages of the trace (see Chapter 11).

The method used for the spectral decomposition first reduces the matrix to tri-diagonal form using Householder transformations (Martin, Reinsch, and Wilkinson 1968); this is followed by a QL algorithm for finding the eigenvalues and eigenvectors (Bowdler, Martin, Reinsch, and Wilkinson 1968). The two-matrix problem is solved using two spectral decompositions, each computed as for the first problem.

The three options of FLRV control the printing of the results. You use the PRINT option to specify whether you want the roots or vectors to be printed. If you request the roots to be printed, the trace will be printed as well. By default nothing is printed. The NROOTS option governs how many of the roots and vectors are printed, while the SMALLEST option determines whether the largest or smallest roots, and corresponding vectors, are printed.

The INMATRIX parameter lists the matrices for which latent roots and vectors are to be calculated. You can use the LRV parameter to save the latent roots and vectors, and the trace. You must declare these structures in advance if you want to save less than the full number of roots; otherwise, they are defined automatically, as LRVs with n rows.

Example 4.10.2a forms the LRV structure, Ulrv, from a matrix of sums of squares and products (4.10.3), and prints the two smallest roots in order of descending magnitude. The trace is printed together with the latent roots, and the latent roots are printed as percentages of the trace.

Example 4.10.2a

```
1   VARIATE [NVALUE=13] U[1...7]
2   OPEN 'HARVF.DAT'; CHANNEL=2
3   READ [CHANNEL=2] U[]
```

Identifier	Minimum	Mean	Maximum	Values	Missing
U[1]	7.91	10.62	12.71	13	0
U[2]	7.71	10.25	12.15	13	0
U[3]	8.32	10.46	13.16	13	0
U[4]	9.19	10.86	13.06	13	0
U[5]	7.72	10.46	13.08	13	0
U[6]	8.69	10.53	12.82	13	0
U[7]	8.81	10.31	11.99	13	0

```
4   SSPM [TERMS=U[]] Us; SSP=Ussp
5   FSSPM Us
6   FLRV [PRINT=roots,vectors; NROOTS=2; SMALLEST=yes] Ussp; LRV=Ulrv
```

6..

***** Spectral decomposition *****

*** Latent Roots ***

```
            6              7
         6.881          1.122
```

```
***   Percentage variation   ***
                6               7
             4.25            0.69

***   Trace   ***

      161.7

***   Latent vectors   ***
                        6               7
        1          0.4770         -0.4040
        2         -0.4345         -0.1617
        3         -0.0525          0.1007
        4          0.1123          0.4762
        5          0.1425         -0.0629
        6          0.4178         -0.5527
        7         -0.6111         -0.5141
```

You can save a subset of the latent roots and vectors by supplying an LRV structure with fewer columns than rows. However this saves only the largest roots and the corresponding vectors. You cannot save the smallest roots directly, as the SMALLEST option applies only to printing. If you want to save the smallest roots, then you must save the complete set of roots and vectors, and extract the last columns of the matrix, for example using qualified identifiers (4.1.6). These rules are the same as those applied in the directives for multivariate analysis.

For the two-matrix problem, you specify the matrix W using the WMATRIX parameter. As an example we take W to be the diagonal of the matrix A. In this case, the solution is equivalent to the spectral decomposition of the correlation matrix derived from A, although the normalization of the latent vectors will be different. Example 4.10.2b shows the equivalence of the two analyses.

Example 4.10.2b

```
   7   CALCULATE Usspcor = CORRMAT(Ussp)
   8   PRINT Usspcor; FIELDWIDTH=7; DECIMALS=3

   Usspcor

   1   1.000
   2   0.215   1.000
   3   0.179   0.113   1.000
   4   0.294   0.439  -0.002   1.000
   5  -0.137   0.100  -0.049   0.345   1.000
   6  -0.754  -0.013  -0.014   0.065   0.062   1.000
   7   0.177  -0.112   0.021   0.419   0.258  -0.359   1.000

           1       2       3       4       5       6       7

   9   DIAGONALMATRIX Dusp
  10   SYMMETRICMATRIX Sdusp
  11   CALCULATE Sdusp = (Dusp = Ussp)
  12   LRV [ROWS=7; COLUMNS=7] Uclrv,Usclrv
  13   FLRV [PRINT=roots,vectors; NROOTS=4] Usspcor; LRV=Uclrv
```

```
13.........................................................................
```

***** Spectral decomposition *****

*** Latent Roots ***

```
            1              2              3              4
          2.114          1.593          1.244          0.936
```

*** Percentage variation ***

```
            1              2              3              4
         30.20          22.76          17.78          13.37
```

*** Trace ***

```
          7.000
```

*** Latent vectors ***

```
                  1              2              3              4
      1        0.5558        -0.3518        -0.1574         0.1238
      2        0.2649         0.2787        -0.6400         0.2819
      3        0.1227        -0.1055        -0.4073        -0.8902
      4        0.4249         0.5042        -0.1067         0.0873
      5        0.1510         0.5429         0.2671        -0.1397
      6       -0.4809         0.4648        -0.1942        -0.1236
      7        0.4138         0.1497         0.5284        -0.2651
```

```
   14   & Ussp; LRV=Uclrv; WMATRIX=Sdusp
```

```
14.........................................................................
```

***** Two-matrix latent decomposition *****

*** Latent Roots ***

```
            1              2              3              4
          2.114          1.593          1.244          0.936
```

*** Percentage variation ***

```
            1              2              3              4
         30.20          22.76          17.78          13.37
```

*** Trace ***

```
          7.000
```

*** Latent vectors ***

```
                  1              2              3              4
      1        0.09616        0.06086       -0.02723       -0.02142
      2        0.06482       -0.06821       -0.15664       -0.06899
      3        0.02494        0.02144       -0.08279        0.18093
      4        0.09889       -0.11735       -0.02484       -0.02033
      5        0.02429       -0.08736        0.04298        0.02248
      6       -0.10395       -0.10046       -0.04198        0.02672
      7        0.13848       -0.05011        0.17682        0.08871
```

A similar use of the two-matrix problem is when *W* is obtained from previous samples of the same set of variables as those in *A*.

For a symmetric matrix *A*, you can use FLRV to form an inverse of *A* in much the same way as the singular value decomposition. If *A* is singular, this forms the Moore-Penrose inverse (pseudo inverse). Example 4.10.2c follows the lines of the SVD example for the generalized inverse of a matrix (4.10.1).

Example 4.10.2c

```
15   SYMMETRICMATRIX [ROWS=3; VALUES=10,13,17,17,22,29] Smx
16   LRV [ROWS=3; COLUMNS=3] Lsmx; VECTORS=Vsmx; ROOTS=Rts
17   FLRV [PRINT=roots,vectors] Smx; LRV=Lsmx
```

17...

***** Spectral decomposition *****

*** Latent Roots ***

```
         Rts
          1              2              3
        55.80           0.20           0.00
```

*** Percentage variation ***

```
         Rts
          1              2              3
        99.65           0.35           0.00
```

*** Trace ***

```
Lsmx['Trace']
     56.00
```

*** Latent vectors ***

```
                 Vsmx
                  1              2              3
       1        0.4233        -0.0512         0.9045
       2        0.5500        -0.7788        -0.3015
       3        0.7199         0.6251        -0.3015
```

```
18   " The value 1.E-6 is to check for roots which,
-19    but for numerical round-off, would be zero.
-20    This might need to be changed in another example. "
21   CALCULATE [ZDZ=zero] Irts = (Rts > 1.E-6) / Rts
22   CALCULATE Ismx = Vsmx *+ Irts *+ TRANSPOSE(Vsmx)
23   PRINT Ismx; FIELDWIDTH=8; DECIMALS=2
```

```
         Ismx
           1        2        3

       1   0.02     0.21    -0.16
       2   0.21     3.08    -2.46
       3  -0.16    -2.46     1.99
```

The relationship between the singular value decomposition of a rectangular matrix A and the spectral decompositions of $A'A$ and AA' is as follows. If $A = USV'$ is the singular value decomposition for A, then $A'A = VSU'USV' = VS^2V'$ and $AA' = USV'VSU' = US^2U'$, since $U'U = V'V = I$. The rank of matrix A is q and $q \le \min(m,n)$, which is p in our earlier notation (4.10.1); q corresponds to the number of non-zero singular values, and the diagonal matrix S consists of the q non-zero singular values followed by $(p-q)$ zero values. This shows that the squares of the q singular values of A are equivalent to the non-zero latent roots of the two symmetric matrices, $A'A$ and AA', derived from A. It also shows that the matrices U and V contain the first p latent vectors of AA' and $A'A$, respectively. For further details, see Rao (1973, Chapter 1) or Digby and Kempton (1987, Appendix A.8).

4.10.3 Forming sums of squares and products (FSSPM)

FSSPM forms the values of SSPM structures.

Options

PRINT = *strings*	Printed output required (correlations, wmeans, SSPM); default * i.e. no printing
WEIGHTS = *variate*	Weightings for the units; default * i.e. all units with weight one
SEQUENTIAL = *scalar*	Used for sequential formation of SSPMs; a positive value indicates that formation is not yet complete (see READ directive); default * i.e. not sequential
SAVE = *DSSP*	To save results in a special DSSP structure (see TERMS directive); default *

Parameter

SSPMs	Structures to be formed

FSSPM forms the values for the component parts of SSPM structures, based on the information supplied when the SSPM structures were declared (2.7.2). You can use an SSPM as input to the regression directive TERMS (8.2.2), or the multivariate directives PCP and CVA (11.2). The method used to form the SSPM is based on the updating formula for the means and corresponding corrected sums of squares and cross products (Herraman 1968).

FSSPM has one parameter which lists the SSPM structures whose values are to be formed. Genstat takes account of restrictions on any of the variates or factors forming the terms of the SSPM, or on the weights variate or grouping factor if you have specified them. If any of these vectors has a missing value, the corresponding unit is excluded from all the means and all the sums of squares and products. You can also exclude units by setting their weights to zero.

In Example 4.10.3a, units 1, 5, and 7 are omitted. Notice that the wmean setting of the PRINT option is ignored, as the GROUPS option of the SSPM directive has not been set.

Example 4.10.3a

```
1   VARIATE [NVALUE=10] Va[1...6]
2   OPEN 'HARVFB.DAT'; CHANNEL=2
3   READ [CHANNEL=2] Va[]
```

Identifier	Minimum	Mean	Maximum	Values	Missing	
Va[1]	15.70	36.86	47.10	10	0	
Va[2]	32.30	37.91	55.60	10	0	Skew
Va[3]	29.40	37.47	53.00	10	0	
Va[4]	26.20	33.66	44.00	10	0	
Va[5]	13.20	38.06	51.90	10	0	
Va[6]	12.70	36.74	54.60	10	0	

```
4   VARIATE Weight; VALUES=!(0,1,1,1,0,1,0,1,1,1)
5   SSPM [TERMS=Va[]] Ssva
6   FSSPM [PRINT=wmean,correlation,sspm; WEIGHT=Weight] Ssva
```

*** Degrees of freedom ***

Sums of squares: 6
Sums of products: 5
Correlations: 5

*** Sums of squares and products ***

Va[1]	1	482.54			
Va[2]	2	91.37	88.11		
Va[3]	3	248.40	141.25	559.24	
Va[4]	4	-82.84	-105.96	-75.79	270.99
Va[5]	5	-305.37	-52.51	-142.92	248.19
Va[6]	6	122.76	-30.11	248.49	43.17

		1	2	3	4

Va[5]	5	983.05	
Va[6]	6	-593.52	799.23

		5	6

*** Means ***

Va[1]	1	33.54
Va[2]	2	35.39
Va[3]	3	38.34
Va[4]	4	34.67
Va[5]	5	34.17
Va[6]	6	34.27

*** Sum of weights ***

        7.000

*** Correlation matrix ***

Va[1]	1	1.000				
Va[2]	2	0.443	1.000			
Va[3]	3	0.478	0.636	1.000		
Va[4]	4	-0.229	-0.686	-0.195	1.000	
Va[5]	5	-0.443	-0.178	-0.193	0.481	1.000

Va[6]	6	0.198	-0.113	0.372	0.093	-0.670	1.000
		1	2	3	4	5	6

When you have very many units, you may not be able to store them all at the same time within Genstat. You can then use the SEQUENTIAL option of READ (3.1.8) to read the data in conveniently sized blocks, and the SEQUENTIAL option of FSSPM to control the accumulation of the sums of squares and products. The SSPM is updated for each block of data in turn until the end of data is found.

Example 4.10.3b

```
 7   OPEN 'HARV.DAT'; CHANNEL=3
 8   SCALAR Sseq; 0
 9   VARIATE [NVALUE=10] V[1...5]
10   SSPM [TERMS=V[]] Vssp
11   FOR [NTIMES=999]
12      READ [CHANNEL=3; SEQUENTIAL=Sseq] V[]
13      FSSPM [SEQUENTIAL=Sseq; PRINT=SSPM] Vssp
14      EXIT Sseq <= 0
15   ENDFOR
```

Identifier	Minimum	Mean	Maximum	Values	Missing
V[1]	8.52	10.13	11.75	10	0
V[2]	8.910	9.829	10.800	10	0
V[3]	8.69	10.95	13.08	10	0
V[4]	7.71	10.01	11.65	10	0
V[5]	9.29	10.50	12.34	10	0

Identifier	Minimum	Mean	Maximum	Values	Missing
V[1]	8.32	10.29	12.71	10	0
V[2]	7.72	11.00	13.16	10	0
V[3]	8.93	10.79	12.66	10	0
V[4]	7.91	11.02	13.06	10	0
V[5]	8.81	10.90	13.07	10	0

Identifier	Minimum	Mean	Maximum	Values	Missing
V[1]	10.67	10.67	10.67	10	9
V[2]	12.15	12.15	12.15	10	9
V[3]	10.67	10.67	10.67	10	9
V[4]	13.06	13.06	13.06	10	9
V[5]	12.89	12.89	12.89	10	9

*** Degrees of freedom ***

Sums of squares: 20
Sums of products: 19

*** Sums of squares and products ***

V[1]	1	27.75			
V[2]	2	4.72	39.99		
V[3]	3	-4.55	-3.45	34.06	
V[4]	4	4.01	16.81	7.59	55.32
V[5]	5	6.76	24.19	10.79	16.81

		1	2	3	4
V[5]	5	34.71			
		5			

*** Means ***

V[1]	1	10.23
V[2]	2	10.50
V[3]	3	10.86
V[4]	4	10.64
V[5]	5	10.80

*** Number of units used ***

        21

Notice that the PRINT option has no effect until the last set of values is processed, when READ sets the scalar indicator to a negative value (3.1.8).

If you use an SSPM as input to the TERMS directive (8.2.2), Genstat will copy the information into a special structure called a DSSP. This uses extra precision for storage on computers where real numbers are not represented by enough bits to guard against unacceptable inaccuracies of round-off during the regression calculations. The extra precision is also used while the information is calculated by FSSPM, but it will be lost when the values are placed in the SSPM. The SAVE option allows you to save the DSSP formed internally by FSSPM, so that you can put it straight into TERMS.

4.11   Operations on tables

A table is a structure that stores numerical summaries of data that are classified into groups. The TABULATE directive lets you form tables from a variate, given also factors to define the groups; you can form and print tables containing counts, means, medians and other quantiles, totals, minima, maxima, or variances of the observations in each group (4.11.1). You can also use tables to save means, effects, and numbers of replications from an analysis of variance (9.6.1), or predictions from regression and generalized linear models (8.3.4 and 8.5.6).

You can do numerical calculations on the values in tables, using the CALCULATE directive (4.1.1, 4.1.4, and 4.2.5). You can re-form a table to omit or combine levels of any of the classifying factors (4.11.4); you can include margins, or omit them, or recalculate them (4.11.2); and you can express the body of a table as percentages of one of its margins (4.11.3).

4.11.1 Tabulation (**TABULATE**)

TABULATE forms summary tables of variate values.

Options

PRINT = *strings*	Printed output required (counts, totals, nobservations, means, minima, maxima,

	variances, quantiles); default * i.e. no printing
CLASSIFICATION = *factors*	Factors classifying the tables; default * i.e. these are taken from the tables in the parameter lists
COUNTS = *tables*	Saves a table counting the number of units with each factor combination; default *
SEQUENTIAL = *scalar*	Used for sequential formation of tables; a positive value indicates that formation is not yet complete (see READ); default *
MARGINS = *string*	Whether the tables should be given margins if not already declared (yes, no); default no
IPRINT = *string*	Whether to print the identifier of the table or the identifier of the (associated) variate that was used to form it (identifier, extra, associatedidentifier); default iden
WEIGHTS = *variate*	Weights to be used in the tabulations; default * indicates that all units have weight 1
PERCENTQUANTILES = *scalar* or *variate*	
	Percentage points for which quantiles are required; default 50 (i.e. median)
OWN = *scalar* or *variate*	Specifies option settings for the OWNTAB subroutine and indicates that this is to supply the data values instead of the variates in the DATA list; default *
OWNFACTORS = *factors*	Factors whose values are to be read by OWNTAB (must include the factors of the classification set); default *
OWNVARIATES = *variates*	Variates whose values are to be read by OWNTAB (must include the DATA variates); default *
INCHANNEL = *scalar*	Channel number of the file from which the OWNTAB subroutine is to read the data (previously opened by an OPEN statement)
INFILETYPE = *string*	Type of the OWN data file (input, unformatted); default inpu

Parameters

DATA = *variates*	Data values to be tabulated
TOTALS = *tables*	Tables to contain totals
NOBSERVATIONS = *tables*	Tables containing the numbers of non-missing values in each cell
MEANS = *tables*	Tables of means
MINIMA = *tables*	Tables of minimum values in each cell
MAXIMA = *tables*	Tables of maximum values in each cell
VARIANCES = *tables*	Tables of cell variances
QUANTILES = *tables* or *pointers*	Table to contain quantiles at a single PERCENTQUANTILE or pointer of tables for several

PERCENTQUANTILEs (not available for sequential or OWN tabulation)

TABULATE allows you to produce the various types of tabular summary listed in the settings of its PRINT option. The variates whose values are to be summarized are listed with the DATA parameter. If you want to save the summaries in tables, for manipulating or for printing later on, you should list identifiers of the tables in the appropriate parameter list: for example, you would save the totals in a table T by including T in the list for the TOTALS parameter. The other parameters similarly give the other kinds of summary: numbers of non-missing values, means, minima, maxima, variances, and quantiles.

The simplest quantile, and the one produced by default, is the median (50% quantile), but the PERCENTQUANTILE option allows you to request any percentage point (between 0 and 100, of course). Moreover, by specifying a variate as the setting for PERCENTQUANTILE, you can obtain several quantiles at the same time. However, if you then want to save the results the setting of the QUANTILE parameter must be a pointer with length equal to the required number of quantiles, instead of a single table.

If you merely want to print the summaries, you do not usually need to list any tables; you need only specify the PRINT option. The only exception to this is with sequential tabulation, described at the end of this subsection.

Any table that you have not declared in advance will be declared implicitly. If you have not declared any of the tables, the classifying factors are taken from the CLASSIFICATION option, which in that case you must have set; likewise, the MARGINS option determines whether or not the tables will have margins. Otherwise these two options are ignored, and the undeclared tables are defined to have the same classifying factors and status for margins as the tables that you have declared previously; all these previously declared tables must have the same set of classifying factors, and must be all with margins or all without margins.

Example 4.11.1a concerns goods of two different types dispatched to four different towns. In the print of the data you will notice that the book-keeping has been rather slack. There is one consignment (in line 6) where the type has not been recorded. With such observations, Genstat cannot find out what the group should be because one of the factor values is missing; so they are ascribed to the *unknown* cell associated with the table (2.5). In the declaration in line 9, the scalar that stores this value has been named so that it can be referred to in later calculations. After the tabulation (line 10), table Totdisp stores the total number of items of each type dispatched to each town, and the scalar Udisp summarizes the observations with unknown type or destination.

Example 4.11.1a

```
 1   VARIATE [NVALUES=15] Quantity,Charge
 2   FACTOR [NVALUES=15; LABELS=!T(A,B)] Type
 3   & [LABELS=!T(London,Manchester,Birmingham,Bristol)] Town
 4   READ [PRINT=data,errors] Town,Quantity,Type; FREPRESENTATION=labels

 5        London 10 A  Manchester    5 B  Birmingham  10 B      Bristol 25 A
 6   Manchester 10 *   Birmingham 100 B       London 200 B   Manchester 25 A
 7      Bristol 50 A   Birmingham  25 A       Bristol  25 B       London 25 A
```

```
 8       London 50 B  Manchester  25 B       London  50 A  :
 9    TABLE [CLASSIFICATION=Town,Type] Totdisp; UNKNOWN=Udisp
10    TABULATE Quantity; TOTALS=Totdisp
11    PRINT Totdisp; DECIMALS=0
```

	Totdisp	
Type	A	B
Town		
London	85	250
Manchester	25	30
Birmingham	25	110
Bristol	75	25

Unknown   Totdisp   10

Example 4.11.1b illustrates what happens when a value of the data variate is missing. Variate Charge stores the charge to be made for the transport of each consignment, and you will see that three of the values are missing (because these invoices have not yet been prepared). In the tables listed with the parameters, missing data values are ignored. For example, the table Invoices is declared automatically by the NOBSERVATIONS parameter to hold the number of invoices sent to each destination; it excludes the observations where Charge has a missing value. Similarly Payment contains the total charge to be paid on behalf of each destination, ignoring the missing values. You can however obtain a count of the numbers of units that would have contributed to each group if no values had been missing: you use the COUNTS option if you want to save the table, or put PRINT=counts if you want to print it. So table Nconsign contains the total number of consignments made to each destination (regardless of whether the corresponding charge is missing or not). The data variates are irrelevant for counts, and so you need not list any if counts are all that you require.

If there are no observations in one of the groups, the corresponding cell will be zero in a table of numbers of observations or counts; in a table of totals, means, minima, maxima, or variances, the cell will contain a missing value.

Example 4.11.1b

```
12   READ [PRINT=data,errors] Charge

13   10 20 15 15 * 60 80 30 25 15 25 15 40 * *  :
14   TABULATE [CLASSIFICATION=Town; COUNTS=Nconsign] DATA=Charge; \
15     TOTALS=Payment; NOBSERVATIONS=Invoices
16   PRINT Nconsign,Invoices,Payment; DECIMALS=0,0,2
```

	Nconsign	Invoices	Payment
Town			
London	5	4	145.00
Manchester	4	2	50.00
Birmingham	3	3	90.00
Bristol	3	3	65.00

Weighted tables can be obtained by setting the WEIGHT option to a variate of weights. You can, in general, think of weights as a set of multipliers which are applied to the data before any operations are performed. Thus, for most aspects of weighted tabulation you can replace x by wx and 1 by w (that is, n by Σw) in the standard formulae (see below). This is not quite what happens in the case of variances, but it is certainly true for all other functions (including counts).

	Unweighted	Weighted
Count	n	$\Sigma\, w$
Total	$\Sigma\, x$	$\Sigma\, wx$
Nobservations	n	$\Sigma\, w$ (x not missing)
Mean	$\Sigma\, x/n$	$\Sigma\, wx\, /\, \Sigma\, w$
Minimum	Min$(\ x\ )$	Min$(\ wx\ )$
Maximum	Max$(\ x\ )$	Max$(\ wx\ )$
Variance	$\Sigma\, (\ x - (\Sigma x/n)\ )^2\, /\, n{-}1$	$\Sigma\, w\, (\ x - (\Sigma x/n)\ )^2\, /\, \Sigma\, w{-}1$

A quick look at the formula used for weighted variance will show that it breaks down for $\Sigma w < 1$, and, in fact, is valid only for integer values of w. If an invalid weight is found during the calculation of a variance it will be reported and the tabulation halted. Temporary tables will be deleted, but named tables may contain partial results. Non-integer weights are allowed in contexts other than variances.

If you have many observations to summarize, there may be insufficient space within Genstat for you to read them all and then form the tables. To cater for such situations, Genstat allows you to process the data in sections, using the SEQUENTIAL option of TABULATE in conjunction with the SEQUENTIAL option of READ (3.1.8). After READ, the absolute value of the option indicates the number of units that have been read in this particular section; the value is positive during interim sections and negative or zero once the terminator at the end of the data is reached. TABULATE will not print any tables until the final section has been processed. If you want to see the intermediate tables, you can include a PRINT statement after the TABULATE statement. To allow Genstat to keep contact with the working tables in which the results are accumulating, you must save at least one out of the various types of table for every DATA variate. Genstat can then link the working tables to this named table during the course of the sequential tabulation, so that the information is not lost between the successive uses of TABULATE.

This is illustrated in Example 4.11.1c, which also shows how to use the IPRINT option to print the identifier of the variate from which the table was formed, instead of the identifier of the table. Also notice that this time the table formed has a margin (2.5). As there is only one type of table being printed, Genstat has labelled the margin appropriately (as "Mean"). If several types of table were printed, Genstat would label the margins as "Margin". For tables of quantiles, the margin label is either "Median" (for the 50% quantile) or, say, "25%" for the 25% quantile. These labels are associated with the tables for later use, for example by PRINT.

The printed table summarizes the amount of excess baggage per person for the passengers on a particular flight. There are 77 passengers. The factors and variates are declared to have length 20, so the data are read in three sections of size 20 and a final section of size 17. The setting of the SEQUENTIAL option is the scalar S: it has the value 20 for the first three times that the loop is executed, and −17 on the final time. Notice that the variate Baggage is given missing values in units 18 to 20 in the final section: the value −17 in S tells Genstat that these units are not to be included in the tabulation. The loop construct FOR-ENDFOR is described in 5.2.1, and the EXIT directive in 5.2.4.

Example 4.11.1c

```
 1   UNITS [NVALUES=20]
 2   FACTOR [LABELS=!T(UK,EEC,other)] National
 3   FACTOR [LABELS=!T(male,female)] Sex
 4   VARIATE Baggage
 5   VARIATE Excess; EXTRA=' baggage per person in Kilograms'; DECIMALS=3
 6   OPEN 'FLIGHT.DAT'; CHANNEL=2
 7   SCALAR S
 8   FOR [NTIMES=999]
 9     READ [CHANNEL=2; SEQUENTIAL=S] Baggage,Sex,National; \
10       FREPRESENTATION=labels
11     CALCULATE Excess=(Baggage>20)*(Baggage-20)
12     TABULATE [PRINT=mean; CLASSIFICATION=Sex,National; SEQUENTIAL=S; \
13       MARGINS=yes; IPRINT=associatedidentifier] \
14       Excess; NOBSERVATIONS=Ntemp; MEANS=Mtemp
15     EXIT S <= 0
16   ENDFOR
```

Identifier	Minimum	Mean	Maximum	Values	Missing
Baggage	15.00	20.50	28.00	20	0

Identifier	Values	Missing	Levels
Sex	20	0	2
National	20	0	3

Identifier	Minimum	Mean	Maximum	Values	Missing
Baggage	17.00	22.45	35.00	20	0

Identifier	Values	Missing	Levels
Sex	20	0	2
National	20	0	3

Identifier	Minimum	Mean	Maximum	Values	Missing
Baggage	15.00	20.65	30.00	20	0

Identifier	Values	Missing	Levels
Sex	20	0	2
National	20	0	3

Identifier	Minimum	Mean	Maximum	Values	Missing
Baggage	15.00	20.35	28.00	20	3

Identifier	Values	Missing	Levels
Sex	20	3	2
National	20	3	3

National	Excess UK	EEC	other	Mean
Sex				
male	1.292	2.364	3.857	2.265
female	1.200	1.400	1.250	1.250

Mean   1.256   2.063   2.909   1.896

The final five options of TABULATE (OWN, OWNFACTORS, OWNVARIATES, INCHANNEL, and INFILETYPE) allow you to link your own Fortran subroutine, G5XZIT, to Genstat to allow you to handle complicated arrangements of data, as can occur for example in hierarchical surveys. Details are given in 13.4.3.

4.11.2 Forming margins of tables (MARGIN)

MARGIN forms and calculates marginal values for tables.

Option

CLASSIFICATION = *factors*	Factors classifying the margins to be formed; default * requests all margins to be formed

Parameters

OLDTABLE = *tables*	Tables from which the margins are to be taken or calculated
NEWTABLE = *tables*	New tables formed with margins
METHOD = *strings*	Way in which the margins are to be formed for each table (totals, means, minima, maxima, variances, medians, deletion, or a null string to indicate that the marginal values are all to be set to the missing value); default tota

---

You can use MARGIN to extend a table to contain marginal values, or to change the marginal values of a table that already has margins, or to delete the margins from a table. The tables whose margins are to be changed are specified by the OLDTABLES parameter. If you specify only this parameter, the new values replace those of the original tables. For example, in 2.5, the statement

```
MARGIN Classnum,Schoolnm
```

formed margins of totals over all the classifying factors for the tables, Classnum and Schoolnm; the new values, including the margins, replaced the original values of Classnum and Schoolnm.

However, if you want to retain the original values, you can specify new tables to contain the amended values, using the NEWTABLES list. These tables will be declared automatically, if you have not declared them already.

Example 4.11.2a creates the new tables, Classt and Schoolt, with margins of totals, using the values in the tables Classnum and Schoolnm.

Example 4.11.2a

```
1   FACTOR [LABELS=!T(boy,girl)] Sex
2   FACTOR [LEVELS=5] Class
3   FACTOR [LEVELS=2] School
4   TABLE [CLASSIFICATION=Class,Sex; \
5     VALUES=15,17,29,31,34,30,33,35,28,27] Classnum
6   TABLE [CLASSIFICATION=School,Class,Sex; VALUES=15,17,29,31,34, \
7     30,33,35,28,27,18,16,33,31,35,36,34,33,31,32] Schoolnm
8   MARGIN Classnum,Schoolnm; NEWTABLE=Classt,Schoolt
9   PRINT Classt,Schoolt; DECIMALS=0
```

		Classt		
	Sex	boy	girl	Margin
Class				
1		15	17	32
2		29	31	60
3		34	30	64
4		33	35	68
5		28	27	55
Margin		139	140	279

		Schoolt		
	Sex	boy	girl	Margin
School	Class			
1	1	15	17	32
	2	29	31	60
	3	34	30	64
	4	33	35	68
	5	28	27	55
	Margin	139	140	279
2	1	18	16	34
	2	33	31	64
	3	35	36	71
	4	34	33	67
	5	31	32	63
	Margin	151	148	299
Margin	1	33	33	66
	2	62	62	124
	3	69	66	135
	4	67	68	135
	5	59	59	118
	Margin	290	288	578

You can form other types of margin by setting the METHOD parameter. The next example forms the tables Classno and Schoolno with margins of means and maxima respectively.

Example 4.11.2b

```
10   MARGIN Classnum,Schoolnm; NEWTABLE=Classno,Schoolno; \
11     METHOD=means,maxima : PRINT Classno,Schoolno; DECIMALS=0
```

```
                    Classno
       Sex           boy           girl         Margin
       Class
         1            15            17            16
         2            29            31            30
         3            34            30            32
         4            33            35            34
         5            28            27            28

       Margin         28            28            28
```

```
                               Schoolno
                  Sex           boy           girl         Margin
       School     Class
         1          1            15            17            17
                    2            29            31            31
                    3            34            30            34
                    4            33            35            35
                    5            28            27            28

                  Margin         34            35            35

         2          1            18            16            18
                    2            33            31            33
                    3            35            36            36
                    4            34            33            34
                    5            31            32            32

                  Margin         35            36            36

       Margin       1            18            17            18
                    2            33            31            33
                    3            35            36            36
                    4            34            35            35
                    5            31            32            32

                  Margin         35            36            36
```

All the examples so far have been of adding margins. But you can delete them too: if you set METHOD=deletion, all the margins of the tables are deleted but the body of the table is retained.

The CLASSIFICATION option specifies the list of factors for which you want to form marginal values. Example 4.11.2c forms a margin of totals for the factor Class in the table Classnum.

Example 4.11.2c

```
   12   MARGIN [CLASSIFICATION=Sex] Classnum
   13   PRINT Classnum; DECIMALS=0

                   Classnum
       Sex          boy           girl         Margin
       Class
         1           15            17            32
         2           29            31            60
         3           34            30            64
         4           33            35            68
```

	5	28	27	55
Margin		0	0	0

Genstat puts missing values in the margins that are excluded if the METHOD parameter is set to maxima or minima; for other settings of METHOD, Genstat puts in zeroes. The classifying sets for each table can be different, but all the factors in the CLASSIFICATION option must be in the classifying sets of each OLDTABLE. So, for example,

    MARGIN [CLASSIFICATION=Sex,School] Classnum, Schoolnm

would fail because the factor School is not in the classifying set of Classnum.

4.11.3 Forming tables of percentages

The PERCENT procedure allows you to express the body of a table as percentages of the values in one of its margins. It has two parameters, OLDTABLE and NEWTABLE, and three options CLASSIFICATION, METHOD, and HUNDRED.

The table is specified using the OLDTABLE parameter. A table to store the new values can be specified using the NEWTABLES parameter, otherwise these replace the values of the original table. The margin is indicated by listing the factors that define it using the CLASSIFICATION option; the default is the final margin (the grand total, or grand mean &c). If the original table has no margins, option METHOD defines how these are to be calculated (totals, means, minima, maxima, variances, medians); the default is to form margins of totals. The values originally in the margin will be left unchanged. If you would prefer these to be replaced by values of 100%, you should set option HUNDRED=yes.

In Example 4.11.3 the contents of the table Totdisp, formed in Example 4.11.1a, are expressed as percentages of the overall margin – which, as option HUNDRED is not set, is left with its original value 625. The unknown cell is not included in the calculations, and option setting PUNKNOWN=never in PRINT suppresses it from being printed with Totdisp.

Example 4.11.3

```
  17  LIBHELP [options,parameters] 'PERCENT'

    Procedure     PERCENT

Help['options']
CLASSIFICATION = factors Factors classifying the margin over which the
                        percentages are to be calculated
METHOD         = string  Method to use to calculate the margin if not already
                        present (totals, means, minima, maxima, variances,
                        medians); default t
HUNDRED        = string  Whether to put 100% values into the margin instead
                        of the original values (no, yes); default n

Help['parameters']
OLDTABLE       = tables  Tables containing the original values
NEWTABLE       = tables  Tables to store the percentage values; if any of
                        these is unset, the new values replace those in the
                        original table
```

```
18   PERCENT Totdisp
19   PRINT [PUNKNOWN=never] Totdisp; DECIMALS=2
```

```
              Totdisp
     Type        A          B        Margin
     Town
   London      13.60      40.00       53.60
Manchester      4.00       4.80        8.80
Birmingham      4.00      17.60       21.60
   Bristol     12.00       4.00       16.00

   Margin      33.60      66.40      625.00
```

4.11.4 Combining or omitting slices of tables and matrices (COMBINE)

COMBINE   combines or omits "slices" of a multi-way data structure (table, matrix, or variate).

Options

OLDSTRUCTURE = *identifier*   Structure whose values are to be combined; no default i.e. this option must be set

NEWSTRUCTURE = *identifier*   Structure to contain the combined values; no default i.e. this option must be set

Parameters

OLDDIMENSION = *factors* or *scalars*

Dimension number or factor indicating a dimension of the OLDSTRUCTURE

NEWDIMENSION = *factors* or *scalars*

Dimension number or factor indicating the corresponding dimension of the NEWSTRUCTURE; this can be omitted if the dimensions are in numerical order, while zero settings (each in conjunction with a single OLDPOSITION) allows a slice of an old table to be mapped into a new table with fewer dimensions

OLDPOSITIONS = *pointers*, *texts*, *variates* or *scalars*

These define positions in each OLDDIMENSION: pointers are appropriate for matrices whose rows or columns are indexed by a pointer; texts are for matrices indexed by a text, variates with a textual labels vector, or tables whose OLDDIMENSION factor has labels; and variates either refer to levels of table factors or numerical labels of matrices or variates, if these are present, otherwise they give the (ordinal) number of the position. If omitted, the positions are assumed to be in (ordinal)

	numerical order. Margins of tables are indicated by missing values
NEWPOSITIONS = *pointers*, *texts*, *variates* or *scalars*	These define positions in each NEWDIMENSION, specified similarly to OLDPOSITIONS; these indicate where the values from the corresponding OLDDIMENSION positions are to be entered (or added to any already entered there)
WEIGHTS = *variates*	Define weights by which the values from each OLDDIMENSION coordinate are to be multiplied before they are entered in the NEWDIMENSION

Sometimes you may wish to reclassify a table to have factors different from those that you used in its declaration. COMBINE allows you to omit or to combine levels of the classifying factors. Furthermore, if you want to take just one level of a factor, you can copy the values into a table with one less dimensions.

You specify the original table using the OLDSTRUCTURE option, and a table to contain the reclassified values using the NEWSTRUCTURE option; if you have not already declared the new table, it will be declared implicitly. You must specify both of these options.

You can modify several of the classifying factors at a time. You list the factors of the original table with the OLDDIMENSION parameter, and the equivalent factors of the new table with NEWDIMENSION. An alternative way of doing this is to give a dimension number, specifying the position of the factor in the classifying set of the table (2.5); for the NEWDIMENSION list, this requires that you have already declared the new table. You can even omit the list of dimensions if they would be in ascending numerical order.

In Example 4.11.4a, the table Sales contains the number of items of some product sold by a retailer with shops in nine towns, in the years 1979 to 1984. Lines 20 to 23 declare and form a table Csales in which the sales are classified by the country where the sale was made, instead of the town; so there is one OLDDIMENSION, the factor Town, and a corresponding NEWDIMENSION, Country.

Example 4.11.4a

```
  1   TEXT [VALUES= Aberdeen,Birmingham,Cardiff,Dundee,Edinburgh, \
  2    Liverpool,Manchester,Sheffield,Swansea] Townname
  3   VARIATE [VALUES=1979,1980,1981,1982,1983,1984] Yearnum
  4   FACTOR  [LABELS=Townname] Town
  5   FACTOR  [LEVELS=Yearnum]  Year; DECIMALS=0
  6   TABLE [CLASSIFICATION=Town,Year] Sales
  7   READ Sales
```

Identifier	Minimum	Mean	Maximum	Values	Missing
Sales	343.0	676.3	1158.0	54	0

```
 17   PRINT Sales; FIELDWIDTH=8; DECIMALS=0
```

```
              Sales
      Year    1979    1980    1981    1982    1983    1984
      Town
   Aberdeen    608     635     672     692     685     723
 Birmingham    618     601     784     720     863     921
    Cardiff    757     743     785     816     783     737
     Dundee    343     391     358     366     418     470
  Edinburgh    714     751     710     763     788     830
  Liverpool    816     859     820     938    1007    1158
 Manchester    662     632     758     721     893     837
  Sheffield    531     569     615     624     607     593
    Swansea    416     461     478     462     497     520
```

```
18   FACTOR [LABELS=!T(England,Wales,Scotland)] Country
19   " Form a table Csales, classified by country instead of town."
20   COMBINE [OLDSTRUCTURE=Sales; NEWSTRUCTURE=Csales] \
21      OLDDIMENSION=Town; NEWDIMENSION=Country; \
22      OLDPOSITIONS=!(2,6,7,8,1,4,5,3,9); \
23      NEWPOSITIONS=!T(4(England),3(Scotland),2(Wales))
24   PRINT Csales; FIELDWIDTH=8; DECIMALS=0
```

```
              Csales
      Year    1979    1980    1981    1982    1983    1984
   Country
   England    2627    2661    2977    3003    3370    3509
     Wales    1173    1204    1263    1278    1280    1257
  Scotland    1665    1777    1740    1821    1891    2023
```

Each of the levels of Country is a combination of several levels of Town. You use the OLDPOSITIONS and NEWPOSITIONS parameters to specify how this combining is to be done. These parameters specify a pair of vectors for each pair of old and new dimensions, listing positions within the old dimension and the corresponding positions to which they are mapped in the new dimension. The positions can be defined in terms of either the levels or the labels of the factor that classifies the dimension. In the example, the vector for the old dimension Town is an unnamed variate !(2,6,7,8,1,4,5,3,9) whose values refer to the levels (1 to 9); the vector for Country is an unnamed text !T(4(England),3(Scotland),2(Wales)) whose values are labels of Country. The correspondence between the two sets of values is:

Town level	Town label	Country label	Country level
2	Birmingham	England	1
6	Liverpool	England	1
7	Manchester	England	1
8	Sheffield	England	1
1	Aberdeen	Scotland	2
4	Dundee	Scotland	2
5	Edinburgh	Scotland	2
3	Cardiff	Wales	3
9	Swansea	Wales	3

Thus, as you can see, the values in the original table for the English towns (Birmingham, Liverpool, Manchester, and Sheffield) are allocated to Country England in the new table, the

Scottish towns (Aberdeen, Dundee, and Edinburgh) are allocated to Scotland, and Cardiff and Swansea are allocated to Wales.

If you omit the vector for one of the dimensions, it is assumed to contain each value once only, taken in the order in which they occur in the levels vector of the factor. Thus the OLDPOSITIONS variate could be omitted in

```
COMBINE[OLDSTRUCTURE=Sales; NEWSTRUCTURE=Csales] \
    OLDDIMENSION=Town; NEWDIMENSION=Country; \
    OLDPOSITIONS=!(1...9); NEWPOSITIONS= \
        !T(Scotland,England,Wales,2(Scotland),3(England),Wales)
```

You indicate a margin of the table by a missing value in a variate, or by a null string in a text.

Values in the original table can be allocated to more than one place. Also, as we have mentioned already, you can modify more than one dimension at a time. In Example 4.11.4b, the Years dimension is modified as well as the Town dimension: years 1979 and 1980 are omitted, while the other years are allocated to two summary lines as well as to themselves in the new dimension Ysummary. Thus the new table Salesum has lines giving sales for the individual years, interspersed with bi-annual totals. Note that the interspersing of the summary lines is ensured by the order in which the FACTOR declaration specifies the labels of the factor Yearsum.

Example 4.11.4b

```
25  " Form a table classified by country and year including biannual totals."
26  FACTOR [LABELS=!T('1981','1982','1981-2','1983','1984','1983-4')] Yearsum
27  TABLE   [CLASSIFICATION=Yearsum,Country] Salesum
28  COMBINE [OLDSTRUCTURE=Sales; NEWSTRUCTURE=Salesum] \
29    OLDDIMENSION=Town,Year; NEWDIMENSION=Country,Yearsum; \
30    OLDPOSITIONS=!(2,6,7,8,1,4,5,3,9),!V((1981...1984)2); \
31    NEWPOSITIONS=!T(4(England),3(Scotland),2(Wales)), \
32    !T('1981','1982','1983','1984',2('1981-2','1983-4'))
33  PRINT Salesum; FIELDWIDTH=8; DECIMALS=0
```

```
             Salesum
   Country England   Wales Scotland
   Yearsum
      1981     2977    1263     1740
      1982     3003    1278     1821
    1981-2     5980    2541     3561
      1983     3370    1280     1891
      1984     3509    1257     2023
    1983-4     6879    2537     3914
```

The final use of the sales data shows how to extract a single slice of a table into a table with fewer dimensions. In Example 4.11.4c, the OLDPOSITIONS parameter specifies a single level England of the OLDDIMENSION Country, the NEWDIMENSION is set to 0. The new table is thus classified by only the factor Yearsum, and contains information about sales in England.

Example 4.11.4c

```
34   COMBINE [OLDSTRUCTURE=Csales; NEWSTRUCTURE=Esales] \
35     OLDDIMENSION=Country; NEWDIMENSION=0; OLDPOSITIONS='England'
36   PRINT Esales
```

```
            Esales
    Year
    1979     2627
    1980     2661
    1981     2977
    1982     3003
    1983     3370
    1984     3509
```

In parallel with the vectors of positions, you can also specify a variate of weights by which the values are multiplied before being entered into the new table. Thus, for example, forming summary lines of means instead of totals would require an extra parameter list

```
    WEIGHTS=*,!(1,1,1,1,0.5,0.5,0.5,0.5)
```

Although the main way in which you will use COMBINE is likely to be for tables, you can also use it on rectangular matrices and even variates. For these, the dimensions can only be numbers: number 1 refers to the rows of a matrix, and 2 to the columns; number 1 refers to the rows (or units) of a variate. The position vectors refer to the labels vectors of matrices (2.4.1), which can be variates, texts, or pointers; or they refer to the unit labels of a variate (2.3.1), which can be held in either a variate or a text. If a dimension has no labels vector, you use a variate to specify its positions; then each value of the variate gives the number of a row, column, or unit. You can do the same also if the labels vector is something other than a variate: that is, a text or a pointer.

P.K.L.
R.W.P.
R.P.W.

5   Programming in Genstat

In this chapter we tell you more about the structure of a Genstat program (5.1). We also describe the directives that allow you to loop over sequences of statements, or to choose between alternative sets of statements (5.2). These give you the flexibility to write general programs, where the exact analysis to be performed depends on information derived from the data when the program is run. Programs that are frequently required can be stored in procedures, as described in 5.3. This not only makes them simpler for you to use; it also means that you can make them easily available to other users, by means of libraries (5.3.1) or by articles in the Genstat Newsletter.

5.1   Genstat programs

A Genstat program is a sequence of statements involving either standard Genstat directives or procedures (1.1). You may often wish to examine several different sets of data within the same program: for example, you may want to be able to collect several analyses in one batch run (1.1.2), or to be able to end one analysis and start a different one, with different data, when you are running Genstat interactively (1.1.1).

The JOB and ENDJOB directives can be used to partition a Genstat program into separate *jobs*. A job is a self-contained subsection of a program. All data structures and procedures are lost at the end of each job. Any setting defined by a UNITS statement (2.3.4) is deleted, as are the special structures set up by analyses like regression and analysis of variance (2.8). The graphics environment is also reset to the initial default (6.5). Thus, in many ways, it is as though Genstat was starting again for each new job.

However, any files that have been attached to Genstat retain their current status from job to job. So, for example, Genstat will continue to add output to the end of an output file, or will continue reading from the current point of an input file.

The JOB directive also has options that allow you to modify some aspects of the Genstat environment: for example what prompt will be used for input and whether input lines are reprinted in an output file. The default settings of the options will leave these aspects unchanged so, if any aspect is modified, it will remain in that form (unless modified again) in any subsequent job. The initial settings, which apply at the outset of a program, are described in 5.1.1; however, remember that it is possible to arrange for Genstat to run commands from a *start-up* file (13.1.4) before it executes the first statement of a program, so the initial environment can differ from machine to machine.

5.1.1   The JOB directive

JOB   starts a Genstat job.

Options

INPRINT = *strings*   Printing of input as in PRINT option of INPUT

	(statements, macros, procedures, unchanged); default unch
OUTPRINT = *strings*	Additions to output as in PRINT option of OUTPUT (dots, page, unchanged); default unch
DIAGNOSTIC = *strings*	Defines the least serious class of Genstat diagnostic which should still be generated (messages, warnings, faults, extra, unchanged); default unch
ERRORS = *scalar*	Limit on number of error diagnostics that may occur before the job is abandoned; default * i.e. no limit
PROMPT = *text*	Characters to be printed for the input prompt

Parameter

| *text* | Name to identify the job |

The JOB directive is used to start a new job. It has a parameter which can be set to a text to identify the job (for example in the message at the end of the job), and options to control some aspects of the Genstat "environment". However, Genstat will automatically start a job at the beginning of a program, or after an ENDJOB statement, so you do not need to give a JOB statement unless you wish to define an identifying text or to modify the environment.

The INPRINT option specifies which pieces of input from the current input channel will be recorded in the current output file. (The current input channel may be a file or, in an interactive run, it may also be the keyboard.) The settings correspond to three types of input:

statements	statements that are typed explicitly on the keyboard or which occur explicitly in an input file,
macros	statements or parts of statements that have been supplied in macros, using the ## notation (1.9.2), and
procedures	statements occurring within procedures.

The initial default is to record only statements for input from a file, or to record nothing if input is from the keyboard. The recording of input can be modified also by the INPRINT option of the SET directive (13.1.1), or by the PRINT option of INPUT (3.4.1).

The OUTPRINT option controls the way in which the output from many Genstat directives will start: page ensures that output to a file will start at the head of a page, and dots produces a line of dots beginning with the line number of the statement that has generated the analysis. The initial default is to give a new page and a line of dots if output is to a file, but neither if output is to the screen. This can be modified also by the OUTPRINT option of the SET directive (13.1.1), or by the PRINT option of OUTPUT (3.4.3).

The DIAGNOSTICS option controls the reporting of errors and possible mistakes. In order of increasing seriousness there three classes of diagnostic: messages, warnings, and faults. Messages are comments that are made to draw your attention to things that might need closer investigation, like large residuals in an analysis of variance or a regression. Warnings are definite errors, but ones that are not sufficiently serious to prevent Genstat from continuing; an example would be an attempt to print a data structure with no values. Faults are the most serious type of error. A fault in a batch run will cause Genstat to stop executing the current job. However, Genstat will continue to read and interpret the statements so that it can find the start of the next job (if any); at the same time it will report any further errors that it finds, up to the number specified by the ERRORS option.

The setting of DIAGNOSTICS indicates the level of stringency to be adopted. Thus, if DIAGNOSTICS=warnings, Genstat will report faults and warnings (but not messages), while DIAGNOSTICS=messages ensures that all three classes are reported. The setting extra is similar to messages but will also generate a dump of system information (2.10.1) after any fault. You can prevent the output of any diagnostics by putting DIAGNOSTICS=*. The initial default is to set DIAGNOSTICS=messages.

5.1.2 The **ENDJOB** directive

ENDJOB ends a Genstat job.

No options or parameters

The ENDJOB directive terminates a job, printing a message summarizing how much workspace has been used. For example:

Example 5.1.2

```
   1   JOB 'Example of ENDJOB message'
   2   PRINT 'This job just prints this message.'

 This job just prints this message.

   3   ENDJOB

******** End of Example of ENDJOB message.  Maximum of 4932 data units used
at line 2 (379054 left)
```

You do not need to give an ENDJOB statement before a JOB statement, as JOB will automatically end any existing job before it starts another. Thus you can begin a new job by specifying either JOB or ENDJOB (or both).

5.1.3 The **STOP** directive

STOP ends a Genstat program.

No options or parameters

The STOP directive indicates the end of a Genstat program, thus telling the computer that you have finished using Genstat. It also ends the existing job, so there is no need to give an ENDJOB statement beforehand. Any input that follows a STOP statement is ignored.

5.2   Program control in Genstat

Usually the statements in a Genstat job are executed in sequence, until either ENDJOB or STOP is reached. But, as with most programming languages, you may sometimes want to control the order in which the statements are executed.

If you have several sets of data that are all to be analysed in the same way, you may want to repeat the necessary series of statements for each set. You can do this by preceding the series with a FOR statement, and ending it with an ENDFOR statement. The FOR directive also allows you to specify dummy structures (2.2.2) which point in turn to the data structures of the successive sets.

To be able to write general programs, you may need to be able to choose between alternative sets of statements, according to the exact form of a particular set of data. There are two ways in which you can do this. The directives IF, ELSIF, ELSE, and ENDIF allow you to define *block-if* structures (5.2.2). Alternatively, the directives CASE, OR, ELSE, and ENDCASE allow you to choose between sets of statements according to an integer value (5.2.3).

The EXIT directive (5.2.4) allows you to abandon any of these control structures while the program is being executed. The exit can be dependent on a condition, for example on an invalid data value or even on a Genstat diagnostic. The Genstat language is designed in accordance with the principles of structured programming: there is no way of "labelling" a statement and no equivalent of the Fortran "GO TO" construct.

5.2.1   FOR loops

FOR   introduces a loop; subsequent statements define the contents of the loop, which is terminated by the ENDFOR directive.

Options

NTIMES = *scalar*   Number of times to execute the loop; default is to execute as many times as the length of the first parameter list or once if the first list is null

Parameters

   dummies   Are set up implicitly by the statement; each dummy appears to be a parameter

---

ENDFOR   indicates the end of the contents of a loop.

No options or parameters

The FOR loop is a series of statements, or a *block*, that is repeated several times. The FOR directive introduces the loop and indicates how many times it is to be executed. In its simplest form FOR has no parameters, and the number of times is indicated by the NTIMES option. Thus the iterative calculation of a square root in the example in 1.9.2 can be specified in a FOR loop like this:

```
FOR [NTIMES=3]
   CALCULATE Previous = Root
   & Root = (X/Previous + Previous)/2
   PRINT Root,Previous; DECIMALS=4
ENDFOR
```

The sequence of CALCULATE and PRINT statements is repeated three times (exactly as in 1.9.2).

The parameters of FOR allow you to write a loop whose contents apply to different data structures each time it is executed. Unlike other directives, the parameter names of FOR are not fixed for you by Genstat: you can put any valid identifier before each equals sign. Each of these then refers to a Genstat dummy structure, as described in 2.2.2; so you must not have declared them already as any other type of structure. The first time that the loop is executed, they each point to the first data structure in their respective lists, next time it is the second structure, and so on. The list of the first parameter must be the longest; other lists are recycled as necessary. You can specify as many parameters as you need. For example

```
FOR Ind=Age,Name,Salary; Dir='descending','ascending'
   SORT [INDEX=Ind; DIRECTION=#Dir] Name,Age,Salary
   PRINT Name,Age,Salary
ENDFOR
```

is equivalent to the sequence of statements

```
SORT [INDEX=Age; DIRECTION='descending'] Name,Age,Salary
PRINT Name,Age,Salary
SORT [INDEX=Name; DIRECTION='ascending'] Name,Age,Salary
PRINT Name,Age,Salary
SORT [INDEX=Salary; DIRECTION='descending'] Name,Age,Salary
PRINT Name,Age,Salary
```

printing the units of the text Name, and variates Age and Salary, first in order of descending ages, then in alphabetic order of names, and finally in order of descending salaries.

You can put other control structures inside the loop. So, for example, you can have loops within loops.

When you are using loops interactively, you may find it helpful to use the PAUSE option of SET to requests Genstat to pause after every so many lines of output (13.1.1). Another useful directive is BREAK, which specifies an explicit break in the execution of the loop (5.4.1).

5.2.2 Block-if structures

The component parts of a *block-if* structure are delimited by IF, ELSIF, ELSE, and ENDIF statements.

IF   introduces a block-if control structure.

No options

Parameter

    expression   Logical expression, indicating whether or not to execute
                                            the first set of statements.

ELSIF   introduces a set of alternative statements in a block-if control structure.

No options

Parameter

    expression   Logical expression to indicate whether or not the set of
                                            statements is to be executed.

ELSE   introduces the default set of statements in block-if or in multiple-selection control structures.

No options or parameters

ENDIF   indicates the end of a block-if control structure.

No options or parameters

A *block-if* structure consists of one or more alternative sets of statements. The first of these is introduced by an IF statement. There may then be further sets introduced by ELSIF statements. Then you can have a final set introduced by an ELSE statement, and the whole structure is terminated by an ENDIF statement. Thus the general form is:
first

      IF expression
        statements

then either none, one, or several blocks of statements of the form

      ELSIF expression
        statements

then, if required, a block of the form

      ELSE
        statements

and finally the statement

```
ENDIF
```

Each expression must evaluate to a single number, which is treated as a logical value: a zero value is treated as *false* and non-zero as *true* (4.1.1). Genstat executes the block of statements following the first true expression. If none of the expressions is *true*, the block of statements following ELSE (if present) is executed.

You can thus use these directives to built constructs of increasing complexity. The simplest form would be to have just an IF statement, then some statements to execute, and then an ENDIF. For example:

```
IF MINIMUM(Sales) < 0
   PRINT 'Incorrect value recorded for Sales.'
ENDIF
```

If the variate Sales contains a negative value, the PRINT statement will be executed. Otherwise Genstat goes straight to the statement after ENDIF.

To specify two alternative sets of statements, you can include an ELSE block. For example

```
IF Age < 20
   CALCULATE Pay = Hours*1.75
ELSE
   CALCULATE Pay = Hours*2.5
ENDIF
```

calculates Pay using two different rates: 1.75 for Age less than 20, and 2.5 otherwise.

Finally, to have several alternative sets, you can include further sets introduced by ELSIF statements. Suppose that we want to assign values to X according to the rules:

X=1 if Y=1
X=2 if Y ≠ 1 and Z=1
X=3 if Y ≠ 1 and Z=2
X=4 if Y ≠ 1 and Z ≠ 1 or 2

This can be written in Genstat as follows:

```
IF Y == 1
   CALCULATE X = 1
ELSIF Z == 1
   CALCULATE X = 2
ELSIF Z == 2
   CALCULATE X = 3
ELSE
   CALCULATE X = 4
ENDIF
```

If Y is equal to 1, the first CALCULATE statement is executed to set X to 1. If Y is not equal to 1, Genstat does the tests in the ELSIF statements, in turn, until it finds a *true* condition; if none of the conditions is *true*, the CALCULATE statement after ELSE is executed to set X to 4. Thus, for Y=99 and Z=1, Genstat will find that the condition in the IF statement is *false*. It will then test the condition in the first ELSIF statement; this produces a *true* result, so X is set to 2. Genstat then continues with whatever statement follows the ENDIF statement. Block-if structures can be nested to any depth, to give conditional constructs of even greater flexibility.

5.2.3 The multiple-selection control structure

The directives CASE, OR, ELSE, and ENDCASE allow you to specify alternative blocks of statements, to be selected according to the value of an expression yielding a single integer value.

CASE introduces a "multiple-selection" control structure.

No options

Parameter

    expression   Expression which is evaluated to an integer, indicating which set of statements to execute

---

OR introduces a set of alternative statements in a "multiple-selection" control structure.

No options or parameters

ELSE introduces the default set of statements in block-if or in multiple-selection control structures.

No options or parameters

ENDCASE indicates the end of a "multiple-selection" control structure.

No options or parameters

A *multiple-selection* control structure consists of several alternative blocks of statements. The first of these is introduced by a CASE statement. This has a single parameter, which is an expression that must yield a single number. Subsequent blocks are each introduced by an OR statement. There can then be a final block, introduced by an ELSE statement, as in the block-if structure (5.2.2). The whole structure is terminated by an ENDCASE statement. Thus the general form is: first

    CASE expression
      statements

then either none, one, or several blocks of statements of the form

    OR
      statements

then, if required, a block of the form

      ELSE

statements

and finally the statement

```
ENDCASE
```

Genstat rounds the expression in the CASE expression to the nearest integer, k say, and then executes the kth block of statements. If there is no kth block present (as for example if k is negative) the block of statements following the ELSE statement is executed, if there is such a block; otherwise an error diagnostic is given. The next example prints the salient details about each day in the song *The twelve days of Christmas*. The scalar Day indicates which day it is.

```
CASE Day
  PRINT 'a partridge in a pear tree'
OR
  PRINT 'two turtle doves and a partridge in a pear tree'
OR
  PRINT 'three French hens, two turtle doves \
    and a partridge in a pear tree'
OR
  PRINT 'four calling birds, three French hens ...'
OR
  PRINT 'five gold rings ...'
OR
  PRINT 'six geese a-laying ...'
OR
  PRINT 'seven swans a-swimming ...'
OR
  PRINT 'eight maids a-milking ...'
OR
  PRINT 'nine drummers drumming ...'
OR
  PRINT 'ten pipers piping ...'
OR
  PRINT 'eleven ladies dancing ...'
OR
  PRINT 'twelve lords a-leaping ...'
ELSE
  PRINT 'sorry, no delivery today'
ENDCASE
```

CASE statements can be nested to any depth.

5.2.4 Exit from control structures

Sometimes you may want simply to abandon part of a program: you may be unable to do any further calculations or analyses. For example, if you are examining several subsets of the units, you would wish to abandon the analysis of any subset that turned out to contain no observations. Another example would be if you wanted to abandon the execution of a procedure whenever an error diagnostic has appeared. The EXIT directive allows you to exit from any control structure.

EXIT exits from a control structure.

Options

NTIMES = *scalar*	Number of control structures, *n*, to exit; default 1. If *n* exceeds the number of control structures of the specified type that are currently active, the exit is to the end of the outer one; while for *n* negative, the exit is to the end of the −*n*th structure (in order of execution)
CONTROL = *string*	Type of control structure to exit (job, for, if, case, procedure); default for
REPEAT = *string*	Whether to go to the next set of parameters on exit from a FOR loop or procedure (yes, no); default no
EXPLANATION = *text*	Text to be printed if the exit takes place; default *

Parameter

expression	Logical expression controlling whether or not an exit takes place

---

In its simplest form EXIT has no parameter setting, and the exit is unconditional: Genstat will always exit from the control structure or structures concerned. You are most likely to use this as part of an ELSE block of a block-if or multiple-selection structure. For example

```
IF N.GT.0
   CALCULATE Percent = R * 100 / N
ELSE
   PRINT [IPRINT=*] 'Incorrect value ',N,'  for N.'
   EXIT [CONTROL=procedure]
ENDIF
```

prints an appropriate warning message for a zero or negative value of N, and then exits from a procedure.

If the warning message is simply a text or string, the EXPLANATION option can be used to print it on exit. For example

```
EXIT [CONTROL=procedure; EXPLANATION='Incorrect value for N.'] \
   N.LE.0
CALCULATE Percent = R * 100 / N
```

has the same effect except that the actual value of N is no longer printed.

The CONTROL option specifies the type of control structure from which to exit. The default setting is for, causing an exit from a FOR loop (5.2.1). For the other settings: if causes an exit from a block-if structure (5.2.2), case exits from a multiple-selection structure (5.2.3), procedure exits from a procedure (5.3), and job causes the entire job to be abandoned. Sometimes, to exit from one type of control structure, others must be left too. To exit from the procedure in the above example, requires Genstat to exit also from the block-if structure. Generally, Genstat does these nested exits automatically, as required. However, inside a procedure, you can exit only from FOR loops and block-if or multiple-selection structures that are within the procedure. You cannot put, for example,

```
EXIT [CONTROL=if]
```

within a part of the procedure where there is no block-if in operation, and then expect Genstat to exit both from the procedure and from a block-if structure in the outer program from which the procedure was called. Genstat regards a procedure as a self-contained piece of program.

The NTIMES option indicates how many control structures of the specified type to exit from. If you ask Genstat to exit from more structures than are currently in operation in your program, it will exit from as many as it can and then print a warning. If NTIMES is set to zero or to missing value no exit takes place. If NTIMES is set to a negative value, say $-n$, the exit is to the end of the nth structure of the specified type, counting them in the order in which their execution began. Consider this example:

```
FOR I=A[1...3]
  FOR J=B[1...3]
    FOR K=C[1...3]
      FOR L=D[1...3]
        "contents of the inner loop, including:"
        EXIT [NTIMES=Nexit]
        "amongst other statements"
      ENDFOR "end of the loop over D[]"
    ENDFOR "end of the loop over C[]"
  ENDFOR "end of the loop over B[]"
ENDFOR "end of the loop over A[]"
```

If the scalar Nexit has the value 2, the exit is to the end of the loop over C[]; so the two exits are from the loop over D[] and the loop over C[]. But if Nexit has the value −2 the exit is to the end of the loop over B[], as this is the second loop to have been started.

A further possibility when EXIT is used within a FOR loop is that you can choose either to go right out of the loop and continue by executing the statement immediately after the ENDFOR statement, or to go to ENDFOR and then repeat the loop with the next set of parameter values. To repeat the loop, you need to set option REPEAT=yes. For example, suppose that variates Height and Weight contain information about children of various ages, ranging from five to 11. The RESTRICT statement causes the subsequent GRAPH statement to plot only those units of Height and Weight where the variate Age equals Ageval (4.4.1). The EXIT statement ensures that the graph is not plotted if there are no units of a particular age; the program then continues with Ageval taking the next value in the list.

```
FOR Ageval=5,6,7,8,9,10,11
  RESTRICT Height,Weight; CONDITION=Age.EQ.Ageval
  EXIT [REPEAT=yes] NVALUES(Height).EQ.0
  GRAPH Height; X=Weight
ENDFOR
```

The REPEAT option can also be used within a procedures to ask Genstat to call the procedure with the next set of parameter settings.

The example of the heights and weights of children also illustrates the use of the parameter of EXIT, to make the effect conditional. The parameter is an expression which must evaluate to a single number which Genstat interprets as a logical value. If the value is zero, the condition is *false* and no exit takes place; for other values the condition is *true* and the exit takes effect as specified. This is particularly useful for controlling the convergence of iterative processes: for example

```
CALCULATE Clim = X/10000
FOR [NTIMES=999]
  CALCULATE Previous = Root
  & Root = (X/Previous + Previous)/2
  PRINT Root,Previous; DECIMALS=4
  EXIT ABS(Previous-Root) < Clim
ENDFOR
```

will calculate the square root of X to four significant figures.

5.3   Procedures

Once you start to write programs for complicated tasks, you may wish to keep them to use again in future. The most convenient way of doing this is to form them into procedures. You may also wish to use procedures written by other people.

The use of a Genstat procedure looks exactly the same as the use one of the standard Genstat directives. The only difference is that the name of the procedure must not be abbreviated beyond eight letters, whereas directive names can be abbreviated to four. Thus, you simply give the name of the procedure, and then specify options and parameters as required.

When Genstat meets a statement with a name that it does not recognize as one of the standard Genstat directives, it first looks to see whether you have a procedure of that name already stored in your program. Then it looks in any procedure library that you may have attached explicitly to your program, taking these in order of their channel number (5.3.3). The people that manage your computer can define a special *site* library and arrange for this to be attached to Genstat automatically when it is run. If they have done so, this library will be examined next. Finally Genstat looks in the official Genstat procedure library (5.3.1), which is also attached automatically to your program. After locating the required procedure, Genstat reads it in, if necessary, and then executes it. So you do not have to do any more than you would to use a Genstat directive.

The official library thus allows new facilities to be offered to all users. Or your computer manager can make procedures available that cover the special needs of the users at your site, and these will over-ride any procedures of the same name in the official library. Or you can form your own libraries of the procedures that you find particularly useful, and these will always be taken in preference to procedures in the site or the official library. Note however that a procedure cannot have the same name as any of the Genstat directives (6.3.1).

Information is transferred to and from a procedure only by means of its options and parameters. Otherwise the procedure is completely self-contained. Anyone who uses it does not need to know how the program inside operates, what data structures it contains, nor what directives it uses. The data structures inside the procedure are local to the procedure and cannot be accessed from outside.

The first part of this section describes the Genstat Procedure Library. Later we describe how to write your own procedures, and how to form and access procedure libraries of your own.

5.3.1 The Genstat Procedure Library

The Genstat Procedure Library contains procedures contributed not only by the writers of Genstat but also by knowledgeable Genstat users from many application areas – and countries. It is controlled by an Editorial Board, who check that the procedures are useful and reliable, and maintain standards for the documentation. Guidelines for Authors were published in Genstat Newsletter 20, or can be obtained from the Secretary of the Genstat Procedure Library, at Rothamsted. The Library initially available with Release 3 of Genstat is known as Release 3[1], and the procedures that it contains are listed in Appendix 2.

The Library is regularly extended and updated, independently to the releases of Genstat itself, and these revised versions are distributed automatically to all supported Genstat sites. The contents and the exact details of individual procedures will thus change as the Library continues to develop. Consequently, the information about the Library is available from within the Library itself. There is a procedure called LIBHELP which gives you general information about the form and contents of the Library, and tells you how to find out about the facilities and syntax of the individual procedures. The default output from LIBHELP tells you about the syntax of LIBHELP itself, and so to get started you should type just the statement LIBHELP. Example 5.3.1 shows how could then continue and find out about the other help procedures: LIBEXAMPLE, LIBINFORM, and LIBMANUAL.

Example 5.3.1

```
  1  LIBHELP

LIBHELP provides information about procedures in the Genstat 5 Procedure
Library. The information is stored in a backing-store file whose name is
defined by Library procedure LIBFILENAME; there must be a free backing-store
to which the file can be attached.
LIBHELP has one parameter, called PROCEDURE, which you use to indicate the
procedures for which you want information; if PROCEDURE is not specified,
information is given about LIBHELP itself. The names of the procedures
should be given in quotes: for example
  LIBHELP 'LIBINFORM'
will obtain information about the procedure LIBINFORM (you can use LIBINFORM
to find out what procedures and modules are in the Library).
LIBHELP has a single option, called PRINT, with which you specify a list
of strings to indicate what information you want about each procedure.
The possible values, with explanations in brackets, are as follows:
'index' (one-line description), 'description' (full description),
'options' (syntax of the options), 'parameters' (syntax of the parameters),
'method' (description of the method used), 'restrict' (action when arguments
 are restricted), 'calls' (list of procedures called by this procedure),
'similar' (procedures with similar facilities), 'authors' (list of authors),
'references' (relevant publications), 'module' (the Library module to
 which the procedure belongs), 'history' (when accepted, modified &c.),
'errors' (details of any reported errors).
Option: PRINT.  Parameter: PROCEDURE.

  2  & [PRINT=options,parameters,similar]
```

    Procedure   LIBHELP

Help['options']
PRINT   = strings Indicates what information is required about each
                      procedure (index, description, options, parameters,
                      method, restrict, calls, similar, authors, references,
                      module, history, errors); default description

Help['parameters']
PROCEDURE = texts   Single-valued texts indicating the procedures about
                     which the information is required; if this is not set,
                     information is given about LIBHELP itself

Help['similar']
Procedure LIBINFORM can be used to obtain a list of the modules in the
library, to list the names of the procedures in any module of the Library,
to print the index lines of the procedures in any module, and to find out
the procedures for which errors have been reported.
Procedure LIBEXAMPLE allows you to obtain an example of the use of any
Library procedure, or a copy of the source code of the procedure.
Procedure LIBMANUAL produces a "Manual" containing Help information about
all the procedures in the Library.

    3 & [PRINT=description] 'LIBINFORM'

    Procedure   LIBINFORM

Help['description']
LIBINFORM provides information about the Genstat 5 Procedure Library.
The information is stored in a backing-store file whose name is defined by
Library procedure LIBFILENAME; there must be a free backing-store channel
to which the file can be attached.

The MODULE parameter allows you to specify that information is required only
for a specified list of modules of the Library; if MODULE is not set, the
information is given for the whole Library. The name of each module should
be given in a quoted string: for example
    LIBINFORM [PRINT=index] 'AOV','MVA'

The PRINT option specifies what information is required about each module.
The possible values are as follows:
    'contents' list of procedures in the module or in the Library (see MODULE),
    'index'   index lines for the procedures in the module/Library,
    'errors'   list of procedures in the module/Library for which errors have
               been reported,
    'modules' list of modules in the Library; given only if MODULE is not set.

Option: PRINT. Parameter: MODULE.

    4 & [PRINT=description] 'LIBEXAMPLE'

    Procedure   LIBEXAMPLE

Help['description']
LIBEXAMPLE allows you to obtain an example of the use of any procedure in the
Genstat 5 Procedure Library, also to access the source code of any procedure,
so that you can see how it works, or modify it. The examples and source are
stored in a backing-store file whose name is defined by Library procedure
LIBFILENAME; there must be a free backing-store channel to which the file can

be attached.

The names of procedures for which examples or source code are required should be listed, in quotes, using the PROCEDURE parameter. The EXAMPLE parameter can be used to specify the identifier of a text to store each example, and the SOURCE parameter to specify texts to store the source code. The examples can then be run (as macros) using the operator ##. Thus,

```
LIBEXAMPLE 'PERCENT'; EXAMPLE=%Ex
##%Ex
```

would put an example of how to use PERCENT into the text %Ex, and then run it.

No options. Parameters: PROCEDURE, EXAMPLE, SOURCE.

```
5  & [PRINT=description] 'LIBMANUAL'

Procedure   LIBMANUAL


Help['description']
```
LIBMANUAL prints a manual containing information about procedures in the Genstat 5 Procedure Library. There is first a header page, with title and list of index lines giving brief details about the procedures. Then Help information is printed about each of the procedures in turn. LIBMANUAL takes account of the current environment (as controlled by the OUTPRINT option of SET) to decide whether to start each procedure on a new page. The information is stored in a backing-store file whose name is defined by Library procedure LIBFILENAME; there must be a free backing-store channel to which the file can be attached.

Unless otherwise specified, the manual will contain every procedure in the library. However, there is a parameter, MODULES, which can be set to a text to indicate that procedures only in a particular set of modules should be included. Details of the modules in the library can be obtained using procedure LIBINFORM, and some procedures may belong to more than one. In particular, there are modules with the prefix NEWHELP to indicate the procedures whose help information has changed since the last release of the library, NEWSOURCE to indicate those whose source code has been modified to improve efficiency or correct errors, and NEW to denote additions in the current release. Thus, a manual for the procedures that were new in Release 3[1] can be obtained by putting

```
LIBMANUAL 'NEW3_1'
```

By default, the manual is printed to the current output channel. The CHANNEL option allows the Manual to be printed to some other channel; however, the OLDCHANNEL option must then also be specified, to allow LIBMANUAL to reset the output channel afterwards.

The REFERENCE option allows just a reference summary to be obtained, instead of the full information each procedure. Finally, the INDENTATION option can be used to indent the information by a specified number of columns, so that the manual can conveniently be put into a folder or binder.

Options: CHANNEL, OLDCHANNEL, REFERENCE, INDENTATION. Parameter: MODULES.

Another procedure in the Library is called NOTICE. This has a PRINT option with settings: errors for information about any aspects of Genstat that do not work correctly, library for information about recent developments concerning the Procedure Library, and news for general Genstat news (conferences, new issues of the Newsletter, and so on).

5.3.2   Forming a procedure

To write your own procedures, you start by giving a PROCEDURE statement.

PROCEDURE   introduces a Genstat procedure.

Option

PARAMETER = *string*	Whether to process the structures in each parameter list of the procedure sequentially using a dummy to store each one in turn, or whether to put them all into a pointer so that the procedure is called only once (dummy, pointer); default dumm
RESTORE = *strings*	Which aspects of the Genstat environment to store at the start of the procedure and restore at the end (inprint, outprint, diagnostic, errors, pause, prompt, newline, case, run, units, blockstructure, treatmentstructure, covariate, asave, dsave, rsave, tsave, vsave); default *
SAVE = *text*	Text to save the contents of the procedure (omitting comments and some spaces)

Parameter

text	Name of the procedure

This has a single parameter which defines the name of the procedure. The name can be up to eight characters with the same rules as for the identifiers of data structures: the first character must be a letter, the second to the eighth can be either letters or digits, and characters beyond the eighth are ignored. However the name cannot be suffixed, neither must it be the same as the name of any of the standard Genstat directives, nor any of their valid abbreviations. Thus you could have a procedure with the name CALCULUS but not CALC or CALCUL as these are abbreviations of the directive name CALCULATE. Similarly, the procedure TRANSFORM in Example 5.3.2 has a name that differs from the directive name TRANSFERFUNCTION only in the seventh and eighth characters.

When you use the procedure, you must give the name in full – all eight characters, or as many as were defined for the name if that was less than eight.

The PARAMETER option indicates whether the settings in any list specified for the parameters of the procedure are to be taken one at a time, or whether they need to be processed together. The difference between these alternatives can be illustrated by considering some of the Genstat directives. For example, with

```
ANOVA Height,Weight; RESIDUALS=Hres,Wres
```

Genstat will first do analysis with the values in the Height variate and store the resulting residuals in the variate Hres; it then analyses Weight and stores the residuals in Wres. This action corresponds to the default setting PARAMETER=dummy; inside the procedure, each parameter will then be a dummy data structure which will point to each item of the list in turn, in the same way as the parameters of a FOR loop (5.2.1). Conversely, in the statement

      PRINT Height,Hres

the values of Height and Hres are printed together down the page, and this is possible only if PRINT is able to access both variates simultaneously. In a procedure this would require the setting PARAMETER=pointer; each parameter is then a pointer, storing the whole list.

You may change some aspects of the Genstat environment within a procedure (13.1). This may be the intended purpose of the procedure; but if it is an unwanted side effect, you should reset them afterwards. The RESTORE option allows you to list aspects that would like Genstat to reset automatically when it finishes executing the procedure. Alternatively, you can save and restore the environment explicitly using the SET and GET directives, but this is usually less efficient.

Finally, the SAVE option allows you to store the contents of the procedure, up to and including ENDPROCEDURE, in a text so that you can edit and redefine it or, for example, print it to a file or save it on backing store. The saved version is a modified form of the original input. Each line of the text contains a single statement; thus, where a statement spans several lines of input, these are concatenated into a single line in the text (deleting the continuation characters). Any line that contains several statements is split. Comments are removed, and any occurrence of several contiguous spaces is replaced by a single space. Also, a colon is placed at the end of each line.

After the PROCEDURE statement, you must define what options and parameters the procedure is to have; this is done by the directives OPTION and PARAMETER respectively. Only one of each of these should be given, and they must appear immediately after the PROCEDURE statement, but it does not matter which of the two you give first. They have very similar syntaxes, except that OPTION has an extra parameter which allows you to indicate whether a list of values or of identifiers is allowed. If you do not wish to define options or parameters for a procedure you can simply omit these directives; alternatively you can use OPTION or PARAMETER but with none of their parameters set, which has precisely the same effect. The OPTION and PARAMETER directives are also used when extending the Genstat language (13.4.1).

OPTION defines the options of a Genstat procedure with information to allow them to be checked when the procedure is executed.

No options

Parameters

NAME = *texts*	Names of the options
MODE = *strings*	Mode of each option (e, f, p, t, v, as for unnamed structures); default p
NVALUES = *scalars* or *variates*	Specifies allowed numbers of values
VALUES = *variates* or *texts*	Defines the allowed values for a structure of type

	variate or text
DEFAULT = *identifiers*	Default values for each option
SET = *strings*	Indicates whether or not each option must be set (yes, no); default no
DECLARED = *strings*	Indicates whether or not the setting of each option must have been declared (yes, no); default no
TYPE = *texts*	Text for each option, whose values indicate the types allowed (ASAVE, datamatrix {i.e. pointer to variates of equal lengths as required in multivariate analysis}, diagonalmatrix, dummy, expression, factor, formula, LRV, matrix, pointer, RSAVE, scalar, SSPM, symmetricmatrix, table, text, TSAVE, TSM, variate); default * meaning no limitation
COMPATIBLE = *texts*	Defines aspects to check for compatibility with the first parameter of the directive or procedure (nvalues, nlevels, nrows, ncolumns, type, levels, labels {of factors or pointers}, mode, rows, columns, classification, margins, associatedidentifier, suffixes {of pointers}, restriction)
PRESENT = *strings*	Indicates whether or not each structure must have values (yes, no); default no
LIST = *strings*	Whether to allow a list of identifiers (MODE=p) or of values (MODE=v or t) instead of just one (yes, no); default no

---

PARAMETER defines the parameters of a Genstat procedure with information to allow them to be checked when the procedure is executed.

No options

Parameters

NAME = *texts*	Names of the parameters
MODE = *strings*	Mode of each parameter (e, f, p, t, v, as for unnamed structures); default p
NVALUES = *scalars* or *variates*	Specifies allowed numbers of values
VALUES = *variates* or *texts*	Defines the allowed values for a structure of type variate or text
DEFAULT = *identifiers*	Default values for each parameter
SET = *strings*	Indicates whether or not each parameter must be set (yes, no); default no
DECLARED = *strings*	Indicates whether or not the setting of each parameter must have been declared (yes, no); default no

TYPE = *texts*	Text for each parameter whose values indicate the types allowed (`scalar`, `factor`, `text`, `variate`, `matrix`, `diagonalmatrix`, `symmetricmatrix`, `table`, `expression`, `formula`, `dummy`, `pointer`, `LRV`, `SSPM`, `TSM`, `ASAVE`, `RSAVE`, `TSAVE`, `datamatrix` i.e. pointer to variates of equal lengths as required in multivariate analysis)
COMPATIBLE = *texts*	Defines aspects to check for compatibility with the first parameter of the directive or procedure (`nvalues`, `nlevels`, `nrows`, `ncolumns`, `type`, `levels`, `labels` {of factors or pointers}, `mode`, `rows`, `columns`, `classification`, `margins`, `associatedidentifier`, `suffixes` {of pointers}, `restriction`)
PRESENT = *strings*	Indicates whether or not each structure must have values (`yes`, `no`); default `no`

The NAMES parameter of OPTION and PARAMETER defines the names of the options and parameters of the procedure. Each name also defines the identifier of a data structure that will be used, within the procedure itself, to refer to the information transmitted by the relevant option or parameter. When you use the procedure, you have the choice of typing each name in capital letters, or in small letters, or in any mixture of the two; this corresponds to the rules for the names of options and parameters of directives. However, to avoid ambiguity, Genstat automatically converts the corresponding identifiers so that they are all in capital letters, and it is in this form that you must use them in the statements within the procedure. Thus, in Example 5.3.2, the identifiers are METHOD for the option of the procedure, PERCENT and RESULT for the parameters.

The MODE parameter tells Genstat whether the setting of each option or parameter of the procedure is to be a number (v), or an identifier of a data structure (p), or a string (t), or an expression (e), or a formula (f). These codes are exactly the same as those that indicate the mode of the values to appear within the brackets containing an unnamed structure (1.6.3).

The type of the structure used to represent an option of the procedure depends on the MODE and LIST parameters of the OPTION directive.

For anything other than mode p, the structure will be a dummy. This will point to an expression for mode e, a formula for mode f, and a text for mode t. With mode v, it will point to a scalar if the corresponding setting of the LIST parameter is no, and a variate if LIST=yes.

For mode p and LIST=no, the structure is a dummy, which will point to whichever structure is supplied for the option when the procedure is called; alternatively, when LIST=yes, it is a pointer which will store the list of structures that are supplied. For example, suppose that procedure ALLPOSS which contains the option definitions

```
OPTION NAMES='EXP','FORM','VLN','VLY','TLN','TLY','PLN','PLY'; \
       MODE=  e,     f,     v,    v,    t,    t,    p,    p; \
       LIST=  no,    no,    no,   no,   yes,  yes,  no,   yes
```

is called with these options settings:

```
ALLPOSS [EXP=LOG10(X+1); FORM=Variety*Nitrogen; VLN=2; \
    VLY=1,3,5,7; TLN=oneval; TLY=one,two,three; PLN=A; PLY=B,C,D]
```

Inside the procedure it will be as though the identifiers had been defined as follows:

```
DUMMY  [VALUE=!E(LOG10(X+1))] EXP
&      [VALUE=!F(Variety*Nitrogen)] FORM
&      [VALUE=2] VLN
&      [VALUE=!(1,3,5,7)] VLY
&      [VALUE='oneval'] TLN
&      [VALUE=!T(one,two,three)] TLY
&      [VALUE=A] PLN
POINTER [VALUE=B,C,D] PLY
```

For parameters, the structures are either all dummies or all pointers, according to the setting of the PARAMETER option of the PROCEDURE directive. If they are pointers, they store all the settings, and the procedure is called only once; if they are dummies, the procedure is called once for every item in the lists. In Example 5.3.2, the PARAMETER option is not set, and so it retains the default of dummy. Thus, in line 24, the procedure is called three times; firstly with the dummy PERCENT set to 25 and the dummy RESULT set to Ang25, then with PERCENT set to 50 and RESULT set to Ang50, and finally with PERCENT set to 75 and RESULT set to Ang75. However, if the PROCEDURE statement had been

```
PROCEDURE [PARAMETER=pointer] 'TRANSFORM'
```

PERCENT for example would have been the pointer !P(25,50,75).

The other parameters of OPTION and PARAMETER allow the settings that are supplied, when the procedure is called, to be checked automatically.

The NVALUES parameter indicates how many values the structures that are supplied for an option or parameter of mode p may contain. For example,

```
OPTION NAME='X','Y'; NVALUES=3,!(3,4); TYPE='variate'
```

indicates that the variates supplied for X must be of length 3, while those supplied for Y can be of length 3 or 4.

The VALUES parameter can be used with modes t and v to specify an allowed set of values against which those supplied for the option or parameter will be checked. In line 4 of Example 5.3.2, the OPTION statement lists the values that are allowed for METHOD, namely LOGIT, COMPLOGL, or ANGULAR. The allowed values for mode t can be up to eight characters in length; characters 9 onwards are ignored and the values are converted to upper case. When the TRANSFORM procedure is used, Genstat will check the string specified for METHOD against those in the VALUES list, using the same abbreviation rules as for options or parameters of the ordinary Genstat directives (1.8.1). Thus, for example, to request an angular transformation we need merely put METHOD=A as the first letter A is sufficient to distinguish ANGULAR from LOGIT and COMPLOGL. Within the procedure, Genstat then sets METHOD to the full string, in capitals, ANGULAR and this greatly simplifies its subsequent use (see lines 11, 13, and 15). As an example of mode v, this specification would ensure that the numbers supplied for an option NV were all odd integers between one and nine

```
OPTION NAME='NV'; MODE=v; VALUES=!(1,3,5,7,9)
```

The DEFAULT parameter specifies default values to be used if the option or parameter or option is not set. In Example 5.3.2, METHOD will be set by default to 'LOGIT'.

The SET parameter indicates whether or not an option or parameter must be set. In the PARAMETER statement in line 7 of Example 5.3.2, we have put SET=yes and so Genstat will check that the parameters of the procedure, PERCENT and RESULT, are both set whenever the procedure is used. The default is SET=no.

The DECLARED parameter specifies whether or not the structures to which options or parameters of mode p are set must already have been declared. For the PERCENT parameter of TRANSFORM they must have been declared, but for the RESULTS parameter they need not have been. (Any undeclared RESULTS structures will be declared automatically by the CALCULATE statements within the procedure.)

The TYPE parameter can be used to specify a text to indicate the allowed types of the structures to which an option or parameter of mode p is set. The parameters of TRANSFORM can be either scalars, variates, tables, or any type of matrix (rectangular, symmetric, or diagonal). In Example 5.3.2 the COMPATIBLE parameter is then used to specify that the type and number of values of each RESULTS structure must be compatible with those of the equivalent PERCENT structure; this parameter is also available in the OPTION directive, but with both options and parameters, the compatibility checks are against the first parameter of the procedure.

Finally, the PRESENT parameter allows you to indicate that the structure to which an option or parameter is set must have values. The PERCENT parameter must have values, but the RESULTS parameter need not (its values will be calculated within the procedure).

After the OPTION and PARAMETER statements, you then list the statements that are to be executed when the procedure is called: these statements are the sub-program that makes up the procedure. Any data structures defined within the procedure are local to the procedure and cannot be accessed from outside. So you can use any identifiers for the structures, without having to worry about whether they may also be used outside by someone who may later use the procedure. You end these statements making up the procedure by an ENDPROCEDURE statement.

ENDPROCEDURE indicates the end of the contents of a Genstat procedure.

No options or parameters

Once you have defined a procedure, its subsequent use is very easy. This example shows a procedure to do various transformations of percentages.

Example 5.3.2

```
1   PROCEDURE 'TRANSFORM'
2   " Define the arguments of the procedure."
3     OPTION NAME='METHOD'; MODE=t; \
4       VALUES=!t(LOGIT,COMPLOGL,ANGULAR); \
```

```
 5          DEFAULT='LOGIT'
 6        PARAMETER NAME='PERCENT','RESULT'; \
 7          MODE=p; SET=yes; DECLARED=yes,no; \
 8          TYPE=!t(scalar,variate,matrix,symmetric,diagonal,table);\
 9          COMPATIBLE=*,!t(type,nvalues); \
10          PRESENT=yes,no
11        IF METHOD .EQS. 'LOGIT'
12          CALCULATE RESULT = LOG( PERCENT / (100-PERCENT) )
13        ELSIF METHOD .EQS. 'COMPLOGL'
14          CALCULATE RESULT = LOG( -LOG((100-PERCENT)/100) )
15        ELSIF METHOD .EQS. 'ANGULAR'
16          CALCULATE RESULT = ANGULAR(PERCENT)
17        ENDIF
18      ENDPROCEDURE
19
20      VARIATE    [VALUES=10,20...90] Every10%
21      " default setting 'logit' for METHOD "
22      TRANSFORM Every10%; RESULT=Logit10%
23      PRINT      Every10%,Logit10%; DECIMALS=0,3
```

```
Every10%     Logit10%
      10       -2.197
      20       -1.386
      30       -0.847
      40       -0.405
      50        0.000
      60        0.405
      70        0.847
      80        1.386
      90        2.197
```

```
24      TRANSFORM [METHOD=A] 25,50,75; RESULT=Ang25,Ang50,Ang75
25      PRINT      Ang25,Ang50,Ang75
```

```
    Ang25        Ang50        Ang75
    30.00        45.00        60.00
```

There are several functions that you may find useful when writing procedures. You might use these either in CALCULATE (4.1.1), or in the program-control directives (5.2). Some of the functions enable you to access information about the structures that have been supplied in the options or parameters of the procedure. For example: the function NVALUES allows you to find out the length of a structure, NROWS enables you to find out the number of rows of a matrix, and so on (4.2.2). Alternatively you can use the GETATTRIBUTE directive (2.10.2). You might want to use this information to check that the supplied structures are suitable for the operations that the procedure is to carry out; or you might use it in the definition of the local structures required within the procedure.

You can use the UNSET function (4.2.6) to check whether the user has set a particular option or parameter. If this option or parameter is necessary for some particular section of the procedure to be executed you might want to use a block-if structure (5.2.2), or you might use the EXIT directive to leave the procedure altogether. The ASSIGN directive (4.9.1) provides a convenient way of setting the dummies to some default structure within the procedure (or even to structures outside the procedure). For example, the following statements assign PERCENT (if unset) to one of two different variates, according to whether this is a batch or an interactive run; then RESULT is assigned, if necessary, to Res.

```
IF UNSET(PERCENT)
  GET [ENVIRONMENT=Env]
  IF Env['run'].eqs.'batch'
    ASSIGN PERCENT; !(1,2.5,5,7.5,(10,15...90),92.5,95,97.5,99)
  ELSE
    ASSIGN PERCENT; !(1,5,10,25,50,75,90,95,99)
  EXIT [CONTROL=procedure]
ENDIF
ASSIGN [METHOD=preserve] RESULT; Res
```

The setting METHOD=preserve in ASSIGN will preserve any existing setting of RESULT but assign it to Res if it is unset.

The CONCATENATE directive can be very useful for checking strings up to a specified number of characters. For example, in LIBHELP, the name of the procedure (specified by the PROCEDURE parameter) needs to be exactly eight characters to match the way in which the names have been stored in the backing-store file that holds the Help information for the Library; the statement below abbreviates a setting of PROCEDURE that is longer than eight characters, or pads out with spaces one that is shorter.

```
CONCATENATE [NEWTEXT=Procname] PROCEDURE; WIDTH=8
```

You can use other procedures from within a procedure; in fact you can even call the procedure itself, allowing you to write recursive programs. However, these auxiliary procedures must already exist when the procedure is defined: they must either have been defined already within your program or be available within one of the libraries attached to your job; you cannot define a procedure within another procedure or within any other control structure. The Utility module of the procedure library contains several procedures useful to procedure writers, including CHECKARGUMENT which can be used to check various aspects of option and parameter settings.

You are allowed to redefine an existing procedure if you wish to change any of the statements that it contains. To do this you specify the PROCEDURE statement, as usual, followed by the statements making up the new version of the procedure, and then an ENDPROCEDURE statement. However, you are not allowed to change the option or parameter definitions, and if there are any changes in the OPTION or PARAMETER statements, Genstat will give an error diagnostic.

If you are running short of workspace, remember that you can use the DELETE directive to delete any procedures that are no longer required, or which can be accessed again from a library if they should be needed. For example

```
DELETE [PROCEDURE=yes] TRANSFORM
```

to delete procedure TRANSFORM or

```
DELETE [PROCEDURE=yes; LIST=all]
```

to delete all the procedures that are currently in store (see 2.9.1 for further details).

The GET directive (13.1.2) allows you to obtain details about the current state of the Genstat environment or the settings of special structures like the model formula most recently specified by the TREATMENTSTRUCTURE directive, and the SET directive (13.1.1) allows you to modify any of these. However, unless the changes are part of the intended purpose of the procedure, they should be reset at the end of the procedure. This can be done by using GET to store the information, and then SET to reset it; alternatively you can use the RESTORE option of the PROCEDURE directive.

The GET directive also allows you to discover the current value of Genstat's internal fault indicator, and the SET directive allows you to set the indicator to some other value. Together with the DISPLAY directive, these allow a procedure to perform some of its own error handling. You can use the FAULT option of DISPLAY to print any particular diagnostic; if this is not set, the default is to reprint the most recent diagnostic.

DISPLAY prints, or reprints, diagnostic messages.

Options

PRINT = *string*	What information to print (diagnostic); default diag
CHANNEL = *identifier*	Channel number of file, or identifier of a text to store output; default current output file
FAULT = *text*	Specifies the fault message to print (for example, FAULT='VA 4' prints the message "Values not set"); default is to print the last diagnostic message

No parameters

5.3.3   Using a procedure library

A *procedure library* is a particular kind of backing-store file that is used to store procedures. It can be used like any other backing store file: you can store procedures in the file, then retrieve them later for further use, using the methods described in 3.5. However you will usually find a library more convenient to use when it is attached to one of the input channels reserved just for procedure libraries. You can then only read procedures from the file and you cannot add new procedures; but the procedures are retrieved from the library automatically, as described at the start of this section.

Several libraries can be attached to a Genstat job. The standard Genstat procedure library is attached automatically, and you may have a local site library that is also attached automatically; your local documentation should give details of this. To attach your own libraries, you can use the OPEN directive (3.3.1). For example:

```
OPEN 'graphicslib'; CHANNEL=2; FILETYPE=procedurelibrary
```

Alternatively, you may be able to open the file in the command that you use to run Genstat; this should also be described in your local documentation.

See Installation S.2. (handwritten annotation)

5.3.4 Forming a procedure library

Procedure libraries are backing-store files that are formed using the normal backing-store directives, as described in full in 3.5. Individual procedures are best stored in separate subfiles; to make constructing and maintaining these libraries easier, you should give the subfiles the same names as the procedures. Using the same name also means that the automatic retrieval works more efficiently. To store procedures you use the STORE directive, for example:

```
STORE [CHANNEL=1; SUBFILE=Jacknife; PROCEDURE=yes] Jacknife
```

Some procedures may contain references to auxiliary procedures for performing particular parts of an analysis; in this case the searching of the library is more efficient if the additional procedures are contained in the same subfile as the main procedure: for example

```
STORE [CHANNEL=1; SUBFILE=Plot; PROCEDURE=yes] \
   Plot,Scalex,Scaley
```

While you are developing a procedure library you will need to use it like any other backing-store file, retrieving any procedures that are required by using RETRIEVE. To edit a procedure library you can use either the STORE or MERGE directives. You can display the contents of a library and subfiles using CATALOGUE.

5.3.5 Macros

There are two ways in which you can insert the contents of a text as a macro during the statements inside a procedure. With the ## operator, the contents are inserted at the time that the procedure is defined. For example

```
TEXT [VALUES='  CALCULATE V = VARIANCE(X)',\
   '  IF V>0',\
   '     CALCULATE X = (X - MEAN(X))/V',\
   '  ELSE',\
   '     CALCULATE X = CONSTANT(''missing'')',\
   '  ENDIF'] Calcs
PROCEDURE 'STANDARD'
PARAMETER NAME='X'
##Calcs
ENDPROCEDURE
```

will define the procedure STANDARD as

```
PROCEDURE 'STANDARD'
PARAMETER NAME='X'
  CALCULATE V = VARIANCE(X)
  IF V>0
    CALCULATE X = (X - MEAN(X))/V
  ELSE
    CALCULATE X = CONSTANT('missing')
  ENDIF
ENDPROCEDURE
```

(Notice that the quotes around missing need to be given twice to tell Genstat that these are a part of the string and are not intended to mark the end of the string; see 1.5.2.)

  Alternatively, you may need to take the contents of the text and execute them only at the same time as the procedure is executed. This facility is provided by the EXECUTE directive.

EXECUTE executes the statements contained within a text.

No options

Parameters

      texts   Statements to be executed

Example 5.3.5 shows a rather simple use of EXECUTE, to execute different statements on each pass through a loop.

Example 5.3.5

```
1   TEXT [VALUES='SCALAR X; VALUE=12'] T1
2   &    [VALUES='DELETE [REDEFINE=yes] X','TEXT [VALUE=Twelve] X'] T2
3   FOR T=T1,T2
4      EXECUTE T
5      PRINT X
6   ENDFOR


        X
    12.00



        X
    Twelve
```

5.4 Debugging Genstat programs

If you are writing a general program in the Genstat language (as in any other high-level language) you may often find that your program is syntactically correct and can be executed by Genstat, but nevertheless produces the wrong answers: somewhere in the logic of your program you have made a mistake. To allow such errors to be identified and corrected, Genstat has two directives, BREAK and DEBUG, that allow you to interrupt the execution of your program. You can then execute other statements, for example to examine the contents of data structures or modify their values, or even to exit from a control structure. This is particularly useful inside a procedure: the data structures used by the procedure are local and cannot normally be accessed from outside; during a break you remain within the procedure and so all the local data structures can be accessed. The BREAK directive allows you to insert breakpoints explicitly; so you must plan its use in advance when you are writing the code. Alternatively you can use DEBUG to insert breakpoints implicitly. This allows you for example to debug an existing procedure without having to edit and redefine it.

5.4.1   Breaking into the execution of a program

BREAK suspends execution of the statements in the current channel or control structure
and takes subsequent statements from the channel specified.

Option

CHANNEL = *scalar*   Channel number; default 1

Parameter

    expression   Logical expression controlling whether or not the break
                                    takes place

The BREAK directive allows you to halt the execution of the current set of statements
temporarily so that you can execute some other statements. If the parameter is not set, the
break will always take place. Alternatively, you can specify a logical expression and then the
break will take place only if this produces a *true* (i.e. non-zero) result.

The CHANNEL option determines where the statements to be executed during the break are
to be found. Usually (and by default) they are in channel 1. The statements are read and
executed, one at a time, until an ENDBREAK statement is reached, at which point control
returns to the statements originally being executed.

ENDBREAK returns to the original channel or control structure and continues execution.

No options or parameters

BREAK also provides a convenient way of interrupting a loop or a procedure so that you can
read one set of output before the next is produced. For example:

Example 5.4.1

```
    1   VARIATE [NVALUES=13] X,Y,LogY
    2   READ X,Y

    Identifier   Minimum      Mean   Maximum     Values    Missing
            X       6.00     30.00     60.00         13          0
            Y      72.50     95.42    115.90         13          0
   16   CALCULATE LogY = LOG(Y)
   17   FOR Dum=Y,LogY
   18      MODEL Dum
   19      TERMS X
   20      FIT [PRINT=summary] X
   21      BREAK
   22      RDISPLAY [PRINT=estimates]
   23      BREAK
   24   ENDFOR

24..................................................................
```

```
***** Regression Analysis *****

*** Summary of analysis ***

              d.f.          s.s.          m.s.          v.r.
Regression      1         1831.9        1831.90        22.80
Residual       11          883.9          80.35
Total          12         2715.8         226.31

Change         -1        -1831.9        1831.90        22.80

Percentage variance accounted for 64.5

* MESSAGE: The following units have large standardized residuals:
              10         2.04

* MESSAGE: The following units have high leverage:
               1         0.34

***** break at statement 5 in For Loop
" RDISPLAY [PRINT=estimates]"
   25  ENDBREAK

25......................................................................

***** Regression Analysis *****

*** Estimates of regression coefficients ***

              estimate          s.e.            t
Constant       117.57           5.26         22.34
X              -0.738           0.155        -4.77

***** break at statement 7 in For Loop
"ENDFOR"
   26  ENDBREAK

26......................................................................
***** Regression Analysis *****

*** Summary of analysis ***

              d.f.          s.s.          m.s.          v.r.
Regression      1         0.21732       0.217322       24.07
Residual       11         0.09931       0.009028
Total          12         0.31663       0.026386

Change         -1        -0.21732       0.217322       24.07

Percentage variance accounted for 65.8

* MESSAGE: The following units have high leverage:
               1         0.34

***** break at statement 5 in For Loop
" RDISPLAY [PRINT=estimates]"
   27  ENDBREAK

27......................................................................

***** Regression Analysis *****
```

```
*** Estimates of regression coefficients ***

                  estimate          s.e.          t
Constant            4.7876        0.0558      85.83
X                 -0.00804       0.00164      -4.91

***** break at statement 7 in For Loop
"ENDFOR"
  28   ENDBREAK
```

5.4.2   Putting automatic breaks into a program

DEBUG puts an implicit BREAK statement after the current statement and after every
   NSTATEMENTS subsequent statements, until an ENDDEBUG is reached.

Options

CHANNEL = *scalar*	Channel number; default 1
NSTATEMENTS = *scalar*	Number of statements between breaks; default 1
FAULT = *string*	Whether to invoke DEBUG only at the next fault (yes, no); default no

No parameters

ENDDEBUG cancels a DEBUG statement.

No options or parameters

The straightforward use of DEBUG causes an immediate break, and then further breaks at
regular intervals until you issue an ENDDEBUG statement. Alternatively, by setting option
FAULT=yes, you can arrange for Genstat to continue until the next fault diagnostic, and then
break.

The interval before each further break is specified by the NSTATEMENTS option; by default,
breaks take place after every statement.

During the breaks, Genstat takes statements from the channel specified by the CHANNEL
option; by default they are taken from channel 1.

Each individual break is terminated by an ENDBREAK, exactly like a break invoked explicitly
by the BREAK directive (5.4.1).

For example:

Example 5.4.2

```
  1   PROCEDURE 'POLAR'
  2      PARAMETER 'X','Y','R','THETA'
  3      " Takes (x,y) and returns (r,theta) "
  4      CALCULATE R = SQRT(X*X + Y*Y)
  5      CALCULATE THETA = ARCCOS(X/R)
```

```
   6        CALCULATE THETA = THETA + 2*(3.14159 - THETA)*(Y < 0)
   7    ENDPROCEDURE
   8    SCALAR Xpos,Ypos; VALUE=3,4
   9    DEBUG
  10    POLAR Xpos; Y=Ypos; R=Radius; THETA=Angle
***** break at statement 1 in Procedure POLAR
"  CALCULATE R = SQRT(X*X + Y*Y)"
  11    ENDBREAK
***** break at statement 2 in Procedure POLAR
" CALCULATE THETA = ARCCOS(X/R)"
  12    PRINT R

       Radius
        5.000


  13    ENDBREAK
***** break at statement 3 in Procedure POLAR
" CALCULATE THETA = THETA + 2*(3.14159 - THETA)*(Y < 0)"
  14    PRINT THETA

        Angle
       0.9273


  15    ENDBREAK
***** break at statement 4 in Procedure POLAR
"ENDPROCEDURE"
  16    CALCULATE Deg = THETA*180/3.14159
  17    PRINT Deg

          Deg
        53.13


  18    ENDDEBUG
  19    PRINT Xpos,Ypos,Radius,Angle

        Xpos         Ypos        Radius        Angle
       3.000        4.000         5.000       0.9273
```

S.A.H.
R.W.P.

6 Graphical display

Genstat can produce graphical output in two distinctively different styles. These are *line-printer* graphics and *high-resolution* graphics. As the name suggests, line-printer graphics are designed for printing on ordinary printers, and are also suitable for display on terminals and 4PC screens. The standard character set, made up of letters, digits, and punctuation characters, is used to produce a graphical representation of the data. This will be of *low resolution*, typically 24 rows by 80 columns for screen display, 132 by 48 or 80 by 60 for a printer; but this is quite adequate for many uses within Statistics, such as producing a scatterplot or histogram of raw data or checking the distribution of residuals. The advantage of this form of output is that no special equipment, such as a graphics terminal or plotter, is required; also the graphics form an integral part of the Genstat output, and can thus be interspersed with other results during the analysis of your data. Histograms, graphs, and contour plots can be produced in this basic style.

On the other hand, high-resolution graphics, utilizing the full resolution of your display, can provide a far more attractive representation of the data. Lines and points are plotted with far greater precision and various plotting symbols and character fonts can be used to enhance the output. As well as directives for graphs, histograms, and contour plots there are also facilities for generating pie charts and displaying three-dimensional surfaces. Graphical input can be used to read information from plots, for example to allow interactive identification of outliers. There is also a great deal of control over the plotting directives, allowing basic elements to be combined into more complex displays; this can be implemented in procedures. High-resolution graphics can be produced interactively on graphics terminals, workstations, or PC screens. Alternatively, they can be saved in files using standard formats that are suitable for plotters or laser printers and can also be incorporated in documents by some word-processing packages. The exact details of what is provided in particular implementations of Genstat is discussed further in 6.5.1.

6.1 Line-printer graphics

There are three directives that provide line-printer output: GRAPH, HISTOGRAM, and CONTOUR. You can use options and parameters to modify the annotation, the symbols used, the size of plot, and so on. Several options apply generally to all three directives and are described now. Others are more specific and are left until the descriptions of the relevant directives.

Normally, output goes to the current output channel, but you can use the CHANNEL option to direct it to another (see 3.3). For example, when you are working interactively, you might want to send a graph to a secondary output file so that you can print it later. Unlike some directives (for example, PRINT) you cannot save the output in a text structure.

The TITLE option lets you set an overall title for the output; graphs and contour plots can also have individual axis titles, specified by the YTITLE and XTITLE options. You can supply the text settings of these options directly, in a string, or give them as the identifier of a predefined text structure. For example:

```
GRAPH [XTITLE='Nitrogen Applied (kg/ha)'] Yield; Nitrogen
```

or

```
TEXT Experiment
READ [CHANNEL=2; SERIAL=yes; SETNVALUES=yes] Experiment,Data
HISTOGRAM [TITLE=Experiment] Data
```

Genstat prints the y-axis title as a column of characters down the left-hand side of a graph or contour plot. New lines are ignored, so that strings within a text are concatenated. Genstat truncates the title if necessary: the maximum possible number of characters is the number of rows of the frame plus 4. The x-axis title is printed below the graph; the maximum number of characters is the number of columns of the frame plus four: long strings are truncated whereas short strings are centred.

6.1.1 The GRAPH directive

GRAPH produces scatter and line graphs on the terminal or line printer.

Options

CHANNEL = *scalar*	Channel number of output file; default is current output file
TITLE = *text*	General title; default *
YTITLE = *text*	Title for y-axis; default *
XTITLE = *text*	Title for x-axis; default *
YLOWER = *scalar*	Lower bound for y-axis; default *
YUPPER = *scalar*	Upper bound for y-axis; default *
XLOWER = *scalar*	Lower bound for x-axis; default *
XUPPER = *scalar*	Upper bound for x-axis; default *
MULTIPLE = *variate*	Numbers of plots per frame; default * i.e. all plots are on a single frame
JOIN = *string*	Order in which to join points (ascending, given); default asce
EQUAL = *string*	Whether/how to make bounds equal (no, scale, lower, upper); default no
NROWS = *scalar*	Number of rows in the frame; default * i.e. determined automatically
NCOLUMNS = *scalar*	Number of columns in the frame; default * i.e. determined automatically
YINTEGER = *string*	Whether y-labels integral (yes, no); default no
XINTEGER = *string*	Whether x-labels integral (yes, no); default no

Parameters

Y = *identifiers*	Y-coordinates

X = *identifiers*	X-coordinates
METHOD = *strings*	Type of each graph (point, line, curve, text); if unspecified, poin is assumed
SYMBOLS = *factors* or *texts*	For factor SYMBOLS, the labels (if defined), or else the levels, define plotting symbols for each unit, whereas a text defines textual information to be placed within the frame for METHOD=text or the symbol to be used for each plot for other METHOD settings; if unspecified, * is used for points, with integers 1-9 to indicate coincident points, ′ and . are used for lines and curves
DESCRIPTION = *texts*	Annotation for key

The simplest form of the GRAPH directive produces a point plot (or scatterplot as it is sometimes called). It can also be used to plot lines and curves, and text can be added for extra annotation. The data are supplied as y- and x-coordinates in separate parameter lists.

In Example 6.1.1a, the identifiers Y and X are variates of equal length; Genstat uses their values in pairs to give the coordinates of the points to be plotted.

Example 6.1.1a

```
1   VARIATE [VALUES=-16,-7,9,16,7,-8,-12,-5,0,10,4,-4,-3,3,16] X
2   & [VALUES=0,-14,-12.5,0,14,0,12,0,-10,-9,5,6,-6,-1.5,16] Y
3   GRAPH Y; X
```

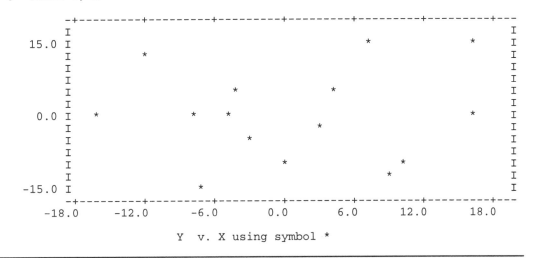

```
        Y   v. X using symbol *
```

By default, if you specify several identifiers, Genstat plots them all in the same frame a pair at a time; for example

```
    GRAPH Y[1...3]; X[1,2]
```

superimposes plots of Y[1] against X[1], Y[2] against X[2], and Y[3] against X[1]. The usual rules governing the parallel expansion of lists apply here: the length of the Y parameter list determines the number of plots within the frame, and the x parameter list is recycled if it is shorter. To generate several frames from one GRAPH statement you can use the MULTIPLE option, described below.

The identifiers supplied by the Y and X parameters need not be variates, but can be any numerical structures: scalars, variates, tables, or matrices. The only constraints are that the pairs of structures must have the same numbers of values, and that tables must not have margins. You cannot use a factor in the Y or X parameter lists; you must first copy its values into a variate, for example by using CALCULATE (4.1.1).

There are four types of graph available, controlled by the METHOD parameter: point (the default), line, curve, and text.

A line plot is one in which each point is joined to the next by a straight line. Alternatively, using the curve method, cubic splines are used to produce a smoothed curve through the data points. This does not represent any model fitted in the statistical sense, but as long as the data points are not too widely spaced (especially where the gradient changes quickly) the plotted curve should be a good representation of the underlying function.

By default, Genstat sorts the data so that the x-values are in ascending order before any line or curve is drawn through the points. However, if you set option JOIN=given, the points are joined in the order in which they occur in the data; if there are then any missing values there will be breaks in the line at each missing unit.

Plots produced with METHOD set to either line or curve do not include markings for the data points themselves; you should plot these separately if they are required, as shown in Example 6.1.1b.

Example 6.1.1b

```
 4   VARIATE [VALUES=-0.1,0.1...0.9] V
 5   & [VALUES=5.5,9.9,8.7,2.3,1.3,5.5] W
 6   GRAPH [TITLE='Point and curve plot'; NROWS=16; NCOLUMNS=61] W,W; V; \
 7     METHOD=curve,point; SYMBOLS=*,'X'; DESCRIPTION='Fitted curve    ...',*
```

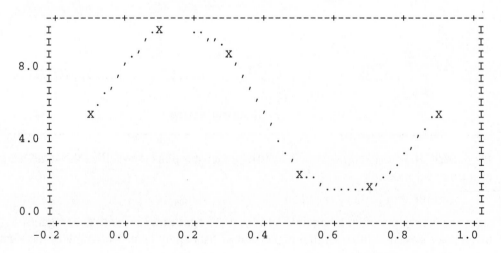

Point and curve plot

```
Fitted curve    ...
W  v. V using symbol X
```

Here W is plotted against V twice, first with the curve method and then with the point method. It is best to plot the line first, so that the symbols for individual points will overwrite those used for the line or curve.

The fourth plotting method is text. You can use this to place an item of text within a graph as extra annotation. For example:

```
SCALAR Xt,Yt; VALUE=20,10
TEXT [VALUES='Y=aX+b'] T
GRAPH Y,Yt; X,Xt; METHOD=line,text; SYMBOLS=*,T
```

This plots a line, defined by the variates Y and X, as described above. In addition, the text T is printed within the frame starting at the coordinates defined by the scalars Yt and Xt. As these statements show, the SYMBOLS parameter then specifies the text that is to be plotted. The text is truncated as necessary, if positioned too close to the edge of the graph.

With other methods SYMBOL defines the plotting symbol to be used to mark either points or lines on the graph. The default symbol for points is the asterisk, and for lines is a combination of dots and single quotes: you can see these in the earlier examples. If several points coincide, Genstat replaces the asterisk by a digit between 2 and 9, representing the number of coincidences, with 9 meaning nine or more. For point plots, the SYMBOLS parameter can be set to either a text or a factor. If you specify a text with a single string, the string is used to label every point; otherwise, the text must have one string for each point.

By default, Genstat automatically calculates the extent of the axes from the data to be plotted, in such a way that all the data are contained within the frame. You can set one or more of the bounds for the axes by options YLOWER, YUPPER, XLOWER, and XUPPER. By setting the upper bound of an axis to a value that is less than the lower bound, you can reverse the usual convention for plotting in which the y-values increase upwards and the x-values increase to the right. Setting the options YINTEGER and XINTEGER constrains the axis markings to be integral, if possible.

The EQUAL option allows you to place constraints on the bounds for the axes. The default setting no (meaning no constraint) uses the boundary values as set by the options or calculated from the data. The settings lower and upper constrain the lower or upper bounds of the two axes to be equal: for example, to plot the line *y=x* along with the data, setting EQUAL=lower will ensure that it will pass through the bottom left-hand corner of the frame. The scale setting adjusts the y-bounds and x-bounds so that the physical distance on one axis corresponds as closely as possible to physical distance on the other: for example, so that one centimetre will represent the same distance along each axis.

Normally each GRAPH statement produces one frame, and Genstat sets the size so that it will fill one screen or line-printer page, based on the settings of WIDTH and PAGE from OPEN or OUTPUT (3.3.1 and 3.4.3), or their defaults if these have not been specified. When output is to a file the graph will be placed on a new page, unless this has been disabled using OUTPUT, JOB, or SET (3.3.1, 5.1.1, and 13.1.1). The size of the graph is defined in terms of the number of characters in each row and the number of rows in the frame, a row being one line of output. You can adjust the size of the frame by using the NROWS and NCOLUMNS options; the minimum allowed is three rows and three columns, and the maximum number of columns is 17 characters less than the width of the output channel (to leave room for axis markings and titles). There is no maximum on the number of rows. By default, the number of columns is 101, subject to the maximum above, and the number of rows is the number of lines per page, less 8, to allow room for annotation. By defining the page size in advance you can avoid having to specify the numbers of rows and columns when you wish to plot many graphs.

The automatic axis scaling aims to find axis markings that are at reasonable values, but because the markings appear at fixed character positions this may not always be possible. If both upper and lower axis bounds are set, or EQUAL is set in conjunction with axis bounds, or you have requested integral axis markings, there may be conflicting constraints on the axis scaling. If the resultant axis markings then require several decimal places, you may be able to obtain better values by slight adjustments to the numbers of rows or columns.

The MULTIPLE option lets you generate several frames (separate graphs) from one statement. If there is room, the graphs can be printed alongside each other, for example to produce a two-by-two array of plots on a line-printer page. The option should be set to a variate whose elements define the number of graphs to plot in each frame and the number of values in the variate determines the number of frames to be output. For example,

        GRAPH [MULTIPLE=!(2,1,2)] A,B,C,D,E; X[1...3]

will produce three frames; the first containing A against X[1] and B against X[2], the second containing C against X[3] and the third containing D against X[1] and E against X[2]. The sum of the values in the MULTIPLE list gives the total number of structures required to form the plots, which must therefore be equal to the length of the Y parameter list. The X list will be recycled if necessary, as here.

By default, each graph will fit the page (as if it had been produced by an individual GRAPH statement). However, if you set the NCOLUMNS option to a suitably small value, Genstat may be able to fit more than one frame across the page. The MULTIPLE option will then produce the graphs side by side. Remember that 17 columns are automatically added to provide annotation, and five blank columns are used to separate multiple graphs in parallel. This means that, for example, setting NCOLUMNS=20 will produce two graphs in parallel on a screen of width 80, and three graphs when output to a file of width 121 or more.

You can annotate the graph by using the TITLE, XTITLE, and YTITLE options described at the beginning of this section. If none of these are set, a simple key will be produced below the graph, as in Example 6.1.2a, which lists the identifiers and plotting symbols for each pair of Y and X structures. You can obtain your own key by setting the DESCRIPTION parameter, which supplies a line of text for each plot, as in Example 6.1.1b.

6.1.2 The **HISTOGRAM** directive

HISTOGRAM produces histograms of data on the terminal or line printer.

Options

CHANNEL = *scalar*	Channel number of output file; default is the current output file
TITLE = *text*	General title; default *
LIMITS = *variate*	Variate of group limits for classifying variates into groups; default *
NGROUPS = *scalar*	When LIMITS is not specified, this defines the number of groups into which a data variate is to be classified; default is the integer value nearest to the square root of the number of values in the variate
LABELS = *text*	Group labels
SCALE = *scalar*	Number of units represented by each character; default 1

Parameters

DATA = *identifiers*	Data for the histograms; these can be either a factor indicating the group to which each unit belongs, a variate whose values are to be grouped, or a one-way table giving the number of units in each group
NOBSERVATIONS = *tables*	One-way table to save numbers in the groups
GROUPS = *factors*	Factor to save groups defined from a variate
SYMBOLS = *texts*	Characters to be used to represent the bars of each histogram
DESCRIPTION = *texts*	Annotation for key

Histograms provide quick and simple visual summaries of data values. The data are divided into several groups, which are then displayed as a histogram consisting of a line of asterisks for each group. The number of asterisks in each line is proportional to the number of values assigned to that group; this figure is also printed at the beginning of each line.

Example 6.1.2a

```
  1  VARIATE Data
  2  READ Data

   Identifier   Minimum      Mean    Maximum    Values   Missing
         Data     0.000     3.960      9.000        25         0

  4  HISTOGRAM Data

Histogram of Data
```

```
          -    2   9  *********
      2 -      4   7  *******
      4 -      6   5  *****
      6 -      8   2  **
      8 -          2  **
```

Scale: 1 asterisk represents 1 unit.

You can form histograms from data stored in variates, factors, or tables. If a histogram is to be formed from a variate, Genstat sorts its values into groups as defined by upper and lower bounds.

You can specify a list of variates, to obtain a parallel histogram. For each group one row of asterisks is printed for each variate, labelled by the corresponding identifier.

Example 6.1.2b

```
    5   VARIATE Data2
    6   READ Data2

     Identifier   Minimum      Mean    Maximum     Values    Missing
          Data2     0.000     3.225      8.000         40          0

    9   HISTOGRAM Data,Data2

Histogram of Data and Data2


              - 1.5    Data    5  *****
                       Data2   9  *********

          1.5 - 3.0    Data    6  ******
                       Data2  14  **************

          3.0 - 4.5    Data    5  *****
                       Data2   7  *******

          4.5 - 6.0    Data    5  *****
                       Data2   8  ********

          6.0 - 7.5    Data    1  *
                       Data2   1  *

          7.5 -        Data    3  ***
                       Data2   1  *
```

Scale: 1 asterisk represents 1 unit.

As shown in Example 6.1.2b, the variates are sorted according to the same intervals; there is no need for them all to have the same numbers of values,

You can use the NGROUPS option to specify the number of groups in the histogram; Genstat will then work out appropriate limits, based on the range of the data, to form intervals of equal width. For example:

```
HISTOGRAM [NGROUPS=5] Data
```

Alternatively, you can define the groups explicitly, by setting the LIMITS option to a variate containing the group limits.

Example 6.1.2c

```
10   VARIATE [VALUES=1,2,3,5,7,8,10] Glimits
11   HISTOGRAM [LIMITS=Glimits] Data
```

Histogram of Data grouped by Glimits

```
            -   1.00   5 *****
      1.00  -   2.00   4 ****
      2.00  -   3.00   2 **
      3.00  -   5.00   7 *******
      5.00  -   7.00   4 ****
      7.00  -   8.00   1 *
      8.00  -  10.00   2 **
     10.00  -          0
```

Scale: 1 asterisk represents 1 unit.

In Example 6.1.2c, Limits is a variate with seven values producing a histogram in which the data is split into eight groups; ≤1, 1-2, 2-3, 3-5, 5-7, 7-8, 8-10, >10. The upper limit of each group is included within that group, so the group 3-5, for example, contains values that are greater than 3 and less than or equal to 5. The values of the limits variate are sorted into ascending order if necessary, but the variate itself is not changed.

You can use the LABELS option to provide your own labelling for the groups of the histogram. It should be set to a text vector of length equal to the number of groups. If neither NGROUPS nor LIMITS has been set, the number of groups is determined from the number of values in the LABELS structure. If LABELS is also unset, the default number of groups is chosen as the integer value nearest to the square root of the number of values (as in Example 6.1.2a where 25 values are sorted into five groups), up to a maximum of 10. Alternatively, procedure AKAIKEHISTOGRAM provides a more sophisticated method of generating histograms, using Akaike's Information Criterion (AIC) to generate an optimal grouping of the data.

The data for the histogram can also be specified as a factor (which defines the assignment of each unit to a group of the histogram), or as a one-way table supplying the group counts directly.

To form a histogram from a factor, Genstat counts the number of units that occur with each level of the factor; thus the number of groups of the histogram is the number of levels of the factor and the value for each group is the corresponding total. The labels of the factor (if present) are used to label the groups, otherwise Genstat uses the factor levels.

Example 6.1.2d

```
12   TEXT [VALUES=apple,banana,peach,cherry,pear,orange] Name
13   FACTOR [LEVELS=6; LABELS=Name; NVALUES=32] Fruit
```

```
  14  READ Fruit

    Identifier     Values    Missing     Levels
         Fruit         32          0          6

  16  HISTOGRAM Fruit

Histogram of Fruit


  apple  3 ***
 banana  2 **
  peach  8 ********
 cherry  5 *****
   pear  8 ********
 orange  6 ******

Scale:  1 asterisk represents 1 unit.
```

When Genstat plots the histogram of a one-way table, the number of groups is the number of levels of the factor classifying the table and the values of the table indicate the number of observations in each group. The labels or levels of the classifying factor are again used to label the histogram.

The LABELS option can also be used when producing a histogram from a factor or table. It should be set to a text of length equal to the number of levels of the factor or classifying factor.

When producing a parallel histogram the data structures must all be of the same type: variate, factor, or table. Variates and factors may be restricted, in which case only the subset of values specified by the restriction will be included in the histogram; however, unlike many directives, restrictions do not carry over to the other structures listed by the DATA parameter. If parallel histograms are to be formed from several factors, they must all have the same number of levels, and the labels or levels of the first factor will be used to identify the groups. Likewise, if you are forming parallel histograms from several tables, they must all have the same number of values, and the classifying factor of the first table will define the labelling of the histogram.

The SYMBOLS parameter can specify alternative plotting characters to be used instead of the asterisk. For example:

    HISTOGRAM Variate; SYMBOLS='+'

You can specify a different string for each structure in a parallel histogram. If you specify strings of more than one character, Genstat uses the characters in order, recycled as necessary, until each histogram bar is of the correct length.

Example 6.1.2e

```
  17  HISTOGRAM Data; SYMBOLS='X-O-'

Histogram of Data
```

```
     -   2   9  X-O-X-O-X
 2   -   4   7  X-O-X-O
 4   -   6   5  X-O-X
 6   -   8   2  X-
 8   -       2  X-
```

Scale: 1 character represents 1 unit.

You can use the DESCRIPTION parameter to provide a text for labelling the histogram instead of the identifiers of the DATA structures.

Normally one asterisk will represent one unit. However, if there are many data values and the groups become large, Genstat may not be able to fit enough asterisks into one row. It will then alter the scaling so that one asterisk represents several units. You can set the scaling explicitly using the SCALE option; the value specified is rounded to the nearest integer, and determines how many units should be represented by each asterisk.

HISTOGRAM has two output parameters that allow you to save information that has been generated during formation of the histogram. The NOBSERVATIONS parameter allows you to save a one-way table of counts that contains the number of observations that were assigned to each group; the missing-value cell of this table will contain a count of the number of units that were missing and that therefore remain unclassified. When producing a histogram from a variate, you can use the GROUPS parameter to specify a factor to record the group to which each unit was allocated.

6.1.3   The CONTOUR directive

CONTOUR produces contour maps of two-way arrays of numbers (on the terminal/printer).

Options

CHANNEL = *scalar*	Channel number of output file; default is current output file
INTERVAL = *scalar* or *variate*	Contour interval for scaling (scalar) or positions of the contours (variate); default * i.e. determined automatically
TITLE = *text*	General title; default *
YTITLE = *text*	Title for y-axis; default *
XTITLE = *text*	Title for x-axis; default *
YLOWER = *scalar*	Lower bound for y-axis; default 0
YUPPER = *scalar*	Upper bound for y-axis; default 1
XLOWER = *scalar*	Lower bound for x-axis; default 0
XUPPER = *scalar*	Upper bound for x-axis; default 1
YINTEGER = *string*	Whether y-labels integral (yes, no); default no
XINTEGER = *string*	Whether x-labels integral (yes, no); default no
LOWERCUTOFF = *scalar*	Lower cut-off for array values; default *
UPPERCUTOFF = *scalar*	Upper cut-off for array values; default *

Parameters

GRID = *identifiers*	Pointers (of variates representing the columns of a data matrix), matrices, or two-way tables specifying values on a regular grid
DESCRIPTION = *texts*	Annotation for key

A contour plot provides a way of displaying three-dimensional data in a two-dimensional plot. The data values are supplied as a rectangular array of numbers that represent the values of the variable in the third dimension, often referred to as *height* or the *z-axis*. The first two dimensions (x and y) are the rows and columns indexing the array; the complete three-dimensional data set is referred to as a *surface* or *grid*. Contours are lines that are used to join points of equal height, and usually some form of interpolation is used to estimate where these points lie. The resulting contour plot is not necessarily very "realistic" when compared to perspective plots (6.3.2), but it has the advantage that the entire surface can easily be examined, without the danger of some parts being obscured by high points or regions.

You might use contour plots for example when you have data sampled at points on a regular grid, such as the concentrations of a trace element or nutrient in the soil. Contours are also very useful when fitting nonlinear models (8.7), when they can be used to study two-dimensional slices of the likelihood surface, to help find good initial estimates of the parameters.

The CONTOUR directive produces output for a line printer by using cubic interpolation between the grid points to estimate a z-value for each character position in the plot. Each value is reduced to a single digit in the range 0 ... 9, according to the rules described below. To produce the contour plot only the even digits are printed: you can then see the contours as the boundaries between the blank areas and the printed digits.

In Example 6.1.3a, a function of two variables is calculated, and the shape of the function is displayed with CONTOUR. Titles have been given to the x-axis and the y-axis, and there is an overall title giving the algebraic form of the function.

Example 6.1.3a

```
  1    MATRIX [ROWS=5; COLUMNS=7] X,Y; VALUES=!((1...7)5),!(7(1...5))
  2    CALCULATE Zvalues = (X-2.5)*(X-6)*X - 10*(Y-3)*(Y-3)
  3    CONTOUR [TITLE='Z(x,y) = x*(x-2.5)*(x-6) - 10*(y-3)**2';\
  4      YTITLE='Y values'; XTITLE='X values'] Zvalues

Contour plot of Zvalues at intervals of 8.400

** Scaled values at grid points **
 -3.8690    -4.2857    -5.2976    -6.1905    -6.2500    -4.7619    -1.0119
 -0.2976    -0.7143    -1.7262    -2.6190    -2.6786    -1.1905     2.5595
  0.8929     0.4762    -0.5357    -1.4286    -1.4881     0.0000     3.7500
 -0.2976    -0.7143    -1.7262    -2.6190    -2.6786    -1.1905     2.5595
 -3.8690    -4.2857    -5.2976    -6.1905    -6.2500    -4.7619    -1.0119

             Z(x,y) = x*(x-2.5)*(x-6) - 10*(y-3)**2
```

```
            0.000      0.167      0.333      0.500      0.667      0.833      1.000
                '          '          '          '          '          '          '
        1.000-66666            4444444444                4444444    666   888-
              666666666666666        4444444444444444444            666    88
                  6666666666            6666666666          66666     88    0
              8888888            6666666666            666666    888    00
              88888888888888        666666666666666666666666     888    00
                  88888888888        666666666666            8888    00   2
        0.750-        888888888                      8888    000   22-
                  0000            8888888888                888888   000   22
    Y         0000000000        88888888888            8888888   000   22
              0000000000000        888888888888888888888888     000   222
    v         00000000000000        888888888888888888888     000   22
    a         000000000000000        88888888888888888888     000   22
    l   0.500-000000000000000000        8888888888888888888     0000   22   -
    u         000000000000000        8888888888888888888     000   22
    e         00000000000000        888888888888888888888     000   22
    s         0000000000000        888888888888888888888888     000   222
              0000000000        88888888888            8888888   000   22
                  0000            8888888888                888888   000   22
        0.250-        888888888                      8888    000   22-
                  88888888888                6666666666      8888    00   2
              88888888888888        666666666666666666666666     888    00
              8888888            6666666666            666666    888    00
                  6666666666            66666     88    0
              666666666666666        4444444444444444444            666    88
        0.000-66666            4444444444                4444444    666   888-
                '          '          '          '          '          '          '

                                    X values
```

The GRID parameter can be set to a matrix, a two-way table (with the first factor defining the rows), or a pointer to a set of variates each containing a column of data. We explain the conventions in terms of a matrix as input, but similar rules apply to the other structures. When reading or printing a matrix the origin of the rows and columns (row 1, column 1) appears at the top left-hand corner. However, in forming the contour plot the rows are reversed in order so that the first row of the matrix is placed at the bottom of the contour; thus the origin of the contour is located, according to the usual conventions, at the bottom left-hand corner of the plot. The DCONTOUR directive (6.3.1) also reverses the rows of the grid in the same way.

CONTOUR scales the grid values by dividing by the contour interval. The scaled grid values are then converted to single digits by taking the remainder modulo 10 and truncating the fractional part. In Example 6.1.3a, the first grid value is −32.5, which is divided by the interval size (8.4) to obtain −3.869; this becomes 6.131 when taken modulo 10, and then 6 after truncation. To aid interpretation of the plot, the array of scaled values is printed out.

The INTERVAL option allows you to set the interval between contour lines. For example, if the grid values range from 17 to 72 and the interval is set to 10, contour lines (the boundaries between blank space and printed digits) will occur at grid values of 20, 30, 40, 50, 60, and 70. By default, the interval is determined from the range of the data in order to obtain 10 contours.

The UPPERCUTOFF and LOWERCUTOFF options can be used to define a window for the grid values that will form the contours. All values above or below these are printed as X. Setting either UPPERCUTOFF or LOWERCUTOFF will change the default contour interval, as the range of data values is effectively curtailed.

You can use the TITLE, YTITLE, and XTITLE option to annotate the contour plot. If you specify several grids, these will be plotted in separate frames and the text of the TITLE option will appear at the top of each one. You should thus use TITLE only to give a general description of what the contours represent. The DESCRIPTION parameter can be used to add specific descriptions to be printed at the bottom of each individual plot.

The YUPPER and YLOWER options allow you to set upper and lower bounds for the y-axis; thus generating axis labels that reflect the range of values over which the grid was observed or evaluated. Setting YINTEGER=yes will ensure the labels are printed as integers, if possible. The default axis bounds are 0.0 and 1.0. The options XLOWER, XUPPER, and XINTEGER similarly control labelling of the x-axis.

Example 6.1.3b shows how a contour plot can be produced from a set of variates. In line 17, the values of the variates are inverted, using the REVERSE function, and y-axis labelling set up so that *depth* increases as you read down the plot.

Example 6.1.3b

```
   5   "
  -6     Data are core samples taken from a wetland rice experiment to
  -7     examine the leaching of ammonium nitrate.  Cores were taken
  -8     centrally and 5 and 10 cm either side of the central core.
  -9     Concentration of ammonium nitrate measured at depths of 4,8...20 cm.
 -10   "
  11   VARIATE [NVALUES=5] Core[1...5]
  12   READ Core[]

      Identifier    Minimum       Mean    Maximum     Values    Missing
         Core[1]      5.000      8.200     11.000          5          0
         Core[2]       6.00      67.60     195.00          5          0
         Core[3]      129.0      940.6     2315.0          5          0
         Core[4]      10.00      36.00      77.00          5          0
         Core[5]      7.000      9.400     15.000          5          0

  15   TEXT [VALUES=' Samples taken 40 days after placement ', \
  16     ' of 2 grams supergranule urea. '] Coredesc
  17   CALCULATE Core[] = LOG10(REVERSE(Core[]))
  18   CONTOUR [YTITLE='Soil depth in cm'; XTITLE='Distance from central core';\
  19     YINTEGER=yes; XINTEGER=yes; YUPPER=4; YLOWER=20; \
  20     XUPPER=10; XLOWER=-10] Core; DESCRIPTION=Coredesc

Contour plot of Core at intervals of 0.267

** Scaled values at grid points **
     3.1704     2.9193     7.9179     3.7515     3.5799
     3.3880     7.1797    12.6222     6.4687     3.1704
     2.6222     8.5911    11.3557     7.0772     3.1704
     3.7515     5.7926    11.3091     5.5415     3.5799
     3.9068     4.8808     8.2790     3.7515     4.4121
```

```
        -10          -5          0          5         10
         '           '          '          '          '
      4- 2222222222  4  66         66   44                    -
          222      44 66  88 888  66 444
              444   6  88   0   8  66 4444
              444  66 88  00 00   88 66   444
  s         4444  66 88 00      00  8  6    444
  o          444   66 88 00   2   00  8 66   4444
  i      8- 444    66   8  0  2222  00 8  66  444         -
  l         444  66  88 00   2222  00 88 66   44
            44  66   88  00  2222  00 88 666   444
  d      2 44  66   88 000   222   00 88  66   444
  e      2 44  6   88 000        000 88  66   44
  p      2 44 66   88 000         00  88 66   44
  t     12-2 44 66  88   000       000  88 66   44        -
  h      2 44  66  88 0000        000  88 666  444
         2 44  666  88 000        00  88 66   444
  i        44  666  88  00        00  88 66  4444
  n        444  666 88  000       00  88 66   444
           4444   66 88  00       00  8  6    4444
  c     16-  4444    6  8  00   000 88 66   4444        -
  m         44444  66 88  00 00   8   6  4444
            44444   6   88  000   88 66 4444
            44444  66 88   0   88  6  444
            444444 66   88      88  66 444           4
            444444  66  888888  66  44             4
     20- 4444444  666    88   66  44         44-
          '           '          '          '          '
```

Distance from central core

Samples taken 40 days after placement
of 2 grams supergranule urea.

6.2   High-resolution graphics

The DGRAPH directive is used in this section to introduce many of the features of high-resolution graphics available in Genstat. DGRAPH is then covered in full in 6.2.1, followed by details of the other types of display: histograms, contour plots, surface plots, and pie charts. Many interactive graphical devices also allow *graphical input*, whereby a mouse or cursor can be used to identify points on a graph. This is provided by the DREAD directive, described in 6.4.1.

Before producing any high-resolution graphics you must first select an appropriate output *device*. This can either be screen-based, for interactive use, or it may send the output to a file in one of a number of standard formats suitable for plotters or printers. Different types of device are supported by different versions of Genstat; inevitably there are minor differences in the details of their operation, and these are discussed further in the description of the DEVICE directive (6.5.1). You should also read the additional documentation provided with each version of Genstat (the *Users' Note*). The default graphics device is chosen to be the most appropriate for each version and, if this is suitable, no explicit action is required before using graphics. The examples in this chapter (and the remainder of the manual) have all been generated using the PostScript device, available in all versions of Genstat. This stores output in a file which can then be sent to a printer or incorporated into other documents, as has been done with this manual.

In its simplest form DGRAPH looks very similar to GRAPH, the corresponding line-printer directive. For example, the statement:

```
DGRAPH [TITLE='Scatter Plot'] Y1,Y2; X1,X2
```

is used to generate the graph shown in Figure 6.2a. There are separate parameter lists for the y- and x-coordinates, which are processed in parallel so that the graph contains plots of Y1 versus X1 and Y2 versus X2. The TITLE option is used to include a title for the graph, which is drawn at the top of the plot. It can be up to 80 characters in length, and it can consist of one line of text only. However, there are many more aspects of the output that can be controlled when producing a graph and it is not feasible to allow all of these to be specified by the options or parameters of DGRAPH. The syntax would have become very complicated, and you would have had to specify all the relevant settings every time that DGRAPH was used.

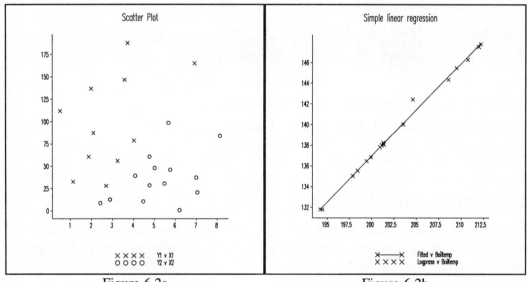

Figure 6.2a   Figure 6.2b

Instead additional directives are used to set up and modify a graphical *environment* which contains most of the information required when plotting. Each time DGRAPH is used it accesses the relevant information from this environment in order to determine how to construct the graph. Thus, to make a simple modification to a graph, for example to change the colour of the plotted symbols, you need make only that change; any other information that you have supplied previously will remain in force. This section illustrates some of the settings that can be used to control or modify the appearance of graphical output, and other examples are shown in Section 6.3. The complete description of the various elements of the environment and the directives that can be used to define them is in Section 6.5.

All the elements of graphical output, such as symbols, lines, axes, titles, labels, annotation, and filled polygons are drawn by *pens*, which have associated definitions covering various attributes, like colour, font, and symbol type. The pen also indicates the plotting method, that is, what kind of plot is to be drawn. For example, the following statements can be used to plot a straight line that has been fitted to the data used in Example 8.1:

```
PEN 1,2; METHOD=line,point; SYMBOL=0,1
DGRAPH [TITLE='Simple Linear Regression'] \
   Fitted,Logpress; Boiltemp; PEN=1,2
```

This means that *pen* 1 will be used to plot a line through the points specified by `Fitted` and `Boiltemp`, and *pen* 2 will be used to plot the points specified by `Logpress` and `Boiltemp`. The corresponding output is shown in Figure 6.2b.

There are a total of 32 pens, each of which has own attribute settings, thus allowing a wide variety of styles within each plot. You can control which pens are used to plot the data, using the `PEN` parameter of `DGRAPH` as shown above. The `PEN` directive can be used to define the various attributes of each pen, such as the colour, symbol type, line style (whether lines should be full, dotted, or dashed), and the font to be used for text. You can also specify labels to be plotted at each point, and control the size of symbols, text, and thickness of lines. When plotting a line you can switch off the symbols if you do not want to mark individual points, by setting `SYMBOLS=0`. Using the `AXES` directive, described below, it is also possible to control the pens that are used for adding axes and titles to the plot.

If the `PEN` parameter of `DGRAPH` is not specified, pen 1 is used for the first pair of `Y` and `X` structures, pen 2 for the second pair, and so on. So the pens need not have been specified explicitly in the `DGRAPH` statement above. The same convention applies to the pens used for different structures in histograms, pie charts, and contour plots. The default settings for each pen are designed so that they will differ in appearance, for example by using different colours or line styles, depending on the output device. Thus, if you specify several data structures to appear in the same plot, the different sets of points or lines will be clearly distinguished by their different pens; see Figure 6.2a. The pens can be re-defined between `DGRAPH` statements in order to have a different effect each time.

You can also control the position and size of the graph. All graphical output is drawn in individual graphics *windows*. A window is a rectangular area of the screen. The position of the window is defined in terms of its lower and upper bounds in the vertical (y) and horizontal (x) directions using a data-independent coordinate system that ranges from 0.0 to 1.0 in each direction. You can use the default window positions defined by Genstat or you can use the `FRAME` directive (6.5.3) to define your own. The `WINDOW` option of `DGRAPH` indicates which window is to be used for the plot and `KEYWINDOW` specifies the location of the key.

Example 6.2a

```
   1  "
  -2         Wulfer's sunspot numbers, from Yule (1927)
  -3  "
   4  READ Sunspots

   Identifier   Minimum      Mean    Maximum     Values    Missing
    Sunspots      0.00      44.76     154.40        176          0

  22  VARIATE Year; VALUES=!(1749...1924); DECIMALS=0
```

```
 23    "
-24           Set the window size so that the y- and x-axis length are in the
-25           ratio 0.065:1, by calculating the upper y-bound to be
-26           0.065 * x-axis length + y-margin (see 6.5.3).
-27    "
 28    CALCULATE Yup = 0.065*0.9 + 0.1
 29    FRAME 4; YLOWER=0.0; YUPPER=Yup; XLOWER=0.0; XUPPER=1.0;\
 30      YMLOWER=0.1; YMUPPER=0.0; XMLOWER=0.1; XMUPPER=0.0
 31    AXES 4; XLOWER=1749; XUPPER=1924; XMARKS=!(1750,1800,1850,1900)
 32    PEN 1; METHOD=line; SYMBOL=0
 33    DGRAPH [WINDOW=4; KEYWINDOW=0] Sunspots; Year
```

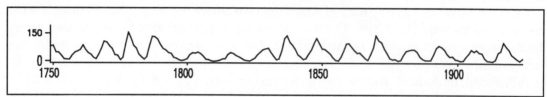

Figure 6.2c

The FRAME statement in line 29 of Example 6.2a defines window 4 to have dimensions that will ensure that the y- and x-axes are suitably in proportion to show the shape of the series (see for example Cleveland and McGill 1987). The setting KEYWINDOW=0 in DGRAPH in line 33 stops the key being displayed. The resulting graph is shown in Figure 6.2c.

Altogether, there are 32 windows, numbered from 1 up to 32. Windows are independent of one another and are allowed to overlap or contain others. So it is possible to build up complex displays with graphs in adjacent windows, or to superimpose plots in different windows as shown in Figure 6.2d. For all the directives that produce graphics, the default window is window 1 and the default key window is window 2. The windows all have initial defaults, as explained in 6.5.3. Generally you would not want the key window to overlap the window containing the picture, unless there are sections of the graph where there are no points. If so, you could first plot just the graph, by setting KEYWINDOW to zero; then use FRAME to define the key window within an area that is clear of plotted data; and then plot just the key, by setting WINDOW to zero (to suppress the graph itself) and KEYWINDOW to the window defined for the key. For example,

```
DGRAPH [WINDOW=9; KEYWINDOW=0] Y; X
FRAME WINDOW=10; YLOWER=0.3; YUPPER=0.5; XLOWER=0.5; XUPPER=0.9
DGRAPH [WINDOW=0; KEYWINDOW=10; SCREEN=keep] Y; X
```

The default key window, window 2, is not very large. If you include too many variates in the plot, the key window may become full and a warning message will be printed. Also, very long identifier names or descriptions will be truncated if the width is not sufficient. In either case you may want to use FRAME to increase the window size.

Each window has an associated definition for the axes that may be drawn in that window. The default definition will often be sufficient, but you can use the AXES directive (6.5.4) to control various aspects of the axes within each window, for example to add axis titles or to specify the spacing of tick marks or the position of labels. You can use AXES also to specify which pens should be used for drawing the axes and adding annotation.

There are also ways in which you can control the output device. There are two options in DGRAPH (and in the other plotting directives) that can be used for this: SCREEN and ENDACTION. By default, when plotting a graph, Genstat will first clear the screen (or, equivalently, start a new page), but you can set option SCREEN=keep to preserve the current display. You can thus add more points to an existing graph, or draw another graph in a different window. The ENDACTION option of DGRAPH is useful when producing graphs on an interactive graphical device; it specifies whether Genstat should pause at the completion of a graph, waiting for the user to press a key before continuing, or should immediately continue to the next statement. Where the screen has to switch between text and graphics modes (as for example when using a PC), you would normally want to pause so that you could look at the graph; whereas using a windowed display, as with X-windows, this is unnecessary unless several graphs are being drawn in succession, for example by a procedure. The default for ENDACTION, in the initial environment, uses the setting most suited to the current device. However, this can be modified by the DEVICE directive (6.5.1). ENDACTION is ignored when output is to a file.

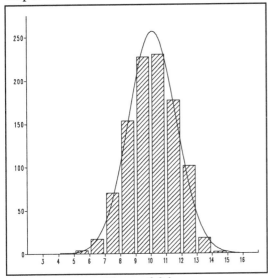

Figure 6.2d

DGRAPH can be used not only as a means of producing a self-contained picture, but also as a basic drawing tool to build up a complicated picture in several stages. This approach has been used in the programming of many of the procedures in the graphics module of the Procedure Library; for example, BIPLOT, DDENDROGRAM, and BOXPLOT. These provide many extensions to the facilities given by the basic Genstat directives.

By defining windows that overlay existing ones, you can also use DGRAPH to superimpose information on other plots such as histograms or contours. Example 6.2b illustrates how a graph and a histogram can be overlaid. These methods are developed further in the library procedure DBARCHART where the two-way table that forms the input to the procedure is converted into a form that can be displayed by DHISTOGRAM, and axes are then added using DGRAPH. The procedure LIBEXAMPLE can be used to extract the source of these graphics procedures from the library, so that you can see how the displays are constructed and obtain ideas about how to construct your own.

Example 6.2b

```
   1   READ Data
```

Identifier	Minimum	Mean	Maximum	Values	Missing
Data	4.39	10.11	14.55	1000	0

```
103   AXES 3; YUPPER=275
104   VARIATE [VALUES=3...16] Limits
105   DHISTOGRAM [LIMITS=Limits; WINDOW=3; KEYWINDOW=0] Data
106   CALCULATE Mu = MEAN(Data)
107   & Sigma = SQRT(VAR(Data))
108   VARIATE X; VALUES=!(2.0,2.1...17.0)
109   CALCULATE Y = 1000/(SQRT(2*C('PI'))*Sigma)*EXP(-0.5*((X-Mu)/Sigma)**2)
110   PEN 1; METHOD=monotonic; SYMBOL=0
111   AXES 4; YLOWER=0.0; YUPPER=275; XLOWER=2.0; XUPPER=17.0; STYLE=none
112   DGRAPH [WINDOW=4; KEYWINDOW=0; SCREEN=keep] Y; X
```

6.2.1   The DGRAPH directive

DGRAPH   Draws graphs on a plotter or graphics monitor.

Options

TITLE = *text*	General title; default *
WINDOW = *scalar*	Window number for the graphs; default 1
KEYWINDOW = *scalar*	Window number for the key (zero for no key); default 2
SCREEN = *string*	Whether to clear the screen before plotting or to continue plotting on the old screen (clear, keep); default clea
KEYDESCRIPTION = *text*	Overall description for the key; default *
ENDACTION = *string*	Action to be taken after completing the plot (continue, pause); default * uses the setting from the last DEVICE statement

Parameters

Y = *identifiers*	Vertical coordinates
X = *identifiers*	Horizontal coordinates
PEN = *scalars* or *variates* or *factors*	Pen number for each graph (use of a variate or factor allows different pens to be defined for different sets of units); default * uses pens 1, 2, and so on for the successive graphs
DESCRIPTION = *texts*	Annotation for key
YLOWER = *identifiers*	Lower values for vertical bars
YUPPER = *identifiers*	Upper values for vertical bars
XLOWER = *identifiers*	Lower values for horizontal bars
XUPPER = *identifiers*	Upper values for horizontal bars

The DGRAPH directive draws high-resolution graphs, containing points, lines, or shaded polygons. The graph is produced on the current graphics device; this can be selected using the DEVICE directive as explained in 6.5.1. The WINDOW option defines the window, within the plotting area, in which the graph is drawn; by default this is window 1.

The Y and X parameters specify the coordinates of the points to be plotted; they must be numerical structures (scalars, variates, matrices, or tables) of equal length. If any of the variates is restricted (4.4.1), only the subset of values specified by the restriction will be included in the graph. The restrictions are applied to the Y and X variates in pairs, and do not carry over to all the variates in a list. For example, suppose the variate Y1 is restricted but the variate Y2 is not. The statement

```
DGRAPH Y1,Y2; X
```

will plot the subset of values of Y1 against X, but all the values of Y2 against X. Conversely, if X were restricted the subset would be plotted for both Y1 and Y2. Any associated structures, like variates specified by the PEN parameter or factors used to provide labels for the points, must be of the same length as Y and X.

Each pair of Y and X structures has an associated pen, specified by the PEN parameter. By default, pen 1 is used for the first pair, pen 2 for the second, and so on. The type of graph that is produced is determined by the METHOD setting of that pen. This can be point, to produce a point plot or scatterplot; line to join the points with straight lines; monotonic, open, or closed to plot various types of curve through the points; or fill to produce shaded polygons. In the initial graphics environment, all the pens are defined to produce point plots. This can be modified using the METHOD option of the PEN directive. Other attributes of the pen can be used to control the colour, font, symbols, and labels as described in 6.5.5.

With METHOD=fill, the points defined by the Y and X variates are joined by straight lines to form one or more polygons which are then filled using the brush style specified for the pen. The JOIN parameter of PEN determines the order in which the points are joined; with the default, ascending, the data are sorted into ascending order of x-values, while with JOIN=given they are left in their original order. There should be at least three points when using this method.

A warning message is printed if the data contain missing values. The effect of these depends on the type of graph being produced, as follows. If the method is point there will be no indication on the graph itself that any points were missing (but obviously none of the points with missing values for either the y- or x-coordinate can be included in the plot). If a line or curve is plotted through the points there will be a break wherever a missing value is found; that is, line segments will be omitted between points that are separated by missing values. When using METHOD=fill missing values will, in effect, define subsets of points, each of which will be shaded separately. Note, however, that the position of the missing values within the data will differ according to whether or not the data values have been sorted; this is controlled by the JOIN parameter of PEN, as described above. If the data are sorted, units with missing x-values are moved to the beginning.

The PEN parameter can also be set to a variate or factor, to allow different pens to be used for different subsets of the units. With a factor, the units with each level are plotted separately, using the pen defined by the level concerned. If PEN is set to a variate, its values similarly define the pen for each unit. For example, if you fit separate regression lines to some grouped data, you can easily plot the fitted lines in just two statements, one to set up the pens and one to plot the data:

```
PEN 1...Ngroups; METHOD=line; SYMBOL=0
DGRAPH Fitted; X; PEN=Groups
```

By default, Genstat calculates bounds on the axes that are wide enough to include all the data; the range of the data is extended by five percent at each end, and the axes are drawn on the left-hand side and bottom edge of the graph. This can all be changed by the AXES directive (6.5.4), using the YLOWER, YUPPER, XLOWER, and XUPPER parameters to set the bounds, and YORIGIN and XORIGIN to control the position of the axes. Other parameters allow you to control the axis labelling and style. If the axis bounds are too narrow, some points may be excluded from the graph, so that *clipping* occurs. If the plotting method is point, Genstat ignores points that are out of bounds. For other settings of METHOD, lines are drawn from points that are within bounds towards points that are out of bounds, terminating at the appropriate edge. Clipping may also occur if the method is monotonic, open, or closed and you have left Genstat to set default axis bounds, because these methods fit curves that may extend beyond the boundaries. If this occurs you should use the AXES directive to provide increased axis bounds. When you use several DGRAPH statements with SCREEN=keep to build up a complex graph, the axes are drawn only the first time, and the same axes bounds are then used for the subsequent graphs. You should thus define axis limits that enclose all the subsequent data. Axes are drawn only if SCREEN=clear, or the specified window has not been used since the screen was last cleared, or the window has been redefined by a FRAME statement.

DGRAPH can also be used to add *error bars* to the plot. You might want to use these, for example, to show confidence limits on points that have been fitted by a regression (Chapter 8). Error bars are requested by setting the YLOWER and YUPPER parameters to variates defining the lower and upper values for the error bar to be drawn at each point. For example, if you know the standard error for each point, you could calculate and plot the bounds as follows:

```
CALCULATE Barlow = Y - 1.96 * Err
& Barhigh = Y + 1.96 * Err
DGRAPH Y; X; YLOWER=Barlow; YUPPER=Barhigh
```

The error bar is drawn from the lower point to the upper point at the associated x-position; the bar will be drawn even if the corresponding y-value (or y-variate) is missing. If the lower value is missing, or the YLOWER parameter is not set, only the upper section of the bar is drawn; likewise if the upper value is missing only the lower section is drawn. The same pen is used to draw the error bars as is used for the y- and x-values. If you want to use a different pen for the error bars you can plot them separately: for example

```
DGRAPH Y,*; X,X; YLOWER=*,Barlow; YUPPER=*,Barhigh; PEN=2,6
```

Similarly, parameters XLOWER and XUPPER allow you to plot horizontal bars at each point.

The KEYWINDOW option specifies the window in which the key appears; by default this is window 2. Alternatively, you can set KEYWINDOW=0 to suppress the key. The key contains a line of information for each pair of Y and X structures, written with the associated pen. This will indicate the symbol used, the line style (for a plotting method of `line` or `curve`) or a shaded block to illustrate the brush style (when `METHOD=fill`), the name of the structure (if any) defined by the `LABELS` parameter of `PEN`, and a description indicating the identifiers of the data plotted (for example `Residuals v Fitted`). Alternatively, you can supply your own key, using the `DESCRIPTION` parameter, and you can specify a title for the key using the `KEYDESCRIPTION` option. If you draw several graphs using `SCREEN=keep` and the same key window, each new set of information is appended to the existing key, until the window is full.

If you have set the `PEN` parameter to a variate or factor in order to plot independent subsets of the data, the key will contain information for each subset. If the `LABELS` parameter of `PEN` has been used to specify labels for the points, each line of the key will contain the label corresponding to the first value of the subset, rather than the identifier of the labels structure itself. In Example 6.2.1 a factor is used to label the points, and the graph is enhanced by using the factor also to specify different pens for different groups of points.

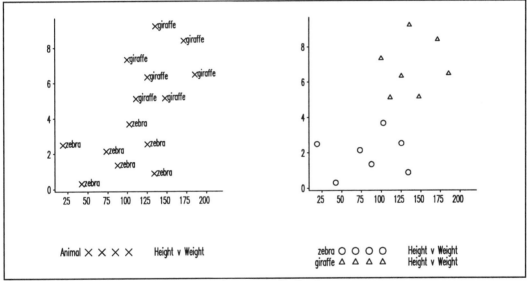

Figure 6.2.1

Example 6.2.1

```
  1  "
 -2         Use of factors for labels and pens
 -3  "
  4  FACTOR [LABELS=!T(zebra,giraffe)] Animal
  5  READ Animal,Height,Weight; FREPRESENTATION=labels
```

Identifier	Minimum	Mean	Maximum	Values	Missing
Height	0.308	4.384	9.228	14	0
Weight	18.7	111.4	185.6	14	0

```
   Identifier     Values    Missing    Levels
     Animal          14         0         2
21  AXES 5,6; XUPPER=220
22  FRAME 7,8; YLOWER=0.3; YUPPER=0.5
23  PEN 1; LABELS=Animal
24  DGRAPH [WINDOW=5; KEYWINDOW=7] Height; Weight
25  PEN 1,2; SYMBOLS=2,7; LABELS=*
26  DGRAPH [WINDOW=6; KEYWINDOW=8; SCREEN=keep] Height; Weight; PEN=Animal
```

The TITLE option can be used to provide a title for the graph. You can also put titles on the axes by using the YTITLE and XTITLE parameters of the AXES directive.

The SCREEN option controls whether the graphical display is cleared before the graph is plotted and the ENDACTION option controls whether Genstat pauses at the end of the plot, as described at the start of this section.

6.2.2   The **DHISTOGRAM** directive

DHISTOGRAM   draws histograms on a plotter or graphics monitor.

Options

TITLE = *text*	General title; default *
WINDOW = *scalar*	Window number for the histograms; default 1
KEYWINDOW = *scalar*	Window number for the key (zero for no key); default 2
LIMITS = *variate*	Variate of group limits for classifying variates into groups; default *
NGROUPS = *scalar*	When LIMITS is not specified, this defines the number of groups into which a DATA variate is to be classified; default is the integer value nearest to the square root of the number of values in the variate
LABELS = *text*	Group labels; default *
APPEND = *string*	Whether or not the bars of the histograms are appended together (yes, no); default no
SCREEN = *string*	Whether to clear the screen before plotting or to continue plotting on the old screen (clear, keep); default clea
KEYDESCRIPTION = *text*	Overall description for the key; default *
ENDACTION = *string*	Action to be taken after completing the plot (continue, pause); default * uses the setting from the last DEVICE statement

Parameters

DATA = *identifiers*	Data for the histograms; these can be either a factor indicating the group to which each unit belongs, a variate whose values are to be grouped, or a one-way

	table giving the height of each bar
NOBSERVATIONS = *tables*	One-way table to save numbers in the groups
GROUPS = *factors*	Factor to save groups defined from a variate
PEN = *scalars*	Pen number for each histogram; default * uses pens 1, 2, and so on for the successive structures specified by DATA
DESCRIPTION = *texts*	Annotation for key

Many of the options and parameters of DHISTOGRAM are the same as those of the HISTOGRAM directive (6.1.1). The data can be specified as variates, which are sorted into groups defined by the LIMITS option or determined automatically if NGROUPS is set. Alternatively, the groups can be defined by factors. In HISTOGRAM the data can also be presented as a one-way table giving the sizes of the groups; the table is then constrained to contain positive integers. In DHISTOGRAM this form of input is extended by allowing the table to contain any numbers, positive or negative, thus allowing bar charts to be drawn, as shown in Example 6.2.2a. As in HISTOGRAM, details of the groups can be saved using the NOBSERVATIONS and GROUPS parameters.

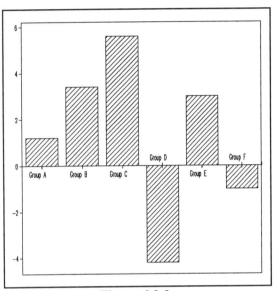

Figure 6.2.2a

Example 6.2.2a

```
1  FACTOR [LABELS=!t('Group A','Group B','Group C','Group D','Group E',\
2    'Group F')] Groups
3  TABLE [CLASSIFICATION=Groups; VALUES=1.2,3.4,5.6,-4.2,3.0,-1] Table
4  AXES 1; STYLE=box
5  DHISTOGRAM [WINDOW=1; KEYWINDOW=0] Table
```

There are also some options and parameters that are specific to high-resolution histograms.

The WINDOW option defines the window where the histogram is plotted, and the KEYWINDOW option similarly specifies where the key should appear. You can set either of these to zero if you want to suppress the corresponding output. Titles can be added to the histogram and key using the TITLE and KEYDESCRIPTION options respectively.

The SCREEN option controls whether the graphical display is cleared before the histogram is plotted and the ENDACTION option controls whether Genstat pauses at the end of the plot, as described at the start of this section.

The APPEND option controls the form of display to be used when the DATA parameter
specifies a list of structures. These parallel histograms can be produced in one of two styles.
By default (APPEND=no), the histogram contains a set of bars for each structure, drawn in
parallel groups, as shown in line 13 of Example 6.2.2b. Alternatively, if you set APPEND=yes,
the bars for the structures are concatenated into a single bars for each group, as in line 14 of
Example 6.2.2b and Figure 6.2.2b. The top portion of each bar then corresponds to the first
structure, and the bottom to the last structure.

Example 6.2.2b

```
  1   READ Y,X

     Identifier    Minimum      Mean    Maximum     Values   Missing
            Y        2.80      14.25      29.01         30         0
            X        4.07      19.33      39.01         30         0

 11   FRAME 7,8; YUPPER=0.45; XLOWER=0.2,0.7
 12   PEN 1,2; BRUSH=2,9
 13   DHISTOGRAM [WINDOW=5; KEYWINDOW=7] Y,X
 14   DHISTOGRAM [WINDOW=6; KEYWINDOW=8; SCREEN=keep; APPEND=yes] Y,X
```

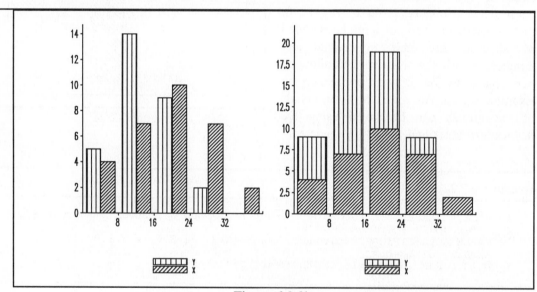

Figure 6.2.2b

The bars for each structure are all shaded according to the pen that has been specified for that
structure, using the PEN parameter. If the PEN parameter is not set, Genstat uses the pens in
turn, pen 1 for the first structure, pen 2 for the second structure, and so on, so that a different
shading is used for each of the structures. The relevant aspects of the pens should be set in
advance, if required, using the BRUSH and COLOUR parameters of the PEN directive, as in line

12 of Example 6.2.2b. Often, however, the default attributes of the pens will be satisfactory (see 6.5.5).

The bars are drawn with equal width and the length of each bar is proportional to the number of values that it represents, irrespective of the width of the corresponding group. Thus, the areas of the bars will represent the relative frequencies of the groups only if the groups are of equal width.

The axes of the histogram are formed automatically from the data. By default, the upper bound of the y-axis is set to be five percent greater than the height of the longest bar. If any of the bars is of negative height the lower bound is adjusted in a similar way, otherwise it is set to zero. When the histogram is formed from a variate, the x-axis markings are set to indicate the limits of each bar or set of bars; when the data are provided in a factor the factor labels or levels are used to label the histogram bars, and when the bar heights are provided directly in a table the classifying factor of the table is used. You can control the form of the axes by using the AXES directive to set the required attributes (see 6.5.4), before the DHISTOGRAM directive is used.

The WINDOW parameter of AXES should be set to the window in which the histogram is to be plotted (controlled by the WINDOW option of DHISTOGRAM). The STYLE parameter then controls which axes are drawn: x-axis only (by specifying x or none), x- and y-axes (y, xy, or grid), or x- and y-axes with a box (box). The YTITLE, YLOWER, YUPPER, YMARKS, and YLABELS parameters control annotation of the y-axis. The YUPPER parameter is particularly useful when you are plotting a series of histograms; by setting YUPPER to a value larger than any of the bars in any of the histograms, you can ensure that they are all plotted on the same scale. However, Genstat ignores the setting of this parameter if the longest bar is greater than the value supplied and, when checking this, you must be careful to allow for the effect of the APPEND option (see above). The x-axis bounds are defined by the data and cannot be altered. However, you can use the LABELS option (of DHISTOGRAM) to specify labels, and XMPOSTION and XLPOSITION to control the positioning of tick marks and labels.

The histogram key consists of the title, if set by KEYDESCRIPTION, followed by a legend for each structure plotted. This consists of a small rectangle that is drawn in the same colour and brush style as that used in the histogram, followed by the identifier name or the piece of text specified by the DESCRIPTION parameter.

6.2.3   The **DPIE** directive

DPIE   draws a pie chart on a plotter or graphics monitor.

Options
TITLE = *text*	General title; default *
WINDOW = *scalar*	Window number for the pie chart; default 1
KEYWINDOW = *scalar*	Window number for the key (zero for no key); default 2
SCREEN = *string*	Whether to clear the screen before plotting or to continue plotting on the old screen (clear, keep); default clea

KEYDESCRIPTION = *text*   Overall description for the key

ENDACTION = *string*   Action to be taken after completing the plot
                                 (continue, pause); default * uses the setting from
                                 the last DEVICE statement

Parameters

SLICE = *scalars*   Amounts in each of the slices (or categories)

PEN = *scalars*   Pen number for each slice; default * uses pens 1, 2, and
                                 so on for the successive slices

DESCRIPTION = *texts*   Description of each slice

A pie chart is formed by taking the values of the scalars in the SLICE parameter, in order, and representing them by segments of a circle starting at "three o'clock" and working in an anti-clockwise direction. The angle subtended by each segment (and thus the area of the segment) is proportional to the value of the corresponding scalar. The values may be raw data or can be expressed as percentages (by ensuring they total 100).

The brush style used for each segment can be controlled using the PEN parameter. By default, pen 1 is used for the first segment, pen 2 for the second segment, and so on. The default attributes of the pens are device specific, so that on a colour display the segments will be solid-filled using different colours, and on a monochrome device different hatching styles will be used. These can be modified using the PEN directive, as described in 6.5.5.

Line 4 of Example 6.2.3 plots a pie chart with four slices, as shown in Figure 6.2.3.

Example 6.2.3

```
1   FRAME 1,2,3; YLOWER=0.6; YUPPER=1.0,0.9,1.0; \
2     XLOWER=0.0,0.4,0.6; XUPPER=0.4,0.6,1.0
3   PEN 1,2,3,4; BRUSH=1,3,2,4
4   DPIE [WINDOW=1; KEYWINDOW=2; KEYDESCRIPTION='1993 Summary'] \
5     24.7,98.8,74.1,49.4; \
6     DESCRIPTION='Administration','Sales','Marketing','Overheads'
7   DPIE [WINDOW=3; KEYWINDOW=0; SCREEN=keep] 10,40,30,-20; \
8     DESCRIPTION='Administration','Sales','Marketing','Overheads'
```

Individual segments can be displaced outwards from the centre, to obtain an "exploded" pie chart, as in Figure 6.2.3. The chosen segments are indicated by setting the corresponding scalars in the SLICE parameter list to negative values (see line 7 of Example 6.2.3).

The WINDOW and KEYWINDOW options specify the windows in which the pie chart and key are to be displayed. The shape of the pie chart is determined by the dimensions of the window; if it is not square the resulting pie chart will be elliptical.

Titles can be added using the TITLE and KEYDESCRIPTION options. The key produced for the pie chart is similar to that produced by the DHISTOGRAM directive: a shaded block is drawn for each segment, followed by the identifier name or the piece of text specified by the DESCRIPTION parameter.

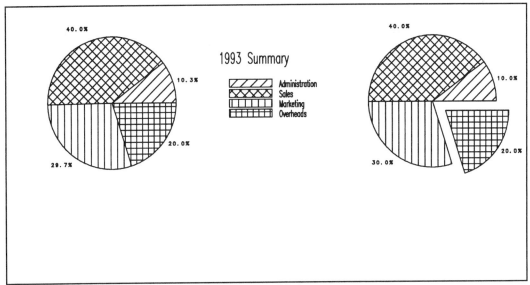

Figure 6.2.3

The SCREEN option controls whether the graphical display is cleared before the histogram is plotted and the ENDACTION option controls whether Genstat pauses at the end of the plot, as described at the start of this section.

6.3   Plotting three-dimensional surfaces in high-resolution

Three-dimensional surfaces are represented in Genstat by a grid of z-values or heights, as for the CONTOUR directive described in 6.1.3. The grid can be a rectangular matrix, a two-way table, or a pointer to a set of variates; the y-dimension is represented by the rows of the structure and the x-dimension by the columns. In each case there must be at least three rows and three columns of data (after allowing for any restrictions on a set of variates). Missing values are not permitted; that is, only complete grids can be displayed. If the grid is supplied as a table with margins, these will be ignored when plotting the surface.

Genstat provides three methods for plotting three-dimensional data. A contour plot (DCONTOUR) can be used to form a two-dimensional representation of the surface, in which contour lines are drawn to link points of equal height. Alternatively, a three-dimensional representation of the surface, or *perspective view*, can be drawn using the DSURFACE directive, to display more fully the three-dimensional nature of the data. The grid can be viewed from any angle, allowing the investigation of features such as maxima, minima, valleys, and plateaux. Finally, when the grid contains discrete data, a three-dimensional (or bivariate) histogram may be appropriate. This is produced using the D3HISTOGRAM directive, which forms the display by drawing cuboid blocks of the appropriate height at each (x,y) position.

6.3.1　The DCONTOUR directive

DCONTOUR draws contour plots on a plotter or graphics monitor.

Options

INTERVAL = *scalar* or *variate*	Contour interval for scaling (scalar) or positions of the contours (variate); default * i.e. determined automatically
TITLE = *text*	General title; default *
WINDOW = *scalar*	Window number for the plots; default 1
KEYWINDOW = *scalar*	Window number for the key (zero for no key); default 2
LOWERCUTOFF = *scalar*	Lower cut-off for array values; default *
UPPERCUTOFF = *scalar*	Upper cut-off for array values; default *
SCREEN = *string*	Whether to clear the screen before plotting or to continue plotting on the old screen (clear, keep); default clea
KEYDESCRIPTION = *text*	Overall description for the key
ENDACTION = *string*	Action to be taken after completing the plot (continue, pause); default * uses the setting from the last DEVICE statement

Parameters

GRID = *identifiers*	Pointers (of variates representing the columns of a data matrix), matrices, or two-way tables specifying values on a regular grid
PEN = *scalars* or *variates*	Pen number to be used for the contours of each grid (use of a variate allows every nth contour to be highlighted; the first $n-1$ units should contain the number of the standard pen, and the nth unit the number of the highlighting pen); default * uses pens 1, 2, and so on for the successive grids
DESCRIPTION = *texts*	Annotation for key

The orientation of the contour plot is the same as that used by the CONTOUR directive, which ensures that element (1,1) of the grid is plotted at the point (1,1); that is, the bottom left-hand corner of the plot. Normally the data will lie on a regular grid but you can also specify an irregular grid as shown in line 10 of Example 6.3.1c; the ROWS and COLUMNS options of the MATRIX directive are set to variates containing the appropriate x- and y- values when the matrix Function is declared. By specifying a list of structures in the GRID parameter you can produce several superimposed contour plots.

The WINDOW option defines the window where the histogram is plotted, and the KEYWINDOW option similarly specifies where the key should appear. The grid axes are scaled so that the y- and x-dimensions (rows and columns respectively) will match the dimensions of the specified window: if you wish to preserve the "shape" of the grid you should use the FRAME directive (6.5.3) to define a window whose y- and x-dimensions are in the same proportions as the grid dimensions, as shown in Example 6.3.1c. Titles can be added to these windows using the TITLE and KEYDESCRIPTION options. The SCREEN option controls whether the graphical display is cleared before the histogram is plotted and the ENDACTION option controls whether Genstat pauses at the end of the plot, as described at the start of Section 6.2.

Example 6.3.1a

```
23    "    ... continuation from Example 6.1.3c    "
24    FRAME 2; YLOWER=0.0; YUPPER=0.9; XLOWER=0.75; XUPPER=1.0
25    DCONTOUR Core
```

By default, the contour lines are plotted at heights that are determined automatically from the range of the data. A constant interval is used, such that the contour heights are round numbers, up to a maximum of ten contours lines. This can be changed by setting the INTERVAL option to a scalar containing the required interval between contours; the number of contours will then depend on the range of the data. Alternatively, you can specify the actual contour heights by setting INTERVAL to a variate containing the required values. If the resulting number of contours is less than two, no contours are drawn. However, if the number is very large, the contours may be difficult to interpret and take a long time to plot. You can also set the LOWERCUTOFF and UPPERCUTOFF options to truncate the grid values; the default interval and contour heights are then adjusted accordingly.

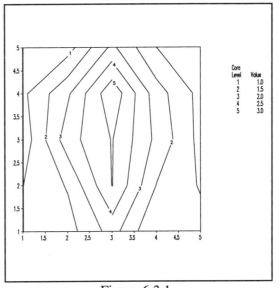

Figure 6.3.1a

The contour lines are labelled by integers, and the translation from contour number to the actual height is provided in the key. Contour lines that are very short will not be labelled but their height can be determined from adjacent contours. Each line of the key occupies a space of height 0.02 (in normalized device coordinates; see 6.5.3), and the key window by default has room for a heading and nine contour levels. If necessary, the size of the window can be redefined using the FRAME directive (6.5.3).

The way in which the contour lines are drawn for each grid is determined by the pen that has been defined for that grid, using the PEN parameter of DCONTOUR. If the PEN parameter is not set, Genstat uses the pens in turn, pen 1 for the first grid, pen 2 for the second grid, and so on, so that the different grids can easily be distinguished. The relevant aspects of the pens should be set in advance, if required, using the METHOD, COLOUR, LINESTYLE, and THICKNESS parameters of the PEN directive (6.5.5).

If the PEN directive is not used, the plotting method will be line, so that individual contours are made up of straight line segments. If curves are required, METHOD should be set to monotonic to use the method of Butland (1980), or open (or closed) to use the method of McConalogue (1970). Both these methods produce curves that are fitted to independent sets of interpolated points and can thus produce contour lines that cross, particularly if the supplied grid of data is coarse or in a region where the contour height is changing rapidly. If METHOD is set to other values, straight lines will be used to draw the contours.

The PEN parameter of DCONTOUR can also be set to a variate containing a list of pen numbers, which allows highlighting of particular contours. The first value specifies the pen for the first (lowest) contour, second value for the second contour, and so on. The list is recycled if there are too few values for the number of contours to be plotted. For example, the statement

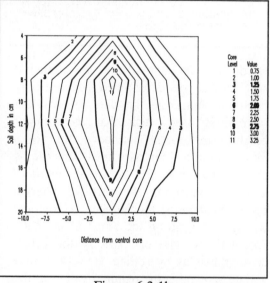

Figure 6.3.1b

```
DCONTOUR Matrix; PEN=!(1,1,2)
```

will produce a contour plot where every third contour is drawn by pen 2. The contours drawn by pen 2 may be highlighted in various ways. The default attributes of the pens, which will be in place unless the PEN directive has been used to specify otherwise, will often be satisfactory. By default, on a colour device, the pens will be defined to use different colours, while on a monochrome device they will use different line styles. In line 31 of Example 6.3.1b, the PEN directive specifies that pen 2 is to use a solid line style (like pen 1), and the THICKNESS is increased to produce the required highlighting.

Example 6.3.1b

```
26    " ... continued from Example 6.3.1a "
27    AXES 1; YTITLE='Soil depth in cm'; XTITLE='Distance from central core';\
28      YUPPER=4; YLOWER=20; XUPPER=10; XLOWER=-10
29    PEN 2; LINESTYLE=1; THICKNESS=3
30    DCONTOUR [INTERVAL=0.25] Core; PEN=!(1,1,2)
```

By default, the axis bounds are determined from the grid. Normally the lower bound for each axis will be 1.0 and the upper bound will be the number of rows of the grid for the y-axis, and the number of columns for the x-axis. If a matrix is used to specify the grid, its row and column labels can be set to variates whose values will then be used to determine the axis bounds. If more than one grid is specified, the axes are derived from the first grid and subsequent grids are plotted relative to these axes. The AXES directive can be used to control how the axes are drawn (see Example 6.3.1b) or, by setting STYLE=none, to suppress them altogether.

In Example 6.3.1c, a matrix of function values is calculated over a regular range of y- and x-values, to produce the contour plot on the left-hand side of Figure 6.3.1c. The function is then recalculated on an irregular grid with the y- and x-values closest where the function is changing most rapidly, and the plot on the right-hand side is produced.

Example 6.3.1c

```
 1  VARIATE Rows,Columns; VALUES=!(0.0,0.2...2.0),!(0.0,0.2...1.0)
 2  MATRIX [ROWS=Rows; COLUMNS=Columns] Function,Y,X
 3  EQUATE Columns; X
 4  CALCULATE Y$[*; 1...6] = Rows
 5  & Function = COS(1/(X+0.1)**2) + SIN(Y**2)
 6  FRAME 1; YLOWER=0.25; YUPPER=1; XLOWER=0; XUPPER=0.375
 7  DCONTOUR [TITLE='Regular Grid'] Function
 8  VARIATE Irr_Cols; VALUES=!(0.0,0.1...0.4,0.6,0.8,1.0)
 9  MATRIX [ROWS=Rows; COLUMNS=Irr_Cols] Function,Y,X
10  EQUATE OLD=Irr_Cols; NEW=X
11  CALCULATE Y$[*; 1...8] = Rows
12  & Function = COS(1/(X+0.1)**2) + SIN(Y**2)
13  FRAME 3,4; YLOWER=0.25,0.0; YUPPER=1.0,0.25; XLOWER=0.375; XUPPER=0.750
14  DCONTOUR [TITLE='Irregular Grid'; WINDOW=3; KEYWINDOW=4; SCREEN=keep] \
15     Function
```

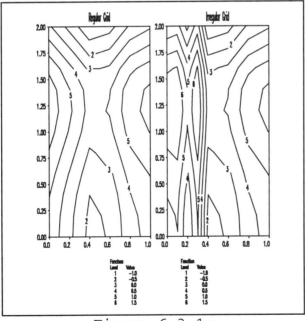

Figure 6.3.1c

6.3.2 The DSURFACE directive

DSURFACE produces perspective views of a two-way arrays of numbers.

Options

TITLE = *text*	General title; default *
WINDOW = *scalar*	Window number for the plots; default 1
ELEVATION = *scalar*	The elevation of the viewpoint relative to the surface; default 25 (degrees)
AZIMUTH = *scalar*	Rotation about the horizontal plane; the default of 225 degrees ensures that, with a square matrix M, the element M\$[1;1] is nearest to the viewpoint
DISTANCE = *scalar*	Distance of the viewpoint from the centre of the grid on the base plane; default * gives a distance of 25 times the number of y-points in the grid
ZORIGIN = *scalar*	Defines the origin of the diagram along the z-axis; default * takes the value defined by LOWERCUTOFF
ZSCALE = *scalar*	defines the scaling of the z-axis relative to the horizontal (x-y) axes; default 2
LOWERCUTOFF = *scalar*	Lower cut-off for array values; default *
UPPERCUTOFF = *scalar*	Upper cut-off for array values; default *
SCREEN = *string*	Whether to clear the screen before plotting or to continue plotting on the old screen (clear, keep); default clea
ENDACTION = *string*	Action to be taken after completing the plot (continue, pause); default * uses the setting from the last DEVICE statement

Parameters

GRID = *identifier*	Pointer (of variates representing the columns of a data matrix), matrix, or two-way table specifying values on a rectangular grid
PEN = *scalar*	Pen number to be used for the plot; default 1

The DSURFACE directive produces a perspective (or *conical*) projection of a surface, showing the view from a particular viewpoint. The position of this viewpoint is specified in polar coordinates, using the options ELEVATION, DISTANCE, and AZIMUTH. These define the angle of elevation, in degrees, above the base plane of the surface, distance from the centre of this plane, and angular position relative to the vertical z-axis, respectively. This is illustrated in Figure 6.3.2a.

The default settings of ELEVATION, DISTANCE, and AZIMUTH have been chosen to produce a reasonable display of most surfaces; but if, for example, some parts of the surface are obscured by high points they can be modified to obtain a better view. Altering the value of

AZIMUTH will, in effect, rotate the surface in the horizontal plane about a vertical axis drawn through the centre of the grid; the default value of 225 degrees ensures that the element in the first row and column of the grid is at the corner nearest the viewpoint.

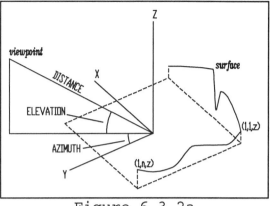

Figure 6.3.2a

As in the DCONTOUR directive, the LOWERCUTOFF and UPPERCUTOFF options specify lower and upper bounds for the z-axis. You can use these to truncate the grid or, alternatively, you can set them to values outside the range of the data to obtain compatible scales when you are plotting several grids. The ZORIGIN option allows the origin of the z-axis to be specified. By default, LOWERCUTOFF and UPPERCUTOFF will be set to the minimum and maximum grid values, and ZORIGIN to the value of LOWERCUTOFF.

The ZSCALE option specifies a scaling factor for the z-axis (or vertical axis) of the plotted surface. Generally values between 0.5 and 2.0 are most successful; large values result in a flatter surface, while smaller values produce a steep surface, accentuating changes in the data.

The TITLE, WINDOW, SCREEN, and ENDACTION options are used to specify a title, the plotting window, whether the screen should be cleared first, and whether there should be a pause once the plotting is finished; as in other graphics directives.

The PEN parameter specifies the pen to be used to plot the surface (by default, pen 1). The PEN directive can be used to modify the colour and the thickness of the pen, but the other attributes of the pen are ignored.

Simple axes are drawn to indicate the directions in which x and y increase. The YTITLE and XTITLE parameters of the AXES directive can be used to add further annotation, as shown in line 7 of Example 6.3.2 which produced the plots in Figure 6.3.2b.

Example 6.3.2

```
 1   MATRIX [ROWS=21; COLUMNS=21] Grid,Y,X
 2   EQUATE OLD=!(0.0,0.05...1.0); NEW=X
 3   CALCULATE Y = TRANSPOSE(X)
 4   & Fy = EXP(-0.5*((Y-0.3)/0.07)**2) + 0.5*EXP(-0.5*((Y-0.7)/0.12)**2)
 5   & Fx = EXP(-0.5*((X-0.3)/0.07)**2) + 0.5*EXP(-0.5*((X-0.7)/0.12)**2)
 6   & [PRINT=summary] Grid = Fx*Fy+0.1

     Identifier   Minimum      Mean   Maximum     Values   Missing
           Grid    0.1000    0.1960    1.1039        441         0      Skew

 7   AXES 5,6; YTITLE='The Y axis'; XTITLE='The X axis'
 8   DSURFACE [WINDOW=5; TITLE='Default option settings'] Grid
 9   DSURFACE [WINDOW=6; TITLE='Effect of various options'; SCREEN=keep; \
10      ZORIGIN=0.0; UPPERCUTOFF=0.6; AZIMUTH=120] Grid
```

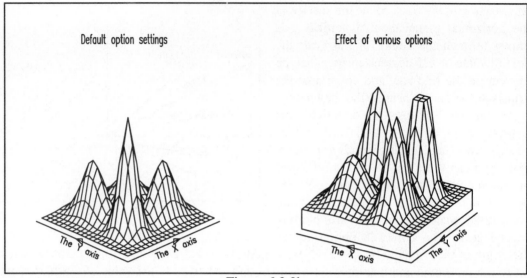

Default option settings   Effect of various options

Figure 6.3.2b

6.3.3   Three-dimensional histograms (**D3HISTOGRAM**)

D3HISTOGRAM produces three-dimensional histograms.

Options

WINDOW = *scalar*	Window number for the plots; default 1
TITLE = *text*	General title; default *
ELEVATION = *scalar*	The elevation of the viewpoint relative to the surface; default 25 (degrees)
AZIMUTH = *scalar*	Rotation about the horizontal plane; the default of 225 degrees ensures that, with a square matrix M, the element M$[1;1] is nearest to the viewpoint
DISTANCE = *scalar*	Distance of the viewpoint from the centre of the grid on the base plane; default * gives a distance of 25 times the number of y-points in the grid
LOWERCUTOFF = *scalar*	Lower cut-off for array values; default *
UPPERCUTOFF = *scalar*	Upper cut-off for array values; default *
SCREEN = *string*	Whether to clear the screen before plotting or to continue plotting on the old screen (clear, keep); default clea
ENDACTION = *string*	Action to be taken after completing the plot (continue, pause); default * uses the setting from the last DEVICE statement

Parameters

GRID = *identifier*	Pointer (of variates representing the columns of a data matrix), matrix, or two-way table specifying values on a regular grid
PEN = *scalar*	Pen number to be used for the plot; default 1

The preceding subsection described how the DSURFACE directive can be used to produce a perspective view of a surface. D3HISTOGRAM provides an alternative way of displaying such data, which may be more appropriate for example if the grid contains counts. D3HISTOGRAM has the same options and parameters as DSURFACE, and so the details are not repeated here.

Example 6.3.3 illustrates the use of D3HISTOGRAM by displaying the table Sales formed in Example 4.11.4. The AZIMUTH and ELEVATION options are used to obtain a clearer view of the surface, and LOWERCUTOFF is used to set the minimum z-value to zero. Note that when the grid is not square, as in this example, the y- and x-axes are scaled appropriately. This is also the case when using DSURFACE.

The axis labelling is derived from the grid, using the classifying factors if it is a table or the row and column labels if it is a matrix. Alternative labels can be supplied using the YLABELS and XLABELS parameters of the AXES directive. If axis labels are not available, either from the grid or from an AXES statement, plain axes will be drawn in the style used by DSURFACE; these can be labelled using the YTITLE and XTITLE parameters of AXES.

Example 6.3.3

```
 37   "
-38           ... continuation of Example 4.11.4
-39   "
 40   PRINT Sales; FIELDWIDTH=8; DECIMALS=0
```

	Sales					
Year	1979	1980	1981	1982	1983	1984
Town						
Aberdeen	608	635	672	692	685	723
Birmingham	618	601	784	720	863	921
Cardiff	757	743	785	816	783	737
Dundee	343	391	358	366	418	470
Edinburgh	714	751	710	763	788	830
Liverpool	816	859	820	938	1007	1158
Manchester	662	632	758	721	893	837
Sheffield	531	569	615	624	607	593
Swansea	416	461	478	462	497	520

```
 41   D3HISTOGRAM [AZIMUTH=150; ELEVATION=40; LOWERCUTOFF=0.0] Sales
```

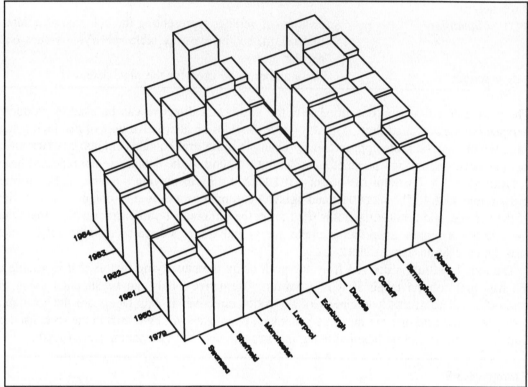

Figure 6.3.3

6.4   Graphical Input

6.4.1   The DREAD directive

DREAD reads the locations of points from an interactive graphical device.

Options

PRINT = *strings*	What to print (data, summary); default summ
CHANNEL = *scalar*	Number of the graphics device from which to read; default * takes the current graphics device
WINDOW = *scalar*	Window from which to read; default 1
CURSORTYPE = *scalar*	Type of cursor; default 1
SETNVALUES = *string*	Whether to set number of values of structures from the number of values read (yes, no); default no causes the number of values to be set only for structures whose lengths are not defined already
ENDACTION = *string*	Action to be taken after completing the plot (continue, pause); default * uses the setting from

the last DEVICE statement

Parameters

Y = *variates*	Variate to receive the y-values that have been read
X = *variates*	Variate to receive the x-values that have been read
YGIVEN = *variates*	Y-coordinates of points that may be located on the graph
XGIVEN = *variates*	X-coordinates of points that may be located
SAVESET = *variates*	Unit numbers of the located points
PEN = *scalars*	Pen number to use to echo points; default 0
YSAVE = *variates*	Variate to receive the y-coordinates of the located points
XSAVE = *variates*	Variate to receive the x-coordinates of the located points

The DREAD directive allows you to input information about the positions of points on interactive graphical terminals. The exact details of how this directive operates will vary slightly from one system to another, so this section attempts to outline the basic principles involved. If you encounter any difficulties using DREAD you should refer to the *Users' Note* supplied with your version of Genstat.

When you type DREAD, a cursor should appear on the graphics screen. This can be moved to the chosen position by using the cursor keys or a mouse; the coordinates of this point can then be read by pressing a key or mouse button (normally the left hand mouse button). The cursor can then be moved to another position to read the next point. You can use graphical input within any window that contains a graph or contour plot, but you cannot input data from an "empty" window or one containing other forms of graphical output. In addition you can identify particular points from those plotted on an existing graph and you can mark the points that you have read.

The CHANNEL and WINDOW options are used to specify the device and the window from which the information is to be read; the default is to read from window 1 of the current device. The values that are read are converted to the scale of the data that was previously plotted in that window, and are then stored in the pair of variates specified by the Y and X parameters.

Any number of points may be read in one DREAD statement. If the required number of points is known in advance, the Y and X variates can be declared with the appropriate length, and the input will terminate automatically when sufficient points have been read. Alternatively, if the lengths of the variates have not been defined in advance, points are read until you terminate the input, and the variates are defined accordingly. This action can be requested explicitly by setting option SETNVALUES=yes; the existing variate lengths are then ignored and points are read until the input is terminated. Graphical input can usually be terminated in two ways, either by pressing a mouse button (usually the right button) or a key that has been specifically defined for this purpose, or by attempting to read a point lying outside the current axes. In case of difficulty you should refer to the *Users' Note* which will explain how to terminate the input on specific devices. The final point read as a terminator is not included in the Y and X variates. If you try to terminate input prematurely when a set number of values is to be read, the corresponding Y and X values are set to missing values.

The PRINT option of DREAD is similar to the PRINT option of READ. Putting PRINT=data lists the y- and x-values of the points that have been read, while PRINT=summary generates the usual summary of mean, minimum, maximum, and number of values.

Several types of cursor may be available; again this will depend on the graphics device. The cursor is selected by setting the CURSORTYPE option to an integer between 1 and 10. Normally cursors 1, 2, and 3 are different graphics cursors; for example, large cross-hair, arrow, and small cross. Cursors 4 and 5 may be set up to provide special functions called *rubber-band* and *rubber-rectangle*.

A rubber-band cursor works by reading one point in the normal way (as if CURSORTYPE was set to 1). This defines an anchoring point for a line whose other end is attached to the cursor. As you move the cursor, the line will change direction and contract or expand, but always linking the fixed point to the current cursor position: hence the term "rubber-band". When you read the next point this will become the anchor point for a new rubber-band segment which you use whilst locating a third point, and so on until the required number of points have been read.

The rubber-rectangle works in a similar way, with the first point being read with a normal cursor. This defines the fixed point and the cursor is now regarded as being attached to the diagonally opposite corner of a rectangle which will contract and expand as you move the cursor around the screen. Reading the second point terminates the input; with a rubber-rectangle cursor Genstat will always read exactly two values, ignoring the SETNVALUES option and any predefined length of Y and X.

The rubber-band and rubber-rectangle types of cursors may not be available on all devices, in which case setting CURSORTYPE to 4 or 5 will use one of the simpler cursors. However, setting CURSORTYPE to 5 will always read just two points, regarded as being diagonally opposite corners of a rectangle, whether or not the rubber-rectangle appears on the screen.

Some devices may have more than one method of manipulating the graphics cursor, for example by use of a joystick or mouse. In this case, cursor-types 1 to 5 will be set up as described above for the joystick, say, and types 6 to 10 will be the same types of cursor but controlled by the mouse. Usually, however, there will be only one method of control, in which case cursor-types 6 to 10 will be the same as types 1 to 5.

The PEN parameter of DREAD can be used to specify a pen which will be used to plot each point as its position is read. The various attributes of this pen determine how the points are plotted; these can be modified, in the usual way, using the PEN directive (6.5.5). If the pen method is set to line, monotonic, open, or closed, then straight line segments will be drawn between the points; otherwise just the points themselves are plotted. If the points are to be joined by lines and a rubber-rectangle cursor is being used, the rectangle will be drawn rather than the diagonal line. If labels are set for the pen, they will be used in turn to mark the points as they are read; if the number of points exceeds the number of labels the labels will be recycled.

The YGIVEN and XGIVEN parameters allow you to identify points that have been plotted in an existing graph. They should be set to the y- and x-variates that were plotted on the graph. Each point that is read by DREAD is then located within this pair of variates, by finding the original point that is physically nearest to the new point, ignoring any differences in the scales of the y- and x-values. The unit number of the located points can be saved in a variate specified by the SAVESET parameter, and their coordinates in a pair of variates supplied by the YSAVE and XSAVE parameters. The length of the variates is defined in the same way as for the Y and X variates. The variates saved by YSAVE and XSAVE contain the actual coordinates of the plotted points that were selected by DREAD; whereas the Y and X variates contain the coordinates of the exact position of the cursor. The SAVESET variate indicates the unit numbers of the selected points. This information could be used, for example, in CALCULATE or RESTRICT statements to refer to the units that have been identified on the graph. For example,

```
DREAD U; V; YGIVEN=Y; XGIVEN=X; SAVESET=SS
RESTRICT Y,X; .NOT.EXPAND(SS; NVALUES(X))
```

would have the effect of excluding the points identified by DREAD; in this example the exact cursor locations recorded in U and V are not of interest.

Example 6.4.1 demonstrates how DREAD can be used to mark some points on a graph. It also shows the output stored in the structures specified by the various parameters. Figure 6.4.1 illustrates the appearance of the screen during the process, and after it has terminated.

Example 6.4.1

```
> VARIATE Ydata; VALUES=!(171,149,159,108,101,166,122)
> VARIATE Xdata; VALUES=!(6.7,5.6,7.4,7.4,5.6,4.4,4.7)
> TEXT Point; VALUES=!T(A,B,C)
> DGRAPH Ydata; Xdata; PEN=2
> PEN 1; LABELS=Point; SYMBOL=3
> DREAD Yread; Xread; YGIVEN=Ydata; XGIVEN=Xdata; SAVESET=Position; \
>   YSAVE=Yselect; XSAVE=Xselect; PEN=1
```

Identifier	Minimum	Mean	Maximum	Values	Missing
Yread	119.0	130.8	153.3	3	0
Xread	4.813	5.877	6.999	3	0

```
> PRINT Label,Yread,Xread,Position,Yselect,Xselect
```

Label	Yread	Xread	Position	Yselect	Xselect
A	120.2	4.813	7	122.0	4.700
B	153.3	5.819	2	149.0	5.600
C	119.0	6.999	4	108.0	7.400

When the PEN parameter is being used to mark the points that are read, you may want to pause at the end of the read so that you can inspect the modified graph. This is controlled by the ENDACTION parameter, as explained in 6.2.

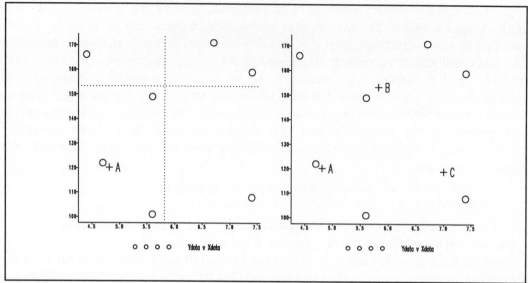

Figure 6.4.1

6.5   The environment for high-resolution graphics

The directives described in Sections 6.2 and 6.3 can display data in various ways. Implicit in all the discussion is the idea of a *graphics environment*, in which the displays are generated. This consists of a choice of graphics devices and a large number of parameters which control the appearance of the output. When you start Genstat an initial environment is created which contains default settings that are designed to be appropriate for the more common types of plot. This section describes the directives that allow you to modify the graphical environment in order to obtain more control over the appearance of your output. The description of each directive in Sections 6.2 and 6.3 indicates how the output will appear by default, and how it is affected by changes to the environment. The examples were chosen to illustrate the default display and some of the ways in which it can be modified by directives such as PEN and AXES.

When you produce a high-resolution plot, the pictures are drawn on a *graphical device*, in a *graphical window*, using a *graphical pen*. Output can be produced on only one type of device at any time; however you can switch between different devices during a Genstat session so that, for example, you can experiment with various displays on the screen before sending some output to a file for printing as hard-copy. The device is selected using the DEVICE directive. The associated directives DDISPLAY and COLOUR may be used to re-display an existing graph and alter the *colour table* of the device.

A graphics window is an area of the screen (or page on a plotter) that is used for plotting output. Many such windows can be used within a sequence of statements, so that several graphs may be plotted on a single screen. The position and size of the windows is defined using the FRAME directive. Associated with each window are the attributes of its axes. These control how axes are drawn by directives such as DGRAPH, DHISTOGRAM, and DCONTOUR. The AXES directive can be used to control the various aspects of the axes associated with any specific window.

Each part of the display is drawn using pens, each of which has attributes such as colour, font, line style, and symbol type. In addition, the pen may be used to control how data is plotted, for example by requesting a straight line or a curve. The PEN directive is used to set attributes of the different pens to be used in each graph.

The Genstat HELP system (1.3) can be used to display information about the graphics environment. If you type

```
HELP environment,pictures,current
```

(or simply HELP e,p,c) you can discover the current settings of the frame, axis, and pen attributes. You can also obtain information about the other available settings of environment directives by typing

```
HELP environment,pictures,possible
```

The directives that define the environment change only the parameters that are mentioned explicitly; unspecified parameters retain their previous values (which may be the initial defaults). When you start a new job (5.1), the environment is reset to the initial default values. On the other hand, when you use RESUME (3.6.2) to re-start an earlier session, the graphics environment will be loaded from the resume file. However, this does not affect the choice of output device (and associated file) which is preserved in both situations.

As the effects of these directives are additive, you need to keep aware of the current settings, and avoid unwanted side-effects which may occur, for example, if you use a pen that has earlier been modified in a way that is incompatible with its current use. This should not cause problems under ordinary circumstances. However, if you are using graphics in a general program or procedure there are various things you can do to make the graphics self-contained, and avoid side-effects. Each directive that modifies the environment includes a SAVE parameter that enables you to save the current settings of its particular aspect of the environment (frame, axes, or pen) after making any modifications specified in the current statement. This enables you to check the current settings and reset particular attributes to their original values after a plot has been produced. The DKEEP directive can be used to obtain additional general information about the graphics devices and environment. The GET and SET directives (13.1) allow the entire graphics environment to be stored in a pointer and later restored to its original state. For example, in a graphics procedure you might have the following statements:

```
GET [SPECIAL=Special]
FRAME 1; YLOWER=0.3; YUPPER=0.6; XLOWER=0.3; XUPPER=0.6
AXES 1; YLOWER=0; YUPPER=100; YTITLE='Percentages'
PEN 1...4; METHOD=line; LINESTYLE=1...4; SYMBOL=0; COLOUR=1,2,1,2
DGRAPH Percent[1...4]; X; PEN=1...4
SET [DSAVE=Special['dsave']]
ENDPROCEDURE
```

This can also be done automatically using the RESTORE option of PROCEDURE (5.3.2).

6.5.1   The DEVICE directive

DEVICE switches between (high-resolution) graphics devices.

No options

Parameters

NUMBER = *scalar*   Device number
ENDACTION = *string*   Action to be taken after completing each plot
                                         (continue, pause)

High-resolution graphics can be generated principally in two forms by Genstat: either on a screen that can operate in graphics mode or by sending output to a file. The screen-based operation is for use in interactive sessions, whereas file output is designed for later use outside Genstat: either to produce hard-copy on a plotter or laser-printer, or to re-display graphics on the screen, if appropriate software is available. Usually there is a choice of various kinds of screen type or file format. Each type of output, whether screen or file, is referred to as a *device*; thus, the first step in producing graphical output is selecting a device within Genstat that is appropriate for the hardware that you have available. Genstat has built-in interfaces to a number of different graphics devices and the *Users' Note* will contain a list of those included for any particular version. This information is also available within the HELP directive, by typing the statement

        HELP environment,pictures,possible

(or HELP e,p,p for short).

   Source code for additional interfaces is provided to allow Genstat to be linked to standard graphical subroutine libraries, such as GKS, to provide access to a wider variety of devices where these are available. Further details of use of these interfaces to extend the graphical facilities are contained in the *Installers' Note* and *Graphical Extension Guide*.

   The output device is selected by the DEVICE statement. The devices are numbered, so that for example

        DEVICE 4

will select the fourth available device. The device numbers to use for the different kinds of output will depend on the version of Genstat that you are using, and how it has been configured. The *Users' Note* contains details of the available devices; in addition the Genstat help system can be used to display this list as described above.

   If you have selected a file-based device you will also have to open a file to receive the output, using the OPEN directive (3.3.1). This can be done before or after selecting the device, so long as the file has been opened before any output is generated. You can close the file when the graphics are complete; if you want to store separate items of graphical output in individual files you can use a sequence of OPEN and CLOSE statements. When opening or closing files for graphical output the CHANNEL parameter of the OPEN and CLOSE statements should be set to the device number specified by the DEVICE statement. For example:

```
OPEN 'PLOT.HPGL'; CHANNEL=4; FILETYPE=graphics
DEVICE 4
DGRAPH Y; X
CLOSE 4; FILETYPE=graphics
```

The default device, selected automatically when you start Genstat, is device 1: sometimes you may be able to specify an alternative device number and associated output file on the command line used to start Genstat (the *Users' Note* should explain if this is possible).

You may get strange results if you try to generate graphics on a screen that is not designed for displaying graphics, or if you specify the wrong device type, as Genstat is not always able to detect the type of device or screen.

This is intended to be a general description of the facilities available and so it is inevitable that some details may vary according to the capabilities of the equipment you are using. For example, when using a PC, the screen will switch from text mode to graphics mode in order to display a graph, whereas in a windowed environment, such as a Unix workstation running X-windows, the graphical output will appear in a window that is independent of the one that contains the Genstat input and text output. There may also be differences in the way that the keyboard or special keys are defined to control the graphics. To obtain the best results from some terminals particular modifications may be required to their settings. For this reason some *device-specific* information (6.5.7) is contained in the *Users' Note*.

Other than this, there should be little difference in the use of Genstat graphics on different machines, as all the plotting symbols, brush styles, and character output are *software-generated* by default, using built-in graphics definitions and font files that are supplied with Genstat. The aspects of graphical output that may depend on particular capabilities of the graphics device are identified in the later parts of this section; for example, different defaults may apply to colour and monochrome devices. It may sometimes be advantageous to use particular features of the hardware or additional graphics software (like GKS); for example, other fonts may be available. These device-specific features are usually selected by negative parameter settings, (for example, SYMBOL=-3) and are described in 6.5.7. Naturally, selection of device-specific attributes may lead to some differences in appearance of the output on different devices.

The ENDACTION parameter, with settings continue and pause, controls the action taken by default at the end of each plot. When using a graphics terminal interactively it may be convenient to pause at the end of a plot to examine the screen. When you are ready to continue, pressing carriage-return or some equivalent key will switch the terminal back to text mode and the Genstat prompt will appear. The precise details will vary according to the device and underlying graphical package; the *Users' Note* should provide the full information. For some interactive devices, for example workstations with separat' graphics windows, it may not be necessary to pause. Each device is initialized to either pause or continue when you start Genstat, according to the particular implementation. If you are running in batch mode the default will always be to continue.

You can repeat the DEVICE statement and set ENDACTION to pause or continue at any time that you wish to change the default action. Alternatively, each graphical directive has an ENDACTION option that controls the device at the end of that directive, without altering the general default setting. For example, if you wish to build up a complex display using several DGRAPH statements with option SCREEN=keep, you could set ENDACTION=continue in the DEVICE statement, then put ENDACTION=pause in the final DGRAPH statement.

6.5.2 Re-displaying the Graphics screen

DDISPLAY redraws the current graphical display.

Option

DEVICE = *scalar*	Device on which to redraw the display (on some systems it may only be possible to redisplay the picture on an interactive graphics device)
ENDACTION = *string*	Action to be taken after completing the plot (continue, pause); default * uses the setting from the last DEVICE statement

No parameters

This directive is provided to allow additional control of some interactive devices. In some of these, such as PC's, the screen can operate in either text mode or graphics mode. Genstat will automatically switch the screen into the appropriate mode when starting or finishing a graph. Having returned to text mode after examining a graph you may later wish to have another look at the graph that was plotted. DDISPLAY will switch the screen back to graphics mode, thus re-displaying the graph. The ENDACTION option controls what happens after re-displaying the graph; normally with this type of device you would want to pause. The default action for DDISPLAY is the setting specified by the most recent DEVICE statement.

This directive has no effect when output is directed to a graphics file. For devices that do not operate in this dual-mode fashion, for example a graphics window under X-windows, DDISPLAY has no effect on the graphical display itself. It will however generate a pause if ENDACTION is set to request one.

Note that DDISPLAY does not actually re-plot the graphical output; it merely switches the screen into graphics mode, and assumes that your system has preserved the graphics image.

6.5.3 The FRAME directive

FRAME defines the positions of windows within the frame of a high-resolution graph. The positions are defined in normalized device coordinates ([0,1]×[0,1]).

No options

Parameters

WINDOW = *scalars*	Window numbers

YLOWER = *scalars*	Lower y device coordinate for each window
YUPPER = *scalars*	Upper y device coordinate for each window
XLOWER = *scalars*	Lower x device coordinate for each window
XUPPER = *scalars*	Upper x device coordinate for each window
YMLOWER = *scalars*	Size of bottom margin (for x-axis labels)
YMUPPER = *scalars*	Size of upper margin (for overall title)
XMLOWER = *scalars*	Size of left-hand margin (for y-axis labels)
XMUPPER = *scalars*	Size of right-hand margin
BACKGROUND = *scalars*	Specifies the colour to be used for the background in each window (where allowed by the graphics device)
SAVE = *pointers*	Saves details of the current settings for the window concerned

You can define up to 32 different windows in which to plot graphics. Each window is a rectangular area of the screen which is defined using *normalized device coordinates* (NDC). These have a range from 0.0 to 1.0 in both Y and X directions, thus defining a square which represents the available plotting area on any device. The mapping from NDC to physical coordinates on the current output device is performed internally, so the window definitions are independent of the choice of device. The actual size of a particular window on different devices will vary according to their relative physical sizes. The NDC system used for window definition is also completely independent of the values of the data that are to be plotted.

To define a window, the upper and lower bounds are required in both y- and x-directions; thus defining both the position and the size of the window. For example

```
FRAME WINDOW=1; YLOWER=0.25; YUPPER=0.75; XLOWER=0; XUPPER=0.5
```

defines window 1 to be a square of size 0.5, whose bottom left corner is at the point (0.0,0.25) and whose top right corner is (0.5,0.75). This does not define the exact size of a graph plotted in this window, as margins may be required for the annotation and titles (see below).

If you do not specify all four values in the FRAME statement, existing values from the previous definition are used. A check is then made on the validity of the window bounds. The settings of YLOWER and XLOWER must be strictly less than those of YUPPER and XUPPER respectively; also, none of the bounds can be outside the permitted range (usually [0.0,1.0], but see 6.5.7). You cannot use * to reset a bound to the default value; if you try to do so, Genstat will produce an error diagnostic.

All the windows have a default size set up when you start Genstat. Window 1 is the default window used for plots by DGRAPH, DCONTOUR, and so on, and is set up to be a square of size 0.75. The default key window is window 2, which is a rectangle of height 0.25 and width 0.75 located immediately below window 1. Windows 3 and 4 are the unit square [0,1]x[0,1] and windows 5, 6, 7, and 8 are the top-left, top-right, bottom-left, and bottom-right quarters respectively of the unit square. The remaining windows, from 9 up to 32, also default to the unit square. You can use FRAME to modify the size or position of any of these windows.

Usually, a margin is provided around each plot so that there is room for the axes to be drawn, along with labelling and titles as specified by AXES. By default, the margin size is designed to allow sufficient room for annotation to be added using the standard character size, as described in the SIZE parameter of PEN (6.5.5). If you use the AXES directive to control the plotting of axes explicitly you may wish to alter the size of the margins, either to increase the space used for the axes or, alternatively, to maximize the space available for the graph itself. For example, if you alter the size of the labelling, by explicitly defining the relevant axis pens, more space may be required for the axes; otherwise the labels may be clipped at the window bounds. The parameters YMLOWER, YMUPPER, XMLOWER, and XMUPPER can be used to set the space (in NDC) for the bottom, top, left-hand, and right-hand margins respectively, and have initial default settings of 0.10, 0.07, 0.12, and 0.05.

On some output devices, mainly colour graphic screens, the background colour for the window may be modified by setting the BACKGROUND parameter; see 6.5.7.

The current FRAME settings for a particular window can be saved in a pointer supplied by the SAVE parameter. The elements of the pointer are labelled to identify the components, as shown in Example 6.5.3.

Example 6.5.3

```
1   FRAME 1; YLOWER=0.0; XUPPER=1.0; SAVE=Win1
2   PRINT [ORIENTATION=across; RLWIDTH=18] Win1[]


    Win1['ylower']          0.00
    Win1['yupper']          1.00
    Win1['xlower']          0.00
    Win1['xupper']          1.00
    Win1['ymlower']         0.10
    Win1['ymupper']         0.07
    Win1['xmlower']         0.12
    Win1['xmupper']         0.05
Win1['background']             0
```

6.5.4   The **AXES** directive

AXES   Defines the axes in each window for high-resolution graphics.

Option

EQUAL = *string*	Whether/how to make axes equal (no, scale, lower, upper); default no

Parameters

WINDOW = *scalars*	Numbers of the windows
YTITLE = *texts*	Title for the y-axis in each window
XTITLE = *texts*	Title for the x-axis in each window
YLOWER = *scalars*	Lower bound for y-axis
YUPPER = *scalars*	Upper bound for y-axis
XLOWER = *scalars*	Lower bound for x-axis

XUPPER = *scalars*	Upper bound for x-axis
YMARKS = *scalars* or *variates*	Distance between each tick mark on y-axis (scalar) or positions of the marks (variate)
XMARKS = *scalars* or *variates*	Distance between each tick mark on x-axis (scalar) or positions of the marks (variate)
YMPOSITION = *strings*	Position of the tick marks across the y-axis (left, right, centre)
XMPOSITION = *strings*	Position of the tick marks across the x-axis (above, below, centre)
YLABELS = *texts*	Labels at each mark on y-axis
XLABELS = *texts*	Labels at each mark on x-axis
YLPOSITION = *strings*	Position of the labels for the y-axis (left, right)
XLPOSITION = *strings*	Position of the labels for the x-axis (above, below)
YORIGIN = *scalars*	Position on y-axis at which x-axis is drawn
XORIGIN = *scalars*	Position on x-axis at which y-axis is drawn
STYLE = *strings*	Style of axes (none, x, y, x, box, grid)
PENTITLE = *scalar*	Pen to use for the title
PENAXES = *scalar*	Pen to use for the axes and their labelling
PENGRID = *scalar*	Pen to use for the grid
SAVE = *pointers*	Saves details of the current settings for the axes concerned

Associated with each window is a definition of axes. This specifies how the axes are to be drawn when graphical output is produced in that window. The default definition for each set of axes requires some of the features to be determined from the data, as described below. Others have fixed defaults that are independent of the data. The AXES directive can be used to override the default action and specify explicitly how particular parts of the axes are drawn. All parameters of AXES are relevant when using DGRAPH, but for other directives only some of the parameters are used. The documentation of the other graphics directives (for example, DCONTOUR and DHISTOGRAM) in earlier sections indicates which aspects of the axes can be controlled by the user, and therefore which parameters of AXES are relevant to those directives.

The WINDOW parameter specifies the window whose axes definition is to be altered. Only those aspects specified by subsequent parameter lists are modified; any parameters that are not set will retain their current settings. WINDOW can be set to a list of window numbers, in which case the other parameter lists are cycled in the usual way.

The YLOWER and YUPPER parameters specify the lower and upper bounds for the y-axis. By default, Genstat derives suitable axis bounds from the data, as described for the appropriate directive. You can set the lower bound to a value greater than the upper bound, to obtain an inverted data scale, but the bounds must not be equal. The XLOWER and XUPPER parameters set bounds for the x-axis in a similar way. The values specified with these parameters are on the scale of the data values that are plotted, and are independent of the normalized device coordinates used to define the window size in FRAME (6.5.3). The EQUAL option can be used to ensure that equal upper or lower bounds are used for the y- and x-axes. For example, if

EQUAL=lower, lower bounds for both axes will be set to the lower of the values determined automatically from the data. The bounds obtained when using the EQUAL option may be constrained by settings of other parameters: for example, if YUPPER is set and EQUAL=upper, the upper bounds of both axes are set to the value specified by YUPPER; but if XUPPER is also set, EQUAL will be ignored. You can set EQUAL=lower,upper to constrain both upper and lower bounds, and EQUAL=scale can be used to ensure physical distance is equal on both axes, for example the y-axis could range from 0 to 100 and the x-axis from 100 to 200.

The YORIGIN parameter determines the value on the y-axis through which the x-axis is drawn. If its value is outside the y-axis bounds, the upper or lower bound is adjusted so that the axis will extend up to the specified origin. This applies whether you have set the bounds explicitly or have left Genstat to calculate them from the data. The XORIGIN parameter sets the origin for the x-axis in a similar way. By default, the lower bounds of each axis are used, so that the axes are drawn on the bottom and left-hand sides of the plot. In line 3 of Example 6.5.4a the origin is set to the position (0,0) so that the axes cross in the centre of the window.

Titles can be added to the axes using the YTITLE and XTITLE parameters. In each case, the title is limited to a single line of characters.

Each axis is marked with a scale, determined automatically so that tick marks are evenly spaced and positioned to give "round" numbers for the scale values. For each axis, you can specify either the increment between tick marks or their actual positions. You can also specify labels to use for scale markings instead of their numerical values.

To specify the increment on the y-axis, the YMARKS parameter should be set to a scalar. For example, YMARKS=1.5 with bounds 10 and 2 causes tick marks to appear at 2, 3.5, 5, 6.5, 8, and 9.5. The interval must be a positive number, irrespective of the values of the bounds. Alternatively, you can set YMARKS to a variate (with more than one value) to specify the actual positions of the tick marks on the y-axis. Any values that lie outside the axis bounds are ignored. The scale values printed next to the tick marks use a format that is determined automatically from the values, but if you have set YMARKS to a variate it will use the number of decimals specified in the variate declaration. When you have set YMARKS, you can also use the YLABELS parameter to specify a set of labels to mark the axis scale. For example,

```
TEXT [VALUES=Mon,Tues,Wed,Thur,Fri,Sat,Sun] Day
VARIATE [VALUES=1...31] Month
AXES 1; YMARKS=Month; YLABELS=Day
```

The strings within the text are cycled if necessary; hence, the number of strings can be less than the number of tick marks.

The tick marks can be drawn to the left or to the right of the axis, or can be centred (that is, across the axis). By default, the tick marks are drawn towards the "outside" of the plot; that is, to the left if the y-axis is to the left of the centre of the plot, or to the right if the y-axis is drawn to the right of centre. The aim is to position the tick marks away from the main part of the plot, so that they interfere with the plotted points as little as possible. You can control the positioning of the tick marks by setting the YMPOSITION parameter to either left, right, or centre. A similar rule governs the default positioning of the scale markings or labels, but you can again control this by setting the YLPOSITION parameter to either left or right. Setting YMARKS=* will return to the default positioning of the tick marks; YLABELS=* will switch off any labels previously specified; and YMPOSITION=* and YLPOSITION=* will switch off tick marks or labels altogether.

Annotation of the x-axis can be controlled in a similar way using the XMARKS, XLABELS, XMPOSITION, and XLPOSITION parameters, except that the settings left and right are replaced by above and below.

The STYLE parameter controls the type of axes that are drawn. By default STYLE=xy, so both y- and x-axes are plotted. Alternative settings allow the axes to be completed by drawing a box around the graph, with an overlaid grid if required. The settings STYLE=x and STYLE=y can be used if only one axis is required. Finally, STYLE=none inhibits the plotting of axes completely, although some other parameters, such as YLOWER, may still have an effect on the plotted data. Example 6.5.4a shows the different settings of STYLE, while Example 6.2b illustrated the use of STYLE=none when DGRAPH was used to add an additional plot to a histogram that already had axes, and thus did not require a second set to be plotted.

There are three parameters that control the pens to be used when drawing the axes. These are PENTITLE, PENAXES, and PENGRID, specifying the pen for the title, the axes and annotation, and the grid, respectively. The initial default is to use pens 30, 31, and 32 in every window. These pens are in turn set up to use colour 1, line style 1, thickness 1, size 1, and font 1, as described in 6.5.5. You can thus control which pens are used for drawing the axes in each window, and the attributes of those pens. For example, if no AXES statement has yet been given,

        PEN 32; LINESTYLE=4; COLOUR=2

will request that the grids in every window should be drawn in line style 4 and colour 2; while

        PEN 29; LINESTYLE=3; COLOUR=4
        AXES 1; PENAXES=29

will change the appearance of just the axes in window 1, as pen 29 is not used for the other windows. Control of the grid pen is particularly useful as a combination of colour and line style can be chosen to ensure that the grid does not obscure the plotted points, as in lines 4-5 of Example 6.5.4a.

Example 6.5.4a

```
 1   AXES 5,6,7,8; YLOWER=-1; YUPPER=1; XLOWER=-1; XUPPER=1; \
 2     STYLE=xy,x,box,grid
 3   AXES 7; YORIGIN=0; XORIGIN=0
 4   AXES 8; PENGRID=29
 5   PEN 29; LINESTYLE=3
 6   PEN 5,6,7,8; SYMBOL=0; LABEL='STYLE=xy','STYLE=x','STYLE=box',\
 7     'STYLE=grid'
 8   DGRAPH [WINDOW=5; KEYWINDOW=0] 0.3; 0.3; PEN=5
 9   DGRAPH [WINDOW=6; KEYWINDOW=0; SCREEN=keep] 0.3; 0.3; PEN=6
10   DGRAPH [WINDOW=7; KEYWINDOW=0; SCREEN=keep] 0.3; 0.3; PEN=7
11   DGRAPH [WINDOW=8; KEYWINDOW=0; SCREEN=keep] 0.3; 0.3; PEN=8
```

You should of course be careful of side-effects when modifying these pens or changing the pen numbers. For example, pen 29 may also have been modified for use in a DGRAPH

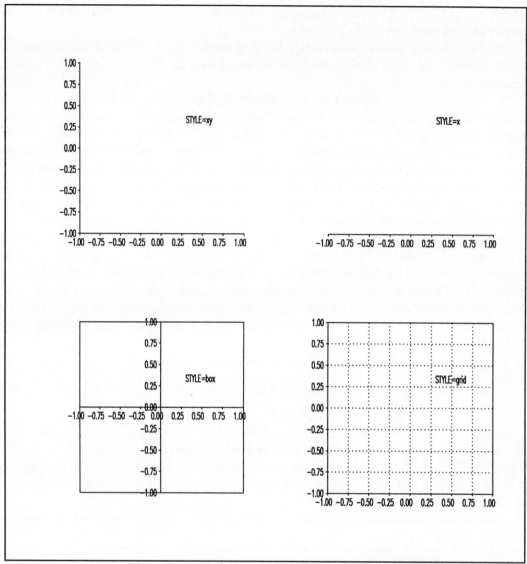

Figure 6.5.4

statement and other attributes may have been set that are not wanted when drawing the axes.

Axis annotation is plotted in the margins specified by the FRAME directive. You may wish to reduce the size of these margins if you have defined axes that use less space, for example by keeping within the area of the graph itself, or by omitting titles or labels. Space can thus be regained and used for plotting data. However, if the margins are too small the axis annotation may be "clipped" at the boundaries of the margins; if this happens, you can use FRAME to increase the margin size. The margins are used by DGRAPH, DHISTOGRAM, and DCONTOUR, but they are ignored by other directives.

The current settings of the axes for a particular window can be saved in a pointer supplied by the SAVE parameter. The elements of the pointer are labelled to identify the components, as shown in Example 6.5.5b.

Example 6.5.4b

```
12   AXES 7; SAVE=Axes7
13   PRINT [SQUASH=yes; RLWIDTH=19] Axes7[]
     Axes7['equal']              *
    Axes7['ytitle']
    Axes7['xtitle']
    Axes7['ylower']          -1.00
    Axes7['yupper']           1.00
    Axes7['xlower']          -1.00
    Axes7['xupper']           1.00
    Axes7['ymarks']             *
    Axes7['xmarks']             *
 Axes7['ymposition']           *
 Axes7['xmposition']           *
    Axes7['ylabels']            *
    Axes7['xlabels']            *
 Axes7['ylposition']           *
 Axes7['xlposition']           *
    Axes7['yorigin']         0.00
    Axes7['xorigin']         0.00
     Axes7['style']           box
   Axes7['pentitle']           30
   Axes7['penaxes']            31
   Axes7['pengrid']            32
```

This facility is of most use within procedures, where it may be necessary to check or modify particular AXES settings before constructing complicated graphs. Also, the DKEEP directive (6.5.8) allows you to extract the actual bounds used when plotting; these will be the bounds determined from the data if none have been defined explicitly by AXES.

6.5.5   The PEN directive

PEN   defines the properties of "pens" for high-resolution graphics.

No options

Parameters

NUMBER = *scalars*	Numbers associated with the pens
COLOUR = *scalars*	Number of the colour used with each pen
LINESTYLE = *scalars*	Style for line used by each pen when joining points
METHOD = *strings*	Method for determining line (point, line, monotonic, closed, open, fill)
SYMBOLS = *scalars, pointers, factors* or *texts*	
	Plotting symbols – scalar for special symbols, pointer

	for user defined symbols, text or factor for character symbols
LABELS = *texts* or *factors*	Define labels that will be printed alongside the plotting symbols, provided these consist of a single character
ROTATION = *scalars* or *variates*	Rotation required for the plotting symbols (in degrees)
JOIN = *strings*	Order in which points are to be joined by each pen (ascending, given)
BRUSH = *scalars*	Number of the type of area filling used with each pen when drawing pie charts or histograms
FONT = *scalars*	Font for to be used for any text written by each pen
THICKNESS = *scalars*	Thickness with which any lines are drawn by each pen
SIZE = *scalars* or *variates*	Multiplier used in the calculation of the size in which to draw characters and symbols by each pen
SAVE = *pointers*	Saves details of the current settings for the pen concerned

Graphical displays are drawn using graphical *pens*. Certain pens are used by default, or you can specify other pens, as described in the preceding sections. The attributes of each pen, such as colour, font, and symbol-type, determine how they are used to generate output. The initial defaults for each pen are device-specific; see 6.5.7. The PEN directive can be used to change these attributes so that you can modify the resulting display. Different attributes are relevant for different types of output, for example symbols and labels are used only within DGRAPH.

The NUMBER parameter lists the numbers of the pens, in the range 1 to 32, that you wish to redefine. For many of the directives that produce pictures, pens 1 to 32 are used in turn for the different structures being plotted, while the default pens used for axis annotation are 30, 31, and 32 (see 6.5.4). Thus, if you modify the attributes of the pens you should be aware of possible side-effects, for example, when the axes are drawn.

The COLOUR parameter specifies a colour in the range 0 to 32 to be used by the pen. The initial settings of this attribute and the effect of this parameter are device-specific, as discussed in 6.5.7.

The SYMBOLS parameter determines what symbol is drawn at each point by DGRAPH. The numbers 1 to 9 correspond to graphical markers, as shown in Figure 6.5.5a.

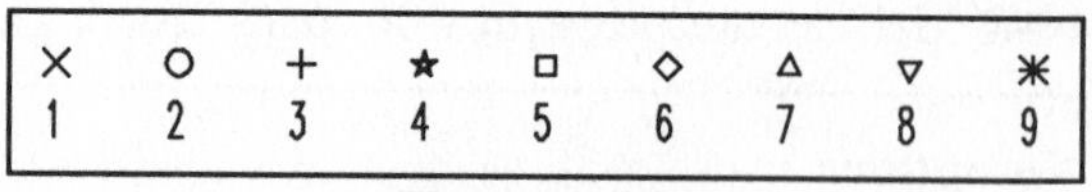

×	○	+	★	□	◇	△	▽	✳
1	2	3	4	5	6	7	8	9

Figure 6.5.5a

The initial default symbols are device specific. For colour displays, symbol 1 is used for all pens but in different colours. On monochrome displays, the pens all use colour 1 and symbols 1 to 9 are used in turn: symbol 1 for pen 1, symbol 2 for pen 2, and so on.

You can also use any standard character to mark the points (for example you could set SYMBOLS='+' to use the plus character), or you can request device-specific symbols as described in 6.5.7. If you do not want to plot symbols at the data points, for example when drawing a line through the points, you can set SYMBOLS=0. You can also set SYMBOLS to a pointer containing a pair of variates, to define your own symbol. The variates contain the coordinates of a set of points to be joined by straight line segments; these points should be within a notional square with bounds −1.0 to 1.0 in each direction. The square is centred on the data point and scaled to the same size as the standard symbols. Example 6.5.5a illustrates two simple user-defined symbols.

Example 6.5.5a

```
 1  FRAME 1...4; YLOWER=0.75; YUPPER=1.0; XLOWER=0.0,0.25,0.5,0.75; \
 2    XUPPER=0.25,0.5,0.75,1.0; YMLOWER=0.05; YMUPPER=0.01; XMLOWER=0.05; \
 3    XMUPPER=0.01
 4  PEN 29; LINESTYLE=3
 5  AXES 1,3; STYLE=grid; PENGRID=29; \
 6    YLOWER=-1.1; YUPPER=1.1; XLOWER=-1.1; XUPPER=1.1
 7  AXES 2,4; YLOWER=0.9; YUPPER=3.1; XLOWER=0.9; XUPPER=3.1
 8  VARIATE Diamond[1]; VALUES=!(-1,0,1,0,-1)
 9  & Diamond[2]; VALUES=!(0,-0.5,0,0.5,0)
10  PEN 1; METHOD=line; SYMBOL=0; JOIN=given
11  & 2; SYMBOL=Diamond
12  DGRAPH [WINDOW=1; KEYWINDOW=0] Diamond[1]; Diamond[2]; PEN=1
13  & [WINDOW=2; SCREEN=keep] 1,2,3; 1,3,2; PEN=2
14  VARIATE Arrow[1]; VALUES=!(0.0,1.0,0.75,*,1.0,0.75)
15  & Arrow[2]; VALUES=!(0.0,0.0,-0.25,*,0.0,0.25)
16  PEN 3; SYMBOL=Arrow; SIZE=!(2,2.5,3,2); ROTATION=!(0,45,90,180)
17  DGRAPH [WINDOW=3; KEYWINDOW=0; SCREEN=keep] Arrow[1]; Arrow[2]; PEN=1

* MESSAGE: There are missing values in the plot

18  & [WINDOW=4] !(1.0,2.7,2.0,1.6); !(1.4,1.8,2.2,2.6); PEN=3
```

The definition of the arrow symbol illustrates how missing values can be included in the definition so that separate pen strokes are used draw line segments. The plot produced by this example is shown in Figure 6.5.5b.

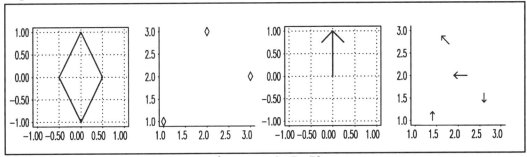

Figure 6.5.5b

You can mark different points with different symbols (for example to indicate groupings in the data) by setting the PEN parameter of DGRAPH to a variate or factor specifying a pen with the appropriate symbol for each point.

You can also label each point with a string or number, as illustrated in Example 6.5.5b, which produced Figure 6.5.5c. The LABELS parameter is set to a text structure specifying the strings to be plotted at each point. You can specify a single string to be plotted at every point, otherwise the text must have the same number of values as the Y and X variates that are being plotted. LABELS can also be specified as a factor; the factor labels are then used, if available, otherwise the levels. This provides another means of representing grouped data.

Note that the graphical symbols are drawn so that they are centred at the specified position. If LABELS are specified they are aligned as shown in Figure 6.5.5c, unless you have set SYMBOLS=0 to suppress the markers, in which case the labels are drawn so that the bottom left point of the first character is at the specified (x,y) position. Figure 6.5.5c illustrates the alignment of various SYMBOLS and LABELS settings. For compatibility with previous releases of

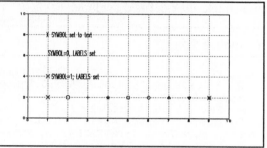

Figure 6.5.5c

Genstat you can also set SYMBOLS to a factor or text, which has the same effect as setting LABELS with SYMBOLS=0.

Example 6.5.5b

```
 1   FRAME 3; YLOWER=0.5; YUPPER=1; XLOWER=0; XUPPER=1
 2   AXES 3; STYLE=grid; PENGRID=29; YLOWER=0; YUPPER=10; \
 3      XLOWER=0; XUPPER=10; XMARK=1; YMARK=2
 4   PEN 29; LINESTYLE=3
 5   DGRAPH [WINDOW=3; KEYWINDOW=0] 9(2); 1...9
 6   PEN 1; LABELS='SYMBOL=1; LABELS set'
 7   DGRAPH [WINDOW=3; KEYWINDOW=0; SCREEN=keep] 4; 1
 8   PEN 2; SYMBOL=0; LABELS='SYMBOL=0, LABELS set'
 9   DGRAPH [WINDOW=3; KEYWINDOW=0; SCREEN=keep] 6; 1; PEN=2
10   PEN 3; SYMBOL='X'; LABELS='SYMBOL set to text'
11   DGRAPH [WINDOW=3; KEYWINDOW=0; SCREEN=keep] 8; 1; PEN=3
```

The METHOD parameter specifies the type of graph to be plotted: points, lines, or filled polygons. The initial default for every pen, METHOD=point, will result in points being plotted using the corresponding symbols, labels, colours, and fonts. Various types of line can be drawn through the plotted points; either straight lines (line) or smooth curves (monotonic, open, and closed). The monotonic setting specifies that a smooth single-valued curve is to be drawn through the data points. The name is derived from the requirement that the x-values (rather than the fitted curve) must be strictly monotonic, so that there is only one y-value for each distinct x-value. To ensure this, a copy of the data is made and sorted before the curve is fitted. This setting is recommended for plotting curves fitted to data, for example with FITCURVE. You should ensure that the points are close enough for the plotted line to be a reasonable approximation. When you know the functional form of the curve, it may be advantageous to calculate extra points. The open and closed settings specify that a smooth, possibly multi-valued, curve is to be drawn through the data points, using the method of McConalogue (1970); the resulting curve is rotationally invariant, although it is not invariant under scaling. The closed setting connects the last point to the first. McConalogue's method (open or closed) is more suited to the situation where the plotted curve is intended to represent the shape of an object. The setting METHOD=fill joins the data points by straight lines to produce one or more polygons. Each polygon is then shaded in the style specified by BRUSH (see below). The plotting method also determines how contours will be drawn, as described in 6.3.1. Also, the combination of SYMBOLS=0 and METHOD=point will produce no plotting at all (and no warning) within DGRAPH.

If the requested plotting method produces a line through the points, the LINESTYLE parameter will specify what sort of line is drawn (for example a solid, dotted, or dashed line). The type of line style is denoted by a number in the range 1 up to 10. Figure 6.5.5d illustrates some of the line styles available. Note that the exact appearance of the different line styles is

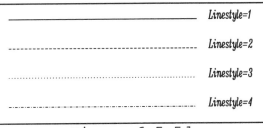

Figure 6.5.5d

device-specific (6.5.7), and there are not necessarily 10 different line styles available on a particular device, but line style 1 should always produce a solid line.

The JOIN parameter controls the order in which points are connected when lines are to be drawn or the points define a polygon to be shaded. Given requests that the data are to be plotted in the order in which they are stored, whereas ascending implies that the data are copied and sorted so that the x-values are in ascending order before plotting. This parameter is ignored when METHOD=monotonic, as this requires that the data must always be sorted.

The BRUSH parameter controls how areas are shaded when METHOD is set to fill, or when plotting histograms and pie charts. There are 16 available patterns indicated by the integers 1 to 16, as shown in Figure 6.5.5e. In general, the higher the number, the denser the hatching, and the longer such areas take to plot. The device-specific brush styles (6.5.7) are generally faster, and produce smaller output files; however results are not guaranteed to be the same on every type of device.

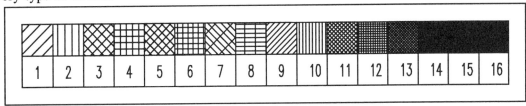

Figure 6.5.5e

The THICKNESS parameter allows you to specify an amount by which the standard thickness of plotted lines is to be multiplied. This allows you to increase the thickness of lines, perhaps to highlight some feature of a plot, as illustrated in the contour plot in Figure 6.3.1b. You can also use thickness to emphasize the axes, by redefining the appropriate pen. Note, however, that some devices do not allow the thickness to be altered.

The default size of characters and symbols is determined from the dimensions of the current window. The SIZE parameter can be used to modify the size, by specifying a value by which this default size is to be multiplied. For example when plotting a graph in a small window you may wish to increase the size of annotation in order to make it legible. SIZE can be set to a scalar, or to a variate to allow the different points to be scaled in different ways.

The ROTATION parameter controls the angle (in degrees) at which to plot text or user-defined symbols. The initial setting of zero will produce text "conventionally" orientated. You can set ROTATION to a scalar value that will apply to all points, or to a variate that allows a different angle to be used at each point. ROTATION is used in line 14 of Example 6.5.5a to plot a user-defined symbol at different angles.

The FONT parameter can be set to an integer between 1 and 25 to select different fonts for text appearing as titles, axis annotation, plotting symbols, and key information. This allows you to control the appearance of textual information and also use other character sets, for example the character 'a' will appear as α when one of the Greek fonts is selected. The available fonts are as follows.

01 Simplex Roman	14 Cyrillic
02 Duplex Roman	15 Triplex Roman
03 Complex Roman	16 Triplex Italic
04 Simplex Greek	17 Map Symbols
05 Complex Greek	18 Astronomical Symbols
06 Complex Italic	19 Music Symbols
07 Mathematical Symbols	20 Monospace Typewriter
08 Meteorological Symbols	21 Typewriter
09 Gothic English	22 Simplex
10 Simplex Script	23 Italic
11 Complex Script	24 Complex
12 Gothic Italian	25 Complex Cyrillic
13 Gothic German	

Sample output from the fonts based on the standard character set and the mappings from the standard character set to other fonts are illustrated in Figure 6.5.5f. Note that some of the symbols produced by fonts 7, 8, 17, and 19 are of non-standard size. Device-specific fonts can also be used where available (6.5.7).

You can also use in-line typesetting commands to change font or character size part-way through plotting text, when specified either as a title or label. This allows you to insert Greek characters in an equation, for example, and also to use subscripts, superscripts, and mathematical symbols. The escape character '!' is used to signal a change of font or character size and must be followed immediately by a code indicating the required action. For a simple change of font the code is just the new font number, for example '!07' will switch to the mathematical symbols font. For fonts 1 up to 9 the leading zero may be omitted, so that '!7' may be used instead, but you should be careful of ambiguities; for example '!021' will plot the character '1' in font 2, whereas '!21' will just switch to font 21. The mnemonics 'G' for (Simplex) Greek, 'M' for mathematical, and 'W' for simplex script can be used as

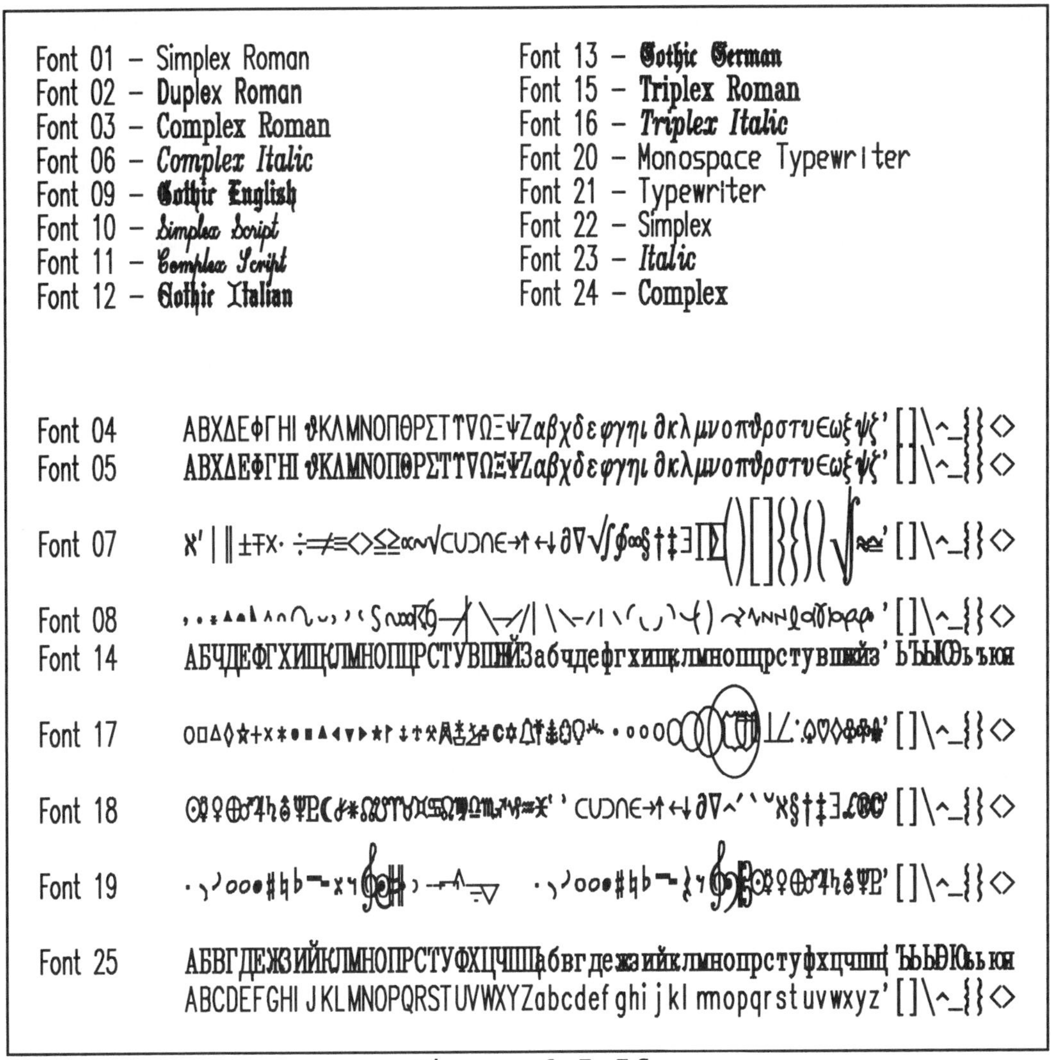

Figure 6.5.5f

well as the font numbers. The additional codes below specify other in-line commands.

A shift above the fraction line　　　　I move to index level
B shift below the fraction line　　　　L move to lower subscript level
U move up to superscript level　　　　N move to normal base line
D move down to subscript level　　　　S save current position and size on stack
E move to exponent level　　　　R restore position and size from stack

In-line commands can be specified in upper or lower-case. To print the escape character, !, it should be entered twice; for example the string `'Outlier!!'` could be used to label a point. If an invalid sequence of characters is specified the remainder of the string will not be plotted and a warning will be printed. Example 6.5.5c contains some examples of the use of in-line commands. The resulting output is shown in Figure 6.5.5g.

Example 6.5.5c

```
    1   PEN 1...20; SIZE=2; SYMBOL=0
    2   AXES 3; YLOWER=0; YUPPER=10; XLOWER=0; XUPPER=10; STYLE=none
    3   "
   -4            Pen 1 - plot the string   y = alpha + beta*exp(kx)
   -5               switch to Greek to obtain alpha, beta, then font 23 for
   -6               'e'; finally font 1 then superscript for 'kx'
   -7   "
    8   PEN 1; LABELS='y = !Ga + b!23e!1!Ukx'
    9   "
  -10            Pen 2 - switch to Maths font to get large sigma (letter n)
  -11               then font 2 for 'x', duplex Greek for mu, finally
  -12               superscript 2
  -13   "
   14   PEN 2; LABELS='!Mn!02(x - !05m)!U2'
   15   "
  -16            Pen 3 - plot x(i) minus y(j)
  -17               !23x!Di - italics, x subscript i,
  -18               !N  - return to normal baseline (from subscript)
  -19               y!Dj - y subscript j
  -20   "
   21   PEN 3; LABELS='!23x!Di!N - y!Dj'
   22   "
  -23            Pen 4 - a**p + b**q
  -24               note !1+ to avoid italic plus sign
  -25   "
   26   PEN 4; LABELS='!23a!Up!N !1+ !23b!Uq'
   27   "
  -28            Pen 5 - plot fraction 1/2
  -29               need to plot 1, then 2, then fraction line
  -30               so must return to start position twice, therefore:
  -31               !S!S - stack current position twice
  -32               !A1  - plot 1 'above'
  -33               !R!B2 - return to start then plot 2 'below'
  -34               !R!8X - return to start then plot fraction line (X in font 8)
  -35   "
   36   PEN 5; LABELS='!S!S!A1!R!B2!R!8X'
   37   "
  -38            Pen 6 - combine items 3 and 4 as a fraction
  -39               !Mo  - large open bracket
  -40               !S!S - save position twice (as for pen 5)
  -41               !23!A!S - italics, move to 'above fraction',
  -42                      save again (thus setting new baseline)
  -43               a!Up!N !1+ !23b!Uq - a**p+b**q, as before. N uses new baseline.
  -44               !R!R - restore twice, to go back to original start
  -45               !B!S - move below, save
  -46               !x!Di!N - y!Dj - x(i)-y(j)
  -47               !R!R - restore twice, to go back to original start
  -48               !8XXXXX!Mp - fraction line, large close bracket
  -49   "
   50   PEN 6; LABELS= \
   51   '!Mo!S!S!23!A!Sa!Up!N !1+ !23b!Uq!R!R!B!Sx!Di!N - y!Dj!R!R!8XXXXX!Mp'
   52   DGRAPH [WINDOW=3; KEYWINDOW=0] 6(5); 1,3,4.5,6,7.5,8; PEN=1...6
```

$$y = \alpha + \beta e^{kx} \qquad \sum(x - \mu)^2 \qquad x_i - y_j \qquad a^p + b^q \qquad \frac{1}{2} \left(\frac{a^p + b^q}{x_i - y_j} \right)$$

Figure 6.5.5g

The current settings of each pen can be saved in a pointer supplied by the SAVE parameter. The elements of the pointer are labelled to identify the components, as shown in Example 6.5.5d.

Example 6.5.5d

```
    1   PEN 8; LABELS='observation'; SYMBOL=8; JOIN=given; SAVE=Pen8
    2   PRINT [RLWIDTH=18; SQUASH=yes] Pen8[]; FIELDWIDTH=18
    Pen8['colour']                   8
Pen8['linestyle']                    *
    Pen8['method']               point
    Pen8['symbols']                  8
    Pen8['labels']         observation
    Pen8['rotation']              0.00
      Pen8['join']               given
     Pen8['brush']                   *
      Pen8['font']                   1
 Pen8['thickness']               1.00
      Pen8['size']               1.00
```

Note that the saved values for line style and brush style are missing value. This is how the initial default settings are represented; the actual values used for these attributes when plotting will depend on the output device, as described in 6.5.7, unless they are set explicitly (as with SYMBOL in this example).

6.5.6   The COLOUR directive

COLOUR defines the red, green, and blue intensities to be used for the Genstat colours; this affects only certain graphics devices (see HELP).

No options

Parameters

NUMBER = *scalars*	Numbers of the colours to be set
RED = *scalars*	Red intensity of each colour (between 0 and 1)
GREEN = *scalars*	Green intensity of each colour (between 0 and 1)

| BLUE = *scalars* | Blue intensity of each colour (between 0 and 1) |
| MATCH = *scalars* | Number of a Genstat colour to define any unset values of RED, GREEN or BLUE; default is to restore the original values of the colour |

The COLOUR directive allows you to redefine the colour map stored internally. Genstat uses the RGB colour system to define each colour (0 up to 32) in terms of its red, green, and blue components. These are specified as values in the range [0,1]. Thus black is represented by (0,0,0), white by (1,1,1), red by (1,0,0), and so on. The COLOUR directive can be used in three ways. Firstly you can define a colour in RGB terms. For example, you could put

```
COLOUR 1; RED=0.5; BLUE=0.5; GREEN=0.0
```

to define colour 1 as yellow. Points plotted in colour 1 would then appear as yellow. Alternatively, the MATCH parameter allows a colour to take its RGB values from the current settings of another colour. For example,

```
COLOUR 2; MATCH=1
```

will set colour 2 also to be yellow. Note that if colour 1 is changed again, colour 2 will not be altered. Finally a colour can be returned to its initial default settings by specifying only the colour number. For example,

```
COLOUR 1,2
```

will set colours 1 and 2 back to their original values. The background colour may be altered by changing the definition of colour 0.

The exact effects of the COLOUR directive will vary for different graphics devices. In some cases, mainly plotters and monochrome terminals, it will be ignored. With some devices, using COLOUR will not affect existing displays but the modified colour definitions will be used for subsequent plots. For other devices it will take effect immediately, allowing plots to be modified dynamically without having to redraw the graphs. For example, typing

```
COLOUR 1; MATCH=0
```

will make all parts of the graph plotted in colour 1 disappear, by changing colour 1 to the background colour. If this is followed by

```
COLOUR 1
```

the points will reappear. This can be achieved only when it is possible to alter the colour table of the terminal dynamically, and where the underlying graphical software allows it. The *Users' Note* should contain details of how COLOUR is implemented in your version of Genstat.

6.5.7   Device-specific information

The high-resolution graphics in Genstat are designed so that, as far as possible, graphical output will be identical on every device. There are, however, two ways in which differences may arise. Firstly, there may be minor differences due to the physical limitations of the chosen device. For example, the resolution of a display will, ultimately, affect the quality of the plotted output. Secondly, it is not feasible to use colour on monochrome displays. Where possible, the default settings are designed to cater for these differences, for example by having default pen settings that depend on whether the output is monochrome or colour. However, if you request a specific setting, that will be used irrespective of the device.

There are also device-specific settings for some aspects. These are designed to take advantage of features built into the hardware or software of particular devices, in order to improve the efficiency of graphical output and perhaps to improve the display itself. An example is the use of *solid-fill* in histograms and pie charts. This can be generated by software, using brush style 16, on any device, but in many cases this can be performed in a fraction of the time by using the hardware instead. By their very nature, of course, these settings cannot be guaranteed to generate identical output on different devices.

Below we identify the areas of graphical output that are likely to be affected. The *Users' Note* should document the information relevant to all the devices available in your version of Genstat. You can also use the procedure DHELP to display the device-specific settings.

When using FRAME to define windows for your plots, the area [0,1]×[0,1] is always available, but for some devices the plotting area may extend further in either the y- or x-direction. By keeping within the [0,1] range you can ensure that the window is always valid, whatever output device is selected. However, you may wish to use the extended area where possible on a particular device.

Pen colours can be set in the range from 0 to 32. If fewer colours are available the colours are used in turn, and then recycled. Where possible, colour 0 is interpreted as the background colour, thus allowing points to be easily erased from a plot. On some output devices, mainly colour graphic screens, the background colour for the window may be specified using the BACKGROUND parameter of FRAME. This has a default of zero, the normal background colour. If you choose another colour as the background, the window will be drawn in this colour, before any graph is plotted. Any points to be plotted in this colour will then use colour 0 instead, and any points to be plotted in colour 0 will use the specified background colour. On devices where areas cannot be filled by hardware in different colours this parameter is ignored. The association of colour numbers with actual colours depends on the particular device. Using a colour graphics screen, the colours should be as defined by the COLOUR directive. However, for many plotters, the colour number relates only to the physical position in which the pen is mounted in the plotter, so the actual colours may vary between devices.

The standard text fonts, graphical symbols, and brush styles are software generated. However, you can set negative values for these parameters of the PEN directive to select device-specific alternatives. For each parameter, the device-specific settings have the same range as the standard settings; thus you can select symbols −1 to −9, fonts −1 to −25, and brush styles −1 to −16. Figure 6.5.7a illustrates the device-specific values available with the standard Genstat PostScript device. If fewer device-specific settings are actually available, the settings are taken in turn, and then recycled. Where a feature has no device-specific settings on a particular device, the standard form is used instead (for example, font −3 appearing as font 3). Device-specific font numbers cannot be used within the in-line typesetting system described in 6.5.5; Genstat will use either the standard fonts or the corresponding device-

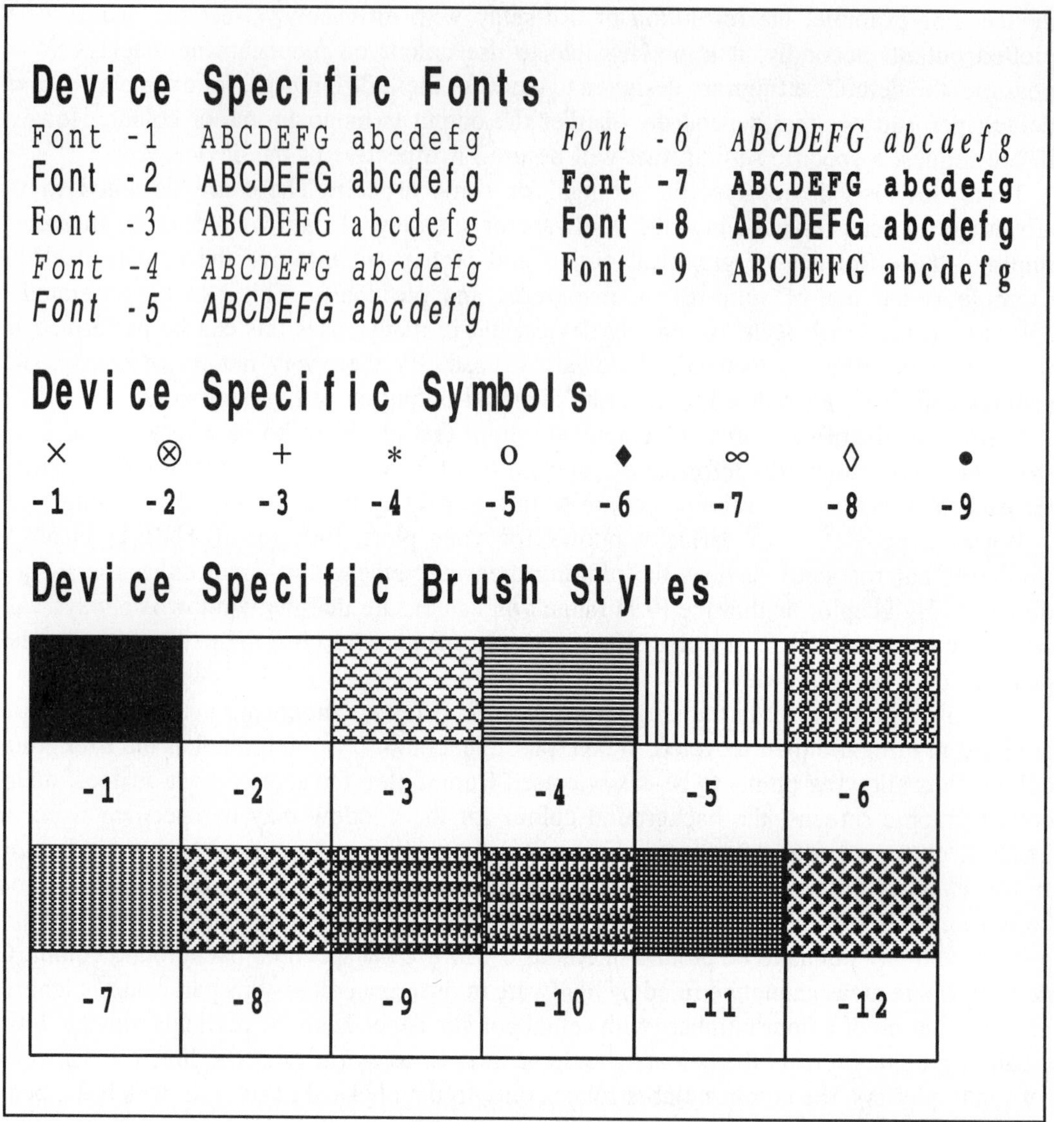

Device Specific Fonts

Font -1 ABCDEFG abcdefg	Font -6 ABCDEFG abcdefg
Font -2 ABCDEFG abcdefg	Font -7 ABCDEFG abcdefg
Font -3 ABCDEFG abcdefg	Font -8 ABCDEFG abcdefg
Font -4 ABCDEFG abcdefg	Font -9 ABCDEFG abcdefg
Font -5 ABCDEFG abcdefg	

Device Specific Symbols

×	⊗	+	*	o	♦	∞	◊	•
-1	-2	-3	-4	-5	-6	-7	-8	-9

Device Specific Brush Styles

-1	-2	-3	-4	-5	-6
-7	-8	-9	-10	-11	-12

Figure 6.5.7

specific fonts depending on the base font originally specified by the PEN directive. In some cases, device-specific symbols or fonts may be of fixed size; the SIZE parameter will then have no effect, and some of the typesetting commands may not function correctly. The *Users' Note* should indicate when device-specific fonts are of fixed size. Although the device-specific settings are likely to be different from device to device, they are arranged to be consistent where possible, so that for example brush style −1 will select solid fill, if available.

By default, Genstat uses software generated symbols and fonts. For colour displays, by default symbol 1 is used for all pens but in different colours. On monochrome displays, the pens all use colour 1 and symbols 1 to 9 are used in turn: symbol 1 for pen 1, symbol 2 for pen 2, and so on. When solid fill and colour are available, the default brush style is −1, in different colours for each pen; otherwise software-generated brushes are used by default.

For some devices, it is not possible to control the thickness of plotted lines; the THICKNESS parameter of PEN is then ignored.

6.5.8   Accessing details of the graphics environment

DKEEP saves information from the last plot on a particular device.

No options

Parameters

DEVICE = *scalars*	The devices for which information is required, if the scalar is undefined or contains a missing value, this returns the current device number
WINDOW = *scalars*	Window about which the information is required; default * gives information about the last window
YLOWER = *scalars*	Lower bound for the y-axis in last graph in the specified device and window
YUPPER = *scalars*	Upper bound for the y-axis in last graph in the specified device and window
XLOWER = *scalars*	Lower bound for the x-axis in last graph in the specified device and window
XUPPER = *scalars*	Upper bound for the x-axis in last graph in the specified device and window
FILE = *scalars*	Returns the value 1 or 0 to indicate whether a file is required for this device
DESCRIPTION = *texts*	Description of the device
DREAD = *scalars*	Returns the value 1 or 0 to indicate whether graphical input is possible from this device
ENDACTION = *texts*	Returns the current ENDACTION setting ('continue' or 'pause')

DKEEP provides information that can be used in general programs and procedures to control the graphical output. For the specified device you can determine whether it generates screen output or uses a file, whether graphical input is possible, a description of the device (as printed by HELP), the current ENDACTION setting, and details of the axis bounds.

The device for which the information is required is specified by the DEVICE parameter. If you specify a scalar containing a missing value, this will be set to the number of the current graphics device. You can then test whether an output file is needed and open one accordingly, as shown in Example 6.5.8a.

Example 6.5.8a

```
 1   READ Y,X,Y2

     Identifier   Minimum     Mean   Maximum   Values   Missing
              Y     46.46    68.11     89.95       20        0
              X     0.940    4.867     8.877       20        0
             Y2     38.00    62.37     82.65       20        0

13   DKEEP DEVICE=Device; FILE=File; DESCRIPTION=Name
14   PRINT Name,Device,File

                   Name       Device        File
              PostScript          13           1

15   IF File
16     OPEN 'GRAPH.OUT'; CHANNEL=Device; FILETYPE=graphics
17   ENDIF
```

When writing a procedure you can find out if axes bounds have been set explicitly, using the SAVE parameter of AXES. This information may then be used when setting up the axes for other graphs. However, if the bounds were not set, but have been evaluated from the data (or if the axes have subsequently been redefined) the information in the save structure will not be of any use. The actual values used when plotting are recorded internally, for each window of each device, and can be accessed using the YLOWER, YUPPER, XLOWER, and XUPPER parameters of DKEEP.

Example 6.5.8b

```
18   DGRAPH [WINDOW=5; KEYWINDOW=7] Y; X
19   "
-20          Now set up window 6 to have the same bounds as window 5,
-21          so that Y2 is plotted on the same scale as Y
-22   "
23   DKEEP Device; WINDOW=5; YLOWER=Ymin; YUPPER=Ymax; XLOWER=Xmin; \
24     XUPPER=Xmax
25   PRINT Ymin,Ymax,Xmin,Xmax

        Ymin        Ymax        Xmin        Xmax
       44.29       92.12      0.5432       9.274

26   AXES 6; YLOWER=Ymin; YUPPER=Ymax; XLOWER=Xmin; XUPPER=Xmax
27   DGRAPH [WINDOW=6; KEYWINDOW=8; SCREEN=keep] Y2; X
```

S.A.H.

7   Basic Statistics

The subject of *statistics* includes a wide range of techniques for dealing with data. In the early days, statisticians were mainly concerned with collecting and summarizing data. Later the subject was widened to include methods for finding and fitting quantitative models and, more recently, methods have been developed for exploring data and searching for patterns.

Genstat can be used in all of these areas: for example, you can obtain tabular summaries of data, such as means, totals, and medians, using the TABULATE directive (4.11.1); you will find a wide variety of graphical displays provided by the directives in Chapter 6; and the more general and powerful methods for quantitative modelling are described in Chapters 8 to 12.

This chapter describes some basic techniques provided by Genstat for exploring data, producing straightforward summaries, and performing simple tests. Some of these techniques use procedures in the standard Procedure Library, automatically available to any Genstat user (see 5.3.1). Others simply use basic options of more powerful methods of analysis provided by Genstat directives. Many can also be found in the Menu System (1.1.3). The aim of this chapter is to provide an introduction that makes these simple methods readily accessible, with references pointing to the more sophisticated methods that are available if required.

Section 7.1 describes methods for exploring the distribution of data using histograms and summary statistics either for quantitative measurements (7.1.1) or for categorical observations (7.1.2). It then gives information on the facilities for studying probability distributions in Genstat: procedure GRANDOM is available for generating pseudo-random numbers from a wide range of probability distributions (7.1.3), and the DISTRIBUTION directive can be used to estimate parameters of distributions (7.1.4).

In Section 7.2 we present a range of simple hypothesis tests for comparing grouped data, ranging from parametric t-tests and non-parametric equivalents (7.2.1 and 7.2.2), to one-way analysis of variance (7.2.3).

Section 7.3 then presents methods of describing relationships between variables. Graphical methods are a powerful tool for exploring relationships (7.3.1). More quantitative methods include the calculation of correlations (7.3.2), modelling and testing relationships between qualitative variables (7.3.3), and regression (7.3.4).

7.1   Distributions

Many probability distributions are used in statistics to describe random variation. An extensive survey of these distributions is given by Johnson and Kotz (1969, 1970a, 1970b). For all the common distributions, functions are available in Genstat to provide probabilities, integrals, and equivalent deviates (see 4.2.9). Other distributions can be handled by programming the necessary equations using the CALCULATE directive (4.1).

In this section, we show ways to display the distribution of a set of data for both continuous measurements (7.1.1) and categorical observations (7.1.2). You can compare these with data simulated from theoretical distributions, generated using the GRANDOM procedure (7.1.3), or you can fit a theoretical distribution to data using the DISTRIBUTION directive (7.1.4).

7.1.1 Displaying the distribution of data

A diagram can be an informative way of
displaying the various features of a set of
observations. One of the easiest diagrams to
construct and interpret is the histogram,
which can be drawn with the DHISTOGRAM
directive (6.2.2). This produces a high-
resolution picture, but if you are constrained
by a non-graphical terminal, a line-printer-
style diagram can be drawn instead by using
the HISTOGRAM directive (6.1.2). A histogram
is formed by splitting the range of the data
into contiguous categories. It displays the
number of observations falling into successive
categories, thus showing whether they are
tightly-packed or spread-out, symmetrically
distributed or skew, and whether there are
observations separated, or outlying, from the
mass of the data.

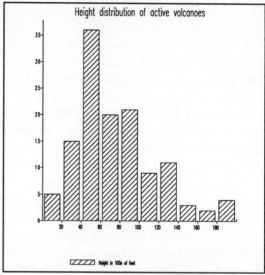

Figure 7.1.1a

Example 7.1.1a shows the statements needed to read some data and draw a histogram. This
set of data is used extensively in this chapter – it contains the heights of active volcanoes,
together with their names, latest eruption dates, and geographical regions.

Example 7.1.1a

```
 1   "  Heights of active volcanoes. Data from Hoffman (1992);
-2       previous version of data also displayed in Tukey (1977) p.40."
 3
 4   OPEN 'VOLCANO.DAT'; CHANNEL=2
 5   TEXT Volcano
 6   TEXT [VALUES=America,'Asia/Oceania',Elsewhere] Regname
 7   FACTOR [LABELS=Regname] Region
 8   READ [CHANNEL=2] Volcano,Year,Height,Region
```

Identifier	Minimum	Mean	Maximum	Values	Missing
Volcano				126	0
Year	1960	1983	1991	126	0
Height	10.00	77.19	199.00	126	0

Identifier	Values	Missing	Levels
Region	126	0	3

```
 9   TEXT [VALUES='Height distribution of active volcanoes'] Head
10   & [VALUES='Height in 100s of feet'] Scale
11   DHISTOGRAM [TITLE=Head] Height; DESCRIPTION=Scale
```

The DHISTOGRAM statement in this example draws the picture in Figure 7.1.1a, showing that
the distribution of heights is positively skewed. The statement automatically chooses the

number of classes into which to divide the observations, and uses the default colours, brush-types, and so on. These details can be changed by setting options in the DHISTOGRAM statement (6.2.2), or by explicitly setting the graphical environment (6.5). For example, to specify a more spread-out picture, the NGROUPS option of DHISTOGRAM could have been set to get 20 groups instead of the 10 produced by default:

```
DHISTOGRAM [NGROUPS=20] Height
```

An alternative diagram for studying the distribution of observations is the boxplot, or box-and-whisker plot. This consists of a simple box, drawn so that its upper and lower edges are the upper and lower quartiles of the data-set. A line through the box shows the median, and whiskers extend out in either direction to the maximum and minimum. This diagram often gives a good indication of the skewness of a sample, as well as its location and spread.

You can draw boxplots in Genstat using the BOXPLOT procedure. Because BOXPLOT is a procedure, it is not described fully in this Manual, but details can be found in the Procedure Library Manual. An option controls whether a line-printer-style or high-resolution picture is produced, and options are also used

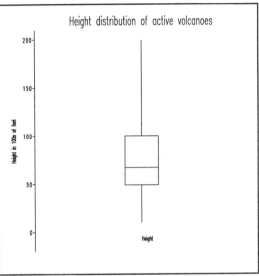

Figure 7.1.1b

to supply titles. Thus, the following statement draws the high-resolution picture shown in Figure 7.1.1b.

```
BOXPLOT [GRAPHICS=high; TITLE=Head; AXISTITLE=Scale] Height
```

Both the histogram and the boxplot can be used to compare the distributions of several sets of data. With DHISTOGRAM, you should specify a list of variates instead of just one (here Height); see 6.2.2. The same method can be used with BOXPLOT; alternatively, you can specify a single variate holding all the values and set the GROUPS parameter to a factor indicating the data-set to which each value belongs. Here is a statement that draws parallel boxes:

```
BOXPLOT [GRAPHICS=high; TITLE=Head; AXISTITLE=Scale; \
    METHOD=schematic; BOXWIDTH=variable] Height; \
    GROUPS=Region; UNITLABELS=Volcano
```

The GROUPS parameter now specifies that separate boxes are to be drawn for volcanoes in three geographic regions, defined by the factor Region read from the datafile. In addition, we have set the METHOD option and UNITLABELS parameter to get *schematic diagrams* rather than simple boxplots: they identify individual outlying points (Tukey 1977). We have also set the BOXWIDTH option so that the widths of the boxes indicate the numbers of volcanoes in each group. The picture drawn by this statement is in Figure 7.1.1c; it shows that all the tallest

volcanoes are in America, and identifies two
volcanoes in the Asia/Oceania region that
seem particularly high compared to the others
in that region.

A third kind of picture – a rugplot – can
be used to represent each observation
separately on a graph. This is easy to draw in
Genstat, although there is no single command
to do it. It is best to draw just an x-axis, and
then simply plot a constant value (here zero)
against the variate holding the data values, so
that the data appear along a horizontal line.
The vertical-line character is a good plotting
symbol to use, since it represents the
individual observations clearly. The following
statements draw the picture in Figure 7.1.1d.

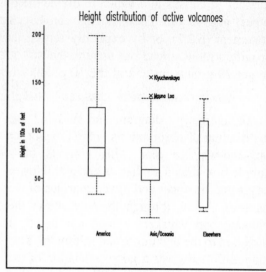

Figure 7.1.1c

```
AXES 3; XTITLE=Scale; STYLE=x
PEN 1; SYMBOL='|'
DGRAPH [WINDOW=3; KEY=0] !(126(0)); Height
```

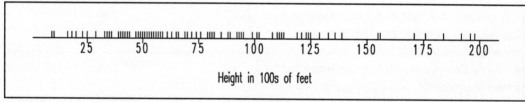

Figure 7.1.1d

Pictures are not the only way to study distributions. It can often be helpful to look at
descriptive statistics, such as the mean, variance, median, and number of missing values. You
can display these and many other statistics with the DESCRIBE procedure. Example 7.1.1b
shows the default set of statistics produced by DESCRIBE for the volcano heights.

Example 7.1.1b

```
  22   DESCRIBE Height

Summary statistics for Height

       Number of observations = 126
     Number of missing values = 0
                       Mean = 77.190
                     Median = 67.000
                    Minimum = 10.000
                    Maximum = 199.000
             Lower quartile = 49.000
             Upper quartile = 100.000
```

You can set the SELECT option of DESCRIBE to pick out individual statistics; for example,

```
DESCRIBE [SELECT=mean,sem] Y
```

will produce just the mean and standard error of the mean. The available settings for SELECT are listed below:

nval number of values	nobs number of non-missing values
nmv number of missing values	sum total of the values
mean arithmetic mean	sem standard error of mean
median median	range range
min minimum	max maximum
q1 lower quartile	q3 upper quartile
sd standard deviation	var variance
ss corrected sum of squares	uss uncorrected sum of squares
%cv coefficient of variation	skew coefficient of skewness
seskew standard error of coefficient of skewness	
kurtosis kurtosis	sekurtosis standard error of kurtosis

7.1.2 Displaying data classified into groups

The measurements displayed in the previous section were all quantitative: that is, they were numbers that came from an essentially continuous, interval-based scale. In fact, the heights of volcanoes were recorded in the data file only to the nearest 100 feet, but it is clear that the difference in height between volcanoes of 12,400 and 11,200 feet is directly comparable to that between volcanoes of 3,400 and 2,200 feet.

There are several other types of measurement scale as well as the continuous interval-based scale. One possibility is an *ordinal* scale, where measurements are ranked. For example, the severity of a disease might be measured on a scale with five levels: extreme, severe, moderate, slight, absent. These levels are qualitative, but they could also be represented by the numbers 4, 3, 2, 1, 0, for example, and so be treated as quantitative. However, this would not alter the fact that the underlying scale is not continuous, and there is no reason why the difference between levels 4 and 3 should be treated as quantitatively the same as that between 1 and 0. Another possibility is a purely *nominal* scale, where the distinct values, or levels, of the scale are not even considered to be ordered. An example is the Region classification in the previous section: each volcano is "measured" to be in one of the regions, but there is no natural ordering of the continents. Data of this kind are nearly always purely qualitative.

When data are not measured on an interval-based scale, statistical methods appropriate to the actual measurement scale should be used. In Genstat, it is usually best to store such measurements in a factor data structure rather than a variate; many of the standard statistical techniques in Genstat make a deliberate distinction between factors and variates. However, data measured on any scale can be represented in histograms, with each bar corresponding to one level or distinct value of the scale. The following statement draws such a histogram, as shown in Figure 7.1.2a.

```
DHISTOGRAM [TITLE=\
  'Active volcanoes in three regions of the world'] Region
```

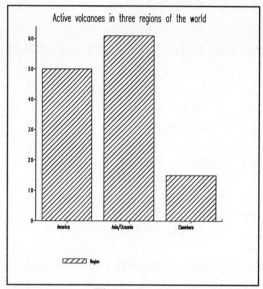

Figure 7.1.2a

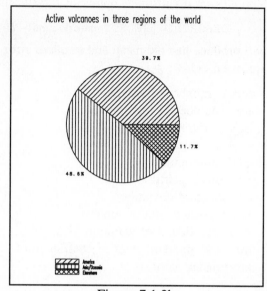

Figure 7.1.2b

An alternative way of displaying the division of data into levels or categories is provided by the pie chart. The DPIE directive in Genstat (6.2.3) requires a list of numbers indicating the size of each group. These can be generated, for example by the TABULATE directive (4.11.1), as in the following statements that draw Figure 7.1.2b.

```
TABULATE [CLASSIFICATION=Region; COUNTS=Nvolcano]
DPIE [TITLE='Active volcanoes in three regions of the world'] \
    #Nvolcano; DESCRIPTION=#Regname
```

The TABULATE statement here has produced a one-way table Nvolcano containing the number of volcanoes in each region. TABULATE can produce a wide range of tabular summaries which can all be printed using its PRINT option. Example 7.1.2a prints the one-way summary of the division of volcanoes between continents.

Example 7.1.2a

```
16    TABULATE [PRINT=counts; CLASSIFICATION=Region]

                      Count
          Region
          America        50
    Asia/Oceania         61
       Elsewhere         15
```

The tables of summaries produced by the TABULATE directive can be classified by up to nine factors at a time. They need not be counts as here, but can be means or other statistics derived from continuous measurements (stored in Genstat in variates). The tables can be given margins, and you can use the PERCENT procedure to express the values inside the tables as percentages of the margins – which often gives a more easily digestible form of summary. Example 7.1.2d show how PERCENT can be used to re-express the table in Example 7.1.2a. Notice that the "Margin" line of the table contains the total number of volcanoes on which the percentages are based.

Example 7.1.2b

```
18   PERCENT OLD=Nvolcano; NEW=%volcano
19   PRINT %volcano

                   %volcano
        Region
        America     39.68
   Asia/Oceania     48.41
      Elsewhere     11.90

        Margin     126.00
```

7.1.3   Generating random numbers from probability distributions

Genstat contains various facilities for generating pseudo-random numbers. The function URAND generates random numbers from the uniform distribution with range (0,1). The statement

```
CALCULATE X = URAND(75123; 100)
```

for example, generates 100 uniform random numbers based on the *seed* 75123, and stores them in the variate X. The seed specifies how the generating process should begin. If the seed is zero, then the first time that the generator is used in a program, Genstat will use a seed constructed from the computer's internal clock; after that, Genstat will continue generating from the point at which it last stopped.

Random numbers from a range of other distributions can be generated by the GRANDOM procedure. It has a DISTRIBUTION option which can request any of the following distribution names: uniform, Normal, logNormal, chisquare, t, F, gamma, beta, Weibull, Poisson, or binomial. You can specify the mean and variance of the distribution with the MEAN and VARIANCE options. These must obey any restrictions applicable to the distribution concerned; for example, the mean and variance of the Poisson distribution must be equal and integral. Alternatively, for some of the distributions, you can set individual defining parameters; for example, the gamma distribution's defining parameters can be set by options AGAMMA and BGAMMA. Thus the following two statements will generate the identical random numbers:

```
GRANDOM [DISTRIBUTION=gamma; MEAN=2.5; VARIANCE=5; \
   SEED=553616; NVALUES=250] Gamma1
GRANDOM [DISTRIBUTION=gamma; AGAMMA=1.25; BGAMMA=2; \
   SEED=553616; NVALUES=250] Gamma2
```

You can also generate random sets of integers, this time using the SAMPLE procedure. Example 7.1.3 shows how to select a random sample of 10 integers from the set 1 to 100, sampling without replacement.

Example 7.1.3

```
  1   SAMPLE [NVALUES=100] NSAMPLE=10; SAMPLE=Selected
* MESSAGE: Default seed for random number generator used with value 744934
  2   PRINT [ORIENTATION=across] Selected; FIELDWIDTH=3; DECIMALS=0

   Selected  8 80 60 53 30 75 62 49 39 11
```

Yon can set both NVALUES and NSAMPLE to *n* if you just want a random ordering of the first *n* integers. The SAMPLE procedure can also generate stratified samples based on classifications stored in factor structures. Full details of procedures SAMPLE and GRANDOM can be found in the Procedure Library Manual or from the procedure LIBHELP (5.3.1).

More generally, if you want to randomize any set of values, you can use the RANDOMIZE directive (9.8.3). For example, these statements copy the volcano heights in Example 7.1.1a into a new variate Nheight, and then put them into a random order.

```
        VARIATE Nheight; VALUES=Height
        RANDOMIZE Nheight
```

7.1.4   Fitting distributions

DISTRIBUTION estimates the parameters of continuous and discrete distributions.

Options

PRINT = *strings*	Printed output required from each individual fit (parameters, samplestatistics, fittedvalues, proportions, monitoring); default para, samp, fitt
CBPRINT = *strings*	Printed output required from a fit combining all the input data (parameters, samplestatistics, fittedvalues, proportions, monitoring); default *
DISTRIBUTION = *string*	Distribution to be fitted (Poisson, geometric, logseries, negativebinomial, NeymanA, PolyaAeppli, PlogNormal, PPascal, Normal, dNvequal, dNvunequal, logNormal, exponential, gamma, Weibull, b1, b2, Pareto); default * i.e. fit nothing
CONSTANT = *string*	Whether to estimate a location parameter for the gamma, logNormal, Pareto, or Weibull distributions (estimate, omit); default omit
LIMITS = *variate*	Variate to specify or save upper limits for classifying the data into groups; default *
NGROUPS = *scalar*	When LIMITS is not specified, this defines the number

	of groups (of approximately equal size) into which the data are to be classified; default is the integer value nearest to the square root of the number of data values
XDEVIATES = *variate*	Variate to specify points up to which the CUMPROPORTIONS are to be estimated
JOINT = *string*	Requests joint estimates from the combined fit to be used for a re-fit to the separate data sets (dispersion, variancemeanratio, Poissonindex); default *
PARAMETERS = *variate*	Estimated parameters from the combined fit
SE = *variate*	Standard errors for the estimated parameters of the combined fit
VCOVARIANCE = *symmetric matrix*	
	Variance-covariance matrix for the estimated parameters of the combined fit
CUMPROPORTIONS = *variate*	Estimated cumulative proportions of the combined distribution up to the values specified by the XDEVIATES option
MAXCYCLE = *scalar*	Maximum number of iterations; default 30
TOLERANCE = *scalar*	Convergence criterion; default 0.0001

Parameters

DATA = *variates* or *tables*	Data values either classified (table) or unclassified (variate)
NOBSERVATIONS = *tables*	One-way table to save the data classified into groups
RESIDUALS = *tables*	Residuals from each (individual) fit
FITTEDVALUES = *tables*	Fitted values from each fit
PARAMETERS = *variates*	Estimated parameters from each fit
SE = *variates*	Standard errors of the estimates
VCOVARIANCE = *symmetric matrices*	
	Variance-covariance matrix for each set of estimated parameters
CUMPROPORTIONS = *variates*	Estimated cumulative proportions of each distribution up to the values specified by the XDEVIATES option
CBRESIDUALS = *tables*	Residuals from the combined fit
CBFITTEDVALUES = *tables*	Fitted values from the combined fit
STEPLENGTH = *variates*	Initial step lengths for each fit
INITIAL = *variates*	Initial values for each fit

The DISTRIBUTION directive is used to fit an observed sample of data to a theoretical distribution function, in order to estimate the parameters of the distribution and test the goodness of fit. The data consists of observations x_i of a random variable X, which has a distribution function $F(x)$ defined by $F(x)=\Pr(X \leq x)$. A selection of both discrete and continuous distributions are available, and full details are given later in this section.

For discrete distributions X may take non-negative integer values only, except for the log-series distribution where only positive integer values are allowed. For continuous distributions the random variable X may take any values, subject to constraints for certain distributions, for example, data values must be strictly positive in order to fit a log-Normal distribution. Constraints are detailed with the individual distributions described below.

The data can be supplied to DISTRIBUTION as a variate or as a one-way table of counts. If the raw data are available, then these should be supplied (as a variate), since the raw data contains more information than grouped data.

If raw data are not available, then a one-way table of counts, or frequencies, should be given. The factor classifying the table must have its levels vector declared explicitly, since the levels are used to indicate the boundary values of the raw data used to create the grouping. For example, if the discrete variable X takes the values 0...8, with numbers of observations 2,6,7,4,2,1,0,1,0 respectively, a table of counts can be declared by

```
FACTOR [LEVELS=!(0...8)] F
TABLE [CLASSIFICATION=F; VALUES=2,6,7,4,2,1,0,1,0] T
```

The factor levels do not have to specify single data values: often it will be desirable to group certain values together, and indeed for continuous data this is the only sensible way to proceed. In general, for a classifying factor with levels $l_1, l_2, \ldots , l_f$, the count n_k for the kth cell of the table will be the number of observations x_i such that

$$
\begin{aligned}
x_i \le l_1, && k=1 \\
l_{k-1} < x_i \le l_k, && 2 \le k \le f-1 \\
l_{f-1} < x_i, && k=f
\end{aligned}
$$

This means that for all except the last cell of the table, the factor level represents the upper limit on values in that cell. The final class of the table is termed the *tail*; it is formed by combining the frequencies for all values of X greater than l_{f-1}, and the upper limit on values in the tail is infinity. For continuous distributions with no lower bound, the first class will be the lower tail. You will often want to form the tail(s) by amalgamating groups with low numbers of counts. In the example above, you might amalgamate the groups for values 6-8:

```
FACTOR [LEVELS=!(0...5,99)] F2
TABLE [CLASSIFICATION=F2; VALUES=2,6,7,4,2,1,1] T2
```

Note that the final factor level, for the tail, can be given a dummy value of 99 to indicate that it has no upper limit, since this value is never used in calculations.

When data is supplied as a table instead of as a variate, the computed log-likelihood is only an approximation to the full log-likelihood and the solution obtained will depend to some extent on the choice of class limits. More reliable results will be achieved with a larger number of classes, since this gives more information on the data distribution, so only classes with very few observations should be amalgamated. In general, care should be taken to choose class limits that give a reasonable number of counts in each class, but with none of the individual classes holding a disproportionately large number of observations.

The DISTRIBUTION option should be set to indicate which distribution is to be fitted to the data. The following distributions are available:

Discrete	Continuous
Binomial (as a special case of the negative binomial)	Normal
	Double Normal (equal variances)
Poisson	Double Normal (unequal variances)
Geometric	Log-Normal
Log-series	Exponential
Negative binomial	Gamma
Neyman type A	Weibull
Pólya-Aeppli	Beta type I
Poisson-log-Normal	Beta type II
Poisson-Pascal	Pareto

The first step of the fitting process is to compute and print various sample statistics. Examining these may help in the selection of appropriate distributions for fitting – properties of the various distributions are listed at the end of this section. The setting `DISTRIBUTION=*` can be used to produce this output without any model fitting. The following sample statistics are calculated:

Sample size	n	
Sample mean	$m = \Sigma\, x_i/n$	
Sample variance	$s^2 = \Sigma\, x_i^2/n - m^2$	discrete distributions
	$s^2 = \Sigma\, (x_i - m)^2 / (n-1)$	continuous distributions
Sample skewness	$g_1 = \Sigma\, (x_i - m)^3 / (n-1)s^3$	
	$= m_3/s^3$	
Sample kurtosis	$g_2 = \Sigma\{(x_i - m)^4/(n-1)s^4\} - 3$	continuous distributions only
Sample quartiles	$x_p;\ F(x_p) = p$	(see 7.1.1)
Poisson index	$(s^2 - m)/m^2$	discrete distributions only
Negative binomial index	$m(m_3 - 3s^2 + 2m)/(s^2 - m)^2$	discrete distributions only

If the original data are not available, the sample statistics are calculated by substituting class mid-points in place of the data. For the lower tail, the class "mid-point" is taken to be $l_1 - \frac{1}{2}(l_2 - l_1)$ and for the upper tail, $l_{f-1} + \frac{1}{2}(l_{f-1} - l_{f-2})$. No corrections are made for groupings. When a distribution has been fitted to data, the relevant theoretical statistics of that distribution are printed for comparison with the sample statistics, as a check on the appropriateness of the model for the data.

If a distribution has been specified, it is then fitted to the data to obtain maximum-likelihood estimates of the parameters, as in Example 7.1.4a below, which fits a log-Normal distribution to the volcano data introduced in 7.1.1.

Example 7.1.4a

```
21  VARIATE Limits; VALUES=!(20,40...180)
22  DISTRIBUTION [DISTRIBUTION=lognormal; LIMITS=Limits] Height

22...............................................................
```

```
***** Fit continuous distribution *****

*** Sample Statistics ***

     Sample Size        126
     Mean             77.19      Variance     1702.91
     Skewness          0.96      Kurtosis        0.62

     Quartiles:          25%        50%        75%
                        49.0       67.0      100.0

*** Summary of analysis ***

Observations: Height
              Parameter estimates from individual data values
Distribution: Lognormal
              Log(x) distributed as Normal(m,s**2)

     Deviance: 10.82 on 7 d.f.

*** Estimates of parameters ***

          estimate        s.e.      Correlations
m          4.1967       0.0513      1.0000
s          0.5760       0.0363      0.0000  1.0000

*** Fitted quartiles ***

            25%           50%           75%
          45.071        66.468        98.024

*** Fitted values (expected frequencies) and residuals ***

     x              Number        Number     Weighted
                   Observed      Expected     Residual
 < 20.0               5             2.33         1.51
 < 40.0              15            21.47        -1.48
 < 60.0              36            30.30         1.00
 < 80.0              20            24.79        -0.99
 < 100.0             21            16.97         0.94
 < 120.0              9            10.91        -0.60
 < 140.0             11             6.88         1.45
 < 160.0              3             4.33        -0.68
 < 180.0              2             2.74        -0.47
 > 180.0              4             5.27        -0.58
```

A summary is given of the fit: the parameter estimates are printed with their standard errors and correlations, including the *working parameters*, which are *stable* functions of the parameters defining the distribution and are used in the internal algorithm. The goodness of fit to the chosen distribution is indicated by the residual deviance which has an asymptotic chi-squared distribution with the specified degrees of freedom. The deviance is also the preferred statistic for comparison of nested models, for example the double Normal distribution with equal and unequal variances. This is followed by a table of observed and fitted values (expected frequencies), together with weighted residuals. If raw data are supplied, by default this table is formed by dividing the data into $\sqrt{n}$ groups of approximately equal observed frequency, which are therefore likely to be of unequal widths. The NGROUPS option may be used to set the number of groups for this table. If data are supplied as a table, as in Example 7.1.4b, the fitted values use the classification from that table. In either case the LIMITS option may be used to supply a different set of limits; with the constraint that if tabulated data are analysed these limits should be a subset of the original limits so that the new groups are formed by aggregation. In Example 7.1.4a, evenly spaced limits were specified.

Example 7.1.4b shows the analysis of some tabulated data: this is disease data, indicating the number of leaves on which zero, one, up to seven red mites were found. A further cell, containing 0, for eight mites is included in the table as the tail. A negative binomial distribution is fitted to investigate the distribution of mites on leaves.

Example 7.1.4b

```
  1   "
 -2         Negative binomial fit to counts of European red mites on apple
 -3         leaves, Bliss (1953). The data are recorded as the number of leaves
 -4         having no mites, number with one mite, and so on.
 -5   "
  6   FACTOR [LEVELS=!(0...8)] Mites; DECIMALS=0
  7   TABLE [CLASSIFICATION=Mites] Leaves; DECIMALS=0
  8   READ [PRINT=*] Leaves
 10   PRINT [ACROSS=Mites] Leaves; FIELDWIDTH=6
```

```
        Leaves
Mites     0     1     2     3     4     5     6     7     8
         70    38    17    10     9     3     2     1     0
```

```
 11   DISTRIBUTION [DISTRIBUTION=negativebinomial] Leaves
```

```
11.........................................................................

***** Fit discrete distribution *****

*** Sample Statistics ***

    Sample Size        150
    Mean              1.15    Variance                   2.26
    Skewness          1.53
    Poisson Index     0.85    Negative Binomial Index    0.66

*** Summary of analysis ***

Observations: Leaves
              Parameter estimates from tabulated data values
Distribution: Negative Binomial
              Pr(X=r) = (r+k-1)C(k-1).(m/(m+k))**r.(1+m/k)**(-k)

    Deviance: 4.22 on 6 d.f.

*** Estimates of working parameters ***

          estimate       s.e.    Correlations
mean        1.1467     0.1273    1.0000
variance    2.4301     0.5379    0.7663  1.0000

*** Estimates of defining parameters ***
```

```
          estimate          s.e.        Correlations
m          1.1467         0.1273        1.0000
k          1.0246         0.2758        0.0001  1.0000
Poisson Index
1/k        0.9760         0.2628
```

*** Fitted values (expected frequencies) and residuals ***

r	Number Observed	Number Expected	Weighted Residual
0	70	69.49	0.06
1	38	37.60	0.07
2	17	20.10	-0.71
3	10	10.70	-0.22
4	9	5.69	1.28
5	3	3.02	-0.01
6	2	1.60	0.30
7	1	0.85	0.16
8+	0	0.95	-1.38

The NOBSERVATIONS, RESIDUALS, and FITTEDVALUES parameters can be used to save the number of observations in each cell, the fitted number, and the residual respectively (all in tables). The parameter estimates and their standard errors can be saved in variates specified by PARAMETERS and SE. The variance-covariance matrix for the estimated parameters can be saved as a symmetric matrix using the VCOVARIANCE parameter.

Having fitted the required distribution, the estimated cumulative distribution function (CDF) can be evaluated at specified values of *X*. These are defined using the XDEVIATES option. The values of the CDF can be printed (by selecting PRINT=proportions) or saved in a variate by setting the CUMPROPORTION parameter.

If you have several sets of data you may be interested in fitting the distribution individually to each set; this can be done by setting the DATA parameter to a list of identifiers. A separate analysis is then performed for each set of data, but of course any option settings are common to all the data sets. The data sets should all be specified in the same way, either as raw data or as tabulated counts. For tabulated counts, the same categories must be used for defining every table. You can also carry out one final fit to the combined data set, in order to investigate whether the data can be adequately modelled as coming from a single population. This combined fit is produced if any of the options relating to the combined fit have been set (that is, options CBPRINT, PARAMETERS, SE, VCOVARIANCE, or CUMPROPORTION which print or save information from the combined analysis). For each individual data set you can also save fitted values and residuals based on the parameters estimated from the combined data set, using the CBRESIDUALS and CBFITTEDVALUES parameters. The JOINT option can be used to specify that certain parameters should be held constant at their estimated values from the combined analysis during refits to the individual data sets. For continuous distributions only, a common dispersion parameter can be requested; for discrete distributions a common value can be requested for either the Poisson index or the ratio of variance to mean. An analysis of deviance is printed to compare the nested models.

If the original data is available, the full log-likelihood is used in the optimization algorithm. Otherwise, an approximate log-likelihood is optimized, using representative values for each class. For some distributions, it is necessary to use stable *working parameters* in the optimization algorithm (Ross 1990), and the *defining parameters* for the distribution are then evaluated by a simple transformation.

The deviance and corresponding degrees of freedom that are printed as part of the model summary are based on the table of fitted values, and thus may be affected by the choice of limits. The residuals computed are deviance residuals (McCullagh and Nelder 1989), and the deviance is therefore the sum of squared residuals. The degrees of freedom are $n-p-1$, where n is the number of cells in the table of fitted values and p is the number of parameters estimated in the model. The default limits for grouping the raw data are designed to avoid small expected frequencies (for example in the tail cells) which can have an inflationary affect on the deviance; however, if the tails are important, because of the origin of the data, it may be important to specify the limits explicitly.

An iterative Gauss-Newton optimization method is used to estimate the parameters of the distribution. The parameterization is chosen for each model so that the optimization is stable, but if there are any problems with particular data sets it may be necessary to control this process. The MAXCYCLE and TOLERANCE options allow you to increase the number of iterations and alter the convergence criterion for data sets that fail to converge. You can also specify initial values and step lengths for the parameters for each set of data using the STEPLENGTH and INITIAL parameters. These parameters should be set to variates of length appropriate for the distribution being fitted; for example, if DISTRIBUTION=Poisson they should have just one value. Another use of INITIAL and STEPLENGTH is to constrain a parameter to a particular value; for example when fitting a double Normal the proportion parameter p could be fixed at 0.5 by setting the initial value to 0.5 and the steplength to 0, thus fitting a double Normal in equal proportions. Note that the degrees of freedom are not adjusted to take account of this. Optimization problems are discussed further in 8.6 and 8.7.

We now discuss the distributions that can be fitted, looking first at the discrete and then the continuous distributions. A summary of the theoretical properties of the discrete distributions is given in Table 7.1.4a.

The *negative binomial* distribution is applicable in many different situations, and can be derived in several ways. For example: waiting times for the rth success in a sequence of Bernoulli trials have a negative binomial distribution; random sampling from a heterogeneous population described by a mixture of Poisson distributions with means varying according to a gamma distribution will produce negative binomial data; and the distribution can also describe the number of events per unit interval given underlying Poisson and log-series distributions. Further explanation can be found in Ross (1987 and 1990) and Johnson and Kotz (1969). The negative binomial distribution can be defined in terms of the expansion of $(q-p)^{-n}$ with $q-p=1$. In Genstat, it is specified in the form obtained by setting $\mu=np$ and $k=n$, so that the probability of observing the value $X=r$ is given by:

$$p_r \;=\; \Pr(X{=}r) \;=\; \binom{r+k-1}{k-1}\left[\frac{\mu}{\mu+k}\right]^r\left[1+\frac{\mu}{k}\right]^{-k} \qquad r=0,1...$$

The parameters estimated are the mean (μ) and the variance ($V=\mu+\mu^2/k$) from which the defining parameters μ and k are derived.

When the sample variance is less than the mean (indicated by a negative value of the Poisson index), the usual (positive) binomial distribution will be fitted where

$$p_r = \binom{N}{r} p^r (1-p)^{N-r} \qquad r=0,1...N$$

In this case, a negative value of k will be estimated, and the index of the binomial distribution, N, will be estimated to be $-k$, where $-k$ exceeds the largest value present in the sample. The probability of a success, p, is derived from $\mu=Np$.

The negative binomial distribution also generates the Poisson distribution (as $k\rightarrow\infty$), the geometric distribution (with $k=1$) and log-series distribution (as $k\rightarrow0$) as special cases. Although the estimated parameters μ and k are independent, estimated standard errors for k are not reliable since the confidence interval for k is skew: the deviance should therefore be used to compare the fit of the negative binomial distribution with nested models for particular values of k.

Table 7.1.4a: Theoretical properties of discrete distributions

	Mean (μ)	Variance (V)	Parameters estimated	Poisson index	Neg.bin. index
Poisson	μ	$V=\mu$	μ	0	–
Geometric	$\mu=(1-p)/p$	$V=(1-p)/p^2$	μ	1	2
Log-series	$\mu=\theta/z(1-\theta)$	$V=\mu[(1-\theta)^{-1}-\mu]$	$z=-\log(1-\theta)$	$z-1$	–
Negative binomial	μ	$V=\mu+\mu^2/k$	μ,V	$1/k$	2
Neyman type A	$\mu=\mu_1\mu_2$	$V=\mu_1\mu_2(1+\mu_2)$	μ,V	$1/\mu_1$	1
Pólya–Aeppli	$\mu=\mu_1/p$	$V=\mu_1(2-p)/p^2$	μ,V	$2(1-p)/\mu_1$	1.5
Poisson-log-Normal	$\mu=\exp(\mu_1+\tfrac{1}{2}\sigma^2)$	$V=\mu+\mu^2(e^{\sigma^2}-1)$	μ,V	$\exp(\sigma^2)-1$	2
Poisson-Pascal	$\mu=\lambda pk$	–	μ $(k+1)/\lambda k$ $(k+2)/(k+1)$	$(k+1)/\lambda k$	$(k+2)/(k+1)$

The *Poisson* distribution with mean μ arises as the number of events per unit time, assuming that events are distributed randomly and independently in time (or space), with mean number of events per unit interval equal to μ. The probability of observing r independent events in a unit interval is then:

$$p_r = \frac{\mu^r}{r!} e^{-\mu} \qquad r=0,1...$$

The distribution is described by the single parameter μ, equal to the mean and variance. The skewness is $g_1=1/\sqrt{\mu}$. For a sample from a Poisson distribution with mean μ, the expected value of the Poisson index is 0, with variance $2/n\mu^2$.

The *geometric* distribution is a discrete analogue of the continuous exponential distribution described later in this section, and can be interpreted as the waiting time in a series of Bernoulli trials before an event occurs. The probability that r trials occur before an event is given by:

$$p_r = p (1-p)^r \qquad 0<p<1, \ r=0,1...$$

where p is the probability that the event occurs in a single trial. The parameter estimated is the mean ($\mu=(1-p)/p$), from which the defining parameter p is derived.

The *logarithmic series* (or *log-series*) distribution is applicable when there is no zero cell, for example when events are not reported unless they occur at least once. This might occur when a crop survey records numbers of parasites per host for infected plants only. The series is also important in the study of species diversity. The distribution is given by

$$p_r = \frac{\theta^k}{zk} \qquad \text{where} \quad z=-\log(1-\theta), \ r=1,2..., \quad 0<\theta<1$$

The parameter estimated is z, from which θ is derived.

The *Neyman type A* distribution is a *contagious* distribution; that is, one allowing for heterogeneity, in which events are aggregated into groups. The number of groups per unit interval has a Poisson distribution (with mean μ_1), and the number of events per group has an independent Poisson distribution with mean μ_2. The Neyman type A distribution is generated by compounding the two Poisson distributions. The probabilities, p_r, can be described by the recurrence relation:

$$p_0 = \exp(-\mu_1(1-\exp(-\mu_2)))$$

$$p_r = \frac{\mu_1}{r} e^{-\mu_2} \sum_{j=1}^{r} \mu_2^j \frac{p_{r-j}}{(j-1)!} \qquad r>0$$

This distribution is less skew than the negative binomial, and cannot be fitted if the variance is less than the mean. When μ_2 tends to zero the distribution becomes a simple Poisson; if μ_2 tends to infinity whilst the mean $\mu=\mu_1\mu_2$ remains constant the distribution tends to a Poisson with added zeroes. The distribution is fitted by estimating the mean and the variance from which the defining parameters μ_1 and μ_2 are obtained. These parameters may be highly negatively correlated, since the mean $\mu=\mu_1\mu_2$ is usually well-defined.

The *Pólya-Aeppli* distribution is a contagious distribution where the number of groups per unit interval has a Poisson distribution with mean μ_1 and the number of events per group has a geometric distribution with parameter p. The probabilities are generated by the recurrence relation:

$$p_0 = e^{-\mu_1}$$

$$p_r = \frac{p\mu_1}{r} \sum_{j=0}^{r-1} (r-j)(1-p)^{r-j-1} p_j \qquad r>0$$

As p tends to 1 the distribution becomes Poisson. As μ_1 tends to 0 the distribution becomes geometric with added zeroes. The distribution is fitted by estimating the mean (μ_1/p) and variance ($\mu_1(2-p)/p^2$), from which estimates of the defining parameters μ_1 and p are obtained.

The *Poisson-Pascal* distribution is a more general three-parameter contagious distribution in which the number of groups per unit interval has a Poisson distribution (with mean λ) and the number of events per group has a negative binomial (or Pascal) distribution. The distribution is defined by the parameters k, p (with $q=1+p$), and λ in the following recurrence relations:

$$p_0 = \exp(-\lambda(1-q^{-k}))$$

$$p_r = \frac{\lambda p k}{r\, q^{k+1}} \sum_{j=1}^{\infty} \left(\frac{p}{q}\right)^{j-1} \binom{k+j-1}{k} p_{r-j} \qquad r>0$$

The distribution is fitted by estimating the mean, the Poisson index and the negative binomial index: the defining parameters can then be derived. This distribution contains several others as special cases:

	k	Negative binomial index
Neyman type A	∞	1
Pólya-Aeppli	1	1.5
Negative binomial	0	2

The *Poisson-log-Normal* distribution is an aggregated distribution which is more skew than the negative binomial. It is generated as a mixture of Poisson distributions whose means are log-Normally distributed with mean μ_1 and variance σ^2. Then the probabilities are obtained as follows:

$$p_r = \frac{1}{r!\sigma\sqrt{2\pi}} \int_0^\infty e^{-z} z^{r-1} \exp(-(\log(z)-\mu_1)^2/2\sigma^2)\, dz$$

The mean and variance of the distribution are fitted, from which the defining parameters μ and σ^2 are obtained. The probabilities are computed by numerical integration when r is small and by an approximation formula when r is large.

Other discrete distributions could be fitted to data using the facilities for fitting nonlinear models: see 8.7 for more details. We now go on to look at the continuous distributions in more detail. For these the density function $f(x)=F'(x)$ is used instead of point probabilities.

Several of the continuous distribution functions available are based around the *Normal* distribution, which has density function:

$$f(x) = \frac{1}{\sigma\sqrt{2\pi}}\, e^{-\frac{1}{2}\left(\frac{x-\mu}{\sigma}\right)^2} = \Phi(x;\mu;\sigma)$$

The parameters to be estimated are μ and σ. The sample skewness is 0 with variance $6/n$ and the sample kurtosis is 0 with sampling variance $24/n$.

The *double Normal* distribution can be used when an observation may come from either of two Normal populations with different means. If a proportion p of the population is Normally distributed with mean μ_1 and variance σ_1^2 and a proportion $(1-p)$ is Normally distributed with mean μ_2 and variance σ_2^2 the density function is:

$$f(x) = p\Phi(x;\mu_1;\sigma_1) + (1-p)\Phi(x;\mu_2;\sigma_2)$$

with mean $p\mu_1+(1-p)\mu_2$ and variance $p\sigma_1^2+(1-p)\sigma_2^2+p(1-p)(\mu_1-\mu_2)^2$. There may be one mode or two, depending on the separation of μ_1 and μ_2. There are two cases of the Double Normal that can be fitted. The variances can be constrained to be equal, by setting `DISTRIBUTION=dNvequal`, so that four parameters (p, μ_1, μ_2, and σ) are fitted. As p tends to 0 or 1 the limiting case of a single Normal is reached, and as μ_1 tends to μ_2, p becomes indeterminate. The more general five-parameter model (p, μ_1, μ_2, σ_1, and σ_2) can be fitted by setting `DISTRIBUTION=dNvunequal`. Unless there is good separation between the two underlying distributions, local maxima may cause problems during the fitting process.

The *log-Normal* distribution assumes that $\log(X)$ (the natural logarithm) is Normally distributed with mean μ and variance σ^2. An additional location parameter a can be included in the model so that the Normal distribution is fitted to $\log(X-a)$, by setting `CONSTANT=estimate`. By default, the constant is omitted and the two-parameter model is fitted (that is, with $a=0$). The density function is

$$f(x) \;=\; \frac{1}{x-a}\; \Phi(\,\log(x-a)\,;\mu\,;\sigma\,)\qquad x>a$$

with mean $a+\exp(\mu+\sigma^2/2)$ and variance $\exp(2\mu+\sigma^2)(\exp(\sigma^2)-1)$. The distribution must have positive skewness; if the sample skewness is negative an automatic switch is made to the Normal distribution, which is the limit as a tends to minus infinity.

The *exponential* (or *negative exponential*) distribution can be used to model *lifetime* distributions, for example the time to failure of a process or death of an organism, where the failure rate can be assumed constant. The density function is

$$f(x) \;=\; b\,e^{-bx}\qquad x>0,\; b>0$$

where b is the failure rate per unit time. The mean is $1/b$, the variance is $1/b^2$, and the median is $\log(2)/b$.

The *Weibull* distribution is a generalization of the exponential distribution in which the failure rate can vary monotonically with time. It can be derived using a power transformation, so that X^c is assumed to have an exponential distribution. The density function is given by

$$f(x) \;=\; cb^{\,c}x^{\,c-1}\exp(\,-(bx)^c\,)\qquad x>0,\; b,c>0$$

which has mean $(1/b)\Gamma(\,(c+1)/c\,)$ and median $(1/b)(\log2)^{1/c}$. For $0<c<1$ the failure rate decreases with time and has a single mode at 0. If $c>1$ the failure rate increases with time and the mode is at $(1/b)(1-c^{-1})^c$. The skewness decreases as c increases, until $c=3.6$ when the skewness is 0, then becomes negative. The Weibull distribution is fitted by holding the median fixed to the sample estimate, whilst obtaining an initial estimate of c; the full model is then fitted. If the option CONSTANT=estimate is set, an additional location parameter is estimated, so that the Weibull is fitted to $(X-a)$. By default, CONSTANT=omit.

The *gamma* distribution is useful as a general empirical distribution. It is similar in form to the Weibull, and is closely related to other standard distributions. By default, it is fitted with two parameters. An additional location parameter can be fitted by setting CONSTANT to estimate, and you must do this if X can take negative values. The density function for the two-parameter model is

$$f(x) \;=\; b^{\,k}\,x^{\,k-1}\,\frac{e^{-bx}}{\Gamma(k)}\qquad \text{where}\quad \Gamma(k) \;=\; \int_0^{\infty} x^{\,k-1}\,e^{-x}\,dx\,,\qquad x>0.$$

with mean $\mu=k/b$ and variance $V=kb^{-2}$. If $k=1$ then this is the exponential distribution; if $b=\frac{1}{2}$ it is a χ^2 distribution with $2k$ degrees of freedom and, if $b=1$, the gamma tends to the standard Normal as k tends to infinity. The gamma distribution is fitted using the sample median, approximately $(k+1)/b$, to provide initial estimates for the parameters before the full model is fitted.

The *beta* distribution is suitable for fitting proportions and ratios. Two forms are available in Genstat, denoted *type I* and *type II*. The type I distribution is a two-parameter model restricted to values in the range $0<x<1$ and is thus used to fit proportions. The density function is

$$f(x) \;=\; \frac{1}{B(p,q)}\, x^{p-1}\,(1-x)^{q-1} \qquad 0 < x < 1, \;\; p,q > 0$$

$$\text{where} \quad B(p,q) \;=\; \int_{0}^{1} x^{p-1}\, x^{q-1}\, dx, \qquad B(p,q) \;=\; \frac{\Gamma(p)\,\Gamma(q)}{\Gamma(p+q)}.$$

This distribution has mean $\mu=p/(p+q)$ and variance $pq/\{(p+q)^2(p+q+1)\}$. If $p>1$ and $q>1$ then there is a single mode at $x=(p-1)/(p+q-2)$; whilst if $p<1$ and $q<1$ there is a minimum at this point. For large values of p and q the distribution is approximately Normal.

The type II beta distribution is suitable for any positive continuous data, and has density

$$f(x) \;=\; \frac{b^{\,p}\, x^{p-1}}{(1+bx)^{p+q}\, B(p,q)} \qquad x > 0$$

now with three parameters b, p and q. The distribution has mean $\mu=1/\{b(q-1)\}$ and variance $V=p(p+q-1)/\{b^2(q-1)^2(q-2)\}$. The mode is at 0 for $p<1$ and at $(p-1)/(q+1)$ otherwise. For large values of q the distribution tends to a gamma distribution with index p. For $p=m/2$, $q=n/2$ and $b=m/n$ we have the F distribution with m and n degrees of freedom.

For either form of the beta distribution it is possible to include an additional location parameter, so that the distribution is fitted to $(X-a)$. This is specified by setting the CONSTANT option to estimate. By default, CONSTANT=omit, so no location parameter is fitted.

The *Pareto* distribution originates in economics where it is used for modelling the distribution of incomes in a population. Like the log-Normal it is suitable for data with very long upper tails; it provides a better fit to the tail but performs less well over the whole range. The "Pareto distribution of the first kind" is defined by its distribution function, only for positive data greater than a minimum value c:

$$F(x) \;=\; 1 - \left(\frac{c}{x}\right)^{b} \qquad x \geq c, \;\; b,c > 0.$$

An additional location parameter can be requested, by setting CONSTANT=estimate. This fits a Pareto distribution "of the second kind", which has the distribution function

$$F(x) \;=\; 1 - \left(\frac{c-a}{x-a}\right)^{b} \qquad x \geq c > a, \;\; b,c > 0.$$

The mean is $\mu = a + (c-a)b/(b-1)$ if $b>1$, and the variance is $b(c-a)^2(b-1)^{-2}(b-2)^{-1}$ if $b>2$.

7.2   Comparison of groups of data

The aim of many statistical studies is to compare different groups of observations. These groups may differ because they have been selected from separate populations, or perhaps because they have received different experimental treatments. It is often possible to fit statistical models containing different parameters for each group, which enable you to answer questions like "How much better is treatment A than treatment B?". These methods of estimation generally also provide extra information such as standard errors, sums of squares, or perhaps deviances to allow you to check statistically whether the treatments genuinely do differ in their effects. Many of the estimation procedures in Genstat provide formal probability levels associated with these *hypothesis tests* (see for example 8.1.2 or 9.1).

However, it is not always possible to make sensible assumptions about the models or the probability distributions from which the observations have been generated. Genstat therefore also contains a range of procedures for performing non-parametric or distribution-free tests that require only relatively simple assumptions.

In this section we start with the simplest situation of the one-sample test, where the purpose is to compare the properties of a single sample against some known characteristic (7.2.1). The DISTRIBUTION directive described earlier (7.1.4) can of course be used to see whether the data come from various probability distributions, but more often the interest is mainly in the location of the mean of the sample. The one-sample t-test (provided by procedure TTEST) allows you to check whether there is evidence that the mean differs from a particular value, under the assumption that the data come from a Normal distribution. The Wilcoxon test (procedure WILCOXON) provides a non-parametric alternative.

Equivalent methods for comparing the locations of two samples, the two-sample t-test (again procedure TTEST) and the Mann-Whitney U test (procedure MANNWHITNEY), are illustrated in 7.2.2, together with the Kolmogorov-Smirnov test (procedure KOLMOG2) which provides a distribution-free test for similarity of the cumulative distribution functions of the two samples.

Finally, when there are more than two samples (7.2.3), you can use the non-parametric Kruskal-Wallis test (procedure KRUSKAL), or the powerful technique of analysis of variance; the latter is covered in depth in Chapters 9 and 10.

7.2.1   One-sample tests

Here we describe ways of testing whether the mean of a set of observations differs from some target value. If you are willing to assume that the observations are Normally distributed, you can use the TTEST procedure. The observations are specified, in a variate, using the first parameter of TTEST; there is also an option, NULL, to set the target (by default zero). Example 7.2.1a applies this test to a set of diffusion data discussed further in 7.2.2.

Example 7.2.1a

```
   1  "   Rates of diffusion of carbon dioxide through a fine soil.
  -2       Data from Smith and Brown (1933); also analysed by
  -3       Snedecor & Cochran (1989) p94, (who give wrong reference for data)."
   4
```

```
 5   VARIATE [VALUES=20,31,18,23,23,28,23,26,27,26,12,17,25] Fine
 6   TTEST [NULL=20] Fine
```

***** One-sample T-test *****

```
     Sample      Size      Mean      Variance
     Fine        13        23.00     26.50
```

*** Test for evidence that distribution mean is different to 20.00 ***

     Test statistic t = 2.10 on 12 df.

     Probability level (under null hypothesis) p = 0.057

The one-sample t-test is based upon the statistic

$$t = (\bar{x} - \mu) / (s / \sqrt{n})$$

where μ is the target value, n is the number of observations in the sample, $\bar{x} = \Sigma\, x_i / n$ is the sample mean and $s^2 = \Sigma\, (x_i - \bar{x})^2 / (n - 1)$ is the usual unbiased estimate of the sample variance. For Normally distributed observations, the statistic t is distributed as Student's t distribution on $n-1$ degrees of freedom. The probability level quoted is the theoretical probability of getting a result as extreme as the value calculated, given that the null hypothesis is true (that is, that the sample mean equals the target value). For this example, the probability of getting a value as large as 2.10 is 0.057 under the null hypothesis. Since this probability is small, there is evidence that the sample mean is different from 20, but (since the probability is greater than 0.05) there is not enough evidence to reject the null hypothesis at the 5% level. By default, the probability level quoted is for a two-sided test, but the METHOD option allows you to specify a one-sided alternative hypothesis, for example that the sample mean is greater than the target value. It is also possible to produce a 95% confidence interval for the sample mean by setting option CIPROB=0.95. Further details on hypothesis testing can be found in any book covering basic statistical methods, such as Snedecor and Cochran (1989).

The Wilcoxon Matched-Pairs Signed-Rank test is a non-parametric equivalent to the one sample t-test, and can be performed using the WILCOXON procedure. This procedure is based upon the ranked data values, and so depends only on the order, not on the actual distribution of the data. It does not have a NULL option like TTEST, so you need to use the CALCULATE directive first to form the differences with the target value.

Example 7.2.1b

```
 8   CALCULATE Fine20 = Fine-20
 9   WILCOXON Fine20
```

Wilcoxon Matched-Pairs Test

```
Test Statistic   =         15.00 (rank sum positive)
Sample size:                  12 (zero values have been excluded)
Normal Approximation =     1.883 (p=0.06 two-sided test)
```

By default, WILCOXON prints the test statistic and sample size, excluding zero values. In this case, the sample size is 12 since one of the original data values was 20 which then gave a zero value in Fine20. There is a function of this statistic which has an asymptotic Normal approximation (that is, it is Normally distributed for very large samples). In practice, this approximation is good enough for sample sizes larger than 8, so it is printed then with its probability under the null hypothesis. In Example 7.2.1b the conclusions are the same as from the t-test.

The SIGNTEST procedure could also have been used to perform a one-sample sign test to test the location of the sample median for this data. This procedure is applicable also in the two sample situation discussed in 7.2.2, and is documented in the Procedure Library Manual.

Most of the non-parametric procedures described in this chapter follow the same basic structure. By default the test statistic is printed. For large samples, they also present an asymptotic approximation and probability value under the null hypothesis while, for small samples, exact probability levels are given where possible. Adjustments for tied data values are made where necessary. By default (PRINT=test) the test statistics are displayed, but it is also possible to print out the ranks on which the test is based (PRINT=ranks), and to save the results in Genstat data structures. Further details of the non-parametric procedures are given in the Procedure Library Manual, and the theoretical background for these tests is given by Siegel (1956) and Conover (1971).

7.2.2   Two-sample tests

If you want to compare two samples of measurements it is important to establish first how the samples were taken. All the tests in this subsection are based on the assumption that individual measurements have been made independently. In particular, these tests are not appropriate for measurements taken in a series where it is likely that neighbouring measurements are more correlated than measurements further apart in the series; in this situation you could use the time-series methods in Chapter 12, or procedures in the repeatedmeasures module of the procedure library (5.3.1). Likewise, if the two samples have been taken in a paired way, so that each measurement in one sample is matched with a measurement in the other, the test procedure must reflect this structure, for example by treating the pairs of observations as blocks in an analysis of variance or by subtracting one set of values from the other and then doing a one-sample test. This structure often arises when several samples are taken from a single set of individuals.

Examples 7.2.2a and 7.2.2b deal with measurements of diffusion of carbon dioxide through two soils of different porosity (data from Smith and Brown 1933, also analysed by Snedecor and Cochran 1989). There is no pairing of the measurements – indeed, there are different numbers of measurements in each sample – and we assume that the measurements are independent. If we can assume also that the measurements in the samples are Normally distributed, or at least approximately Normally distributed, then we can use the t-test to test the difference between the means of the samples. (It is possible to use the DISTRIBUTION directive (7.1.4) to assess the distribution of a sample; however, with small samples like this one, there is rarely enough evidence to clearly determine the distribution.) Example 7.2.2a shows how to use the TTEST procedure to compare these two samples assuming Normally distributed observations. The TTEST procedure performs a test based on the assumption of equal variance within the two samples, and performs an F test for homogeneity of variances to check this assumption. If there is evidence of unequal variation, a warning is printed out and the test may not be valid. The t-test statistic calculated is

$$t = (\bar{x}_1 - \bar{x}_2) / (s / \sqrt{\{ (n_1 + n_2) / n_1 n_2 \}})$$

where n_1 and n_2 are the numbers of observations in the two samples, $\bar{x}_1$ and $\bar{x}_2$ are the two sample means, $s_1{}^2$ and $s_2{}^2$ are unbiased estimates of the sample variances (see 7.2.1), and $s^2 = \{ (n_1 - 1)s_1{}^2 + (n_2 - 1)s_2{}^2) \} / (n_1 + n_2 - 2)$ is a combined estimate of the variance. For Normally distributed data, t is distributed as Student's t on n_1+n_2-2 degrees of freedom.

Example 7.2.2a

```
   1   "  Rates of diffusion of carbon dioxide through two soils.
  -2      Data from Smith and Brown (1933); also analysed by
  -3      Snedecor & Cochran (1989) p94."
   4   VARIATE [VALUES=20,31,18,23,23,28,23,26,27,26,12,17,25] Fine
   5   &       [VALUES=19,30,32,28,15,26,35,18,25,27,35,34] Coarse
   6   TTEST   [CIPROB=0.95] Fine; Coarse

***** Two-sample T-test *****

      Sample      Size      Mean      Variance
      YY1         13        23.00     26.50
      YY2         12        27.00     46.00

*** Test for evidence that the distribution means are different ***

      Test statistic t = -1.67 on 23 df.

      Probability level (under null hypothesis) p = 0.109

      95.0% Confidence Interval for difference in means: ( -8.957 , 0.9567 )
```

If you would prefer not to make any assumptions of Normality, you can use the Mann-Whitney U test, as provided by the MANNWHITNEY procedure.

Example 7.2.2b

```
   7   MANNWHITNEY Fine; Coarse

Mann-Whitney U (Wilcoxon Rank-Sum) Test

Value of U               =       47.0   (second variate has highest rank score )
Normal Approximation =           1.686  (p=0.09)
Adjusted for ties        =       1.690  (p=0.09)

Sample sizes:    13      12
```

For this data set, the results of the two tests are similar, indicating a probability of 9-11% of obtaining a result this extreme under the null hypothesis of no difference between sample means. So there is some, but not strong, evidence that the mean of the second sample is higher; it is conventional, however, to reject the hypothesis of no difference only at the 5% level. Indeed, the 95% confidence interval for the difference between the two means, produced from TTEST by setting option CIPROB=0.95, includes zero, showing that a zero difference is not inconsistent with the data at this significance level.

With paired samples, you can subtract one set of observations from the other and then analyse these differences using one of the one-sample tests described in 7.2.1. Example 7.2.2c uses this method to compare counts of lesions on halves of eight tobacco leaves, where the two halves of each leaf were treated with two different preparations of a virus (data from Youden and Beale 1934, also analysed by Snedecor and Cochran 1989).

Example 7.2.2c

```
   10   " Numbers of lesions on halves of eight tobacco leaves
  -11     treated with two virus preparations. Data from Youden & Beale (1934);
  -12     also analysed by Snedecor & Cochran (1989) p86."
   13   VARIATE [VALUES=31,20,18,17, 9, 8,10, 7] Prep1
   14   &        [VALUES=18,17,14,11,10, 7, 5, 6] Prep2
   15   CALCULATE Diffprep = Prep1-Prep2
   16   TTEST Diffprep

***** One-sample T-test *****

      Sample      Size     Mean      Variance
      Diffprep    8        4.000     18.57

*** Test for evidence that distribution mean is different to 0 ***

      Test statistic t = 2.63   on 7 df.

      Probability level (under null hypothesis) p = 0.034

   17   WILCOXON Diffprep

Wilcoxon Matched-Pairs Test

Test Statistic =          2.000  (rank sum positive)
Sample size:                 8
Normal  Approximation =   2.240  (p=0.03 two-sided test)
```

Again, the results from the t-test and the non-parametric test agree, showing strong evidence that more lesions arise from the first virus preparation than from the second. In this case it is important to use a one-sample test on the differences: we would expect two halves of the same leaf to respond similarly to the virus, and whereas analysing the differences preserves the connection between the two halves of each leaf, a two-sample test assumes no connection between the observations.

There are of course other aspects of two samples that can be compared besides their mean or median values. The KOLMOG2 procedure provides a distribution-free test of overall similarity between the distributions of two samples. In Example 7.2.2d we apply this to the diffusion data of Example 7.2.2a and find no significant difference between the cumulative distribution functions (CDFs) of the two samples.

Example 7.2.2e

```
  20   KOLMOG2 Fine; Coarse

Kolmogorov-Smirnov Two-Sample Test

Maximum Difference    =    0.3526
Chi-squared           =    3.103    (p=0.21)

Sample Sizes:    13    12

Signed Differences between the CDFs

   0.0769   -0.0064    0.0705    0.0641   -0.0192    0.0577    0.2885    0.2821
   0.3526    0.3462    0.3397    0.2564    0.3333    0.2500    0.1667    0.0000
```

7.2.3   Testing differences between more than two groups

Analysis of variance provides a structured approach to the analysis of data that are multiply classified: treatment factors generally describe aspects that are controlled by the experimenter, while block factors are used to specify the underlying structure of the design (for example, randomized blocks, Latin square, split-plot, and so on) and thus the sources of random variation. Chapter 9 describes the powerful facilities in Genstat for analysing balanced designs, with one or more error terms. Chapter 10 describes the REML algorithm for analysing unbalanced designs with several error terms, and for estimating variance components. Both of these chapters assume that the measurements are approximately Normally distributed. The generalized linear models framework, in Chapter 8, caters for other distributions but with the constraint that there can be only one error term; however, procedure GLMM implements one way of extending this methodology to data with several error terms.

In this subsection we introduce only the straightforward one-way analysis. This can be seen simply as an extension of the two-sample problems in 7.2.2, allowing more than two samples to be compared. In Examples 7.2.3a and 7.2.3b, we analyse some measurements on fat absorbance of doughnuts during cooking (from Snedecor and Cochran 1989). Firstly, here is the analysis of variance of these four samples each of six measurements, using the standard Library procedure AONEWAY. The factor indicating the groups is specified using the GROUPS option, and the data variate is specified as the first parameter of the procedure.

Example 7.2.3a

```
   1   "  Absorbance of four types of fat while cooking doughnuts. Data
  -2       from Lowe (1935); also analysed by Snedecor & Cochran (1989) p217."
   3
   4   VARIATE [VALUES=64,72,68,77,56,95,  78,91,97,82,85,77, \
   5                   75,93,78,71,63,76,  55,66,49,64,70,68] Absorb
   6   FACTOR [LEVELS=4; VALUES=6(1...4)] Fat
   7   AONEWAY [GROUPS=Fat; HOMOGENEITY=yes] Absorb

7.................................................................................

***** Analysis of variance *****
```

```
Variate: Absorb

Source of variation      d.f.        s.s.       m.s.     v.r.   F pr.
Fat                        3       1636.5      545.5     5.41   0.007
Residual                  20       2018.0      100.9
Total                     23       3654.5

***** Tables of means *****

Variate: Absorb

Grand mean  73.8

        Fat          1         2         3         4
                  72.0      85.0      76.0      62.0

*** Standard errors of differences of means ***

Table                 Fat
rep.                    6
s.e.d.               5.80

   *** Bartlett's Test for homogeneity of variances ***

       Chisq          d_f
        1.75            3
```

The procedure packages the analysis-of-variance directives available in Genstat to provide an appropriate one-way analysis. The variance ratio (v.r.) can be used to construct an F test of the null hypothesis that there are no differences between the groups, here the different types of fat. In this example, the probability of the statistic under the null hypothesis (F pr.) is 0.007, indicating differences between the fat types which can be seen in the table of means: fat type 2 tends to have a higher absorbance and fat type 4 has a lower absorbance. See Chapter 9 for further details about the output, or if you want to analyse more complex designs. In this example we have also set the HOMOGENEITY option of the AONEWAY procedure to carry out Bartlett's test for homogeneity of the variances in the four groups. This statistic is small compared to a chi-squared distribution on three degrees of freedom, and so there is no evidence against the assumption of equal variation across the groups.

An alternative to this distribution-based approach is to use the Kruskal-Wallis one-way analysis of variance, a non-parametric method based on the ranks of the data. This method can be used only for a one-way classification; that is, with only a single grouping factor. Example 7.2.3b shows the use of the procedure KRUSKAL to analyse the doughnut data.

Example 7.2.3b

```
   9  KRUSKAL [GROUPS=Fat] Absorb

Kruskal-Wallis One-Way Analysis of Variance

Value of H           =     11.81
Adjusted for ties    =     11.83

Sample Sizes:          6         6         6         6
```

```
Mean Ranks            11.3   19.5   13.6    5.7

Degrees of freedom =           3
Chi-square p-value =        0.01
```

In this example, the input is given in the same format as for AONEWAY. However, in KRUSKAL it is also possible to specify a list of variates, containing the data for each of the groups. The Kruskal-Wallis test ranks the whole data set, and then compares ranks across the different groups. The chi-square test indicates that differences do exist between groups, and the mean ranks show which samples tend to have higher or lower scores: in this case sample 2 tends to have higher and group 4 lower scores, as in the analysis of variance in Example 7.2.3a.

7.3   Relationships between variables

Many scientific studies involve establishing and understanding relationships between variables, and statistics offers many useful techniques for this purpose. In this chapter we discuss only some of the simpler methods. Subsection 7.3.1 introduces the graphical displays available in Genstat for exploring relationships, referring to Chapter 6 for more detail. We then outline some of the simplest methods for quantifying relationships: correlation (7.3.2), tests for independence of categorical variables (7.3.3), and regression analysis (7.3.4). These are all covered in more detail in Chapter 8.

Chapter 8 describes techniques for fitting nonlinear models as well as linear ones: there is a range of standard curves which can be fitted automatically, as well as the ability to specify general nonlinear relationship between the variables. Genstat also provides the class of *generalized linear models*, which extend the available probability distributions beyond the usual Normal distribution, and allow for a *link* transformation to relate the fitted values to the linear model. Nonparametric relationships can be included using smoothing techniques (see 8.4.3); the models are then described as *additive* or *generalized additive models*.

The linear and nonlinear models described in Chapter 8 are all based on the assumption of uncorrelated errors. To analyse series of measurements for which this assumption cannot be made, such as successive observations in time, you can use the methods of *time-series analysis* described in Chapter 12.

There are also many techniques for searching for underlying variables to describe relationships; these *multivariate methods* are covered in Chapter 11. *Principal components analysis* is the most common method used in the reduction of the dimensionality of a set of multivariate data. Genstat also provides *canonical variates analysis*, which is a method of linear *discriminant analysis*, and both hierarchical and non-hierarchical clustering.

7.3.1   Displaying relationships

The scatterplot is a powerful graphical method for displaying the relationship between two variables (see Tufte 1983, page 47). In Genstat, the scatterplot and many variants are easily produced using the DGRAPH directive (6.2.1), with the GRAPH directive providing line-printer output when high-resolution graphics are not available. Example 7.3.1 shows the statements needed to draw a scatterplot showing cancer death rates and cigarette consumption, discussed and drawn by Tufte (1983, page 47). The picture produced is shown in Figure 7.3.1.

Example 7.3.1

```
 1   " Display the relationship between death rates from lung cancer and
-2       per capita cigarette consumption. Data from Tufte (1983).
-3   "
 4
 5   TEXT Country
 6   READ [PRINT=data] Country,Deaths,Cigarettes

 7   AUSTRALIA        172   452
 8   CANADA           151   508
 9   DENMARK          168   379
10   FINLAND          353  1113
11   'GREAT BRITAIN'  468  1145
12   HOLLAND          244   468
13   ICELAND           60   226
14   NORWAY            95   258
15   SWEDEN           116   315
16   SWITZERLAND      252   540
17   U.S.A.           194  1290
18   :
19   " Fit a linear regression line to the relationship."
20   MODEL Deaths; FITTED=Fitdeath
21   FIT [PRINT=*] Cigarettes
22   " Set axis and labelling details."
23   PEN 1; LABELS=Country
24   PEN 2; METHOD=line; LINESTYLE=1; SYMBOLS=0
25   AXES WINDOW=1; YTITLE='DEATHS PER MILLION'; \
27      XTITLE='CIGARETTE CONSUMPTION'
27   DGRAPH [KEY=0; TITLE=\
28      'Lung cancer deaths 1950 vs cigarette consumption 1930'] \
29        Deaths,Fitdeath; Cigarettes; PEN=1,2
```

A single DGRAPH statement is all that would have been necessary to produce a simple unlabelled scatterplot; the other statements here are included to provide labelling for the axes and the points, and to draw the fitted regression line. If further statements are given to modify the graphical environment (see 6.5) it is possible to reproduce Tufte's picture exactly.

When more than two variables are to be investigated, their inter-relationships can be studied by multiple scatterplots. There are also many methods for looking for patterns in the data and for reducing the dimensionality needed to display the patterns. Chapter 11 describes these *multivariate methods*, and the methods of *cluster analysis* which allow you to look for groupings of the subjects based on the various available measurements.

Genstat also provides graphical displays specifically for examining the way in which one variable changes with two other variables, namely contour plots (6.3.1), perspective views of surfaces (6.3.2), and three-dimensional histograms (6.3.3).

7.3.2 Correlation

Correlation is a measure of the association between two variables. The most commonly used correlation coefficient is the product-moment correlation coefficient which measures linear association. This is provided in Genstat by the CORRELATE directive, described in 12.1. CORRELATE can also provide information, either as tables or graphs, about the autocorrelation of a series of measurements, and about cross-correlation between two matched series. These and other methods of analysing time series are covered in Chapter 12. Example 7.3.2a shows how to use the CORRELATE directive simply to display a matrix of correlation coefficients between quantitative variables, in this case three measures of phosphorus in soil.

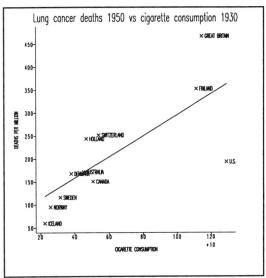

Figure 7.3.1

Example 7.3.2a

```
   1   " Correlations between inorganic phosphorus, organic phosphorus,
  -2     and estimated plant-available phosphorus. Data from Eid et al. (1954);
  -3     also analysed by Snedecor & Cochran (1989) p335."
   4   READ [PRINT=data] Inorg_P,Org_P,Plant_P

   5    0.4 53 64    0.4 23 60    3.1 19 71    0.6 34 61    4.7 24 54
   6    1.7 65 77    9.4 44 81   10.1 31 93   11.6 29 93   12.6 58 51
   7   10.9 37 76   23.1 46 96   23.1 50 77   21.6 44 93   23.1 56 95
   8    1.9 36 54   29.9 51 99 :

   9   " Display product-moment correlations."
  10   CORRELATE [PRINT=correlations] Inorg_P,Org_P,Plant_P

*** Correlation matrix ***

      Inorg_P    1.000
        Org_P    0.399    1.000
      Plant_P    0.720    0.212    1.000

                Inorg_P    Org_P   Plant_P
```

An alternative measure of association is the non-parametric coefficient formed from ranked data, called Spearman's rank correlation coefficient. This measures the association between two variables in terms only of the ranks, or ordering, of the units. Example 7.3.2b shows the output of the SPEARMAN procedure which calculates these coefficients, using the same data as in Example 7.3.2a.

Example 7.3.2b

```
11   " Display rank correlations."
12   SPEARMAN [PRINT=correlations] Inorg_P,Org_P,Plant_P

Spearman Rank Correlation

Sample size:                    17

*** Correlation matrix (adjusted for ties) ***

   1    1.000
   2    0.360   1.000
   3    0.663   0.245   1.000

            1       2       3
```

If the PRINT option is set to correlations, SPEARMAN prints just the correlation matrix between the three variables. The default, PRINT=test, prints both the correlation matrix and Student's t approximations to test for zero correlation. You can see that the rank and product-moment correlations are quite similar for these data.

Another non-parametric measure of rank correlation is Kendall's coefficient of concordance, which measures the overall level of association between several different sets of measurements taken on a single set of subjects. This can be calculated by procedure CONCORD. Example 7.3.2c shows the overall concordance between the three different measures of phosphorus in soil, indicating evidence of association between the orderings of the three variables.

Example 7.3.2c

```
15   " Calculate coefficient of concordance. "
16   CONCORD [PRINT=test] Inorg_P,Org_P,Plant_P

Kendall Coefficient of Concordance

Coefficient         =   0.612
Adjusted for ties   =   0.615
Sample size:            17
Number of samples:       3
Chi-Squared         =   29.5     (p=0.02)
Degrees of freedom  =   16
```

7.3.3   Tests for independence in two-way tables

When measurements are qualitative or categorical, a different approach is needed to establish relationships than when they are quantitative. One way is to analyse the counts of individuals with each combination of levels of the categorical variables: a set of counts like this is most conveniently stored in a table, and is usually referred to as a *contingency table*. The analysis can be done in an analogous way to the analysis of quantitative measurements, fitting what is known as a *log-linear model*, in which the counts are assumed to have a multinomial or

Poisson distribution. But before illustrating how to do this in Genstat, we shall describe a similar technique for testing for independence between classifications, known as the chi-squared test for independence.

In a contingency table there are two or more classifying factors. The table will contain several cells and the interest is in analysing the number of observations in each cell. The tables can be computed from the raw data using the TABULATE directive, as described in 4.11.1. Alternatively, values can be assigned to the table by the TABLE directive or read by the READ directive. In Example 7.3.3a, a two-way table is declared, called simply Counts, and assigned the results from a survey of smoking habits. The classifying factors both have two levels, so the table has four cells.

Example 7.3.3a

```
  1   " Relationship between smoking habits and mortality in Canada.
 -2     Data from Snedecor and Cochran (1989) p124."
  3
  4   FACTOR [LABELS=!t(Dead,Alive)] Mortality
  5   & [LABELS=!t(Nonsmoker,'Pipe smoker')] Smoking
  6   TABLE [CLASSIFICATION=Mortality,Smoking; VALUES=117,54,950,348] Counts
  7   PRINT Counts; DECIMALS=0

               Counts
     Smoking  Nonsmoker Pipe smoker
     Mortalit
        Dead      117          54
        Alive     950         348
```

You can test the independence of these two classifications using the CHISQUARE procedure. This is simply a test of whether the distribution of subjects between the two categories of one factor appears to change according to the categories of the other factor. Example 7.3.3b shows the output from the procedure for these data.

Example 7.3.3b

```
  9   " Perform Pearson chi-square test of independence of classifications."
 10   CHISQUARE Counts

Pearson chi-square value is  1.73  with   1 df.

Probability level (under null hypothesis) p = 0.189
```

The test statistic here indicates that the two classifying factors, Smoking and Mortality, are independent; that is, that there is no evidence of association between the two factors.

The CHISQUARE procedure can also be used with one-way tables, to test whether there is any pattern in the distribution of the counts between the levels of a classifying factor. However, it cannot deal with three-way or more complicated tables.

The alternative method for the analysis of contingency tables, log-linear analysis, is provided in Genstat as part of the regression facilities. This is because the log-linear model is a *generalized linear model* which can be treated in a very similar way to the linear model of classical regression. This produces a statistic known as the *deviance*, equivalent (although calculated differently) to the Pearson chi-squared statistic shown in Example 7.3.3b.

To use the regression facilities, you need to copy the counts from the table into a variate, and define factors to indicate the cell of table from which each value was taken. The MODEL and FIT directives can then be used to calculate the test statistic as shown in Example 7.3.3c.

Example 7.3.3c

```
12   " Calculate maximum-likelihood chi-square using regression directives."
13   VARIATE Vcounts; VALUES=Counts
14   FACTOR [MODIFY=yes; NVALUES=4] Mortality,Smoking
15   GENERATE Mortality,Smoking
16   MODEL [DISTRIBUTION=Poisson] Vcounts
17   FIT [PRINT=deviance] Mortality,Smoking
```

17...

Residual d.f. 1, deviance 1.685

```
18   " Display the test statistic with approx. probability."
19   RKEEP DEVIANCE=Chisquare
20   CALCULATE Chiprob = CUCHI(Chisquare; 1)
21   PRINT Chisquare,Chiprob
```

```
    Chisquar    Chiprob
       1.685     0.1943
```

The test of independence is provided by the *residual deviance*, so you can see that the two methods give a very similar result for these data. Log-linear models can be used with any number of classifying factors, and more detailed models can be fitted than simple independence between classifications. See 8.5 for details.

The analysis by either of these two methods is approximate: the test statistics are only approximately distributed as chi-squared statistics with one degree of freedom. The approximation improves as the number of observations increases; the numbers in this example are large enough for the approximation to be good. However, Genstat also provides an exact method to test for independence in this simple case of a two-way table with only four cells. This is called Fisher's exact test, and is carried out by the FEXACT2X2 procedure. The result for the smoking data is shown in Example 7.3.3c. The two-tailed significance values are equivalent to the probabilities given by the chi-squared tests. The remainder of the output is explained under procedure FEXACT2X2 in the Procedure Library Manual.

Example 7.3.3c

```
23   " Use Fisher's exact test."
24   FEXACT2X2 Counts
```

```
One-tailed significance level =          0.1115
Mid-P value =        0.09635
```

Two-tailed significance level
    Two times one-tailed significance level =   0.2229
    Mid-P value =   0.1927
    Sum of all outcomes with Prob<=Observed =   0.2015
    Mid-P value =   0.1864

7.3.4   Quantifying relationships

Regression, and its extensions already mentioned, provide a good way of quantifying the relationship between a response variable and other, explanatory, variables. To perform the simple linear regression of Y on X, you need only the following two statements:

```
MODEL Y
FIT X
```

The MODEL statement indicates the response (or dependent) variable, while the FIT statement specifies the explanatory (or independent) variables that are to explain the variation of the response. Once the response variable has been specified, you can explore a series of sets of explanatory variables in a search for a suitable model (8.2).

Example 7.3.4 fits the line that was shown in Figure 7.3.1, representing the relationship between lung cancer and smoking.

Example 7.3.4

```
  1   " Fit a linear regression of the death rates from lung cancer on
 -2     per capita cigarette consumption. Data as in Example 7.3.1a"
  3
  4   TEXT Country
  5   READ [PRINT=*] Country,Deaths,Cigarettes

 18   UNITS Country
 19   MODEL Deaths
 20   FIT Cigarettes

20.............................................................
```

***** Regression Analysis *****

Response variate: Deaths
    Fitted terms: Constant, Cigarett

*** Summary of analysis ***

	d.f.	s.s.	m.s.	v.r.
Regression	1	77881.	77881.	11.05
Residual	9	63434.	7048.	
Total	10	141315.	14131.	

Percentage variance accounted for 50.1
Standard error of observations is estimated to be 84.0
* MESSAGE: The following units have large standardized residuals:
         U.S.A.   -2.62
* MESSAGE: The following units have high leverage:
         U.S.A.   0.41

```
*** Estimates of regression coefficients ***

                 estimate           s.e.        t(9)
Constant            66.7            49.1        1.36
Cigarett          0.2299          0.0692        3.32
```

This analysis highlights the data from the U.S.A. as being different from the rest: not only is the cigarette consumption the highest, making it *influential* in the analysis, but also the death rate is much lower than would be expected on the basis of the linear model fitted for all the countries. Another drawback is that the death rates are probably more variable between countries that have high rates than between those that have low ones, so it would be better to assume a gamma distribution for the response, rather than a Normal distribution with constant variance as is assumed in linear regression (8.5.1).

Ignoring these anomalies, the analysis estimates that the death rate from lung cancer seems to be higher by 23 men per million (s.e. 7) for each increase of 100 cigarettes smoked on average in a country per man per year.

Further details of the regression facilities in Genstat, and many other examples, are given in Chapter 8.

S.A.H.
P.W.L.
G.W.M.
S.J.W.

8 Regression analysis

The simplest meaning of the word *regression* is the technique for fitting a straight line that relates one quantitative variable to another. The *response variable* is supposed to be dependent on the *explanatory variable*. We describe how to do this simple linear regression with Genstat in 8.1.

In later sections we use the word regression to cover a much wider class of relationships. We look at more than two variables, at qualitative variables, and at nonparametric and nonlinear relationships. But the common feature is that we shall always be modelling the dependence of one variable on others.

The word linear here does not mean linear in terms of the explanatory variables, but rather linear in terms of the parameters or coefficients that have to be estimated. Thus the regression
$$y_i = \alpha + \beta x_i + \gamma x_i^2 + \varepsilon_i$$
is in fact linear: it is linear in terms of the parameters α, β, and γ, even though it is not linear in terms of the explanatory variable X.

In the model for simple linear regression, it is usually assumed that the response variable has a Normal distribution with constant variance. But other distributions can be used, and the variance need not be constant. For example, the distribution could be Poisson in which the variance is equal to the mean. These extensions are provided by generalized linear models, as described in 8.5.

In all the models in this chapter, we assume that there is only one component of variation: that is, they contain only one error term like ε in the equation above. When there are more components, some results can be obtained by the methods described here: for example, you could analyse the effects of treatment factors after eliminating some blocking of the units, by treating the blocking factor as if it were another treatment factor. But it is usually more convenient, and more efficient, to use the methods of Chapter 9 if the design is balanced, or those of Chapter 10 otherwise.

We assume in this chapter that you know which is the response variable and which are explanatory variables. There are more general methods of investigating relationships between variables, in which no single variable is treated as a response; see Chapter 11. We also assume that the relationship between the response variable and explanatory variables relates the mean of the response to given explanatory values. The methods of regression analysis are not applicable to law-like relationships, with values of both the response and the explanatory variables subject to error; for more details, see Sprent (1969).

Finally, we assume in this chapter that the errors in the regression models are uncorrelated. For example, the quantities ε_i in the equation above are assumed to be independently distributed. When there is some correlation between the errors, the methods of Chapter 9 may be suitable, particularly if the correlation is constant within some groups of the data and zero between the groups. Alternatively, if there is a serial pattern of correlation, where the order of the observations is important, the methods of Chapter 12 may be used.

Throughout this chapter we assume that you already know about regression. We give references to standard books, where you can find the theory explained in full.

8.1   Simple linear regression

The word *simple* here does not mean easy, but rather that there is only one explanatory variable. Suppose you have observations $\{y_i: i = 1...N\}$ of a response variable Y, and $\{x_i: i = 1...N\}$ of an explanatory variable X. Then the model for simple linear regression is:

$$y_i = \alpha + \beta x_i + \varepsilon_i$$

where α and β are unknown *parameters*: that is, they are numerical characteristics of the model that determine the precise nature of the relationship. The values $\{\varepsilon_i: i = 1...N\}$ are *errors* which are random variables, identically and independently distributed with a Normal distribution. For details of this model, see the books by Seber (1977), Draper and Smith (1981), or Weisberg (1985), or indeed any other standard statistical text.

The model can alternatively be written in matrix form:

$$y = X\beta + \varepsilon$$

where the vector $\beta = (\alpha,\beta)'$, and X is an $N\times2$ matrix whose first column consists just of 1's, called the *design matrix*. (This is standard terminology although, of course, regression is often used when it has not been possible to use any special design.)

Example 8.1 shows how to fit a simple linear regression. The model here is a linear relationship between the logarithm of barometric pressure and the boiling point of water. Forbes (1857) collected these measurements at the tops of mountains with the intention that, on any other mountain, he would be able to predict barometric pressure (and hence the height of the mountain) by boiling water at the summit.

Example 8.1

```
  1   " Simple linear relationship between boiling point and barometric
 -2     pressure.  Data from Forbes (1857); analysed by Weisberg (1985) p3."
  3
  4   READ [PRINT=data] Boiltemp,Pressure

  5   194.50 20.79   194.25 20.79   197.90 22.40   198.43 22.67   199.45 23.15
  6   199.95 23.35   200.93 23.89   201.15 23.99   201.35 24.02   201.30 24.105
  7   203.55 25.14   204.60 26.57   209.47 28.49   208.57 27.760 210.72 29.040
  8   211.95 29.879 212.18 30.064 :
  9   CALCULATE Logpress = 100*LOG10(Pressure)
 10   DGRAPH [TITLE='Forbes data'] Logpress; Boiltemp
 11   MODEL Logpress
 12   FIT Boiltemp

12.............................................................................


***** Regression Analysis *****

 Response variate: Logpress
     Fitted terms: Constant, Boiltemp


*** Summary of analysis ***

                d.f.          s.s.          m.s.          v.r.
Regression         1       425.350      425.3497       3000.13
Residual          15         2.127        0.1418
```

```
Total            16       427.476        26.7173
```

Percentage variance accounted for 99.5
Standard error of observations is estimated to be 0.377
* MESSAGE: The following units have large standardized residuals:
                12   3.71

*** Estimates of regression coefficients ***

```
              estimate        s.e.      t(15)
Constant       -42.10         3.32      -12.68
Boiltemp       0.8953        0.0163      54.77
```

The first two statements set up variates storing the values of the two variables to be analysed and the DGRAPH statement displays the scatterplot in Figure 8.1; the next two statements fit the regression.

It is often necessary to give CALCULATE statements before the regression statements. Though the model is linear, it can be fitted to a transformation of the response variable, as here, or of the explanatory variable, or both. This can be done to get variables that are expected to be linearly related, or to get a response variable with an approximately Normal distribution with constant variance. Unfortunately, both of these conditions are needed for the regression analysis to be valid; when one set of transformations does not achieve both – as is usually the case with a

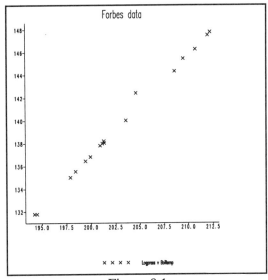

Figure 8.1

response variable of counts or proportions, for example – then it is best to fit a generalized linear model, 8.5, or a nonlinear model, 8.6 and 8.7. Additive models, 8.4, can be used when there is no predetermined form of a relationship.

You can fit models to subsets of the data by using the RESTRICT directive (2.4.1). The regression directives also automatically exclude any unit that contains a missing value for either variate. However, if only the response is missing, Genstat does give you some information about the unit (8.1.2).

Most of the directives in this section are relevant also to multiple regression and to nonlinear regression. But you can understand their main features most readily by seeing them in the simplest case.

8.1.1   The MODEL directive

MODEL   defines the response variate(s) and the type of model to be fitted for linear, generalized linear, generalized additive, and nonlinear models.

Options

DISTRIBUTION = *string*	Distribution of the response variable (normal, poisson, binomial, gamma, inversenormal, multinomial, calculated); default norm
LINK = *string*	Link function (canonical, identity, logarithm, logit, reciprocal, power, squareroot, probit, complementaryloglog, calculated); default cano i.e. iden for DIST=norm or calc, loga for DIST=pois, logi for DIST=bino or mult, reci for DIST=gamm, powe for DIST=inve
EXPONENT = *scalar*	Exponent for power link; default −2
DISPERSION = *scalar*	Value of dispersion parameter in calculation of s.e.s etc; default * for DIST=norm, gamm, inve, or calc, and 1 for DIST=pois, bino, or mult
WEIGHTS = *variate*	Variate of weights for weighted regression; default *
OFFSET = *variate*	Offset variate to be included in model; default *
GROUPS = *factor*	Absorbing factor defining the groups for within-groups linear or generalized linear regression; default *
RMETHOD = *string*	Type of residuals to form, if any, after each model is fitted (deviance, Pearson); default devi
FUNCTION = *scalar*	Scalar whose value is to be minimized by calculation; default *
YRELATION = *string*	Whether to analyse the y-variates separately, as in ordinary regression, or to analyse them cumulatively as counts in successive categories of a multinomial distribution (separate, cumulative); default sepa
DCALCULATION = *expressions*	Calculations to define the deviance contributions and variance function for a non-standard distribution; must be specified when DIST=calc
LCALCULATION = *expressions*	Calculations to define the fitted values and link derivative for a non-standard link; must be specified when LINK=calc
SAVE = *identifier*	To name regression save structure; default *

Parameters

Y = *variates*	Response variates; only the first is used in nonlinear models and in generalized linear models except when DIST=mult, when they specify the numbers in each

	category of an ordinal response model
NBINOMIAL = *variate* or *scalar*	Total numbers for DIST=bino
RESIDUALS = *variates*	To save residuals for each y variate after fitting a model
FITTEDVALUES = *variates*	To save fitted values, and provide fitted values if no terms are given in FITNONLINEAR
LINEARPREDICTOR = *variate*	Specifies the identifier of the variate to hold the linear predictor
DERIVATIVE = *variate*	Specifies the identifier of the variate to hold the derivative of the link function at each unit
DEVIANCE = *variate*	Specifies the identifier of the variate to hold the contribution to the deviance from each unit
VFUNCTION = *variate*	Specifies the identifier of the variate to hold the value of the variance function at each unit

In most applications, you will need only a simple form of the directive:

    MODEL identifier_of_response_variate

Notice that MODEL does not actually fit anything: it simply sets up some structures inside Genstat that are used when you give a FIT statement later on (8.1.2). So when you are doing regression, MODEL will always be accompanied by at least one other regression statement to fit a model, like FIT.

The Y parameter allows a list of variates; if you put more than one for linear regression, then you will get an analysis for each. This is a more efficient way of doing many linear regressions with the same explanatory variables, than separate pairs of MODEL and FIT statements. With additive models, generalized linear models, and nonlinear models (8.4 to 8.7), only the first variate will be analysed (with the exception of multinomial response models, 8.5.5); the others will be ignored.

The NBINOMIAL parameter is relevant only for the binomial setting of the DISTRIBUTION option (8.5.1).

The RESIDUALS and FITTEDVALUES parameters allow you to specify variates to contain the residuals and fitted values for each response variable. For example, you could change the MODEL statement above to ensure that each subsequent FIT statement will put the residuals into a variate R and fitted values into a variate F:

    MODEL Logpress; RESIDUALS=R; FITTEDVALUES=F

The residuals are the "unexplained" component of the response variable, standardized in some way according to the RMETHOD option (see below). The fitted values are the "explained" component: that is, the combination of parameters and explanatory variables fitted in the model. You can get access to these sets of values in a different way through the RKEEP directive (8.1.4).

The remaining parameters are of relevance only with generalized linear models (8.5).

The DISTRIBUTION, LINK, and EXPONENT options are also for generalized linear models, allowing you to go beyond linear regression; they are described in 8.5.1.

The DISPERSION option controls how the variance of the distribution of the response values is calculated. By default, the variance is estimated from the residual mean square (8.1.2), and standard errors and standardized residuals are calculated from the estimate. If you use DISPERSION to supply a value for the variance of the Normal distribution, or for the dispersion parameter of other distributions (8.5), then standard errors and residuals are based on this given value instead.

The WEIGHTS option allows you to specify a variate holding weights for each unit. Suppose, for example, you have assigned values to a weights variate W earlier in the program; then the option takes the form: WEIGHTS=W. If the weight for unit i is w_i, the regression directives will weight by w_i the contribution to the estimate of dispersion from the ith unit. In simple linear regression, the estimate of dispersion is then the weighted residual mean square:

$$\Sigma\{w_i \varepsilon_i^2\}/(N-2)$$

Thus, if the variance of the response variable is not constant, and you know the relative size of the variance for each observation, you can set the weight to be proportional to the inverse of the variance of an observation. Alternatively, if the variance is related in a simple way to the mean, you may just need to specify a different distribution for the response.

The OFFSET option allows you to include in the regression a variable with no corresponding parameter:

$$y_i = \alpha + o_i + \beta x_i + \varepsilon_i$$

where o_i is the ith value of the offset variable, O say. Linear regression analysis of Y with offset O is just the same as analysis of $Y-O$, but the offset has non-trivial applications in generalized linear models (8.5.1).

The GROUPS option specifies a factor whose effects you want to eliminate before any regression is fitted. The factor must already have been defined. (The effects of factors on regression are discussed in 8.3.) This method of elimination is sometimes called *absorption*; you might want to use it when data from many different groups are to be modelled. Use of GROUPS gives less information than you would get if you included the factor explicitly in the model (leverages, predictions, and some parameter correlations cannot be formed), but it saves space and time in fitting the model. You can use GROUPS only with linear and generalized linear regression.

The RMETHOD option controls how residuals are formed. By default, residuals are *deviance residuals* standardized by their estimated variance. For linear regression, these are:

$$r_i = (y_i - f_i) \sqrt{(w_i / v_i)}$$

In this, f_i is the ith fitted value, and v_i is the variance of an unstandardized residual:

$$v_i = (1 - l_i) s^2$$

Here, s^2 is the estimate of dispersion and l_i is the *leverage* (diagonal of the projection matrix), defined in terms of the design matrix X and the diagonal matrix of weights W by

$$l_i = w_i \{X(X'W X)^{-1}X'\}_{ii}$$

The alternative *Pearson residuals* are defined in exactly the same way if the distribution is Normal. For regression models with distributions other than Normal, the two kinds of residual are different (8.5.6).

If you do not want residuals, you can set the option to a missing value (*) to save space within Genstat. However, you will then not be able to get residuals, fitted values, or leverages, and the automatic checks on the fit of a model will not be done (8.1.2).

The FUNCTION option is relevant only when you want to optimize a general function (8.7.4). It is ignored unless no response variates are specified by the Y parameter.

The YRELATION option is relevant only for ordinal response models (8.5.5), and the DCALCULATION and LCALCULATION options only for generalized linear models that you define yourself (8.5.4).

The SAVE option allows you to specify an identifier for the regression save structure. This structure stores the current state of the regression model, and can be used explicitly in the directives RDISPLAY (8.1.3), RKEEP (8.1.4), PREDICT (8.3.4), and RFUNCTION (8.6.5). If the identifier in SAVE is of a regression save structure that already has values, those values are deleted. You can reset the current regression save structure at any point in a program by using the SET directive (13.1.1). Then, later regression statements would use the model stored in this save structure.

8.1.2 The **FIT** directive

FIT fits a linear, generalized linear, or generalized additive model.

Options

PRINT = *strings*	What to print (model, deviance, summary, estimates, correlations, fittedvalues, accumulated, monitoring); default mode,summ,esti
CONSTANT = *string*	How to treat the constant (estimate, omit); default esti
FACTORIAL = *scalar*	Limit for expansion of model terms; default as in previous TERMS statement, or 3 if no TERMS given
POOL = *string*	Whether to pool ss in accumulated summary between all terms fitted in a linear model (yes, no); default no
DENOMINATOR = *string*	Whether to base ratios in accumulated summary on rms from model with smallest residual ss or smallest residual ms (ss, ms); default ss
NOMESSAGE = *strings*	Which warning messages to suppress (dispersion, leverage, residual, aliasing, marginality); default *
FPROBABILITY = *string*	Printing of probabilities for variance ratios (yes, no); default no
TPROBABILITY = *string*	Printing of probabilities for t-statistics (yes, no); default no

Parameter

formula	List of explanatory variates and factors, or model formula

A FIT statement must always be preceded by a MODEL statement, though not necessarily immediately. You can give several FIT statements after a single MODEL statement: for example, you might want to try out different explanatory variables.

The parameter of the FIT directive specifies the explanatory variables in the model. In the simple linear regression above, it consists of the identifier of the explanatory variate alone:

        FIT Boiltemp

If you omit the parameter, Genstat fits a *null model*; that is, a model consisting of just one parameter, the overall mean:

$$y_i = \alpha + \varepsilon_i$$

The PRINT option controls output. You can give several settings at the same time, to provide reports on several aspects of the analysis.

The model setting gives a description of the model, including response and explanatory variates. Here is a repeat of this aspect of the analysis in Example 8.1; model gives the first lines in this output:

Example 8.1.2a

```
  14  FIT [PRINT=model,summary; FPROBABILITY=yes] Boiltemp

14...............................................................................


***** Regression Analysis *****

 Response variate: Logpress
     Fitted terms: Constant, Boiltemp


*** Summary of analysis ***

                d.f.          s.s.         m.s.        v.r.    F pr.
Regression         1       425.350     425.3497     3000.13    <.001
Residual          15         2.127       0.1418
Total             16       427.476      26.7173

Percentage variance accounted for 99.5
Standard error of observations is estimated to be 0.377
* MESSAGE: The following units have large standardized residuals:
                12         3.71
```

The output from the summary setting is also reproduced here: this gives a summary analysis of variance, which subdivides the total sum of squares, corrected for the mean, between that explained by the regression (Regression), and that which is not explained (Residual). The table has the standard form with columns for the degrees of freedom (d.f.), the sums of squares (s.s.), the mean squares (m.s.), and for the variance ratio (v.r.). In addition, because we have set the FPROBABILITY option, there is a column giving the probability that the variance ratio would be as large as this under the null hypothesis of no relationship; this probability is based on the F-distribution, which is valid only if the distribution of the response is indeed Normal. By default, as seen in Example 8.1, this probability does not appear.

Following the analysis of variance is some more information about the fit of the model. The percentage variance accounted for is the *adjusted R^2 statistic*, expressed as a percentage:

Percentage variance accounted for = $100 \times (1 - $ (Residual m.s.)/(Total m.s.))

This statistic is usually a better guide to the fit of a model than the unadjusted version, but you should remember that neither version is an absolute measure of fit, and both depend on the range of response and explanatory values as well as on the goodness of fit (Seber, 1977). If this statistic had a negative value, indicating a very poorly fitting model, the message `Residual variance exceeds variance of Y variate` would be printed instead.

The standard error of the observations is estimated simply by the square root of the residual mean square.

The next message in the output is produced as a result of several checks made by Genstat on the adequacy of the model. Here, the only report concerns an apparently extreme observation in the data. This report appears for any standardized residuals whose values are particularly large: the criterion is to list residuals greater than that value c corresponding to probability $1/d$ of being exceeded by a standard Normal deviate, where d is the number of residual degrees of freedom. However, the value $c=2.0$ is used instead of any smaller value when there are less than 20 residual degrees of freedom, and the value 4.0 is used instead of any larger value when there are more than 15,773 degrees of freedom. Thus, a message should appear for any extreme outlier, but messages should not appear too often just as a result of random variation.

Genstat makes four other checks on the model that can generate messages in the summary of the analysis. Examples of these can be seen in the other examples of this chapter. One check is for particularly large values of the leverage, using the criterion ck/N, where k and N are the number of parameters and number of units used in the regression model, and c is as used in the check on residuals. The sum of the leverages is always k, so this criterion brings to your attention those observations with more than about twice the average influence. Unlike the other checks, this one does not indicate a potential violation of assumptions, but rather that the analysis may be greatly affected by some observations.

If there are at least 20 observations, two checks are made on the constancy of the variance of the response variable. The fitted values are ordered into three roughly equal-sized groups; Levene tests (Snedecor & Cochran, 1989) are carried out to compare the variance of the standardized residuals in the bottom group with those in the top group, and then the middle group is compared with the other two groups combined. Each test will generate a message if the test statistic is significant at the 2.5% level, indicating that the assumption of constant variance may not be tenable. Finally, a "runs" test is carried out on the standardized residuals, ordered according to the fitted values. A message is generated if the sign of successive residuals does not change often enough (again using a 2.5% significance level), indicating that there is still some systematic pattern in the residuals.

These messages are intended to warn you about potential problems in interpreting the analysis, but cannot be relied on to detect all problems. See Cook and Weisberg (1982) for more information about these and other model-checking techniques; the RCHECK procedure (8.1.5) provides some further techniques.

You can prevent these messages appearing by using the NOMESSAGE option. They will not appear in any case if you have set option RMETHOD=* in the MODEL statement.

The estimates setting produced the last section of output in Example 8.1:

Example 8.1.2b

```
16  FIT [PRINT=estimates; TPROBABILITY=yes] Boiltemp

16...........................................................................


***** Regression Analysis *****

*** Estimates of regression coefficients ***

                    estimate        s.e.       t(15)    t pr.
Constant              -42.10        3.32      -12.68    <.001
Boiltemp              0.8953       0.0163       54.77    <.001
```

The standard errors of the estimates are based here on the residual mean square. Alternatively, you can supply an estimate of variance by using the DISPERSION option of MODEL; if you do this, Genstat will print a reminder about the basis of the standard errors. You can prevent this reminder appearing by setting the NOMESSAGE option. The t-statistics allow you to test whether each parameter differs significantly from zero, keeping the other parameters fixed. The number of degrees of freedom for such a test is the number of residual degrees of freedom reported in the summary analysis of variance, and this number appears in the column heading. If the estimate of variance is supplied, then the "t-statistics" actually have a standard Normal distribution, indicated by the column heading "t(*)". By default, as in Example 8.1, probabilities are not printed, but if the TPROBABILITY option is set (as in Example 8.1.2b), the corresponding probabilities are displayed.

You can use the deviance setting if you want only an abbreviated output.

Example 8.1.2c

```
18  FIT [PRINT=deviance] Boiltemp

18...........................................................................


Residual d.f. 15, s.s. 2.127
```

The other settings that you can use in the PRINT option are correlations, fitted, accumulated, and monitoring. The first two of these are illustrated in Example 8.1.2d. There is a correlation matrix of the parameter estimates, followed by a table of unit labels, values of response variate, fitted values, standardized residuals, and leverages. For the unit labels, Genstat will take those associated with the response variate using the NVALUES option of the VARIATE directive (2.3.1), if available, or the values of the units structure (2.3.4). If neither is available, the integers 1...N are printed. If you have weighted the regression by setting the WEIGHTS option of the MODEL directive, the weights are also listed. The accumulated and monitoring settings are discussed later, in 8.2.1 and 8.5.6 respectively.

Example 8.1.2d

```
 20  FIT [PRINT=correlations,fitted] Boiltemp

20.........................................................................

***** Regression Analysis *****

*** Correlations between parameter estimates ***

estimate        ref     correlations

Constant         1      1.000
Boiltemp         2     -1.000  1.000
                          1      2


*** Fitted values and residuals ***

                                    Standardized
            Unit     Response Fitted value residual Leverage
             1       131.785     132.044    -0.76     0.19
             2       131.785     131.820    -0.10     0.20
             3       135.025     135.088    -0.18     0.11
             4       135.545     135.562    -0.05     0.10
             5       136.455     136.476    -0.06     0.08
             6       136.829     136.923    -0.26     0.08
             7       137.822     137.801     0.06     0.07
             8       138.003     137.998     0.02     0.06
             9       138.057     138.177    -0.33     0.06
            10       138.211     138.132     0.22     0.06
            11       140.037     140.146    -0.30     0.06
            12       142.439     141.086     3.71     0.06
            13       145.469     145.447     0.06     0.14
            14       144.342     144.641    -0.85     0.12
            15       146.300     146.566    -0.78     0.17
            16       147.537     147.667    -0.39     0.21
            17       147.805     147.873    -0.21     0.22

Mean                 139.614     139.614    -0.01     0.12
```

In the table, units are omitted according to any restriction in force or to any missing values of explanatory variates (8.1). Fitted values are shown, however, for units with zero weight or in which only the response variate is missing. Residuals are standardized as described in 8.1.1. You can use the RCHECK procedure to provide unstandardized residuals.

The CONSTANT option controls whether the constant parameter is included in the model. In simple linear regression, this parameter is the intercept, in other words the estimate of the response variable when the explanatory variable is zero. By setting CONSTANT=omit, you can prevent the constant parameter being estimated, so that the simple linear regression becomes

$$y_i = \beta x_i + \varepsilon_i$$

This model is particularly useful when y_i and x_i are measurements of the same attribute of a unit, as in calibration, and when you know that they are zero together. However, you need to be careful here: you must be sure that the relationship remains linear right down to zero.

When you omit the constant, the analysis of variance produced by PRINT=summary will not be corrected for the mean, so that the model will be compared with the null model $y_i=0$. (However, if the effects of factors are present in the model (8.3), setting CONSTANT=omit merely affects how the model is parameterized, and so the analysis will still be corrected for the mean.) The percentage variance accounted for will still be expressed as a percentage of the variance of the response variable about the mean.

The FACTORIAL option is described in 8.3.1, and the POOL and DENOMINATOR options in 8.2.1.

The NOMESSAGE option controls printing of messages. The aliasing setting is discussed in 8.2.1 and 8.3.2, and the marginality setting in 8.3.3. The leverage setting prevents messages about large leverages, and residual prevents messages about large residuals or non-constant variance or systematic pattern in the residuals. (These messages are those that are associated with the summary setting of the PRINT option.) You use the dispersion setting to prevent reminders appearing about the basis of the standard errors (as would be produced by the estimates setting of the PRINT option).

The FPROBABILITY and TPROBABILITY options are described above with PRINT=summary and PRINT=estimates.

8.1.3   The **RDISPLAY** directive

RDISPLAY displays the fit of a linear, generalized linear, generalized additive, or nonlinear model.

Options

PRINT = *strings*	What to print (model, deviance, summary, estimates, correlations, fittedvalues, accumulated); default mode, summ, esti
CHANNEL = *identifier*	Channel number of file, or identifier of a text to store output; default current output file
DENOMINATOR = *string*	Whether to base ratios in accumulated summary on rms from model with smallest residual ss or smallest residual ms (ss, ms); default ss
NOMESSAGE = *strings*	Which warning messages to suppress (dispersion, leverage, residual, vertical, df); default *
FPROBABILITY = *string*	Printing of probabilities for variance ratios (yes, no); default no
TPROBABILITY = *string*	Printing of probabilities for t-statistics (yes, no); default no

| SAVE = *identifier* | Specifies save structure of model to display; default * i.e. that from latest model fitted |

No parameters

The PRINT option has the same settings as in the FIT directive, except that no monitoring is available. The CHANNEL option selects the output channel to which the results are output, as in the PRINT directive (3.2); this may be a text structure, allowing output to be stored prior to display. The DENOMINATOR (8.2.1) and NOMESSAGE, FPROBABILITY, and TPROBABILITY options are also as in the FIT directive.

The SAVE option lets you specify the identifier of a regression save structure; the output will then relate to the most recent regression model fitted with that structure.

8.1.4 The RKEEP directive

RKEEP stores results from a linear, generalized linear, generalized additive, or nonlinear model.

Options

EXPAND = *string*	Whether to put estimates in the order defined by the maximal model for linear or generalized linear models (yes, no); default no
DISTRIBUTION = *text*	Saves the distribution of the response variable (either normal, poisson, binomial, gamma, inversenormal, multinomial, or calculated)
LINK = *text*	Saves the link function (either identity, logarithm, logit, reciprocal, power, squareroot, probit, complementaryloglog, or calculated)
EXPONENT = *scalar*	Saves the exponent of a power link
DISPERSION = *scalar*	Saves the fixed dispersion parameter
WEIGHTS = *dummy*	Saves the identifier of the weight variate
OFFSET = *dummy*	Saves the identifier of the offset variate
GROUPS = *dummy*	Saves the identifier of the absorbing factor
RMETHOD = *text*	Saves the type of the residuals
Y1 = *dummy*	Saves the identifier of the first response variate
NBINOMIAL = *dummy*	Saves the identifier of the binomial totals
SAVE = *identifier*	Specifies save structure of model; default * i.e. that from latest model fitted

Parameters

| Y = *variates* | Response variates for which results are to be saved; default takes the response variates from the most recent |

	MODEL statement
RESIDUALS = *variates*	Standardized residuals for each Y variate
FITTEDVALUES = *variates*	Fitted values for each Y variate
LEVERAGES = *variate*	Leverages of the units for each Y variate
ESTIMATES = *variates*	Estimates of parameters for each Y variate
SE = *variates*	Standard errors of the estimates
INVERSE = *symmetric matrix*	Inverse matrix from a linear or generalized linear model, inverse of second derivative matrix from a nonlinear model
VCOVARIANCE = *symmetric matrix*	
	Variance-covariance matrix of the estimates
DEVIANCE = *scalars*	Residual ss or deviance
DF = *scalar*	Residual degrees of freedom
TERMS = *pointer* or *formula*	Fitted terms (excluding constant)
ITERATIVEWEIGHTS = *variate*	Iterative weights from a generalized linear model
LINEARPREDICTOR = *variate*	Linear predictor from a generalized linear model
YADJUSTED = *variate*	Adjusted response of a generalized linear model
EXIT = *scalar*	Exit status from a generalized linear or nonlinear model
GRADIENTS = *pointer*	Derivatives of fitted values with respect to parameters in a nonlinear model
GRID = *variate*	Grid of function or deviance values from a nonlinear model
DESIGNMATRIX = *matrix*	Design matrix whose columns are explanatory variates and dummy variates
PEARSONCHI = *scalar*	Pearson chi-squared statistic from a generalized linear model
STERMS = *pointer*	Saves the identifiers of the variates that have been smoothed in the current model
SCOMPONENTS = *pointer*	Saves a pointer to variates holding the nonlinear components of the variates that have been smoothed

RKEEP allows you to copy information from a regression analysis into Genstat data structures. You do not need to declare the structures in advance; Genstat will declare them automatically to be of the correct type and length.

The Y parameter specifies the response variates for which the results are to be saved. Unusually for the first parameter of a directive, this has a default: if you leave it out, Genstat assumes that results are to be saved for all the response variates, as given in the previous MODEL statement.

The RESIDUALS, FITTEDVALUES, and LEVERAGES parameters allow you to save the standardized residuals, the fitted values, and the leverages. For example, RESIDUALS=R puts the residuals in a variate R. You cannot save these values if you had set RMETHOD=* in the MODEL statement. Unstandardized residuals are available from the RCHECK procedure.

The ESTIMATES and SE parameters save the parameter estimates and their standard errors;

RKEEP puts them in variates, using the same order as in the display produced by the PRINT option of FIT. However, you can use an alternative order by setting the EXPAND option (8.2.4). The variates saving these values are set up with labels (2.3); thus, you can refer to individual values in expressions using the labels as displayed by the FIT directive. For example, to get the estimate of the constant into a scalar, you could use:

```
RKEEP ESTIMATES=Esti
SCALAR Const
CALCULATE Const = Esti$['Constant']
```

The INVERSE parameter allows you to save the inverse matrix as a symmetric matrix: that is, $(X'X)^{-1}$ where X is the design matrix. This matrix is the same for all response variates.

The VCOVARIANCE parameter saves the variance-covariance matrix of the estimates for each response variate: these are formed by multiplying the inverse matrix by the relevant variance estimate based on the estimated dispersion, or on the dispersion that you have supplied.

The DEVIANCE parameter lets you save the residual sum of squares, or the *deviance* for distributions other than Normal (8.5). The DF parameter saves the residual degrees of freedom.

The ITERATIVEWEIGHTS, LINEARPREDICTOR, and YADJUSTED parameters are discussed in 8.5.6, the EXIT and GRADIENTS parameters in 8.6.4, and the GRID parameter in 8.7.1.

The DESIGNMATRIX parameter allows you to save the matrix X. The columns correspond to the parameters of the model, ordered as for the ESTIMATES parameter. For simple linear regression with a constant this has only two columns, the first containing ones and the second containing the values of the explanatory variate.

The PEARSONCHI parameter provides the Pearson chi-squared statistic for dispersion, which is the same as the residual sum of squares for the Normal distribution, but is different to the deviance for other distributions (8.5.6). The STERMS and SCOMPONENTS parameters are discussed in 8.4.3.

The options, apart from EXPAND and SAVE mentioned above, allow you to find out details of the current regression model as defined by a MODEL statement. They correspond directly to the options of MODEL (8.1.1) with the addition of Y1 to allow reference to the first response variate and NBINOMIAL for the variate of binomial totals (8.5.1). These options are most likely to be of use in general procedures that work with a fitted model.

8.1.5   The RGRAPH and RCHECK procedures

There are two procedures in the Library that are particularly useful in regression analysis. RGRAPH draws a picture of the current regression, showing the observed points and fitted model, while RCHECK can produce a variety of diagnostic pictures, such as half-Normal plots of residuals, and calculate diagnostic quantities such as modified Cook's statistics. Details can be found in the Procedure Library Manual, or from the procedure LIBHELP (5.3.1). Figure 8.1.5 was produced by the following two statements given after the FIT statement of Example 8.1.2d.

```
RGRAPH [GRAPHICS=high; WINDOW=5]
RCHECK [GRAPHICS=high; WINDOW=6; SCREEN=keep] residual;\
   halfnormal
```

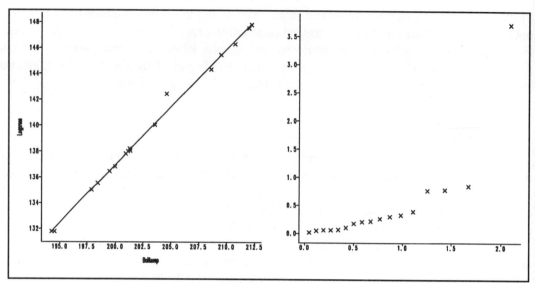

Figure 8.1.5

8.2 Multiple linear regression

The model for simple linear regression can be extended by adding the effects of further explanatory variables. It is then called *multiple linear regression* and can be written:

$$y_i = \alpha + \beta_1 x_{1i} + \beta_2 x_{2i} + \dots + \beta_k x_{ki} + \varepsilon_i$$

or in matrix form:

$$y = X \beta + \varepsilon$$

where the design matrix X has $k+1$ columns. The errors ε_i will be assumed in this section to be Normally distributed, as in 8.1.

You can fit a multiple linear regression with the MODEL and FIT directives as before; the only change is that you now give a list of explanatory variates in FIT. In Example 8.2, data is read from a file attached to the second input channel and a multiple linear regression is fitted for the response variable Heat on the four explanatory variables X[1...4]; the RESTRICT directive is used to confine the analysis to those samples that have 3.2% gypsum.

Example 8.2

```
   1   "  Multiple linear regression of the heat given out by setting cement
  -2       on four chemical constituents. Data from Woods, Steinour and Starke
  -3       (1932); analysed by Draper and Smith (1981) p629."
   4
   5   OPEN 'CEMENT.DAT'; CHANNEL=2
   6   READ [PRINT=data; CHANNEL=2] X[3,1,4,2],%gypsum,Heat

       1   6   7 60 26 3.2   78.5     15   1 52 29 3.2    74.3
       2   8  11 20 56 3.2  104.3      8  11 47 31 3.2    87.6
       3   6   7 33 52 3.2   95.9      9  11 22 55 3.2   109.2
       4   9  11 22 55 4.3  108.0      9  11 22 55  *    110.2
```

```
 5  17   3   6 71 3.2 102.7    22   1 44 31 3.2  72.5
 6  18   2  22 54 3.2  93.1     4  21 26 47 3.2 115.9
 7   4  21  26 47 6.5 114.0    23   1 34 40 3.2  83.8
 8   9  11  12 66 3.2 113.3     8  10 12 68 3.2 109.4
 9  18   1  61 17 3.2   *
 7  " Analyse only those samples with 3.2% gypsum."
 8  RESTRICT Heat; %gypsum==3.2
 9  MODEL Heat
10  " Constituents are: X[1]   tricalcium aluminate
-11                     X[2]   tricalcium silicate
-12                     X[3]   tetracalcium aluminoferrite
-13                     X[4]   beta-dicalcium silicate"
14  FIT X[]

14.........................................................................


***** Regression Analysis *****

Response variate: Heat
     Fitted terms: Constant, X[1], X[2], X[3], X[4]


*** Summary of analysis ***

               d.f.         s.s.        m.s.        v.r.
Regression       4      2667.90      666.975      111.48
Residual         8        47.86        5.983
Total           12      2715.76      226.314

Percentage variance accounted for 97.4
Standard error of observations is estimated to be 2.45


*** Estimates of regression coefficients ***

               estimate       s.e.       t(8)
Constant          62.4        70.1       0.89
X[1]             1.551       0.745       2.08
X[2]             0.510       0.724       0.70
X[3]             0.102       0.755       0.14
X[4]            -0.144       0.709      -0.20
```

One thing you might want to do with multiple regression is find the subset of explanatory variables that gives the most satisfactory fit. You can search for this subset by using the ADD, DROP, SWITCH, TRY, and STEP directives. Each of these makes and reports changes to the current regression model. You must use the TERMS directive before any of these, to define a common set of units for the regression and to carry out initial calculations efficiently.

The directives described in this section let you supervise the search for suitable sets of explanatory variables. Automatic searches can be carried out by incorporating statements into a loop (5.2.1); for example, a suitable STEP statement in a loop can carry out the methods of forward selection, backward elimination, or stepwise regression (8.2.6).

8.2.1   Extensions to the FIT and RDISPLAY directives

You would usually want to divide the explained variation between explanatory variables. The summary analysis of variance from the PRINT options of FIT and RDISPLAY does not do this, but there is a further setting accumulated. This divides the variation according to the order in which you listed the variables in the parameter of the FIT directive: therefore, the sum of squares for each variable ignores the effects of variables fitted later and eliminates the effect for variables already fitted. This contrasts with the t-statistics from PRINT=estimates which can be used to test the effect of each variable after eliminating the effects of all the other variables. You will find the accumulated setting useful also for summarizing changes in the regression model that you might make by the directives described later in this section. Here is the accumulated summary produced after Example 8.2.

Example 8.2.1

```
   16   RDISPLAY [PRINT=accumulated]

16...........................................................................


***** Regression Analysis *****

*** Accumulated analysis of variance ***

Change              d.f.          s.s.          m.s.          v.r.
+ X[1]                 1      1450.076      1450.076        242.37
+ X[2]                 1      1207.782      1207.782        201.87
+ X[3]                 1         9.794         9.794          1.64
+ X[4]                 1         0.247         0.247          0.04
Residual               8        47.864         5.983

Total                 12      2715.763       226.314
```

The table shows the sum of squares and degrees of freedom attributable to each individual change in the model. As for the summary analysis of variance, if you set the FPROBABILITY option at the same time as PRINT=accumulated you will get an extra column in the table with F-probabilities. By default the variance ratios are obtained by dividing the mean squares by the mean square corresponding to the smallest residual sum of squares in the table; that is from the model with fewest residual degrees of freedom.

If you do not want the sum of squares and the degrees of freedom to be subdivided between changes to the explanatory variables that you make within a statement, you should set option POOL=yes. There would then be just one entry in the table for each statement. The main use of POOL is with the ADD, DROP, and SWITCH directives (8.2.3). With FIT, the POOL option merely gives the same table as you would get using the summary setting of the PRINT option.

The DENOMINATOR option of the FIT and RDISPLAY directives can be set to produce variance ratios in the summary based on the smallest residual mean square, rather than on the mean square corresponding to the smallest residual sum of squares. You might, for example, know in advance of doing the regression that certain variables are unlikely to have a relationship with the response variable. So you would want to be able to include the sum of squares for these variables in the residual sum of squares for the other explanatory variables. You can do that by listing the interesting variables first, and these potentially uninteresting variables last, and setting DENOMINATOR=ms.

Sometimes you will find that the effect of an explanatory variable turns out to be exactly zero. This is no problem if it happens because the correlation of the explanatory variable with the response variable is itself zero. But it is a problem if it happens because the explanatory variable is a linear combination of other explanatory variables. We call this *collinearity* or *aliasing* of the explanatory variables. There is then no unique set of parameter estimates, and the method of computing information about the regression would break down, since it involves inverting a singular matrix $X'X$. The method also becomes unstable if the explanatory variables are nearly linearly related. Therefore Genstat tests for such a linear relationship, and will not include an explanatory variable that fails the test (8.2.2). A warning message is displayed, telling you which variable is not being included and the form of the linear relationship that has been found (see Examples 8.3.4f and 8.5.1). You can prevent the message appearing by using the aliasing setting of the NOMESSAGE option of the FIT directive.

If you then change the model, Genstat will continue to try to include this problem variable unless it is explicitly dropped. This is because the changes in the model may cause the original collinearity to disappear. If the variable is successfully included, a message is printed; again you can prevent the message appearing by the aliasing setting of the NOMESSAGE option.

8.2.2 The **TERMS** directive

TERMS specifies a maximal model, containing all terms to be used in subsequent linear, generalized linear, generalized additive, and nonlinear models.

Options

PRINT = *strings*	What to print (correlations, wmeans, SSPM); default *
FACTORIAL = *scalar*	Limit for expansion of model terms; default 3
FULL = *string*	Whether to assign all possible parameters to factors and interactions (yes, no); default no
SSPM = *SSPM*	Gives sums of squares and products on which to base calculations; default *
TOLERANCE = *scalar*	Criterion for testing for linear dependence; default is 10ε or 10000ε, depending on the computer's precision, where ε is the smallest real value such that $1+\varepsilon$ is greater than 1 on the computer

Parameter

formula	List of explanatory variates and factors, or model formula

You use the TERMS directive before starting to explore different subsets of explanatory variables, so that Genstat can define a common set of units for the regression and carry out some initial calculations. The directives that allow you to search through the different subsets, ADD, DROP, SWITCH, TRY, and STEP, are described later in this section.

TERMS thus initializes Genstat ready for the exploration. It overrules any model that has already been fitted with FIT, and resets the current model to be the null model.

The formula specified by the parameter of TERMS should contain all the explanatory variables that you may wish to use in the subsets; if you later need to include others, you should give another TERMS statement. For multiple regression, the formula is a simple list of variates; it may include the response variates, but need not. Here is an example.

Example 8.2.2

```
   18   TERMS [PRINT=correlated] X[]

*** Degrees of freedom ***

Correlations:      11


*** Correlation matrix ***

Heat            1   1.000
X[1]            2   0.731  1.000
X[2]            3   0.816  0.229  1.000
X[3]            4  -0.535 -0.824 -0.139  1.000
X[4]            5  -0.821 -0.245 -0.973  0.030  1.000

                    1      2      3      4      5
```

The TERMS directive actually fits a model: the null model containing only the constant term (in this case a mean). It also calculates the sums of squares and products and the means (SSPM) of the variates, including any response variates: the matrix of SSPMs is $X'X$, augmented by rows and columns for response variables, and is the basis of the regression calculations. The matrix is weighted if you have specified weights in the MODEL statement, and the calculations are made within groups if you have specified a grouping factor. All units of the variates are used unless there are restrictions or missing values. You are not allowed to have different restrictions on the different vectors. Thus you can define the set of units that Genstat uses in the calculations by putting a restriction on any one of: a response variate, an explanatory variate, the weight variate, the offset variate, or the groups factor. A missing value in any of these structures except a response variate will also exclude the corresponding unit. You should not alter the restriction applied to the vectors between the TERMS statement and subsequent fitting statements.

The model containing all the terms specified by the parameter of TERMS, excluding the response variates, is called the *maximal model*.

The PRINT option allows you to display the calculated sums of squares and products, and means, together with the degrees of freedom. It can also display the corresponding matrix of correlations between variables, as above, and group means if the regression is within groups.

The FACTORIAL and FULL options are relevant only if there are factors in the model (8.3.1 and 8.3.2).

The SSPM option lets you use values that you have already calculated for an SSPM or DSSP structure (2.7.2 and 4.10.3). You might find this especially useful when you are analysing very large sets of data: you can accumulate a DSSP sequentially to avoid storing all the data at one time (4.10.3). Later regression calculations will be based on the supplied values of the DSSP, though no fitted values, residuals, or leverages will be available. The values of a supplied SSPM or DSSP are accepted without checking by the TERMS directive: Genstat simply assumes you are giving it something sensible. On most computers, however, regression should not be based on sums of squares and products stored with single-precision accuracy. All standard data structures in Genstat except for the DSSP structure are in single precision. The TERMS directive will print a warning if you supply an SSPM structure on a computer that has less than 48 bits for single-precision storage.

The TOLERANCE option controls the detection of aliasing in subsequent model fitting. By default, a parameter in a linear or generalized linear model will be deemed to be aliased if the ratio between the original diagonal value of the SSPM corresponding to this parameter and the current diagonal value of the partially inverted SSPM is less than 10ε. The quantity ε depends on the computer and is defined to be the smallest number such that the computer recognizes $1.0 + \varepsilon$ as greater than 1.0 in single precision. On computers for which single and double precision are equivalent in Genstat, ε is a much smaller quantity; so the criterion 10000ε is used by default. Any positive value can be supplied by the TOLERANCE option to replace this default criterion in subsequent linear regression and generalized linear regression.

8.2.3   The ADD, DROP, and SWITCH directives

The directives ADD, DROP, and SWITCH all have identical options and parameters.

ADD   adds extra terms to a linear, generalized linear, generalized additive, or nonlinear model.

DROP   drops terms from a linear, generalized linear, generalized additive, or nonlinear model.

SWITCH   adds terms to, or drops them from a linear, generalized linear, generalized additive, or nonlinear model.

Options

PRINT = *strings*	What to print (model, deviance, summary, estimates, correlations, fittedvalues, accumulated, monitoring); default mode,summ,esti
NONLINEAR = *string*	How to treat nonlinear parameters between groups (common, separate, unchanged); default unch

CONSTANT = *string*	How to treat the constant (estimate, omit, unchanged); default unch
FACTORIAL = *scalar*	Limit for expansion of model terms; default * i.e. that in previous TERMS statement
POOL = *string*	Whether to pool ss in accumulated summary between all terms fitted in a linear model (yes, no); default no
DENOMINATOR = *string*	Whether to base ratios in accumulated summary on rms from model with smallest residual ss or smallest residual ms (ss, ms); default ss
NOMESSAGE = *strings*	Which warning messages to suppress (dispersion, leverage, residual, aliasing, marginality, vertical, df); default *
FPROBABILITY = *string*	Printing of probabilities for variance ratios (yes, no); default no
TPROBABILITY = *string*	Printing of probabilities for t-statistics (yes, no); default no

Parameter

| *formula* | List of explanatory variates and factors, or model formula |

You use the directives ADD, DROP, and SWITCH to change the current model. Broadly, ADD lets you add extra explanatory variables, DROP lets you remove variables, and SWITCH lets you simultaneously add and remove variables.

The directives have a common syntax, which is also much the same as the syntax of the FIT directive. They modify the current regression model, which may be linear, generalized linear, generalized additive, standard curve, or nonlinear. You must give a TERMS statement before using any of the three directives, in order to define a set of units and carry out basic calculations. If no model is fitted after the TERMS statement before an ADD, DROP, or SWITCH statement, the current model is taken to be the null model.

Here is some output that continues the example from the beginning of this section:

Example 8.2.3

```
  20   ADD [PRINT=deviance,estimates] X[1,2,4]

20................................................................................


***** Regression Analysis *****

Residual d.f. 9, s.s. 47.97; Change d.f. -3, s.s. -2667.79

*** Estimates of regression coefficients ***

                  estimate          s.e.        t(9)
Constant              71.6          14.1        5.07
X[1]                 1.452         0.117       12.41
```

```
X[2]                    0.416         0.186          2.24
X[4]                   -0.237         0.173         -1.37

   21  DROP [PRINT=deviance,estimates] X[4]

21.....................................................................
```

***** Regression Analysis *****

Residual d.f. 10, s.s. 57.90; Change d.f. 1, s.s. 9.93

*** Estimates of regression coefficients ***

	estimate	s.e.	t(10)
Constant	52.58	2.29	23.00
X[1]	1.468	0.121	12.10
X[2]	0.6623	0.0459	14.44

```
   22  SWITCH [PRINT=estimates,accumulated] X[2,4]

22.....................................................................
```

***** Regression Analysis *****

*** Estimates of regression coefficients ***

	estimate	s.e.	t(10)
Constant	103.10	2.12	48.54
X[1]	1.440	0.138	10.40
X[4]	-0.6140	0.0486	-12.62

*** Accumulated analysis of variance ***

Change	d.f.	s.s.	m.s.	v.r.
+ X[1]	1	1450.076	1450.076	272.04
+ X[2]	1	1207.782	1207.782	226.59
+ X[4]	1	9.932	9.932	1.86
Residual	9	47.973	5.330	
- X[4]	-1	-9.932	9.932	1.86
- X[2]	-1	-1207.782	1207.782	226.59
+ X[4]	1	1190.925	1190.925	223.43
Total	12	2715.763	226.314	

The formula specified by the parameter of each of these directives indicates the terms that are to be added or dropped, as appropriate, from the model; you must have included all of these in the formula of the previous TERMS statement. The terms in the formula (variates in the case of multiple linear regression) are compared with those in the current regression model to form the new model.

For the ADD directive, the new model consists of all terms in the current model together with any terms in the formula; terms may appear in both the current model and the formula, in which case they will remain in the new model.

In Example 8.2.3, remember that the TERMS statement has reset the current model to be the null model. The ADD statement in line 20 thus has the same effect as the statement

        FIT [PRINT=deviance,estimates] X[1,2,4]

If the ADD statement were followed by another, for example

        ADD X[3,4]

then the variate X[3] would be added to the model, which would then be the same as in Example 8.2.

For the DROP directive, the new model consists of all terms in the current model excluding any that are in the formula: terms in the formula that are not in the current model are ignored. You can see this at line 21 of Example 8.2.3. If the DROP statement had instead been

        DROP [PRINT=deviance,estimates] X[3,4]

it would still have had the same effect, since X[3] does not appear in the current model as defined by the previous statements.

Terms in the formula for the SWITCH directive are dropped from the current model if they are already there, and added to it if they are not. For example, if the current model consists of R and S, the effect of

        SWITCH S,T

is to make a new model consisting of R and T (assuming that T was included in the previous TERMS statement).

The options of the ADD, DROP, and SWITCH directives are the same as those of the FIT directive, but with the extra NONLINEAR option (see 8.6.3). The output from the summary and accumulated settings of the PRINT option is modified when a TERMS statement has been given, and is described in 8.2.4. The model fitted by ADD, DROP, or SWITCH will include a constant term if the previous model included one, and will not include one if the previous model did not. You can, however, change this using the CONSTANT option.

8.2.4   Extensions to output and the RKEEP directive following TERMS

Following a TERMS statement, extra output is produced by the PRINT option of the FIT, ADD, DROP, SWITCH, and RDISPLAY directives. The summary analysis of variance produced by the summary setting includes an extra line called "Change". This shows the change in the Residual line since the last model. If no previous model has been fitted, the change refers to the null model.

The accumulated summary produced by the accumulated setting of the PRINT option shows all changes made to the model since the last TERMS or FIT statement, including those made by the FIT statement. You can see this after the SWITCH statement in Example 8.2.3: three terms are added, then X[4] is removed, and then X[2] is removed and X[4] reinstated. Notice the two very different sums of squares for X[4]: the smaller is the sum of squares after eliminating X[1] and X[2] while the larger is the sum of squares after eliminating X[1] but ignoring X[2]. The large difference implies that X[2] and X[4] are highly correlated after elimination of X[1]; in fact, the correlation matrix from the TERMS statement shows that they are also highly correlated ignoring X[1].

The variance ratios from the setting PRINT=accumulated are calculated either from the smallest residual mean square, or from the residual mean square corresponding to the smallest residual sum of squares, depending on how the DENOMINATOR option has been set in the statement that prints the accumulated summary. In Example 8.2.3, DENOMINATOR has its default value and so the variance ratios are calculated from the residual mean square corresponding to the smallest residual sum of squares.

You can use the EXPAND option of the RKEEP directive to re-order the parameters of a regression model when they are stored by RKEEP, so that they correspond to the order of the terms in the maximal model defined by the previous TERMS statement. So, after the ADD statement in Example 8.2.3, you could put the statement

    RKEEP [EXPAND=yes] ESTIMATE=E

This would cause E to have parameters in the following order:

```
*
X[1]
X[2]
*
X[4]
Constant
```

The first missing value corresponds to the response variate, which is added at the start of the list of items if not included explicitly, and the second missing value corresponds to X[3] which has not been fitted. The constant always appears last, if fitted. So, if you set EXPAND=yes in RKEEP, all the variates and matrices that are stored by RKEEP will contain information for every parameter in the maximal model. That is, for each term in the maximal model there will be one value in the variate of parameter estimates and in the variate of standard errors, and there will be one row and one column in the inverse matrix and in the variance-covariance matrix, and one column in the design matrix. These values or rows or columns of values will be set to missing by Genstat if the corresponding parameter is not in the current model.

8.2.5 The **TRY** directive

TRY displays results of single-term changes to a linear, generalized linear, or generalized additive model.

Options

PRINT = *strings*	What to print (model, deviance, summary, estimates, correlations, fittedvalues, accumulated, monitoring); default mode,summ,esti
FACTORIAL = *scalar*	Limit for expansion of model terms; default * i.e. that in previous TERMS statement
POOL = *string*	Whether to pool ss in accumulated summary between all terms fitted in a linear model (yes, no); default no

DENOMINATOR = *string*	Whether to base ratios in accumulated summary on rms from model with smallest residual ss or smallest residual ms (ss, ms); default ss
NOMESSAGE = *strings*	Which warning messages to suppress (dispersion, leverage, residual, aliasing, marginality); default *
FPROBABILITY = *string*	Printing of probabilities for variance ratios (yes, no); default no
TPROBABILITY = *string*	Printing of probabilities for t-statistics (yes, no); default no

Parameter

| *formula* | List of explanatory variates and factors, or model formula |

The essential difference between TRY and SWITCH is that TRY makes no permanent change to the current model. Explanatory variables are added or removed only temporarily.

The current regression model is modified by each term in the formula specified by the parameter of TRY, one term at a time, dropping terms that are in the current model and adding terms that are not. After each change, the output requested by the PRINT option is displayed; then Genstat reverts to the original model before going on to the next change. The current model is also reinstated after the last change.

In Example 8.2.5, TRY is used to study the effect of X[3].

Example 8.2.5

```
  24   TRY [PRINT=summary] X[3]

24..............................................................................


***** Regression Analysis *****

*** Summary of analysis ***

                d.f.           s.s.           m.s.         v.r.
Regression         3        2664.93        888.309       157.27
Residual           9          50.84          5.648
Total             12        2715.76        226.314

Change            -1         -23.93         23.926         4.24

Percentage variance accounted for 97.5
Standard error of observations is estimated to be 2.38
* MESSAGE: The following units have high leverage:
                 12           0.69
```

The only circumstances in which TRY does make a permanent change is when the current model includes a term that had been found to be aliased before this TRY statement was reached. If the aliased term can be fitted after dropping one of the terms in the TRY formula, then that is indeed done. The term that was dropped will be aliased thereafter.

The options are as in the FIT directive, except that there is no CONSTANT option. The accumulated setting of the PRINT option will show only one change at a time. Accumulated summaries produced by later statements will not have any entries for a TRY statement.

8.2.6 The STEP directive

STEP selects a term to include in or exclude from a linear, generalized linear, or generalized additive model according to the ratio of residual mean squares.

Options

PRINT = *strings*	What to print (model, deviance, summary, estimates, correlations, fittedvalues, accumulated, monitoring, changes); default mode, summ, esti, chan
FACTORIAL = *scalar*	Limit for expansion of model terms; default * i.e. that in previous TERMS statement
POOL = *string*	Whether to pool ss in accumulated summary between all terms fitted in a linear model (yes, no); default no
DENOMINATOR = *string*	Whether to base ratios in accumulated summary on rms from model with smallest residual ss or smallest residual ms (ss, ms); default ss
NOMESSAGE = *strings*	Which warning messages to suppress (dispersion, leverage, residual, aliasing, marginality); default *
FPROBABILITY = *string*	Printing of probabilities for variance ratios (yes, no); default no
TPROBABILITY = *string*	Printing of probabilities for t-statistics (yes, no); default no
INRATIO = *scalar*	Criterion for inclusion of terms; default 1.0
OUTRATIO = *scalar*	Criterion for exclusion of terms; default 1.0

Parameter

formula	List of explanatory variates and factors, or model formula

---

Example 8.2.6 shows how you can use STEP to pick the "best" change to make to the set of explanatory variables at any stage.

Example 8.2.6

```
26  FIT [PRINT=*] X[1]
27  STEP [INRATIO=4; OUTRATIO=4] X[1...4]
```

27..

*** Residual mean squares ***

```
        5.790    Adding   X[2]
        7.476    Adding   X[4]
      115.062    No change
      122.707    Adding   X[3]
      226.314    Dropping X[1]
```

***** Regression Analysis *****

Response variate: Heat
     Fitted terms: Constant, X[1], X[2]

*** Summary of analysis ***

	d.f.	s.s.	m.s.	v.r.
Regression	2	2657.86	1328.929	229.50
Residual	10	57.90	5.790	
Total	12	2715.76	226.314	
Change	-1	-1207.78	1207.782	208.58

Percentage variance accounted for 97.4
Standard error of observations is estimated to be 2.41
* MESSAGE: The following units have high leverage:
              12   0.55

*** Estimates of regression coefficients ***

	estimate	s.e.	t(10)
Constant	52.58	2.29	23.00
X[1]	1.468	0.121	12.10
X[2]	0.6623	0.0459	14.44

Example 8.2.6 starts by fitting a model containing just the term X[1]. Then the STEP statement tries, one at a time, to drop X[1] and to add X[2], X[3], and X[4]. After each of these it reverts to the original model. Thus far, therefore, it is like a TRY statement. But then STEP, unlike TRY, permanently modifies the current model according to the change that was most successful. This means (putting it loosely at the moment) that if, for example, dropping X[1] "improves" the model, then X[1] is permanently removed; or, when no removals are worthwhile, if adding X[2] gives the biggest "improvement", then X[2] is permanently included. We see in fact that the latter happened, and so the current model is now as displayed at the end of Example 8.2.6.

We now define what constitutes an "improvement" in the model. The current model is modified by each term in the formula specified by the parameter of STEP, one term at a time, as with TRY (8.2.5). For each term, the residual sum of squares and the residual degrees of freedom are recorded; then Genstat reverts to the original model before trying the next term.

The current model is finally modified by the best term, according to a criterion based on the variance ratios. Suppose that the residual sum of squares and residual degrees of freedom of the current model are s_0 and d_0, and of the model after making a one-term change are s_1 and d_1. If the variance ratio for any term that is dropped is greater than the value of the setting of the OUTRATIO option, then the term that most reduces the residual mean square is dropped. That is, a term will be dropped only if at least one term has

$$\{(s_1-s_0) \ / \ (d_1-d_0)\} \ / \ \{s_0/d_0\} \ > \ \text{OUTRATIO}$$

If you have set OUTRATIO=*, then no term is dropped. Note that, though the criteria are ratios of variances, you should not interpret them as F-statistics with the usual interpretation of significance. The probability levels would need be adjusted to take account of correlations between the explanatory variables concerned, and the number of changes being considered.

If no term satisfies the criterion for dropping, then the term that most reduces the residual mean square will be added to the model if its variance ratio is greater than the setting of the INRATIO option. That is, if

$$\{(s_0-s_1) \ / \ (d_0-d_1)\} \ / \ \{s_1/d_1\} \ > \ \text{INRATIO}$$

Likewise, if you have set INRATIO=*, no term will be added.

If neither criterion is met, the current model is left unchanged.

The effect of the STEP directive is to make one change of a stagewise regression search. You can make STEP do forward selection by repeating a statement with option OUTRATIO=*: for example,

```
TERMS X[]
FOR [NTIMES=4]
   STEP [OUTRATIO=*] X[]
ENDFOR
```

Similarly, you can make STEP do backward elimination, by setting option INRATIO=*.

You might want to avoid the loop repeatedly making no change if some terms have little effect. To do that, you should cause an exit from the loop (5.2.4) whenever the residual d.f. does not change; you can obtain the residual d.f. from the RKEEP directive. For example, with backward elimination:

```
TERMS X[]
FIT X[]
RKEEP DF=D0
FOR [NTIMES=4]
   STEP [INRATIO=*; OUTRATIO=4] X[]
   RKEEP DF=D1
   EXIT D1.EQ.D0
   CALCULATE D0 = D1
ENDFOR
```

The changes setting of the PRINT option produces a list of terms with the corresponding residual mean squares and residual degrees of freedom, ordered according to the sizes of the residual mean squares; you can see this in Example 8.2.6. Note that this list is not available for display later by the RDISPLAY directive. The INRATIO and OUTRATIO options are explained above. The rest of the options are as in the FIT directive, except that there is no CONSTANT option.

8.3 Linear regression with grouped or qualitative data

You can incorporate the effects of grouped variables into a regression model. These are sometimes called qualitative variables to distinguish them from the quantitative ones that we have discussed so far in this chapter. For example, you could fit a separate constant term for each level of some classification: you would then get a series of parallel regression lines of the response variable on the quantitative variable. You might also want to fit separate slopes for the quantitative variable at each level of the classification.

In Example 8.3, the data from a cloud-seeding experiment include two qualitative variables, referred to as A and E; their effects are included in a linear model along with the effects of four quantitative variables referred to as D, S, C, and Lp.

Example 8.3

```
 1    "  Comparison of multiple linear regressions of rainfall on associated
-2        variables in the presence and absence of cloud seeding.
-3        Data from Woodley et al. (1977); analysed by Weisberg (1985) p169."
 4
 5    OPEN 'CLOUD.DAT'; CHANNEL=2
 6    FACTOR A,E
 7    READ [PRINT=data; CHANNEL=2] A,D,S,C,P,E,Y; \
 8       FREPRESENTION=labels,4(*),levels,*

      1  NS  0 1.75 13.4 0.274 2 12.85     S  1 2.70 37.9 1.267 1  5.52
      2   S  3 4.10  3.9 0.198 2  6.29    NS  4 2.35  5.3 0.526 1  6.11
      3   S  6 4.25  7.1 0.250 1  2.45    NS  9 1.60  6.9 0.018 2  3.61
      4  NS 18 1.30  4.6 0.307 1  0.47    NS 25 3.35  4.9 0.194 1  4.56
      5  NS 27 2.85 12.1 0.751 1  6.35     S 28 2.20  5.2 0.084 1  5.06
      6   S 29 4.40  4.1 0.236 1  2.76     S 32 3.10  2.8 0.214 1  4.05
      7  NS 33 3.95  6.8 0.796 1  5.74     S 35 2.90  3.0 0.124 1  4.84
      8   S 38 2.05  7.0 0.144 1 11.86    NS 39 4.00 11.3 0.398 1  4.45
      9  NS 53 3.35  4.2 0.237 2  3.66     S 55 3.70  3.3 0.960 1  4.22
     10  NS 56 3.80  2.2 0.230 1  1.16     S 59 3.40  6.5 0.142 2  5.45
     11   S 65 3.15  3.1 0.073 1  2.02    NS 68 3.15  2.6 0.136 1  0.82
     12   S 82 4.01  8.3 0.123 1  1.09    NS 83 4.65  7.4 0.168 1  0.28
 9    " Variables are: A   Action (NS not seeded, S seeded)
-10                     D   Days after first day of experiment
-11                     S   Suitability for seeding (from model)
-12                     C   Percent cloud cover
-13                     P   Previous rainfall (in 10**7 cubic m)
-14                     E   Type of cloud (1 or 2)
-15                     Y   Subsequent rainfall (in 10**7 cubic m)"
 16   CALCULATE Lp,Ly = LOG10(P,Y)
 17   MODEL Ly
 18   TERMS A*(D+S+C+Lp+E)
 19   FIT [PRINT=model,estimates] A+S+D+C+Lp+E
```

19...

```
***** Regression Analysis *****

  Response variate: Ly
        Fitted terms: Constant + A + S + D + C + Lp + E

*** Estimates of regression coefficients ***

                           estimate          s.e.       t(17)
Constant                      1.030         0.381        2.70
A S                           0.274         0.149        1.84
S                           -0.0817        0.0966       -0.85
D                          -0.00604       0.00359       -1.68
C                           -0.0049        0.0119       -0.41
Lp                            0.348         0.240        1.45
E 2                           0.340         0.195        1.74

   20   ADD [PRINT=model,estimates,accumulated] A.S


20...........................................................................


***** Regression Analysis *****

  Response variate: Ly
        Fitted terms: Constant + A + S + D + C + Lp + E + S.A

*** Estimates of regression coefficients ***

                           estimate          s.e.       t(16)
Constant                      0.614         0.368        1.67
A S                           1.670         0.559        2.99
S                             0.107         0.111        0.96
D                          -0.00925       0.00336       -2.76
C                           -0.0128        0.0108       -1.18
Lp                            0.379         0.208        1.82
E 2                           0.470         0.177        2.66
S.A S                        -0.430         0.167       -2.57


*** Accumulated analysis of variance ***

Change                        d.f.           s.s.         m.s.       v.r.
+ A                              1        0.19498      0.19498       2.20
+ S                              1        0.38967      0.38967       4.39
+ D                              1        0.86460      0.86460       9.75
+ C                              1        0.00214      0.00214       0.02
+ Lp                             1        0.08127      0.08127       0.92
+ E                              1        0.35882      0.35882       4.05
+ S.A                            1        0.58560      0.58560       6.60
Residual                        16       1.41926      0.08870

Total                           23       3.89635      0.16941
```

Before we go into details, look at the FIT statement in line 19. A is a factor with two levels labelled NS and S, and E also has two levels, 1 and 2. This statement fits a multiple linear regression of the variate Ly on the variates S, D, C, and Lp; the model also includes the *main effects* of the factors A and E. This means that for each factor an additive constant is estimated, representing the mean difference between the responses at the two levels of the factor. In other words, a set of parallel linear regressions is fitted, one for each combination of levels of the two factors.

Now look at the ADD statement. Here the interaction between the factor A and the variate S is included too. This means that different effects of the variate S are estimated for each level of A. In other words, separate linear regressions are fitted as before, except that the fitted relationships between Ly and S for each level of A are not constrained to be parallel.

We now make some more formal definitions, after which we shall return to this example.

You store data from qualitative variables in factors (2.3.3). After factors have been declared and assigned values, their effects can be included in regression models. You do this by putting their identifiers in directives such as FIT and TERMS, along with the identifiers of variates storing the values of quantitative explanatory variables.

You represent the *main effect* of a factor by its identifier as a single term: a model including such a main effect has a separate constant or intercept for each level of the factor.

Interactions between factors allow more detailed modelling of the constant term for combinations of levels of more than one factor. They are represented by terms consisting of the dot operator between factor identifiers in formulae.

Interactions between factors and variates allow modelling of the changes in the regression coefficient of the variate between combinations of levels of factors. They too are represented by terms including dot operators.

"Interactions" between quantitative variables can also be expressed in this way. They represent simply the product of two or more variates.

8.3.1   Formulae in parameters of regression directives

Formulae are described in 1.7.3, and further details are given in 9.1.1. In regression directives you cannot use the // operator, nor the functions POLND and REGND. The functions POL, REG, and SSPLINE can be used to represent polynomial effects, general sets of contrasts, and nonparametric smoothed effects respectively; these are described in 8.4. The basic operators are those of summation (+) and dot product (.), and if you want you can write all formulae using just these two. The other operators provide a shorthand for representing complicated formulae. Of particular use in regression are the cross-product operator (*)

        A*B = A + B + A.B

and the nesting operator (/)

        A/B = A + A.B

For more complicated formulae, remember that the nesting operator is not distributive (see 1.7.3 and 9.1.1): for example,

        (A + B)/C = A + B + A.B.C

Terms are ignored if they are put in an invalid order. For example the formula A.B + A becomes just A.B, since A is *marginal* to A.B. Genstat takes care to avoid fitting uninterpretable models that violate the principles of marginality, and will not accept any model where a term is specified before any of its margins (8.3.3).

If a formula contains commas, they are treated in the same way as + operators together with pairs of brackets. For example, X,Y*A is the same as (X+Y)*A, which is X+Y+A+X.A+Y.A.

The expansion of formulae into constituent terms is controlled in all regression directives by the FACTORIAL option. The default setting is 3, which excludes all interactions involving more than three identifiers. For example,

        FIT [FACTORIAL=2] A*B*C

will fit a model that includes the terms A, B, C, A.B, A.C, and B.C, but excludes A.B.C. However, following a TERMS statement, the default of FACTORIAL in other regression statements is whatever was set or implied by default in TERMS.

8.3.2 Parameterization of factors

A regression model that includes the main effect of a single factor and omits the constant (8.1.2), contains one parameter for each level of the factor: this parameter represents the constant term for that level. If an explicit constant term is also included in the model, then some constraint must be applied to the parameters for the factors. In Genstat, the parameter corresponding to the first level of the factor is set to zero. For example, in the first model fitted by line 19 of Example 8.3, the parameter estimates are:

Example 8.3.2a

```
*** Estimates of regression coefficients ***
```

	estimate	s.e.	t(17)
Constant	1.030	0.381	2.70
A S	0.274	0.149	1.84
S	-0.0817	0.0966	-0.85
D	-0.00604	0.00359	-1.68
C	-0.0049	0.0119	-0.41
Lp	0.348	0.240	1.45
E 2	0.340	0.195	1.74

No parameter estimate is shown for "A NS" or for "E 1". You can interpret the constant term here as the constant when both these factors are at level 1, that is, on days when there was no seeding and the cloud was of Type 1. Thus the parameter labelled "A S" is the difference between the constant for days with and without seeding. The same is true for factors with more than two levels: the parameters all represent differences from the first level.

This form of parameterization makes it easy to compare each level of a factor with a base level. In the example, the t-statistic of the estimate for "A S" shows that the difference between the constants for the levels of A is not quite significant at the 5% level.

However, you will not necessarily find this parameterization very convenient for summarizing the effect of a factor, especially when there are several levels, or several factors in a model. Instead you may wish to use the PREDICT directive to produce summaries (8.3.4), unless the methods of Chapter 9 or 10 are relevant.

You can obtain other parameterizations by modifying the definition of the model. For example, you can fit a constant for each level of factor A by setting option CONSTANT=omit in FIT:

Example 8.3.2b

```
22   MODEL Ly
23   FIT [PRINT=estimates; CONSTANT=omit] A+S+D+C+Lp+E
```

23..

***** Regression Analysis *****

*** Estimates of regression coefficients ***

	estimate	s.e.	t(17)
A NS	1.371	0.432	3.17
A S	1.645	0.480	3.43
S	-0.0817	0.0966	-0.85
D	-0.00604	0.00359	-1.68
C	-0.0049	0.0119	-0.41
Lp	0.348	0.240	1.45
E 1	-0.340	0.195	-1.74
E 2	0	*	*

Since there is no constant term in this model, no constraint needs to be imposed on the parameters representing factor A. However, the parameterization of factor E must still be constrained as before. Genstat always chooses to parameterize the first factor in the model fully when the constant is omitted; so to get E fully parameterized you should put E before A in the FIT statement.

If you want to fit a sequence of models and use any form of parameterization other than the standard one (including the constant), you must set option FULL=yes in the TERMS statement. This is because TERMS allocates the number of parameters for each term in the model, and automatically imposes constraints when there is over-parameterization. The setting FULL=yes specifies that a parameter is to be associated with every level of each factor, regardless of the presence of a constant term. If you include a constant term in a model as well as some factors, you will again find that one of the parameters of each factor will be aliased. Similarly, if you omit the constant and fit more than one factor, each factor other than the first will also have an aliased parameter.

Example 8.3.2c

```
25   TERMS [FULL=yes] A*(D+S+C+Lp+E)
26   FIT [PRINT=estimates; CONSTANT=omit] A+S+D+C+Lp+E
```

26..

***** Regression Analysis *****

*** Estimates of regression coefficients ***

	estimate	s.e.	t(17)
A NS	1.371	0.432	3.17
A S	1.645	0.480	3.43
S	-0.0817	0.0966	-0.85
D	-0.00604	0.00359	-1.68
C	-0.0049	0.0119	-0.41
Lp	0.348	0.240	1.45
E 1	-0.340	0.195	-1.74
E 2	0	*	*

The last level of the factor E is aliased in both Example 8.3.2b and Example 8.2.3c since this is the last parameter to be fitted, and its estimate is left as 0. Notice that no reports are given on partial aliasing of terms involving factors when the constant is omitted or when FULL=yes, regardless of the setting of the NOMESSAGE option of the FIT directive.

Factor effects are also fully parameterized if an SSPM or a DSSP structure, supplied through the SSP option of the TERMS directive, was declared by an SSPM statement (2.7.2) with option FULL set to yes.

8.3.3 Parameterization of interactions, and marginality

The parameters representing interactions in a model are also constrained to remove over-parameterization.

For example, if A and B are factors with two and three levels respectively and the model A*B is fitted (including a constant), the parameters will be: Constant, A2, B2, B3, A2.B2, and A2.B3. No parameter is assigned to A1 because there is a constant, and none to B1 or A1.B1. Similarly, no parameter is assigned to A2.B1 because the main effect of A is included, and none to A1.B2 nor A1.B3 because the main effect of B is included. The terms A and B are described as being *marginal* to the term A.B. The constant term is also marginal to A and B, and to the term A.B.

In general, one term is marginal to a second if the second can be written as an interaction between the first term and a third term involving factors only; for example, A is marginal to A.B and to A.B.C.D. Whenever one term is marginal to a second, some parameters of the full set of the second term are aliased with the first term. Genstat will automatically constrain selected parameters to be zero to avoid aliasing. The automatic constraint can be removed by setting the FULL option of the TERMS directive.

In the analysis fitted in line 20 of Example 8.3, the fitted model is A+S+D+C+Lp+E+S.A. The term S.A is an interaction between a factor and a variate, and so represents variations in the effect of the variate between levels of the factor: that is, the regression lines of Ly on S are allowed to have separate slopes for the days with and without seeding, as well as separate intercepts. The linear model is

$$y_{ijk} = \alpha + \gamma_{1i} + \beta_1 x_{1ijk} + \beta_2 x_{2ijk} + \beta_3 x_{3ijk} + \beta_4 x_{4ijk} + \gamma_{2j} + \delta_i x_{1ijk} + \varepsilon_{ijk}$$
$$\text{for } i = 1, 2; j = 1, 2; k = 1 \ldots N_{ij}$$

where α represents the constant term, and is the intercept for "A NS" and "E 1". The parameters γ_{1i} and γ_{2j} represent the main effects of A and E: γ_{1i} is the difference between the intercept for the *i*th level of A and that for the first level (labelled "A NS"), so that γ_{11} is zero. The parameter β_1 represents the variate S, and is the slope for "A NS" and "E 1". Lastly, the parameters δ_i represent the interaction term S.A; δ_i is the difference between the slope for the *i*th level of A and that for the first level, so that δ_1 is zero. In this model, the constant is marginal to the terms A and E, and S is marginal to A.S.

Again, you can present the results differently, either using the PREDICT directive (8.3.4), or by modifying the model. The parameters can be made to be the actual slopes by omitting S from the model, as long as you have set option FULL=yes in TERMS:

Example 8.3.3a

```
 28   FIT [PRINT=estimates; CONSTANT=omit] A+D+C+Lp+E+A.S
```

```
28.............................................................................
```

```
***** Regression Analysis *****

*** Estimates of regression coefficients ***
```

	estimate	s.e.	t(16)
A NS	1.084	0.391	2.77
A S	2.754	0.600	4.59
D	-0.00925	0.00336	-2.76
C	-0.0128	0.0108	-1.18
Lp	0.379	0.208	1.82
E 1	-0.470	0.177	-2.66
E 2	0	*	*
S.A NS	0.107	0.111	0.96
S.A S	-0.323	0.126	-2.57

If option FULL had been left at its default setting no, the FIT statement would fail:

Example 8.3.3b

```
 30   TERMS A*(D+S+C+Lp+E)
 31   FIT [PRINT=*; CONSTANT=omit] A+D+C+Lp+E+A.S
```

```
31.............................................................................
```

```
* MESSAGE: Term A cannot be added
  because term Constant is marginal to it and is not in the model

* MESSAGE: Term E cannot be added
  because term Constant is marginal to it and is not in the model

* MESSAGE: Term S.A cannot be added
  because term S is marginal to it and is not in the model
```

The messages about marginality can be suppressed by using the marginality setting of the NOMESSAGE option of the FIT directive.

As an alternative to setting FULL=yes, you could omit the marginal terms from the TERMS statement as well; above you would need to omit the effect of S. However, the constant cannot be omitted in TERMS.

8.3.4 The **PREDICT** directive

PREDICT forms predictions from a linear or generalized linear model.

Options

PRINT = *string*	What to print (description, predictions, se); default desc,pred,se
CHANNEL = *scalar*	Channel number for output; default * i.e. current output channel
COMBINATIONS = *string*	Which combinations of factors in the current model to include (all, present); default all
ADJUSTMENT = *string*	Type of adjustment (marginal, equal); default marg
WEIGHTS = *tables*	Weights classified by some or all of the factors in the model; default *
METHOD = *string*	Method of forming margin (mean, total); default mean
ALIASING = *string*	How to deal with aliased parameters (fault, ignore); default faul
BACKTRANSFORM = *string*	What back-transformation to apply to the values on the linear scale, before calculating the predicted means (link, none); default link
PREDICTIONS = *tables* or *scalars*	To save tables of predictions for each y variate; default *
SE = *tables* or *scalars*	To save tables of standard errors of predictions for each y variate; default *
VCOVARIANCE = *symmetric matrices*	To save variance-covariance matrices of predictions for each y variate; default *
SAVE = *identifier*	Specifies save structure of model to display; default * i.e. that from latest model fitted

Parameters

CLASSIFY = *vectors*	Variates and/or factors to classify table of predictions
LEVELS = *variates* or *scalars*	To specify values of variates, levels of factors

---

The PREDICT directive provides a convenient way of summarizing the results of a regression, by using the fitted relationship to predict the values of the response variate at particular values of the explanatory variables. In simple or multiple linear regression, the parameters of the model may be sufficient summaries in themselves, but these may not provide a very clear description when the model contains factors and their interactions. PREDICT can also be used to answer "what-if" questions, effectively predicting what fitted values would have been obtained if the data had been balanced in some way.

The simplest use of PREDICT is to make estimates from a simple linear regression for specific values of the explanatory variable. For example, if we had regressed Ly on just S in the example above, we could get the predicted value of Ly at S = 3.5 (say) by putting

```
PREDICT S; LEVELS=3.5
```

If we wanted the predicted values at 3.5 and 4, we would have to put these into a variate. The easiest way to do that is to use an unnamed variate (1.6.3):

```
PREDICT S; LEVELS=!(3.5,4)
```

Suppose now that we had regressed Ly on both S and C, and wanted to predict the value of Ly at S = 3.5 and 4 and C = 4, 8, and 12. We would then put 3.5 and 4 into one variate, and 4, 8, and 12 into another:

```
PREDICT S,C; LEVELS=!(3.5,4),!(4,8,12)
```

This would give six predicted values, one for each combination of 3.5 and 4 with 4, 8, and 12.

If we had also included the factor E in the regression, we might want to predict Ly for S equal to 3.5 at both levels 1 and 2 of E:

```
PREDICT E,S; LEVELS=!(1,2),3.5
```

This would produce two predicted values, classified by the levels of A. Since C is not mentioned in the PREDICT statement, the predictions will be based on the mean value of C by default. It is not actually necessary to list the levels of E if predictions are wanted for all of them; we could thus have put:

```
PREDICT E,S; LEVELS=*,3.5
```

If the factor A was also in the model, we could still use either of the previous two statements to get a summary of the effects of E. Since there is no mention of A, the predictions would automatically be averaged over the levels of A, as described later in this section.

For more complicated structures the rules are more intricate, as we shall see. But the basic ideas remain the same as in the simpler cases. In Example 8.3.4a, we summarize the model fitted at line 20 of Example 8.3, for each combination of levels of the two factors.

Example 8.3.4a

```
  33  FIT [PRINT=*] A+S+D+C+Lp+E+A.S
  34  PREDICT A,E
```

```
34.......................................................................
```

```
*** Predictions from regression model ***
```

```
These predictions are fitted values.
```

```
The predictions are based on fixed values of some variates:
```

```
     Variate    Fixed value    Source of value
         D            35.33    Mean of variate
         S            3.169    Mean of variate
         C            7.246    Mean of variate
        Lp          -0.6489    Mean of variate

    Table contains predictions followed by standard errors

Response variate: Ly
       E            1.00                        2.00
       A
      NS          0.2883        0.0995        0.7582        0.1583
       S          0.5958        0.0918        1.0656        0.1799
```

The four values are estimates, based on the fitted model, of the mean logged rainfall at the mean values of the four explanatory variates.

By using the LEVELS parameter, we can ask for the summary to be calculated for cloud-type 2 only, for a range of suitability values (variate S), and as if all observations were made on the first day of the experiment (D=0).

Example 8.3.4b

```
 36   PREDICT S,A,E,D; LEVELS=!(1...4),*,2,0


 36................................................................................


*** Predictions from regression model ***

These predictions are fitted values.

The predictions are based on fixed values of some variates:

     Variate    Fixed value    Source of value
         D              0.     Supplied
         C            7.246    Mean of variate
        Lp          -0.6489    Mean of variate

The predictions are calculated at fixed levels of some factors:

     Factor   Fixed level
       E      2

    Table contains predictions followed by standard errors

Response variate: Ly
       A            NS                           S

      1.00        0.852         0.212         2.093         0.387
      2.00        0.960         0.163         1.770         0.285
      3.00        1.067         0.182         1.447         0.211
      4.00        1.174         0.254         1.124         0.199
```

The first parameter, CLASSIFY, specifies those variates or factors in the current regression model whose effects you want to summarize. Any variate or factor in the current model that you do not include will be standardized in some way, as described below.

The LEVELS parameter specifies values at which the summaries are to be calculated, for each of the structures in the CLASSIFY list. For factors, you can select some or all of the levels, while for variates you can specify any set of values. A single level or value is represented by a scalar; several levels or values must be combined into a variate (which may of course be unnamed). A missing value in the LEVELS parameter is taken by Genstat to stand for all the levels of a factor, or for the mean value of a variate.

You can best understand how Genstat forms predictions by regarding its calculations as consisting of two steps. The first step, referred to below as Step A, is to calculate the full table of predictions, classified by every factor in the current model. For any variate in the model, the predictions are formed at its mean, unless you have specified some other values using the LEVELS parameter; if so, these are then taken as a further classification of the table of predictions. The second step, referred to as Step B, is to average the full table of predictions over the classifications that do not appear in the CLASSIFY parameter: you can control the type of averaging using the COMBINATIONS, ADJUSTMENT, and WEIGHTS options.

Printed output is controlled by the PRINT option. The description setting produces a summary of what standardization policies are used when forming the predictions, the predictions setting prints the predictions, and se produces predictions and standard errors; by default all these components of output are printed. The standard errors are relevant for the predictions when considered as means of those data that have been analysed (with the means formed according to the averaging policy defined by the options of PREDICT). The word *prediction* is used because these are predictions of what the means would have been if the factor levels been replicated differently in the data; see Lane and Nelder (1982) for more details. The standard errors are not augmented by any component corresponding to the estimated variability of a new observation.

You can send the output to another channel, or to a text structure, by setting the CHANNEL option.

The COMBINATIONS option specifies which cells of the full table in Step A are to be filled for averaging in Step B. By default all the cells are used. Alternatively, you can set COMBINATIONS=present to exclude cells for factor combinations that do not occur in the data, as shown in Example 8.3.4i below. In the examples above, however, this would make no difference because all four cells in the A by E table contain some values.

Setting COMBINATIONS=present overrules the LEVELS parameter. Any subsets of factor levels in the LEVELS parameter are ignored, and predictions are formed for all the factor levels that occur in the data. Likewise, the full table cannot then be classified by any sets of values of variates; the LEVELS parameter must then supply only single values for variates.

The ADJUSTMENT and WEIGHTS options define how the averaging is done in Step B. Values in the full table produced in Step A are averaged with respect to all those factors that you have not included in the settings of the CLASSIFY parameter. By default, the levels of any such factor are combined with what we call *marginal weights*: that is, by the number of occurrences of each of its levels in the whole dataset. Line 38 of Example 8.3.4c uses the TABULATE directive (4.11.1) to display the occurrences of combinations of levels of the factors

A and E, and then line 40 produces a summary of the effects of A alone, averaging over E.

Example 8.3.4c

```
38   TABULATE [PRINT=counts; CLASSIFICATION=A,E; MARGINS=yes]
```

	Count		
E	1.00	2.00	Count
A			
NS	9	3	12
S	10	2	12
Count	19	5	24

```
39
40   PREDICT A
```

```
40..........................................................................
```

```
*** Predictions from regression model ***
```

These predictions are fitted values
adjusted with respect to some factors as specified below.

The predictions are based on fixed values of some variates:

Variate	Fixed value	Source of value
D	35.33	Mean of variate
S	3.169	Mean of variate
C	7.246	Mean of variate
Lp	-0.6489	Mean of variate

The predictions have been standardized by averaging
over the levels of some factors:

Factor	Weighting policy	Status of weights
E	Marginal weights	Constant over levels of other factors

Table contains predictions followed by standard errors

Response variate: Ly

A		
NS	0.3862	0.0890
S	0.6937	0.0909

In forming the averages for A, the data from the two levels of E have been combined with weights 19 and 5, since these are the frequencies with which they occur in all the data. Because we are using the default settings of ADJUSTMENT and WEIGHTS, these weights are constant over the levels of the other factors: that is, the same weights are used when forming the prediction for each level of A, even though the levels of E occurred with different frequencies at the different levels of A. The effect, therefore, is to *standardize* the prediction for the estimated effects of E.

The ADJUSTMENT and WEIGHTS options allow you to change the weights. The setting ADJUSTMENT=equal specifies that the levels are to be weighted equally, when the predictions are averaged over the standardizing factors. The weights would then be 1 and 1 instead of 19 and 5, as shown in Example 8.3.4d.

Example 8.3.4d

```
  42   PREDICT [ADJUSTMENT=equal] A


42.......................................................................


*** Predictions from regression model ***

These predictions are fitted values
adjusted with respect to some factors as specified below.

The predictions are based on fixed values of some variates:

        Variate    Fixed value   Source of value
              D          35.33    Mean of variate
              S          3.169    Mean of variate
              C          7.246    Mean of variate
             Lp        -0.6489    Mean of variate

The predictions have been standardized by averaging
over the levels of some factors:

        Factor   Weighting policy  Status of weights
             E      Equal weights   Constant over levels of other factors


      Table contains predictions followed by standard errors

  Response variate: Ly

          A
         NS       0.5232         0.0984
          S       0.8307         0.1122
```

The WEIGHTS option is more powerful than the ADJUSTMENT option, allowing you to specify an explicit table of weights. This table can be classified by any, or all, of the factors over whose levels the predictions are to be averaged; the levels of remaining factors will be weighted according to the ADJUSTMENT option. Moreover, you can classify the weights by the factors in the CLASSIFY parameter as well, to provide different weightings for different combinations of levels of these factors. If you supply explicit weights in the WEIGHTS option, any setting of the COMBINATIONS option is ignored.

You will find explicit weights useful in particular when you have population estimates of the proportions of each level of a factor – proportions which may not be matched well in the available data. For example, you might know that these proportions for Type of cloud are in the ratio 2:1 rather than the 19:5 observed in the data. You might then specify these weights with the WEIGHTS option, as shown in Example 8.3.4e.

Example 8.3.4e

```
44   TABLE [CLASSIFICATION=E; VALUES=2,1] Wte
45   PREDICT [WEIGHTS=Wte] A
```

45...

*** Predictions from regression model ***

These predictions are fitted values
adjusted with respect to some factors as specified below.

The predictions are based on fixed values of some variates:

```
      Variate   Fixed value   Source of value
            D        35.33     Mean of variate
            S        3.169     Mean of variate
            C        7.246     Mean of variate
           Lp      -0.6489     Mean of variate
```

The predictions have been standardized by averaging
over the levels of some factors:

```
      Factor   Weighting policy  Status of weights
           E   Supplied weights  Constant over levels of other factors
```

```
      Table contains predictions followed by standard errors
```

Response variate: Ly

```
      A
     NS      0.4449       0.0896
      S      0.7524       0.0973
```

If a model contains any aliased parameters, predicted values cannot be formed for some cells of the full table without assuming a value for the aliased parameters. If the aliased parameters simply represent effects of variates that are correlated with other explanatory variables in the model, it may be sufficient just to ignore them. This can be done by setting the ALIASING option to ignore. The aliased parameters are then taken to be zero, and fitted values are calculated for all cells of the table from the remaining parameters in the model.

Aliasing can also occur if there are some combinations of factors that do not occur in the data, and here it may be more sensible to set option COMBINATIONS=present so that these cells are all excluded from the calculation of predictions.

To illustrate the action of the ALIASING and COMBINATIONS options, we fit a new model to the cloud-seeding data. The factor Sf is formed by grouping the values of the variate S; it happens that there were no days in the experiment when the suitability S was 2 or less and seeding was done, so one parameter of the interaction between A and Sf cannot be fitted.

Example 8.3.4f

```
53   GROUPS S; FACTOR=Sf; LIMITS=!(2,3,4)
54   TERMS A*Sf
55   FIT [PRINT=estimates] A*Sf
```

```
55..........................................................................
```

```
* MESSAGE: Term A.Sf cannot be fully included in the model
  because 1 parameter is aliased with terms already in the model
  (A S .Sf 4.175) = (A S) - (A S .Sf 2.525) - (A S .Sf 3.350)

***** Regression Analysis *****

*** Estimates of regression coefficients ***
```

	estimate	s.e.	t(17)
Constant	0.446	0.236	1.89
A S	0.369	0.354	1.04
Sf 2.525	0.348	0.373	0.93
Sf 3.350	−0.054	0.298	−0.18
Sf 4.175	−0.398	0.373	−1.07
A S .Sf 2.525	−0.362	0.500	−0.72
A S .Sf 3.350	−0.192	0.448	−0.43
A S .Sf 4.175	0	*	*

When the model is fitted, the last parameter of the interaction term A.Sf is aliased, and the form of the aliasing relationship is shown in the message. This relationship appears complicated because it is the first level of Sf that has no observations when A takes level 'S', and parameters are usually differences from the first level. When this happens, the parameters become differences with the last level instead, and the parameter for the last level becomes aliased.

If we now try to form predictions based on this model as before, a message appears.

Example 8.3.4g

```
57   PREDICT [PRINT=prediction] A,Sf
```

```
******** Warning (Code RE 36). Statement 1 on Line 57
Command: PREDICT [PRINT=prediction] A,Sf
Predictions cannot be formed
Option ALIAS is set to 'fault' and 1 parameter is aliased.
```

We can overcome this in one of three ways, depending on what we are prepared to assume about the aliased parameter. If we are not concerned about the prediction for the missing factor combination, or want to assume that the missing parameter (the difference between the first and last levels of Sf when A is 'S') is actually zero, then we can just set ALIASING=IGNORE.

Example 8.3.4h

```
 59   PREDICT [PRINT=prediction; ALIASING=ignore] A,Sf

59.......................................................................

*** Predictions from regression model ***

Response variate: Ly
          Sf       1.60        2.53        3.35        4.18
          A
          NS      0.446       0.794       0.392       0.048
          S       0.815       0.801       0.569       0.417
```

Alternatively, we can specify that predictions are to be formed only for the cells of the full table in Step A that have observations.

Example 8.3.4i

```
 61   PREDICT [PRINT=prediction; COMBINATIONS=present] A,Sf

61.......................................................................

*** Predictions from regression model ***

Response variate: Ly
          Sf       1.60        2.53        3.35        4.18
          A
          NS      0.446       0.794       0.392       0.048
          S          *        0.801       0.569       0.417
```

The third way to overcome aliasing is to supply explicit weights using the WEIGHTS option.

The use of COMBINATIONS=present has consequences on any averaging that is done. Here there is none, but if we were to give the statement

        PREDICT [COMBINATIONS=present] A

the averages for the two levels of A would not be formed with the same weights for Sf: that for level NS would include a contribution from level 1 of Sf, whereas that for level S would not. This must be borne in mind when interpreting the results.

We have assumed in this section that averaging is the appropriate way of combining predicted values over levels of a factor. But sometimes summation is needed, for example in the analysis of counts by log-linear models (8.5.1). You can achieve this by setting the METHOD option to total. The rules about weights and so on still apply. The BACKTRANSFORM option is also relevant to generalized linear models (8.5.6).

The PREDICTIONS, SE, and VCOVARIANCE options let you save the results of PREDICT as well as, or instead of, printing them. We use them in Example 8.3.4j to produce 95% confidence limits for predictions of the amount of rainfall at each level of A, transformed back to the natural scale of the original data. (The EDT function is described in 4.2.9.)

Example 8.3.4j

```
47   PREDICT [PRINT=*; PREDICTION=Pa; SE=Sa] A
48   RKEEP DF=df
49   CALCULATE High,Low = Pa + 1,-1*Sa*EDT(0.95; df)
50   & Low,Pa,High = 10**Low,Pa,High
51   PRINT Low,Pa,High
```

```
              Low        Pa        High
   A
   NS        1.702      2.433      3.479
    S        3.427      4.939      7.118
```

The SAVE option allows you to specify the regression save structure of the analysis on which the predictions are based. If SAVE is not set, the most recent regression model is used.

8.4   Polynomials and additive models

This section describes how to fit regression models containing functions of explanatory variables. The POL function allows you to specify polynomials representing quadratic, cubic, or quartic curves. The REG function allows you to specify your own functions of an explanatory variate, as long as they are linear in the parameters (nonlinear models are described in later sections of this chapter). REG can also be used to fit orthogonal polynomials. The SSPLINE function, or S for short, provides general smoothing splines. These are actually cubic splines with constraints to ensure smoothness, but they are usually regarded as nonparametric effects of variables. Models containing smoothing splines are referred to as *additive models*.

In Release 3.1 of Genstat, these three functions can be used only with explanatory variates. However, in later releases it may be possible to use them also with factors. In the meantime, it is easy to use CALCULATE to form a variate from a factor, as in

       CALCULATE V = F

or

       CALCULATE V = NEWLEVELS(F; W)

and then use a function of the variate in the regression model.

Another restriction in Release 3.1 is that you cannot include interactions between variables in a model if any one of the variables is in a function; again, this restriction may be relaxed in later releases. The TERMS parameter of the RKEEP directive (8.1.4) does not store information about functions, and it is not displayed in accumulated summaries (8.2.1).

8.4.1   Polynomial regression

You can fit a polynomial model simply by using the CALCULATE statement before FIT. For example, the following statements fit the quadratic regression of Y on X:

```
CALCULATE X2 = X**2
MODEL Y
FIT X,X2
```

However, you can do this more quickly, and using less storage space, with the POL function:

```
MODEL Y
FIT POL(X; 2)
```

The latter method also has the advantage that the PREDICT directive can produce predictions for specific values for X: with the former method, PREDICT treats X and X2 as if they varied separately rather than having a fixed relationship.

Example 8.4.1a shows the fitting of a cubic relationship between two variables measured on children with diabetes. The fitted polynomial curve is plotted by RGRAPH in Figure 8.4.1.

Example 8.4.1a

```
   1   "  Relationship between serum C-peptide and measured variables in
  -2       children with diabetes. Data from Sochett et al. (1987);
  -3       analysed by Hastie and Tibshirani (1990) p304."
   4
   5   OPEN 'DIABETES.DAT'; CHANNEL=2
   6   READ [CHANNEL=2; PRINT=data] Age,Base,Cpep

    1     5.2   -8.1 4.8     8.8 -16.1 4.1    10.5   -0.9 5.2    10.6   -7.8 5.5
    2    10.4  -29.0 5.0     1.8 -19.2 3.4    12.7  -18.9 3.4    15.6  -10.6 4.9
    3     5.8   -2.8 5.6     1.9 -25.0 3.7     2.2   -3.1 3.9     4.8   -7.8 4.5
    4     7.9  -13.9 4.8     5.2   -4.5 4.9    0.9  -11.6 3.0    11.8   -2.1 4.6
    5     7.9   -2.0 4.8    11.5   -9.0 5.5   10.6  -11.2 4.5     8.5   -0.2 5.3
    6    11.1   -6.1 4.7    12.8   -1.0 6.6   11.3   -3.6 5.1     1.0   -8.2 3.9
    7    14.5   -0.5 5.7    11.9   -2.0 5.1    8.1   -1.6 5.2    13.8  -11.9 3.7
    8    15.5   -0.7 4.9     9.8   -1.2 4.8   11.0  -14.3 4.4    12.4   -0.8 5.2
    9    11.1  -16.8 5.1     5.1   -5.1 4.6    4.8   -9.5 3.9     4.2  -17.0 5.1
   10     6.9   -3.3 5.1    13.2   -0.7 6.0    9.9   -3.3 4.9    12.5  -13.6 4.1
   11    13.2   -1.9 4.6     8.9  -10.0 4.9   10.8  -13.5 5.1
   7   MODEL Cpep
   8   FIT POL(Age; 3)
```

8. . . . .

***** Regression Analysis *****

Response variate: Cpep
    Fitted terms: Constant + Age
      Submodels: POL(Age; 3)

*** Summary of analysis ***

	d.f.	s.s.	m.s.	v.r.
Regression	3	7.95	2.6515	7.46
Residual	39	13.85	0.3552	
Total	42	21.81	0.5192	

```
Percentage variance accounted for 31.6

Standard error of observations is estimated to be 0.596

* MESSAGE: The following units have large standardized residuals:
                7         -2.58
               22          2.94

* MESSAGE: The following units have high leverage:
                8          0.37
               15          0.32
               24          0.29
               29          0.34


*** Estimates of regression coefficients ***

                         estimate          s.e.        t(39)
Constant                    2.740         0.508         5.39
Age Lin                     0.665         0.250         2.66
Age Quad                  -0.0641        0.0339        -1.89
Age Cub                   0.00198       0.00134         1.47

    9   RGRAPH [GRAPHICS=high] Age
```

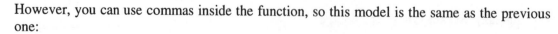

The FIT statement in Example 8.4.1a fits a cubic curve relating the response to a single explanatory variate. You can also use POL functions in multiple regression models with some or all the explanatory variates, and with different orders (quadratic, cubic, and so on). The maximum order for POL is 4. This limit is used because polynomial models with high orders can be very unstable; higher orders are allowed for orthogonal polynomials with the REG function.

When using POL, or the other functions, you must follow the syntax of model formulae (1.7.3). This means that you cannot use commas between functions: for example,

    FIT POL(X; 3),POL(Z; 3),F

would be faulted. Instead, you should use the plus operator, as in

    FIT POL(X; 3)+POL(Z; 3)+F

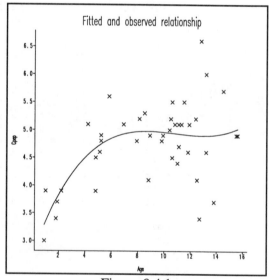

Figure 8.4.1

However, you can use commas inside the function, so this model is the same as the previous one:

    FIT POL(X,Z; 3)+F

The models specified by POL are simple polynomials: they are not orthogonalized. Thus, the parameter estimates are simply the linear coefficients of powers of an explanatory variate. This can result in computational problems with some data, when successive polynomial effects can be highly correlated; this would be evidenced in Genstat by a report of linear dependence and the omission of some of the effects. For this reason, it can be better to use the REG function to fit orthogonal polynomials, though the estimated parameters are then not so easy to interpret. Example 8.4.1b shows the correlations between the estimated parameters.

Example 8.4.1b

```
  11  RDISPLAY [PRINT=correlations]

11..............................................................................


***** Regression Analysis *****

*** Correlations between parameter estimates ***

estimate               ref     correlations

Constant                1      1.000
Age Lin                 2     -0.899  1.000
Age Quad                3      0.799 -0.975  1.000
Age Cub                 4     -0.721  0.928 -0.986  1.000
                               1      2      3      4
```

Functions can also be used in the TERMS directive. If a variable appears in a POL function in the model formula of TERMS, then the fitting statements that follow will fit the function of the variable rather than just its ordinary (linear) effect, whether or not the function name and parentheses are given. If a particular variate or factor has already been fitted in the model, the default order for the POL or REG function is the order already fitted; otherwise it is the order used in the TERMS directive. The order specified by TERMS cannot be exceeded (unless a new TERMS statement is given). It may be changed to a lower value whenever the variate is added to the model, or in a FIT statement. Attempts to change the order of a function already in the model by any other directive apart from FIT and SWITCH are ignored. For example, you can give the following statements to compare a quadratic with a cubic model:

```
TERMS POL(X; 3)
FIT POL(X; 2)
SWITCH POL(X; 3)
```

8.4.2 Orthogonal polynomials and general functions

The REG function can be used in exactly the same way as the POL function to fit polynomial effects. The difference is that REG will fit orthogonalized effects. It is also possible to fit orthogonalized effects by calculating them in advance with the ORTHPOL procedure, as in the following statements:

```
ORTHPOL [ORDER=4] X; POLYNOMIAL=P
FIT P[1...4]
```

The same model can be fitted using REG as follows:

```
FIT REG(X; 4)
```

The calculation of orthogonal polynomials is not trivial for a general variate, because in regression there is not necessarily any regular replication. By contrast, in a balanced design as in the examples of Chapter 9, orthogonal polynomials of factors are relatively easy to calculate, and so are formed automatically when the POL function is used. Use of the REG function as above will result in the automatic calculation of orthogonal polynomials internally, by the same method as used in the procedure ORTHPOL. Consequently REG uses more storage space than POL.

In Example 8.4.2, we fit the same model as in Example 8.4.1a but using orthogonalized polynomials for comparison; note that there is now no correlation between the parameter estimates.

Example 8.4.2

```
13   FIT [PRINT=estimates,correlations] REG(Age; 3)

13...........................................................................
```

```
***** Regression Analysis *****

*** Estimates of regression coefficients ***

                            estimate         s.e.        t(39)
Constant                    4.7465         0.0909        52.22
Age Reg1                    0.0831         0.0229         3.63
Age Reg2                   -0.01490        0.00562       -2.65
Age Reg3                    0.00198        0.00134        1.47

*** Correlations between parameter estimates ***

estimate                ref       correlations

Constant                1       1.000
Age Reg1                2       0.000   1.000
Age Reg2                3       0.000   0.000   1.000
Age Reg3                4       0.000   0.000   0.000   1.000
                                  1       2       3       4
```

The REG function can also be used to specify general functions of a variate. You must form these functions yourself, for example by using the CALCULATE directive, and put the results into a matrix for use in the third argument of REG. This matrix must have as many columns as there are values of the variate. The number of rows is the maximum order of the function and it must be greater than or equal to the setting of the second parameter of REG. For example, the following statements use the orthogonal polynomials formed by ORTHPOL and combine them into the matrix Ox used in REG; this would give the same result as using REG with no third argument.

```
ORTHPOL [ORDER=4] X; POLYNOMIAL=P
MATRIX [ROWS=4; COLUMNS=X; VALUES=£P[]] Ox
FIT REG(X; 3; Ox)
```

The values of the variate X are not actually used in the analysis, but must nonetheless be present.

This mechanism can be used to specify the whole design matrix of a model. Suppose that the matrix M contains the design matrix, as would be the case if it were formed by RKEEP after an analysis:

```
RKEEP DESIGNMATRIX=M
```

Then the model corresponding to this design matrix can be fitted by the statements

```
CALCULATE Mt = TRANSPOSE(M)
& N = NCOLUMNS(M)
FIT [CONSTANT=omit] REG(X; N; Mt)
```

where X is any of the explanatory variates fitted in the model.

The use of REG functions in regression directives other than FIT is the same as described for POL in 8.4.1, apart from the PREDICT directive: after fitting a model that includes a REG function, it is not possible to form predictions.

8.4.3 Cubic smoothing splines and additive models

The SSPLINE function, or S for short, specifies a cubic smoothing spline for the effect of a variate. Smoothing splines are complicated functions, constructed from segments of cubic polynomials between the distinct values of the variate, and constrained to be "smooth" at the junctions. Models that contain such a function are no longer linear, but are described as *additive models* because the effects of separate explanatory variates are still combined additively. Another way of describing the effects of a variate that has been smoothed in this way is *nonparametric*: in fact, there is a complicated parameterization of the fitted smooth curve, but it is unlikely to be of use for interpretation. See Hastie and Tibshirani (1990) for further details of these models. The main uses of smoothed terms in regression are to investigate the shape of a relationship with a view to later parametric fitting, and to remove the effect of nuisance variables so as to concentrate on the variables of interest.

The degree of smoothness can be controlled, effectively increasing or relaxing the constraints. For example,

```
FIT S(X; 4)
```

would fit a spline for X that has four effective degrees of freedom. This curve will be similar to the curve fitted by

```
FIT REG(X; 4)
```

However, the smoothing spline does not exhibit the awkward end-effects of the polynomial, where the curve by its parametric nature tends to bend much more sharply than the observed data would suggest. The smoothing spline with one degree of freedom has the same effect as a linear fit, although the iterative fitting process may not give exactly the same results. At the other extreme, if the variate X has precisely N values, all distinct, then the statement

```
     FIT S(X; N)
```

would fit a curve that actually passes through each data point (and so would be of little practical use). By default, if the second parameter of S is omitted, four effective degrees of freedom are assigned. For an explanation of effective degrees of freedom, see Hastie & Tibshirani (1990).

Example 8.4.3a shows a smoothing spline fitted to the relationship in the previous examples. The resulting fit is displayed in Figure 8.4.3a using the procedure RGRAPH.

Example 8.4.3a

```
   15  FIT S(Age; 3)

15.........................................................................


***** Regression Analysis *****

Response variate: Cpep
   Link function: Identity
   Fitted terms: Constant + Age
      Submodels: SSPLINE(Age; 3)


*** Summary of analysis ***

               d.f.        s.s.        m.s.        v.r.
Regression        3        7.88      2.6269        7.36
Residual         39       13.93      0.3571
Total            42       21.81      0.5192

Percentage variance accounted for 31.2

Standard error of observations is estimated to be 0.598

* MESSAGE: The following units have large standardized residuals:
               7       -2.66
              22        2.87

* MESSAGE: The following units have high leverage:
               8        0.29
              15        0.26
              24        0.24
              29        0.27


*** Estimates of regression coefficients ***

                     estimate        s.e.      t(39)
Constant                3.996       0.226      17.66
Age                    0.0831      0.0229       3.62

   16  RGRAPH [GRAPHICS=high] Age
```

Note that the linear component of the smoothing spline is reported in the same way as when just the linear effect of a variate is fitted and, in fact, has the same value. No other parameters of the smoothed effect are available from Genstat.

If a TERMS statement is given before fitting a smoothed variate, the same function must be defined for the variate in TERMS. Again the default number of degrees of freedom is four, but if a number is given in the second argument of the S function in TERMS it becomes the default for subsequent fitting statements, until another number is specified in a fitting statement, as with POL (8.4.1). The order of S can be either increased or decreased in subsequent SWITCH statements, and whenever the variate is re-introduced into the model after being dropped. Unlike POL and REG, there is no theoretical maximum number of degrees of freedom; the number available, however, is one less than the

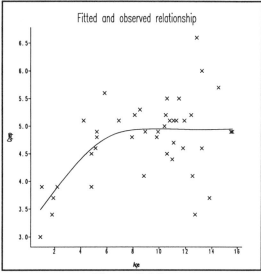

Figure 8.4.3a

number of distinct values in the variate. After you have used the S function to fit a smooth function of a variate, you revert to fitting just the linear effect by specifying the variate, without the function, in either SWITCH or FIT. However, as with POL and REG, attempts to change the order of an S function already in the model, by any directive other than SWITCH or FIT, will be ignored.

Example 8.4.3b shows the effect of a second smoothed variable Base being added to Age, after first giving a TERMS statement.

Example 8.4.3b

```
  18   TERMS S(Age,Base)
  19   FIT [PRINT=model,deviance] S(Age)

19...........................................................................................


***** Regression Analysis *****

 Response variate: Cpep
    Link function: Identity
     Fitted terms: Constant + Age
        Submodels: SSPLINE(Age; 4)


Residual d.f. 38, s.s. 13.72; Change d.f. -4, s.s. -8.08
  20   ADD S(Base)

20...........................................................................................


***** Regression Analysis *****
```

```
Response variate: Cpep
   Link function: Identity
   Fitted terms: Constant + Age + Base
      Submodels: SSPLINE(Age; 4)
                 SSPLINE(Base; 4)
```

*** Summary of analysis ***

	d.f.	s.s.	m.s.	v.r.
Regression	8	12.356	1.5445	5.56
Residual	34	9.451	0.2780	
Total	42	21.807	0.5192	
Change	-4	-4.273	1.0682	3.84

Percentage variance accounted for 46.5

Standard error of observations is estimated to be 0.527

```
* MESSAGE: The following units have large standardized residuals:
              22        2.70

* MESSAGE: The following units have high leverage:
               5        0.85
              10        0.52
```

*** Estimates of regression coefficients ***

	estimate	s.e.	t(34)
Constant	4.494	0.243	18.46
Age	0.0617	0.0208	2.97
Base	0.0374	0.0117	3.19

Finally, Figure 8.4.3b is produced by the statements

```
SWITCH S(Age; 20)
RGRAPH [GRAPHICS=high] Age
```

This shows how SWITCH can be used to change the order of the smoothing function, in this case increasing it to a point where the curve follows individual fluctuations too closely to be of much practical use.

After fitting spline functions, you can access the fitted effects with the RKEEP directive (8.1.4). The STERMS parameter can be used to store a pointer to those variates whose effects in the model are smoothed. The SCOMPONENTS parameter stores a pointer to variates, one for each smoothed variate in the same order as in STERMS, containing the

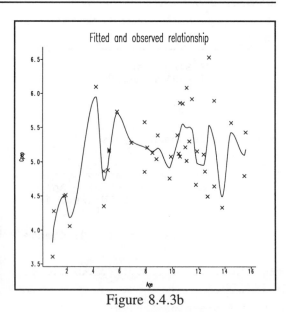

Figure 8.4.3b

fitted nonlinear component of each smoothed variate – this does not include the linear component or the constant term.

The PREDICT directive cannot be used to form predictions at specific values of a variate that has been smoothed. If predictions are formed for other explanatory variates or factors in the model, only the linear effect of the smoothed variate will be incorporated in the predictions.

Regression models may contain any combination of POL, REG, and S functions and linear terms, except that no interactions can be fitted involving a variate appearing in a function.

When a spline function is included in the model, it has to be fitted iteratively using a technique known as *backfitting*. This iterative process can be monitored if required in the same way as the iterative process for generalized linear models (8.5.6). Because an iterative method is needed, Genstat will analyse only the first response variate, even if several have been listed in the MODEL statement. Similarly, it is not possible to fit additive models based on sequentially accumulated SSPM structures (8.2.2), nor can individual changes to the model be summarized separately in an accumulated analysis of variance (8.2.4).

When splines are fitted within generalized linear models themselves, the models are called *generalized additive models* which Genstat fits by nesting the iterative process for smoothing within that for generalized linear models. For details of this technique, see Hastie and Tibshirani (1990) or Lane and Hastie (1992).

8.5 Generalized linear models

Generalized linear models extend the ordinary regression framework to situations where the data do not follow a Normal distribution, or where a transformation (known as the *link function*) needs to be applied before a linear model can be fitted. Subsection 8.5.1 contains a brief account of the essential concepts, but for more information see Dobson (1990) or McCullagh and Nelder (1989).

Example 8.5a shows a probit analysis (Finney, 1971). This is a particular type of generalized linear model which models the relationship between a stimulus, like a drug, and a quantal response (recorded simply as success or failure). In probit analysis it is assumed that for each subject there is a certain level of stimulus below which it will be unaffected, but above which it will respond. This level of stimulus, known as the tolerance, will vary from subject to subject within the population. The assumption in Example 8.5a is that the tolerance of the mice to the logarithm of the dose will have a Normal distribution; so, if we were to plot the proportion of the population with each tolerance against log dose, we would obtain the familiar bell-shaped curve. Likewise, if we plotted the probability that a randomly-selected individual will respond, against the logarithm of dose, we would obtain the sigmoid (S-shaped) cumulative-Normal curve limited below by zero and above by one. To make the relationship linear, then, we could transform the y-axis to Normal equivalent deviates or *probits* (see 8.5.1). Thus, in this example, we need a probit link function in order to fit a linear model.

The data in Example 8.5a consist of observations, in each of which a particular dose of one of the drugs was applied to a group of mice, and the number that responded was counted. The data can thus be assumed to follow a binomial distribution, instead of the Normal distribution assumed for the examples earlier in this chapter.

As Example 8.5a shows, you can fit generalized linear models using exactly the same directives as for linear regression: the only difference is that you need to set extra options in the MODEL directive to specify the distribution and the link function, and, for binomial data, an extra parameter to define the total number of subjects at each observation.

Example 8.5a

```
   1   " Comparison of the effectiveness of three analgesic drugs to a standard
  -2     drug, morphine. Data from Grewal (1952), analysed by Finney (1971)
  -3     p103. Four drugs were compared at several doses for their effect on
  -4     groups of mice; the numbers of mice that responded were recorded."
   5
   6   FACTOR [LABELS=!T(Morphine,Amidone,Phenadoxone,Pethidine)] Drug
   7   READ [PRINT=data] Drug,Dose,Ntest,Nrespond

   8   1 1.50 103 19   1 3.00 120 53   1 6.00 123 83
   9   2 1.50  60 14   2 3.00 110 54   2 6.00 100 81
  10   3 0.75  90 31   3 1.50  80 54   3 3.00  90 80
  11   4 5.00  60 13   4 7.50  85 27   4 10.00 60 32
  12                   4 15.00 90 55   4 20.00 60 44 :
  13
  14   " Fit standard probit models, relating the number of responses to the
 -15     logarithm of the dose.  The probit model is a generalized linear
 -16     model, assuming a binomial distribution for the number of responses
 -17     and a 'probit' link function (cumulative Normal distribution
 -18     function) between the number of responses and the logarithm of
 -19     the dose."
  20   CALCULATE Logdose = LOG10(Dose)
  21   MODEL [DISTRIBUTION=binomial; LINK=probit] Nrespond; NBINOMIAL=Ntest
  22   TERMS Logdose*Drug
  23   " Fit a model ignoring the types of drug used."
  24   FIT [NOMESSAGE=leverage,residual] Logdose

 24............................................................................

***** Regression Analysis *****

 Response variate: Nrespond
 Binomial totals: Ntest
     Distribution: Binomial
    Link function: Probit
     Fitted terms: Constant + Logdose


*** Summary of analysis ***

                                      mean   deviance
                d.f.    deviance    deviance    ratio
Regression        1        39.4      39.41     39.41
Residual         12       210.6      17.55
Total            13       250.0      19.23

Change           -1       -39.4      39.41     39.41
* MESSAGE: ratios are based on dispersion parameter with value 1


*** Estimates of regression coefficients ***
```

```
                               estimate        s.e.        t(*)
Constant                       -0.2976       0.0663       -4.49
Logdose                         0.5972       0.0959        6.23
* MESSAGE: s.e.s are based on dispersion parameter with value 1

  25
  26   " Fit parallel responses (on the probit scale) for the drugs; morphine
 -27     has been assigned as the first level of the factor so that Genstat
 -28     will automatically compare the other drugs to it."
  29   ADD Drug

29.............................................................................
```

***** Regression Analysis *****

```
 Response variate: Nrespond
  Binomial totals: Ntest
     Distribution: Binomial
    Link function: Probit
     Fitted terms: Constant + Logdose + Drug
```

*** Summary of analysis ***

	d.f.	deviance	mean deviance	deviance ratio
Regression	4	246.090	61.5225	61.52
Residual	9	3.868	0.4298	
Total	13	249.958	19.2275	
Change	-3	-206.682	68.8940	68.89

```
* MESSAGE: ratios are based on dispersion parameter with value 1
```

*** Estimates of regression coefficients ***

```
                               estimate        s.e.        t(*)
Constant                        -1.379        0.114      -12.08
Logdose                          2.468        0.173       14.30
Drug Amidone                     0.238        0.108        2.20
Drug Phenadoxone                 1.360        0.130       10.49
Drug Pethidine                  -1.180        0.133       -8.87
* MESSAGE: s.e.s are based on dispersion parameter with value 1

  30
  31   " Fit separate models for the different drugs"
  32   ADD [PRINT=accumulated] Logdose.Drug

32.............................................................................
```

***** Regression Analysis *****

*** Accumulated analysis of deviance ***

Change	d.f.	deviance	mean deviance	deviance ratio
+ Logdose	1	39.4079	39.4079	39.41
+ Drug	3	206.6821	68.8940	68.89
+ Logdose.Drug	3	1.5336	0.5112	0.51

```
Residual                      6        2.3344        0.3891

Total                        13      249.9579       19.2275
* MESSAGE: ratios are based on dispersion parameter with value 1

  33
  34   " There is no evidence of non-parallelism, so return to the parallel
 -35     model and display it with procedure RGRAPH."
  36
  37   DROP [PRINT=*] Logdose.Drug
  38   RGRAPH [GRAPHICS=high]
```

The graph of the fitted model drawn by the RGRAPH procedure is shown in Figure 8.5.

This analysis does not include information on LD50s or similar quantities that are usually used to characterize the effectiveness of drugs (see Finney, 1971). These can be produced by the FIELLER procedure. FIELLER can calculate either the LD50s themselves for each drug, or *relative potencies* which compare one drug with another. Example 8.5b shows the relative potencies of the drugs compared with the standard one, Morphine.

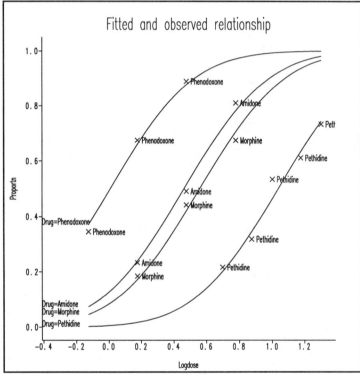

Figure 8.5

Example 8.5b

```
  40   " Estimate the relative potencies of the drugs using the
 -41     procedure FIELLER."
  42   FIELLER [RELATIVE=yes] SLOPE=2; TREATMENT=3,4,5; \
  43      VALUE=relp[2...4]; LOWER=low[2...4]; UPPER=up[2...4]
  Relative potency    Lower 95%    Upper 95%
          0.09636      0.01032       0.1861
  Relative potency    Lower 95%    Upper 95%
          0.5508       0.4615        0.6465
  Relative potency    Lower 95%    Upper 95%
         -0.4780      -0.5578       -0.3970
```

```
44   " The results are the log-potency and 95% fiducial limits:
-45      transform these to the natural scale."
 46   CALCULATE relp[],low[],up[] = 10**relp[],low[],up[]
 47   PRINT relp[2],low[2],up[2] & relp[3],low[3],up[3] & relp[4],low[4],up[4]
```

```
     relp[2]      low[2]        up[2]
       1.248       1.024        1.535


     relp[3]      low[3]        up[3]
       3.554       2.894        4.431


     relp[4]      low[4]        up[4]
      0.3327      0.2768       0.4008
```

8.5.1 Introduction to generalized linear models

Generalized linear models are natural generalizations of ordinary linear regression models. The ordinary regression model can be written as:

$$y_i = \mu_i + \varepsilon_i , \qquad\qquad i=1...N$$
$$= \alpha + \Sigma\{\beta_j\, x_{ji}\} + \varepsilon_i$$

where x_{ji} is the ith observation of the jth explanatory variable and y_i is the ith observation of the response variable; and

$$\mathrm{Var}(y_i) = \sigma^2$$

where σ^2 is constant for all observations. The residuals ε_i are assumed to be uncorrelated, and usually the model is specialized further by assuming the observations y_i to be Normally distributed.

The generalized linear model is still

$$y_i = \mu_i + \varepsilon_i , \qquad\qquad i=1...N$$

but now the linear model describes η_i, the *linear predictor*,

$$\eta_i = \alpha + \Sigma\{\beta_j\, x_{ji}\}$$

and η_i is related to μ_i by

$$\eta_i = G(\mu_i)$$

where $G()$ is a monotonic and differentiable function called the *link function*. Also,

$$\mathrm{Var}(y_i) = \phi\, V(\mu_i) , \qquad\qquad i=1...N$$

where ϕ is a *dispersion parameter*, known or unknown. Again the model is usually specialized further, now so that the observations y_i have some distribution such as the Normal, Poisson, binomial, or gamma from the exponential family. $V()$ is a differentiable function, called the *variance function*.

The model could equally well be expressed using the inverse of the link function:

$$\mu_i = G^{-1}(\eta_i)$$

However, the convention is to use G rather than G^{-1} because the model is then similar to fitting a linear model to the link transformation of the response. For example, if G is the log function,

$$\log(\mathbf{E}(y_i)) = \alpha + \Sigma\{\beta_j\, x_{ji}\}$$

is similar to

$$E(\log(y_i)) = \alpha + \Sigma\{\beta_j \, x_{ji}\}$$

but they are not identical – the logarithm of the expectation of a random variable is not the same as the expectation of the logarithm.

Ordinary linear regression is in the class of generalized linear models, with $G()$ being the identity function, ϕ being σ^2, and $V()$ being constant. Many other familiar statistical models are in this class too.

(a) The model used in the probit analysis of proportions is a generalized linear model with $G(\mu)=\Phi^{-1}(\mu/n)$, where Φ is the cumulative Normal distribution function, $\phi=1$, and $V(y)=\mu(1-\mu/n)$, n being the number of trials of which y respond. The distribution is usually assumed to be binomial. Example 8.5a shows such an analysis.

(b) The log-linear model for contingency tables is a generalized linear model with $G(\mu)=\log(\mu)$, $\phi=1$, and $V(y)=\mu$. The distribution for the counts is usually stipulated to be Poisson or multinomial. This is illustrated in Example 8.5.1.

(c) Logistic regression models are very similar to models used in probit analysis, except that they use the logit link function, $G(\mu)=\log(\mu/(n-\mu))$, rather than the probit. Data can often be analysed in two equivalent forms: units may correspond to individuals that are tested, so that the response is always 0 or 1; alternatively, groups of individuals with common values of explanatory variables may be treated as units, so that each data value is the number responding out of the number in the group. Example 8.5.2 shows an analysis using the latter form.

(d) Dilution assays are usually analysed by a model that has $G(\mu)=\log(-\log(1-\mu/n))$, $\phi=1$, and the binomial distribution. The logarithm of the dilution is included in the model as an offset variable, similarly to the offset in Example 8.5.1, as described later in this subsection.

(e) The proportional-odds and proportional-hazards models for ordinal response variables can be treated as generalized linear models with a multinomial distribution for the response. The link functions for the two models are the logit as in (c) and the complementary-log-log as in (d); the former is shown in Example 8.5.5.

(f) Inverse polynomial models are generalized linear models with $G(\mu)=1/\mu$. They are usually used for response variables with constant coefficient of variation, $V(y)=\mu^2$, rather than constant variance, so the distribution is taken to be gamma.

You can fit these and other models using the options of the MODEL directive. The DISTRIBUTION option specifies the characteristic form of the variance function $V()$, according to these rules:

Distribution	Variance function, V
Normal	1
Poisson	μ
binomial	$\mu(1-\mu/n)$
multinomial	$\mu(1-\mu/n)$
gamma	μ^2
inverse Normal	μ^3

If you use the binomial distribution, you must put the number of successes (or the number of failures) into the response variate, and supply the total numbers (that is successes plus failures) in another variate using the NBINOMIAL parameter of the MODEL directive. For example:

```
VARIATE [VALUES=3,5,6] Nsuccess
&        [VALUES=5,9,17] Ntrial
MODEL [DISTRIBUTION=binomial] Nsuccess; NBINOMIAL=Ntrial
```

Alternatively, if you have units for each individual, the total numbers will all be 1 and the above statements would be replaced by:

```
VARIATE [VALUES=3(1),2(0), 5(1),4(0), 6(1),11(0)] Nsuccess
MODEL [DISTRIBUTION=binomial] Nsuccess; NBINOMIAL=1
```

The multinomial distribution can be used only for ordinal response models (8.5.5). A list of response variates is required, one for each category of the response; the number of trials for each unit is determined automatically by adding the values of each response.

When you use the Normal, gamma, or inverse Normal distribution, the dispersion parameter ϕ is usually unknown and is assumed to be constant over all observations. For the Normal distribution this is the constant variance, usually written as σ^2, and for the gamma distribution it is the reciprocal of the index, written either as σ^2 or as v^{-1}. Sometimes, however, you may know a value for the dispersion parameter. For example, you may know that the response variable has a Normal distribution with a variance that you can estimate from previous experiments or surveys. In this case, you can fix the value of the dispersion parameter using the DISPERSION option of MODEL (8.1.1). The effect of this is that standard errors and other measures of variability for the fit of the model will be based on the given fixed value rather than on a value estimated from the data.

The Poisson and binomial distributions do not have any dispersion parameter, so Genstat fixes it at 1.0. This has the effect described above: the variance of an observation is a function only of its mean, and so no estimator of variance is required from the observations as a whole. Another example arises when you want to use the exponential distribution; this can be specified as a gamma distribution, fixing the dispersion parameter at 1.0.

You may sometimes want to include a dispersion parameter even though you are using the binomial, multinomial, or Poisson distributions. An example is the *heterogeneity factor* of probit analysis: the distribution of the observations is taken to be "superbinomial", in the sense that the variance is greater than what would be expected for a binomial distribution; specifically, $V(y)=\theta\mu(1-\mu/n)$, where θ is the heterogeneity factor (Finney 1971). This can be achieved by setting the DISPERSION option to *:

```
MODEL [DISTRIBUTION=binomial; DISPERSION=*] Nsuccess; \
  NBINOMIAL=Ntrial
```

Data for which a "superbinomial", "supermultinomial", or "superPoisson" distribution incorporating such a heterogeneity factor are needed are called *overdispersed*, or *underdispersed* if θ is less than 1; see McCullagh and Nelder (1989) for more details.

Using a heterogeneity factor means formally that the method of analysis is no longer based on maximum likelihood, because there is no probability distribution in the exponential family to provide a likelihood to be maximized. Instead, the method requires a *quasi-likelihood*, which relies solely on the description of the relationship between variance and mean. However, the model can still be analysed and interpreted in the same way as with a given distribution; see McCullagh & Nelder (1989).

The link function is specified by the LINK option of the MODEL directive. The link functions available in Genstat are as follows:

Link function	$G(\mu)$	$G^{-1}(\eta)$
identity	μ	η
logarithm	$\log(\mu)$	$\exp(\eta)$
logit	$\log(\mu/(n-\mu))$	$n \exp(\eta)/(1+\exp(\eta))$
reciprocal	$1/\mu$	$1/\eta$
power	μ^{power}	$\eta^{(1/power)}$
square root	$\mu^{1/2}$	η^2
probit	$\Phi^{-1}(\mu/n)$	$n*\Phi(\eta)$
complementary log-log	$\log(-\log(1-\mu/n))$	$n\,(1-\exp(-\exp(\eta)))$

In the original definition the probit was equal to the Normal equivalent deviate Φ^{-1} plus five but, for simplicity, in Genstat the five is omitted. Similarly, the logit transformation is sometimes defined with a multiplier of ½, but this too is omitted in Genstat.

By default, the power setting uses the exponent −2; you can specify other values using the EXPONENT option, for example:

```
MODEL [DISTRIBUTION=gamma; LINK=power; EXPONENT=1.5] Y
```

For each of the available distributions, one of the links is known as the *canonical link*. This has special properties. In particular, a model with its canonical link always provides a unique set of parameter estimates, whereas with other models this may not be so. There are often practical scientific reasons for using the canonical link, but there may sometimes also be very good reasons for using a non-canonical link. If you do not set the LINK option, the default is the canonical link of the chosen distribution:

Normal	Identity
Poisson	Log
Binomial	Logit
Gamma	Reciprocal
Inverse Normal	Power, with exponent −2
Multinomial	Logit

The MODEL directive also allows you to specify your own distributions or link functions or both. There is an example in 8.5.4.

When the binomial distribution is used, it is usually natural to choose the logit, probit, or complementary-log-log link function; and vice versa. If another link is chosen with the binomial distribution, it is assumed to relate the expected proportion of responses (rather than the expected number of responses) to the linear predictor. Similarly, if one of the above three links is chosen with a distribution other than the binomial, the number of trials is assumed to be 1.

Only the logit or complementary-log-log links can be used with the multinomial distribution.

An *offset* variable is a variable that appears in the linear predictor without a parameter. It provides for each observation a fixed offset, o_i say, from the estimated constant:

$$G(\mu_i) = o_i + \alpha + \Sigma\{\beta_j \, x_{ji}\}$$

You set an offset by the OFFSET option of the MODEL directive. Offsets arise naturally in the standard analysis for dilution assay, involving a complementary-log-log link function. The model then takes the form:

$$E(y_i) = n_i \exp(-d_i \exp(\alpha)) = n_i \exp(-\exp(\log(d_i)+\alpha))$$

where y_i is the number of positives out of n_i samples tested at dilution d_i, and α is the unknown concentration. So the logarithm of the dilution is an offset. This model contains no explanatory variables other than the dilution, but the concentration can sometimes be expressed as a linear function of variables such as time. Dilution assays can conveniently be analysed in Genstat using the DILUTION procedure.

Offset variables also occur naturally in log-linear models for rates where each cell has a different exposure time. Example 8.5.1 shows an analysis of data of this kind.

Example 8.5.1

```
   1   " Analysis of the damage caused by waves to forward sections of
  -2     cargo-carrying ships. The data, from McCullagh & Nelder (1989) p204,
  -3     are counts of damage incidents for each combination of three risk
  -4     factors: the type of ship, the year of construction, and the
  -5     period of operation."
   6
   7   UNITS [NVALUES=40]
   8   FACTOR [LABELS=!T(A,B,C,D,E)] Type
   9   & [LABELS=!T('1960-64','1965-69','1970-74','1975-79')] Constrct
  10   & [LABELS=!T('1960-74','1975-79')] Operatn
  11   GENERATE Type,Constrct,Operatn
  12   " Read the number of months service and number of damage incidents."
  13   OPEN 'ship.dat'; CHANNEL=2
  14   READ [CHANNEL=2] Service,Damage
```

```
   Identifier   Minimum      Mean   Maximum   Values   Missing
      Service          0      4674     44882       40         5    Skew
       Damage       0.00     10.17     58.00       40         5    Skew
```

```
  15
  16   " Use the log of the number of months of service as an offset in the
 -17     model; CALCULATE turns zeroes into missing values, which will then
 -18     be excluded by TERMS as required for a correct analysis."
  19   CALCULATE Logserv = LOG(Service)
```

```
******* Warning (Code CA 7). Statement 1 on Line 19
Command: CALCULATE Logserv = LOG(Service)
Invalid value for argument of function
The first argument of the LOG     function in unit 34 has the value      0.0000
```

```
  20   MODEL [DISTRIBUTION=poisson; LINK=log; OFFSET=Logserv] Damage
  21   TERMS [FACTORIAL=2] Type * Constrct * Operatn
  22   " Fit the main effects."
  23   FIT Type + Constrct + Operatn
```

```
23............................................................
```

```
***** Regression Analysis *****

 Response variate: Damage
     Distribution: Poisson
```

```
      Link function: Log
   Offset variate: Logserv
     Fitted terms: Constant + Type + Constrct + Operatn


*** Summary of analysis ***

                                      mean  deviance
               d.f.     deviance    deviance   ratio
Regression       8      107.63      13.454    13.45
Residual        25       38.70       1.548
Total           33      146.33       4.434

Change          -8     -107.63      13.454    13.45
* MESSAGE: ratios are based on dispersion parameter with value 1


* MESSAGE: The following units have large standardized residuals:
                21        3.01
                22       -2.29
                30        2.30
                36        2.15
* MESSAGE: The following units have high leverage:
                 9        0.70
                11        0.64
                12        0.65
                14        0.59
                16        0.56
                38        0.56


*** Estimates of regression coefficients ***

                          estimate        s.e.      t(*)
Constant                   -6.406        0.217     -29.46
Type B                     -0.543        0.178      -3.06
Type C                     -0.687        0.329      -2.09
Type D                     -0.076        0.291      -0.26
Type E                      0.326        0.236       1.38
Constrct 1965-69            0.697        0.150       4.66
Constrct 1970-74            0.818        0.170       4.82
Constrct 1975-79            0.453        0.233       1.94
Operatn 1975-79             0.384        0.118       3.25
* MESSAGE: s.e.s are based on dispersion parameter with value 1

   24
   25   " Try adding the two-factor interactions."
   26   TRY [PRINT=accumulated] Type.Constrct + Type.Operatn + Constrct.Operatn

26.........................................................................


***** Regression Analysis *****

*** Accumulated analysis of deviance ***

Change
                                                  mean   deviance
                    d.f.      deviance         deviance    ratio
+ Type
+ Constrct
+ Operatn            8        107.633           13.454    13.45
+ Type.Constrct     12         24.108            2.009     2.01
```

```
Residual              13        14.587       1.122

Total                 33       146.328       4.434
* MESSAGE: ratios are based on dispersion parameter with value 1


***** Regression Analysis *****

*** Accumulated analysis of deviance ***
```

Change	d.f.	deviance	mean deviance	deviance ratio
+ Type				
+ Constrct				
+ Operatn	8	107.633	13.454	13.45
+ Type.Operatn	4	4.939	1.235	1.23
Residual	21	33.756	1.607	
Total	33	146.328	4.434	

```
* MESSAGE: ratios are based on dispersion parameter with value 1


* MESSAGE: Term Constrct.Operatn cannot be fully included in the model
  because 1 parameter is aliased with terms already in the model
  (Constrct 1975-79 .Operatn 1975-79) = (Constrct 1975-79)

***** Regression Analysis *****

*** Accumulated analysis of deviance ***
```

Change	d.f.	deviance	mean deviance	deviance ratio
+ Type				
+ Constrct				
+ Operatn	8	107.633	13.454	13.45
+ Constrct.Operatn	2	1.787	0.894	0.89
Residual	23	36.908	1.605	
Total	33	146.328	4.434	

```
* MESSAGE: ratios are based on dispersion parameter with value 1
```

8.5.2 The deviance

You can assess how well a linear regression fits by doing an analysis of variance. Based on the assumption that the residuals have independent Normal distributions with equal variances, the variance ratio (mean square due to the regression divided by the residual mean square) has an F distribution.

With generalized linear models, there is no similarly simple exact distributional property. However, you can get approximate assessments of the quality of the fit from a statistic called the *scaled deviance*. This is defined as minus twice the log-likelihood ratio between the model you have fitted and a full model that explains all the variation in the data. The scaled deviance has approximately a χ^2_d distribution, d being the number of residual degrees of freedom. The approximation is better for large numbers of observations than for small numbers, and is poor when there are many extreme observations (such as zeroes for the Poisson distribution). In particular, in the special case of a binary response variable (with values 1 and 0), the scaled deviance is absolutely uninformative about the fit of the model.

The scaled deviance is a function of the dispersion parameter, and so its distribution depends also on any estimate of that parameter. Usually you would obtain the estimate from a model that you believe explains all systematic variation – a *maximal model*, as in the analysis of variance for linear regression. You can assess the importance of a term in any generalized linear model by considering the difference between the scaled deviances of that model and the model excluding the term. The difference in scaled deviances also has an approximate χ^2_t distribution, where t is the number of degrees of freedom of the term; in fact this approximation is better than that for the scaled deviance itself.

Alternatively, you can consider ratios of mean scaled deviances between competing models, one of which is nested inside the other. (The mean scaled deviance is the scaled deviance divided by the corresponding number of degrees of freedom.) The resulting ratios do not involve the dispersion parameter. Such a ratio has approximately an F distribution – exact for linear regression models with Normal errors.

Genstat reports the *deviance* of the data for each type of model, which is equivalent to the scaled deviance multiplied by the dispersion parameter. The deviance is otherwise known as the log-likelihood ratio statistic.

You can summarize the fit of a sequence of nested models by an *analysis of deviance*, which you interpret in much the same way as an analysis of variance (but do not forget that the distributions have only approximate χ^2 distributions).

Here are the formulae for the deviance for each distribution; the ith response is represented by y_i, and the corresponding fitted value by f_i:

Normal	$\Sigma(y_i - f_i)^2$
Poisson	$2 \Sigma\{y_i \log(y_i/f_i) - (y_i - f_i)\}$
Binomial	$2 \Sigma\{y_i \log(y_i/f_i) + (n_i - y_i) \log((n_i - y_i)/(n_i - f_i))\}$
Gamma	$2 \Sigma\{(y_i - f_i)/f_i - \log(y_i/f_i)\}$
Inverse Normal	$\Sigma\{(y_i - f_i)^2/(y_i f_i^2)\}$
Multinomial	$2 \Sigma\Sigma\{y_{ij} \log(y_{ij}/f_{ij})\}$

Sometimes parameter estimates cannot be obtained. The commonest cause with models using the binomial or Poisson distribution is the presence of observations at the extremes (0 for Poisson, 0 or n for binomial). One or more of the parameters may then need to be infinite to maximize the likelihood: in practice, approximate convergence will usually be achieved with the parameters large but finite (the meaning of "large" being dependent on the link function).

This is illustrated in Example 8.5.2: all subjects at level 1 of the factor Li responded positively (that is, they were disease-free for three years). Hence, on the logit scale which is the default link function for the binomial distribution, the difference between the two levels is infinite. Genstat achieves convergence here, so the only indications of the problem are the large estimates and standard errors for the constant and "Li 2". The PREDICT statement shows what is happening: all the predicted proportions at level 1 of Li are almost exactly 1.0.

Example 8.5.2

```
    1    " Logistic regression including a factor with a 100% response rate.
   -2      Data from Goorin et al. (1987).
   -3      46 patients were studied, to determine predictors of non-metastatic
   -4      sarcoma: this analysis uses Li (Lymphocytic infiltration), Sex,
   -5      and Aop (any osteoid pathology). The response variable is the number
   -6      disease free for three years."
    7
    8    FACTOR [NVALUES=8; LEVELS=2] Li,Sex,Aop
    9    GENERATE Li,Sex,Aop
   10    VARIATE [VALUES=3,2,4,1,5,3,5,6] Nfree
   11    &        [VALUES=3,2,4,1,5,5,9,17] Nstudy
   12    MODEL [DISTRIBUTION=binomial] Nfree; NBINOMIAL=Nstudy
   13    TERMS Sex,Aop,Li
   14    ADD [PRINT=*] Sex
   15    & Aop
   16    & [PRINT=estimates,accumulated] Li
```

```
16.....................................................................
```

```
***** Regression Analysis *****
```

```
*** Estimates of regression coefficients ***
```

	estimate	s.e.	t(*)
Constant	13.9	90.4	0.15
Sex 2	-1.636	0.912	-1.79
Aop 2	-1.220	0.771	-1.58
Li 2	-11.8	90.4	-0.13

`* MESSAGE: s.e.s are based on dispersion parameter with value 1`

```
*** Accumulated analysis of deviance ***
```

Change	d.f.	deviance	mean deviance	deviance ratio
+ Sex	1	5.8795	5.8795	5.88
+ Aop	1	5.0105	5.0105	5.01
+ Li	1	6.9148	6.9148	6.91
Residual	4	1.6279	0.4070	
Total	7	19.4327	2.7761	

`* MESSAGE: ratios are based on dispersion parameter with value 1`

```
   17    PREDICT Sex,Aop,Li
```

```
17.....................................................................
```

```
*** Predictions from regression model ***
```

```
These predictions are fitted proportions.
```

```
        Table contains predictions followed by standard errors
```

```
  Response variate: Nfree
                          Li           1                    2
          Sex         Aop
```

1	1	0.9999990	0.0000853	0.8916535	0.0934515
	2	0.9999968	0.0002890	0.7083464	0.1752989
2	1	0.9999952	0.0004380	0.6157480	0.1517264
	2	0.9999836	0.0014840	0.3210746	0.1091177

```
* S.e.s are approximate, since model is not linear
* S.e.s are based on dispersion parameter with value 1
```

Occasionally, the iterative process may converge only very slowly when a parameter needs to be infinite: you can increase the limit on the number of cycles with the RCYCLE directive (8.5.3), though this may not always help. Very rarely you may even get divergence; this can also happen when the initial guesses for the fitted values are very bad, and the deviance appears to increase after the first cycle. But usually in such cases, the model would not fit the data satisfactorily anyway.

Failure to find a solution may occur when estimates from a fit take impossible values. For example, the gamma distribution is defined in the range $(0, \infty)$, but some sets of data may produce an estimated mean that is negative. In such cases, you should consider a different link, or try a new fit omitting those explanatory variables whose parameters were estimated as negative.

8.5.3   The RCYCLE directive

RCYCLE controls iterative fitting of generalized linear, generalized additive, and nonlinear models, and specifies parameters, bounds etc for nonlinear models.

Options

MAXCYCLE = *scalar*	Maximum number of iterations; default * gives 15 for generalized linear and generalized additive models, 30 for nonlinear models
TOLERANCE = *scalar*	Convergence criterion; default 0.0001
FITTEDVALUES = *variate*	Initial fitted values for generalized linear model; default *
METHOD = *string*	Algorithm for fitting nonlinear model (GaussNewton, NewtonRaphson, FletcherPowell); default Gaus, but Newt for scalar minimization

Parameters

PARAMETER = *scalars*	Nonlinear parameters in the model
LOWER = *scalars*	Lower bound for each parameter
UPPER = *scalars*	Upper bound for each parameter
STEPLENGTH = *scalars*	Initial step length for each parameter
INITIAL = *scalars*	Initial value for each parameter

The parameters of the RCYCLE directive are ignored when generalized linear models are fitted; see 8.6.5 and 8.7.1 for their use in nonlinear models.

The MAXCYCLE option allows you to change the limit on the number of cycles in the iterative estimation process. Usually, the algorithm converges in four or five cycles, but when there are many extreme observations more cycles may be needed; however, the resulting fit is then often uninformative.

The TOLERANCE option controls the criterion for convergence. The iteration stops when the absolute change in deviance in successive cycles is less than the tolerance times the current value of the deviance.

When additive terms are included in the model (8.4.3) the resulting *generalized additive model* is fitted by nested iteration. This means that at each cycle of the iterative fit required by the presence of a non-identity link function or non-Normal distribution or both, the iterative search described in 8.4.3 will take place. This can, of course, be a time-consuming operation, particularly if the number of units is large. The two iterative processes are both controlled by the settings of MAXCYCLE and TOLERANCE.

The algorithm has to start by estimating an initial set of fitted values. Genstat usually obtains these by a simple transformation of the observed responses. It may be that better estimates are available, for example from a previously fitted model; if so, you can supply these by the FITTEDVALUES option.

The METHOD option is relevant only for nonlinear models, as described in 8.7.1.

8.5.4 Non-standard distributions and link functions

If you want a non-standard distribution for the response variable or a non-standard link function, you can specify your own. It will then be up to you to ensure that the iterative process is suitable and to decide how to interpret the resulting fit (if convergence is achieved). Formally, the methods for generalized linear models are suitable only for distributions in the exponential family, and for a monotonic differentiable link function.

To specify your own distribution, you need to set DISTRIBUTION=calculated in the MODEL statement. You must then supply expression structures with the DCALCULATION option to calculate the deviance and the variance function for each unit of the response variate, using the current values of the fitted-values variate. You must also set the FITTEDVALUES, DEVIANCE, and VFUNCTION parameters of the MODEL statement to indicate which identifiers are used to represent these in the expressions.

For example, the following statements specify the calculations for the gamma distribution (though it would be more efficient of course just to set DISTRIBUTION=gamma). The deviance is calculated by expression Dc[1] and placed into the scalar D, and the variance function V is defined by expression Dc[2].

```
EXPRESSION Dc[1]; VALUES=!e(D=2*((Y-F)/F-log(Y/F)))
& Dc[2]; VALUES=!e(V=F*F)
MODEL [DISTRIBUTION=calculated; LINK=reciprocal; \
  DCALCULATION=Dc[]] Y; FITTED=F; VFUNCTION=V; DEVIANCE=D
FIT X
```

To specify your own link, you need to set LINK=calculated and provide expressions for two other calculations to form the fitted values and the derivative of the link function for each unit of the response variate, using the current values of the linear predictor. You must also set the FITTEDVALUES, LINEARPREDICTOR, and DERIVATIVE parameters to specify the identifiers used to represent these in the calculations. In addition, you must provide initial values for the linear predictor, so that the iterative process can get started: often this can be done just by applying the link function to the response variate itself, but it may be necessary to modify extreme values such as 0 that may be mapped to infinity by the link function.

Example 8.5.4 defines a link function for a probit model, incorporating a known control mortality. (If the control mortality is not known, the model cannot be treated as a generalized linear model, but the PROBITANALYSIS procedure can be used instead.) The inverse of the link function here takes the form

$$\mu = n(c + (1-c)\Phi(\eta))$$

where c is the control mortality, and the derivative of the link is

$$d = \sqrt{(2\pi)}\exp(\eta^2/2)/(n(1-c))$$

Example 8.5.4

```
   1  " Analysis of toxicity of derris roots to grain beetle, using
  -2     probit analysis with allowance for control mortality.
  -3      Data from Martin (1940), analysed by Finney (1971) p131."
   4  READ [PRINT=data] Conc,Nspray,Ndead

   5  1480 142 142   1000 127 126    480 128 115    120 126  58
   6   619 125 125    458 117 115    310 127 114    149  51  40
   7  37.1 132  37  :
   8  FACTOR [LABELS=!t(w213,w214); VALUES=4(1),5(2)] Root
   9  CALC Logconc = LOG10(Conc)
  10  " Estimate of control mortality is 17% "
  11  SCALAR [VALUE=0.17] Cm
  12  " Give calculations for probit link with control mortality."
  13  EXPRESSION [VALUE=Fv1=Nspray*(Cm+(1-Cm)*NORMAL(Lp1))] E[1]
  14  & [VALUE=Ld1=SQRT(2*C('pi'))*EXP(Lp1**2/2)/Nspray/(1-Cm)] E[2]
  15  MODEL [DISTRIBUTION=binomial; LINK=calculated; LCALCULATION=E[1,2]] \
  16    Ndead; NBINOMIAL=Nspray; LINEARPRED=Lp1; FITTED=Fv1; DERIVATIVE=Ld1
  17  " Initialize the linear predictor."
  18  CALCULATE Lp1 = NED((Ndead+0.5)/(Nspray+1))
  19  FIT Logconc,Root

  19.............................................................................
```

```
***** Regression Analysis *****

 Response variate: Ndead
 Binomial totals: Nspray
     Distribution: Binomial
    Link function: Calculated
    Fitted terms: Constant, Logconc, Root


*** Summary of analysis ***

                                      mean    deviance
               d.f.    deviance   deviance      ratio
Regression        2     450.778    225.389     225.39
```

```
Residual            6          7.391           1.232
Total               8        458.169          57.271
* MESSAGE: ratios are based on dispersion parameter with value 1


* MESSAGE: The following units have high leverage:
                  4          0.70


*** Estimates of regression coefficients ***

                  estimate          s.e.        t(*)
Constant            -6.222         0.462       -13.46
Logconc              2.795         0.184        15.16
Root w214            0.668         0.146         4.57
* MESSAGE: s.e.s are based on dispersion parameter with value 1
```

8.5.5 Models for ordinal response

The models in this section may be relevant when a response variable can take one out of a fixed set of possible values. A response variable of this kind is called *polytomous*, and the possible values are called *response categories.*

If the categories are purely nominal – that is, with no concept of an ordering – it may be appropriate to fit log-linear models. The response variable is represented as a factor with one level for each category, and included in the model along with its interactions with the explanatory factors, while the response variate stores the numbers of units with each combination of levels of the factors. Effects of the explanatory factors are then represented by their interactions with the "response" factor. This approach can be seen in Example 8.5.1 above, where the number of damage incidents was analysed, though in that example there is no specific "response" factor.

If the categories are on an interval scale, so that differences between categories can be compared quantitatively, the response variable can be analysed as for a continuous variable, using linear regression or some generalized linear model with an appropriate distribution.

If the categories are ordinal, so that there is a known ordering of the categories but no concept of distance between them, Genstat provides two possible models for the relationship between explanatory variables and the division into categories. These are both cumulative models, describing the relationship between numbers of observations up to a particular category and the explanatory values. They are both described in Chapter 5 of McCullagh & Nelder (1989), where they are called the *proportional-odds* model and the *proportional-hazards* model. They have the following form:

$$G(\gamma_{ij}) = \theta_j - \Sigma\{\beta_i\, x_{ij}\}$$

where $G()$ is the logit or complementary-log-log link function, respectively, and γ_{ij} is the probability that the response for unit i is in category j or lower. The quantities θ_j are referred to as the *cut-points*, and provide a quantification of the difference between successive categories on the scale of the chosen link function. It is conventional to have the minus sign in this model, rather than the plus sign that would be expected in a multiple linear model: this convention ensures that as the linear predictor increases, the probability of the response lying in the higher categories also increases.

Example 8.5.5 uses the proportional odds model. Note that it is necessary to set the option YRELATION=ordinal in the MODEL statement, as well as DISTRIBUTION=multinomial; this is to allow for further models using the multinomial distribution in the future.

Example 8.5.5

```
   1  " Analysis of a tasting experiment with ordinal response categories.
  -2      Data from McCullagh & Nelder (1989) p175.
  -3      Four types of cheese were rated by 52 panellists on a nine-point
  -4      'hedonic scale' for taste, ranging from 'strong dislike' (1) to
  -5      'excellent taste' (9)."
   6  OUTPUT [WIDTH=80] 1
   7  READ [PRINT=data] Taste[1...9]

   8   0  0  1  7  8  8 19  8  1
   9   6  9 12 11  7  6  1  0  0
  10   1  1  6  8 23  7  5  1  0
  11   0  0  0  1  3  7 14 16 11 :
  12  FACTOR [LABELS=!t(A,B,C,D); VALUES=1...4] Cheese
  13  " Specify the proportional-odds model (LINK=logit is the default)
 -14      and ask for Pearson residuals rather than deviance residuals,
 -15      since these are reported by McCullagh and Nelder."
  16  MODEL [DISTRIBUTION=multinomial; YRELATION=cumulative; \
  17      RMETHOD=Pearson] Taste[]
  18  " Use full parameterization to get differences with Cheese D, as in
 -19      McCullagh & Nelder, rather than with Cheese A."
  20  TERMS [FULL=yes] Cheese
  21  FIT Cheese

21........................................................................
```

```
***** Regression Analysis *****

Response variates: ordinal model for categories defined by
                   Taste[1], Taste[2], Taste[3], Taste[4], Taste[5],
                   Taste[6], Taste[7], Taste[8], Taste[9]
       Distribution: Multinomial
      Link function: Logit
      Fitted terms: Cheese


*** Summary of analysis ***

                                       mean   deviance
                  d.f.    deviance   deviance    ratio
Regression           3     148.45    49.4846    49.48
Residual            21      20.31     0.9671
Total               24     168.76     7.0318

Change              -3    -148.45    49.4846    49.48
* MESSAGE: ratios are based on dispersion parameter with value 1


  Response variate: Taste[4]
* MESSAGE: The following units have large standardized residuals:
                1        2.23
```

```
Response variate: Taste[6]
* MESSAGE: The following units have large standardized residuals:
                2          2.30


*** Estimates of regression coefficients ***

                        estimate         s.e.        t(*)
Cut-point 0/1            -7.080         0.562       -12.59
Cut-point 1/2            -6.025         0.475       -12.67
Cut-point 2/3            -4.925         0.427       -11.53
Cut-point 3/4            -3.857         0.390        -9.88
Cut-point 4/5            -2.521         0.343        -7.35
Cut-point 5/6            -1.569         0.309        -5.08
Cut-point 6/7            -0.067         0.266        -0.25
Cut-point 7/8             1.493         0.331         4.51
Cheese A                -1.613         0.378        -4.27
Cheese B                -4.965         0.474       -10.47
Cheese C                -3.323         0.425        -7.82
Cheese D                     0            *            *
* MESSAGE: s.e.s are based on dispersion parameter with value 1
```

8.5.6 Modifications to output and the RKEEP and PREDICT directives

Some aspects of the results of fitting generalized linear models differ from those described for linear regression, because of the iterative process that is involved. We call any generalized linear model other than linear regression an *iterative* model.

Genstat will analyse only one response variate if the model is iterative, except for models for ordinal response where several response variates are involved in each set of data (8.5.5). If the Y parameter of the MODEL statement contains more than one variate, Genstat will analyse only the first. This is because the fitting process involves weights that depend on the fitted values, which would thus differ from response variate to response variate (see ITERATIVEWEIGHTS below).

Genstat does not report the percentage variance accounted for by models with distributions other than Normal.

The standard errors of parameter estimates produced by the estimates setting of the PRINT option are only approximate for iterative models; the same applies to the t-statistics, and to the correlations produced by the correlations setting. The adequacy of the approximation depends on the model and the context, so you should use these values as a guide only: for example, a "t-statistic" greater than 3.0 will usually be significant and one less than 1.0 will not be significant. You can get a better test of the corresponding parameter by dropping it from the model and then assessing the change in the deviance.

Genstat displays leverages with the fittedvalues setting of the PRINT option and allows them to be stored by the LEVERAGE parameter of the RKEEP directive. With iterative models, the formula for the *i*th leverage is:

$$l_i = u_i \, w_i \, \{X(X'UWX)^{-1}X'\}_{ii} \,, \qquad i = 1...N$$

where U is a diagonal matrix consisting of the iterative weights u_i (defined below). These values are also used in the standardization of residuals, according to the formula given in 8.1.1. However, no leverages are formed for ordinal response models, because there is no analogous quantity for assessing influence in these effectively multivariate generalized linear models; the standardized residuals, therefore, contain no adjustment for relative influence.

By default, the residuals are deviance residuals, as described in 8.1.1: each residual is the signed square root of the contribution to the deviance. (See 8.5.2 for the definition of deviance for each distribution.) The standardization of the residuals uses the leverages, l_i, described above, and the weights, w_i, if specified; by default, if the WEIGHTS option of MODEL is not set the weights are 1.0 . The ith residual is

$$r_i = \text{sign}(y_i - f_i) \sqrt{\{w_i d_i / (s^2(1 - l_i))\}}$$

where d_i is the contribution to the deviance from unit i, and s^2 is the estimated or fixed dispersion. For example, the deviance residuals for a model with the Poisson distribution are given by:

$$r_i = \text{sign}(y_i - f_i) \sqrt{\{2 (y_i \log(y_i / f_i) - (y_i - f_i)) / (1 - l_i)\}}$$

If you set the RMETHOD option of the MODEL directive to Pearson then Genstat forms the residuals by adjusting the ordinary residuals for their estimated variance:

$$r_i = (y_i - f_i) \sqrt{\{w_i / (V(f_i) s^2 (1 - l_i))\}}$$

With the binomial distribution, the table produced by the fittedvalues setting includes a column for the binomial totals specified by the NBINOMIAL parameter of the MODEL directive. For the multinomial distribution, a separate table is printed for each category.

The accumulated setting of the PRINT option produces an accumulated analysis of deviance for iterative models, just as for linear models except that all contributions from one statement are pooled. The POOL option of the RDISPLAY directive, and of directives like FIT, has no effect with iterative models. Thus you cannot calculate the change in deviance attributable to each individual term unless you add the terms into the model individually. For example, these statements would provide a full analysis of deviance for two factors A and B and their interaction:

```
TERMS A*B
ADD [PRINT=*] A
& B
& [PRINT=accumulated] A.B
```

The monitoring setting of the PRINT option provides a report on the progress of the fit. Example 8.5.6 shows how convergence was achieved in Example 8.5.2 above.

Example 8.5.6

```
  19  FIT [PRINT=monitoring] Sex,Aop,Li

19.................................................................................
```

```
*** Convergence monitoring ***

Cycle           Deviance     Current parameters
    1          12.618057     -0.695419      -0.801977      -0.470358
    2          4.2247982     -1.22439       -1.03232       -1.58180
    3          2.4242139     -1.54368       -1.18072       -2.66966
    4          1.9038448     -1.62716       -1.21515       -3.72636
    5          1.7275255     -1.63560       -1.21979       -4.75119
```

6	1.6642329	-1.63614	-1.22030	-5.76057
7	1.6411533	-1.63620	-1.22037	-6.76404
8	1.6326904	-1.63620	-1.22038	-7.76531
9	1.6295807	-1.63620	-1.22038	-8.76578
10	1.6284373	-1.63620	-1.22038	-9.76595
11	1.6280167	-1.63620	-1.22038	-10.7660
12	1.6278620	-1.63620	-1.22038	-11.7660

Convergence at cycle 12

Three of the parameters of the RKEEP directive are relevant only for saving results of iterative models. The LINEARPREDICTOR parameter lets you save the linear predictor; that is

$$p_i = a + o_i + \Sigma\{b_j\, x_{ij}\}\,, \qquad\qquad\qquad i = 1...N$$

where a and b_j are estimates of α and β_j. The values of the linear predictor are the same as the fitted values if the link function is the identity function.

The ITERATIVEWEIGHTS parameter saves a variate containing the iterative weights used in the last cycle of the iteration. The weight for unit i is

$$\{\,V(f_i\,)\,\}^{-1}\,\{\,p_i'\,\}^{-2}$$

where $V()$ is the variance function (8.5.1) and p_i' is the derivative of the linear predictor with respect to the mean. The iterative weights do not contain any contribution from the weights that can be specified whether or not the model is iterative by the WEIGHTS option of the MODEL directive. The iterative weights are 1.0 for ordinary linear regression.

The YADJUSTED parameter saves the adjusted response variate Z that was used in the last cycle of the iteration:

$$z_i = p_i + (y_i{-}f_i)p_i'$$

With the identity link function this is the same as the response variate.

The Pearson chi-squared statistic can be saved using the PEARSONCHI parameter of RKEEP. It is calculated as the sum of the squared Pearson residuals, defined above. This can be used as an alternative to the deviance for testing goodness of fit; see Nelder and McCullagh (1989).

The EXIT parameter of RKEEP provides a code that indicates the success or type of failure when fitting a generalized linear model (codes for nonlinear models are given in 8.6.4).

0   Successful fitting
8   Data incompatible with model
9   Predicted mean or linear predictor out of range
10 Invalid calculation for calculated link or distribution
11 All units have been excluded from the analysis
12 Iterative process has diverged
13 Failure due to lack of space or data access

The PREDICT directive forms summaries of the fit of an iterative model as for a linear model. However, note that averaging is done by default on the scale of the original response variable, not on the scale transformed by the link function. In other words, linear predictors are formed for all the combinations of factor levels and variate values specified by PREDICT, and then transformed by the link function back to the natural scale. This back transformation may be useful when you are reporting results, since the tables from PREDICT can then be interpreted as natural averages of means predicted by the fitted model. You can set option BACKTRANSFORM=none if you want the averaging to be done on the scale of the linear predictor; PREDICT will then form averages and report predictions on the transformed scale.

PREDICT calculates the standard errors of predictions from iterative models by using first-order approximations that allow for the effect of the link function. Thus you should interpret them only as a rough guide to the variability of individual predictions.

8.6   Standard nonlinear curves

This section describes various standard nonlinear curves that can be fitted using the FITCURVE directive. These standard curves have been found useful in many applications of statistics. They are fitted by a modified Newton method of maximizing the likelihood, using stable forms of parameterization (Ross 1990). Facilities for fitting other user-defined curves are described in 8.7.

The method Genstat uses to fit curves is iterative, using a search procedure to find parameter values that maximize the likelihood. The search is much quicker when Genstat knows the shape of the curve; thus, fitting a curve by the methods in this section is more efficient than using those in 8.7. With standard curves you will not usually need to supply starting values for the search, nor to control the course of the search; in contrast, you will nearly always have to do these things when you are fitting non-standard curves. For more information about nonlinear curve fitting, see Ratkowsky (1983, 1990), Ross (1990), or Seber and Wild (1989).

Example 8.6 fits the exponential curve

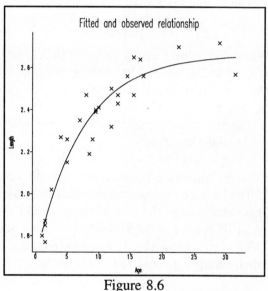

Figure 8.6

$$y_i = \alpha + \beta \, \rho^{x_i} + \varepsilon_i$$

to the relationship between length and age of dugongs. At line 8 the RGRAPH procedure is used to produce the graph of the fitted curve shown in Figure 8.6.

Example 8.6

```
    1   " Asymptotic regression (exponential curve) of length on age
   -2       of dugongs.   Data from Ratkowsky (1983) p101. "
    3
    4   OPEN 'DUGONG.DAT'; CHANNEL=2
    5   READ [PRINT=data; CHANNEL=2] Age,Length

        1     1.0 1.80    1.5 1.85    1.5 1.87    1.5 1.77    2.5 2.02
        2     4.0 2.27    5.0 2.15    5.0 2.26    7.0 2.35    8.0 2.47
        3     8.5 2.19    9.0 2.26    9.5 2.40    9.5 2.39   10.0 2.41
```

```
    5  15.5 2.65  15.5 2.47  16.5 2.64  17.0 2.56  22.5 2.70
    6  29.0 2.72  31.5 2.57
 6   MODEL Length
 7   FITCURVE [CURVE=exponential] Age
```

```
7.................................................................
```

```
***** Nonlinear regression analysis *****

 Response variate: Length
      Explanatory: Age
     Fitted Curve: A + B*R**X
      Constraints: R < 1


*** Summary of analysis ***

              d.f.        s.s.         m.s.        v.r.
Regression      2       1.7745      0.887258      114.02
Residual       24       0.1868      0.007782
Total          26       1.9613      0.075434

Percentage variance accounted for 89.7
Standard error of observations is estimated to be 0.0882
* MESSAGE: The following units have high leverage:
                    1          0.26
                   26          0.26
                   27          0.30


*** Estimates of parameters ***

              estimate          s.e.
R               0.8735        0.0223
B              -0.9725        0.0647
A               2.6666        0.0579

 8   RGRAPH [GRAPHICS=high]
```

8.6.1   The **FITCURVE** directive

FITCURVE fits a standard nonlinear regression model.

Options

PRINT = *strings*	What to print (`model, deviance, summary, estimates, correlations, fittedvalues, accumulated, monitoring`); default `mode,summ,esti`
CURVE = *string*	Type of curve (`exponential, dexponential, cexponential, lexponential, logistic, glogistic, gompertz, ldl, qdl, qdq, fourier, dfourier, gaussian, dgaussian`); default `expo`
SENSE = *string*	Sense of curve (`right, left`); default `righ`

ORIGIN = *scalar*	Constrained origin; default *
NONLINEAR = *string*	How to treat nonlinear parameters between groups (common, separate); default comm
CONSTANT = *string*	How to treat the constant (estimate, omit); default esti
FACTORIAL = *scalar*	Limit for expansion of model terms; default as in previous TERMS statement, or 3 if no TERMS given
POOL = *string*	Whether to pool ss in accumulated summary between all terms fitted in a linear model (yes, no); default no
DENOMINATOR = *string*	Whether to base ratios in accumulated summary on rms from model with smallest residual ss or smallest residual ms (ss, ms); default ss
NOMESSAGE = *strings*	Which warning messages to suppress (dispersion, leverage, residual, aliasing, marginality, vertical); default *
FPROBABILITY = *string*	Printing of probabilities for variance ratios (yes, no); default no

Parameter

formula	Explanatory variate, list of variate and factor, or variate*factor

The parameter of FITCURVE can be set just to the variate that supplies the x-values for the curve, if you simply want to fit a single curve. You can also include a factor if you want to fit separate curves for different groups of the observations: these facilities for *parallel curve analysis* are described in 8.6.3.

The CURVE option specifies which of the standard curves is to be fitted. For some of these, the SENSE option lets you choose between alternative forms. Figure 8.6.1 shows the shapes of representative curves of each type, although you should be aware that several of the curves, particularly the rational functions, can exhibit a wide variety of shapes as their parameters vary. Before describing the curves in detail, here is a list for convenient reference:

Exponential

exponential $\quad y_i = \alpha + \beta \rho^{x_i} + \varepsilon_i$

dexponential $\quad y_i = \alpha + \beta \rho^{x_i} + \gamma \sigma^{x_i} + \varepsilon_i$

cexponential $\quad y_i = \alpha + (\beta + \gamma x_i) \rho^{x_i} + \varepsilon_i$

lexponential $\quad y_i = \alpha + \beta \rho^{x_i} + \gamma x_i + \varepsilon_i$

Logistic

logistic $\quad y_i = \alpha + \dfrac{\gamma}{1 + \exp(-\beta(x_i - \mu))} + \varepsilon_i$

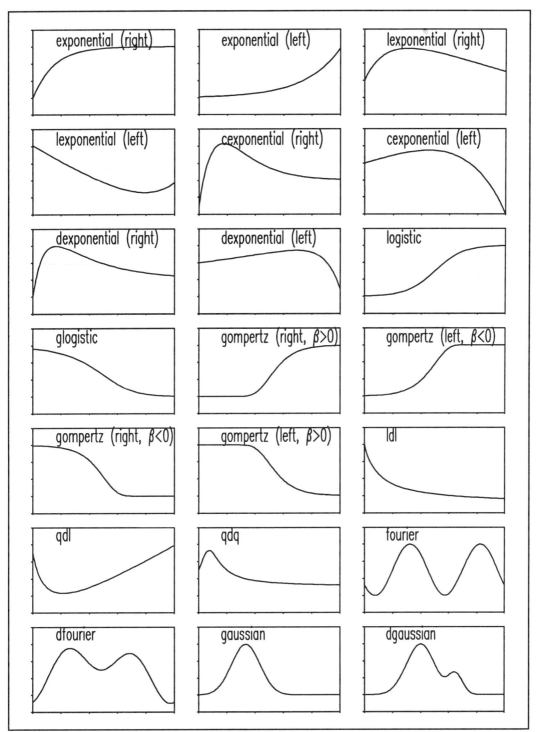

Figure 8.6.1

glogistic $\qquad$ $y_i = \alpha + \dfrac{\gamma}{(1 + \tau \, \exp(-\beta(x_i - \mu)))^{\tau^{-1}}} + \varepsilon_i$

gompertz $\qquad$ $y_i = \alpha + \gamma \, \exp(-\exp(-\beta(x_i - \mu))) + \varepsilon_i$

Rational functions

ldl $\qquad$ $y_i = \alpha + \dfrac{\beta}{1 + \delta \, x_i} + \varepsilon_i$

qdl $\qquad$ $y_i = \alpha + \dfrac{\beta}{1 + \delta \, x_i} + \gamma \, x_i + \varepsilon_i$

qdq $\qquad$ $y_i = \alpha + \dfrac{\beta + \gamma \, x_i}{1 + \delta \, x_i + \eta \, x_i^2} + \varepsilon_i$

Fourier

fourier $\qquad$ $y_i = \alpha + \beta \, \sin\!\left(\dfrac{2\pi(x_i - \eta)}{\omega}\right) + \varepsilon_i$

dfourier $\qquad$ $y_i = \alpha + \beta \, \sin\!\left(\dfrac{2\pi(x_i - \eta)}{\omega}\right) + \gamma \, \sin\!\left(\dfrac{4\pi(x_i - \phi)}{\omega}\right) + \varepsilon_i$

Gaussian

gaussian $\qquad$ $y_i = \alpha + \dfrac{\beta}{\sqrt{2\pi}} \, \exp\!\left(\dfrac{-(x_i - \mu)^2}{2\sigma^2}\right) + \varepsilon_i$

dgaussian $\qquad$ $y_i = \alpha + \dfrac{\beta}{\sqrt{2\pi}} \, \exp\!\left(\dfrac{-(x_i - \mu)^2}{2\sigma^2}\right) + \dfrac{\gamma}{\sqrt{2\pi}} \, \exp\!\left(\dfrac{-(x_i - \nu)^2}{2\sigma^2}\right) + \varepsilon_i$

The four exponential curves each arise as solutions of linear ordinary differential equations. These represent processes that increase exponentially with time, for example, or that increase with a law of diminishing returns (that is, for which the rate of increase decreases with time).

The default setting of the CURVE option is exponential, corresponding to the "asymptotic regression" or Mitscherlich curve. An equivalent form of the equation shown above for this curve is

$$y_i = \alpha + \beta \, \exp(-\kappa \, x_i) + \varepsilon_i$$

where $\rho = \exp(-\kappa)$. The form involving ρ is used in Genstat to avoid problems with large values of κ. The model has only one nonlinear parameter, ρ, which defines the rate of exponential increase or decrease. FITCURVE estimates the other parameters by linear regression at each stage of an iterative search for the best estimate of ρ. The values of the explanatory variate are automatically scaled to avoid any computational problems near the boundary of the allowed values of ρ. By default, ρ is restricted to the range $0 < \rho < 1$, giving a curve corresponding to the law of diminishing returns. The alternative is $\rho > 1$, which can be requested by setting the SENSE option to left: for all the exponential curves, SENSE=left corresponds to a curve whose asymptote is to the left – that is, as X decreases to $-\infty$. If

Genstat finds that a better fit is obtained by the opposite sense to the one specified, the sense is reversed and a warning is printed. The parameter α is the asymptote – to the right if $\rho<1$ and to the left if $\rho<1$; β is the range of the curve between the value at $X=0$ and the asymptote.

The double exponential curve also has two forms: you can choose either $0<\rho<1$ and $0<\sigma<1$ or $\rho>1$ and $\sigma>1$, by using the SENSE option as for the exponential curve. The fitting process is unlikely to find a satisfactory solution for this curve unless there are enough data to estimate both components separately: there should be at least four points for which the fast component is larger than the slow component; the fast component corresponds to the smaller of ρ and σ when SENSE=right, or to the larger of ρ and σ when SENSE=left.

Two limiting cases of the double exponential are provided as special curves. The critical exponential curve can take a variety of shapes like the double exponential, whereas the line-plus-exponential curve is an exponential curve with a non-horizontal asymptote. Again here, the constraint on the parameter ρ depends on the setting of the SENSE option as for the exponential curve.

Another type of standard curve is sigmoid and monotonic, and is often used to model the growth of biological subjects. There are three types of these growth curves in Genstat, each a logistic of some sort. The first type is the generalized logistic without any constraints. In the equation above, α is the lower asymptote, μ is the point of inflexion for the explanatory variable, β is a slope parameter, τ is a power-law parameter, and $\alpha+\gamma$ is the upper asymptote. To fit this curve you need data for the steep central part and for both flat parts.

There are two special cases of the generalized logistic. The ordinary logistic curve is sometimes known as the autocatalytic or inverse exponential curve. The same curve can be rewritten in several different forms, so you should be alert for concealed equivalences of apparently different curves: otherwise you might be tempted to use the methods in 8.7, which would be less efficient. The other special case is the Gompertz curve. It is non-symmetrical about the inflexion, $X=\mu$, and has asymptotes at $Y=\alpha$ and $Y=\alpha+\gamma$.

You can also fit these three growth curves to data in which Y decreases as X increases. For the logistic and generalized logistic curves, you are not allowed to constrain the sense of the curve by the SENSE option. This is because the sense depends on both the parameters β and γ. In fact, the logistic curve with parameters α, β, γ, and μ is the same as the logistic curve with parameters $(\alpha+\gamma)$, $-\beta$, $-\gamma$, and μ; Genstat will report only one of the two possible versions. For the Gompertz curve, you can set SENSE=left to specify the upside-down Gompertz curve corresponding to $\gamma<0$; otherwise γ is constrained to be positive. When the sign of γ is changed for a response Y that increases with X, the sign of β will also change so that the curve remains an ascending one, and similarly for descending curves. All four possible shapes are shown in Figure 8.6.1. The interpretation of SENSE=left thus depends on the shape of the data; for ascending curves it means that the asymptote is reached more slowly to the left than to the right, but for descending curves it means the opposite.

The three rational functions are ratios of polynomials. The linear-divided-by-linear curve is a rectangular hyperbola, which occurs for example as the Michaelis-Menten law of chemical kinetics. The quadratic-divided-by-linear curve is a hyperbola with a non-horizontal asymptote. The quadratic-divided-by-quadratic curve is a cubic curve having an asymmetric maximum falling to an asymptote. The SENSE option is ignored for all three rational functions.

Fourier curves are trigonometric functions, involving the sine function in Genstat's implementation, used to model periodic behaviour. Sometimes the wavelength or period ω is a known constant, such as 2π radians (or 360 degrees), 24 hours, or 12 months; the models are then linear and should be fitted by linear regression using the FIT directive, instead of by FITCURVE. For example, the simple Fourier curve with fixed ω can be expressed in the form:

$$y_i = \alpha + \beta \sin\left(\frac{2\pi x_i}{\omega}\right) + \gamma \cos\left(\frac{2\pi x_i}{\omega}\right) + \varepsilon_i$$

and so can be fitted by statements like the following.

```
CALCULATE X1 = SIN(2*C('pi')*X/W)
& X2 = COS(2*C('pi')*X/W)
FIT X1,X2
```

The parameters β and γ are the amplitudes of the components of the curve. The SENSE option is ignored for Fourier curves.

The Gaussian curve is a bell-shaped curve like the Normal probability density. The double Gaussian is a sum of two overlapping curves of this type, and arises for example in spectography. The parameter α is usually called the *background*, and the parameters μ and ν are the peaks. The parameter σ is the standard deviation: for the double Gaussian, Genstat can deal only with the case of equal standard deviation for the two components. The parameters β and γ represent the strength of a spectrographic signal in each component, excluding the background. The SENSE option is ignored for Gaussian curves.

The PRINT, FACTORIAL, POOL, DENOMINATOR, NOMESSAGE, and FPROBABILITY options are as for FIT. The ORIGIN and CONSTANT options are described in 8.6.2, and the NONLINEAR option in 8.6.3.

8.6.2 Distributions and constraints in curve fitting

The curves available with FITCURVE can be fitted in Genstat only with the Normal likelihood. If you set some other distribution in the MODEL statement, you will get a warning message and the distribution will automatically be reset to Normal. However, you can specify a weighted Normal likelihood by providing weights in the WEIGHTS option of the MODEL directive, as for linear regression, and hence mimic other distributions.

You can set the DISPERSION option if you want Genstat to use a known variance for the distribution of the response variate (8.1.1).

FITCURVE ignores the LINK and EXPONENT options of the MODEL directive, and you are not allowed to set the GROUPS option.

You can constrain the exponential and rational curves to pass through a given point. The ORIGIN option of the FITCURVE directive specifies a value for the response variate corresponding to a zero value of the explanatory variate; to specify the response for another value of the explanatory variate you would need to modify the explanatory variate beforehand. For all these standard curves except the double exponential, the supplied origin corresponds to the expression $(\alpha+\beta)$; in the double exponential it is $(\alpha+\beta+\gamma)$. If you constrain the origin in this way, you should probably use some form of weighting, because points near the constraint are likely to vary less than points further away. You can get approximately log-

Normal weighting by using a weight variate with values $1/(Y-\text{origin})^2$.

Another way of constraining the curves is by omitting the constant term – the parameter α in each case. This parameter represents the asymptote: for growth curves with parameter $\beta>0$ it represents the asymptote as $X \to -\infty$, and for those with $\beta<0$ it represents the asymptote as $X \to +\infty$. To constrain the asymptote to be other than 0, you should put the value that you require into every element of the variate in the OFFSET option of the MODEL directive. An example is the exponential curve

$$y_i = o + \beta\, \rho^{x_i} + \varepsilon_i$$

where o is the constant value to be supplied by the offset variate. Note that the constant cannot be omitted from the Gompertz fitted with SENSE=left.

8.6.3 Parallel curve analysis

When data are grouped, a common requirement in curve fitting is to compare curves fitted to each group. The curves can be constrained to be similar to each other to some degree, governed by restricting some of the parameters to be common to all groups. Genstat provides four levels of similarity to be specified for a single grouping factor.

If you give just a variate in the parameter of the FITCURVE directive, a single curve is fitted to all groups defined by the factor. Thus, for the data in Example 8.6.3 below, the statements

```
FACTOR [LEVELS=4; VALUES=16(1...4)] Solution
MODEL Density
FITCURVE [CURVE=logistic] Log
```

fit the model

$$y_i = \alpha + \frac{\gamma}{1 + \exp(-\beta(x_i-\mu))} + \varepsilon_i\,, \qquad j=1...4$$

in which x_i stands for the explanatory variable (the logarithm of the dilution), y_i stands for the response variable (the optical density of the solution), and j stands for the solution number.

If you specify a variate and a factor, separate curves are fitted for each group, constrained to be parallel: that is, they differ only by a constant (the analogy of what in linear regression would be called the intercept). The statement

```
FITCURVE [CURVE=logistic] Log,Solution
```

fits

$$y_i = \alpha_j + \frac{\gamma}{1 + \exp(-\beta(x_i-\mu))} + \varepsilon_i\,, \qquad j=1...4$$

If you include the interaction between the variate and the factor, the curves are constrained to have common nonlinear parameters, but all linear parameters are estimated separately for each group. So the statement

```
FITCURVE [CURVE=logistic] Log*Solution
```

fits

$$y_i = \alpha_j + \frac{\gamma_j}{1 + \exp(-\beta(x_i-\mu))} + \varepsilon_i , \qquad j=1...4$$

If you set the NONLINEAR option to separate when the model includes the variate, the factor, and the interaction, Genstat estimates all the parameters independently; only the information about variability is pooled:

      FITCURVE [CURVE=logistic; NONLINEAR=separate] Log*Solution

fits

$$y_i = \alpha_j + \frac{\gamma_j}{1 + \exp(-\beta_j (x_i-\mu_j))} + \varepsilon_i , \qquad j=1...4$$

You can modify a model fitted by FITCURVE by using the ADD, DROP, or SWITCH directives as for linear models, provided you have given an appropriate TERMS statement before the FITCURVE statement. The alterations must, however, produce a model that would be allowed in the FITCURVE directive: that is, it must contain one variate, or one variate and one factor, or one variate and one factor and their interaction. The NONLINEAR options of the ADD, DROP, and SWITCH directives have the same effect as the NONLINEAR option of FITCURVE. Thus you can compare curves between groups of a factor, assessing for example whether they are parallel. The accumulated setting of the PRINT option of these directives allows you to summarize the results. Example 8.6.3 shows such *an analysis of parallelism*.

Example 8.6.3

```
 1   " Modelling the relationship between dilution and optical density
-2      for four solutions.
-3      Data from Bouvier et al (1985) page 129."
 4
 5   READ [PRINT=data] Density

 6   1.914 1.878 1.717 1.195 0.587 0.264 0.099 0.114
 7   1.891 1.887 1.703 1.158 0.599 0.277 0.106 0.069
 8   1.876 1.830 1.608 1.099 0.513 0.236 0.096 0.074
 9   1.913 1.847 1.622 1.109 0.536 0.227 0.100 0.086
10   1.873 1.859 1.707 1.191 0.611 0.262 0.111 0.082
11   1.877 1.873 1.696 1.185 0.617 0.259 0.122 0.041
12   1.897 1.800 1.495 0.915 0.417 0.203 0.068 0.047
13   1.869 1.780 1.500 0.922 0.396 0.165 0.096 0.035 :
14   FACTOR [LEVELS=4; VALUES=16(1...4)] Solution
15   VARIATE [VALUES=(30,90,270,810,2430,7290,21870,65610)8] Dilution
16   VARIATE Log; EXTRA=' dilution'
17   CALCULATE Log = LOG10(Dilution)
18   MODEL Density
19   TERMS Log*Solution
20   FITCURVE [PRINT=model,estimates; CURVE=logistic] Log

20.........................................................................

***** Nonlinear regression analysis *****

Response variate: Density
        Explanatory: Log dilution
```

```
Fitted Curve: A + C/(1 + EXP(-B*(X - M)))
```

*** Estimates of parameters ***

	estimate	s.e.
B	-2.816	0.139
M	2.9973	0.0184
C	1.8633	0.0329
A	0.0658	0.0184

```
    21  ADD [PRINT=model,estimates] Solution
```

21..

***** Nonlinear regression analysis *****

```
 Response variate: Density
        Explanatory: Log dilution
   Grouping factor: Solution, constant parameters separate
        Fitted Curve: A + C/(1 + EXP(-B*(X - M)))
```

*** Estimates of parameters ***

		estimate	s.e.
B		-2.8156	0.0671
M		2.99732	0.00889
C		1.863	
A	Solution 1	0.1069	
A	Solution 2	0.06406	
A	Solution 3	0.1012	
A	Solution 4	-0.008880	

```
    22  ADD [PRINT=model,estimates] Log.Solution
```

22..

***** Nonlinear regression analysis *****

```
 Response variate: Density
        Explanatory: Log dilution
   Grouping factor: Solution, all linear parameters separate
        Fitted Curve: A + C/(1 + EXP(-B*(X - M)))
```

*** Estimates of parameters ***

		estimate	s.e.
B		-2.8164	0.0681
M		2.99798	0.00903
C	Solution 1	1.884	
A	Solution 1	0.09700	
C	Solution 2	1.859	
A	Solution 2	0.06584	
C	Solution 3	1.870	
A	Solution 3	0.09773	
C	Solution 4	1.839	
A	Solution 4	0.001892	

```
    23   ADD [PRINT=model,summary,estimates,accumulated; NONLINEAR=separate]
  23.............................................................................
```

***** Nonlinear regression analysis *****

```
  Response variate: Density
         Explanatory: Log dilution
   Grouping factor: Solution, all parameters separate
       Fitted Curve: A + C/(1 + EXP(-B*(X - M)))
```

*** Summary of analysis ***

	d.f.	s.s.	m.s.	v.r.
Regression	15	34.95314	2.3302090	4742.60
Residual	48	0.02358	0.0004913	
Total	63	34.97672	0.5551860	
Change	6	-0.08345	-0.0139087	-28.31

Percentage variance accounted for 99.9
Standard error of observations is estimated to be 0.0222
* MESSAGE: The following units have large standardized residuals:
                12   -2.39

*** Estimates of parameters ***

		estimate	s.e.
B	Solution 1	-2.9175	0.0979
M	Solution 1	3.0491	0.0122
C	Solution 1	1.8622	0.0217
A	Solution 1	0.0825	0.0127
B	Solution 2	-2.8285	0.0967
M	Solution 2	2.9924	0.0127
C	Solution 2	1.8572	0.0227
A	Solution 2	0.0693	0.0126
B	Solution 3	-2.8693	0.0968
M	Solution 3	3.0783	0.0125
C	Solution 3	1.8560	0.0220
A	Solution 3	0.0655	0.0131
B	Solution 4	-2.7433	0.0951
M	Solution 4	2.8610	0.0134
C	Solution 4	1.8822	0.0245
A	Solution 4	0.0487	0.0121

*** Accumulated analysis of variance ***

Change	d.f.	s.s.	m.s.	v.r.
+ Log	3	34.7306061	11.5768690	23562.03
+ Solution	3	0.1363831	0.0454610	92.53
+ Log.Solution	3	0.0026935	0.0008978	1.83
+ Separate nonlinear	6	0.0834520	0.0139087	28.31
Residual	48	0.0235841	0.0004913	
Total	63	34.9767189	0.5551860	

8.6.4   Modifications to regression output and the RKEEP directive

The output produced by the PRINT options of the FITCURVE and RDISPLAY directives for fitted curves is much like that for iterative generalized linear models with a Normal distribution (8.5.6). In particular, only one response variable is analysed, standard errors are approximate, and the accumulated summary contains pooled contributions for all the terms fitted in one statement.

You cannot get standard errors and correlations for linear parameters in models where you have constrained some parameters of the curve to be equal for all the groups defined by a fitted factor. When you fit separate curves for the groups of a factor, correlations between parameters in different groups are zero and are not shown.

Neither can you get leverages for models in which parameters are constrained to be equal across groups. Genstat therefore does not standardize residuals with respect to the leverages in these models. For other models, the leverages are defined as:

$$l_i = \{D'CD\}_{ii}$$

where D is the matrix of derivatives of the fitted values with respect to the parameters, and C is the variance-covariance matrix of the parameters.

You can display intermediate results of the iteration by the monitoring setting of the PRINT option of the FITCURVE directive. At each cycle, the current parameter values are displayed together with the total number of times the likelihood function has been evaluated (Nfun) and an indication of the state of the search (Move). The possible states are:

Move
0   The current step is acceptable
1   Preconvergence; small adjustments are being made
2   The function is concave in at least one direction
3   Convergence is being approached, but there is distinct curvature
4   A bound has been violated
5   The current step is too large relative to the steplengths
6   Convergence
7   A step has been taken within a boundary plane

The steplengths used in the search are also reported whenever they are changed, and information is given about any temporary scaling used to simplify the search. Example 8.6.4 shows the progress of the search for the curve fitted in Example 8.6.

Example 8.6.4

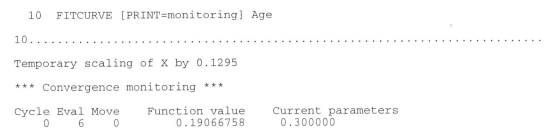

```
  10   FITCURVE [PRINT=monitoring] Age

10..............................................................................

Temporary scaling of X by 0.1295

*** Convergence monitoring ***

Cycle Eval Move    Function value    Current parameters
    0    6    0        0.19066758    0.300000
```

			Steps	0.0100000
1	9	0	0.18676342	0.350165
			Steps	0.00250000
2	12	1	0.18675923	0.351720
3	16	6	0.18675919	0.351886

The search may not converge, particularly if the model to be fitted is unsuitable for the data. Genstat will give a warning message to indicate why convergence has not been achieved; often it will also suggest a limiting form of the curve that might be a more suitable description of the data than the one you have specified. You can find out about the final status of the search by the EXIT parameter of the RKEEP directive. It takes a value according to the following key:

Exit
0   Successful convergence
1   Limit on number of cycles has been reached without convergence
2   Parameter out of bounds
3   Likelihood appears constant
4   Failure to progress towards solution
5   Some standard errors are not available because the information matrix
    is nearly singular
6   Calculated likelihood may be incorrect because of missing fitted values
7   Curve is close to a limiting form

With code 7, the limiting form of the curve is described by the warning diagnostic.

Further messages warn you about vertical asymptotes of rational curves. You can use the summary setting of the PRINT option to display the value or values of the explanatory variate for which the fitted curve is infinite. A warning is also printed if an asymptote occurs within the range of the data.

The derivatives of the fitted values with respect to each parameter can be stored in variates using the GRADIENTS parameter of the RKEEP directive. You can use these quantities to assess the relative influence of each observation on a parameter; you can also construct a measure of leverage by summing the gradients for all the parameters.

The RGRAPH procedure can be used to display a fitted curve, as shown in Figure 8.6; it can also display a set of curves fitted for each level of a factor (8.6.3). The RCHECK procedure cannot be used to produce diagnostic information or pictures after curve fitting.

8.6.5   Functions of parameters: the RFUNCTION directive

RFUNCTION estimates functions of parameters of a nonlinear model.

Options

PRINT = *strings*	What to print (estimates, se, correlations); default esti,se
CALCULATION = *expressions*	Calculation of functions involving nonlinear and/or linear parameters; no default

SE = *variate*	To save approximate standard errors; default *
VCOVARIANCE = *symmetric matrix*	
	To save approximate variance-covariance matrix; default *
SAVE = *identifier*	Specifies save structure of regression model; default * i.e. that from last model fitted

Parameter

scalars	Identifiers of scalars assigned values of the functions by the calculations

The RFUNCTION directive provides estimates of functions of parameters in nonlinear models, together with approximate standard errors and correlations. It can be used after any of the models fitted by the FITCURVE or FITNONLINEAR directives; information about the latter is in 8.7.2. However, if there are any linear parameters in a general nonlinear model for which standard errors have not been estimated, standard errors and correlations cannot be estimated for functions that depend on those parameters (see 8.7.2). In addition, it is not possible to use the RFUNCTION directive after fitting standard curves with separate nonlinear parameters for each level of a factor (option NONLINEAR=separate in FITCURVE, ADD, DROP and SWITCH).

The functions are defined by the expressions supplied by the CALCULATION option of RFUNCTION; these define how to calculate the function from the values of the parameters. Unless initial values have been specified (8.6.6), the parameters in standard curves usually have no identifiers associated with them. If this is the case, you should refer to each parameter by using a text structure containing the name of the parameter as displayed, for example, by the option PRINT=estimates of the FITCURVE directive. The text structure can, of course, just be a string, for example 'R'.

In Example 8.6.5, we use RFUNCTION to provide us with an alternative parameterization of the exponential model fitted in Example 8.6, using the parameter κ (8.6.1) instead of ρ, and reporting $-\beta$ instead of β.

Example 8.6.5

```
  12   " Get estimates of parameters in the form
 -13       Y = A - Bneg*EXP(-K*X) "
  14   EXPRESSION e[1,2]; VALUE=!e(Bneg = -'B'),!e(K = -LOG('R'))
  15   RFUNCTION [CALCULATION=e[]] Bneg,K

15...........................................................................

***** Estimates of functions of parameters *****

*** Estimates and standard errors ***

               estimate        s.e.
Bneg            0.9725        0.0647
K               0.1352        0.0256
```

The parameter of RFUNCTION provides a list of scalars that are to hold the estimated values of the functions. These need not be declared in advance, but will be defined automatically if necessary. The CALCULATION option specifies a list of one or more expressions to define the calculations necessary to evaluate the functions from the parameters of the nonlinear model, and place the results into the scalars. Note that when parameters are referred to by their names, these must match exactly, including case, the names as displayed by FITCURVE.

The PRINT option controls output as usual. By default, the estimates of the function values are formed – as could be done simply by a CALCULATE statement using the expressions if the parameters were available in scalars. In addition, approximate standard errors are calculated, using a first-order approximation based on difference estimates of the derivatives of each function with respect to each parameter. Approximate correlations can also be requested.

The SE and VCOVARIANCE options allow standard errors and the approximate variance-covariance matrix of the functions to be stored; the estimates of the functions themselves are automatically available in the scalars listed by the parameter of RFUNCTION. The SAVE option specifies which fitted model is to be used, as in the RDISPLAY and RKEEP directives.

8.6.6   Controlling the start of the search with the **RCYCLE** directive

You can use the RCYCLE directive to supply initial values for the nonlinear parameters: you might do this, for example, to improve efficiency if you are fitting a standard curve and already have good prior knowledge of the likely values o the parameters. Usually, FITCURVE determines a reasonable starting value for each parameter by a short grid search, or by some manipulation of the data values: this will not be done if you supply initial values. For example

```
RCYCLE PARAMETER=Rate; INITIAL=0.62
FITCURVE [CURVE=exponential] X
```

You must usually give an identifier (here Rate) and an initial value for each nonlinear parameter in the model to be fitted. For logistic curves, however, you must include all the parameters – both nonlinear and linear. The parameters must be listed in the same order as Genstat uses to print them. The RCYCLE directive defines the identifiers as scalars holding the initial values that you have supplied; after the model has been fitted they contain the estimated values of the parameters.

The other parameters of RCYCLE are ignored by FITCURVE: bounds are set up automatically according to the curve to be fitted and the way in which it is parameterized by Genstat (over which you have no control).

You can use the MAXCYCLE option to reset the limit on the number of iterations, but Genstat ignores the METHOD and TOLERANCE options. For all standard curve fitting Genstat uses a modified Newton method (8.7.1).

8.7   General nonlinear regression, and minimizing a function

You can use the methods described in this section to fit any kind of regression. However, you should check first that the model does not belong to any of the categories described earlier in this chapter, for the appropriate directives are then much more efficient. These categories are linear models, generalized linear models, and the standard curves provided by FITCURVE.

Because the methods described here are very general, they are neither as robust nor as automatic as, for example, the method that is used for fitting linear models. Nonlinear methods make use of iterative optimization algorithms, designed to search for the minimum value of a function as the parameters vary; for nonlinear regression models, the function involved is the deviance, or minus twice the log-likelihood ratio, so the algorithm searches for the maximum-likelihood solution. It is often necessary to provide the algorithm with good starting values, to set bounds on the parameter values, and sometimes even to define the initial direction of search.

Optimization is easiest with few parameters, approximately quadratic functions, small correlations between parameters, and good initial parameter estimates.

Where possible, you can effectively reduce the number of parameters to be optimized by separating linear and nonlinear ones: that is, you can first fit the linear parameters, and treat the resulting residual sums of squares as functions of the nonlinear parameters alone (8.7.2).

Problems with optimization methods are most likely to arise if you neglect the parameterization of the function. You can often transform the parameters to make the function nearly quadratic; after finding a solution, you can then use the RFUNCTION directive (8.6.5) to estimate the original parameters. Another source of difficulty is if you try to fit inappropriately many parameters.

You can usually find descriptive statistics based on the data that will provide initial estimates reasonably close to the final parameter estimates. For example, suitably spaced ordinates provide parameters for curve fitting that give much the same likelihood surface whatever curve is being fitted.

For advice on reformulating functions to speed up optimization, see Ross (1990). The methods used for optimization in Genstat are the same as those in MLP, the Maximum Likelihood Program. The MLP Manual (Ross 1987) contains further useful advice on alternative ways of specifying models.

Example 8.7 shows the fitting of a nonlinear model with four parameters. The model has the form

$$y_i = \frac{\theta_1 \, \theta_3 \left(x_{2i} - \dfrac{x_{3i}}{1.632} \right)}{1 + \theta_2 \, x_{1i} + \theta_3 \, x_{2i} + \theta_4 \, x_{3i}} + \varepsilon_i$$

which is linear in the parameter θ_1 but nonlinear in θ_2, θ_3, and θ_4. The parameterization of this model is reasonable, and it fits the data well; the algorithm succeeds in finding the solution without requiring the definition of initial values or bounds.

Example 8.7

```
  1   " Nonlinear model for a chemical process, involving four parameters.
 -2     Data from Carr (1960), analysed in Seber & Wild (1989) p78.
 -3     The response R is the rate of disappearance of n-pentane by catalytic
 -4     isomerization to i-pentane, and the three associated variables
 -5     X1, X2 and X3 are the partial pressures of hydrogen, n-pentane and
 -6     i-pentane. Fit the unweighted model (Seber & Wild, p83)."
  7
```

```
 8   OPEN 'REACTION.DAT'; CHANNEL=2
 9   READ [PRINT=data; CHANNEL=2] X1,X2,X3,R

     1   205.8  90.9  37.1  3.541    404.8  92.9  36.3  2.397
     2   209.7 174.9  49.4  6.694    401.6 187.2  44.9  4.722
     3   224.9  92.7 116.3  0.593    402.6 102.2 128.9  0.268
     4   212.7 186.9 134.4  2.797    406.2 192.6 134.9  2.451
     5   133.3 140.8  87.6  3.196    470.9 144.2  86.9  2.021
     6   300.0  68.3  81.7  0.896    301.6 214.6 101.7  5.084
     7   297.3 142.2  10.5  5.686    314.0 146.7 157.1  1.193
     8   305.7 142.0  86.0  2.648    300.1 143.7  90.2  3.303
     9   305.4 141.1  87.4  3.054    305.2 141.5  87.0  3.302
    10   300.1  83.0  66.4  1.271    106.6 209.6  33.0 11.648
    11   417.2  83.9  32.9  2.002    251.0 294.4  41.5  9.604
    12   250.3 148.0  14.7  7.754    145.1 291.0  50.2 11.590
10   " Change units from psia to atmospheres."
11   CALCULATE X1,X2,X3 = X1,X2,X3/14.7
12   " Specify how to form the nonlinear component of the model from
-13    the parameters and associated variables."
14   EXPRESSION e1; VALUE=!e(Z = T3*(X2-X3/1.632)/(1+T2*X1+T3*X2+T4*X3))
15   MODEL R
16   " List the nonlinear parameters: attempt optimization from
-17    default starting values of 1 with no bounds."
18   RCYCLE T2,T3,T4
19   " Fit the model, estimating the linear parameter (called theta1 by
-20    Seber & Wild) by linear regression with no additional constant."
21   FITNONLINEAR [CALCULATION=e1; CONSTANT=omit; SELINEAR=yes] Z

21.........................................................................
```

***** Nonlinear regression analysis *****

 Response variate: R

*** Summary of analysis ***

	d.f.	s.s.	m.s.	v.r.
Regression	4	637.254	159.3135	985.09
Residual	20	3.234	0.1617	
Total	24	640.489	26.6870	

Percentage variance accounted for 98.5
Standard error of observations is estimated to be 0.402

*** Estimates of parameters ***

	estimate	s.e.
T2	1.05	2.68
T3	0.56	1.60
T4	2.47	6.47
* Linear		
Z	35.9	11.4

8.7.1 Fitting nonlinear models

This subsection describes the preliminary things that you must do before fitting a general nonlinear model. It also gives information about the algorithms that Genstat uses.

Before using the FITNONLINEAR directive to fit a nonlinear model, you must use the MODEL directive to specify either the response variate, or the scalar that is to store the value of a general function (8.7.4). You must also use the RCYCLE directive to specify the nonlinear parameters. The TERMS directive can be used as in linear regression, to list the explanatory variables to be used in modelling. The model calculations themselves are provided in expression structures which are supplied by the CALCULATION option of FITNONLINEAR; in Example 8.7, a single expression called e1 is used. If you have used TERMS you can modify the model using the ADD, DROP, and SWITCH directives, as in the previous sections. You can use the RDISPLAY and RKEEP directives to display or save the results. The RCHECK procedure does not work with nonlinear models, but RGRAPH can be used to display the fit of a nonlinear model with respect to some specified variate.

Genstat fits nonlinear regression models by maximum likelihood. The likelihood is usually from a distribution in the exponential family; this is specified using the DISTRIBUTION option of the MODEL directive. With the Normal and the Poisson distribution you can take advantage of linear parameters that the model contains; see 8.7.2. The fitting of models with the other settings of DISTRIBUTION, or with no linear parameters, is described in 8.7.3. To use other forms of likelihood, you should specify how it is to be calculated and set the FUNCTION option of the MODEL directive to a scalar whose value is assigned by the calculation (8.7.4). You can use this same device to minimize a general function with respect to its parameters.

The settings of the LINK and EXPONENT options of the MODEL directive are ignored, and you are not allowed to set the GROUPS option; other options and parameters are as in linear regression.

Genstat provides three algorithms for fitting general nonlinear models; they work with numerical differences and so do not require you to specify derivatives. The default algorithm is a modified Gauss-Newton method. This takes advantage of the fact that the likelihood function can be expressed as a sum of squares. However, you cannot use it for minimizing a general function (8.7.4). The second algorithm, a modified Newton method, is requested by setting option METHOD=Newton in the RCYCLE statement (8.5.3). This can be used for any nonlinear model. The third algorithm is a modified Fletcher-Powell method, specified by setting METHOD=Fletcher. In fact, this is similar to the Newton method, with an occasional step in the search being determined by the Fletcher-Powell algorithm rather than by the Newton algorithm.

The modification in all these methods is to use estimated numerical differences instead of evaluating derivatives. In nonlinear regression problems, particularly ones with separable linear parameters, specification of the derivatives would be very complex, and so it is much more convenient to estimate them numerically.

You can change the limit on the number of iterations by the MAXCYCLE option of the RCYCLE directive, as for the FITCURVE directive.

You must set the PARAMETER parameter of the RCYCLE directive to the identifiers of scalars that will be used to represent the nonlinear parameters in the model calculations (8.7.2). There must be at least one nonlinear parameter. There is no formal upper limit on the number of nonlinear parameters, but the greater the number of parameters the longer the time required for the search and the smaller the chance of finding a satisfactory solution.

You can set the LOWER and UPPER parameters of RCYCLE to provide fixed bounds for each parameter. By default, the values $\pm 10^9$ are used. Where possible you should always set bounds, particularly to avoid such problems as attempting to take the log of a negative number. You can incorporate more general constraints as logical functions within the calculations. For example you could compute an extra term

        (Constr > 0) * K * Constr

to impose a penalty on exceeding the constraint, controlled by setting different values of K. Often, the best way to impose a constraint is to reparameterize. For example, if a parameter α must be positive, you could replace α by $\exp(\beta)$, and allow β to take any value.

The STEPLENGTH parameter of RCYCLE can be used to provide initial steplengths for the search. By default the steplength is 0.05 times the initial value of the corresponding parameter, or precisely 1.0 if the initial value is zero. If you set a steplength to zero, Genstat treats the corresponding parameter as being fixed at its initial value. This allows complex problems in many dimensions to be tackled in stages, optimizing some parameters with others fixed, and then optimizing the others in the turn.

By default, the initial value of a parameter is taken to be the current value of the scalar that represents it in the calculation, or 1.0 if the value is missing. Other values can be specified using the INITIAL parameter of RCYCLE.

If you can calculate a range within which you expect a parameter to lie, you should choose a steplength of about 1% of the width of the range. If the steps are too small, numerical differencing may not work; if they are too large, gradients may be unreliable and you may get premature convergence. Genstat tests convergence by the relationship of final adjustments to step lengths.

The more parameters there are to estimate, and the more scattered are the data, the more iterations are required to find the optimum. The maximum number of iterations is set to 30 by default, but you can reset this with the MAXCYCLE option of RCYCLE (8.5.3). However, if convergence fails with a given setting of MAXCYCLE, you should check the data and consider reparameterizing the model before you indiscriminately increase the number of iterations.

Genstat prints a warning when convergence fails. The only sections of output that are then available are the residual degrees of freedom, the residual deviance, the fitted values, and the parameter estimates (without standard errors) for the current cycle. The EXIT parameter of the RKEEP directive (8.6.4) allows you to obtain a numerical code indicating why convergence failed.

For any nonlinear model, you can choose just to evaluate the likelihood for a range of combinations of parameter values, rather than to maximize the likelihood with respect to the parameters. You do this by setting the NGRIDLINES option of FITNONLINEAR (8.7.2). The calculated values of the likelihood can be stored in a variate using the GRID parameter of the RKEEP directive (8.1.4), and used to produce pictures of the surface for example with the DCONTOUR or DSURFACE directives (6.3). This is illustrated in Example 8.7.4b.

8.7.2 Nonlinear regression for models with some linear parameters

---

FITNONLINEAR fits a nonlinear regression model or optimizes a scalar function.

Options

PRINT = *strings*	What to print (model, deviance, summary, estimates, correlations, fittedvalues, accumulated, monitoring, grid); default mode, summ, esti or grid if NGRIDLINES is set
CALCULATION = *expressions*	Calculation of fitted values or of explanatory variates involving nonlinear parameters; default * (valid only if OWN set)
OWN = *scalar*	Option setting for OWN directive if this is to be used rather than CALCULATE; default * requests CALCULATE to be used
CONSTANT = *string*	How to treat the constant (estimate, omit); default esti
FACTORIAL = *scalar*	Limit for expansion of model terms; default as in previous TERMS statement, or 3 if no TERMS given
POOL = *string*	Whether to pool ss in accumulated summary between all terms fitted in a linear model (yes, no); default no
DENOMINATOR = *string*	Whether to base ratios in accumulated summary on rms from model with smallest residual ss or smallest residual ms (ss, ms); default ss
NOMESSAGE = *strings*	Which warning messages to suppress (dispersion, leverage, residual, aliasing, marginality, vertical, df); default *
FPROBABILITY = *string*	Printing of probabilities for variance ratios (yes, no); default no
NGRIDLINES = *scalar*	Number of values of each parameter for a grid of function evaluations; default *
SELINEAR = *string*	Whether to calculate s.e.s for linear parameters (yes, no); default no
INOWN = *identifiers*	Setting to be used for the IN parameter of OWN if used in place of CALCULATE; default *
OUTOWN = *identifiers*	Setting to be used for the OUT parameter of OWN if used in place of CALCULATE; default *

Parameter

formula	List of explanatory variates and/or one factor to be used in linear regression, within nonlinear optimization

---

If the model is linear in some of the parameters, it may be fitted more efficiently using the methods described in this subsection. To use these the data must either be Normally distributed, or they must follow a Poisson distribution and the model must contain only one explanatory variable and no constant term.

The linear parameters are fitted by a linear regression of the response variate (specified by the parameter of the MODEL statement) on the variates listed by the parameter of FITNONLINEAR. At least one of these variates must depend on the nonlinear parameters in the model but they need not all do so. You can define how to calculate the variates from the nonlinear parameters either by the CALCULATION option or by the OWN, INOWN, and OUTOWN options of FITNONLINEAR. If the parameter of FITNONLINEAR is not set, Genstat uses the methods described in either 8.7.3 or 8.7.4.

In Example 8.7, the linear parameter (θ_1 in the equation) is estimated by a regression of the response variate R on the variate Z; expression e1 defines how to form Z from the values of the parameters T2, T3, and T4 (θ_2, θ_3, and θ_4 in the equation) and from the variates X1, X2, and X3. The setting CONSTANT=omit in the FITNONLINEAR statement ensures that there is no constant term.

As already mentioned, the parameter of FITNONLINEAR may include variates that are not changed by the calculations as well as those that are. One factor may also be included so that a separate constant is fitted for each level. Thus

```
FACTOR [LEVELS=3; VALUES=8(1...3)] F
FITNONLINEAR Z,F,X2
```

would fit the model of Example 8.7 modified to include a constant for each of the three levels of F and an additional linear effect of the variable X2. The effect of including the factor is to fit a set of parallel nonlinear regressions. You cannot include interactions between a variate and a factor, as is allowed with FITCURVE; nor can you include POL, REG, or SSPLINE functions, nor interactions between variates as allowed with FIT. However, procedure FITPARALLEL allows you to assess the various ways in which nonlinear models can be non-parallel (see 8.6.3 for an explanation of analysis of parallelism with FITCURVE).

If there is a constant in the linear regression, as specified by the CONSTANT option, the factor will be parameterized in terms of differences from the first level – as in linear regression. If you set CONSTANT=omit, the actual constants are fitted; there is no need to set option FULL of the TERMS which is ignored in nonlinear models.

If you specify an offset variate (8.1.1), its values can also be modified by the calculations, and depend on the parameters.

The PRINT option is as for the FIT directive except for the grid setting: see below, with the NGRIDLINES option.

You must set one of the CALCULATION and OWN options to define how the nonlinear parameters are included in the model. The CALCULATION option does this by a list of one or more expressions. The expressions are evaluated in turn at every step of the estimation process, just as if they had been given in a sequence of CALCULATE statements. For example:

```
EXPRESSION Diffuse[1]; \
  VALUES=!E(X1,Xr=NORMAL((H+1,-1*X)/SQRT(2*D*T)))
& Diffuse[2]; VALUES=!E(Z=X1+Xr-1)
FITNONLINEAR [CALCULATION=Diffuse[1,2]] Z
```

Here, the CALCULATION option is set to the two expressions Diffuse[1] and Diffuse[2], to define a model for one-dimensional diffusion.

Alternatively, you can set the OWN option to specify that the calculation is to be done by executing your own source code, called by a version of the subroutine G5XZXO, as for the OWN directive (13.4.2). Generally, using OWN is likely to be worthwhile only when calculations are very extensive, or when a particular function is needed often. The setting of the OWN option will be passed to G5XZXO in the same way as the setting of the SELECT option of the OWN directive is passed to G5XZXO.

The CONSTANT, FACTORIAL, POOL, DENOMINATOR, NOMESSAGE, and FPROBABILITY options are as for the FIT directive.

If you set the NGRIDLINES option to n, say (with $n \geq 2$), the FITNONLINEAR directive evaluates the likelihood at a grid of values of the nonlinear parameters, and does not search for an optimum. For each parameter, the distance between the upper and lower bounds (set by the RCYCLE directive) will be divided into $(n-1)$ equal parts, defining a rectangular grid with n gridlines in each dimension. By setting some upper and lower bounds equal, you can look at the behaviour of the function with respect to a few parameters at a time. The default setting of the PRINT option is grid in this case, and produces a display of the function values. Other settings of the PRINT option are ignored. The calculated grid of values is available from the GRID parameter of the RKEEP directive. This is illustrated in 8.7.4.

By default, standard errors are calculated only for nonlinear parameters. To obtain standard errors for the linear parameters as well, you can set option SELINEAR=yes. Then, after the optimum has been found, Genstat increases the number of dimensions to include the linear parameters and estimates the rate of change of the likelihood in all the dimensions.

The INOWN and OUTOWN options are relevant only when the OWN option is set; see 13.4.2.

8.7.3 Nonlinear regression models with no linear parameters

If there are no linear parameters in the model, or if the distribution is not one of those that can be handled by the method described in 8.7.2, you should no longer use the parameter of FITNONLINEAR. Instead you should set the FITTEDVALUES parameter in the MODEL statement to the identifier of a variate that is to contain the fitted values for any set of values of the nonlinear parameters. Then define how to calculate the fitted values from the nonlinear parameters and the explanatory variates, using either the CALCULATION or the OWN options of FITNONLINEAR, as in 8.7.2.

Example 8.7.3a shows how to refit the model of Example 8.7 without taking advantage of the linearity of parameter θ_1. Expression e2 in line 25 calculates the variate of fitted values F as T1 (θ_1) multiplied by the variate Z (calculated by the expression e1 used in Example 8.7). F is identified as the fitted-value variate in line 28, initial values are specified for the parameters in line 32, and then the model can be fitted, to obtain the same answers as before.

Example 8.7.3a

```
  23   " Specify how to form the fitted values from Z and the linear
 -24     parameter theta 1."
```

```
  25    EXPRESSION e2; VALUE=!e(F=T1*Z)
  26    " Supply the name of the variate that will hold fitted values
 -27      calculated by the expressions."
  28    MODEL R; FITTED=F
  29    " Include theta1 with the list of nonlinear parameters;
 -30      use initial values of 1 as before, except for theta 1
 -31      (if this is not done, FITNONLINEAR will not converge)."
  32    RCYCLE T1,T2,T3,T4; INITIAL=36,1,1,1
  33    " Fit the model, with no linear regression involved."
  34    FITNONLINEAR [CALCULATION=e1,e2]
```

```
  34...............................................................
```

```
***** Nonlinear regression analysis *****

Response variate: R

*** Summary of analysis ***

                d.f.          s.s.          m.s.          v.r.
Regression        4        637.254      159.3135        985.09
Residual         20          3.234        0.1617
Total            24        640.489       26.6870

Percentage variance accounted for 98.5
Standard error of observations is estimated to be 0.402

*** Estimates of parameters ***

               estimate              s.e.
T1                 35.9              11.4
T2                 1.05              2.69
T3                 0.56              1.61
T4                 2.47              6.50
```

The output from the monitoring setting of the PRINT option, not displayed here, shows that solution takes 18 iterations involving 164 function evaluations compared to 13 and 123 when θ_1 is treated as linear. Moreover, convergence is not achieved here without supplying an initial value for θ_1. So clearly you should exploit linearity where possible.

With the methods described in this section, the distribution can be any of those available from the DISTRIBUTION option of the MODEL directive, with the exception of the inverse-Normal distribution. Thus, the deviance will be based on the likelihood function of either the Normal, Poisson, binomial, gamma, or multinomial distributions, taking account of the settings of the DISPERSION and WEIGHTS options of the MODEL directive. The first four of these distributions were discussed in 8.4.1 and 8.4.2.

The multinomial distribution is used rather differently from the others: it is for fitting distributions. The DISTRIBUTION directive (7.1.4) provides a wide range of standard distributions, and is more convenient and efficient than FITNONLINEAR for these; but FITNONLINEAR allows you to fit other distributions. (Despite the terminology "multinomial", this setting is thus not for fitting models to response variables that take one of a finite set of values for each unit; these can be fitted using generalized linear models as described in 8.5.5.)

To specify and fit your own distribution, you should supply as response variate a set of counts of observations falling into a series of groups; the fitted values should then be a set of expected counts for the groups, calculated from the distribution being considered. The resulting multinomial likelihood is the same as that of the Poisson distribution, but with the constraint $\Sigma f_i = M$, where M is the sum of the counts.

Example 8.7.3b fits a Normal distribution to a set of observations produced by the Genstat pseudo-random number generator. It would be much easier to use the DISTRIBUTION directive for this, but use of this familiar distribution here should make it clear how FITNONLINEAR can be used in more complicated situations.

Example 8.7.3b

```
   1    " Fit a Normal distribution to pseudo-random numbers in the range (0,1)
  -2      generated by the functions URAND and EDNORMAL."
   3
   4    CALCULATE Random = EDNORMAL(URAND(25384; 50))
   5    " Define bounds to subdivide the observations."
   6    SCALAR  Limit[1...8]; VALUE=-100,-1,-0.6,-0.2,0.2,0.6,1,100
   7    " Form response variate: counts of numbers within specified bounds."
   8    CALCULATE S[1...7] = SUM(Random<=Limit[2...8] .AND. Random>Limit[1...7])
   9    VARIATE [VALUES=S[1...7]] Count
  10    " Set up expression to calculate expected counts for a Normal variable."
  11    & [VALUES=Limit[2...8]] L1
  12    & [VALUES=Limit[1...7]] L2
  13    EXPRESSION [VALUE=P=50*(NORMAL((L1-Mean)/SD)-NORMAL((L2-Mean)/SD))] \
  14      Normal
  15    MODEL [DISTRIBUTION=multinomial] Count; FITTED=P
  16    RCYCLE Mean,SD; STEPLENGTH=0.02,*; LOWER=*,0.5; INITIAL=0,1
  17    FITNONLINEAR [CALCULATION=Normal]
```

17...

***** Nonlinear regression analysis *****

Response variate: Count
        Distribution: Multinomial

*** Summary of analysis ***

	d.f.	deviance	mean deviance	deviance ratio
Regression	2	*	*	
Residual	4	1.904	0.4760	
Total	6	*	*	

* MESSAGE: ratios are based on dispersion parameter with value 1

*** Estimates of parameters ***

	estimate	s.e.
Mean	0.068	0.151
SD	1.024	0.137

*MESSAGE: s.e.s are based on dispersion parameter with value 1

8.7.4 General nonlinear models

The earlier parts of this section have dealt with two methods of calculating the likelihood at each step of the iterative search: performing linear regression of the response variate on calculated explanatory variates, and directly comparing the response variate with a calculated variate of fitted values. A third method is to calculate the likelihood explicitly. You can also use this to minimize the value of a function that is not a likelihood at all. Remember, however, that the methods described earlier in this chapter actually maximize the likelihood function by minimizing the deviance, which is minus twice the log-likelihood ratio.

To use the regression directives to minimize a function, you need to start with a MODEL statement that has no response variate, but where the FUNCTION option is set to a scalar. You then specify the parameters with the RCYCLE directive as before, and perform the minimization with FITNONLINEAR, supplying an expression that calculates the function from the parameters and places the result into the scalar. Example 8.7.4a shows the minimization of an awkward two-dimensional test function.

Example 8.7.4a

```
  1    " Finding the minimum of a function of two parameters:
 -2      Rosenbrock's steep-sided valley."
  3
  4    EXPRESSION Rbrock; VALUE=!e(F = 100*(P2-P1*P1)**2+(1-P1)**2)
  5    MODEL [FUNCTION=F]
  6    RCYCLE P1,P2; STEPLENGTH=0.01; INITIAL=-1.2,1
  7    FITNONLINEAR [PRINT=summary,estimates,correlation,monitoring; \
  8        CALCULATION=Rbrock]
```

8..

*** Convergence monitoring ***

Cycle	Eval	Move	Function value	Current parameters	
0	1	0	24.200005	-1.20000	1.00000
			Steps	0.0100000	0.0100000
			Steps	0.00622704	0.0160590
1	10	0	4.7307487	-1.17502	1.38004
2	22	5	4.0673242	-0.946587	0.843290
3	31	0	3.2409029	-0.777289	0.575517
4	40	0	2.7767313	-0.513074	0.193435
			Steps	0.00954871	0.0104726
5	49	0	2.0038674	-0.411870	0.159395
6	61	5	1.7240570	-0.295930	0.0664510
7	70	0	1.4788369	-0.0473323	-0.0595603
8	79	0	0.94260949	0.0310666	-0.00518106
			Steps	0.0248445	0.00402503
9	91	5	0.76754379	0.139210	0.00307479
10	100	0	0.58812839	0.334366	0.0737138
11	109	0	0.35515174	0.406305	0.159909
12	121	5	0.26724744	0.495740	0.234370
			Steps	0.00995933	0.0100408
13	130	0	0.16906521	0.628462	0.377350
14	139	0	0.089376554	0.707711	0.494575
15	148	0	0.051455472	0.830482	0.674627
16	157	0	0.017500430	0.868503	0.752852
			Steps	0.00757936	0.0131937

17	166	0	0.7883231E-02	0.956477	0.907109
18	175	0	0.9507093E-03	0.969209	0.939204
19	184	0	0.1537610E-03	0.988128	0.976039
			Steps	0.00177713	0.00351691
20	193	1	0.1294126E-03	0.988624	0.977377
21	202	0	0.1743590E-05	0.999366	0.998616
22	212	6	0.3698422E-06	0.999392	0.998784
			Steps	0.0699250	0.139938
1	222	0	0.3698422E-06	0.999392	0.998784

***** Results of optimization *****

*** Minimum function value ***

 0.36984E-06

*** Estimates of parameters ***

	estimate	"s.e."
P1	0.999	0.820
P2	1.00	1.64

*** Scaled inverse of second derivatives ***

estimate	ref	scaled inverse of 2nd derivatives	
P1	1	1.000	
P2	2	0.998	1.000
		1	2

The FUNCTION option of the MODEL statement defines the scalar to be F, and the expression Rbrock in the CALCULATION option of FITNONLINEAR sets F to the value of the function.

When you are minimizing a general function in this way, some of the output from FITNONLINEAR is different. Genstat ignores the accumulated and fittedvalues settings, and the deviance and summary settings display only the minimum function value. The correlation setting displays the inverse of the estimated matrix of second derivatives of the function with respect to the parameters, scaled by the diagonal values. Similarly, in place of the standard errors usually displayed by the estimates setting, Genstat prints the square roots of the diagonal values of twice the inverse of the second-derivative matrix. These can give a useful indication of the form of the function near the minimum. As indicated by their title in the output, if the function is a deviance you can interpret these as asymptotic standard errors and correlations (not scaled by an estimate of dispersion). For a general function, the "s.e." can be interpreted as the approximate change in a parameter required to increase the function by 1.0 starting from the minimum.

Genstat ignores the CONSTANT option of the FITNONLINEAR directive for general functions, and you must not set the parameter. Similarly, the WEIGHTS and OFFSET options of the MODEL directive are ignored, and the GROUPS option must not be set. The only parameters of the RKEEP directive that are available are ESTIMATES, SE, INVERSE, EXIT, GRADIENTS, and

GRID. The minimum value of the function is of course available in the scalar specified by the FUNCTION option of the MODEL directive.

You will usually want to inspect the shape of the function near the minimum. So next we form a grid of function values using the NGRIDLINES option of FITNONLINEAR; to save space in the output, we do not display the values with the option setting PRINT=grid, but just extract them with the GRID parameter of RKEEP, and display them with the DSURFACE directive (6.3.2). The picture is in Figure 8.7.4.

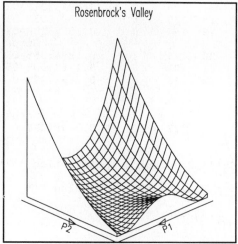

Figure 8.7.4

Example 8.7.4b

```
10   " Draw a contour map of the function with P1 and P2 in (-1.4,1.4)."
11   RCYCLE P1,P2; LOWER=-1.4,-1.4; UPPER=1.4,1.4
12   FITNONLINEAR [PRINT=*; NGRIDLINE=21; CALCULATION=Rbrock]
13   RKEEP GRID=Vgrid
14   MATRIX [ROWS=21; COLUMNS=21] Mgrid; VALUES=Vgrid
15   AXES 3; YTITLE='P2'; XTITLE='P1'; YLOWER=-1.4; YUPPER=1.4;\
16      XLOWER=-1.4; XUPPER=1.4
17   DSURFACE [TITLE='Rosenbrock''s Valley'; WINDOW=3; AZIMUTH=45] Mgrid
```

P.W.L.

9 Design and analysis of experiments

This chapter first describes the Genstat directives for estimating the effects of treatments and doing an analysis of variance with data from a designed experiment. Then, in Section 9.8, it describes the facilities for designing experiments.

In a designed experiment, each treatment is applied to several units, such as plots of land, or animal or human subjects, or samples of material. Usually the treatments are allocated randomly, since the units might not be absolutely identical. This guards against any treatment systematically getting more than its fair share of the best units, which might cause it to appear to be better than the treatments on the less favourable units. It is also one form of justification for the statistical analysis. For a more detailed discussion of why randomization is important, see for example Chapter 5 of Cox (1958).

In the simplest type of investigation, the treatments do not have any particular structure. In a field experiment, for example, they may be several varieties of a crop; in an industrial experiment they could be different types of catalyst. In Genstat you represent treatments like these by a factor. The factor has a level for each treatment; the values of the factor indicate which treatment was applied to each unit.

More complicated are factorial experiments. Here there are several different types of treatment, each represented by a different factor. For example, in an investigation of animal diets, you might wish to vary the amounts both of protein and of carbohydrates; in a fertilizer trial, you might have different levels of both nitrogen and phosphorus. Then the set of treatments is the set of all combinations of the levels of the different factors. Thus if there were a levels of nitrogen and b of phosphorus, there would be $a \times b$ treatments altogether.

The advantage of factorial experiments is that you can look not only at the overall effects of each factor, but also at *interactions* which show how the effects of one factor differ according to other factors (9.1). The overall effects are often called *main effects* (though that does not mean that they have to be the main thing that you are interested in). An interaction would be, for example, nitrogen having a large effect in the absence of phosphorus, but only a small effect in its presence.

You specify which main effects, interactions and other treatment terms are to be included in the model using the TREATMENTSTRUCTURE directive (9.1.1). You can also do more sophisticated modelling of the effects of factors, by partitioning them (and their interactions) into polynomial or other contrasts (9.5): for example, the yield of a crop might increase linearly with the amount of nitrogen.

There can also be structure in the units themselves. In a simple experiment, they are unstructured: that is, they are assumed to come from a single homogeneous population. The treatments can then be allocated to the units at random, without the need to consider any other groupings of the units. This is called a *completely randomized design* (see 9.1). The analysis of experiments where the units do have an underlying structure is described in 9.2. For example, you might expect there to be less variation among animals from the same litter than among different litters. You specify the structure of the units by the BLOCKSTRUCTURE directive; if you omit to do this, Genstat assumes that the units are unstructured.

In an experiment, various measurements will be made to assess how the treatments affect the units. These may be made at the end of the experiment, or while it is still in progress. For example, in a field experiment on potatoes, you might be interested in the yield from each plot, the number of potatoes from each plot, estimates of the percentage areas of potato skin affected by particular diseases, and so on. Analysis of variance allows you to examine only one such measurement at a time. The value measured on each unit (or plot) should be entered into a variate and analysed by the ANOVA directive (9.1.2). After the analysis you can produce further output using the ADISPLAY directive (9.1.3), or you can save some of the quantities that have been calculated during the analysis using the AKEEP directive (9.6.1). You can also use procedures APLOT or DAPLOT to produce plots of residuals.

Sometimes measurements are made before the experiment. For example, the initial blood pressures and other attributes of human subjects might be recorded before the treatments are given. You would want to allow for these baseline readings (or *covariates*) when analysing the effects of the treatments. You specify the variates that are to act as covariates using the COVARIATE directive (9.3.1). By default, the model is assumed to contain no covariates.

The analysis can cope with missing values, either in the variates to be analysed, or in the covariates (9.4). But no factor values should be missing.

In summary, then, the model to be fitted is specified by the BLOCKSTRUCTURE, COVARIATE, and TREATMENTSTRUCTURE directives; the analysis is done by the ANOVA directive; further output can be obtained by the ADISPLAY directive; and information from the analysis can be saved by the AKEEP directive.

There are several other directives and procedures that you may find useful. You can use the GET directive to obtain the current model settings specified by BLOCKSTRUCTURE, COVARIATE, and TREATMENTSTRUCTURE (13.1.2), and you can change them by the SET directive (13.1.1); however, this will mainly be useful if you are writing procedures. You can use the RESTRICT directive (4.4.1) to restrict the analysis to only a subset of the units. You can specify how many decimal places will be used in the output of tables of means, effects, contrasts and residuals by setting the DECIMALS parameter in the declaration of the variate to be analysed (2.1.2). Procedure VHOMOGENEITY can be used to check the homogeneity of variances, and procedure REPMEAS can check the validity of the ordinary analysis of variance if you have repeated measurements; if ordinary anova cannot be used, alternatives are provided by procedures MANOVA (multivariate analysis of variance; 11.2.4), and ANTORDER and ANTTEST (antedependence structure). Other relevant procedures include LVARMODEL (spatial analysis of field trials using the linear variance neighbour model), CENSOR (pre-processing of censored data before analysis by ANOVA), and ABIVARIATE (bivariate analysis of variance).

The designs that can be analysed by Genstat are said to be *balanced* or, more accurately, to have the property of *first-order balance* defined by Wilkinson (1970) and James and Wilkinson (1971). A brief explanation of the property is given in 9.7, where the method of analysis is explained, but you do not need to understand this in order to use Genstat. Virtually all the standard designs can be analysed, including all the generally-balanced designs of Nelder (1965 a,b). Here are some examples.

(a)   All orthogonal designs, whether with a single error term or with several: for example, completely randomized designs, randomized blocks, split plots, Latin and Graeco-Latin squares, split-split plots, and fractional replicates.

(b)   All designs with balanced confounding: for example, balanced incomplete blocks, balanced lattices, and Youden squares.

(c)   Designs with partial balance, provided the pattern of balance can be specified by pseudo-factors (9.7.3).

Amongst the worked examples available on your computer is a data file showing how to analyse all the worked examples in Cochran and Cox (1957); this should cover most of the designs that you are likely to encounter. Genstat itself detects whether or not your design is balanced, by a process known as the *dummy analysis* (9.7.5). So, if you are unsure about whether or not a particular design can be analysed, try it and see what happens. Unbalanced designs with a single error term can be analysed by the regression directives described in Chapter 8 or by procedures AUNBALANCED and AUDISPLAY, while those with several error terms can be analysed by the REML directive (Chapter 10). Also, procedure GLMM provides a method for fitting a generalized linear model to data from a stratified experiment. However, the output from these directives and procedures may be less fully comprehensive than ANOVA.

The facilities in Genstat for designing experiments are described in Section 9.8. The GENERATE directive provides an easy way of generating blocking factors or any other factors whose values occur in a systematic order. You can also use it to form values of treatment factors in an experimental design, using the design-key method, or to define values for the pseudo-factors required to specify partially balanced experimental designs. Alternatively, procedure AKEY combines both these uses of GENERATE, allowing you to generate the block factors in systematic order, and then the treatment factors using a design key; it can also call procedure PDESIGN to display the design. Guidance for the selection of suitable keys will be provided by extensions to the conversational system and by further procedures in the Design module of the procedure library. The procedure library also contains procedure AFALPHA to generate alpha designs, and AFCYCLIC for cyclic designs.

The RANDOMIZE directive can be used to randomize any of these designs.

9.1   Designs with a single error term

Suppose that you have done an experiment to examine v different treatments, and that the value measured on the jth unit out of r receiving treatment i is y_{ij}. For each treatment i, we suppose that there is an underlying mean value of y that we wish to estimate; we shall write this as m_i. This will not be the value observed because there will be measurement error, there may be uncontrolled differences in the way the different units have been dealt with, and the units themselves may not be uniform. So y_{ij} is assumed to follow the linear model

$$y_{ij} = m_i + \varepsilon_{ij}$$

where ε_{ij}, termed the *residual* for the ijth unit, represents the difference between the true value m_i and the value actually observed. The residuals are assumed to be independently distributed: that is, the size of the residual on one unit is assumed to be unaffected by the residuals on other units. They are also assumed to have a zero mean and a constant variance (so the expected value for the ijth unit is m_i). For some of the properties of analysis of variance, it is necessary to assume also that the residuals each have a Normal distribution.

The process by which values for the parameters m_i are estimated from the observed measurements y_{ij} is known as *least squares*. The estimators $\hat{m}_i$ are chosen to minimize the sum of squares of the estimated residuals:

$$RSS = \sum_{i=1}^{v} \sum_{j=1}^{r} (y_{ij} - \hat{m}_i)^2$$

You can find details of this process in any standard statistical textbook. For a simple design like this one, the estimate of each mean, $\hat{m}_i$, is simply the average of the values observed on the units with treatment i. However this may not be so in more complicated experiments, for example where there is non-orthogonality (9.7) or where there are covariates (9.3). In such cases Genstat uses the term *mean* to denote the prediction of the mean value for a treatment, rather than its crude average, and we follow the same convention in this chapter.

Analysis of variance also estimates the uncertainty attached to the estimates of the parameters, allowing you to assess whether the treatments genuinely differ in their effects. In simple cases, this involves assessing whether the variation between the units with different treatments is genuinely greater than that between units with the same treatment. To help investigate this, a more common form of the linear model is

$$y_{ij} = \mu + e_i + \varepsilon_{ij}$$

where μ is known as the *grand mean*, and e_i as the effect of treatment i. So:

$$m_i = \mu + e_i$$

If the treatments do not differ, the effects (e_i, $i = 1 \dots v$) will all be zero. To assess this we would fit first a model containing just the grand mean (and residuals), and then a model with the effects as well. The difference between the residual sums of squares of these two models measures whether the treatments differ: this difference is called the sum of squares due to treatments. Conventionally the different sums of squares are presented in a table known as the analysis-of-variance table.

The example below shows the analysis-of-variance table for a rather more complicated experiment, details of which can be found in Snedecor and Cochran (1980, page 305); further output is shown in 9.1.3 and 9.5. The experiment studies the effect of diet on the weight gains of rats. There were six treatments arising from two treatment factors: the source of protein (beef, pork, or cereal), and its amount (high or low). The 60 rats that provided the experimental units were allocated at random into six groups of ten rats, one group for each treatment combination. The model to be fitted in the analysis contains three terms to explain the effects of the treatments: s_i ($i = 1,2,3$) the main effects of the source of protein (beef, pork, or cereal); a_j ($j = 1,2$) the main effects of the amount of protein (high or low); and sa_{ij} the interaction between source and amount of protein.

$$y_{ijk} = \mu + s_i + a_j + sa_{ij} + \varepsilon_{ijk}$$

The parameters a_j make the same contribution to the model irrespective of the source of the protein received by the rat. So they represent the overall effects of the amount of protein. Similarly, the parameters s_i represent the overall effects of the source of protein. If the interaction effects were all zero, we would have a model in which the difference between high and low amounts of protein was the same whatever the source of the protein. Also, the difference between sources of protein would be identical whether at high or low amounts. So the parameters sa_{ij} indicate whether or not these two factors interact: whether we can determine the best source of protein without regard to its amount; likewise whether we can decide the best amount without considering the source. The estimates of the parameters are included in the output under the heading "Tables of effects" (9.1.3).

Genstat prints the analysis-of-variance table in the conventional form, which you can find in statistical textbooks: there is a line for each treatment term, a line for the residual, and a final "Total" line recording the total sum of squares after fitting the grand mean. The first column, "d.f." standing for *degrees of freedom*, records the number of extra independent parameters included when each term is added into the model; thus with the source of protein, there are three parameters (s_1, s_2, s_3) but, since the grand mean μ has already been fitted, they sum to zero and so the degrees of freedom are two. (A full explanation of this too can be found in statistical textbooks.) The second column "s.s." contains the sums of squares. The column "m.s.", standing for *mean square*, has sums of squares divided by numbers of degrees of freedom. You can assess whether a particular treatment term has had an effect by comparing its mean square with the residual mean square: if there has been an effect, then the mean square for the treatment term will be large compared to the residual. The column denoted "v.r." (for *variance ratio*) helps you make these comparisons: it contains the ratio of each treatment mean square to the residual mean square. If the residuals do indeed have independent Normal distributions with zero mean and equal variance, then each such ratio has an F distribution with t and r degrees of freedom, where t is the number of degrees of freedom of the treatment term and r is the number of degrees of freedom of the residual. The corresponding probabilities can be looked up in statistical tables, or you can ask Genstat to calculate them for you, by setting option FPROBABILITY=yes in the ANOVA or ADISPLAY directives. However you should not interpret these probabilities too rigidly, as the assumptions are rarely more than approximately satisfied; for this reason, Genstat does not print probabilities less than 0.001, but will put "<.001" instead. Also, you should not merely report that a term in an analysis is significant; you should also study its means or its effects to see what their biological (or economic) importance may be, whether their pattern can be explained scientifically, and so on.

Example 9.1

```
   1   " 3x2 factorial experiment.
  -2      (Snedecor and Cochran 1980, page 305)."
   3   UNITS [NVALUES=60]
   4   FACTOR [LABELS=!T(beef,cereal,pork); VALUES=(1...3)20] Source
   5   & [LABELS=!T(high,low); VALUES=3(1,2)10] Amount
   6   READ Gain
```

Identifier	Minimum	Mean	Maximum	Values	Missing
Gain	49.00	87.87	120.00	60	0

```
  17   TREATMENTSTRUCTURE Source*Amount
  18   ANOVA [PRINT=aovtable] Gain
```

```
18...........................................................................
```

```
***** Analysis of variance *****

Variate: Gain

Source of variation        d.f.       s.s.       m.s.      v.r.
Source                        2       266.5      133.3      0.62
Amount                        1      3168.3     3168.3     14.77
Source.Amount                 2      1178.1      589.1      2.75
Residual                     54     11586.0      214.6
Total                        59     16198.9
```

Before you can do the analysis you must set up factors to define the treatment that was applied to each unit. Here there are two factors, for source and for amount of protein. Also you must form a variate containing the data values y_{ijk} that are to be analysed. The ways in which you can do this (as shown in lines 3 to 6) are described in earlier chapters of this manual.

For the analysis of variance, you must first define the model to be fitted. Here we have a single error term ε_{ijk}: the units have no structure. Consequently you need not give a BLOCKSTRUCTURE statement (9.2.1) but can let it take its default value. If you have already defined some other structure (perhaps for an earlier analysis), you should cancel it by giving either a BLOCKSTRUCTURE statement with a null formula, or else one with a single factor indexing the units (9.2.1). Provided you have no covariates (9.3), the only statement that you need give is TREATMENTSTRUCTURE.

9.1.1   The **TREATMENTSTRUCTURE** directive

TREATMENTSTRUCTURE specifies the treatment terms to be fitted by subsequent ANOVA
        statements.

No options

Parameter
    formula   Treatment formula, specifies the treatment model terms
                                    to be fitted by subsequent ANOVAs

---

The single unnamed parameter of the TREATMENTSTRUCTURE directive is a formula known as the *treatment formula*. Formulae (1.5.3) are composed of identifier lists and functions, separated by the operators:

      .   +   *   /   //   −   −*   −/

In the formulae for analysis of variance, the identifier lists can only be of factors. Variates and matrices can appear in the functions (to fit polynomials, for example); these are described in 9.5. Here we describe the first four operators, which are those that are used most often. The pseudo-factorial operator //, which occurs only in treatment formulae, is described in 9.7.3. The final three operators are for deletion. Full definitions of all the operators are in 1.7.3.

Genstat expands a formula into a series of model terms, linked by the operator plus (+). Each model term consists of one or more elements, separated from one another by the operator dot (.); in analysis of variance the elements are either factors or functions. You can always specify a formula in this expanded form: the other operators simply provide a more succinct way of writing long formulae. For the formulae defined by TREATMENTSTRUCTURE and by BLOCKSTRUCTURE (9.2.1), this expansion does not take place until the analysis is being done (by ANOVA). TREATMENTSTRUCTURE and BLOCKSTRUCTURE merely store the formulae in their original form. Consequently there are some syntactic errors that will not be found until the ANOVA statement. When Genstat does the expansion, the FACTORIAL option of ANOVA sets a limit on the number of elements in a model term from the treatment formula: any terms with more elements are deleted.

Each model term in the treatment formula corresponds to a treatment term in the linear model. The expanded version of the formula in line 17 of the example is

```
Source + Amount + Source.Amount
```

(So you could have specified this instead of Source*Amount.) Terms with a single factor represent main effects of the factor: for example Source corresponds to the main effects of the source of protein, s_i. Terms with several factors define higher-order effects: for example Source.Amount corresponds to the interaction effects between source and amount of protein, sa_{ij}. However the meaning of a higher-order term depends on the context: in general, it refers to all those joint effects of the factors in the term that have not been accounted for by preceding terms in the model. So Source.Amount, above, is an interaction because the main effects of source and amount have both been fitted already. But, in the formula

```
A + A.B
```

there are no main effects of B, merely a_i and ab_{ij}, so A.B denotes the fitting of different B effects for each level of A; these are usually called the *B-within-A* effects.

Any redundant terms in a formula are deleted. So, for example,

```
A + B + A
```

becomes

```
A + B
```

Also, as A.B is defined to include all the joint effects of A and B that are not yet accounted for, the formula

```
A.B + B
```

becomes just A.B, which already includes the B main effects. Thus the order in which you specify the terms is important.

The operators * and / are termed the crossing and nesting operators respectively. For example,

```
Source * Amount
```

defines the factors Source and Amount to have a crossed relationship: that is, we wish to examine the effects of each factor individually, and then their interaction. Models containing only crossing are often called factorial models. Another factorial model, but with three factors, is in 9.7.1: the formula is

```
N * K * D
```

which expands to

```
N + K + D + N.K + N.D + K.D + N.K.D
```

including not only two-factor interactions, like N.K, but also the three-factor interaction N.K.D. In general, if L and M are two formulae, the definition (1.7.3) is that

```
L*M = L + M + L.M
```

Nesting (/) occurs most often in block formulae, which are specified by the BLOCKSTRUCTURE directive (9.2). To take an illustration from later in this chapter, for the example analysed in 9.3 the formula is

```
Blocks / Plots
```

indicating that plots are nested within blocks; so the interest is in block effects and the effects of plots within blocks (see 9.2.1). This is exactly what the operator / provides: the expanded form of the formula is

```
Blocks + Blocks.Plots
```

The general definition of the slash operator (1.7.3) is that

```
L/M = L + L.M
```

where L is a model term containing all the factors that occur in L. (The rationale for this is that if M is nested within all the terms in L, it must be nested within all the factors in L.) For example, if you expand the first operator in the formula

```
Blocks/Wplots/Subplots
```

used to specify a split-plot design (9.2.1), you obtain

```
(Blocks + Blocks.Wplots)/Subplots
```

This then expands to

```
Blocks + Blocks.Wplots + Blocks.Wplots.Subplots
```

(which, reassuringly, gives an identical list of terms to those obtained by expanding the second operator before the first operator).

An example of a treatment formula in which there is nesting is the factorial plus added control

```
Fumigant/(Dose*Type) = Fumigant + Fumigant.Dose + Fumigant.Type
                     + Fumigant.Dose.Type
```

in which the factorial combinations of dose and type occur within the 'fumigated' level of the factor Fumigant, as explained in 9.3.

The definition of the operator dot (.) with model formulae L and M is that L.M is the sum of all pairwise combinations of a term in L with a term in M. For example

```
(A + B.C).(D + E) = A.D + A.E + B.C.D + B.C.E
```

After expanding the operators dot (.), star (*), and slash (/), Genstat rearranges the list of model terms so that the numbers of factors in the terms are in increasing order. Where several terms contain the same numbers of factors, the terms are put into lexicographical order according to the order in which the factors first appeared in the formula. For example

```
(A + C.D + B + A.B) * E
```

expands to

```
A + C.D + B + A.B + E + A.E + C.D.E + B.E + A.B.E
```

which is reordered to

```
A + B + E + A.B + A.E + C.D + B.E + A.B.E + C.D.E
```

9.1.2   The ANOVA directive

Once you have defined the model, you can analyse the variates containing the data (the *y-variates*) using ANOVA. All the options and parameters are listed here, although some are relevant only to the more complicated designs and analyses described later in this chapter.

ANOVA   analyses y-variates by analysis of variance according to the model defined by earlier BLOCKSTRUCTURE, COVARIATE, and TREATMENTSTRUCTURE statements.

Options

PRINT = *strings*	Output from the analyses of the y-variates, adjusted for any covariates (aovtable, information, covariates, effects, residuals, contrasts, means, cbeffects, cbmeans, stratumvariances, %cv, missingvalues); default aovt, info, cova, mean, miss
UPRINT = *strings*	Output from the unadjusted analyses of the y-variates (aovtable, information, effects, residuals, contrasts, means, cbeffects, cbmeans, stratumvariances, %cv, missingvalues); default * i.e. no printing
CPRINT = *strings*	Output from the analyses of the covariates, if any (aovtable, information, effects, residuals, contrasts, means, %cv, missingvalues); default * i.e. no printing
FACTORIAL = *scalar*	Limit on number of factors in a treatment term; default 3
CONTRASTS = *scalar*	Limit on the order of a contrast of a treatment term; default 4
DEVIATIONS = *scalar*	Limit on the number of factors in a treatment term for

	the deviations from its fitted contrasts to be retained in the model; default 9
PFACTORIAL = *scalar*	Limit on number of factors in printed tables of means or effects; default 9
PCONTRASTS = *scalar*	Limit on order of printed contrasts; default 9
PDEVIATIONS = *scalar*	Limit on number of factors in a treatment term whose deviations from the fitted contrasts are to be printed; default 9
FPROBABILITY = *string*	Printing of probabilities for variance ratios (yes, no); default no
PSE = *string*	Standard errors to be printed with tables of means, PSE=* requests s.e.'s to be omitted (differences, means); default diff
TWOLEVEL = *string*	Representation of effects in 2^n experiments (responses, Yates, effects); default resp
DESIGN = *pointer*	Stores details of the design for use in subsequent analyses; default *
WEIGHTS = *variate*	Weights for each unit; default * i.e. all units with weight one
ORTHOGONAL = *string*	Whether or not design to be assumed orthogonal (no, yes, compulsory); default no
SEED = *scalar*	Seed for random numbers to generate dummy variate for determining the design; default 12345
MAXCYCLE = *scalar*	Maximum number of iterations for estimating missing values; default 20
TOLERANCES = *variate*	Tolerances for zero in various contexts; default * i.e. appropriate zero values assumed for the computer concerned
NOMESSAGE = *strings*	Which warning messages to suppress (nonorthogonal, residual); default *

Parameters

Y = *variates*	Variates to be analysed
RESIDUALS = *variates*	Variate to save residuals for each y variate
FITTEDVALUES = *variates*	Variate to save fitted values
SAVE = *identifiers*	Save details of each analysis for use in subsequent ADISPLAY or AKEEP statements

Before Genstat does any calculations with the y-variates, it does an initial investigation to acquire all the information that it needs for the analysis. Alternatively, you can supply this from an earlier analysis using the DESIGN option.

During this initial investigation Genstat first generates the model, excluding covariates (9.3), by expanding the block and treatment formulae into a list of model terms (9.1.1). For a design with a single error term, you do not have to define the block formula; its use in the definition of more complicated designs is described in 9.2.1. Genstat also finds out whether the treatment formula contains any functions and, if so, forms the contrasts that they define (9.5).

The treatment terms to be included in the model are controlled by the options FACTORIAL, CONTRASTS, and DEVIATIONS. FACTORIAL sets a limit on the number of factors in a treatment term: terms containing more than that number are deleted. CONTRASTS and DEVIATIONS control the inclusion of contrasts, and of deviations from fitted contrasts (9.5). The maximum number of different factors that you can have in the block and treatment formulae is two less than the number of bits that your computer uses to store integers. On most computers this gives you 30 different factors, which should be sufficient for most sensible purposes.

Genstat then checks whether any of the y-variates is restricted (4.4.1). If several variates are restricted, they must all be restricted to the same set of units. Only these units are included in the analysis of each y-variate.

Next Genstat investigates the design: for example, it checks whether each term can be estimated, whether any are non-orthogonal (9.7), which error term is appropriate for each estimated treatment term if the model contains several, and indeed whether the design has the balance required for ANOVA to analyse it. This process, known as the *dummy analysis*, is described in 9.7.5, but you do not have to understand how it works in order to use ANOVA.

Options WEIGHT, ORTHOGONAL, SEED, and TOLERANCES control various aspects of the analysis.

The WEIGHT option allows you to specify a weight for each unit, to define a weighted analysis of variance. You might want to do this if, for example, different parts of the experiment have different variability; each weight would then be proportional to the reciprocal of the expected variance for the corresponding unit. However unless the weights are fairly systematic, for example to give proportional weighted replication (9.5), the design is unlikely to be balanced.

Genstat has a simplified version of the dummy analysis which you can use to save computing time if all the model terms are orthogonal and if, for every term, all the combinations of its factors were applied to the same number of units (9.7.5). A check is incorporated which will detect non-orthogonality except in particularly complicated designs where terms are aliased. If you set option ORTHOGONAL=yes, Genstat does the simple version unless non-orthogonality is detected, whereupon it gives a warning message and then switches to the full version. The simplified version is done also if ORTHOGONAL=compulsory, but non-orthogonality now causes the analysis to stop altogether, with an error message; this is useful for checking for typing errors in the factor values when you know that the design should otherwise be orthogonal.

Options SEED and TOLERANCES control numerical aspects of the dummy analysis and of the analysis of the y-variates (see 9.7.5).

You can use the DESIGN option to store the details of the model, of the design and of any restrictions of the units, so that Genstat need not recalculate them for future ANOVA statements. The structure in the option is automatically declared as a pointer if you have not declared it already. It points to several other structures which store information about different aspects of the analysis. The only other details that are required for future analyses are the values of the factors in the block and treatment formulae.

If you have not previously declared the design structure, or if it has no values, then the current statement derives and stores the necessary information. If the pointer does already have values, then these are used to do the analysis. In that case, of course, values of the factors in the block and treatment formulae must not have been changed since the design structure was formed. The current settings of options FACTORIAL, CONTRASTS, DEVIATIONS, and WEIGHT are then ignored, as is any change in the restrictions on the y-variates. The DESIGN option is particularly useful with designs where there are many model terms or where there is non-orthogonality, as the dummy analysis may then be time-consuming.

The MAXCYCLE option, which sets a limit on the number of iterations for estimating missing values, is described in 9.4. The other ANOVA options control the printed output, and are described with the ADISPLAY directive (9.1.3).

The first parameter of ANOVA, Y, lists the variates whose values are to be analysed. Genstat examines them all and forms a list of units for which any of the y-variates or any covariate (9.3) has a missing value. These units are treated as missing in all the analyses. (This is necessary to avoid having to re-analyse covariates for each y-variate; analysis of covariance is described in 9.3.) However, if your y-variates have different missing units, you may prefer to analyse them with separate ANOVA statements, while saving details of the model and design with the DESIGN option to improve efficiency (see 9.4).

The RESIDUALS parameter allows you to specify a variate to save the estimated residuals from each analysis. Genstat will declare this variate for you if you have not done so already. In models where there are several error terms, only the final one is included. Others can be obtained using the AKEEP directive (9.6.1).

The fitted values from the analysis are defined to be the data values minus the estimated residuals. These too can be saved, using the FITTEDVALUES parameter. In models where there are several error terms, only the final error term is subtracted. If this is not what you want, you can save the other error terms using AKEEP (9.6.1) and subtract them by CALCULATE (4.1).

The last parameter, SAVE, allows you to save the complete details of the analysis in an *ANOVA save structure*. The ADISPLAY directive lets you use a save structure to produce further output (9.1.3). You can also use it in the AKEEP directive to put quantities calculated from the analysis into data structures which you can then use elsewhere in Genstat (9.6.1). Save structures are special compound structures (2.8), and Genstat declares them automatically. The save structure for the last y-variate analysed is stored automatically, and forms the default for ADISPLAY and AKEEP if you do not provide one explicitly.

Genstat still generates the model and does the dummy analysis even if a y-variate has no values, or if you specify a null entry in the Y list. You then get a *skeleton* analysis-of-variance table, which excludes sums of squares, mean squares and variance ratios; the only other output available is the information summary (9.1.3). You can save a design structure, but no save structure is formed. This is a good way of checking that a design can be analysed, before the experiment is carried out.

9.1.3 The **ADISPLAY** directive

ADISPLAY displays further output from analyses produced by ANOVA.

Options

PRINT = *strings*	Output from the analyses of the y-variates, adjusted for any covariates (aovtable, information, covariates, effects, residuals, contrasts, means, cbeffects, cbmeans, stratumvariances, %cv, missingvalues); default * i.e. no printing
UPRINT = *strings*	Output from the unadjusted analyses of the y-variates (aovtable, information, effects, residuals, contrasts, means, cbeffects, cbmeans, stratumvariances, %cv, missingvalues); default * i.e. no printing
CPRINT = *strings*	Output from the analyses of the covariates, if any (aovtable, information, effects, residuals, contrasts, means, %cv, missingvalues); default * i.e. no printing
CHANNEL = *identifier*	Channel number of file, or identifier of a text to store output; default current output file
PFACTORIAL = *scalar*	Limit on number of factors in printed tables of means or effects; default 9
PCONTRASTS = *scalar*	Limit on order of printed contrasts; default 9
PDEVIATIONS = *scalar*	Limit on number of factors in a treatment term whose deviations from the fitted contrasts are to be printed; default 9
FPROBABILITY = *string*	Printing of probabilities for variance ratios in the aov table (yes, no); default no
PSE = *string*	Standard errors to be printed with tables of means, PSE=* requests s.e.'s to be omitted (differences, means); default diff
TWOLEVEL = *string*	Representation of effects in 2^n experiments (responses, Yates, effects); default resp
NOMESSAGE = *strings*	Which warning messages to suppress (nonorthogonal, residual); default *

Parameters

SAVE = *identifiers*	Save structure (from ANOVA) to provide details of each analysis from which information is to be displayed; if omitted, output is from the most recent ANOVA

The ADISPLAY directive allows you to display further output from one or more analyses of variance, without having to repeat all the calculations. You can store the information from each analysis in a save structure, using ANOVA, and then specify the same structure in the SAVE parameter of ADISPLAY. Several save structures can be listed, corresponding to the analyses of several different variates. They need not all have been produced by the same ANOVA statement nor even be from the same design. Alternatively, if you just want to display output from the last y-variate that was analysed, you need not specify the SAVE parameter in either ANOVA or ADISPLAY: the save structure for the last y-variate analysed is saved automatically, and provides the default for ADISPLAY.

Apart from CHANNEL, all the options of ADISPLAY also occur with ANOVA. CHANNEL can be set to a scalar to divert the output to another output channel. Alternatively, it can specify the identifier of text data structure to store the output (and in fact an undeclared structure will be defined as a text, automatically).

The PRINT option selects which components of output are to be displayed. These are all illustrated in this chapter, as indicated in this list.

aovtable	analysis-of-variance table (9.1, 9.2.1, 9.3, 9.5, and 9.7)
information	information summary, giving details of aliasing and non-orthogonality (9.1.3 and 9.7.1) or of any large residuals (9.2.1 and 9.7.1)
covariates	estimates of covariate regression coefficients (9.3.1)
effects	tables of estimated treatment parameters (9.1.3 and 9.7.1)
residuals	tables of estimated residuals (9.1.3 and 9.2.1)
contrasts	estimated contrasts of treatment effects (9.5)
means	tables of predicted means for treatment terms (9.1.3)
cbeffects	estimated effects of treatment terms combining information from all the strata in which each term is estimated (9.7.1)
cbmeans	predicted means for treatment terms combining information from all the strata in which each term is estimated (9.7.1 and 9.7.3)
stratumvariances	estimated variances of the units in each stratum (9.7.1)
%cv	coefficients of variation and standard errors of individual units (9.1.3 and 9.2)
missingvalues	estimates of missing values (9.4)

The default for PRINT with ADISPLAY is different from that with ANOVA. With ANOVA, the default gives the output that you will require most often from a full analysis: aovtable, information, covariates, means, and missingvalues. You are most likely to use ADISPLAY when you are working interactively, to examine one component of output at a time, and it is not obvious that any one component will then be more popular than any other. So the default for ADISPLAY produces no output (that is, PRINT=*). This also means that you do not need to suppress the output explicitly when you are using UPRINT and CPRINT to examine components of output from analysis of covariance (9.3).

The settings information, covariates, and missingvalues have a slightly different effect with ANOVA than with ADISPLAY. As they are part of the default specified for ANOVA, they will not produce any output unless there is something definite to report. With ADISPLAY you need to request them explicitly, so Genstat will always produce some sort of report. For example, there are no missing values with the variate Gain analysed earlier in this section, there are no covariates, and there is no aliasing or non-orthogonality. The information summary will also contain warnings about any large residuals (see 9.2.1) unless the NOMESSAGES option has been set to residuals to exclude these: for example

        ADISPLAY [PRINT=information; NOMESSAGES=residuals]

The criterion used to decide whether or not to report a residual is the same that used in regression analysis (8.1.2). In this set of data there are none. The other setting, nonorthogonality, of the NOMESSAGES option suppresses the warning produced when there is orthogonality between treatment terms (9.7.4) or covariates (9.3.1).

Example 9.1.3a

```
 19   ADISPLAY [PRINT=information,covariates,missingvalues]

19............................................................................

***** Information summary *****

All terms orthogonal, none aliased.


***** Covariate regressions *****

No covariates


***** Missing values *****

Variate: Gain

No missing values
```

If you had asked for these three pieces of information by ANOVA, you would not have obtained any output, since there is nothing positive to report.

The other default components produced by ANOVA are the analysis-of-variance table, shown earlier in this section, and the tables of means.

Example 9.1.3b

```
  20  ADISPLAY [PRINT=means]

20..................................................................................................

***** Tables of means *****

Variate: Gain

Grand mean  87.9

      Source        beef     cereal         pork
                    89.6       84.9         89.1

      Amount        high        low
                    95.1       80.6

      Source   Amount        high          low
        beef                 100.0         79.2
      cereal                  85.9         83.9
        pork                  99.5         78.7


*** Standard errors of differences of means ***

Table                   Source       Amount       Source
                                                   Amount
rep.                        20           30           10
s.e.d.                    4.63         3.78         6.55
```

A table of means is produced for each term in the treatment model. By using the PFACTORIAL option you can exclude tables for terms containing more than a specified number of factors; Genstat does not allow tables to have more than nine factors, so the default value of nine gives all the available tables.

The means are predicted mean values: estimated expected values for each combination of levels in the table, averaged over the levels of other factors. The table for each term is calculated by taking the table of estimated effects for the term and then adding in the estimated effects of all its margins. The grand mean is a margin, as is every term whose factors are a subset of those in the table. For example, the effects of source of protein have only the grand mean as a margin, and so the table of means for Source is calculated by adding the grand mean to each of the Source effects. Source.Amount has three margins; its table of means is formed by adding the grand mean and the main effects of Source and of Amount to the Source.Amount interaction effects. (You can verify this from the tables of effects printed in the next part of the output, below.)

An assumption of analysis of variance is that the effects of each error term (or residuals) are independently distributed with zero mean and a common variance (see the initial part of 9.1); so they have predicted values of zero. Consequently, even if a term from the block formula (9.2.1) is a margin of a treatment term, its effects will not be included in the table of means. Similarly, if the deviations from fitted contrasts have been ascribed to error (9.5), these effects are also excluded; the table of means is then said to be *smoothed* (9.5).

Usually this process of prediction produces tables of means that are the same as the averages of the observed values: for example, in the common situation where the design is orthogonal and there are no covariates, the only further requirements for this to happen are that the term for the table must have no block terms as margins nor any of its deviations ascribed to error. In an analysis of covariance, the means are all adjusted to correspond to a common value, namely the grand mean of each covariate (9.3.1). Adjusted means are also produced when there is non-orthogonality: they are adjusted for the effects that are non-orthogonal to the term or to its margins (9.7.4).

Below the tables of means, Genstat prints an array of standard errors, usually (as here) one for each table. Unless you request otherwise, each will be an s.e.d. – that is, a standard error for assessing the difference between a pair of means within the table. If you prefer standard errors for the means themselves, you should set the option SE=means. By putting SE=* you can suppress the standard errors altogether. More than one s.e. or s.e.d. will be given when some of the comparisons between the means in a table have different standard errors, as for example in split-plot designs (9.2.1).

The replication of the means in each table is also printed. In an unweighted analysis of variance, like that above, the replication is the number of units that received each combination of the treatments in the table. In a weighted analysis, the weighted replication (wt. rep.) is given: this is the sum of the weights of the units that received each treatment combination. If the replication (or weighted replication) is the same for every combination in the table, it is printed with the standard error; otherwise a table of replications is printed in parallel with the table of means, as illustrated in 9.3.

When the means have different replications, standard errors are presented for three types of comparison: between two means with the minimum replication, between two means with the maximum replication, and between a mean with minimum replication and one with maximum replication. But if, for example, there is only one mean with the minimum replication, the first type of comparison will not arise. If Genstat detects such situations, the appropriate s.e.d. is marked with an X.

In stratified designs (9.2), there may be information on a treatment term in more than one stratum. The setting means uses only the effects from the lowest stratum in which the term is estimated (9.7.1). Alternatively, you can specify cbmeans to obtain means that combine information from all the strata in which the term or its margins are estimated. These will provide more accurate predictions. However, their distributional properties are not well understood, and so it is better to use effects or ordinary means for testing. Combined estimates of means are illustrated in 9.7.1 and 9.7.3, along with the combined estimates of effects (cbeffects) and estimated stratum variances (stratumvariances) from which they are calculated.

Example 9.1.3c

```
  21  ADISPLAY [PRINT=effects]
```

```
21. . . . . . . . . . . . . . . . . . . . . . . . . . . . . . . . . . . . . . . . . . . . . . . . . . . . . . . . . . . . . . .
```

```
***** Tables of effects *****

Variate: Gain

Source effects     e.s.e. 3.28    rep. 20

    Source      beef   cereal      pork
                1.7     -3.0        1.2

Amount response                   -14.5   s.e. 3.78      rep. 30

Source.Amount effects   e.s.e. 4.63     rep. 10

    Source   Amount      high      low
     beef                3.1       -3.1
    cereal               -6.3      6.3
     pork                3.1       -3.1
```

Tables of effects are estimates of treatment parameters in the linear model (9.1). Although effects are less often used than means for summarizing the results of an experiment, they may be useful if you wish to study the model in more detail. The option PFACTORIAL applies to tables of effects in the same way as to tables of means. In this example, there are tables for the Source main effects and the Source.Amount interaction. (The Amount main effects are presented as a *response*, as we explain later.)

Each term is subject to constraints that are generated by the fitting of the terms that come before it in the linear model. The grand mean is fitted first of all. So the sum of the effects, each multiplied by its replication (or weighted replication), is zero within every table. The replication is printed in the header line of the table or, if the replications are unequal, with the table itself. Here the effects within all the tables are equally replicated, and you can check that their sum is zero within each table.

Similarly the table of Source.Amount interaction effects has zero row and column sums because the main effects of Source and Amount have been fitted first.

The header also specifies an e.s.e. or a range of e.s.e.'s for the effects in the table: e.s.e. stands for *effective standard error* – the adjective *effective* denotes that it is appropriate only for comparisons that are unaffected by the constraints within the table. So the e.s.e. for Source is appropriate for obtaining an s.e.d. to assess differences between effects, but not for testing the sum of the effects, nor any individual effect, against zero.

To understand how the e.s.e. arises, we can consider the Source main effects. (If you do not want to know about this piece of theory, skip this paragraph.) These effects are estimated by

$$s_i = 1/20 \sum_{j=1}^{2} \sum_{k=1}^{10} y_{ijk} - 1/60 \sum_{i=1}^{3} \sum_{j=1}^{2} \sum_{k=1}^{10} y_{ijk}$$

and can be shown to have a variance of $\sigma^2(1/20 - 1/60)$ where 20 is the replication of the Source effects, and 60 is the total number of units. The second term in the formula (which is the estimate of the grand mean) is common to all the estimates, and it is because of this that pairs of effects have a non-zero covariance of $-\sigma^2/60$. The variance of the difference between two effects can be calculated by a familiar formula: it is the sum of the variances of the two effects minus twice their covariance, giving an s.e.d. of $\sqrt{(2\sigma^2/20)}$. However an easier way of deriving this s.e.d. is to notice that, when you subtract one estimate from the other, the second term cancels out to leave the difference between two sums of independent random variables, each with variance $\sigma^2/20$. We can thus refer to each estimated effect as having an effective variance of $\sigma^2/20$ and an effective covariance of zero when calculating the variance of a comparison unaffected by the constraint. The general formula for the e.s.e. is:

e.s.e. = $\sqrt{(\sigma^2/((\text{weighted}) \text{ replication} \times \text{efficiency factor} \times \text{covariance efficiency factor}))}$

The efficiency factor is described in 9.7.1; for an orthogonal term its value is one. Likewise, the covariance efficiency factor is one when there are no covariates (9.3). The variance σ^2 is estimated by the residual mean square of the stratum where the effects are estimated. Strata are explained in 9.2. Here there is only one stratum and residual, so σ^2 is estimated by 214.6 and the e.s.e. is $\sqrt{(214.6/20)}$.

When a factor has only two levels, like Amount above, Genstat prints the difference between the two main effects. This difference is called a *response*. For interaction terms whose factors all have only two levels, there are two forms of response. The choice between them is controlled by the TWOLEVEL option. If you leave the default, TWOLEVEL=response, Genstat calculates the response for an interaction between two factors as the difference between the two main-effect responses, and so on; this is the form described in most textbooks. By putting TWOLEVEL=Yates, you can obtain the form as defined by Yates (1937). Alternatively, you can put TWOLEVEL=effects if you prefer not to have responses, but to have the effects themselves, as for factors with more than two levels.

Example 9.1.3d

```
  22  ADISPLAY [PRINT=residuals]

22..............................................................................

***** Tables of residuals *****

Variate: Gain

*Units* residuals    s.e. 13.90    rep. 1

   *units*         1          2          3          4          5          6          7
               -27.0       12.1       -5.5       10.8       23.1      -29.7        2.0

   *units*         8          9         10         11         12         13         14
               -11.9      -20.5       -3.2       11.1        3.3       18.0      -29.9

   *units*        15         16         17         18         19         20         21
                -3.5       10.8       13.1       -5.7        4.0       25.1       -1.5

   *units*        22         23         24         25         26         27         28
               -15.2       -3.9        7.3      -19.0        9.1        2.5        6.8

   *units*        29         30         31         32         33         34         35
                14.1        2.3        7.0        2.1        2.5      -28.2       -9.9
```

units	36	37	38	39	40	41	42
	18.3	0.0	-3.9	8.5	-7.2	-9.9	27.3
units	43	44	45	46	47	48	49
	-13.0	-8.9	-8.5	10.8	-16.9	-8.7	17.0
units	50	51	52	53	54	55	56
	0.1	20.5	15.8	5.1	-17.7	11.0	6.1
units	57	58	59	60			
	5.5	-1.2	-25.9	3.3			

Residuals correspond to the error parameters of the linear model (9.1). Here there is a single error term, and thus a single set of residuals. There is no block model (9.2.1) to define factors to index the units of the design, and so each estimated residual is printed with a unit number, under the heading *units*. The header line shows the replication or weighted replication, and gives a standard error appropriate for comparing any residual with zero. If the replications or weighted replications were unequal, these would be printed in parallel with the residuals, and the range of standard errors would be printed, the lower value being appropriate for residuals with the maximum replication or weighted replication, and the upper value for those with the minimum replication or weighted replication.

Example 9.1.3e

```
 23   ADISPLAY [PRINT=%cv]

23.............................................................................

***** Stratum standard errors and coefficients of variation *****

Variate: Gain

   d.f.          s.e.        cv%
     54         14.65        16.7
```

The setting PRINT=%cv displays the residual number of degrees of freedom, the standard error of a single unit of the design, and the *coefficient of variation* (cv%), which is the standard error of a single unit expressed as a percentage of the grand mean. The coefficient of variation is often used as an index of the variability when comparing several experiments on the yields of the same field crop. However it can be misleading, especially with transformed variables like the logarithm of yield, where the grand mean may even be zero, or with other variables that can take negative values. In designs with several error terms, the same information is presented for each stratum, as shown in 9.2.1. If the units in a stratum have unequal replication or weighted replication, there is no single standard error for a unit; so a missing value is printed instead.

The only component of output that we have not yet mentioned contains the estimates of treatment contrasts, which you can obtain by putting PRINT=contrasts. These are shown in 9.5, together with an explanation of how to control their printing by the options PCONTRASTS and PDEVIATIONS.

With analysis of covariance, you can also print output from the analyses of the covariates and from the analysis of the y-variate ignoring the covariates. This is controlled by options CPRINT and UPRINT respectively, as shown in 9.3.1.

9.1.4 Procedures for examining residuals

Procedure APLOT can produce plots of residuals on an ordinary terminal or on a line-printer. There is a SAVE option to indicate the analysis from which the residuals are to be plotted (which, by default, will use those from the last ANOVA analysis) and a parameter called METHOD to select the types of plot that are required. Similarly, you can produce high-resolution plots of residuals using procedure DAPLOT; if this is done interactively, you can also use the cursor to identify any interesting points. Example 9.1.4 shows various plots of the residuals in Example 9.1.3d.

Example 9.1.4

```
 24   APLOT fitted,normal

             -
         I          *
   r     I                    *      *
   e     I         **                              *2
   s     I          3         3    2                *
   i  6. I         3*         *    *               22
   d     I         **              2               32
   u     I         22         *    2                3
   a     I                    2    *                *
   l     I         **         *                    **
   s     I                    *
     -30. I         **              *               *
         -+---------+---------+---------+---------+---------+---------+---
          76.      80.       84.       88.       92.       96.      100.

                                     fitted values
```

Histogram of residual grouped by !(-28.03,...,21.02)

```
            - -28.03    3 ***
    -28.03 - -21.02    2 **
    -21.02 - -14.01    5 *****
    -14.01 -  -7.01    8 *******
     -7.01 -   0.00    9 ********
      0.00 -   7.01   14 **************
      7.01 -  14.01   10 *********
     14.01 -  21.02    6 ******
     21.02 -           3 ***
```

Scale: 1 asterisk represents 1 unit.

25   DAPLOT fitted,normal,halfnormal,histogram

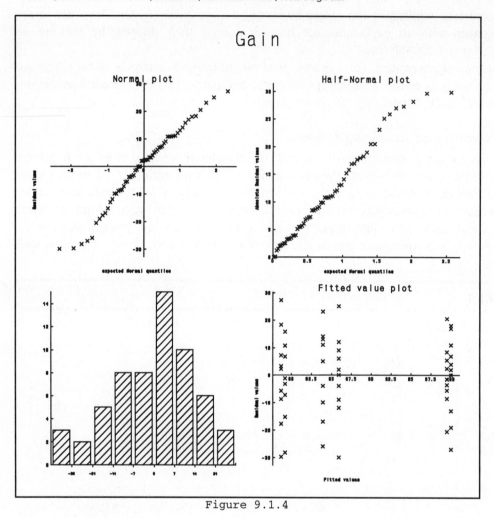

Figure 9.1.4

9.2   Designs with several error terms

The units in the designs covered in 9.1 had no structure: they were assumed to be from a single homogeneous population. The randomization was over the design as a whole, without taking account of any groupings of the units, and there was thus a single error term. Often, however, the population of units is not homogeneous. The rats used to study a set of diets might be grouped according to their litter. An agricultural experiment might involve several different fields, or parts of a field, all with different underlying levels of fertility. An industrial experiment might need to be conducted on several different days, with different batches of material. Or you might wish to impose a structure artificially, by trying to form sets of similar units (and perhaps also subsets) with the aim of decreasing the variability of the experiment.

This structure should then be reflected in the way that you do the randomization and apply the treatments. Some examples are described below. Others can be found in text books on design of experiments: for example, Cochran and Cox (1957), John and Quenouille (1977), and John (1971).

9.2.1 The **BLOCKSTRUCTURE** directive

BLOCKSTRUCTURE defines the blocking structure of the design and hence the strata and the error terms.

No options

Parameter

formula	Block model (defines the strata or error terms for subsequent ANOVA statements)

The BLOCKSTRUCTURE directive specifies the underlying (or *blocking*) structure of the design that is to be analysed. Examples of its use are given below and in 9.3 and 9.7. For unstructured designs with a single error term you can omit this directive, as described in 9.1.

In many designs, the units are nested. The simplest is the randomized block design. Here the units are grouped into sets, known as *blocks*, the aim being that units in the same block should be more similar than those in different blocks. The allocation of the treatments is randomized independently within each block. The design thus has two sources of random variation: differences between blocks as a whole, and differences between the units within each block. An example is in 9.3, where the units are plots of land and the blocks are groupings of nearby plots. The block model is

    Blocks/Plots

indicating that the plots are nested within blocks, and thus that there is no special similarity, for example, between the plot numbered 3 in block 1 and plot 3 of the other blocks. The expanded version of the formula is

    Blocks + Blocks.Plots

giving terms for the differences between blocks as a whole, and the differences between the units within each block, as required.

In the simplest form of the randomized block design, there is a single treatment factor, each of whose levels occurs once in every block. More complicated arrangements are possible, but each treatment combination must still occur exactly the same number of times in every block. This means that any differences found between the blocks cannot be caused by differences between treatments. Thus the treatment terms are all estimated between the plots within the blocks. If the blocks have been chosen successfully, the variation within the blocks should be less than that between blocks, and so the treatment estimates will be less variable than if a completely randomized design had been used.

For the example in 9.3, the treatments have the structure

```
TREATMENTSTRUCTURE Fumigant/(Dose*Type)
```

If you look at the first analysis shown in 9.3, which ignores the covariate discussed later in that section, you can see that the analysis of variance is split into two components called *strata*. The Blocks stratum contains the sums of squares between blocks; this all arises from the variability between the blocks. The Blocks.Plots stratum contains the sum of squares for the plots within the blocks; this is partitioned into the sums of squares due to each of the treatment terms, and a residual against which these can be assessed.

Thus, you can deduce the block model from the structure of the units, which should correspond to the way in which the randomization has been done. Genstat expands the block model to form the list of *block* (or *error*) terms, each of which defines a stratum corresponding to one of the sources of variability in the design. Alternatively, if you prefer to deduce the error terms by some other means, as for example if you follow the philosophy of fixed and random effects, you can specify the block model to be the sum of these terms.

In the analysis, Genstat initially partitions the sums of squares according to the block model alone. This gives the total sum of squares for each of the strata. Then it partitions each stratum sum of squares into sums of squares for those treatment terms estimated in that stratum, and a residual which provides an estimate of variability against which these treatment sums of squares should be compared.

In the randomized block design, the treatments are estimated only in the final (bottom) stratum. You would thus get the same sums of squares if you omitted the BLOCKSTRUCTURE statement and put Blocks at the start of the treatment model. In the example, you would put

```
TREATMENTSTRUCTURE Blocks + Fumigant/(Dose*Type)
```

The effect would also be the same if you specified this treatment model and retained the block model, because any model term that occurs in both the block and treatment models is deleted from the block model. So Blocks would be deleted and there would then be a single stratum Blocks.Plots. You may prefer this specification as it gives an analysis of variance that looks more conventional. However the form in the example better reflects the structure of the design, as it correctly identifies Blocks as an error term. It also allows for the possibility of treatments being estimated between blocks, as in the balanced incomplete-blocks design.

The simplest design in which the treatments are not all estimated in one stratum is the split-plot design. This again is a nested structure. It was originally devised for agricultural experiments where some of the factors can be applied to smaller plots of land than others. However, it also occurs in industrial experiments (for example Cox 1958, page 149), in medical experiments (Armitage 1974), and even in the study of cake mixtures (Cochran and Cox 1957, page 299). A well-known example (Yates 1937, page 74; John 1971, page 99) is shown below. There are two treatment factors: three different varieties of oats (line 8), and four levels of nitrogen (line 9). Because of limitations on the machines for sowing seed, different varieties cannot conveniently be applied to plots as small as those that can be used for the different rates of fertilizer. So the design was set up in two stages. First of all, the blocks were each divided into three plots of the size required for the varieties, and the three varieties were randomly allocated to the plots within each block (exactly as in the randomized blocks design). Then each of these plots, or *whole-plots* as they are usually known, was split into four *sub-plots* (one for each rate of nitrogen), and the allocation of nitrogen was randomized independently within each whole-plot.

To specify the block structure for this design, three factors are required (lines 4 to 6): Blocks to indicate the block (1 to 6) to which each unit belongs, Wplots to indicate the whole-plot (numbered 1 to 3 within each block), and Subplots to identify the sub-plot (numbered 1 to 4 within each whole-plot). You can use the same whole-plot numbers in each block, since the block model (defined below) does not contain any main effect for whole-plots: that is, Genstat will not assume any special similarity between whole-plots with the same numbers. In fact it is best that you do use the same numbering, since otherwise the tables of residuals become very sparse and wasteful of space. In situations like this, it is often convenient to arrange the values of the factors in the block model in a systematic order, for example to reflect positions on the field. This makes patterns in the tables of residuals easier to see. The GENERATE directive (9.8.1) provides a convenient way of specifying their values (line 7).

The design has sub-plots nested within whole-plots, which are themselves nested within the blocks: that is,

```
BLOCKSTRUCTURE Blocks/Wplots/Subplots
```

The block model expands to

```
Blocks + Blocks.Wplots + Blocks.Wplots.Subplots
```

(see 9.1.1 and 1.7.3), giving strata for variation between blocks, between whole-plots within the blocks, and for sub-plots within the whole-plots (within blocks). The treatment model (line 24) specifies terms for the main effects of variety and of nitrogen, and for their interaction (9.1.1).

Just as in the randomized block design, the blocks all contain the same sets of treatments, and so no treatments are estimated in the Blocks stratum. But varieties, which were applied to whole-plots, are estimated in the Blocks.Wplots stratum; in conventional terminology this is called the stratum for whole-plots within blocks. The variance ratio for varieties is calculated by dividing the Variety mean square by the Blocks.Wplots residual mean square. It is easy to see that this is the correct thing to do. When we look to see whether the varieties differ we are really trying to answer the question: "Do the yields from the three sets of whole-plots, on the first of which the variety Victory was grown, on the second Golden rain, and on the third Marvellous, differ by more than the amount that we would expect for any three randomly chosen sets of whole-plots?". Technically, variety is said to be *confounded* with whole plots. The terms for Nitrogen, which was applied to sub-plots, and for the Variety.Nitrogen interaction are both estimated in the stratum for sub-plots within whole-plots (Blocks.Wplots.Subplots).

Variance ratios are also produced for block terms, provided there is an appropriate term lower in the hierarchy of strata with which to compare them. Here Blocks can be compared with Blocks.Wplots, and Blocks.Wplots with Blocks.Wplots.Subplots. Thus, for example, the variance ratio of 5.28 for Blocks indicates that the blocks of land in this experiment are indeed more variable than the plots within each block. However, F probabilities are not produced for variance ratios of block terms. Conversely, in the block formula for replicated Latin squares, discussed later in this section,

        Squares / (Rows * Columns)

which expands to

        Squares + Squares.Rows + Squares.Columns + Squares.Rows.Columns

the term Squares could equally well be compared with either Squares.Rows or Squares.Columns. The ratio of most interest would depend on the exact layout of the trial; for example, if the squares were alongside each other, it might be interesting to see whether the squares were more variable than columns within squares. Genstat has no information about layout, and so leaves you to make these comparisons yourself.

Example 9.2.1a

```
    1   " Split-plot design
   -2      (Yates 1937, page 74; also John 1971, page 99)."
    3   UNITS [NVALUES=72]
    4   FACTOR [LEVELS=6] Blocks
    5   & [LEVELS=3] Wplots
    6   & [LEVELS=4] Subplots
    7   GENERATE Blocks,Wplots,Subplots
    8   FACTOR [LABELS=!T(Victory,'Golden rain',Marvellous)] Variety
    9   & [LABELS=!T('0 cwt','0.2 cwt','0.4 cwt','0.6 cwt')] Nitrogen
   10   VARIATE Yield; EXTRA=' of oats'
   11   READ [SERIAL=yes] Nitrogen,Variety,Yield
```

```
      Identifier    Minimum      Mean    Maximum      Values    Missing
           Yield       53.0      104.0      174.0          72          0
   24   TREATMENTSTRUCTURE Variety*Nitrogen
   25   BLOCKSTRUCTURE Blocks/Wplots/Subplots
   26   ANOVA Yield
```

```
26............................................................................

***** Analysis of variance *****

Variate: Yield of oats
```

Source of variation	d.f.	s.s.	m.s.	v.r.
Blocks stratum	5	15875.3	3175.1	5.28
Blocks.Wplots stratum				
Variety	2	1786.4	893.2	1.49
Residual	10	6013.3	601.3	3.40
Blocks.Wplots.Subplots stratum				
Nitrogen	3	20020.5	6673.5	37.69
Variety.Nitrogen	6	321.8	53.6	0.30
Residual	45	7968.8	177.1	
Total	71	51985.9		

```
* MESSAGE: the following units have large residuals.

Blocks 1          31.4    s.e. 14.8


***** Tables of means *****

Variate: Yield of oats

Grand mean   104.0

   Variety      Victory Golden rain  Marvellous
                  97.6        104.5       109.8

 Nitrogen      0 cwt  0.2 cwt  0.4 cwt  0.6 cwt
                79.4     98.9    114.2    123.4

     Variety Nitrogen    0 cwt  0.2 cwt  0.4 cwt  0.6 cwt
        Victory           71.5     89.7    110.8    118.5
    Golden rain           80.0     98.5    114.7    124.8
    Marvellous            86.7    108.5    117.2    126.8


*** Standard errors of differences of means ***

Table              Variety    Nitrogen    Variety
                                          Nitrogen
rep.                  24          18          6
s.e.d.              7.08        4.44       9.72
Except when comparing means with the same level(s) of
  Variety                                  7.68
```

This shows the default output from ANOVA. Notice that a separate s.e.d. is given for comparisons between means in the variety × nitrogen table when both means are for the same variety. To see why this is necessary, consider how you might calculate the difference between two of the means, using the original data. One way would be to look at each block to find the pairs of sub-plots with these two treatment combinations, and then to calculate the sum of the differences between the values recorded on each pair. If the means are both for the same variety, each pair of sub-plots will be within the same whole-plot; when you take the differences any whole-plot variation then cancels out, to give a smaller s.e.d.

Example 9.2.1a also illustrates the messages that are printed about large residuals. Checking is done for the residuals of every stratum, and the criterion used is the same that used in regression analysis (8.1.2). Here there are no large residuals in either the Blocks.Wplots.Subplots or the Block.Wplots strata, but the residual for block 1 is 31.4 compared to its standard error of 14.8. In this instance, the message can be taken as confirming the success of the choice of blocks: that is, that the yields of the plots in block 1 are consistently higher than those in other blocks. Large residuals in the lower strata might indicate aberrant values, or outliers.

This second section of output shows the tables of residuals and of estimated treatment effects from each stratum, and the coefficients of variation.

Example 9.2.1b

```
  27  ADISPLAY [PRINT=effects,residuals,%cv]

27..........................................................................

***** Tables of effects and residuals *****

Variate: Yield of oats

***** Blocks stratum *****

Blocks residuals   s.e. 14.85  rep. 12

    Blocks       1        2        3        4        5        6
               31.4     -5.8      3.3    -13.1     -8.1     -7.7

***** Blocks.Wplots stratum *****

Variety effects   e.s.e. 5.01   rep. 24

   Variety    Victory Golden rain  Marvellous
               -6.3        0.5         5.8

Blocks.Wplots residuals   s.e. 9.14   rep. 4

    Blocks   Wplots       1        2        3
       1                -11.4     14.0     -2.6
       2                 -9.0     -0.3      9.3
       3                  5.5      8.2    -13.7
       4                -11.5      4.1      7.4
       5                 -9.7     -7.1     16.8
       6                 -0.4     -6.5      6.9

***** Blocks.Wplots.Subplots stratum *****

Nitrogen effects   e.s.e. 3.14   rep. 18

  Nitrogen    0 cwt  0.2 cwt  0.4 cwt  0.6 cwt
             -24.6     -5.1     10.3     19.4

Variety.Nitrogen effects    e.s.e. 5.43   rep. 6

     Variety Nitrogen    0 cwt  0.2 cwt  0.4 cwt  0.6 cwt
        Victory          -1.5     -2.9      3.0      1.5
   Golden rain            0.1     -0.9     -0.1      0.9
    Marvellous            1.5      3.8     -2.9     -2.4

Blocks.Wplots.Subplots residuals    s.e. 10.52  rep. 1

    Blocks   Wplots Subplots      1        2        3        4
       1       1                 9.2    -19.1     11.5     -1.6
               2                -5.9     -5.0     10.1      0.8
               3                 8.3    -13.3     17.6    -12.6
       2       1                 1.6     -1.9     -4.7      5.0
               2                 9.6      8.6      5.5    -23.7
               3                 1.0    -19.5     13.8      4.7
       3       1                 0.7      2.6     15.4    -18.8
               2                 5.7      4.0     -7.6     -2.1
               3                -0.1     -8.1     11.7     -3.5
```

```
  4           1           16.4        -3.3         0.9       -14.0
              2            9.0       -11.7        10.2        -7.5
              3            6.0        -5.2         3.1        -3.9
  5           1          -11.1        -2.2        -7.9        21.2
              2           16.3       -17.4        11.6       -10.5
              3            6.1       -11.5        11.8        -6.4
  6           1           15.3       -10.4         2.6        -7.5
              2           -6.6       -14.4        23.3        -2.2
              3           11.1        -8.7         2.6        -5.0
```

***** Stratum standard errors and coefficients of variation *****

Variate: Yield of oats

Stratum	d.f.	s.e.	cv%
Blocks	5	16.27	15.6
Blocks.Wplots	10	12.26	11.8
Blocks.Wplots.Subplots	45	13.31	12.8

There are some designs where the units have a crossed instead of a nested structure. A simple example is the Latin square. This was devised for agricultural experiments to cater for situations where there are fertility trends both along and across the field, but it can be used whenever there are two independent ways of grouping the units: for example time of testing and batch of material, or the litter of the rat and its order by weight within the litter. In field experiments, the plots are arranged in a square, with blocking factors called Rows and Columns. These each have the same number of levels as there are treatments. Values of the single treatment factor are arranged so that each level occurs once in each row and once in each column. The block structure has rows crossed with columns: that is,

```
BLOCKSTRUCTURE Rows*Columns ( = Rows + Columns + Rows.Columns )
```

The treatments are estimated only in the Rows.Columns stratum. Removing variation between rows and between columns should make these estimates less variable. We do not include output from a Latin square, but recommend that you try an example from one of the books listed earlier in this section.

More complicated designs can involve both crossing and nesting, for example:

```
BLOCKSTRUCTURE Squares/(Rows*Columns)
(= Squares + Squares.Rows + Squares.Columns + Squares.Rows.Columns)
```

which is used for replicated Latin squares (John 1971, page 114), quasi-Latin squares (Cochran and Cox 1957, pages 317-324; John and Quenouille 1977, pages 146-152), and lattice squares (Cochran and Cox 1957, pages 483-506; John and Quenouille 1977, page 192). Another example is

```
BLOCKSTRUCTURE (Rows*Columns)/Subplots
(= Rows + Columns + Rows.Columns + Rows.Columns.Subplots )
```

which is for a Latin square with the plots split into sub-plots (Kempthorne 1952, page 378).

If the factors in the block formula do not provide a unique index for every unit of the experiment, the terms in the block model will not account for all the variation. Genstat must then define a final stratum to contain the variation between the sets of units whose levels are the same for each block factor. At the end of the block model, Genstat therefore sets up an extra term containing all the block factors, together with an extra "factor", denoted *units*, which numbers the units within each set. So, for the randomized block design, you could put just

        BLOCKSTRUCTURE Blocks

which would then become

        BLOCKSTRUCTURE Blocks + Blocks.*units*

Likewise, for the split-plot design,

        BLOCKSTRUCTURE Blocks/Wplots

would become

        BLOCKSTRUCTURE Blocks/Wplots + Blocks.Wplots.*units*

Consequently, if you define no block structure at all, Genstat assumes

        BLOCKSTRUCTURE *units*

giving a single source of variation representing random differences between the units; this defines a completely randomized design, as in 9.1. However, you may prefer to define a more meaningful labelling of the units, for example

        BLOCKSTRUCTURE Rat

The factor Rat would be very easy to set up. To produce a factor equivalent to *units* in more complicated situations, you can use procedure AFUNITS. For example

        AFUNITS [BLOCKSTRUCTURE=Blocks/Wplots] Splot

to generate a factor Splots to index the units within Blocks and Wplots.

9.3   Analysis of covariance

You can do analysis of covariance for any of the designs that can be analysed by ANOVA (9.1). As well as defining the block and treatment models (9.1.1 and 9.2.1), you must also list the covariates, using the COVARIATE directive (9.3.1). Then you can do the analysis by ANOVA (9.1.2), get further output by ADISPLAY (9.1.3), and save information by AKEEP (9.6.1), all exactly as in an ordinary analysis of variance.

   The example used in this section illustrates the treatment structure of a factorial arrangement of several types of treatment, as well as a control. This structure of *factorial plus added control* can be useful when you wish to examine several ways of modifying a preparation, and also wish to see what would happen if you applied nothing at all. This experiment was done at Rothamsted in 1935 to study soil fumigants for decreasing the numbers of nematodes (or eelworms as they were then known). Further details are given in Cochran and Cox (1957, pages 45-46), although there the data are analysed untransformed. There were four types of fumigant, each of which was applied in either a single or a double dose. A randomized block design was used, with four blocks of twelve plots. In each block, four plots were untreated (to act as controls), and there was one plot for each dose of each type of fumigant. This first section of output analyses the logarithm of the numbers of nematode cysts counted in a sample of 400 grammes of soil, taken at the end of the experiment.

Example 9.3

```
   1   "Example of a factorial + added control and analysis of covariance
  -2      (Cochran and Cox 1957, page 46).
  -3      A log transformation has been used, and unit 43 has a missing value
  -4      in the y-variate."
   5   UNITS [NVALUES=48]
   6   FACTOR [LEVELS=4] Blocks
   7   & [LEVELS=12] Plots
   8   FACTOR [LEVELS=5; LABELS=!T(None,CN,CS,CM,CK)] Type
   9   & [LEVELS=3; LABELS=!T(None,Single,Double)] Dose
  10   & [LEVELS=2; LABELS=!T('Not fumigated',Fumigated)] Fumigant
  11   GENERATE Blocks,Plots
  12   READ Dose,Type,Initnem,Finalnem
```

Identifier	Minimum	Mean	Maximum	Values	Missing
Initnem	9.0	128.5	283.0	48	0
Finalnem	80.0	311.7	708.0	48	1

```
  25   CALCULATE Fumigant = NEWLEVELS(Dose; !(1,2,2))
  26   & Initnem,Finalnem = LOG(Initnem,Finalnem)
  27   BLOCKSTRUCTURE Blocks/Plots
  28   TREATMENTSTRUCTURE Fumigant/(Dose*Type)
  29   ANOVA Finalnem
```

29..

***** Analysis of variance *****

Variate: Finalnem

Source of variation	d.f.(m.v.)	s.s.	m.s.	v.r.
Blocks stratum	3	4.0295	1.3432	7.24
Blocks.Plots stratum				
Fumigant	1	0.6918	0.6918	3.73
Fumigant.Dose	1	0.0650	0.0650	0.35
Fumigant.Type	3	0.6656	0.2219	1.20
Fumigant.Dose.Type	3	0.1212	0.0404	0.22
Residual	35(1)	6.4898	0.1854	
Total	46(1)	11.7582		

***** Tables of means *****

Variate: Finalnem

Grand mean   5.618

Fumigant	Not fumigated	Fumigated
	5.788	5.533
rep.	16	32

```
                Fumigant    Dose     None    Single   Double
Not fumigated            5.788
   Fumigated                                 5.488    5.578


                Fumigant    Type     None     CN       CS       CM       CK
Not fumigated            5.788
   rep.            16
   Fumigated                         5.529    5.370    5.763    5.470
   rep.                              8        8        8        8


                Fumigant    Dose     Type     None     CN       CS       CM       CK
Not fumigated            None             5.788
   rep.                16
   Fumigated   Single                      5.483    5.280    5.818    5.371
   rep.                                    4        4        4        4
               Double                      5.575    5.461    5.707    5.570
   rep.                                    4        4        4        4
```

*** Standard errors of differences of means ***

```
Table           Fumigant   Fumigant   Fumigant   Fumigant
                             Dose       Type       Dose
                                                   Type
rep.            unequal        16      unequal    unequal
s.e.d.                                 0.2153     0.3045    min.rep
                0.1318       0.1522    0.1865     0.2407    max-min
                                       0.1522X    0.1522X   max.rep
```

(No comparisons in categories where s.e.d. marked with an X)
(Not adjusted for missing values)

***** Missing values *****

Variate: Finalnem

```
Unit   estimate
 43     5.071
```

Max. no. iterations 3

The block model for this design (line 27) is discussed in 9.2.1. The treatment model requires three factors (lines 8 to 10): Fumigant indicates whether or not the plot has been fumigated with any type of fumigant at all, Type indicates the type of fumigant (if any), and Dose indicates how much was used. If you examine the table of means classified by Fumigant, Dose, and Type, you can see that Dose and Type have a crossed structure within the 'fumigated' level of Fumigant. This suggests a treatment model

        Fumigant/(Dose*Type)

which expands to

        Fumigant + Fumigant.Dose + Fumigant.Type + Fumigant.Dose.Type

As explained in 9.1.1, a term like Fumigant.Dose represents all the joint effects of these two factors, after eliminating any terms that precede it in the model. The main effect Fumigant removes the difference between no fumigant and any positive dose (either single or double).

So `Fumigant.Dose` represents the difference between a single and a double dose. Similarly, `Fumigant.Type` represents differences between types of fumigant, and `Fumigant.Dose.Type` represents the interaction between dose and type of fumigant. Notice that one of the units has a missing value; this aspect of the analysis is explained in 9.4.

The numbers of nematodes were also sampled at the start of the experiment, before any treatments were applied. This gives extra information about the plots, which we can incorporate into the analysis by using the original numbers as a covariate. We have transformed the initial numbers to logarithms, in the same way as the final numbers; so the model to be fitted assumes that the final numbers are related to some power of the original numbers.

You can use covariates to incorporate any quantitative information about the units into the model. In field experiments there may often be linear trends in fertility. These can be estimated and removed by fitting a covariate of the position of the plot along the direction of the trend. For a quadratic trend, you would also include a covariate containing the squares of the positions. In experiments on animals, you may wish to use measurements such as the original weight. However the assumption is always that the y-variate is linearly related to the covariates.

After you have defined variates to contain the measurements that are to act as covariates and done any transformations that may be required, you list them in the COVARIATE directive.

9.3.1 The **COVARIATE** directive

COVARIATE specifies covariates for use in subsequent ANOVA statements.

No options

Parameter

variates	Covariates

Covariates are incorporated into the model as terms for a linear regression. Genstat fits the covariates, together with the treatments, in each stratum. This should explain some of the variability of the units in the stratum, and so decrease the stratum residual mean square.

Each treatment combination will have been applied to units whose mean value for each covariate differs from that of other treatment combinations; so even in the absence of any treatment effects, the y-values recorded for the different combinations would not be identical. A further effect of the analysis is to adjust the treatment estimates for the covariates, to correct for this. This adjustment causes some loss of efficiency in the treatment estimation. The remaining efficiency is measured by the *covariance efficiency factor*, shown for each treatment term in the "cov. ef." column of the analysis-of-variance table. The values are in the range zero to one. A value of zero indicates that the treatment contrasts are completely correlated with the covariates: after the covariates have been fitted there is no information left about the treatments. A value of one indicates that the covariates and the treatment term are orthogonal. Usually the values will be around 0.8 to 0.9. A low value should be taken as a warning: either the measurements used as covariates have been affected by the treatments, which can occur when the measurements on covariates are taken after instead of before the experiment (see for example Cochran and Cox 1957, page 90); or the random allocation of treatments has been unfortunate in that some treatments are on units with generally low values of the covariates while others are on generally high ones. The covariance efficiency factor is analogous to the efficiency factor printed for non-orthogonal treatment terms (see 9.7.1); details of its derivation can be found in Payne and Tobias (1992).

For a residual line in the analysis of variance, the value in the "cov. ef." column measures how much the covariates have improved the precision of the experiment. This is calculated by dividing the residual mean square in the adjusted analysis by its value in the unadjusted analysis (which excludes the covariates).

The covariance efficiency factor is used by Genstat in the calculation of standard errors for tables of effects, as shown by the formula in 9.1.3. So, if you want to calculate the net effect of the analysis of covariance on the precision of the estimated effects of a treatment term, you should multiply the covariance efficiency factor of the term by the value printed in the residual line of the stratum where the term is estimated. Where a term has more than one degree of freedom, the adjustment given by the covariance efficiency factor is an average over all the comparisons between the effects of the term. However this adjustment should not differ by much from those required for any particular comparison unless the randomization has been especially unfortunate. For Fumigant in the example, the calculation is 0.99×2.35. So the e.s.e. of the Fumigant effects from the adjusted analysis is less than that from the unadjusted analysis by a factor of $\sqrt{2.3}$.

In the example we have printed tables of means, but no tables of effects. However, since the table of means for Fumigant is calculated merely by adding the grand mean to each entry in its table of effects (9.1.3), the same factor also applies to the s.e.d. of the Fumigant means. For a table of means classified by several factors, Genstat combines the covariance efficiency factors of the effects from which the means are calculated (9.1.3) into a harmonic mean, weighted according to the numbers of degrees of freedom of each term: for example $4/(1/0.99 + 3/0.92)$ for Fumigant.Type.

The adjusted analysis-of-variance table has an extra line in the analysis of each stratum, giving the sum of squares due to the covariates. This is the extra sum of squares that is removed by the covariates after eliminating all that can be ascribed to the treatments. It lets you assess whether there is any evidence that the covariates are required in the model. If there are several covariates Genstat will also print their individual contributions to that sum of squares, giving first the sum of squares that can be explained by the first covariate in the COVARIATE list, then the extra sum of squares that can be accounted for by fitting the second covariate, and so on. The line for each treatment term contains the sum of squares eliminating the covariates. It indicates whether there is evidence of any effects of that term, after taking account of the differences in the values of the covariates on the units to which each treatment was applied.

As explained in 9.7.4, when an analysis of variance contains non-orthogonal components, the total sum of squares is given by adding the sum of squares for component 1 ignoring component 2 to that for component 2 eliminating component 1, and so on. Here, however, the sums of squares are for covariates eliminating the treatment terms, and for each treatment term eliminating the covariates. So you will find that the values in the s.s. column of the analysis-of-variance table do not add up to the total.

Example 9.3.1

```
 30   COVARIATE Initnem
 31   ANOVA [PRINT=aovtable,covariates,means] Finalnem

31.........................................................................

***** Analysis of variance (adjusted for covariate) *****

Variate: Finalnem
Covariate: Initnem
```

Source of variation	d.f.(m.v.)	s.s.	m.s.	v.r.	cov.ef.
Blocks stratum					
Covariate	1	3.35292	3.35292	9.91	
Residual	2	0.67657	0.33828	4.29	3.97
Blocks.Plots stratum					
Fumigant	1	0.95330	0.95330	12.10	0.99
Fumigant.Dose	1	0.00020	0.00020	0.00	0.98
Fumigant.Type	3	1.43634	0.47878	6.07	0.92
Fumigant.Dose.Type	3	0.11913	0.03971	0.50	0.99
Covariate	1	3.81015	3.81015	48.34	
Residual	34(1)	2.67969	0.07881		2.35
Total	46(1)	11.75815			

```
***** Covariate regressions *****

Variate: Finalnem
```

Covariate	coefficient	s.e.
Blocks stratum		
Initnem	0.48	0.153
Blocks.Plots stratum		
Initnem	0.522	0.0751

```
***** Tables of means (adjusted for covariate) *****

Variate: Finalnem
Covariate: Initnem

Grand mean   5.618
```

Fumigant	Not fumigated	Fumigated
	5.818	5.518
rep.	16	32

Fumigant	Dose	None	Single	Double
Not fumigated		5.818		
Fumigated			5.520	5.515

	Fumigant	Type	None	CN	CS	CM	CK
Not fumigated			5.818				
		rep.	16				
	Fumigated			5.783	5.357	5.692	5.239
		rep.		8	8	8	8

	Fumigant	Dose	Type	None	CN	CS	CM	CK
Not fumigated		None	Type	5.818				
			rep.	16				
	Fumigated	Single			5.703	5.401	5.768	5.209
			rep.		4	4	4	4
		Double			5.864	5.313	5.616	5.269
			rep.		4	4	4	4

*** Standard errors of differences of means ***

Table	Fumigant	Fumigant Dose	Fumigant Type	Fumigant Dose Type	
rep.	unequal	16	unequal	unequal	
s.e.d.			0.1449	0.2024	min.rep
	0.0862	0.0999	0.1255	0.1600	max-min
			0.1025X	0.1012X	max.rep

(No comparisons in categories where s.e.d. marked with an X)
(Not adjusted for missing values)

The method that Genstat uses for analysis of covariance essentially reproduces the method that you would use if you were doing the calculations by hand. First of all, it analyses each covariate according to the block and treatment models. You can print information from these analyses using the CPRINT option of either ANOVA or ADISPLAY. As ADISPLAY (9.1.3) does not constrain you to list save structures that were all produced by the same ANOVA, CPRINT will produce information about the covariate analyses from every save structure that you list; duplicate information will thus be produced if several of the save structures are for analyses involving the same covariates. The output from CPRINT, particularly the analysis-of-variance table, gives you another way of assessing the relationship between treatments and covariates: a large variance ratio for a treatment term in the analysis of one of the covariates would indicate either that the treatment had affected the covariate or that the randomization had been unfortunate (as discussed in the description of cov. ef. above).

Genstat then analyses each y-variate in turn. First of all it does the usual analysis ignoring the covariates. You can control output from this unadjusted analysis by the UPRINT option of ANOVA and ADISPLAY. (So the whole of the output given for the example could have been produced by a single ANOVA statement.) Then the covariates are fitted by linear regression and the full, adjusted, analysis is calculated. Output from the adjusted analysis is controlled by the PRINT option of ANOVA and ADISPLAY. This option has an extra setting, not available for UPRINT and CPRINT: PRINT=covariates prints the regression coefficients of the covariates as estimated in each stratum.

9.4   Missing values

Values from some of the units of an experiment may occasionally fail to be recorded. A laboratory animal may become ill or die during the experiment for reasons unconnected with the treatments. A human subject may withdraw from a clinical trial before it is complete. A plot in a field experiment may become flooded and fail to produce any plants. A value may need to be regarded as missing if a mistake has been made in its recording, or in the way in which the unit was managed during the experiment.

To obtain the exact analysis in such circumstances these units should be excluded, but that would lose the properties such as balance for which the experiment was designed. Consequently techniques have been devised by which missing values are entered for these units, and then estimated during the analysis. The estimates can be printed using the missingvalues setting of the PRINT, CPRINT, or UPRINT options of ANOVA or ADISPLAY:

Example 9.4

```
 32  ADISPLAY [PRINT=missingvalues]

32.............................................................................

***** Missing values (adjusted for covariate) *****

Variate: Finalnem
Covariate: Initnem

 Unit  estimate
   43     5.290

Max. no. iterations 3
```

You can have missing values in the y-variates or the covariates, but not in the block or treatment factors: that is, you should at least know where each missing unit belongs according to the factors of the block model, and what treatments it was scheduled to receive. Genstat regards a unit as missing for all the y-variates listed in an ANOVA statement if it is missing for any one of them, or if it is missing for a covariate. This is because the analysis of covariance requires a missing value in either the y-variate or a covariate to be set missing throughout (Wilkinson 1957); forming the complete list over all the y-variates avoids having to re-analyse the covariates for each y-variate. If you have units where some but not all of the y-variates have missing values, you may prefer to analyse each y-variate separately: for example

```
FOR Y=Weight,Age,Height
  ANOVA [DESIGN=Dsave] Y
ENDFOR
```

instead of

```
ANOVA Weight,Age,Height
```

Use of the DESIGN option (9.1.2) avoids Genstat having to redetermine the structure of the design for each analysis.

Genstat uses the method of Healy and Westmacott (1956). This estimates the missing values by an iterative approach in which they are initially set to the grand mean, then the analysis is repeated with the estimate for each missing unit adjusted each time to set its residual to zero. Genstat also employs the modification discussed by Preece (1971) which over-adjusts each residual to accelerate convergence, but this is discontinued if divergence results instead. Missing cells can occur in higher strata, for example if all the sub-plots in a whole-plot are missing. These missing effects are estimated by a similar iteration of the analysis within the stratum. Likewise missing treatment effects are estimated by minimizing the sum of squares of the treatment term concerned. There is a limit on the number of iterations; by default it is 40, but this can be changed by the MAXCYCLE option of ANOVA. Genstat decides that the process has converged when the residual sum of squares from the previous iteration exceeds the current residual sum of squares by less than 10^{-5} times the current residual sum of squares. This value of 10^{-5} can be changed using the third value of the variate in the TOLERANCES option of ANOVA. Genstat prints the maximum number of iterations required in any of the strata of the design, along with the estimates of the missing values. Convergence is usually fairly rapid: for the example above, only three iterations were required.

In the analysis of variance, as shown in the example in 9.3, the numbers of degrees of freedom are decreased to take account of the missing units and effects; the number subtracted is shown in brackets. The analysis of variance is only approximate. The residual sums of squares are correct (to within the tolerance of convergence) but the treatment sums of squares will be larger than their correct value. (As a result, the sums of squares in the analysis-of-variance table will no longer sum to the total.) If there are few missing values, this increase is unlikely to be large. The estimated effects and means are correct but the calculation of the standard errors does not take account of the missing units. So some standard errors will be too small. For further details, see for example Cochran and Cox (1957, pages 80-82).

If the model has only one error term, you can obtain the exact analysis using regression (Chapter 8). Alternatively you could use the method of Bartlett (1937), in which a dummy covariate is specified for each missing value with minus one in the missing unit and zero elsewhere. The missing units in the y-variates should be set to zero; the regression coefficients of the covariates then estimate the missing values.

9.5   Contrasts between treatments

Sometimes there may be comparisons between the levels of a treatment factor that you particularly wish to assess. With the three sources of protein in 9.1, you might wish to see whether the animal sources (beef and pork) were uniformly better than the cereal source, or you might suspect that the type of meat made little difference and so wish to compare beef with pork. With factors whose levels represent the application of different amounts of some substance like a fertilizer or a drug, you may wish to model the relationship between the effect and the amount. For example, with the nitrogen fertilizer in 9.2, you might wish to see if the yield of oats increases linearly with the amount of fertilizer; you might also include a quadratic term to check for curvature in the response.

Each of these comparisons can be described by specifying a coefficient for each level of the factor. The estimated value of the contrast is obtained by taking the sum of the coefficients each multiplied by the appropriate effect. For example the contrasts for the source of protein are defined by coefficients:

```
                            Source: beef      cereal      pork
      Contrast: animal versus cereal    0.5      -1.0        0.5
                beef versus pork        1.0       0.0       -1.0
```

To compare beef with pork you subtract one effect from the other; while for animal versus cereal sources, you subtract the effect of cereal from the mean of the effects of the animal sources.

As illustrated by this example, to represent a comparison between the levels of the factor, the sum of the coefficients must be zero. These two contrasts are also orthogonal: they represent independent comparisons between the effects. This is shown by the fact that the sum of the pairwise products of the coefficients is zero: $0.5 \times 1.0 + (-1.0) \times 0.0 + 0.5 \times (-1.0)$. With polynomial contrasts it is usual to fit orthogonal polynomials, so that the quadratic term represents the effect of adding a quadratic term into a linear polynomial, the cubic represents the effect of adding a cubic term into a quadratic polynomial, and so on (see for example: John 1971, page 50; John and Quenouille 1977, pages 33-36). The coefficients of the orthogonal polynomials to examine the linear and quadratic effects of the Nitrogen factor in 9.2 are

```
              Nitrogen:  0.0    0.2    0.4    0.6
   Contrast:  linear    -0.3   -0.1    0.1    0.3
              quadratic  0.4   -0.4   -0.4    0.4
```

To examine contrasts like these, you put a function of the factor into the treatment formula, instead of the factor itself. The simplest function, POL, is for fitting polynomial contrasts. It has three arguments: the first specifies the factor, the second is a number or a scalar giving the order of polynomial to be fitted (1 for linear, 2 for quadratic, 3 for cubic, and 4 for quartic), and the third is a variate specifying numerical values for each level of the factor. Genstat calculates the orthogonal polynomials for you. In the Nitrogen example, the levels are equally spaced and in ascending order of magnitude, but this need not be so. You can omit the third argument if the levels already declared with the factor are suitable. For Nitrogen, the declaration (line 9 in the output shown in 9.2.1) specified only labels, and so the levels are the defaults 1 to 4. The variate Nitlev is defined to supply the correct values (line 28).

Example 9.5a

```
  28   VARIATE [VALUES=0,0.2,0.4,0.6] Nitlev
  29   TREATMENTSTRUCTURE POL(Nitrogen; 2; Nitlev) * Variety
  30   ANOVA [PRINT=aov] Yield

  30..........................................................................
```

```
***** Analysis of variance *****

Variate: Yield of oats

Source of variation        d.f.        s.s.       m.s.      v.r.

Blocks stratum                5      15875.3     3175.1      5.28

Blocks.Wplots stratum
Variety                       2       1786.4      893.2      1.49
Residual                     10       6013.3      601.3      3.40

Blocks.Wplots.Subplots stratum
Nitrogen                      3      20020.5     6673.5     37.69
  Lin                         1      19536.4    19536.4    110.32
  Quad                        1        480.5      480.5      2.71
  Deviations                  1          3.6        3.6      0.02
Nitrogen.Variety              6        321.8       53.6      0.30
  Lin.Variety                 2        168.3       84.2      0.48
  Quad.Variety                2         11.1        5.5      0.03
  Deviations                  2        142.3       71.2      0.40
Residual                     45       7968.8      177.1

Total                        71      51985.9
```

In the analysis of variance, the sum of squares for Nitrogen is partitioned into the amount that can be explained by a linear relationship of the yields with nitrogen (the line marked Lin), the extra amount that can be explained if the relationship is quadratic (the line Quad), and the amount represented by deviations from a quadratic polynomial. A cubic term would be labelled as Cub, and a quartic as Quart. You are not allowed to fit more than fourth-order polynomials.

The interaction of nitrogen and variety is also partitioned: Lin.Variety lets you assess the effect of fitting three different linear relationships, one for each variety, instead of a single overall linear contrast; Quad.Variety represents three different quadratic contrasts; and Deviations represents deviations from these three quadratic polynomials. You can print the estimated values of the contrasts by putting PRINT=contrasts in either ANOVA or ADISPLAY.

Example 9.5b

```
  31  ADISPLAY [PRINT=contrasts]

31..................................................................................

***** Tables of contrasts *****

Variate: Yield of oats

***** Blocks.Wplots.Subplots stratum *****

*** Nitrogen contrasts ***

Lin      73.7   s.e. 7.01      ss.div. 3.60
```

```
Quad      -65.   s.e.  39.2      ss.div.  0.115

Deviations     e.s.e.  3.14     ss.div.  18.0

  Nitrogen     0 cwt  0.2 cwt  0.4 cwt  0.6 cwt
                 0.1    -0.3      0.3     -0.1

*** Nitrogen.Variety contrasts ***

Lin.Variety    e.s.e.  12.1    ss.div.  1.20

  Variety      Victory Golden rain  Marvellous
                  7.          2.          -9.

Quad.Variety   e.s.e.  67.9    ss.div.  0.0384

  Variety      Victory Golden rain  Marvellous
                 -1.         13.         -11.

Deviations    e.s.e.  5.43    ss.div.  6.00

  Nitrogen  Variety     Victory Golden rain  Marvellous
    0 cwt                  0.7         0.1         -0.8
  0.2 cwt                 -2.2        -0.2          2.4
  0.4 cwt                  2.2         0.3         -2.4
  0.6 cwt                 -0.7        -0.1          0.8
```

The table of estimated contrasts for `Quad.Variety`, for example, gives the differences between the overall contrast of −65 for `Quad` and the contrasts fitted for the three varieties separately. So the estimated contrast for Golden rain is −65 + 12 = −53. The accompanying "`ss.div`" value is analogous to the replication in a table of effects: it is the divisor used in calculating the estimated values of the contrasts. This is useful mainly where there is a range of e.s.e.'s for a table of contrasts: the contrasts with the smallest values of the ss. div. are those with the largest e.s.e., and vice versa. The ss. div. of each estimated contrast is the sum of squares of the values of the orthogonal polynomial (or other contrast) used to calculate it, weighted according to the replication (or weighted replication in a weighted analysis of variance). The formula for the e.s.e. is similar to that for tables of effects (9.1.3):

$$e.s.e. = \sqrt{(\sigma^2 / (\text{ ss. div. } * \text{ efficiency factor } * \text{ covariance efficiency factor }))}$$

The variance σ^2 is estimated from the residual mean square of the stratum (9.2) where the contrasts are estimated. The efficiency factor (9.7.1) has the value one for terms that are orthogonal, like those in this design. The covariance efficiency factor (9.3.1) equals one when there are no covariates.

To define your own comparisons, you can use the function REG. The first two arguments are the same as those for POL: the first specifies the factor; the second is a number or scalar giving the number of contrasts to be fitted, which must be in the range 1 to 7. The third argument is a matrix, with a column for each level of the factor. Each row of the matrix specifies the coefficients of one of the contrasts. Genstat orthogonalizes these; so the sum of squares, and the estimate, for the second contrast represent the improvement from fitting the second contrast after the first has already been fitted, and so on. If you use a text to label the rows of the matrix (2.4.1), Genstat will use it to annotate the output. Otherwise the contrasts are labelled Reg1 to Reg7.

Example 9.5c shows how to extend the analysis of the example in 9.1, to examine the contrasts between the sources of protein.

Example 9.5c

```
26   MATRIX [ROWS=!T('animal vs cereal','beef vs pork'); COLUMNS=3; \
27      VALUES=0.5,-1,0.5,1,0,-1] Contrasts
28   TREATMENTSTRUCTURE REG(Source; 2; Contrasts) * Amount
29   ANOVA [PRINT=aov,contrasts] Gain
```

29..

***** Analysis of variance *****

Variate: Gain

Source of variation	d.f.	s.s.	m.s.	v.r.
Source	2	266.5	133.3	0.62
animal vs cereal	1	264.0	264.0	1.23
beef vs pork	1	2.5	2.5	0.01
Amount	1	3168.3	3168.3	14.77
Source.Amount	2	1178.1	589.1	2.75
animal vs cereal.Amount	1	1178.1	1178.1	5.49
beef vs pork.Amount	1	0.0	0.0	0.00
Residual	54	11586.0	214.6	
Total	59	16198.9		

***** Tables of contrasts *****

Variate: Gain

*** Source contrasts ***

animal vs cereal   3.0   s.e. 2.67   ss.div. 30.0

beef vs pork   0.2   s.e. 2.32   ss.div. 40.0

*** Source.Amount contrasts ***

animal vs cereal.Amount   e.s.e. 3.78   ss.div. 15.0

Amount	high	low
	6.3	-6.3

beef vs pork.Amount   e.s.e. 3.28   ss.div. 20.0

Amount	high	low
	0.0	0.0

The main effect and interaction are again partitioned to examine the contrasts of interest. The interaction term "beef vs pork.Amount", for example, allows you to examine whether there is any evidence that the difference between beef and pork varies according to the amount of protein fed to the rat.

Where a term has two or more factors partitioned into contrasts, Genstat will also fit interactions between the contrasts. For example `Lin.Lin` looks at the linear change in the linear component of each factor with the other. With two `REG` functions, terms like `Reg1.Reg1` or `Reg2.Reg1` will appear whose interpretation will depend on exactly what comparisons you have defined. If the partitioning of a factor has a component for deviations, there will also be terms like `Dev.Lin`, which represents the interaction between the deviations component of the first factor and the linear part of the second factor. You can suppress the fitting of these interactions by using the function `POLND` instead of `POL`, or `REGND` instead of `REG`. For example, putting `POLND(A; 1)` instead of `POL(A; 1)` ensures that no interactions will be fitted between other contrasts and the `Dev` component of `A`.

The `CONTRASTS` option in the `ANOVA` directive (9.1.2) places a limit on the order of contrast to be fitted. For a term involving a single factor, the orders of successive terms run from one upwards, with the deviations term (if any) numbered highest. So for `Nitrogen` in the example above, the orders are `Lin` 1, `Quad` 2, and `Deviations` 3; while for `Source` they are "animal vs cereal" 1, "beef vs pork" 2. In interactions between contrasts, the order is the sum of the orders of the component parts, so `Lin.Lin` has order 2, `Quad.Lin` has order 3, `Reg1.Quad` has order 3, `Reg1.Reg3` has order 4, and so on. Where the component is a factor, it contributes one to the sum, so `Lin.Variety` has order 2. The default value for `CONTRASTS` is 4.

Option `PCONTRASTS` sets a limit on the order of the contrasts that are printed by either `ANOVA` or `ADISPLAY` (9.1.3); its default value is 9.

If your design has few or no degrees of freedom for the residual, you may wish to regard the deviations from some of the fitted contrasts as error components, and assign them to the residual of the stratum where they occur. You can do this by the `DEVIATIONS` option of `ANOVA` (9.1.2); its value sets a limit on the number of factors in the terms whose deviations are to be retained in the model. For example, by putting `DEVIATIONS=1`, the deviations from the contrasts fitted to all terms except main effects will be assigned to error. The option `PDEVIATIONS` in `ANOVA` or `ADISPLAY` (9.1.3) similarly controls the printing of deviations: to put `PDEVIATIONS=0`, for example, would ensure that no deviations are printed. When deviations have been assigned to error, they will not be included in the calculation of tables of means (9.1.3), which will then be labelled "smoothed". However the associated standard errors of the means are not adjusted for the smoothing.

There are limitations on the models and designs for which Genstat can fit contrasts. In a factorial model, each interaction that is partitioned into contrasts must have equal or proportional replication (or proportional weighted replication in a weighted analysis of variance). Otherwise Genstat gives an error. Here is an example of proportional replication for two factors A and B, giving the numbers of replications for each combination of their levels.

B:	1	2	3	Total over B
A:				
1	4	8	12	24
2	2	4	6	12
Total over A:	6	12	18	36

The fraction of the replication in each cell is the product of the fractions in the marginal total cells: for example the cell for level 1 of A and level 3 of B has 12/36 (= 1/3) of the total replication; the product of the marginal totals for these levels is also 1/3, being 24/36 * 18/36.

An exception to this rule occurs in nested models like the factorial with added control which we discussed in 9.3.1. The table below shows what the replication of the factors Fumigant, Dose, and Type would be if, for illustration, there were also a triple level of dose.

Fumigant:	not fumigated					fumigated				
Type:	none	CN	CS	CM	CK	none	CN	CS	CM	CK
Dose:										
none	16	–	–	–	–	–	–	–	–	–
single	–	–	–	–	–	–	4	4	4	4
double	–	–	–	–	–	–	4	4	4	4
triple	–	–	–	–	–	–	4	4	4	4

The treatment model has

```
Fumigant/(Dose*Type)
= Fumigant + Fumigant.Dose + Fumigant.Type + Fumigant.Dose.Type
```

None of the higher-order terms (such as Fumigant.Dose) has either equal or proportional replication. However, within the 'fumigated' level of Fumigant, there is equal replication. So Genstat can fit any contrast of the nested factors (Type and Dose) provided the level 'none' is excluded. For example, you could estimate linear and quadratic contrasts of Dose using only the non-zero doses by:

```
MATRIX [ROWS=4; COLUMNS=2; VALUES= 0, -1,  0,  1 \
                                    0,  1, -2,  1 ] Quadcon
  TREATMENTSTRUCTURE Fumigant / ( REG(Dose; 2; Quadcon) * Type )
```

But the rows of Quadcon must be specified in orthogonal form. Otherwise the automatic orthogonalization, using the replications of Dose, would produce contrasts involving 'none'.

A further limitation is that contrasts cannot be fitted to terms that involve pseudo-factors (9.7.3). In such situations, the specification of the contrasts is ignored by Genstat.

In nested models, no coherent meaning can be given to contrasts between levels of one of the nested factors if the factor within which it is nested is also partitioned into contrasts. So, for example, the specification

```
POL(A; 1) / POL(B; 2)
```

would generate an error.

Finally there is the limitation that no model term that is to be partitioned into contrasts can contain more factors than one third of the number of bits in the integers on your computer. On most computers, integers contain more than 30 bits, so this limit should not be restrictive.

The contrasts described above, that can be fitted directly by ANOVA, are all linear in their coefficients. Procedure NLCONTRASTS in the Genstat Procedure Library extends this to enable nonlinear contrasts to be fitted to the effects of a quantitative factor and its interaction with another factor. Full details can be found in the Procedure Library Manual, or by using procedure LIBHELP (5.3.1).

9.6   Saving information from an analysis of variance

Most of the quantities calculated during an analysis of variance can be saved in data structures within Genstat. This allows you to write analyses where the analysis of variance itself is only a component part. One example is the multivariate analysis of variance (11.2.4); further examples are in the procedure library (5.3.1). Alternatively, you may wish to save components of the output (such as tables of means) for plotting, or for printing in the form required for a publication.

     You can save variates containing residuals for the final error term of the model, using the RESIDUALS parameter of ANOVA (9.1.2). The FITTEDVALUES parameter similarly allows you to save the fitted values. Other components of the output can be saved using AKEEP.

9.6.1   The **AKEEP** directive

AKEEP copies information from an ANOVA analysis into Genstat data structures.

Options

FACTORIAL = *scalar*	Limit on number of factors in a model term; default 3
STRATUM = *formula*	Model term of the lowest stratum to be searched for effects; default * implies the lowest stratum
SUPPRESSHIGHER = *string*	Whether to suppress the searching of higher strata if a term is not found in STRATUM (yes, no); default no
TWOLEVEL = *string*	Representation of effects in 2^n experiments (responses, Yates, effects); default resp
RESIDUALS = *variate*	To save residuals from the final stratum (as in the RESIDUALS parameter of ANOVA)
FITTEDVALUES = *variate*	To save fitted values (data values or missing value estimates, minus the residuals from the final stratum – as in the FITTEDVALUES parameter of ANOVA)
CBRESIDUALS = *variate*	To save the sum of the residuals from all the strata
CBCREGRESSION = *variate*	To save the estimates of the covariate regression coefficients, combining information from all the strata
TREATMENTSTRUCTURE=*formula*	To save the treatment formula used for the analysis
BLOCKSTRUCTURE = *formula*	To save the block formula used for the analysis
WEIGHTS = *variate*	To save the weights used in the analysis
SAVE = *identifier*	Defines the Save structure (from ANOVA) that provides details of the analysis; default * gives that from the most recent ANOVA

Parameters

TERMS = *formula*	Model terms for which information is required
MEANS = *tables*	Table to store means for each term (available for

	treatment terms only)
EFFECTS = *tables* or *scalars*	Table or scalar (for terms with 1 d.f. when TWOLEVEL=responses or Yates) to store effects (for treatment terms only)
PARTIALEFFECTS = *tables*	Table or scalar (for terms with 1 d.f. when TWOLEVEL=responses or Yates) to store partial effects (for treatment terms only)
REPLICATIONS = *tables*	Table to store replications
RESIDUALS = *tables*	Table to store residuals (for block terms only)
DF = *scalars*	Number of degrees of freedom for each term
SS = *scalars*	Sum of squares for each term
EFFICIENCY = *scalars*	Efficiency factor for each term
VARIANCE = *scalars*	Unit variance for the effects of each term
CEFFICIENCY = *scalars*	Covariance efficiency factor for each term
CREGRESSION = *variates*	Estimated regression coefficients for the covariates in the specified stratum
CSSP = *symmetric matrices*	Covariate sums of squares and products in the specified stratum
CONTRASTS = *pointers*	Estimates for the fitted contrasts of each treatment term, stored in a pointer to scalars or tables; units of the pointer are labelled by the contrast name (as used in the aov table)
XCONTRASTS = *pointers*	X-variates used to fit contrasts, as orthogonalized by ANOVA, stored in a pointer to tables; units of the pointer are labelled as for CONTRASTS
SECONTRASTS = *pointers*	Standard errors for estimated contrasts, stored in a pointer to scalars or tables; units of the pointer are labelled as for CONTRASTS
DFCONTRASTS = *pointers*	Degrees of freedom for estimated contrasts, stored in a pointer to scalars; units of the pointer are labelled as for CONTRASTS
CBMEANS = *tables*	Table to store estimates of the means, combining information from all the strata (for treatment terms only)
CBEFFECTS = *tables* or *scalars*	Table or scalar (for terms with 1 d.f. when TWOLEVEL=responses or Yates) to store estimates of the effects, combining information from all the strata (for treatment terms only)
CBVARIANCE = *scalars*	Unit variance for the combined estimates of the effects of each term
CBCEFFICIENCY = *scalars*	Covariance efficiency factor for the combined estimates of each term
STRATUMVARIANCE = *scalars*	Estimates of the stratum variances (block terms only)

AKEEP allows you to copy components of the output from an analysis of variance into standard Genstat data structures. You can save the information from the analysis in a save structure, using the SAVE option of ANOVA (9.1.2) and then specify the same structure in the SAVE option of AKEEP. Alternatively, Genstat automatically stores the save structure from the last y-variate that has been analysed, and this is used as a default by AKEEP if you do not specify a save structure explicitly.

Several options are provided to save information about the analysis as a whole. The RESIDUALS and FITTEDVALUES options allow variates to be specified to store the residuals and fitted values, respectively. The residuals, like those saved by the RESIDUALS parameter of ANOVA, are taken only from the final stratum. As an alternative, the CBRESIDUALS option saves residuals that incorporate the variability from all the strata. With an orthogonal design, these are simply the sum of the residuals from every stratum. For a non-orthogonal design, they are the data values minus the combined estimates of the treatment effects (9.7.1). Likewise, the CBCREGRESSION option allows you to save estimates of covariate regression coefficients that combine information from all the strata. (The estimates from each individual stratum can be saved using the CREGRESSION parameter, as described below.) The TREATMENTSTRUCTURE, BLOCKSTRUCTURE, and WEIGHTS options can save the treatment and block formulae, and the weights variate (if any) that were used to specify the analysis.

The parameters of AKEEP save information about particular model terms in the analysis. With the TERMS parameter you specify a model formula, which Genstat expands to form the series of model terms about which you wish to save information. As in ANOVA (9.1.2), the FACTORIAL option sets a limit on the number of factors in each term. Any term containing more than that limit is deleted. The subsequent parameters allow you to specify identifiers of data structures to store various components of information for each of the terms that you have specified. If there are components that are not required for some of the terms, you should insert a missing identifier (*) at that point of the list. For example

```
AKEEP Source + Amount + Source.Amount; MEANS=*,*,Meangain; \
    SS=Ssource,Samount,Ssbya; VARIANCE=Vsource,*,*
```

sets up a table Meangain containing the source by amount table of means; it forms scalars Ssource, Samount, and Ssbya to hold the sums of squares for Source, Amount, and Source.Amount respectively, and scalar Vsource to store the unit variance for the effects of Source.

The structures to hold the information are defined automatically, so you need not declare them in advance. If you have declared any of the tables already, its classification set will be redefined, if necessary, to match the factors in the table that you wish to store. Thus Meangain here would be redefined to be classified by the factors Source and Amount, if it had previously been declared with some other set of classifying factors. Sizes of variates and symmetric matrices will also be redefined if necessary.

Most of the components are self-explanatory. Tables of means and effects are described in 9.1.2; these are relevant only for treatment terms. Partial effects (which are also available only for treatment terms) differ from the usual effects, presented by Genstat, only when there is non-orthogonality. The usual effects of a treatment term are estimated after eliminating the terms that precede it in the model (9.1.1), whereas the partial effects are those that would be estimated after eliminating the subsequent treatment terms as well (9.7.4). The TWOLEVEL option controls what it stored for terms whose factors all have only two levels. The settings response (the default) or Yates generate a scalar response, as described in 9.1.3; whereas TWOLEVELS=effects produces a table of effects.

Replication tables are described in 9.1.3 and appear in the example in 9.3. The replications will be arranged in a table, even if all the values are identical. Tables of residuals, available for block terms, are illustrated in 9.1.3 and 9.2.1.

Four components can be saved in scalars: sums of squares (9.1), numbers of degrees of freedom (9.1), efficiency factors (9.7.1), and unit variances. The unit variance of a treatment term is the residual mean square of the stratum where the term is estimated, divided by its efficiency factor and covariance efficiency factor. Thus you can calculate the estimated variance of any of the effects of the term by dividing its unit variance by the replication of the effect (9.1.3).

The next two parameters allow you to save information about the covariates (9.3). To save the regression coefficients estimated in a particular stratum, you should specify the model term of the stratum with the TERMS parameter and a variate with the CREGRESSION parameter. Genstat defines the variate to have a length equal to the number of covariates, and stores the estimated regression coefficients of the covariates in the order in which they were listed in the COVARIATE statement (9.3.1). For the example in 9.3.1, you could put

```
AKEEP Blocks.Plots; CREGRESSION=B
```

to save the regression coefficient estimated for the covariate in the Blocks.Plots stratum; B will be declared implicitly as a variate of length one, as there was only one covariate. The CSSP parameter allows you to obtain sums of squares and products between the covariates for the specified model term. These are arranged in a symmetric matrix. The value in row i on the diagonal is the sum of squares for the term in the analysis of variance that has as its y-variate the ith covariate listed in the COVARIATE statement. The value in row i and column j is the cross-product between the effects estimated for the term in the analysis of variance of covariate i and those estimated for the same term in the analysis of covariate j.

There are four parameters for saving information about contrasts (9.5). For each treatment term there will generally be several contrasts, so the information is stored in pointers with one element for each contrast. Example 9.6 continues Example 9.5b and shows how to save the estimates, the x-variates, the standard errors, and the degrees of freedom for the contrasts of Nitrogen and of Variety.Nitrogen. The structure Ncontr, for example, is defined as a pointer with three elements, labelled 'Lin', 'Quad', and 'Deviations': Ncontr['Lin'] (that is Ncontr[1]) is a scalar containing the estimated linear contrast of Nitrogen; Ncontr['Quad'] similarly contains the estimated quadratic contrast; while Ncontr['Deviations'] is a one-way table, classified by Nitrogen, containing the deviations from the fitted quadratic polynomial. Lines 34-41 of the program print the information for each contrast, to show the structure of each identifier and what it stores.

Example 9.6

```
32   AKEEP Nitrogen+Nitrogen.Variety; XCONTRASTS=Nxvar,NVxvar; \
33     CONTRASTS=Ncontr,NVcontr; SECONTRASTS=Nse,NVse; DFCONTRASTS=Ndf,NVdf
34   PRINT Ncontr[1],Nse[1],Ndf[1]; FIELD=16
```

Ncontr['Lin']	Nse['Lin']	Ndf['Lin']
73.67	7.014	1.000

```
35   PRINT Ncontr[2],Nse[2],Ndf[2]; FIELD=16
```

Ncontr['Quad']	Nse['Quad']	Ndf['Quad']
-64.58	39.21	1.000

```
36   PRINT Ncontr[3],Nse[3]; FIELD=21   & Ndf[3]
```

Nitrogen	Ncontr['Deviations']	Nse['Deviations']
0 cwt	0.1000	3.137
0.2 cwt	-0.3000	3.137
0.4 cwt	0.3000	3.137
0.6 cwt	-0.1000	3.137

```
Ndf['Deviations']
    1.000
```

```
37   PRINT Nxvar[]; FIELD=20
```

Nitrogen	Nxvar['Lin']	Nxvar['Quad']	Nxvar['Deviations']
0 cwt	-0.3000	0.04000	1.000
0.2 cwt	-0.1000	-0.04000	1.000
0.4 cwt	0.1000	-0.04000	1.000
0.6 cwt	0.3000	0.04000	1.000

```
38   PRINT NVcontr[1],NVse[1]; FIELD=24   & NVdf[1]
```

Variety	NVcontr['Lin.Variety']	NVse['Lin.Variety']
Victory	7.417	12.15
Golden rain	1.667	12.15
Marvellous	-9.083	12.15

```
NVdf['Lin.Variety']
    2.000
```

```
39   PRINT NVcontr[2],NVse[2]; FIELD=24   & NVdf[2]
```

Variety	NVcontr['Quad.Variety']	NVse['Quad.Variety']
Victory	-1.042	67.91
Golden rain	12.500	67.91
Marvellous	-11.458	67.91

```
NVdf['Quad.Variety']
    2.000
```

```
40   PRINT [SERIAL=yes] NVcontr[3],NVse[3],NVdf[3]

             NVcontr['Deviations']
     Variety     Victory Golden rain  Marvellous
     Nitrogen
        0 cwt        0.725      0.083      -0.808
      0.2 cwt       -2.175     -0.250       2.425
      0.4 cwt        2.175      0.250      -2.425
      0.6 cwt       -0.725     -0.083       0.808


             NVse['Deviations']
     Variety     Victory Golden rain  Marvellous
     Nitrogen
        0 cwt        5.433      5.433       5.433
      0.2 cwt        5.433      5.433       5.433
      0.4 cwt        5.433      5.433       5.433
      0.6 cwt        5.433      5.433       5.433

 NVdf['Deviations']
     2.000

41   PRINT NVxvar[1,2]; FIELD=24  & NVxvar[3]

             NVxvar['Lin.Variety']   NVxvar['Quad.Variety']
     Nitrogen
        0 cwt              -0.3000                  0.04000
      0.2 cwt              -0.1000                 -0.04000
      0.4 cwt               0.1000                 -0.04000
      0.6 cwt               0.3000                  0.04000


             NVxvar['Deviations']
     Variety     Victory Golden rain  Marvellous
     Nitrogen
        0 cwt        1.000      1.000       1.000
      0.2 cwt        1.000      1.000       1.000
      0.4 cwt        1.000      1.000       1.000
      0.6 cwt        1.000      1.000       1.000
```

The final five parameters, CBMEANS, CBEFFECTS, CBVARIANCE, CBCEFFICIENCY, and STRATUMVARIANCES save details of estimates that combine information from all the strata of the design. These are explained in 9.7.1.

In designs where there is partial confounding, and treatment terms are estimated in more than one stratum (9.7.1), options STRATUM and SUPPRESSHIGHER allow you to specify the strata from which the information is to be taken. This is relevant to tables of effects and partial effects, sums of squares, efficiency factors, unit variances, sums of squares and products between covariates, and information about contrasts. By default, Genstat searches all the strata, and takes the information from the lowest of the strata where the term is estimated. If you set the STRATUM option, only strata down to the specified stratum are searched. By setting SUPPRESSHIGHER=yes, you can restrict the search to only that stratum. For Example 9.7.1a,

    AKEEP [STRATUM=Blocks] K.D; EFFECTS=EffKD; EFFICIENCY=EfacKD

would take the effects estimated for K.D in the Blocks stratum, and put them into the table EffKD, and it would put their efficiency factor into the scalar EfacKD.

You cannot save tables of means if you have excluded any stratum from the search. Likewise, tables of residuals and residual sums of squares cannot be saved for any of the excluded strata. If a term is not estimated in any of the strata that are searched, the corresponding data structures are filled with missing values.

As explained in Section 9.2.1, Genstat will set up an extra "factor" denoted *Units* if the block formula does not specify the final stratum explicitly. AKEEP allows you to refer to this "factor", if necessary, by putting the string '*Units*' (or '*units*' or '*UNITS*') in the TERMS formula. Thus, to save the residual sum of squares in Example 9.1 you could put

```
AKEEP '*Units*'; SS=RatRSS
```

9.7   Non-orthogonality and balance

So far, all the examples in this chapter have all been orthogonal. Each treatment term has been estimated in only one stratum. Any confounding between block and treatment terms has been complete: for example, in the split-plot design in 9.2.1, differences between varieties were completely confounded with whole-plots, and so were estimated only in that stratum.

The ANOVA directive can also analyse designs where there is partial confounding or where there is non-orthogonality, provided there is still the necessary property of balance. These concepts are discussed in this section.

9.7.1   Efficiency factors

The example below is of a design where there is partial confounding. Full details are given by Yates (1937, page 21) and by John (1971, page 135). This is an experiment to study the effects of three factors N, K, and D on the yields of King Edward potatoes. The factor levels were as follows.

     N: sulphate of ammonia at rates of 0 and 0.45 cwt per acre

     K: sulphate of potash at rates of 0 and 1.12 cwt per acre

     D: dung at rates of 0 and 8 tons per acre

The treatment formula (line 22) is

```
N * K * D  =  N + K + D + N.K + N.D + K.D + N.K.D
```

There were eight treatment combinations, but the blocks each had only four plots. Consequently some of the treatment terms needed to be confounded between blocks. This was done by confounding N.K.D between blocks 1 and 2, N.K between blocks 3 and 4, N.D between blocks 5 and 6, and K.D between blocks 7 and 8. There was thus only partial confounding: the interaction terms could be estimated within some of the blocks but not others. To illustrate how this was done, we can consider N.K: this represents the difference in the effect of N according to the level of K (and vice versa). Representing the treatment combinations as triplets of letters, giving respectively the level of N (– or n), K (– or k) and D (– or d), this can be written as

```
{('n--' + 'n-d')-('---' + '--d')}-{('nk-' + 'nkd')-('-k-' + '-kd')}
= ('n--' + 'n-d' + '-k-' + '-kd') - ('---' + '--d' + 'nk-' + 'nkd')
```

The combinations in the first pair of brackets all occur in block 3, while those in the second pair all occur in block 4. Thus within blocks 3 and 4 there is no information on N.K; but information is available within the other 6 blocks. Thus N.K is estimated with efficiency 6/8 (= 0.75) in the Blocks.Plots stratum. The difference between the mean of the yields of the plots in block 3 and those in block 4 also provides an estimate of N.K; this represents the remaining 1/4 of the efficiency available for estimating N.K.

If a term is orthogonal, its efficiency factor equals one: the term is estimated with full efficiency in the stratum concerned. The efficiency factors of non-orthogonal terms are listed in the Information Summary obtained by setting option PRINT=information in either ANOVA or ADISPLAY (9.1.3). Terms that are aliased with earlier terms in the model (and so cannot be estimated) are also listed: these have zero efficiency factors. You can obtain details of the model terms with which they are aliased, using the ALIAS procedure.

The efficiency factors are not always so easy to derive and interpret as here: the original definition by Yates (1936) was for the balanced incomplete-block design. But they always represent the proportion of the information available to estimate a term.

Example 9.7.1a

```
 1    " A partially confounded factorial experiment
-2      (Yates 1937, page 21; also John 1971, page 135)."
 3    UNITS [NVALUES=32]
 4    FACTOR [LEVELS=8] Blocks
 5    &  [LEVELS=4] Plots
 6    &  [LEVELS=2; LABELS=!T(_,n)] N
 7    &  [LABELS=!T(_,k)] K
 8    &  [LABELS=!T(_,d)] D
 9    GENERATE Blocks,Plots
10    READ [PRINT=data,errors] N,K,D; FREPRESENTATION=labels

11    _ _ _   n k _   n _ d   _ k d     n _ _   _ k _   _ _ d   n k d
12    n _ _   _ k _   n _ d   _ k d     _ _ _   _ _ d   n k _   n k d
13    n _ _   _ _ d   n k _   _ k d     _ _ _   _ k _   n _ d   n k d
14    _ k _   _ _ d   n k _   n _ d     _ _ _   n _ _   _ k d   n k d  :
15    VARIATE Yield
16    READ Yield
```

Identifier	Minimum	Mean	Maximum	Values	Missing
Yield	87.0	291.6	471.0	32	0

```
21    BLOCKSTRUCTURE Blocks/Plots
22    TREATMENTSTRUCTURE N * K * D
23    ANOVA Yield
```

23...

***** Analysis of variance *****

Variate: Yield

Source of variation	d.f.	s.s.	m.s.	v.r.
Blocks stratum				
N.K	1	780.1	780.1	3.02
N.D	1	276.1	276.1	1.07
K.D	1	2556.1	2556.1	9.91
N.K.D	1	112.5	112.5	0.44

Residual	3	774.1	258.0	0.81

Blocks.Plots stratum

N	1	3465.3	3465.3	10.86
K	1	161170.0	161170.0	505.21
D	1	278817.8	278817.8	873.99
N.K	1	28.2	28.2	0.09
N.D	1	1802.7	1802.7	5.65
K.D	1	11528.2	11528.2	36.14
N.K.D	1	45.4	45.4	0.14
Residual	17	5423.3	319.0	
Total	31	466779.7		

***** Information summary *****

Model term	e.f.	non-orthogonal terms
Blocks stratum		
N.K	0.250	
N.D	0.250	
K.D	0.250	
N.K.D	0.250	
Blocks.Plots stratum		
N.K	0.750	Blocks
N.D	0.750	Blocks
K.D	0.750	Blocks
N.K.D	0.750	Blocks

* MESSAGE: the following units have large residuals.

Blocks 6   Plots 4   28.2   s.e. 13.0

***** Tables of means *****

Variate: Yield

Grand mean   291.6

N	$\overline{n}$	n
	281.2	302.0

K	$\overline{k}$	k
	220.6	362.6

D	$\overline{d}$	d
	198.3	384.9

N	K	$\overline{k}$	k
		211.3	351.1
$\overline{n}$		229.9	374.1

N	D	$\overline{d}$	d
		196.5	365.9
$\overline{n}$		200.0	404.0

```
   K        D              _           d
   _                     105.4       335.9
   k                     291.1       434.0


            K            _                      k
   N        D            _           d          _           d
   _                   106.1       316.5      286.9       415.2
   n                   104.6       355.2      295.3       452.8
```

*** Standard errors of differences of means ***

Table	N	K	D	N K	N D	K D	N K D
rep.	16	16	16	8	8	8	4
s.e.d.	6.31	6.31	6.31	8.93	8.93	8.93	13.15
Except when comparing means with the same level(s) of							
N				9.65	9.65		13.64
K				9.65		9.65	13.64
D					9.65	9.65	13.64
N.K							14.12
N.D							14.12
K.D							14.12

(Notice in the output that the underline symbol has been used instead of minus for the zero level, to avoid having to put quotes around the labels when they are read in lines 11 to 14.)

As we explained in 9.1.3, the means produced by setting PRINT=means in ANOVA or ADISPLAY take the effects of each term only from the lowest stratum where it is estimated. Thus it would estimate N.K for example only from the Blocks.Plots stratum. The different efficiency factors for the component terms of the two-way and three-way tables of means in the example lead to different standard errors for some comparisons. For example, the s.e.d. for the N.K.D table is 13.15 when comparing means with different levels of all three factors, it is 13.64 if the level of one of the factors is identical for both means, and it is 14.12 if two of the factors are at identical levels.

The effects from the lowest stratum are usually those that are estimated most precisely; the lower strata generally have smaller mean squares and, in most designs, terms will have higher efficiency factors in the lower strata. Moreover, under the usual assumptions of Normality of residuals, differences between means can be tested by the usual t-statistics. Nevertheless, for prediction you will often want to present means and effects that combine the information about each term from all the strata where it is estimated. Provided the design possesses the condition of *first-order balance* that is required for it to be analysed by Genstat (see 9.7.2), and provided there is no non-orthogonality between treatment terms, you can use the PRINT settings cbeffects and cbmeans to print combined estimates of the effects and the means respectively. (The design is then a *generally-balanced design*; see Payne and Tobias 1992).

Example 9.7.1b

```
  24  ADISPLAY [PRINT=effects,cbeffects,cbmeans]
```

24 . . . .

```
***** Tables of effects *****

Variate: Yield

***** Blocks stratum *****

N.K response                  39.5    s.e. 22.72    rep. 8

N.D response                 -23.5    s.e. 22.72    rep. 8

K.D response                 -71.5    s.e. 22.72    rep. 8

N.K.D response               -30.0    s.e. 45.43    rep. 4

***** Blocks.Plots stratum *****

N response                    20.8    s.e. 6.31     rep. 16

K response                   141.9    s.e. 6.31     rep. 16

D response                   186.7    s.e. 6.31     rep. 16

N.K response                   4.3    s.e. 14.58    rep. 8

N.D response                  34.7    s.e. 14.58    rep. 8

K.D response                 -87.7    s.e. 14.58    rep. 8

N.K.D response               -11.0    s.e. 29.17    rep. 4


***** Tables of combined effects *****

Variate: Yield

N response                    20.8    s.e. 6.36     rep. 16

K response                   141.9    s.e. 6.36     rep. 16

D response                   186.7    s.e. 6.36     rep. 16

N.K response                  12.3    s.e. 12.94    rep. 8

N.D response                  21.6    s.e. 12.94    rep. 8

K.D response                 -84.0    s.e. 12.94    rep. 8

N.K.D response               -15.3    s.e. 25.88    rep. 4


***** Tables of combined means *****

Variate: Yield

        N          _           n
               281.2        302.0

        K          _           k
               220.6        362.6

        D          _           d
               198.3        384.9
```

```
    N         K                      k
    _             213.3     349.1
    n             228.0     376.0


    N         D                      d
    _             193.2     369.1
    n             203.3     400.7


    K         D                      d
    _             106.3     335.0
    k             290.2     434.9


              K                  _             k
    N         D         _             d         _             d
    _             106.2     320.3     280.2     417.9
    n             106.3     349.6     300.2     451.9
```

*** Standard errors of differences of combined means ***

Table	N	K	D	N K	N D	K D	N K D
rep.	16	16	16	8	8	8	4
s.e.d.	6.36	6.36	6.36	9.00	9.00	9.00	12.78

Except when comparing means with the same level(s) of

	N	K	D	N K	N D	K D	N K D
N				9.08	9.08		12.83
K				9.08		9.08	12.83
D					9.08	9.08	12.83
N.K							12.89
N.D							12.89
K.D							12.89

The combined estimates of the effects of any treatment term take the form of a weighted average of the estimates from each of the strata, where the weight for any particular stratum is given by the efficiency factor of the term in that stratum, divided by the variance of the units of the stratum. One common method of estimating the stratum variances simply uses the residual mean squares. However, this method does not make use of all the available information - the differences between the various estimates of each treatment effect also contain information about variability. Moreover, there may sometimes be strata with no residual degrees of freedom, as in the square lattice shown in 9.7.3. Thus, a rather more powerful algorithm is used (Payne and Tobias 1992). This is equivalent to the use of residual maximum likelihood (REML) but, for the generally-balanced designs on which it operates, is very much more efficient particularly in its use of workspace (Payne and Welham 1990). The estimated stratum variances can be printed by setting PRINT=stratumvariance.

Example 9.7.1c

```
 25  ADISPLAY [PRINT=stratumvariances]
```

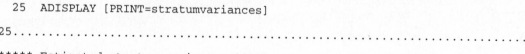

```
25...........................................................................
```

***** Estimated stratum variances *****

```
Variate: Yield

Stratum                      variance  effective d.f.

Blocks                         371.56           6.099
Blocks.Plots                   324.10          17.901
```

9.7.2   Balance

The condition of first-order balance required for a design and its specification to be analysable by the ANOVA directive is explained algorithmically by Wilkinson (1970) and mathematically by James and Wilkinson (1971) and Payne and Tobias (1992). Essentially it is that the contrasts of each term should all have a single efficiency factor, wherever the term is estimated. In the example in 9.7.1, all the terms have only one degree of freedom, and so represent only one contrast. There is thus no difficulty in verifying that the design is balanced.

Suppose instead that the treatment combinations were represented by a single factor T with eight levels:

```
    FACTOR [LABELS=!T('---','--d','-k-','-kd','n--','n-d','nk-','nkd')] T
```

The main effect of T would not be balanced: the comparison of levels

```
    '---' '--d' '-k-' '-kd'
```

with

```
    'n--' 'n-d' 'nk-' 'nkd'
```

has efficiency factor one in the Blocks.Plots stratum and zero in the Blocks stratum (this contrast is equivalent to the main effect of N in the original specification); but the comparison of levels

```
    'n--' 'n-d' '-k-' '-kd'
```

with

```
    '---' '--d' 'nk-' 'nkd'
```

has efficiency 0.25 in the Blocks stratum and 0.75 in the Blocks.Plots stratum (this is equivalent to N.K in the original specification). Thus the main effect of T is not balanced, since in the Block.Plots stratum some of its contrasts have efficiency factor one, while others have efficiency factor 0.75. Genstat can detect this imbalance and will give you an error diagnostic: see later in this section.

For the design to have been balanced for T, a further three pairs of blocks would be required. By confounding the comparison corresponding to the main effect of N between the first pair of extra blocks, that for K between the second pair, and that for D between the third pair, all the contrasts of T would be estimated within twelve of the (now) fourteen blocks, and confounded in the other two. The extended design would thus be balanced - as you may wish to verify!

To analyse the original design with a single treatment term T, a more complicated specification is required involving pseudo-factors.

9.7.3   Pseudo-factors

The pseudo-factorial operator `//` allows you to partition an unbalanced treatment term into pseudo-terms, which are each balanced. In our example, there is a factor T, some of whose contrasts have efficiency one in the `Blocks.Plots` stratum and zero elsewhere, while others have efficiency 0.25 in the `Blocks` stratum and 0.75 in the `Blocks.Plots` stratum. If instead of

        TREATMENTSTRUCTURE T

we specify

        TREATMENTSTRUCTURE T // (N + K + D + N.K + N.D + K.D)

the terms within the brackets that follow the operator `//` are linked to the term T as pseudo-terms. (Without the brackets, only the term immediately after `//` would be linked to T.) When the time comes for T to be fitted, the pseudo-terms N, K, D, N.K, N.D, and K.D are fitted first. All the contrasts wholly estimated in the `Blocks.Plots` stratum are thus removed (by N, K, and D), as well as some of the other contrasts. The remaining contrasts (denoted by T in the information summary) are all estimated with efficiency 0.25 between blocks and 0.75 within blocks. Thus all the pseudo-terms are balanced: those specified explicitly (N, K, D, N.K, N.D, and K.D), and the final pseudo-term which represents the contrasts not accounted for by N, K, D, N.K, N.D, and K.D. So by using the pseudo-factors, the design becomes analysable. In this example all the pseudo-terms represent single degrees of freedom - the final pseudo-term corresponds to the contrast represented earlier by N.K.D - but later we give an example where the pseudo-terms each have several degrees of freedom.

The sums of squares of the pseudo-terms are automatically combined to form the sum of squares for T in the analysis-of-variance table. Similarly the effects are all added together to form the table of means for T.

Example 9.7.3a

```
24   FACTOR [LABELS=!T('_ _ _','_ _ d','_ k _','_ k d', \
25                     'n _ _','n _ d','n k _','n k d')] T
26   READ [PRINT=data,error] T; FREPRESENTATION=labels

27   '_ _ _' 'n k _' 'n _ d' '_ k d'      'n _ _' '_ k _' '_ _ d' 'n k d'
28   'n _ _' '_ k _' 'n _ d' '_ k d'      '_ _ _' '_ _ d' 'n k _' 'n k d'
29   'n _ _' '_ _ d' 'n k _' '_ k d'      '_ _ _' '_ k _' 'n _ d' 'n k d'
30   '_ k _' '_ _ d' 'n k _' 'n _ d'      '_ _ _' 'n _ _' '_ k d' 'n k d' :
31   TREATMENTSTRUCTURE T // (N + K + D + N.K + N.D + K.D)
32   ANOVA Yield

32..........................................................................

***** Analysis of variance *****

Variate: Yield

Source of variation          d.f.      s.s.        m.s.      v.r.

Blocks stratum
T                               4     3724.9       931.2      3.61
```

Residual	3	774.1	258.0	0.81

Blocks.Plots stratum

T	7	456857.4	65265.3	204.58
Residual	17	5423.3	319.0	

Total	31	466779.7		

***** Information summary *****

Model term	e.f.	non-orthogonal terms

Blocks stratum

N.K	0.250	
N.D	0.250	
K.D	0.250	
T	0.250	

Blocks.Plots stratum

N.K	0.750	Blocks
N.D	0.750	Blocks
K.D	0.750	Blocks
T	0.750	Blocks

* MESSAGE: the following units have large residuals.

Blocks 6	Plots 4	28.2	s.e. 13.0

***** Tables of means *****

Variate: Yield

Grand mean 291.6

T	− − −	− − d	− k −	− k d	n − −	n − d	n k −	n k d
N	1	1	1	1	2	2	2	2
K	1	1	2	2	1	1	2	2
D	1	2	1	2	1	2	1	2
	106.1	316.5	286.9	415.2	104.6	355.3	295.3	452.8

*** Standard errors of differences of means ***

Table	T
rep.	4
s.e.d.	13.15

Except when comparing means with the same level(s) of

N	13.64
K	13.64
D	13.64
N.K	14.12
N.D	14.12
K.D	14.12

The basic idea, then, is to use each pseudo-term to pick out a set of contrasts whose efficiency factors are all the same, wherever they are estimated. This should be reasonably straightforward, provided you understand how your design has been constructed. A further example is given below. But first we demonstrate that Genstat can indeed detect an unbalanced design. If we do not include the pseudo-factors, the design would be unbalanced. The error message correctly identifies T as the unbalanced term.

Example 9.7.3b

```
 33   TREATMENTSTRUCTURE T
 34   ANOVA Yield

******** Fault (Code AN 1). Statement 1 on Line 34
Command: ANOVA Yield
Design unbalanced - cannot be analysed by ANOVA
Model term T (non-orthogonal to term Blocks) is unbalanced, in the Blocks.Plots
stratum.
A fatal fault has occurred - the rest of this job will be ignored
```

The traditional example for pseudo-factors is the partially balanced lattice. This has a single treatment factor, with number of levels equal to the square of some integer, k. To form the design, this factor is arbitrarily represented as the factorial combinations of two pseudo-factors, below called A and B, each with k levels. Further details are given by Yates (1937) or Kempthorne (1952). The example below is a simple lattice, taken from Cochran and Cox (1957, page 406). Here the treatment factor, Variety, has 25 levels. The correspondence between levels of Variety and the two pseudo-factors is:

```
    B:    1    2    3    4    5
 A:
 1         1    2    3    4    5
 2         6    7    8    9   10
 3        11   12   13   14   15
 4        16   17   18   19   20
 5        21   22   23   24   25
```

The simple lattice has two replicates, each with k blocks of k plots: the block model is

    Rep/Block/Plot = Rep + Rep.Block + Rep.Block.Plot

The main effect of A is confounded with the blocks in the first replicate: block 1 has the five levels of Variety that correspond to level 1 of A, block 2 has those with level 2, and so on. Similarly, B is confounded with the blocks of the second replicate. Thus A and B are each confounded with blocks in one out of the two replicates. So they have efficiency 0.5 in the Rep.Block (or blocks-within-replicates) stratum, and 0.5 in the Rep.Block.Plot (or plots-within-blocks) stratum. The treatment model is

    Variety//(A + B)

The partially confounded parts of Variety are specified by the two pseudo-terms, A and B, and will be fitted first. The remaining contrasts of Variety correspond to the interaction between A and B, which is all estimated in the Rep.Block.Plot stratum. This final pseudo-term is thus also balanced, so the design can be analysed. The analysis-of-variance table in Example 9.7.3c differs from that presented by Cochran and Cox (1957); they do not present the treatment sums of squares between and within blocks, but merely a sum of squares unadjusted for blocks. Example 9.7.3c also prints the table of means combining information from both the `Rep.Block` and the `Rep.Block.Plot` strata (9.7.1).

Example 9.7.3c

```
   1   " 5x5 Simple lattice
  -2      (Cochran and Cox 1957, page 406)."
   3   UNITS [NVALUES=50]
   4   FACTOR [LEVELS=2] Rep
   5   & [LEVELS=5] Block,Plot,A,B
   6   & [LEVELS=25; VALUES=(1...25),(1,6...21),(2,7...22), \
   7      (3,8...23),(4,9...24),(5,10...25)] Variety
   8   GENERATE Rep,Block,Plot
   9   & [TREATMENTS=Variety; REPLICATES=Rep; BLOCKS=Block] A,B
  10   READ Yield
```

```
    Identifier   Minimum     Mean   Maximum    Values    Missing
       Yield        4.00    13.62     30.00        50          0
  13   BLOCKSTRUCTURE Rep/Block/Plot
  14   TREATMENTSTRUCTURE Variety//(A+B)
  15   ANOVA [PRINT=aovtable,cbmeans] Yield
```

```
15....................................................................
```

***** Analysis of variance *****

Variate: Yield

Source of variation	d.f.	s.s.	m.s.	v.r.
Rep stratum	1	212.18	212.18	
Rep.Block stratum				
Variety	8	350.00	43.75	
Rep.Block.Plot stratum				
Variety	24	711.12	29.63	2.17
Residual	16	218.48	13.66	
Total	49	1491.78		

***** Tables of combined means *****

Variate: Yield

Variety	1	2	3	4	5	6	7
A	1	1	1	1	1	2	2
B	1	2	3	4	5	1	2
	19.07	16.97	14.65	14.77	12.85	13.17	9.07

Variety	8	9	10	11	12	13	14
A	2	2	2	3	3	3	3
B	3	4	5	1	2	3	4
	6.75	8.37	8.45	23.55	12.46	12.63	20.75

Variety	15	16	17	18	19	20	21
A	3	4	4	4	4	4	5
B	5	1	2	3	4	5	1
	19.33	12.62	10.53	10.70	7.32	11.40	11.63

Variety	22	23	24	25
A	5	5	5	5
B	2	3	4	5
	18.53	12.20	17.33	15.40

```
*** Standard errors of differences of combined means ***

Table              Variety
rep.                  2
s.e.d.              4.234
Except when comparing means with the same level(s) of
A                   3.974
B                   3.974
```

Example 9.7.3c also illustrates how to use the GENERATE directive (line 8) to form the values of pseudo-factors; the details are explained in 9.8.1.

9.7.4   Non-orthogonality between treatment terms

The examples earlier in this section illustrate non-orthogonality between treatment and block terms. Balanced designs can also occur where the non-orthogonality is between treatment terms. However the interpretation of the analysis requires more care; indeed there may be information that Genstat is unable to calculate. (Similar difficulties occur in ordinary regression with observational data, see Chapter 8: usually the explanatory variables will not be orthogonal to each other and so their sums of squares, and thus the importance that may be ascribed to them, will depend on the order in which they are fitted.)

Suppose that the treatment model is

    A + B + C

that B is non-orthogonal to A, and that C is non-orthogonal to both A and B. Genstat fits the model sequentially. Thus the sum of squares produced for A is for A ignoring B and C: no account is taken of these two factors, which are still to be fitted. With B, A has already been fitted and thus eliminated, whereas C has not. So the sum of squares produced for B is for B eliminating A and ignoring C. The sum of squares for C, which is fitted last, is eliminating both A and B.

Each sum of squares can be expressed as the difference between the residual sums of squares before and after fitting a particular term. So the sums of squares that are presented by Genstat will automatically add to the total sum of squares. Examining these enables you to check whether any of the terms in the model has an effect. However, to be sure that there is an effect of A, for example, that cannot be explained by B and C requires the sum of squares for A eliminating B and C. To obtain this you could redefine the treatment model as either

    B + C + A

or

```
C + B + A
```

but the design would not necessarily be balanced according to these specifications.

Similarly, the effects estimated for each term are eliminating those terms fitted before it, and ignoring those that are still to be fitted. *Partial effects*, defined as the effects of a term eliminating all the other treatment terms, are calculated during the analysis and can be obtain using AKEEP (9.6.1).

A table of means for A.B, if this were in the model, would require the effects for A eliminating B, those for B eliminating A, and those for the interaction A.B. However, with the treatment model A + B + C, the necessary effects for A are not available. Consequently, no means are presented for terms that contain mutually non-orthogonal margins (like A and B for the table A.B).

A maximum of 10 mutually non-orthogonal terms is allowed. For example, term T[10] may be non-orthogonal to T[9], which is non-orthogonal to T[8], and so on down to term T[2], which is non-orthogonal to term T[1]; but to include an extra term T[11] in the sequence would exceed the limit. This limit should be sufficient for any designed experiment. Data with many non-orthogonal terms are, in any case, analysed more efficiently by the regression directives described in Chapter 8.

Note that, if the terms A, B, and C here had been orthogonal, the sum of squares and effects obtained for any one of them would remain the same irrespective of which of the other two terms had been fitted. For example, the sum of squares for A ignoring B and C would be identical to that for A eliminating B and C. Thus each of these three terms could be assessed independently, without regard to the other two. If two terms are far from orthogonal, you may find that the effects of either term ignoring the other are significant, but that neither set of effects is significant when the other term is eliminated. Deciding which of the terms are important may then be very difficult, and you may have to recommend that another experiment be done. This illustrates that orthogonality between treatment terms is not merely a convenience for making the computations more efficient: it also greatly simplifies the interpretation of the results.

9.7.5 The method of analysis

In this subsection we briefly describe the algorithm that is used to do the analysis of variance. However, for most purposes you will not need this information.

The model formulae defined by the BLOCKSTRUCTURE and TREATMENTSTRUCTURE are interpreted by an extension of the algorithm of Rogers (1973); further details are given by Wilkinson and Rogers (1973).

The method used to do the analysis is described in detail by Payne and Wilkinson (1977), Wilkinson (1970), and Payne and Tobias (1992). It operates on a working vector which initially contains the data values, and finally contains the residuals. The terms in the model are fitted by a series of *sweep* operations. Each sweep estimates the effects of a term, and then subtracts them from the current working vector, which then becomes the working vector for the next sweep. The first sweep is for the grand mean. The block terms are fitted next, to give an initial partitioning into strata. Then the treatments are fitted within each stratum.

If a term is orthogonal, its estimated effects are simply the corresponding table of means calculated from the current working vector. If the term is non-orthogonal to any of the terms already fitted, some of the information about the term is unavailable, and its effects are the totals calculated from the current vector, divided by its replication and efficiency factor. For the term to be balanced, the information still available must be the same for all the contrasts between the effects of the term, so that there is a single efficiency factor for all the contrasts. If the term is orthogonal, the efficiency factor is one. A zero efficiency factor indicates that the term is completely aliased with earlier terms in the model, and so cannot be estimated.

A sweep for a non-orthogonal term reintroduces effects for the terms to which it is non-orthogonal. Before sweeping for the next term in the model, these effects are removed by a sequence of *re-analysis* sweeps for the terms concerned. If any term in the re-analysis sequence is itself non-orthogonal, it must itself be followed by its own re-analysis sequence, and so on. Genstat allows for re-analysis sequences to be nested only ten deep, which is why there is the limit of ten mutually non-orthogonal terms (9.7.4).

When there are several strata, the analysis of each one is introduced by a special sweep known as a *pivot*, in which the value in each unit of the working vector is replaced by the corresponding effect calculated for the block term of the stratum. During the analysis of a stratum, the re-analysis sweeps for its own block term take the form of recalculating the effects and repeating the pivot.

Procedure ASWEEP, which can perform all these types of sweep, is provided in the Genstat Procedure Library for those who wish to study the process further.

The algorithm, unlike multiple regression algorithms, does not distinguish between the individual contrasts of each term (unless you partition it up into pseudo-terms: 9.7.3). This makes the computations more efficient, but it means that only balanced terms can be fitted.

The design can be analysed if all the terms in the model are balanced: that is if they each have a single efficiency factor for their effects, in any stratum where they are estimated. The design is then said to have *first-order balance* with respect to the specified model (Wilkinson 1970, James and Wilkinson 1971, Payne and Tobias 1992): for a brief description, see 9.7.2.

A further consequence of the way in which the effects of each terms are all fitted together is that, if any part of a term is present in a stratum, Genstat must assume that all its effects can be estimated there. Thus if a term is only partially estimatable in a stratum (due to partial aliasing or to partial confounding), the degrees of freedom will be incorrect. In such situations Genstat prints a warning diagnostic. To obtain an analysis with the correct numbers of degrees of freedom you should use pseudo-factors (9.7.3) to identify the parts of a term that are estimated in the different strata.

Genstat determines the structure of the design by a process known as the *dummy analysis* (9.1.2). This is similar to the analysis of the data, but involves extra sweeps to detect whether each term can be estimated in a particular stratum, and to determine its efficiency factor there. In these sweeps, a near-zero sum of squares is taken to indicate that the term cannot be estimated. However the test cannot be against an exact value of zero, because computer calculations always involve errors of round-off. Thus Genstat tests against a number slightly larger than zero; this zero limit is calculated as the total sum of squares in the working variate (after removing the grand mean) multiplied by the first element of the variate specified in the TOLERANCE option of ANOVA (9.1.2). By default, this first element contains the value 10^{-7}. A similar limit checks for zero sums of squares in the analysis of the data, but here the multiplier is given in the second element of the TOLERANCE variate; the default value is 10^{-9}.

The working vector for the dummy analysis contains random values from a Cauchy distribution. The starting value for their generation is set by the SEED option of ANOVA (9.1.2). Thus if you have doubts about a particular dummy analysis, for example if you think that a term is incorrectly listed as aliased, you can change the starting value and repeat the analysis with a different working vector.

A simpler and quicker form of the dummy analysis is available for designs that are orthogonal, and for which all the effects of each term have equal replication. (An orthogonal design is one in which each term has efficiency factor either zero or one in each stratum.) This incorporates a check which will detect any non-orthogonality, unless the design is particularly complicated and terms are aliased. The ORTHOGONAL option of ANOVA (9.1.2) allows you to specify whether non-orthogonality should cause Genstat to switch to the full dummy analysis, or to terminate the analysis with an error diagnostic.

9.8 Design of experiments

The main Genstat facilities for design are provided by the directives GENERATE and RANDOMIZE, and the procedures in the Design module of the procedure library. GENERATE provides an easy way of generating blocking factors or any other factors whose values occur in a systematic order. It can also form the values of treatment factors, using the design-key method, and define values for the pseudo-factors required to specify partially balanced experimental designs (9.7.3). Alternatively, procedure AKEY combines both these uses of GENERATE, allowing you to generate the block factors in systematic order, and then the treatment factors using a design key; it can also print the design. Other procedures in the design module include AFALPHA (generation of alpha designs), AFCYCLIC (for generating cyclic designs), and PDESIGN (for printing a design). The RANDOMIZE directive allows any of these designs to be randomized.

Further facilities, to give help in the selection of a suitable design, will be provided by extensions to the conversational system and by a DESIGN procedure to be added to the library.

9.8.1 Generating factor values (**GENERATE**)

GENERATE generates factor values for designed experiments: with no options set, factor values are generated in standard order; the options allow treatment factors to be generated using the design-key method, or pseudo-factors to be generated to describe the confounding in a partially balanced experimental design.

Options

TREATMENTS = *formula* Model term for which pseudo-factors are to be generated; default *

REPLICATES = *formula*	Factors defining replicates of the design; default *
BLOCKS = *formula*	Block formula (for design-key generation) or term (for generation of pseudo factors); default *
KEY = *matrix*	Key matrix (number of factors in the parameter list by number of factors in the BLOCKS formula) to generate the factors by the design key method; default *
BASEVECTOR = *variate*	Base vector for design key generation; default *

Parameter

| *factors* | Factors whose values are to be generated |

GENERATE is invaluable when you have a set of data that is to be read in a systematic order: for example, you may want to take all the observations within one group, then the same number of observations within the next group, and so on until an equal number of observations has been read for every group. You can then define values of the grouping factor or factors by GENERATE; so the only values that you need to read are the observed data. Designed experiments are the obvious instance where the data are structured in this way: for example, you might have all the data from the first block, then all those from the second block, and so on.

The best way to understand GENERATE is to look at some examples. The values of a set of factors that you have defined by GENERATE are said to be in *standard order*: that is their units are arranged so that the levels of the first factor occur in the same order as in its levels vector then, within each level of the first factor, the levels of the second factor are arranged similarly, and so on. For example

```
FACTOR [NVALUES=24; LEVELS=2] A
& [LEVELS=!(4,1,2)] B
& [LEVELS=4] C
GENERATE A,B,C
```

gives A, B, and C the values

```
A: 1 1 1 1 1 1 1 1 1 1 1 1 2 2 2 2 2 2 2 2 2 2 2 2
B: 4 4 4 4 1 1 1 1 2 2 2 2 4 4 4 4 1 1 1 1 2 2 2 2
C: 1 2 3 4 1 2 3 4 1 2 3 4 1 2 3 4 1 2 3 4 1 2 3 4
```

Placing a number or a scalar in the parameter list has the same effect as if a factor with that number of levels had been listed. Thus to generate values only for A and C, all that you require is

```
GENERATE A,3,C
```

To generate values for just B and C is even simpler since the cycling process is itself recycled until all the units have been covered. Omitting A therefore causes all combinations of a level of B with a level of C to be used twice, in the same pattern as displayed above; so you need specify only

```
GENERATE B,C
```

You get a warning if one of the cycles is incomplete, as would happen for example if B and C had 18 values instead of 24.

This first use of GENERATE, then, is particularly appropriate for generating the blocking factors in an experimental design as can be seen in line 7 of Example 9.2.1a, line 11 of Example 9.3, line 9 of Example 9.7.1a, and line 8 of Example 9.7.3c.

Another use, obtained by setting the BLOCKS, KEY, and BASEVECTOR options, is to form values of treatment factors using the design-key method. This method, described by Patterson (1976) and Patterson and Bailey (1978), provides a very flexible way of specifying the allocation of treatments in an experimental design. The method assumes that the units are identified by a set of what are called "plot" factors. In Genstat terms, these will often be the same as the factors that occur in the block formula of the design (9.2), and they are specified by the BLOCKS option of GENERATE. The setting is a formula, but remember this can be just a list of factors if you do not wish to indicate their inter-relationships; if the setting is more than just a list, Genstat forms the set of plot factors by taking the factors from the block formula in the order in which they occur there. Of course, the factors need not be identical to those in the block formula. For example if one these factors has a non-prime number of levels, it may need to be specified instead as the combination of two or more (pseudo) factors: for example, in a block design with blocks of size eight, the plots might need to be indexed by three factors with two levels.

The treatment factors to be generated are again specified by the parameter of GENERATE.

The KEY option specifies a matrix known as the *design key*, which indicates how the values of each treatment factor are to be calculated from the plot factors. The matrix has a row for each treatment factor and a column for each plot factor; below k_{ij} represents the element in row i and column j. (This is the transpose of the form used by Patterson 1976, but in Genstat it seems more convenient to specify the treatments by rows.) There is also an option called BASEVECTOR, which can specify a variate with an element b_i for each treatment factor to allow the levels of the factor to be shifted cyclically; if this is unset, Genstat assumes $b_i=0$.

The calculation assumes that the values of the plot factors are represented by the integers zero upwards (and GENERATE will perform this mapping automatically if necessary). The value $q[i]_u$ in unit u of treatment factor i is then given by

$$q[i]_u = b_i + k_{i1} \times p[1]_u + k_{i2} \times p[2]_u + ... + k_{in} \times p[n]_u \qquad \text{modulo } t_i$$

where $p[1]_u ... p[n]_u$ are the values of the plot factors in unit u, and t_i is the number of levels of treatment factor i. The calculated values are integers in the range $0, 1 ... t_i-1$, but GENERATE will again map these to the defined levels if necessary.

To illustrate the process, the treatments to be allocated (before randomization) to the plots of an $n \times n$ Latin Square may be calculated as

Latin-factor-value = Row-factor-value + Column-factor-value   modulo n

The values of the extra factor in a Graeco-Latin square can then be formed as

Graeco-factor-value = Row-factor-value + 2 × Column-factor-value   modulo n

The design key thus has rows (1,1) and (1,2); Example 9.8.1 uses this to this generate a 5×5

Graeco-Latin square. Notice also how procedure PDESIGN is used to display the design.

Example 9.8.1

```
1   " Graeco-Latin square "
2   FACTOR      [NVALUES=25; LEVELS=5] Row,Column,A,B
3   GENERATE    Row,Column
4   " specify key matrix (row and column labelling is unnecessary
-5    other than to indicate how the matrix is stored) "
6   MATRIX      [ROWS=!t(A,B); COLUMNS=!t(Row,Column); VALUES=1,1, 1,2] GLkey
7   PRINT       GLkey; DECIMALS=0
```

```
                    GLkey
                    Row        Column

         A           1            1
         B           1            2
```

```
8   GENERATE    [BLOCKS=Row,Column; KEY=GLkey] A,B
9   PDESIGN     [BLOCKSTRUCTURE=Row*Column; TREATMENTSTRUCTURE=A+B]
```

*** Treatment combinations on each unit of the design ***

```
       Column   1      2      3      4      5
         Row
          1     1 1    2 3    3 5    4 2    5 4
          2     2 2    3 4    4 1    5 3    1 5
          3     3 3    4 5    5 2    1 4    2 1
          4     4 4    5 1    1 3    2 5    3 2
          5     5 5    1 2    2 4    3 1    4 3
```

Treatment factors are listed in the order: A B

The design key thus provides a very convenient way of defining treatment factors. Essentially, the key identifies each factor i with the set of contrasts (in the usual terminology)

$$p[1]^{k_{i1}}\ p[2]^{k_{i2}}\ ...\ p[n]^{k_{in}}$$

and the skill when forming a design is in selecting the best set for each factor. Further keys are presented by Patterson & Bailey (1978), and these are used in the example of procedure AKEY; this procedure extends the GENERATE facilities by allowing the block factors to be generated automatically, and the design to be printed after the factors have been generated. Guidance for the selection of suitable keys will be provided by extensions to the conversational system and by further procedures in the Design module of the procedure library.

GENERATE can also be used to form the values of pseudo-factors in partially-balanced designs, as shown in line 8 of Example 9.7.3c:

```
GENERATE [TREATMENTS=Variety; REPLICATES=Rep; BLOCKS=Block] A,B
```

The treatment term to which the pseudo-factors are to be linked is specified by the TREATMENTS option; here this is the main effect of Variety. The factors that identify the replicates are specified by the REPLICATES option, and those that identify the blocks within each replicate are specified by the BLOCKS option. The settings of these two options are model formulae, but Genstat merely scans them to find which factors they contain; so you may again find it easiest simply to give the factors as a list. Here the replicates and blocks are identified by the single factors Rep and Block respectively. The parameter of GENERATE lists the pseudo-factors. These have as many levels as there are blocks within each replicate. The blocks in the first replicate are used to determine which combinations of the factors in the treatment term correspond to each level of the first pseudo-factor, those in the second replicate are used for the second pseudo-factor, and so on. Here the first pseudo-factor is A, and the five blocks of replicate 1 contain Variety levels 1-5, 6-10, 11-15, 16-20, and 21-25. Thus the plots with varieties 1 to 5 are allocated level 1 of A, and so on. If a treatment combination occurs in more than one block within the same replicate, the level of the corresponding pseudo-factor is not determined uniquely and Genstat will report an error.

9.8.2   The **Design** module of the Genstat procedure library

The Design module of the library contains procedures useful for the design of experiments. In Example 9.8.2a, procedure LIBINFORM is used to list the contents of Release 3[1], and then LIBHELP is used to print brief details of the syntax of procedures AFUNITS, AKEY, FACPRODUCT, and PDESIGN.

Example 9.8.2a

```
  1  LIBINFORM [PRINT=index] 'design'

 Index of procedures in Library module design
AFALPHA generates alpha designs
AFCYCLIC generates block and treatment factors for cyclic designs
AFUNITS forms a factor to index the units of the final stratum of a design
AKEY generates values for treatment factors using the design key method
FACPRODUCT forms a factor with a level for every combination of other factors
PDESIGN prints or stores treatment combinations tabulated by the block factors

  2  LIBHELP [PRINT=options,parameters] 'AFUNITS'

  Procedure      AFUNITS

Help['options']
BLOCKSTRUCTURE      = formula    Defines the block factors for the design;
                                 the default is to take those specified by
                                 the BLOCKSTRUCTURE directive
Help['parameters']
UNITS               = factor     Factor to be formed

  3  LIBHELP [PRINT=options,parameters] 'AKEY'

  Procedure      AKEY

Help['options']
PRINT               = string     Allows the generated TREATMENTFACTOR values
                                 to be printed, tabulated by the BLOCKFACTORS
                                 (design); default * i.e. no printing
BLOCKFACTORS        = factors    Defines the block factors for the design, values
                                 of which will be generated if none already exist;
                                 default is to take those in the formula already
```

```
                         specified by the BLOCKSTRUCTURE directive, in
                         the order in which they occur there
KEY              = matrix   Matrix (number of treatment factors
                         x number of block factors) key for the design
BASEVECTOR       = variate  Base vector (length = number of treatment factors)
                         for the design; default is a variate of zeros
Help['parameters']
TREATMENTFACTORS = factors  Defines the treatment factors for the design;
                         default is to take those in the formula already
                         specified by the TREATMENTSTRUCTURE directive,
                         in the order in which they occur there


   Procedure  FACPRODUCT

Help['options']
None.


Help['parameters']
FACTORS   = pointers    factors contributing to each product
          or formulae
PRODUCT   = factors     factors to be formed

   5  LIBHELP [PRINT=options] 'PDESIGN'    "Note: there are no parameters"

   Procedure     PDESIGN

Help['options']
PRINT                 = string   Controls the printing of the design
                                 (design); default design
BLOCKSTRUCTURE        = formula  Defines the block factors for the design;
                                 the default is to take those specified by
                                 the BLOCKSTRUCTURE directive
TREATMENTSTRUCTURE    = formula  Defines the treatment factors for each design;
                                 the default is to take those specified by
                                 the TREATMENTSTRUCTURE directive
TABLES                = pointer  Contains tables to store the tabulated factor
                                 values for printing outside the procedure in
                                 some other format
```

The output from procedure PDESIGN is illustrated in Example 9.8.1. If this is unsuitable, printing can be suppressed by setting option PRINT=* (by default PRINT=design), and the tables of treatment levels can be saved for printing outside the procedure by setting the TABLES option to a pointer. This will be returned with an element for each treatment factor, pointing to a table classified by the block factors and storing the tabulated levels of the treatment. If any of the factors is restricted, only the part of the design not excluded by the restriction will be displayed.

Example 9.8.2b

```
   6  LIBHELP [PRINT=options,parameters] 'AFALPHA'

   Procedure     AFALPHA

Help['options']
PRINT        = string    Whether to print the design (design); default *
                         i.e. no printing
Help['parameters']
```

```
GENERATOR   = matrices    generating array (size number-of-plots-per-block
                          by number-of-reps)
LEVELS      = scalars     Defines the levels of each treatment factor; if this
            or variates   is omitted, the levels of the TREATMENT factor are
                          used, if available, otherwise LEVELS is determined
                          from the generating array on the assumption that the
                          blocks are to be of equal size
SEED        = scalar      Seed to be used to randomize the design, if required
TREATMENTS  = factors     Specifies the treatment factor for each design
REPLICATES  = factors     Specifies the replicate factor
BLOCKS      = factors     Specifies the block factor
UNITS       = factors     Specifies the factor to index the units within
                          each block
```

```
7  MATRIX [ROWS=5; COLUMNS=3; VALUES=0,0,0, 0,1,2, 0,2,3, 0,3,1, 0,3,2] Array
8  PRINT  Array; DECIMALS=0
```

```
            Array
              1           2           3

      1       0           0           0
      2       0           1           2
      3       0           2           3
      4       0           3           1
      5       0           3           2
```

```
 9  FACTOR   [LEVELS=!(0...19)] Treat
10  AFALPHA  [PRINT=design] Array; TREATMENTS=Treat; REPLICATES=Rep; \
11           BLOCKS=Block; UNITS=Plot
```

```
*** Treatment combinations on each unit of the design ***

                    Plot   1    2    3    4    5
        Rep        Block
         1           1     0    4    8   12   16
                     2     3    7   11   15   19
                     3     2    6   10   14   18
                     4     1    5    9   13   17
         2           1     0    5   10   15   19
                     2     3    4    9   14   18
                     3     2    7    8   13   17
                     4     1    6   11   12   16
         3           1     0    6   11   13   18
                     2     3    5   10   12   17
                     3     2    4    9   15   16
                     4     1    7    8   14   19

Treatment factors are listed in the order: Treat
```

Example 9.8.2b shows first the syntax and then an example of procedure AFALPHA. Alpha designs are a very flexible class of resolvable incomplete block designs. A resolvable design is one in which each block contains only a selection of the treatments, but the blocks can be grouped together into subsets in which each treatment is replicated once. The groupings of blocks thus form replicates, and the block structure of the design is

```
    Replicates / Blocks / Units
```

Such designs are particularly useful when there are many treatments to examine and the variability of the units is such that the block size needs to be kept small. Alpha designs were thus devised originally for the analysis of plant breeding trials (Patterson and Williams 1976), where many varieties may need to be evaluated in a single trial, and have the advantage that they can provide effective designs for any number of treatments.

The construction of an alpha design requires a $k \times r$ array of integers between 0 and $s-1$, where r is the number of replicates, and s is the number of blocks per replicate. If the number of treatments, v, is a multiple of the number of blocks per replicate, k will be the number of units in each block, and v will be given by $s \times k$. Otherwise, the design will have some blocks of size k and some of size $k-1$, and v will lie between $s \times (k-1)$ and $s \times k$. Each column of the generating array is used to form $s-1$ further columns by successively adding 1 modulo s. Next, s is added to row 2 of every column, $2s$ to row 3, and so on. Each resulting column then gives one of the blocks of the design, and the replicates are formed by the sets of columns that were all generated from the same initial column.

Clearly, the properties of the design that is formed will be very dependent on the choice of array. Patterson, Williams and Hunter (1978) present 11 basic arrays to generate designs with up to 100 treatments and 2, 3 or 4 replicates when k is greater than 3 and s is greater than or equal to k, while Williams (1975) presents arrays for any sensible values of s and k with up to 100 treatments and 2 to 4 replicates.

Procedure AFALPHA generates the treatment, replicate, block and unit factors for the design. The design can be printed by setting option PRINT=design, and the factors can be saved using the parameters TREATMENTS, REPLICATES, BLOCKS, and UNITS. The generating array for the design must be specified as a $k \times r$ matrix using the GENERATOR parameter, and the number of levels of the treatment factor can be defined by the LEVELS parameter. If LEVELS is omitted, AFALPHA will see whether the TREATMENTS parameter has been set to a factor whose levels have already been defined; if not, AFALPHA will set LEVELS to the scalar value $v = s \times k$. By default the design is unrandomized, but randomization can be requested by setting the SEED parameter.

Example 9.8.2c

```
   12   LIBHELP [PRINT=options,parameters] 'AFCYCLIC'

      Procedure    AFCYCLIC

Help['options']
 PRINT       = string     Whether to print the design (design); default *
                          i.e. no printing
Help['parameters']
 INITIAL     = variates   Defines one (variate) or more (pointer to variates)
               or pointers initial blocks for a treatment factor
 INCREMENT   = scalars    Defines the size of the successive increments (scalar)
               or pointers or increments (pointer to scalars) for each initial
                          block
 LEVELS      = scalars    Defines the levels of each treatment factor; this
               or variates need not be specified if the factor has already been
                          declared
 SEED        = scalar     Seed to be used to randomize each design, if required
 TREATMENTS = factors     Specifies treatment factors
 BLOCKS      = factors     Specifies block factors
```

```
UNITS      = factors      Specifies factors to index the units within each block

13  AFCYCLIC [PRINT=design] !(1,2,4); \
14           TREATMENTS=Treat; Blocks=Block; LEVELS=7

*** Treatment combinations on each unit of the design ***

        Unit   1   2   3
        Block
          1    1   2   4
          2    2   3   5
          3    3   4   6
          4    4   5   7
          5    5   6   1
          6    6   7   2
          7    7   1   3

Treatment factors are listed in the order: Treat
```

Example 9.8.2c illustrates procedure AFCYCLIC. The cyclic method is a very powerful way of constructing incomplete block designs. In its simplest form, it starts with an initial block, containing some subset of the treatments. This subset is then represented by the ordinal number in the range 0 ... m-1 where m is the number of treatment levels. The second and subsequent blocks are then generated by successively addition modulo m of one to the numbers in the subset. Thus, for seven treatments (0...6) and an initial block (0,1,4), the subsequent blocks would contain treatments (1,2,5), (2,3,6), (3,4,0), (4,5,1), (5,6,2) and (6,0,3). As can be seen, if m is a prime number, m blocks are generated with each initial block. However, if m can be expressed as the product of other integers, shorter cycles can occur. For example, for $m=8$ and initial block (0,1,4,5), four blocks are generated altogether, the others being (1,2,5,6), (2,3,6,7) and (3,4,7,0). The procedure allows for all of this. It is also possible to have more than one initial block, and the increment need not be one.

The INITIAL parameter specifies the initial blocks. If the design is to be generated from a single initial block, INITIAL should be set to a variate containing the levels corresponding to the treatments concerned; if there are several, the appropriate variates should be placed into a pointer. Similarly the INCREMENT parameter, which specifies the increment to be used, should be set to a scalar if the same increment is to be used for all the initial blocks, otherwise to a pointer of scalars. The levels of the treatment factor are specified by the LEVELS parameter and the SEED parameter allows the design to be randomized. As is customary in Genstat, if LEVELS is set to a scalar the levels are assumed to be represented by the integers 1 upwards, but LEVELS can be set to a variate to specify other numbers. LEVELS can be omitted if the TREATMENTS parameter is used to supply a factor to store the treatments, provided the levels of that factor have already been defined outside the procedure. The factors for blocks and units within blocks can be saved similarly by the BLOCKS and UNITS parameters respectively. The design can also be printed, by setting option PRINT=design.

The properties of the cyclic designs that can be generated for any particular number of treatments or size of block varies according to the choice of initial block and increment. Tables showing the most efficient combinations have been presented for example by John, Wolock and David (1972), John (1981, 1987) and Lamacraft and Hall (1982).

9.8.3   Randomization

RANDOMIZE randomizes the units of a designed experiment or the elements of a factor or variate.

Options

BLOCKSTRUCTURE = *formula*	Block model according to which the randomization is to be carried out; default * i.e. as a completely-randomized design
EXCLUDE = *factors*	(Block) factors whose levels are not to be randomized
SEED = *scalar*	Seed for the random-number generator; default 12345

Parameter

factors or *variates*	Structures whose units are to be randomized according to the defined block model

---

In its simplest form, RANDOMIZE merely performs a random permutation of the units of a list of factors or variates. You list these structures with the parameter of RANDOMIZE. Genstat gives them all exactly the same permutation, which is produced by a set of random numbers generated from the SEED option. For example

        RANDOMIZE [SEED=144556] X,Y

puts the values of X and Y into an identical random order. The seed can be any positive integer, but only the last six digits of its integer part are used. Thus the seeds 2144556 and 7144556.3 are both equivalent to the seed 144556. If you put SEED=*, or leave it unset, Genstat picks a seed at random.

If you have restricted any of the structures in the parameter list (4.4.1), then all will be treated as though they were restricted; moreover, all the restricted structures must be restricted in exactly the same way.

The main use of RANDOMIZE, however, is to randomize the allocation of treatments to units in a designed experiment. In the analysis of designed experiments, the underlying structure of an experiment is defined by the block formula, as described in 9.2. Provided the only operators in a block formula are the nesting (/) and crossing (*) operators, this also specifies the correct randomization of the experiment.

The nesting operator specifies that one factor is to be randomized within another one. The simplest example is the randomized block design: its block formula is Blocks/Plots; a separate randomization of plots is done for each block. Another example is a split-plot design, the formula for which is Blocks/Wplots/Subplots; this means randomize first the levels of Blocks, then the levels of Wplots within levels of Blocks, and finally the levels of Subplots within the levels of Blocks and Wplots. In other words, there is a separate randomization of Wplots for each Block, and a separate randomization of Subplots for each Wplot. A similar formula and randomization would apply to a resolvable incomplete-block design.

The crossing operator specifies that the factors are to be randomized independently of each other. For example the formula Rows*Cols means randomize the levels of Rows and Cols separately. Thus the same randomization of Cols appears within each Row. This is the block formula associated with a row and column design, for example a Latin square.

You specify the block formula by the BLOCKSTRUCTURE option, which thus defines the way in which the randomization is to be carried out. Genstat does not randomize the factors in the block structure themselves, unless you put them into the parameter list. This is because the original order of the block-factor levels often describes actual positions in the experiment; for example, in a field. So you will be interested in keeping these values, rather than the random ordering of them that is used to allocate treatments.

Example 9.8.3a, shows the randomization of a randomized block design.

Example 9.8.3a

```
1   UNITS [NVALUES=16]
2   FACTOR [LEVELS=4; VALUES=4(1...4)] Blocks
3   & [VALUES=(1...4)4] Plots
4   & [LABELS=!T(A,B,C,D)] Dose
5   PRINT Blocks,Plots,Dose
```

Blocks	Plots	Dose
1	1	A
1	2	B
1	3	C
1	4	D
2	1	A
2	2	B
2	3	C
2	4	D
3	1	A
3	2	B
3	3	C
3	4	D
4	1	A
4	2	B
4	3	C
4	4	D

```
6   RANDOMIZE [BLOCKSTRUCTURE=Blocks/Plots; SEED=556743] Dose
7   PRINT Blocks,Plots,Dose
```

Blocks	Plots	Dose
1	1	C
1	2	B
1	3	D
1	4	A
2	1	C
2	2	B
2	3	A
2	4	D
3	1	B
3	2	C
3	3	D
3	4	A
4	1	A
4	2	C

```
           4              3              D
           4              4              B
```

Notice that the values of the Blocks and Plots factors have not been randomized because they did not appear in the parameter list. Note also that the block formula for this design is Blocks/Plots and not just Blocks. This is because the formula must define each experimental unit by a unique combination of the block factor levels, for example block 1, plot 3. To put a block formula of just Blocks would not give Genstat any information about what to do with the elements of the blocks.

To show a more complicated design, Example 9.8.3b shows the randomization of the Latin square generated in Example 9.8.1. The block formula is now Rows*Cols. Rows and Cols are randomized separately, so the same randomization of Treat appears within each row and column - thus preserving the properties of the Latin square.

Example 9.8.3b

```
   10   RANDOMIZE  [SEED=682747; BLOCKSTRUCTURE=Row*Column] A,B
   11   PDESIGN    [BLOCKSTRUCTURE=Row*Column; TREATMENTSTRUCTURE=A+B]

*** Treatment combinations on each unit of the design ***

        Column    1       2       3       4       5
          Row
            1     1 2     4 3     5 5     3 1     2 4
            2     3 4     1 5     2 2     5 3     4 1
            3     2 3     5 4     1 1     4 2     3 5
            4     4 5     2 1     3 3     1 4     5 2
            5     5 1     3 2     4 4     2 5     1 3

Treatment factors are listed in the order: A B
```

You should use the EXCLUDE option if you want to restrict the randomization so that one or more of the factors in the block formula is not randomized. The most common instance where this is required is when one of the treatment factors is time-order, which cannot be randomized. For example, suppose the main plot treatments in a split-plot experiment were lengths of time between two chemicals being mixed together, and that the analysis is of the amount of gas produced. If all the jars of chemicals needed to be mixed up at the beginning of the day, and the analyses were performed after the appropriate time lapse, the standing times would have to be in the same order in each replicate. A suitable randomization is shown in Example 9.8.3c.

Example 9.8.3c

```
    1   UNITS [NVALUES=18]
    2   FACTOR [LEVELS=3; VALUES=6(1,2,3)] Block
    3   & [LABELS=!T(A,B,C); VALUES=(1...3)6] Method
    4   & [LEVELS=2; LABELS=!T('2 hours','4 hours'); VALUES=3(1,2)3] Time
    5   & [LABELS=*] Mplot
    6   PRINT Block,Time,Method
```

```
       Block          Time       Method
           1       2 hours          A
           1       2 hours          B
           1       2 hours          C
           1       4 hours          A
           1       4 hours          B
           1       4 hours          C
           2       2 hours          A
           2       2 hours          B
           2       2 hours          C
           2       4 hours          A
           2       4 hours          B
           2       4 hours          C
           3       2 hours          A
           3       2 hours          B
           3       2 hours          C
           3       4 hours          A
           3       4 hours          B
           3       4 hours          C
```

```
7    RANDOMIZE [BLOCKSTRUCTURE=Block/Mplot/Method; EXCLUDE=Mplot; \
8      SEED=888667] Time,Method
9    PRINT Block,Time,Method
```

```
       Block          Time       Method
           1       2 hours          C
           1       2 hours          A
           1       2 hours          B
           1       4 hours          A
           1       4 hours          B
           1       4 hours          C
           2       2 hours          C
           2       2 hours          B
           2       2 hours          A
           2       4 hours          C
           2       4 hours          B
           2       4 hours          A
           3       2 hours          C
           3       2 hours          B
           3       2 hours          A
           3       4 hours          B
           3       4 hours          A
           3       4 hours          C
```

In this example we have also used a simplification of the terminology for the block structure: we have used a treatment factor, Method, to specify what is actually a term in the block formula. The strict specification of the structure should have a block factor that is synonymous with Method; but having to specify such duplicate structures can be wasteful, and may not conform to the way in which such experiments are described colloquially. In fact the RANDOMIZE statement in line 7 could be modified further to remove the Mplot factor:

```
RANDOMIZE [BLOCKSTRUCTURE=Block/Time/Method; EXCLUDE=Time; \
   SEED=888667] Method
```

The SEED option determines which randomization Genstat gives. If you use the same seed, you will get the same random numbers, and hence the same randomization (provided the block formula and the block factors are the same as before). If you omit SEED Genstat picks a seed at random, and prints a message to tell you what it is in case you want to reproduce the randomization later.

R.W.P.

10   REML estimation of variance components and analysis of unbalanced designs

This chapter describes the facilities for estimation of variance components and analysis of linear mixed models using the method of residual maximum likelihood (REML), sometimes also known as restricted maximum likelihood.

The REML algorithm estimates the treatment effects and variance components in a linear mixed model: that is, a linear model with both fixed and random effects. Like regression, REML can be used to analyse unbalanced data sets; but, unlike regression, it can also account for more than one source of variation in the data, providing an estimate of the variance components associated with the random terms in the model.

The REML method is applicable in a wide variety of situations. It can be used to obtain information on sources and sizes of variability in data sets. This can be of interest where the relative size of different sources of variability must be assessed, for example to identify the least reliable stages in an industrial process, or to design more effective experiments. REML provides efficient estimates of treatment effects in unbalanced designs with more than one source of error. It can be used to provide estimates of treatment effects that combine information from all the strata of a partially balanced design. It can also be used to combine information over similar experiments conducted at different times or in different places, so that you can obtain estimates that make use of the information from all the experiments, as well as the separate estimates from each individual experiment. Examples from several different areas of application can be found in Robinson (1987).

Section 10.1 examines the large class of linear mixed models that can be fitted using the REML facilities, and describes the underlying methodology. Section 10.2 explains how these models are defined in Genstat using the VCOMPONENTS directive. Section 10.3 describes the REML directive which performs and controls the analysis, and presents two examples to show how to interpret the output. In the first example, the emphasis is on analysis of the fixed model terms, while the second example looks at estimation and interpretation of the variance structure of the data. Section 10.4 goes on to look at the VDISPLAY directive, which is used to display further output from a REML analysis without having to rerun the algorithm. Section 10.5 shows how to use the VKEEP directive to copy results from an analysis into Genstat data structures. Finally, section 10.6 gives some technical information on the way the REML algorithm is implemented within Genstat, and the methods available for reducing the amount of computing time and data space required to fit large or complex models.

10.1   Models for REML estimation

This section describes the linear mixed models that can be fitted using the REML algorithm in Genstat. The fixed and random parts of the model are discussed 10.1.1, before a formal description of the model, is given in 10.1.2. Subsection 10.1.3 then explains the theory behind the residual maximum likelihood method.

10.1.1 Fixed and random effects

Fixed effects are used to describe treatments imposed in an experiment where it is the effect of those specific choices of treatment that are of interest. Random effects are generally used to describe the effects of factors where the values present in the experiment represent a random selection of the values in some larger homogeneous population. It is then possible to make some inference about this population, for example to estimate its variance and to assess the contribution from a factor to the total variation in the data. Predictions of random effects may also be of interest.

For example, consider the split-plot experiment of Section 9.2, used to assess the effects on yield of three oat varieties with four levels of nitrogen application. In this experiment, specific levels of nitrogen application have been used and the aim is to estimate the effects of these levels; so they would be considered as fixed effects in the model, as would the three oat varieties. However, the effects of the actual blocks and plots in the experiment are not of interest in themselves, but they do provide a means of estimating the variability of the more general population of blocks and plots in order to get an estimate of background variation against which to compare the fixed effects. Blocks and plots would therefore be defined as random effects. In this case, the fixed effects correspond to the effects used as treatments in ANOVA and the random effects would correspond to the blocking factors in ANOVA. The REML analysis of this example is shown in 10.3.1.

Another example (from Dempster *et al* 1984) involves an experiment to assess the effect of an experimental compound on maternal performance (see 10.3.3). Twenty-seven female rats (dams) were treated with either a control substance or a high or low dose of an experimental compound in order to examine the effects on their litters. The experimental data were then the weights of each individual pup. The different treatments are specified as fixed effects. Since litter size and the sex of the pup influence weight, these factors must also be included, and as the effects of the specific values of these factors in the experiment are of interest, we define them as fixed effects. Further variation is introduced into the data from the effects of different dams. Since the dams could be considered as a random selection from a wider homogeneous population they are introduced to the model as a random effect. The effect of pups is clearly also a random effect. In fact, since the pups are the units of the experiment, the variation between pups is the error variance component (*units*).

The choice of fixed and random terms is not always determined by the structure of the experiment, but may depend on the information required. For example, variety trials are often carried out over different sites and in several years. If a general assessment of varieties over time is required, then the years present in the trial are considered as a random selection of years, and year would be defined as a random term in the model. On the other hand, if the effect of the specific years present in the trial was to be assessed, year would be defined as a fixed term.

Further discussion of the choice of fixed and random effects can be found in Snedecor and Cochran (1989) and Searle (1971).

In general, both the fixed and random parts of the model are constructed from several factors or variates. The structure of both parts is specified using model formulae, in the same way that models are specified for regression (8.3.1) or analysis of variance (9.1.1). The model for both the fixed and random parts can contain factors and variates and can use the usual crossing and nesting operators (10.2).

In the split-plot example, the fixed part of the model must include the main effects of oat variety and nitrogen application plus their interaction, and is specified as

```
Nitrogen*Variety
```

where `Nitrogen` is a factor indicating nitrogen application on each unit, and `Variety` is a factor indicating the variety grown on each unit. The random part of the model describes the nested blocking structure of subplots within wholeplots within blocks and is specified as

```
Block/Wplot/Subplot
```

where `Block` is a factor indicating which block contains each unit, `Wplot` is a factor indicating which whole plot contains the unit within its block, and `Subplot` is a factor indicating which subplot contains the unit within its whole plot (see 10.3.1).

Similarly, the fixed model in the rat reproductive study described above might be written as `Dose*Sex+Littersz` with random model `Dam/Pup` (10.3.3).

10.1.2 The linear mixed model

Returning to the split-plot example, the model for the yield y_{ijk} from block i, wholeplot j, subplot k is

$$y_{ijk} = m + v_r + a_s + va_{rs} + b_i + w_{ij} + \varepsilon_{ijk}$$

where the fixed part of the model consists of: m the overall constant; v_r the main effect of variety r (where r indicates the variety assigned to unit ijk); a_s the main effect of nitrogen application at level s (where s indicates the nitrogen application on unit ijk); and va_{rs} their interaction. The random model terms are b_i the effect of block i, w_{ij} the effect of wholeplot j within block i, and ε_{ijk} the random error for unit ijk (which here is the same as the subplot effect, since the subplots are the smallest units of the experiment).

This model can be re-written as a general linear mixed model by grouping the fixed and random terms and using matrix and vector notation:

$$y = X\alpha + Z\beta + \varepsilon$$

where

    y is a vector of data (length n)
    α is a vector of fixed effects (length p) with $n \times p$ design matrix X
    β is a vector of random effects (length q) with $n \times q$ design matrix Z
    ε is a vector of random error (length n).

In the split-plot example above, there are 72 units. The vector α contains the fixed effects m, v_1, v_2, v_3, $a_1 \ldots a_4$ and $va_{11} \ldots va_{34}$. The rows of matrix X correspond to the units of the experiment and the columns correspond to the fixed effects. The values in each row of X are 1 or 0 to indicate presence or absence of each effect for that unit. Similarly, the vector β contains the random effects b_i ($i=1\ldots6$) and w_{ij} ($i=1\ldots6$; $j=1\ldots3$) and matrix Z indicates which units occur within each block and wholeplot.

More generally, the random model $Z\beta$ is constructed from c model terms (in this example, it consists of the two random model terms `Block` and `Block.Wplot`). Z and β can then be partitioned as $Z = \{ Z_1 \mid Z_2 \mid \dots \mid Z_c \}$ and $\beta = (\beta_1 \, \beta_2 \, \dots \, \beta_c)'$ where β_i is a vector of length q_i. The model can then be written in terms of the separated random model terms as

$$y = X\alpha + \sum_{i=1}^{c} Z_i \beta_i + \varepsilon$$

It is assumed that the random effects β_i and ε are mutually independent Normally distributed random variables with zero mean, such that $Cov(\varepsilon)=\sigma^2 I_n$, where I_n is the identity matrix of size n, $Cov(\beta_i)=\sigma_i^2 C_i$ where C_i is a symmetric matrix of size q_i, and $Cov(\beta_i \beta_j)=0$ for $i \neq j$. Therefore, effects that occur in different random model terms are independent. This means that the variance-covariance matrix for the whole set of random effects takes a particularly simple form, since $Cov(\beta)=diag\{\sigma_1^2 C_1,\dots \sigma_c^2 C_c\}$ is block diagonal. Furthermore, it is usually assumed that effects within the same random model term are also independent, that is, $Var(\beta_i)=\sigma_i^2 I_{q_i}$. Then $Var(\beta_i)$ is diagonal, which means that units of y are correlated only if they take the same value of some element in β. The variance parameters σ_i^2 associated with the random model terms are called the variance components of the model. The variance parameter σ^2 associated with the random error ε is called the residual variance (or the variance of the factor *units*). The REML algorithm estimates the variance components using residual maximum likelihood, and then uses the variance parameter estimates to form the generalized least squares estimates of the treatment effects and the best linear unbiased predictors (BLUPs) of the random effects.

The general linear model defined above has the properties

$$E(y) = X\alpha$$

$$Cov(y) = V$$

$$= Z Cov(\beta) Z' + \sigma^2 I_n$$

$$= \sum_i \sigma_i^2 Z_i C_i Z_i' + \sigma^2 I_n$$

$$= \sigma^2 (\sum_i \gamma_i Z_i C_i Z_i' + I_n) \qquad \text{where} \quad \gamma_i = \frac{\sigma_i^2}{\sigma^2}$$

$$= \sigma^2 H.$$

If $C_i = I_{qi}$ then $H = Z\Gamma Z'+I_n$ where $\Gamma = diag\{ \gamma_1 I_{q1} \dots \gamma_c I_{qc} \}$.

The expected value of the data is a function of the fixed terms alone, and its variance-covariance matrix can be expressed either as a function of the variance components $\{ \sigma_i^2 ; i=1\dots c \}$ or as a function of σ^2 and the set $\{ \gamma_i ; i=1\dots c \}$ which are ratios of the variance components to σ^2, the residual variance, and are called the "gammas". When the model is defined solely in terms of its expectation and variance-covariance matrix, the components can be interpreted as constituent parts of the variance-covariance matrix. Therefore, so long as the variance-covariance matrix of the data remains positive definite overall, there is no constraint on the individual variance components to remain positive.

10.1.3 REML estimation of variance components and analysis of linear mixed models

The method of residual maximum likelihood (REML) was introduced by Patterson and Thompson (1971). It was developed in order to avoid the biased variance component estimates that are produced by ordinary maximum likelihood estimation: because maximum likelihood estimates of variance components take no account of the degrees of freedom used in estimating treatment effects, they have a downwards bias which increases with the number of fixed effects in the model. This in turn leads to under-estimates of standard errors for fixed effects, which may lead to incorrect inferences being drawn from the data. Estimates of variance parameters which take account of the degrees of freedom used in estimating fixed effects, like those generated by ANOVA in balanced data sets, are more desirable.

The REML method splits the data into two parts: treatment contrasts which contain information only on the fixed effects; and error contrasts (that is, all contrasts with zero expectation) which contain information on the variance components. The error contrasts alone are then used to estimate the variance parameters, since they contain all of the information available on the variance parameters. This is done by projecting the data into the residual space: the vector space of error contrasts, where all the data contrasts have zero expectation. The projected data has log-likelihood RL where

$$-2RL(y) = (n-p^*)\log 2\pi - \log|X'X| + \log|V| + \log|X'V^{-1}X| + (y-X\hat{\alpha})'V^{-1}(y-X\hat{\alpha})$$

with n as the number of data values and p^* as the number of degrees of freedom used in estimating fixed effects; that is, rank(X). Variance components are then estimated by maximizing the log-likelihood function RL of the projected data.

The log-likelihood of the original data is L, where

$$-2L(y) = n\log 2\pi + \log|V| + (y-X\alpha)'V^{-1}(y-X\alpha).$$

Compared with the usual log-likelihood L, the log-likelihood of the residual data, RL, contains several extra terms. The only extra term involving the variance components (which is therefore the only extra term used in estimating the variance components) is $\log|X'V^{-1}X|$ which effectively removes the degrees of freedom used in estimating the fixed effects.

To take the simplest example, the maximum likelihood estimate of the variance of a set of n observations y_i from the same population would be $\Sigma(y_i-\bar{y})^2/n$ which has expectation $(n-1)\sigma^2/n$, whereas the more usual unbiased (REML) estimate is $\Sigma(y_i-\bar{y})^2/(n-1)$.

Similarly, in an orthogonal design, the REML estimates of the variance components are identical to the unbiased estimates that can be produced from residual mean squares in the analysis of variance. However, REML can also be used with unbalanced data to produce estimates of variance components that do not suffer the downward bias associated with maximum-likelihood estimation.

Once the variance components have been estimated, they are used to construct an estimate of the variance-covariance matrix, $\hat{V}$. The fixed effects are then estimated by generalized least squares

$$\hat{\alpha} = (X'\hat{V}^{-1}X)^{-1}X'\hat{V}^{-1}y \quad \text{with} \quad \text{Var}(\hat{\alpha}) = (X'\hat{V}^{-1}X)^{-1}.$$

Predictions of the random effects are given by the best linear unbiased predictors (BLUPs)

$$\beta = (Z'Z+\Gamma^{-1})^{-1}Z'(y-X\hat{\alpha}).$$

10.2   Setting up a model for analysis by REML

The VCOMPONENTS directive is used to set up the linear mixed model to be analysed by REML in a similar way to which an ANOVA model is set up using the TREATMENTSTRUCTURE and BLOCKSTRUCTURE directives (Chapter 9). The syntax of VCOMPONENTS is summarized in 10.2.1, then the rest of the section explains certain aspects in more detail. The parameterization of the fixed model is described 10.2.2, and 10.2.3 shows how the random model is defined. The last two parts of this section, 10.2.4 and 10.2.5, show how to modify the model by setting constraints on the variance parameters and specifying non-standard covariance matrices for random model terms.

10.2.1 The VCOMPONENTS directive

VCOMPONENTS   defines the variance-components model for REML.

Options

FIXED = *formula*	Fixed effects; default *
ABSORB = *factor*	Defines the absorbing factor; default * i.e. none
CONSTANT = *string*	How to treat the constant term (estimate, omit); default esti
RELATIONSHIP = *matrix*	Defines relationships constraining the values of the components; default *

Parameters

RANDOM = *formula*	Random effects
INITIAL = *scalars*	Initial values for each component
CONSTRAINTS = *strings*	How to constrain each component (none, positive, fixrelative, fixabsolute); default none
COVARIANCEMATRIX = *symmetric matrices*	
	Matrix specifying the pattern of covariances between the effects of each random term; default * assumes the identity matrix

---

The VCOMPONENTS directive specifies the linear mixed model to be fitted by subsequent REML statements. The fixed terms in the model are defined by a model formula supplied using the FIXED option, and the random model terms are defined by a model formula supplied by the RANDOM parameter. Thus, for example, the model for the split-plot experiment described in 10.1.1 would be specified by

    VCOMPONENTS [FIXED=Nitrogen*Variety] RANDOM=Block/Wplot/Subplot

where Nitrogen and Variety are factors indicating the treatments applied to each unit, and Block, Wplot, and Subplot are factors indicating the block, wholeplot (within block) and subplot (within wholeplot) to which each unit belongs; see Example 10.3.1.

The model for the rat reproduction experiment would be

```
VCOMPONENTS [FIXED=Dose*Sex+Littersz] RANDOM=Dam/Pup
```

In this case, each pup is a separate unit. The analysis of this experiment is shown in 10.3.3.

If you do not specify the fixed model, the default fixed model consists of just the constant term, which then becomes the grand mean. If the random model is unset, only a single source of variation (the residual component) is used. In this case, REML will produce the same analysis as the regression facilities which, since they take full advantage of the simple variance structure of the model, would be computationally more efficient. Note that any model term found both in both the fixed and the random model will be deleted from the random model and retained in the fixed model only. A complete definition of the operators available in model formulae is given in 9.1.1.

As well as defining the basic model using the FIXED option and RANDOM parameter, the VCOMPONENTS directive can also be used to modify the model or to add extra information.

A constant term is automatically included in the fixed part of the model but this can be omitted by setting option CONSTANT=omit, provided you have also specified a fixed model.

You can supply initial values for the variance components using the INITIAL parameter, and you can impose constraints on the variance components using either the RELATIONSHIP option or the CONSTRAINTS parameter. By default, all the ratios γ_i have initial values of 1. The CONSTRAINTS parameter allows you to request that any variance component should be held positive or fixed at its initial value. The default setting, none, allows the variance components to become negative, provided the overall estimated variance-covariance matrix for the data remains positive definite. The RELATIONSHIP option can be used to define linear relationships between the variance components, for example that component A should be constrained to be twice component B. Full details are given in 10.2.4.

By default, it is assumed that within each random model term the variance-covariance matrix takes the form $\sigma_i^2 I$, that is, equal variation and no correlation between different levels of the factor. You can use the COVARIANCEMATRIX parameter to specify a matrix C_i which defines some other pattern of covariances within a model term (10.2.5).

The ABSORB option allows you to specify a factor from either the fixed or the random model to act as an absorbing factor for the model. The absorbing factor is used to divide the model terms into two groups; this partition is then used in calculations during the fitting process to reduce the size of the matrices that have to be inverted and stored. Use of an absorbing factor can therefore save computing time and data space. However, although exactly the same model is fitted when an absorbing factor is used, some of the standard errors are unavailable (see 10.3.2). A good choice of absorbing factor might be a factor with a large number of levels, or any factor whose effects and standard errors are not of interest. The choice of an absorbing factor is considered in detail in 10.6.2.

10.2.2 The Fixed Model

You define the fixed terms to be included in the linear mixed model using the FIXED option of VCOMPONENTS. The model formula that you specify can include both factors and variates.

Factors are used, as in regression and analysis of variance, to represent qualitative effects.

Consider a simple example where a factor Dose might be used to describe the effect of different doses (none, low or high). This would be specified using the FIXED option of VCOMPONENTS

    VCOMPONENTS [FIXED=Dose]

and would lead to the model

    $y_{ij} = a + b_i + \varepsilon_{ij}$

where a is the overall constant, and the parameters b_i describe the effects of the different doses. As in regression models, unless the constant term is omitted, some form of constraint is needed to avoid over-parameterization. For model terms containing only factors, the parameters corresponding to the first levels of the factors are constrained to be zero, as in Genstat regression (8.3.2). The parameters for other levels of the factor are then comparisons with the first level. Here, for example, b_1 (Dose=none) would be set to zero, and parameter b_2 (Dose=low) would estimate the difference between the low dose and no dose. Similarly, b_3 would estimate the difference between high dose and no dose. Note that the parameter a is not the grand mean, since it contains both the grand mean and the effect of the first level of factor Dose; it is an estimate of the expected value of y when the first level of the factor is applied. The parameters b_i are then the adjustments to be added to a to estimate the expected value for the other levels of Dose.

    Variates can be included in the model to represent a linear relationship between the y-variate and a covariate. For example, if a variate X is added to the FIXED model above

    VCOMPONENTS [FIXED=Dose+X]

the model to be fitted becomes

    $y_{ij} = a + b_i + cx_{ij} + \varepsilon_{ij}$

where c estimates the slope of the linear relationship between the expectation of y and the covariate X. Any covariates are automatically centred for the analysis.

    When interactions of factors and variates are included in the model, then terms are added to fit a different slope parameter (c above) for each level of the factor. Again, constraints are required to avoid over-parameterization, but this time it is parameters associated with the last level of the factors that are set to zero. If the interaction between X and Dose is added to our example

    VCOMPONENTS [FIXED=Dose*X]

then the model becomes

    $y_{ij} = a + b_i + (c + d_i)x_{ij} + \varepsilon_{ij}$

The parameter c is then the slope for the last level of factor Dose (that is, high dose) and the parameters d_i are the differences between the slope for other doses and high dose.

    Whenever a parameter is completely aliased with parameters fitted earlier in the model, the aliased parameter is set to zero.

    The rules above describe how the fixed model is parameterized when the constant term is included, using the default option setting CONSTANT=estimate. If the constant term is omitted, by setting CONSTANT=omit, for example

    VCOMPONENTS [FIXED=Dose*X; CONSTANT=omit]

then, where over-parameterization occurs, the parameters corresponding to the last levels of factors are set to zero. The only exception occurs when some of the variance components have negative estimates. In this case, when the constant is omitted from the model, the parameter estimates for fixed model terms are not easily interpretable and a Genstat prints a warning.

You can set a limit on the number of factors and variates allowed within each term of the fixed model using the FACTORIAL option of the REML directive (see 10.3.1).

10.2.3 The Random Model

The model formula for the random part of the model is specified using the RANDOM parameter of the VCOMPONENTS directive and can also include both factors and variates. Each random model term defines a set of random effects and an associated variance component. For example, the nested block structure of the split-plot experiment is specified by

    VCOMPONENTS [FIXED=Nitrogen*Variety] RANDOM=Block/Wplot/Subplot

so three variance components are included in the model representing the variation due to the blocks, the wholeplots and the subplots.

The random term which corresponds to the residual variation between units, or error variance σ^2, is called the residual component and is considered separately from the rest of the random terms within the algorithm. If the residual component is not specified, it is automatically added onto the end of the random model. Here the Block.Wplot.Subplot term is the residual component since it represents the variation between units at the lowest level of the experiment; that is, the subplots. The same model could have been specified as

    VCOMPONENTS [FIXED=Nitrogen*Variety] RANDOM=Block/Wplot

and the residual component would have been added automatically. Genstat would then refer to it as *units*.

If the model is viewed in terms of random effects, then for a random model term specified by factors, an effect is included for each combination of the levels of the factors. For example, the random model in the split-plot example is

$$y_{ijk} = \{\text{fixed model terms}\} + b_i + w_{ij} + \varepsilon_{ijk}$$

with one parameter b_i ($i=1...6$) for each of the blocks and one parameter w_{ij} ($i=1...6, j=1...3$) for each wholeplot within each block. To avoid over-parameterization, the random effects are constrained so that their sum is zero within each model term. When variance components are estimated as negative values, standard errors are not available for the effects of the corresponding random terms.

You can also include variates in the RANDOM formula to specify random covariates. This may be useful, for example, in specifying models where a linear response to an explanatory variable varies randomly between groups or individuals.

In general, care must be taken not to specify the residual component more than once unless some form of constraint is imposed. If there are no constraints, no estimation will then be possible.

In a very few cases, you may wish to add the residual component onto the end of the random model even though it has already been specified. For example, in some algorithms for fitting generalized linear mixed models, it is necessary to estimate the residual component on the linear predictor scale whilst fixing the variance parameter on the natural scale. You can tell Genstat to add an 'extra' residual component to the model by using the string `'*units*'` at the end of the random model. For example

```
VCOMPONENTS Block/Plot+'*units*'; CONSTRAIN=none,none,fix; \
   INITIAL=1,1,2
REML [WEIGHTS=W] Y
```

will produce the variance structure

$$\sigma_b^2 \, Z_b I_b Z_b' + \sigma_p^2 \, I_n + 2 \, \text{diag}\{ \, w_i \, ; \, i=1...n \, \}.$$

This facility should rarely be needed and should be used with care.

10.2.4 Setting initial values and constraints on variance components

Computing time can be saved by specifying good initial values for some, or all, of the variance parameters. This is especially helpful if the data set is large or many model terms are to be fitted. Since it is the ratios of the variance components to the error variance (the gamma ratios defined in 10.1.2) that are estimated, initial values for all except the residual component should be specified in terms of these ratios. This is done using the INITIAL parameter of the VCOMPONENTS directive. The initial values run in parallel with the expanded form of the RANDOM model, which follows the rules given in 9.1.1. For example, the split-plot model `Block/Wplot/Subplot` expands to `Block + Block.Wplot + Block.Wplot.Subplot` and would require three initial values

```
VCOMPONENTS RANDOM=Block/Wplot/Subplot; INITIAL=7,3,1
```

The random model `Row*Column` would expand to `Row + Column + Row.Column` and would also require three initial values

```
VCOMPONENTS RANDOM=Row*Column; INITIAL=5,8,20
```

As usual, the list of initial values is recycled if it is shorter than the list of terms in the RANDOM model formula. You must remember that the residual component will be added onto the end of the random model (unless it is specified explicitly) and so you must give an initial value for the error variance at the end of the list. For example, if the random model for the split-plot experiment is specified as `Block/Wplot`, then the residual component will be added onto the end of the random model to give three terms in total, so three initial values must be specified

```
VCOMPONENTS RANDOM=Block/Wplot; INITIAL=7,3,1
```

Any gamma for which no initial value is available should be given an initial value of 1, which is the default when no initial values are specified.

By default, the estimates of variance components are allowed to take any non-zero value (positive or negative) such that the variance-covariance matrix of the data (*V*) remains positive definite. However, you may sometimes wish to constrain the components to remain positive. This can be done by setting parameter CONSTRAINTS=positive. You can give a list of strings to specify different constraints for each term in the random model. These again run in parallel with the expanded form of the random model (plus residual component if necessary), and will be recycled if the list is too short. For example,

```
VCOMPONENTS RANDOM=Block/Wplot; CONSTRAINTS=positive
```

would constrain all estimates of components to be positive in the split-plot example. If the `Block` component alone was to be held positive, the command would be

```
VCOMPONENTS RANDOM=Block/Wplot; CONSTRAINTS=positive,none,positive
```

The constraints can be relaxed again to allow negative components by setting `CONSTRAINTS=none`, which is the default.

If the value of a gamma or a variance component is sufficiently well known for there to be no need for further estimation, it can be fixed at its initial value. You can fix the gamma (that is, the ratio of the variance component to the error variance) for a model term by setting `CONSTRAINTS=fixrelative` (=`fix` for short). For example, the command

```
VCOMPONENTS Row*Column; INITIAL=5,8,20; CONSTRAINTS=none,fix,none
```

means that the gamma for the second component will be fixed at 8, that is, the `Column` component will be estimated by $8\sigma^2$.

You can also fix the absolute value of the variance component at its initial value by setting `CONSTRAINTS=fixabsolute`. You must then specify the value of the component (not the gamma ratio) in the list of initial values. Thus the command

```
VCOMPONENTS Row*Column; INITIAL=5,8,20; CONSTRAINTS=none,fixabs,none
```

means that the final estimated value of the `Column` component will be 8.

Note that components that are constrained to be fixed (relative or absolute) at their initial values do not appear in the output as estimated variance components although they are included in the model. Components fixed at zero will be reset to $10^{-3}\sigma^2$ with a warning.

Constraints on the residual component are treated slightly differently to those on the other components. Clearly, the error variance cannot be allowed to become negative, so the default constraint is that the error variance remains positive. The error variance can be fixed at its initial value, using either the `fixabsolute` or `fixrelative` setting. Note that a single parameter setting `CONSTRAINTS=fix` will be recycled, to fix all the gamma ratios and the residual component at their initial values.

Because estimation is carried out in terms of the gamma ratios, the error variance and their inverses, none of these variance parameters are allowed to become zero. No estimation can take place if the error variance is zero, which may happen because the fixed model contains as many parameters as data values. Any random term that is found to be completely aliased with other model terms (so that it cannot be estimated) will be deleted automatically from the random model, and the analysis will be rerun. If any of the gammas becomes very close to zero for any other reason, it will be reset to a small positive value ($10^{-3}\sigma^2$); REML generates a warning diagnostic if this has to be done repeatedly.

If components that have been constrained to be positive are estimated to be negative, they will also be reset to a small positive value ($10^{-3}\sigma^2$). If a component remains negative when the algorithm converges, a warning will be given since the constrained component is being held at an artificial value and may bias other estimates. In this case, it may be wise to estimate the value of the component without constraint to investigate whether the component is effectively zero (see Section 10.3.3) or whether it takes a relatively large negative value, which may indicate some unexpected structure in the variability of the data. Omission of an important term from the fixed model can lead to unexpected negative components, so the structure of the data should also be checked in order to detect any missing terms in the fixed model. Constraining components to be positive does save some data space which may be useful for very large problems.

You can also apply linear equality constraints between the variance components. These are defined by a matrix which is supplied using the RELATIONSHIP option of VCOMPONENTS. The matrix must be square, with one row and one column for each component (including the residual component, even if this is not specified explicitly in the random model). The entries in each row of the matrix define the constraints on the component corresponding to that row, in terms of multiples of the other components.

For example, consider the random model R*C, and suppose we wish to constrain the component for R to be twice the component for C, that is $\sigma_R^2 = 2\sigma_C^2$. This random model has 3 terms, therefore we need a 3×3 matrix. The rows and columns of the matrix correspond to the terms of the expanded model R+C+R.C in order. The first row is used to define constraints on the component for the first random model term, which is the R component. Since this is to be constrained to be twice the C component, the values for this row are 2 for the column corresponding to the C component (the second model term and therefore the second column) and zero elsewhere. The second row is used to define constraints on the second component, C. Since this component is unconstrained, the row has value 1 for the C component (second column), and zeros elsewhere, that is, the C component is constrained to be itself. Similarly R.C, the residual component, is unconstrained and the final row has zeros except for value 1 in the R.C (the third) column. The statements to define this model would then be

```
MATRIX [ROWS=3; COLUMNS=3; VALUES=0,2,0, 0,1,0, 0,0,1] M
VCOMPONENTS [FIXED=F; RELATIONSHIP=M] R*C
```

giving

component	R	C	R.C	Constraint
R	0	2	0	$\sigma_R^2 = 2\sigma_C^2$
C	0	1	0	none
R.C	0	0	1	none

In this case, since the residual component is specified in the model as R.C, there is no need to add an extra row and column to the matrix. However, if the same model had been specified as R+C, a third row and column would still have been needed in the matrix to correspond to the residual component.

If a component is defined to be a multiple of the residual component, it will be treated as if it had been constrained fixed using parameter CONSTRAINTS=fix and will not appear in the list of estimated variance components.

10.2.5 Covariance Matrices for Random Model Terms

You can use the COVARIANCEMATRIX parameter to specify known covariance matrices C_i for random model terms, that is, to introduce correlation between different levels within a random model term. By default $Cov(\beta_i)=\sigma_i^2 I_{qi}$, that is, no correlation is allowed between levels of a factor. However, there are some circumstances where this is not a reasonable assumption. For example, if the data are repeated measurements, data collected from consecutive timepoints might be expected to be more highly correlated than data collected from widely-spaced timepoints. In this case, you can specify $Cov(\beta_i)=\sigma_i^2 C_i$, for a given matrix C_i.

The settings of the COVARIANCEMATRIX parameter run in parallel with the random model, and allow you to specify a variance-covariance matrix C_i for each random model term. The default, *, indicates that the identity matrix is to be used. If a variance-covariance matrix is specified, it must be a symmetric matrix with number of rows equal to the possible number of different factor combinations in the random model term.

For example, consider a field experiment consisting of 3 blocks each containing 6 plots laid out in a row where there is evidence of positive correlation between neighbouring plots which diminishes with distance. Suppose further investigation indicates that an auto-regressive structure of order 1, with parameter 0.5, could model the correlation structure. This structure can be included in the model for REML analysis as shown in Example 10.2.5. The random model is Block/Plot and the covariance structure applies to the Block.Plot model term. This model term has 18 levels (=3×6) and so a symmetric matrix with 18 rows is required. Assuming that the blocks are separated in the field, the correlations will occur only between plots within the same block, so the covariance matrix C contains the covariance matrix for correlation within blocks on the diagonal and zeroes elsewhere.

Example 10.2.5

```
 1   " Declare covariance matrix and initialize values "
 2   SYMMETRICMATRIX [ROWS=18] C
 3   CALCULATE C = 0
 4   " Declare within-blocks submatrix "
 5   SYMMETRICMATRIX [ROWS=6] PLOTVCOV
 6   " Set diagonal values of submatrix "
 7   CALCULATE PLOTVCOV$[ 1...6 ; 1...6 ] = 1
 8   " Set off-diagonal values of submatrix "
 9   CALCULATE PLOTVCOV$[ 2...6 ; 1...5 ] = 0.5
10   &         PLOTVCOV$[ 3...6 ; 1...4 ] = 0.5**2
11   &         PLOTVCOV$[ 4...6 ; 1...3 ] = 0.5**3
12   &         PLOTVCOV$[ 5,6   ; 1,2   ] = 0.5**4
13   &         PLOTVCOV$[ 6     ; 1     ] = 0.5**5
14   PRINT PLOTVCOV
```

PLOTVCOV

1	1.0000				
2	0.5000	1.0000			
3	0.2500	0.5000	1.0000		
4	0.1250	0.2500	0.5000	1.0000	
5	0.0625	0.1250	0.2500	0.5000	1.0000
6	0.0313	0.0625	0.1250	0.2500	0.5000	1.0000
	1	2	3	4	5	6

```
15   " Put submatrix into full matrix "
16   CALCULATE C$[ !(1...6), !(7...12), !(13...18) ] = PLOTVCOV
```

```
17   " Declare model for REML analysis "
18   VCOMPONENTS [FIXED=T] RANDOM=Block/Plot; COVARIANCE=*,C
```

When the residual component is specified in the random model with a covariance matrix, as above, its variance component will be estimated as a gamma ratio, and the usual residual term $\sigma^2 I_n$ will still be added. Example 10.2.5 defines the variance structure

$$V = \sigma^2 (\gamma_b Z_b I_3 Z_b' + \gamma_p C + I_{18})$$

where Z_b is the part of the random model design matrix Z corresponding to the blocks, and C is defined as above. If the difference between the matrix C and the identity matrix I_{18} is very small, then estimation of both γ_p and σ^2 becomes difficult and the algorithm may not converge. It should be noted that the inclusion of the residual term in the random model in this way may greatly increase the size of the matrices that have to be inverted within the algorithm, so it may not be possible to fit models of this type for very large data sets.

There is no direct method of excluding the term $\sigma^2 I_n$ from the variance structure.

Note that if the COVARIANCEMATRIX parameter is used for any of the random terms, no absorbing factor is allowed.

In Release 3.1 of Genstat, it is not possible to estimate the matrices C_i. Parametric covariance matrices will be allowed in future releases, when these facilities will be extended.

10.3   Analysing data using REML

This section explains how to use the REML directive to fit a mixed model, and illustrates the information that is available from the analysis. Subsection 10.3.1 gives details of the REML directive and illustrates it using the split-plot example. This shows how REML produces the same results as ANOVA for orthogonal designs (see 9.2), although the results are presented slightly differently. As for ANOVA, tables of means and effects are available for fixed model terms. REML also provides these tables for random model terms (see 10.3.2).

In general, REML analyses have two purposes: to study fixed effects when there are several sources of variability, and to estimate the variance components and assess the relative importance of the sources of variability. Of course, most analyses will involve both purposes to some extent, but for clarity the rest of this section will look at the two situations separately. Subsection 10.3.3 illustrates the output from a REML analysis to study the fixed effects in the rat reproduction experiment. This is an unbalanced data set where there is more than one source of variation in the data. REML estimation in this situation is more appropriate than a linear regression analysis since it makes use of all the available information on the fixed effects. Subsection 10.3.4 illustrates the output available for assessing the structure of the variability in the data from a factory production process. In this situation REML provides estimates of the variance components and a formal assessment of the random model.

10.3.1   The REML directive

Once you have defined a variance components model using VCOMPONENTS, you can then fit the model to the data (the y-variates) using the REML directive.

REML fits a variance-components model by residual (restricted) maximum likelihood.

Options

PRINT = *strings*	What output to present (`model`, `components`, `effects`, `means`, `stratumvariances`, `monitoring`, `vcovariance`, `deviance`, `Waldtests`); default `mode`,`comp`,`stra`
FACTORIAL = *scalar*	Limit on the number of factors or covariates in each fixed term; default 3
PTERMS = *formula*	Terms (fixed or random) for which effects or means are to be printed; default `*` implies all the fixed terms
PSE = *string*	Standard errors to be printed with tables of effects and means (`differences`, `estimates`, `alldifferences`, `allestimates`, `none`); default `diff`
WEIGHTS = *variate*	Weights for the analysis; default `*` implies all weights 1
MVINCLUDE = *string*	Whether to include units with missing factor values (`yes`, `no`); default `no`
SUBMODEL = *formula*	Defines a sub-model of the fixed model to be assessed against the full model
RECYCLE = *string*	Whether to reuse the results from the estimation when printing or assessing a sub-model (`yes`, `no`); default `no`
RMETHOD = *string*	Whether to use all the random terms or just the final one (`*units*`) when calculating RESIDUALS (`final`, `all`); `final`
MAXCYCLE = *scalar*	Limit on the number of iterations; default 10
TOLERANCES = *variate*	Tolerances for matrix inversion; default `*` i.e. appropriate values assumed for the type of computer concerned

Parameters

Y = *variates*	Dependent variates
RESIDUALS = *variates*	Residuals from each analysis
FITTEDVALUES = *variates*	Fitted values from each analysis
SAVE = *pointers*	Saves the details of each analysis for use in subsequent VDISPLAY and VKEEP directives

---

The REML directive allows control over both the estimation process and the output produced.

The first parameter, Y, lists the variates that are to be modelled. You can restrict any of the y-variates or any of the factors or variates in the fixed and random models to indicate that only a subset of the units are to be used in the analysis (see 4.4.1). If more than one of these vectors is restricted, they must all be restricted to the same set of units.

The parameters FITTEDVALUES and RESIDUALS allow you to store the fitted values and residuals from the fitted model. Parameter SAVE can be used to name the REML save structure for use with later VKEEP and VDISPLAY directives.

The three options PRINT, PTERMS, and PSE all control the printed output. The PRINT option selects the output to be displayed. The different settings are explained in detail in different sections of this chapter, as indicated below:

model	description of model fitted (10.3.1)
components	estimates of variance components (10.3.4)
effects	tables of effects; that is, estimates of parameters α and β (10.3.2)
means	tables of means; that is, predicted means for factor combinations (10.3.2)
stratumvariances	approximate stratum variances from a decomposition of the information matrix for the variance components (10.3.4)
monitoring	monitoring information at each iteration (10.4)
vcovariance	variance-covariance matrix of the estimated components (10.4)
deviance	deviance of the fitted model ($-2 \times$ log-likelihood RL) plus deviance of submodel if specified (10.3.3 and 10.3.4)
waldtests	Wald tests for all fixed terms in model (10.3.3)

The default setting, PRINT=model,components,stratumvariances, gives a description of the model fitted plus estimates of the variance components and the approximate stratum variances. Options PTERMS and PSE control the tables of means and effects that are printed, and their accompanying standard errors (see 10.3.2).

The FACTORIAL option is used to set a limit on the number of factors and variates allowed in each fixed term; any term containing more than that number is deleted from the model.

The MVINCLUDE option allows the inclusion of units that are normally excluded from the analysis. Units for which the y-variate has a missing value are always excluded. Units with missing values in any of the factors or variates in the model terms are usually also excluded, but can be included by setting MVINCLUDE=yes.

The WEIGHTS option can be used to specify a weight for each unit in the analysis. This is useful when it is suspected that the size of the random error varies between units. For example, if the random error for unit i is known to have variance $v_i\sigma^2$, a weight variate should be used containing values $w_i=1/v_i$. Note that if the error component varies between different groups, dummy factors can be used to estimate the error component for each group separately. These dummy factors would be defined for all but one of the groups. For each group, the corresponding dummy factor would take a different value for each observation within the group and have missing values for observations from other groups. The model would then be fitted using the dummy random factors as well as any other random terms, and with option setting MVINCLUDE=yes.

Option SUBMODEL is used to specify a sub-model of the fixed model. This model will be fitted as well as the full fixed model, using a slightly modified version of the algorithm, and the difference in deviances between the full and sub-model can be used as a likelihood-based test to assess the importance of the fixed terms dropped from the full model. This is explained in detail in 10.3.4. Once the full model has been fitted, the RECYCLE option can be used to test a series of sub-models of the fixed model. If option RECYCLE=yes is set, then only the estimation for the sub-model is performed. Information for the full fixed model is picked up from the corresponding save structure. When the RECYCLE option is set, only the deviance and model settings of PRINT can be used. Note that the change in deviance will not be printed unless the setting PRINT=deviance is used.

The RMETHOD option controls the way in which residuals and fitted values are formed. For the default setting RMETHOD=final, the fitted values $\hat{y}$ are calculated from all the fixed and random effects: $\hat{y} = X\hat{\alpha} + Z\hat{\beta}$. The residuals are the difference between the data and the fitted values and, in this case, are estimates of the values of ε, the *units* random error. These residuals can be used to check the Normality and variance homogeneity assumptions for the random error. To get fitted values constructed from the fixed terms alone, omitting all random terms, the setting RMETHOD=all must be used. The fitted values are then $\hat{y} = X\hat{\alpha}$, and the residuals are predictors of $Z\beta+\varepsilon$. The library procedure VPLOT can also be used to produce various diagnostic plots.

The TOLERANCES option controls the tolerances for matrix inversion. Three values can be specified in a variate. The first two values are matrix inversion tolerances for the information matrix and the mixed model equations respectively and take the value 10^{-5} by default. The third value is used to detect zero frequency counts for factor combinations in the mixed model equations: 10^{-6} is used by default.

Option MAXCYCLE can be used to change the maximum number of iterations performed by the algorithm from the default of 10.

Example 10.3.1a shows how to analyse the split-plot design in Section 9.2.1 using REML.

Example 10.3.1a

```
  1   "
 -2     Split-plot design
 -3     (Yates 1937, page 74; also John 1971, page 99).
 -4   "
  5   UNITS [NVALUES=72]
  6   FACTOR [LEVELS=6] Blocks
  7   & [LEVELS=3] Wplots
  8   & [LEVELS=4] Subplots
  9   GENERATE Blocks,Wplots,Subplots
 10   FACTOR [LABELS=!T(Victory,'Golden rain',Marvellous)] Variety
 11   & [LABELS=!T('0 cwt','0.2 cwt','0.4 cwt','0.6 cwt')] Nitrogen
 12   VARIATE Yield; EXTRA=' of oats'
 13   READ [SERIAL=yes] Nitrogen,Variety,Yield
```

Identifier	Minimum	Mean	Maximum	Values	Missing
Yield	53.0	104.0	174.0	72	0

Identifier	Values	Missing	Levels

```
        Nitrogen        72          0          4
        Variety         72          0          3
 26
 27  VCOMPONENTS [FIXED=Nitrogen*Variety] Blocks/Wplots/Subplots
 28  REML Yield
```

28...

***** REML Variance Components Analysis *****

Response Variate : Yield of oats

Random model   : Blocks+Blocks.Wplots+Blocks.Wplots.Subplots
Fixed model   : Constant+Nitrogen+Variety+Nitrogen.Variety

Number of units   : 72
No absorbing factor

*** Estimated Variance Components ***

Random term   Component   S.e.

Blocks   214.5   168.8
Blocks.Wplots   106.1   67.9
Blocks.Wplots.Subplots   177.1   37.3

*** Approximate stratum variances ***

                                              Effective d.f.
Blocks   3175.1   5.00
Blocks.Wplots   601.3   10.00
Blocks.Wplots.Subplots   177.1   45.00

* Matrix of coefficients of components for each stratum *

                   Blocks   12.00   4.00   1.00
            Blocks.Wplots   0.00   4.00   1.00
   Blocks.Wplots.Subplots   0.00   0.00   1.00
```

---

This example shows the default output from REML. First, a summary of the model is given by the default setting PRINT=model; this includes details of the response variate, the fixed and random model terms, the number of units analysed and whether options such as the absorbing factor, weights, or mvinclude are set. The number of units analysed takes account of units excluded because of restrictions, zero weights, or missing values in either the response variate or the factors and variates in the model. After the model description, the estimates of the variance components are printed with their standard errors. Finally, approximate stratum variances, derived from a decomposition of the information matrix for the variance components, are given together with the matrix of coefficients used to construct the stratum variances from the components.

In this orthogonal design, the approximate stratum variances are exactly the same as the
```

residual mean squares from the strata in Example 9.2.1. Note that this will be the case only when the design is orthogonal: that is when the efficiency factors for the treatments are either 1 or 0 in each stratum. Also, under these circumstances, the estimates of variance components are the same as those that can be obtained from the analysis of variance by equating the residual mean squares to their expectations:

$$\text{EMS(Blocks)} = 3175.1 = 12\sigma_b^2 + 4\sigma_{b.w}^2 + \sigma^2$$
$$\text{EMS(Wplots)} = 601.3 = 4\sigma_{b.w}^2 + \sigma^2$$
$$\text{EMS(Subplots)} = 177.1 = \sigma^2$$

then $\sigma_b^2 = 214.5$, $\sigma_{b.w}^2 = 106.1$, $\sigma^2 = 177.1$ as above.

The other PRINT option settings allow further output to be obtained. In Example 10.3.1b, the setting PRINT=wald,means is used to get estimates of means and Wald test statistics for the fixed model terms from the split-plot analysis.

Example 10.3.1b

```
29   REML [PRINT=wald,means] Yield

29.......................................................................

*** Wald tests for fixed effects ***


    Fixed term              Wald statistic        d.f.

    Nitrogen                     113.1              3
    Variety                        3.0              2
    Nitrogen.Variety               1.8              6

* All Wald statistics are calculated ignoring terms fitted later in the model


*** Table of mean effects for Constant ***

                        1
                    104.0
Table has only one entry: standard error        9.107


*** Table of mean effects for Nitrogen ***

    Nitrogen        0 cwt       0.2 cwt        0.4 cwt       0.6 cwt
                     79.4          98.9          114.2         123.4

Standard error of differences:        4.436


*** Table of mean effects for Variety ***

    Variety        Victory  Golden rain    Marvellous
                     97.6         104.5         109.8

Standard error of differences:        7.079
```

```
*** Table of mean effects for Nitrogen.Variety ***

        Variety     Victory  Golden rain   Marvellous
        Nitrogen
           0 cwt       71.5         80.0         86.7
         0.2 cwt       89.7         98.5        108.5
         0.4 cwt      110.8        114.7        117.2
         0.6 cwt      118.5        124.8        126.8

Standard error of differences:        Average          9.161
                                      Maximum          9.715
                                      Minimum          7.683

Average variance of differences:                      84.74

Standard error of differences for same level of factor:

               Nitrogen      Variety
Average          9.715        7.683
Maximum          9.715        7.683
Minimum          9.715        7.683
```

The first section of Example 10.3.1b gives Wald test statistics for each of the fixed model terms. These statistics are explained in detail in 10.3.3. They can be used to test the significance of the fixed model terms as they are added into the model, and have an asymptotic chi-squared distribution with degrees of freedom equal to those of the fixed model term. In Example 10.3.1b, it is clear that only Nitrogen, the overall effect of the nitrogen application, is important. This result agrees with the F-tests in the analysis of variance. For orthogonal designs like this the interpretation is straightforward, but more care must be taken when the fixed model terms are non-orthogonal (see 10.3.3). The second part of the output shows the predicted means for all factor combinations from the fixed model. Again, these agree with the means from ANOVA.

10.3.2 Tables of means and effects for fixed and random terms

This section explains the construction of tables of effects and means for fixed and random terms in a REML analysis. The estimates of parameters α and β in the general linear model are called the effects. Tables of effects generally differ from those obtained from ANOVA since REML uses a different parameterization of the linear model, described in 10.2.2 and 10.2.3.
  The estimates of α and β satisfy the "mixed model equations":

$$\begin{pmatrix} X'X & X'Z \\ Z'X & Z'Z+\Gamma^{-1} \end{pmatrix} \begin{pmatrix} \alpha \\ \beta \end{pmatrix} = \begin{pmatrix} X'y \\ Z'y \end{pmatrix}$$

The fixed effects are estimated by the usual generalized least squares estimators
$$\hat{\alpha} = (X'\hat{V}^{-1}X)^{-1}X'\hat{V}^{-1}y$$
and the random effects are predicted by best linear unbiased prediction (BLUP)
$$\beta = (Z'Z+\Gamma^{-1})^{-1}Z'(y-X\hat{\alpha}).$$
The variance-covariance matrix for the whole set of parameters (α' β') is

$$\mathrm{Var}\begin{pmatrix} \alpha \\ \beta \end{pmatrix} = \sigma^2 \begin{pmatrix} X'X & X'Z \\ Z'X & Z'Z+\Gamma^{-1} \end{pmatrix}^{-1}$$

and the variance matrix for the estimated parameters is obtained by using the estimated values of the variance parameters in $\hat{\Gamma}$. The estimated variance-covariance matrix for the fixed effect parameters can then be shown to be $\mathrm{Var}(\hat{\alpha}) = (X'\hat{V}^{-1}X)^{-1}$.

The difference between estimates of fixed and random parameters can be seen from the form of the estimates. If the matrix Γ^{-1} is zero, the random effects are estimated as though they were fixed effects. For positive Γ, the BLUP estimates β for random effects are smaller than if the effects had been estimated as fixed effects. For this reason, the BLUP random effects estimates are often called "shrunken" parameter estimates. The amount of shrinkage depends both on the values $\{\gamma_i\}$ and on the information available for each element of β. Consider the simple case of a model

$$y_{ij} = \beta_i + \varepsilon_{ij}$$

where y_{ij} measures the jth replicate for the ith group ($i=1...p$; $j=1...n_i$), and there are two variance components σ_1^2 and σ^2. The BLUP estimator for the random effects is

$$\beta_i = \frac{n_i}{n_i + \gamma^{-1}} \bar{y}_{i.} \qquad \text{where } \bar{y}_{i.} = \frac{1}{n_i} \sum_{j=1}^{n_i} y_{ij}$$

The amount of shrinkage increases as $\gamma = \sigma_1^2/\sigma^2$ decreases; that is, shrinkage increases as the variability σ_1^2 of the random effect β decreases relative to the residual variance σ^2. The shrinkage discounts the likely contribution from the random error to the apparent random effect, using a factor that depends on their relative variability. This is intuitively satisfactory since high/low values in β may be due partly to high/low values of ε. Clearly this effect would be expected to decrease as the replication for each element of β increases. In fact, for fixed γ, the shrinkage decreases as the amount of information (here the replication n_i) on each random effect increases. So the random effects for which most information is available, where the estimates are most reliable, are shrunk least.

The BLUP estimates can be interpreted as predictions of the random effects given the data, formed by regressing β on residuals calculated by adjusting the data for the fixed effects only.

Tables of effects are obtained by setting option PRINT=effects, as shown in Example 10.3.3a. The constraints imposed upon the parameters α and β are explained in Subsections 10.2.2 and 10.2.3 respectively.

The setting PRINT=means produces tables of predicted means based on the estimates of parameters α and β. In a generally balanced design, the tables of means produced by REML for fixed model terms are the same as the combined means produced by setting option PRINT=cbmeans in ANOVA, which are the same as the ordinary means when the design is orthogonal (see Examples 10.3.1b and 9.2.1a). There is no such correspondence for unbalanced data. With REML, the means are calculated from a linear transformation of the estimated parameter values, taking no account of the frequency counts for different factor combinations. Therefore, these predicted means will correspond to the averages over the factor combinations only with orthogonal data. In other cases, tables of means can be thought of as mean effects of factor levels adjusted for the mean values of any covariates and for any lack of balance in the other factors: that is, as the means you would have expected if the data had been orthogonal. If there are no random terms in the model, the means from REML are those that would be calculated from fitting a regression model to the fixed terms and then using PREDICT with option settings COMBINATIONS=all and ADJUST=equal (see 8.3.4).

Predicted means are calculated using all the parameter estimates and taking means over the model terms not present in the table. For fixed model terms, means need be taken only over the estimates for fixed model terms, since means over random terms will always be zero. For example, in the split-plot design of Example 10.3.1a above, if c, $v_1...v_3$, $a_1...a_4$ and $va_{11}, va_{12}...va_{34}$ are the estimated parameters for the constant, Variety, Nitrogen and the Variety.Nitrogen interaction respectively, the means for Variety are calculated by

$$\text{mean}\{ \text{Variety } i \} = c + v_i + \text{mean}\{ a_j \} + \text{mean}\{ va_{ij} \}$$

and those for Variety.Nitrogen by:

$$\text{mean}\{ \text{Variety } i, \text{Nitrogen } j \} = c + v_i + a_j + va_{ij}.$$

For random terms, means must be taken over the parameter estimates for all the terms in the model. Since the means are based on the shrunken parameter estimates described above, predicted means for random terms will also be shrunk.

When various parameter combinations do not occur and the calculation of a mean effect involves taking means over any of the missing combinations, then that mean will also be a missing value.

Option PTERMS controls the model terms for which tables of means or effects are produced. By default, if means or effects are requested but option PTERMS is not set, tables are printed for all the fixed model terms and none of the random terms. For covariates in the model, the linear regression parameter associated with the covariate can be printed as an effect, but predicted means are not available. Predicted means for other model terms are adjusted to the mean value of the covariate. If you want tables for terms from the random model, or for only a subset of terms in the fixed model, you can use PTERMS to list exactly which tables you require. The setting of PTERMS can contain the string 'Constant' (in capital or lower-case letters, or any mixture), to obtain details of the constant term.

By default, each table is accompanied by a summary – minimum, mean and maximum – of standard errors of differences (seds) for the entries in the table. This can be changed by option PSE: putting PSE=* suppresses the production of standard errors, the setting estimates gives a summary of the standard errors of individual table entries, while the settings alldifferences and allestimates give the full matrix of standard errors of differences and the table of standard errors respectively, as well as the summary. Only one setting of PSE is allowed at a time.

When an absorbing factor is used, the variance-covariance matrix is not available for the estimated parameters in the absorbing factor model. Therefore standard errors cannot be provided for tables of effects for terms in the absorbing factor model. For tables of means the situation is as follows: for fixed model terms, no errors are available for any term which is in the absorbing factor model or has a fixed interaction in the absorbing factor model; for random model terms, no errors are available for any term which is in the absorbing factor model or has an interaction in the absorbing factor model. No standard errors are available for tables of means if there are fixed effects in the absorbing factor model, although standard errors of differences may be available, subject to the conditions above.

Note that in unbalanced models with more than one source of variation the ratio of an effect to its standard error is not, in general, distributed as Student's t. This happens because the variance of an effect is some linear combination of "stratum variances": that is, a weighted sum of variables proportional to chi-squared distributions, rather than a simple multiple of a single chi-squared variable. For small samples, if the structure of the variances for effects is known, Satterthwaite's formula can be used to calculate approximate degrees of freedom for t-tests. For large samples, Wald tests can be used reliably to test for parameters, or contrasts between parameters, significantly different from zero. The procedure VCONTRAST constructs Wald tests for linear functions of parameters. Wald statistics for assessing fixed model terms are described in 10.3.3.

10.3.3 Analysis of unbalanced data: assessing fixed effects

We now consider in more detail the rat reproduction example described in 10.3.1 (Dempster *et al* 1984). This is an unbalanced design with fixed effects and more than one variance component. In this case, it is the fixed effects, here different doses of the experimental compound and its interactions, that are the primary interest. We describe below how to produce tests for the significance of fixed effects. These tests have only asymptotic distributions and not the exact distributional properties associated with tests from ANOVA and linear regression. Care is therefore needed when making inferences from small samples.

The experiment was designed to compare three doses of an experimental compound for improving maternal performance (control, low and high), so the thirty female rats (dams) were randomly split into 3 groups of 10, and the three groups were randomly assigned to the three different treatments. All the pups in each litter were then weighed. The fixed model is Dose*Sex+Littersz, since the sex of the pup and the size of the litter both affect pup weight, and including the Dose.Sex interaction meant that any differential effect of the compound on male or female pups could be estimated. Three of the litters had to be dropped from the study, which meant that one treatment group had only seven litters. Also, litters contain different numbers of male and female pups, as well as being of different total size. This means that the experiment is not balanced, and so cannot be analysed using ANOVA. If it had only one component of variance, the experiment could be analysed by linear regression. However, further variation is introduced into the data by the effects of different dams. Since the dams could be considered as a random selection from the wider population we use the dams as a random effect. The effect of pups is also a random effect. Since the pups are the units of the experiment, the variation between pups is in fact the error variance component. There are therefore two components of variance, due to dams and to pups within dams. Example 10.3.3a shows the analysis of this experiment.

Since the different doses are applied to different dams, most of the information on the compounds is contained in the differences between the dams. Including dams as a random effect means that REML can make use of the between-dam information when estimating the effects of compounds. The variance component due to dam is also estimated, and used to construct appropriate standard errors for the effects.

Example 10.3.3a

```
  1   UNITS [NVALUES=322]
  2   FACTOR [LEVELS=27] Dam
  3   &       [LEVELS=18] Pup
  4   FACTOR [LEVELS=2; LABELS=!T('M','F')] Sex
  5   FACTOR [LEVELS=3; LABELS=!T('C','Low','High')] Dose
  6   VARIATE Littersz,Weight
  7   OPEN 'RATS.DAT'; CHANNEL=2; FILETYPE=input
  8   READ [CHANNEL=2] Dose,Sex,Littersz,Dam,Pup,Weight; \
        FREPRESENTATION=2(labels),4(levels)
```

Identifier	Minimum	Mean	Maximum	Values	Missing
Littersz	2.00	13.33	18.00	322	0
Weight	3.680	6.084	8.330	322	0

Identifier	Values	Missing	Levels
Dose	322	0	3
Sex	322	0	2
Dam	322	0	27
Pup	322	0	18

```
  9   VCOMPONENTS [FIXED=Dose*Sex+Littersz] RANDOM=Dam/Pup
 10   REML [PRINT=model,components,effects] Weight; FITTED=fit; \
        RESIDUALS=res
10..........................................................................
```

***** REML Variance Components Analysis *****

Response Variate : Weight

Random model   : Dam+Dam.Pup
Fixed model   : Constant+Dose+Sex+Dose.Sex+Littersz

Number of units : 322
No absorbing factor

*** Estimated Variance Components ***

Random term	Component	S.e.
Dam	0.0970	0.0334
Dam.Pup	0.1653	0.0137

*** Table of effects for Constant ***

                6.612

Table has only one entry: standard error   0.1099

*** Table of effects for Dose ***

Dose	C	Low	High
	0.0000	-0.4528	-0.9046

```
Standard error of differences:      Average           0.1815
                                    Maximum           0.1936
                                    Minimum           0.1587

Average variance of differences:                      0.03320


*** Table of effects for Sex ***

     Sex            M               F
                 0.0000         -0.4116

Standard error of differences:     0.07356


*** Table of effects for Dose.Sex ***

          Sex        M               F
          Dose
           C      0.00000         0.00000
          Low     0.00000         0.07008
          High    0.00000         0.10719

Standard error of differences:      Average           0.1210
                                    Maximum           0.1342
                                    Minimum           0.1063

Average variance of differences:                      0.01482


*** Table of effects for Littersz ***

                      1
                   -0.1279

Table has only one entry: standard error       0.01882


11   GRAPH [NROWS=15; NCOLUMNS=50] Y=res; X=fit
```

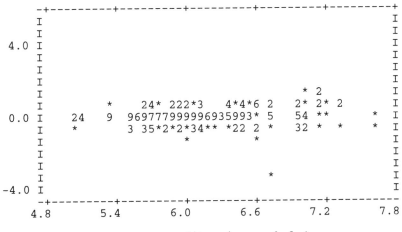

```
         -+---------+---------+---------+---------+---------+
       I                                                    I
       I                                                    I
  4.0  I                                                    I
       I                                                    I
       I                                                    I
       I                                                    I
       I                                            * 2     I
       I        *     24* 222*3    4*4*6 2    2* 2* 2       I
  0.0  I   24   9  969777999996935993* 5    54 **       *   I
       I    *         3 35*2*2*34** *22 2 *  32 *    *    * I
       I                    *            *                  I
       I                                                    I
       I                                                    I
       I                                  *                 I
 -4.0  I                                                    I
         -+---------+---------+---------+---------+---------+
         4.8       5.4       6.0       6.6       7.2       7.8

                 res   v. fit using symbol *
```

Tables of effects contain the values of the estimated parameters $\hat{\alpha}$ and $\hat{\beta}$. By default REML prints estimated effects $\hat{\alpha}$ for the fixed model terms, as shown in Example 10.3.3a. The effects are subject to constraints (as described in 10.2.2) so that parameters corresponding to the first level of a factor are set to zero. The constant term is then not the grand mean, but the mean for a unit with the first level of all the factors: that is, Dose=Control and Sex=male (and mean value for the covariate Littersz). The other parameters represent differences from the first levels of the factors.

As discussed earlier (10.3.2), individual parameter estimates are not in general distributed as Student's t. However, the importance of individual terms in the model can be assessed formally using either Wald statistics or a likelihood-based test.

The Wald statistic to test the null hypothesis $\alpha_1 = 0$ for a fixed model term, denoted by α_1, is defined as $\hat{\alpha}_1'[\mathrm{Var}(\hat{\alpha}_1)]^{-1}\hat{\alpha}_1$, which is the treatment sum of squares for α_1 divided by $\hat{\sigma}^2$. This statistic has an asymptotic chi-squared distribution with degrees of freedom equal to the those of the model term α_1. When there is only one fixed model term, or when the fixed model terms are orthogonal, the interpretation of this statistic is unambiguous. However, for non-orthogonal fixed effects, the Wald statistics will depend on the order in which terms are specified in the fixed model formula. The Wald statistics are calculated from the Cholesky decomposition of the matrix $Y'\hat{V}^{-1}Y$ where Y is the fixed model design matrix X augmented with the data: that is, $(X \mid y)$. The Cholesky decomposition sequentially removes the sum of squares due to each of the fixed effects in turn, ignoring all terms following later in the model. Therefore, the Wald statistics assess the change in fit due to adding the current term to a model containing all the terms listed previously.

For example, for fixed model A*B = A+B+A.B there will be three Wald statistics: the first, due to A, can be used to compare model H_0: $E(y_{ij})=\mu$ with model H_1: $E(y_{ij})=\mu+a_i$; the second, due to fixed model term B, compares model H_1 with model H_2: $E(y_{ij})=\mu+a_i+b_j$; and the third Wald statistic, due to model term A.B compares model H_2 with model H_3: $E(y_{ij})=\mu+a_i+b_j+ab_{ij}$.

Where model terms are non-orthogonal, parameter estimates and Wald statistics depend on the order of the fixed model terms. You may therefore need to specify the model in several different orders to obtain all the required tests. Marginality of nested model terms should still be taken into account: that is, main effects must always be listed before their interactions (see 8.3.3). Problems of interpretation associated with non-orthogonal model terms are discussed further in 9.7.4.

The Wald statistics are obtained by setting the REML option PRINT to waldtests. For some very large models, the formation of the matrix $Y'\hat{V}^{-1}Y$ is not possible, and so the Wald statistics cannot be calculated.

Example 10.3.3b

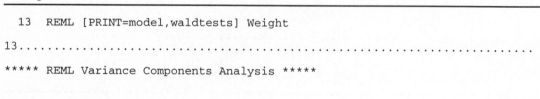

```
   13   REML [PRINT=model,waldtests] Weight

13.......................................................................

***** REML Variance Components Analysis *****

Response Variate : Weight
```

```
Random model      : Dam+Dam.Pup
Fixed model       : Constant+Littersz+Dose+Sex+Dose.Sex

Number of units  : 322
No absorbing factor


*** Wald tests for fixed effects ***


    Fixed term            Wald statistic        d.f.

    Littersz                  28.0               1
    Dose                      24.3               2
    Sex                       58.0               1
    Dose.Sex                   0.8               2
```

* All Wald statistics are calculated ignoring terms fitted later in the model

The Wald statistic for the Dose.Sex factor interaction is 0.8 on 2 degrees of freedom, and is not significant under the asymptotic χ^2 distribution on 2 degrees of freedom. To preserve marginality, we would always fit the interaction after the main effects, so there is no need to recalculate the Wald statistic for the interaction using a different fixed model order. The Dose.Sex interaction can therefore be dropped from the model. To judge which of the main effects should be retained, it is then necessary to fit the model terms in several different orders, as shown in Example 10.3.3c.

Example 10.3.3c

```
  16   VCOMPONENTS [FIXED=Dose+Sex+Littersz] RANDOM=Dam/Pup
  17   REML [PRINT=waldtests] Weight


17.........................................................................

*** Wald tests for fixed effects ***


    Fixed term            Wald statistic        d.f.

    Dose                       9.8               2
    Sex                       54.0               1
    Littersz                  46.4               1
```

* All Wald statistics are calculated ignoring terms fitted later in the model

```
  18   VCOMPONENTS [FIXED=Sex+Dose+Littersz] RANDOM=Dam/Pup
  19   REML [PRINT=waldtests] Weight


19.........................................................................

*** Wald tests for fixed effects ***
```

Fixed term	Wald statistic	d.f.
Sex	55.8	1
Dose	8.0	2
Littersz	46.4	1

* All Wald statistics are calculated ignoring terms fitted later in the model

---

From Example 10.3.3c, it is clear that for any model order, all three remaining fixed model terms are important in explaining the pattern of the data.

For this data set, where the number of degrees of freedom used to estimate variance parameters is large, the asymptotic approximation would be expected to be quite close to the actual distribution of the Wald statistics if the null hypothesis ($\alpha_1=0$) is true. For smaller data sets, the type I error of the Wald statistic tends to increase, so it is more likely to give a significant test statistic than would be expected under the null hypothesis. Also, as mentioned above, for some large models the calculation of Wald statistics is not possible. In these cases, the use of a likelihood-based test statistic may be preferable.

A likelihood ratio test statistic for fixed model terms using REML has been proposed by Welham and Thompson (1992) and can be calculated using the REML directive. Unlike linear regression, the difference in log-likelihoods between two nested fixed models does not give a sensible test statistic. This is because it is the residual likelihood *RL*, the likelihood of the data after projection into the residual space, that is maximized rather than the likelihood of the original data. For the residual likelihood, two different fixed models correspond to two different projections and, hence, effectively to two different data sets on which the same random terms are estimated. The statistic proposed by Welham and Thompson can be used to test a fixed model against a nested sub-model. The method calculates the likelihood for the full fixed model as usual. The same projection is then used for the sub-model and fixed effects to be dropped in the sub-model are constrained to be zero. This gives log-likelihoods calculated from the same projected data-set, using the same random model, but with some fixed effects constrained to zero for the sub-model. The difference in log-likelihoods therefore gives a likelihood ratio test in the usual way, where $-2(RL-RL_0)$ is the test statistic which has an asymptotic chi-squared distribution with degrees of freedom equal to the degrees of freedom of the fixed model terms constrained to be zero in the sub-model.

Simulations have indicated that for small samples this statistic tends to be slightly conservative, that is, it gives a significant test statistic slightly less often than would be expected when the null hypothesis is true.

You can obtain likelihood ratio test statistics by using the SUBMODEL option of REML to define the nested sub-model that is to be fitted and compared to the full fixed model. In other words, the sub-model is the full model with the terms of interest dropped out. For our example above, we would first try dropping the Dose.Sex interaction.

Example 10.3.3d

```
18   VCOMPONENTS [FIXED=Littersz+Dose*Sex] RANDOM=Dam/Pup
19   REML [PRINT=deviance; SUBMODEL=Littersz+Dose+Sex] Weight
```

```
19.............................................................
```

```
*** Deviance: -2*Log-Likelihood ***
```

```
Submodel        :   Constant+Littersz+Dose+Sex
Full fixed model:   Constant+Littersz+Dose+Sex+Dose.Sex
```

```
                    Deviance    d.f.
```

```
Submodel     :      373.8853    315
Full model   :      373.0901    313
Change       :        0.7952      2
```

The inference here is the same as that from the Wald statistic, again suggesting that the Dose.Sex interaction is not important in explaining the pattern of the data. The Dose.Sex interaction can thus be removed from the model, and the other fixed model terms can then be dropped in turn to assess their importance.

The option RECYCLE is very useful for saving computing time when testing a series of sub-models like this. Ordinarily, each time the REML directive is used with the SUBMODEL option set, two runs of the algorithm are made: one to estimate the full model and one to estimate the sub-model. Clearly, for subsequent sub-models, the only new information required is from the sub-model run. The RECYCLE option is used to specify that only the sub-model run is to be made and the remainder of the information is to be picked up from the save structure. If no save structure is specified, the save structure from the most recent REML analysis is used automatically. Note that if you have analysed several y-variates using a single REML statement, then unless you specify a save structure for each y-variate (using the SAVE parameter), only the information from the last y-variate specified will be available. So if the pointer Y held 4 variates to analyse, you would need to use statements of the form

```
REML [PRINT=deviance; SUBMODEL=Sub1] Y[]; SAVE=S[1...4]
  & [RECYCLE=yes; SUBMODEL=Sub2] Y[]; SAVE=S[]
```

to get the test statistics for the two submodels for each of the variates.

In Example 10.3.3e, only one variate is analysed, so there is no need to specify the save structure.

Example 10.3.3e

```
  20   VCOMPONENTS [FIXED=Littersz+Dose+Sex] RANDOM=Dam/Pup
  21   REML [PRINT=deviance; SUBMODEL=Dose+Sex] Weight
```

```
21.............................................................
```

```
*** Deviance: -2*Log-Likelihood ***
```

```
Submodel        :   Constant+Dose+Sex
Full fixed model:   Constant+Dose+Sex+Littersz
```

```
                     Deviance    d.f.

Submodel    :         401.71     316
Full model  :         374.19     315
Change      :          27.52       1
```

    22   & [PRINT=deviance; SUBMODEL=Littersz+Dose; RECYCLE=yes] Weight

22..

*** Deviance: -2*Log-Likelihood ***

```
Submodel       :   Constant+Littersz+Dose
Full fixed model:  Constant+Littersz+Dose+Sex

                     Deviance    d.f.

Submodel    :         426.65     316
Full model  :         374.19     315
Change      :          52.46       1
```

    23   & [PRINT=deviance; SUBMODEL=Littersz+Sex; RECYCLE=yes] Weight

23..

*** Deviance: -2*Log-Likelihood ***

```
Submodel       :   Constant+Littersz+Sex
Full fixed model:  Constant+Littersz+Sex+Dose

                     Deviance    d.f.

Submodel    :         390.02     317
Full model  :         374.19     315
Change      :          15.84       2
```

Again, the results agree with the Wald statistics, and it seems that all the remaining terms in the fixed model are important in explaining the data.

You can specify a sub-model consisting of the constant term alone by using the string 'Constant': that is by putting SUBMODEL='Constant'. The string is case-insensitive: any combination of upper and lower case within the string is accepted.

The use of CONSTRAINTS=positive in a VCOMPONENTS statement may lead to biased results when testing sub-models, since the omission of an important fixed model term often leads to negative estimates of variance components. A warning is given if the constraints have to be enforced when fitting the sub-model, and it is then recommended that the analysis be rerun with parameter setting CONSTRAINTS=none.

Other proposals have been made for the testing of fixed effects using REML estimation procedures. Several of these are based on estimating the full fixed model, fixing the values of the gammas, and then estimating the nested sub-model. The change in residual sum of squares under this procedure is equivalent to the Wald statistic. The change in log-likelihood under this procedure may also give a useful test statistic. These statistics can be constructed by fitting several models and fixing the gammas using the INITIAL and CONSTRAINTS parameters of the VCOMPONENTS directive (10.2.4) then saving the required values from the REML analysis using the VKEEP directive (10.5).

10.3.4 Analysing unbalanced designs: examining sources of variability

Example 10.3.3 shows how REML can be used to estimate variance components in order to form sensible estimates of fixed effects and their standard errors. Sometimes, however, you may be more interested in studying the random effects, in order to gain knowledge about the sources of variability in a data set. The results from REML analyses can help you do this: estimates of the variance parameters are available with their variance covariance matrix; likelihood tests can be used to compare competing random models; and a decomposition of the information matrix for the variance parameters can indicate any underlying structure in the data. Also, library procedure VFUNCTION will calculate functions of variance parameters with standard errors. Some of these facilities are illustrated in Example 10.3.4.

The data in the example were obtained to investigate sources and sizes of variability in an industrial process, the production of car voltage regulators (Example S from Cox and Snell 1981, Snell and Simpson 1991). Within the factory, each regulator was passed from the production line to a setting station where it was adjusted to operate within the correct range of voltages. It would then be passed to a testing station where it would be tested and sent back if outside the acceptable range. An experiment was designed to examine the sources of variability in the voltages produced by the regulators. This experiment used four testing stations, ten setting stations and between four and eight regulators from each setting station. In this situation, small components of variance can be tested for exclusion from the model and the approximate stratum variances can be used to give insight into the structure of the data.

Using factors Teststat and Setstat to indicate the testing and setting stations used for each unit, and factor Regulatr which numbers regulators within each setting station, the random model containing all possible sources of variation is

    Teststat*(Setstat/Regulatr).

The three-way interaction Teststat.Setstat.Regulatr is the residual error component in this model, and there are no fixed effects except the overall mean.

Example 10.3.4a

```
   1   "
  -2      Voltage Regulator Performance
  -3
  -4      Investigation into sources of variability encountered
  -5      during the production of voltage regulators for cars.
  -6
  -7      (Example S from Applied Statistics Principles and Examples,
  -8       D.R.Cox & E.J.Snell, 1981)
  -9   "
  10   UNITS [256]
  11   FACTOR [LEVELS=4; VALUES=(1...4)64] Teststat
  12   FACTOR [LEVELS=10; LABELS=!T(A,B,C,D,E,F,G,H,J,K); \
  13     VALUES=32(1),16(2),28(3),28(4),16(5),28(6), \
  14     32(7),24(8),24(9),28(10)] Setstat
  15   FACTOR [LEVELS=8; VALUES=4( 1...8, 1,2...4, 1,2...7, 1,2...7, \
```

```
16              1,2...4, 1,2...7, 1,2...8, 1,2...6, 1,2...6, 1,2...7)] Regulatr
17   OPEN 'RANDOM.DAT'; CHANNEL=2; FILETYPE=input
18   READ [CHANNEL=2] Voltage

    Identifier   Minimum    Mean   Maximum    Values   Missing
      Voltage     15.30    16.12    17.80       256         0

19   VCOMPONENTS [ABSORB=Setstat] Teststat*(Setstat/Regulatr)
20   REML [PRINT=model,components,stratumvariances,deviance] Voltage

20.........................................................................

***** REML Variance Components Analysis *****


Response Variate : Voltage

Random  model        : Teststat+Setstat+Teststat.Setstat+Setstat.Regulatr+
Teststat.Setstat.Regulatr
Fixed model      : Constant

Number of units  : 256
Absorbing factor : Setstat


******** Warning (Code VC 32). Statement 1 on Line 20
Command: REML [PRINT=model,components,stratumvariances,mean,deviance] Voltage
Insufficient space available to form Wald statistics/check variance matrix


*** Estimated Variance Components ***

Random term              Component        S.e.

Teststat                  0.00350        0.00320
Setstat                   0.01297        0.00902
Teststat.Setstat         -0.00413        0.00139
Setstat.Regulatr          0.02980        0.00851
Teststat.Setstat.Regulatr
                          0.05507        0.00606


*** Approximate stratum variances ***

                                      Effective d.f.
Teststat                     0.24455        3.00
Setstat                      0.47531        8.93
Teststat.Setstat             0.02627       24.03
Setstat.Regulatr             0.17426       54.07
Teststat.Setstat.Regulatr    0.05507      164.97


* Matrix of coefficients of components for each stratum *


                  Teststat   62.40    0.00    7.04    0.00    1.00
                   Setstat    0.00   25.22    6.30    4.00    1.00
          Teststat.Setstat    0.00    0.00    6.98    0.00    1.00
          Setstat.Regulatr    0.00    0.00    0.00    4.00    1.00
 Teststat.Setstat.Regulatr    0.00    0.00    0.00    0.00    1.00
```

```
*** Deviance: -2*Log-Likelihood ***

          Deviance    d.f.

           52.52      250
```

Because a large number of effects are to be fitted in this model (135 parameters), `Setstat` is used as an absorbing factor to reduce the amount of space required. More discussion of the choice of absorbing factor is given in Section 10.6.

In this analysis, a warning is given that Wald statistics cannot be calculated and the estimated variance-covariance matrix of the data, V, cannot be checked for positive-definiteness. The warning for the two events is combined because they are both dependent on the Cholesky decomposition of the matrix $Y'V^{-1}Y$, which cannot be formed here due to lack of space. The likelihood-based test described in 10.3.3 is still available for testing fixed effects. Also, as long as the algorithm has reached convergence, no checking of the variance-covariance matrix should be necessary, as the algorithm should have detected a non-positive-definite variance-covariance matrix at the previous iteration. If the algorithm has not converged, then it is possible that at the last step the components have gone out of bounds, and it is this event that the check is designed to detect. If the components are all positive, then clearly there is no problem. However, if some of the components are negative, it might be advisable to repeat the analysis if possible with extra data space. A further indication that the covariance matrix is not positive definite would occur if REML were also unable to form the stratumvariances. In Example 10.3.4a, the algorithm has converged so there should be no problem, and this can be confirmed by a rerun of the analysis using extra space.

The `Teststat.Setstat` component is estimated as a small negative value. This would mean that the variability due to the testing station and setting station together is less than the variability expected from simply adding the variability of testing stations and setting stations. Rather than assume this to be the case, and since the negative value is small relative to the other components, it might seem more plausible that in reality the `Teststat.Setstat` component is zero.

The list of estimated variance components indicates that two of the components, `Teststat` and `Teststat.Setstat`, are much smaller than the others. They are small compared to their standard errors, but these estimates are based on only four testing stations. (The variance-covariance matrix and standard errors for the components are obtained from the inverse of their information matrix.) In order to decide whether the smaller components are effectively zero, or whether they are really necessary to explain the variation in the data, you can use a likelihood ratio test. You can obtain this by running REML again with the same fixed model but omitting the component from the random model. The test statistic is given by the difference between the deviances of the two models.

Example 10.3.4b

```
21   VCOMPONENTS [ABSORB=Setstat] .Teststat+(Setstat/Regulatr)
22   REML [PRINT=components,deviance] Voltage
```

```
22.......................................................................

*** Estimated Variance Components ***

Random term              Component             S.e.

Teststat                  0.00329             0.00334
Setstat                   0.01193             0.00906
Setstat.Regulatr          0.03078             0.00847
*units*                   0.05114             0.00526


*** Deviance: -2*Log-Likelihood ***


              Deviance    d.f.

               56.60      251
```

The change in log-likelihood of 4.08 is large compared to a chi-squared variable on one d.f. which indicates that the Teststat.Setstat component should be retained in the model.

The original analysis, in Example 10.3.3a, contained estimates of the approximate stratum variances. These can be used to interpret the information on the variance components available from the experiment. They are calculated from a Cholesky decomposition of the information matrix of the variance components E($-\partial^2 RL/\partial\hat{\sigma}^2$), the expected value of the second derivative of the residual likelihood RL, using the vector $\hat{\gamma}$ of estimated variance components. This decomposition is motivated by analogy with the structure of orthogonal designs. Since the decomposition is based on the residual likelihood RL it can give no direct information on the fixed model terms, and therefore effectively gives a decomposition of a random effects model with a grand mean only, ignoring any other fixed model terms.

In an orthogonal design, the information matrix I_ξ for the independent stratum variances { ξ_s } is diagonal with elements $df_s/2\xi_s^2$ where df_s is the degrees of freedom of stratum s. Furthermore, these stratum variances are linear combinations of the variance components which always include the term σ^2, so $\xi=L\sigma$ where L is the matrix mapping the components onto the stratum variances and has value 1 for all elements in the final row (corresponding to σ^2). The information matrix for the variance components I_σ can then be calculated from the information matrix of the stratum variances by $I_\sigma=L'I_\xi L$.

From the results of a REML analysis, the Cholesky decomposition of the information matrix I of the estimated variance components can be written as $I=TDT'$, where T is the lower triangular Cholesky decomposition of I, standardized so that all values in the last row of T are 1 and D is a diagonal matrix containing the squares of the scaling factor for each column. This decomposition gives the information matrix in a form similar to that which occurs naturally in an orthogonal design. T' is then analogous to the matrix of coefficients used to construct the stratum variances from the variance components and D is analogous to the information matrix of the stratum variances.

The components of the decomposition can then be interpreted as if they had arisen from a hypothetical orthogonal experiment which gives information on the variance components equivalent to that available in the actual experiment. In other words, if it was carried out, the hypothetical experiment would be expected to give estimates of the variance components with precision similar to those in the actual experiment. This information can be useful for the planning of future experiments.

For orthogonal experiments, the decomposition will give the stratum variances expected from analysis of variance. As the model becomes non-orthogonal (either through the structure of the fixed or random model) the relationship breaks down, although the decomposition is usually fairly easy to interpret.

It should be remembered that the information matrix I represents the information on the variance components available from the data projected to remove all the treatment contrasts and hence all the information on treatments. There is, however, no information about where the treatment degrees of freedom would have been, and this may lead to a slightly unexpected allocation of degrees of freedom where treatment efficiency factors are not all zero or one. The decomposition can indirectly give information on the fixed model terms. The change in structure when a fixed model term is dropped may give useful information about where the term was estimated and the variation in the data it accounted for.

It should also be noted that the information matrix is evaluated at the estimated value of the variance components, and thus depends on these values. For this reason, two experiments with the same structure may give slightly different decompositions.

In some circumstances the decomposition cannot be interpreted. If any of the variance components has been constrained in a VCOMPONENTS statement, using either CONSTRAINTS or a RELATIONSHIP matrix, there is no information directly available on the constrained components: the information on associated components is pooled, and the approximate stratum variances cannot be related back to the individual random model terms. Also, since the Cholesky decomposition works sequentially, for non-orthogonal random terms the decomposition will depend on the order of the random model. In particular, results may be difficult to interpret if the structure of the random model is non-hierarchical. Occasionally in these circumstances, the Cholesky decomposition yields negative coefficients leading to negative stratum variances which cannot be interpreted.

Example 10.3.4 gives a good illustration of how to interpret the decomposition in terms of the underlying structure of the data. The data for the voltage regulators is not quite balanced, since the number of regulators tested at each setting station varies between four and eight.

The structural information is contained in the matrix of coefficients (T' above) and the degrees of freedom (the diagonal of D above). Within an orthogonal design, the coefficients would indicate the replication of each level of the factors.

In Example 10.3.3a, the Setstat.Regulatr stratumvariance (variation between regulators) has equation $\xi_{S.R}=4\sigma_{S.R}^{2}+\sigma^{2}$ indicating 4 readings for each regulator, which matches the experiment since each regulator was measured on each of the 4 testing stations. Similarly, the equation for the Teststat.Setstat stratum indicates 7 readings for each combination of setting station and testing station. This again matches the structure of the data since 64 regulators were tested on 10 setting stations, giving on average 6.4 regulators at each station. The equation for the Setstat stratum disagrees with this slightly, suggesting 25 readings at each setting station consisting of 6 regulators read 4 times each. Then the Teststat stratum again indicates 7 regulators and 62 readings at each testing station, which implies 9 setting stations. Putting these results together gives a structure consisting of 4 testing stations, 9 setting stations and 6-7 regulators used at each setting station. The degrees of freedom more or less correspond to this structure, suggesting 10 instead of 9 setting stations. This is the structure of a hypothetical orthogonal experiment which would have given the same amount of information on the variance components. Since the original experiment is nearly balanced, this hypothetical experiment is quite similar.

There are no hard and fast rules for interpreting this decomposition. In Example 10.3.3a, there was no fixed model. In general, the removal of treatment contrasts may affect both the coefficients and the degrees of freedom, making interpretation less straightforward.

10.4 The VDISPLAY directive

The VDISPLAY directive allows further output to be produced from one or more REML analyses without having to repeat all the calculations.

VDISPLAY display further output from a REML analysis.

Options

PRINT = *strings*	What output to present (model, components, effects, means, stratumvariances, monitoring, vcovariance, deviance, Waldtests); default mode, comp, stra
CHANNEL = *identifier*	Channel number of file, or identifier of a text to store output; default current output file
PTERMS = *formula*	Terms (fixed or random) for which effects or means are to be printed; default * implies all the fixed terms
PSE = *string*	Standard errors to be printed with tables of effects and means (differences, estimates, alldifferences, allestimates, none); default diff

Parameters

pointers	Save structure containing the details of each analysis; default is to take the save structure from the latest REML analysis

---

You can store the information from a REML analysis using the parameter SAVE in the REML statement, and then specify the same structure with the SAVE parameter of VDISPLAY. Several SAVE structures can be specified, corresponding to the analyses of several different variates. These need not have been analysed using the same REML statement, or even from the same model (as defined by VCOMPONENTS). Alternatively, if you just want to display output from the last y-variate that was analysed, there is no need to use the SAVE parameter in either REML or VDISPLAY: the save structure for the last y-variate analysed is saved automatically, and provides the default for VDISPLAY.

The options of VDISPLAY are the same as those that control output from REML: PRINT, PTERMS and PSE, plus the CHANNEL option which allows output to be directed to another output channel or into a text structure. The available settings of PRINT are identical to those in REML, and are listed in Section 10.3.1.

Example 10.4a illustrates the use of VDISPLAY to display output from the analysis of a variety trial. The data, from Robinson (1984), consist of yields from 16 varieties which have been tested at 6 centres over 4 years. The data is unbalanced, since at most 13 of the varieties are tested at the same time (see the table of counts in Example 10.4a) and different centres are used in different years. Given similar variability within the different trials REML can be used to combine information over the different centres and different years (although the WEIGHTS option could be used if variability changed between trials).

Example 10.4a

```
  1   "
 -2      Variety trial: 16 varieties tested at 6 centres over 4 years
 -3      From the manual for program REML, Robinson (1984).
 -4   "
  5   UNITS [216]
  6   FACTOR [LEVELS=16; LABELS=!T(BRAESIDE, DRAGON, EXPRESS, KAIZEE, \
  7      KEEPWELL, KIERSBG, MAYON, NBYELLOW, SENSGY, TOPKEEP, GLOBE, \
  8      AVANTI, KYU218, ADVANCE, GIONFEST, SEN234)] Variety
  9   FACTOR [LEVELS=6; LABELS=!T(K,SH,EF,LD,AR,R)] Centre
 10   &        [LEVELS=!(77,78,79,80); LABELS=*] Year
 11   &        [LEVELS=!(0,1)] Trial
 12   VARIATE Yield,Pop
 13   READ Year,Trial,Variety,Centre,Yield,Pop; FREPRESENT=2(levels,labels)
```

Identifier	Minimum	Mean	Maximum	Values	Missing
Yield	1.00	28.15	75.10	216	0
Pop	1.00	42.15	116.00	216	0

Identifier	Values	Missing	Levels
Year	216	0	4
Trial	216	0	2
Variety	216	0	16
Centre	216	0	6

```
231   TABULATE [PRINT=counts; CLASS=Centre,Year] Yield
```

	Count			
Year	77.00	78.00	79.00	80.00
Centre				
K	12	12	12	6
SH	12	12	13	0
EF	12	13	13	6
LD	12	13	13	0
AR	12	0	13	6
R	12	12	0	0

```
232   VCOMPONENTS [FIXED=Variety; ABSORB=Year] \
233      RANDOM=Year+Centre+Year.Centre+Variety.Year
234   REML [PRINT=*] Yield
235   VDISPLAY [PRINT=model,components,vcovariance,monitor]
```

235...

```
***** REML Variance Components Analysis *****

Response Variate : Yield

Random model     : Year+Centre+Year.Centre+Year.Variety
Fixed model      : Constant+Variety

Number of units  : 216
Absorbing factor : Year


*** Convergence monitoring ***

Cycle    Deviance   Current gammas and sigma squared
  1      10410.2    0.835660   0.502509   0.966263   0.627500   53.9912
  2      1127.06    0.844899   0.528088   0.960083   0.635791   53.9221
  3      1127.06    0.844015   0.525928   0.960834   0.635512   53.9250


*** Estimated Variance Components ***

Random term                Component        S.e.

Year                         28.36          37.83
Centre                       45.51          41.79
Year.Centre                  51.81          25.28
Year.Variety                 34.27          13.02
*units*                      53.92           6.08


*** Estimated Variance matrix for Variance Components ***

         Year       1430.7
         Centre        37.2     1746.5
   Year.Centre       -140.9     -211.1      638.9
  Year.Variety        -18.8        0.0        0.6      169.5
       *units*          0.6        0.0       -3.4       -7.9       37.0

                     Year      Centre  Year.Centre Year.Variety    *units*


 236    & [PRINT=means; PTERMS=Variety+Centre.Year]

236..............................................................................

*** Table of mean effects for Variety ***

    Variety    BRAESIDE      DRAGON      EXPRESS      KAIZEE     KEEPWELL
                  17.93       26.49        23.52       22.61        33.83


    Variety     KIERSBG       MAYON     NBYELLOW      SENSGY      TOPKEEP
                  17.51       30.44        26.45       36.59        34.02


    Variety       GLOBE      AVANTI       KYU218     ADVANCE     GIONFEST
                  23.31       36.27        24.37       22.80        34.23


    Variety      SEN234
                  34.49
```

Standard error of differences: Average 6.024
            Maximum 6.873
            Minimum 4.805

Average variance of differences: 36.57

*** Table of mean effects for Year.Centre ***

Centre Year	K	SH	EF	LD	AR
77.00	39.83	5.11	22.13	31.87	7.86
78.00	31.53	29.06	21.85	29.95	20.06
79.00	39.32	31.80	34.29	46.12	23.86
80.00	40.16	26.35	23.92	36.49	30.00

Centre Year	R
77.00	20.05
78.00	18.91
79.00	29.78
80.00	27.02

* Standard errors of means cannot be formed for this table as either the
 term or relevant interactions are in the absorbing factor model

---

Example 10.4a illustrates the use of the monitoring setting to print the estimates of the variance ratios, or gammas, and the deviance at each iteration of the algorithm. The deviance printed in the monitoring information differs from the deviance given by PRINT=deviance, since it omits the terms of the residual-log-likelihood, *RL*, which are independent of the variance parameters. The role of the deviance in model testing is covered in Section 10.3.3 for fixed model terms and Section 10.3.4 for random model terms.

10.5 The VKEEP directive

You can use the VKEEP directive to copy results from a REML analysis into Genstat data structures. Genstat automatically stores the save structure for the last y-variate that was analysed using REML, and by default this save structure provides the information for VKEEP. As for VDISPLAY, you can save the information from a REML analysis in a save structure using the SAVE parameter in the REML directive, then access the information by specifying the same structure in the SAVE option of VKEEP.

VKEEP copies information from a REML analysis into Genstat data structures.

Options

RESIDUALS = *variates*	Residuals from each analysis
FITTEDVALUES = *variates*	Fitted values from each analysis
SIGMA2 = *scalar*	Variance component for the lowest stratum

VCOVARIANCE = *symmetric matrix*

    Variance-covariance matrix for the estimates of the variance components

FULLVCOVARIANCE = *symmetric matrix*

    Variance-covariance matrix for the full set of fixed and random effects not associated with the absorbing factor

DEVIANCE = *scalar*   Residual deviance from fitting the full fixed model

DF = *scalar*   Residual degrees of freedom after fitting the full fixed model

SUBDEVIANCE = *scalar*   Residual deviance after fitting the sub-model of the fixed model

SUBDF = *scalar*   Residual degrees of freedom after fitting the sub-model of the fixed model

RSS = *scalar*   Residual sum of squares from fitting the FIXED model by general least squares with a covariance matrix derived from the estimated variance components

SAVE = *pointer*   Save structure from the required analysis; default * takes the save structure from the latest REML statement

Parameters

TERMS = *formula*   Terms for which information is to be saved

COMPONENTS = *scalars*   Estimated variance components

MEANS = *tables*   Table of predicted means for each term

SEDMEANS = *symmetric matrices*   Standard errors of differences between the predicted means

VARMEANS = *symmetric matrices*   Variance-covariance matrix of the means

EFFECTS = *tables*   Table of estimated regression coefficients for each term

SEDEFFECTS = *symmetric matrices*

    Standard errors of differences between the estimated parameters of each term

VAREFFECTS = *symmetric matrices*

    Variance-covariance matrix of the effects of a term

Overall information from the analysis is saved using the options of VKEEP, while the parameters are used to save information for specific model terms. The terms (fixed, random or a mixture) for which you require information are defined by a formula using the TERMS parameter. The other parameters can then be used to specify structures for saving information for each of the model terms.

    Options RESIDUALS and FITTEDVALUES are used to specify variates to hold the residuals and fitted values, which are defined according to the setting of the RMETHOD option of REML that was used when the model was fitted (see 10.3.1). The residual variance can be stored in a scalar using option SIGMA2. The VCOVARIANCE option can supply a symmetric matrix to save the variance-covariance matrix for the estimates of variance components. The

FULLVCOVARIANCE option can be used to store the variance-covariance matrix for the full set of fixed and random effects, excluding those in the absorbing factor model. This matrix will often be very large, and is useful only for looking at covariances between effects associated with different model terms, since the variance-covariance matrices for individual model terms can be stored using the VAREFFECTS parameter. The residual deviance from fitting the full fixed model or the submodel can be saved using options DEVIANCE and SUBDEVIANCE respectively, and the associated residual degrees of freedom can be saved using options DF and SUBDF. The RSS option is used to save the residual sum of squares $(y-X\alpha)'\hat{H}^{-1}(y-X\alpha)$ from fitting the fixed model by generalized least squares, as described in 10.1.3.

The formula given in the TERMS parameter is expanded to give a series of model terms. The other parameters of VKEEP are taken in parallel with these terms. The string 'Constant' can be used within the formula to save structures associated with the constant term. Example 10.5 shows how to save information from the split-plot analysis in Section 10.3.1.

Example 10.5

```
32   VCOMPONENTS [FIXED=Variety*Nitrogen] RANDOM=Blocks/Wplots/Subplots
33   REML [PRINT=*] Yield
34
35   VKEEP TERMS=Variety*Nitrogen; MEANS=Mv,Mn,Mnv; VARMEANS=Vv,Vn,Vnv
36   & [SIGMA2=Sigma2] Blocks/Wplots; COMPONENTS=Cb,Cwp
37
38   PRINT Cb,Cwp,Sigma2

          Cb          Cwp       Sigma2
       214.5        106.1        177.1

39   VKEEP Blocks+Nitrogen; COMPONENTS=Cb,*; EFFECTS=Eb,En; SEDE=SEDb,SEDn
40   PRINT Eb,SEDb

                        Eb
      Blocks
           1        25.422
           2        -4.706
           3         2.657
           4       -10.583
           5        -6.530
           6        -6.260

                        SEDb

  Blocks  1           *
  Blocks  2        9.013        *
  Blocks  3        9.013     9.013        *
  Blocks  4        9.013     9.013     9.013        *
  Blocks  5        9.013     9.013     9.013     9.013        *
  Blocks  6        9.013     9.013     9.013     9.013     9.013        *

              Blocks 1  Blocks 2  Blocks 3  Blocks 4  Blocks 5  Blocks 6
```

The COMPONENTS parameter allows you to save the estimated variance component for each random term in the TERMS list. If you try to save a variance component for a fixed model term it will be given a missing value. Tables of means for each term can be saved using the MEANS parameter, and standard errors of differences between the means are saved by SEDMEANS. You can also save the estimated variance-covariance matrix for the means of each term using parameter VARMEANS. The EFFECTS parameter is used to save tables of estimated parameters, as described in Section 10.4. A symmetric matrix of the standard errors of differences between the effects of each term can be saved using parameter SEDEFFECTS, and the estimated variance-covariance matrix for the parameters can be saved using parameter VAREFFECTS.

The rules for which standard errors of effects or means are available when an absorbing factor has been used are described in Section 10.4. A fault will occur if you try to save standard errors that are not available. You can avoid this by putting a missing value instead of a structure identifier in the parameter list. For example, if Centre is the absorbing factor:

```
VKEEP TERMS=Centre+Variety; MEANS=Mc,Mv; SEDM=*,SEDMv
```

10.6   Technical details

For large data sets and models with many parameters, the REML algorithm may take a large amount of computing time and/or data space. Use of the absorbing factor can help avoid these problems. However, to use the absorbing factor efficiently, it may help to have some idea of the basic structure of the algorithm. This section provides information on the REML algorithm in Genstat, and advice on choice of an absorbing factor for any given model.

10.6.1   The REML algorithm in Genstat

The REML algorithm in Genstat estimates the variance components iteratively using Fisher scoring to solve the normal equations. For the model

$$y = X\alpha + Z\beta + \varepsilon$$

described in detail in Section 10.1, the residual-log-likelihood *RL* can be written as

$$RL(y) \;=\; -\tfrac{1}{2}\log|V| \,-\, \tfrac{1}{2}\log|X'V^{-1}X| \,-\, \tfrac{1}{2}(y-X\hat{\alpha})'\,V^{-1}(y-X\hat{\alpha})$$

ignoring terms independent of the variance parameters. The first derivatives of *RL* with respect to the gammas { γ_i }, where $\gamma_i = \sigma_i^2/\sigma^2$, and the residual variance σ^2 are

$$\frac{\partial RL}{\partial \gamma_i} \;=\; -\tfrac{1}{2}\mathrm{trace}(Z_i'\,PZ_i) \,+\, (\beta'\Gamma^{-1}D_i\Gamma^{-1}\beta)/2\sigma^2$$

$$\frac{\partial RL}{\partial \sigma^2} \;=\; -\tfrac{1}{2}(n-p^*)/\sigma^2 \,+\, (y-X\hat{\alpha})'\,V^{-1}(y-X\hat{\alpha})/2\sigma^4$$

$$\text{where} \quad P = V^{-1} - V^{-1}X(X'V^{-1}X)^{-1}X'V^{-1}\,; \quad \frac{\partial\Gamma}{\partial\gamma_i} = D_i\,; \quad p^* = \mathrm{rank}(X).$$

As well as the unknown variance parameters, these equations involve the estimates of the fixed and random effects. At each iteration, these parameters can be estimated using current estimates of the variance parameters and inverting the mixed model equations

$$\begin{pmatrix} X'X & X'Z \\ Z'X & Z'Z+\hat{\Gamma}^{-1} \end{pmatrix} \begin{pmatrix} \alpha \\ \beta \end{pmatrix} = \begin{pmatrix} X'y \\ Z'y \end{pmatrix}.$$

Then, defining

$$Q = \begin{pmatrix} Q_{11} & Q_{12} \\ Q_{21} & Q_{22} \end{pmatrix} = \begin{pmatrix} X'X & X'Z \\ Z'X & Z'Z+\hat{\Gamma}^{-1} \end{pmatrix}^{-1}$$

It can be shown that trace($Z_i'PZ_i$) = trace(UD_i) where $U = \Gamma^{-1} - \Gamma^{-1}Q_{22}\Gamma^{-1}$. The information matrix I can also be written in terms of U, the estimates of β, and the residual sum of squares. The REML algorithm implemented in Genstat takes the following steps at each iteration:

0) Obtain initial estimates of the variance parameters.
1) Calculate estimates of α and β by inverting the mixed model equations using current estimates of the variance parameters. Form U.
2) Using $\hat{\alpha}$, β and U calculated in step 1, form the first derivatives of the likelihood RL and the information matrix I. Then use Fisher scoring (see equation below) to obtain updated estimates of the variance parameters.

$$\begin{pmatrix} \gamma_{new} \\ \sigma^2_{new} \end{pmatrix} = \begin{pmatrix} \gamma_{old} \\ \sigma^2_{old} \end{pmatrix} + I^{-1} \begin{pmatrix} \dfrac{\partial RL}{\partial \gamma_{old}} \\ \dfrac{\partial RL}{\partial \sigma^2_{old}} \end{pmatrix}$$

3) Check for convergence of variance parameter estimates: exit algorithm on convergence; otherwise, return to step 1.

The inversion of the mixed model equations at step 1 involves inversion of a symmetric matrix with number of rows equal to the number of fixed effects (n_f) plus the number of random effects (n_r) in the model. For models specifying a large number of effects, the inversion of this matrix can be time-consuming and requires $(n_f+n_r)^2$ units of double precision data space.

Since the size of the mixed model equations can limit the speed of the algorithm, it is sensible to try and reduce the size of this matrix. Use of an absorbing factor is one way of tackling the problem. However, it is also important to check that the model is specified succinctly, without unnecessary parameters.

When the mixed model equations are formed, a row of the matrix is allowed for each of the parameters indicated by the model structure. No note is taken at this stage of whether every factor combination occurs, and combinations that do not occur are detected only while inverting the matrix. (Similarly, although some parameters are constrained to be zero, the rows corresponding to these parameters are present in the matrix). For this reason, it is important that model structures do not implicitly specify many unnecessary parameters, since this leads to the inversion of matrices that are much larger than necessary, which can in turn lead to large increases in computing time. This may cause problems for sparse data with a factorial structure, if there are many factor combinations which are not present. In this case, it may be better to fit the model ignoring the factorial structure of the data, and then construct Wald tests from the parameter estimates.

An absorbing factor is a factor from either the fixed or random model, which is used to define a partition of the mixed model equations and hence decrease the size of matrices which must be inverted and stored. However, the information required to calculate estimated errors for some of the tables of means and effects will no longer be available (see Section 10.3.2). When an absorbing factor is specified, the model terms are reordered into two groups: the first contains all the model terms involving the absorbing factor; and the second contains all the other model terms. Each part of the model may include both fixed and random terms. The general mixed model above can be partitioned in this way, so that α_1 and β_1 denote the elements of α and β that are associated with the absorbing factor model, with associated design matrices X_1 and Z_1, and α_2 and β_2 are the remaining fixed and random parameters, with design matrices X_2 and Z_2. The mixed model equations can be reordered to give

$$\begin{pmatrix} U'U + \Gamma_1^{-1} & U'W \\ W'U & W'W + \Gamma_2^{-1} \end{pmatrix} \begin{pmatrix} \theta \\ \phi \end{pmatrix} = \begin{pmatrix} U'y \\ W'y \end{pmatrix}$$

where $U = (X_1 \,|\, Z_1)'$; $W = (X_2 \,|\, Z_2)'$; $\theta' = (\alpha_1' \, \beta_1')$; $\phi' = (\alpha_2' \, \beta_2')$

and Γ_1 and Γ_2 are the parts of Γ relating to β_1 and β_2 respectively, with zero rows added to correspond to α_1 and α_2.

The first set of equations can be *absorbed* into the second set, giving the matrix

$$\begin{pmatrix} (U'U + \Gamma_1^{-1})^{-1} & U'M^{-1}W \\ W'M^{-1}U & W'M^{-1}W + \Gamma_2^{-1} \end{pmatrix} \qquad \text{where} \quad M = I - U'(U'U + \Gamma_1^{-1})^{-1}U$$

It is possible to write most of the expressions in the iterative REML algorithm in terms of the matrices $U'U + \Gamma_1^{-1}$ and $W'M^{-1}W + \Gamma_2^{-1}$ and their inverses. The inversion of the whole set of mixed model equations can be avoided by working with these two matrices separately. Since the inverse sum of squares matrix Q is the estimated variance-covariance matrix for the parameter estimates, this separation means that estimates of covariances between the two sets of parameters are not calculated. By reordering the parameters within the absorbing factor model by level of the absorbing factor, the matrix $U'U + \Gamma_1^{-1}$ becomes block diagonal, which means that any expression involving the matrix $U'U + \Gamma_1^{-1}$ can be calculated using each of these blocks in turn and accumulating the result. This results in a further reduction in the size of matrices that have to be stored, but since the same workspace is used for each block of $U'U + \Gamma_1^{-1}$ and the whole matrix is not stored, the covariances and variance estimates for parameters in the absorbing factor model are not available.

Further details of the REML algorithm and its use of the absorbing factor are given by Thompson (1977).

10.6.2 How to choose the absorbing factor

The calculations for comparing different choices of absorbing factor are quite straightforward.
1) Choose an absorbing factor A with v levels.
2) Split the model terms into two groups and count the number of parameters defined by the factor combinations in each group: (a) model terms containing the absorbing factor (n_1 parameters) and (b) model terms not containing the absorbing factor (n_2 parameters).
3) The matrices that must be inverted using absorbing factor A are then: one matrix of order n_2 plus v matrices of order n_1/v.

As well as considering the numerical advantages of an absorbing factor, it is also important to check that the choice of absorbing factor does not mean that the estimates of error are lost for important comparisons. It should also be noted that the inversion of very many smaller matrices can sometimes take longer than the inversion of a few matrices of intermediate size.

For example, with the variety trial data in Section 10.4, the components of variance for `Year`, `Centre` and some interactions are of interest. Predictions of overall `Variety` means are also required in order to predict individual variety performances. There are 216 units, 4 years, 6 centres and 16 varieties, so to fit `'constant'` + `Variety` + `Year` + `Centre` `Year.Centre` + `Year.Variety` means estimating $1+16+4+6+24+64 = 115$ parameters. This would mean inverting a symmetric matrix with 115 rows. Use of an absorbing factor can reduce this size considerably. Although the `Variety` factor has the highest number of levels, we cannot use `Variety` as the absorbing factor, since we would then lose the standard errors for the effects of interest. We could, however, use either `Year` or `Centre`.

If `Centre` is used as the absorbing factor, the model will be split into two components: first the absorbing factor model which is `Centre+Year.Centre` with 6+24=30 parameters. The second component contains all model terms not involving the absorbing factor, here `'constant'+Variety+Year+Year.Variety` with 1+16+4+64=85 parameters. Although this is still a large matrix, it represents a reduction of more than 25% on the size of the complete matrix, without losing any of the information of interest (that is, the estimates of error for `Variety` means). The parameters in the absorbing factor model are estimated sequentially for each level of the absorbing factor, which here means inverting six 5 × 5 matrices.

In this case, the gain is not as large as it might be by using `Variety` as the absorbing factor (one 35 × 35 matrix plus sixteen 5 × 5 matrices). If computing time is still a problem, it might be worth running the analysis using `Variety` as the absorbing factor, then passing the results into a further REML analysis with a different absorbing factor. Then only two iterations would be required to check convergence, and errors for `Variety` means would become available.

R.T.
S.J.W.

Acknowledgement
These facilities were first included in Genstat 5 Release 2 and have been extended in Release 3. The algorithm was originally taken from the REML program of the Scottish Agricultural Statistics Service (Robinson, Thompson, and Digby 1982); we are very grateful for their permission to adapt their code for Genstat.

11 Multivariate and cluster analysis

11.1 Introduction

In this chapter we are concerned with statistical methods for analysing more than one variable simultaneously. Very often such methods initially combine information on all the given variables into a measure of association, such as a distance or dissimilarity; so, in a sense, they become univariate. Indeed in some fields of application, notably psychology and the social sciences, a single variable of associations may be observed directly, rather than calculated from more basic information. Multivariate analysis is concerned with two forms of data: (a) information on p variables for each of n samples (this can be called the data matrix); or (b) information, usually presented as a symmetric matrix, giving associations between all pairs of samples or all pairs of variables.

In the simplest cases the data matrix has no further structure, and may be regarded as the multivariate generalization of a simple random sample. Genstat does not have a special data structure for a data matrix; generally you must either list the corresponding variables, or collect them in a pointer (2.6). From a data matrix you can calculate the symmetric matrix of sums of squares and products, or alternatively, the correlation matrix of the variables. These are stored in compound data structures, which also contain the means of the variables and other information (2.7.2). However, you can easily extract the basic symmetric matrix from this more general structure (4.10). Genstat has a set of directives for multivariate analyses based on sums of squares and products; these are described in 11.2.

Just as univariate samples may have structure imposed on the units, so may multivariate samples. In canonical variates analysis the units belong to a set of k mutually exclusive groups. For this Genstat lets you calculate the matrix of sums of squares and products, pooled within groups, as well as the means of all the variables in all the groups (4.10.3); these means are held as a set of p variates, each with k values, from which Genstat can calculate a matrix of between-group sums of squares and products. Sums of squares and products arising from more general sample structures are provided by ANOVA (9.6.1).

Correlations and sums of squares and products are elementary examples of how associations can be measured between variables; methods based on such measures are sometimes termed *R-techniques*. Measures of association between units lead to methods known as *Q-techniques*, which are discussed in 11.3 and 11.5. Section 11.4 is concerned with how to calculate symmetric matrices giving similarities or distances between all pairs of samples.

You can think of matrices of distances or dissimilarities as being generated by a cloud of n points in a multidimensional Euclidean space, where the distance between the points representing two samples is or is related to the corresponding distance or dissimilarity in the given matrix. To visualize such a cloud of points is difficult, and much multivariate analysis is concerned with providing approximate graphical representations that are easily interpreted by eye. These representations fall into two main classes: those depending on scatter plots of points in two or, more rarely, three dimensions; and those expressed in the form of networks, especially rooted trees. The plotted distance is usually supposed to approximate to the "true" distance in multidimensional space. Alternatively you may need to examine angle, inner product or area, rather than distance: for example, angles are used to interpret the output from biplots (Gabriel 1971). Apart from the minimum spanning tree given by the HDISPLAY directive, all other standard network-type displays in Genstat are in the form of rooted trees. In particular, you can get dendrograms: these are rooted trees with a scale associated with the nodes. Dendrograms are especially useful for representing hierarchical classifications (11.5).

The directives described in this chapter are for principal components analysis, canonical variates analysis, principal coordinates analysis, multidimensional scaling, orthogonal Procrustes rotation, factor rotation, and various forms of hierarchical and non-hierarchical classification. Other techniques, provided by procedures in the Genstat Procedure Library, include multivariate analysis of variance (11.2.4), canonical correlation analysis (11.2.5), correspondence analysis (11.8.1), biplots (11.8.2), and the analysis of skew-symmetry (11.8.3). The full list of procedures in the MVA module of the current Genstat Procedure Library is given at the start of 11.8.

You can also write your own procedures, taking advantage of CALCULATE (4.1.1) to operate with matrices. You may also find useful the operations of singular value decomposition (4.10.1), spectral (eigenvalue) decomposition (4.10.2), and extracting a sub-matrix from a larger matrix. You can obtain the singular value decomposition for any rectangular real matrix; this is valuable statistically because of its least-squares properties (Eckart and Young 1936). Genstat can calculate the eigenvalues and eigenvectors only of a symmetric matrix, as otherwise the results may be complex numbers. In statistics, non-symmetric matrices often arise in the form $B^{-1}A$, where A and B are both symmetric. Then the eigenvalues of $B^{-1}A$ are real, and are easily found by solving the two-sided algebraic eigenvalue problem $Ax = \lambda Bx$; this case is covered by Genstat (4.10.2). You can extract sub-matrices by CALCULATE, either with qualified identifiers or by using the ELEMENTS function (4.2.8).

For general reading in applied multivariate analysis see for example Manly (1986), Mardia, Kent, and Bibby (1979), Krzanowski (1988), Chatfield and Collins (1986), and Gower (1985a). For work in classification and cluster analysis, see Gordon (1981).

11.2   Analyses based on sums of squares and products

Two of the multivariate methods in Genstat use SSPM structures as input: principal components analysis (11.2.1) and canonical variates analysis (11.2.2). The declaration of SSPM structures is described in 2.7.2, and their formation using the FSSPM directive is described in 4.10.3. Alternatively, if the data contain outliers, you may wish to use procedure ROBSSPM which implements the method of Campbell (1980) for down-weighting units with high Mahalanobis distance from the weighted mean.

Both principal components analysis and canonical variates analysis give loadings of a set of variates: the methods of factor rotation (11.2.3) can sometimes help with the interpretation of loadings.

This section also covers multivariate analysis of variance (11.2.3) and canonical correlation analysis (11.2.4).

11.2.1 Principal components analysis (PCP)

Principal components analysis finds linear combinations of a set of variates that maximize the variation contained within them, thereby displaying most of the original variability in a smaller number of dimensions. Principal components analysis operates on sums of squares and products, or a correlation matrix, or a matrix of variances and covariances, formed from the variates.

PCP performs principal components analysis.

Options

PRINT = *strings*	Printed output required (loadings, roots, residuals, scores, tests); default * i.e. no printing
NROOTS = *scalar*	Number of latent roots for printed output; default * requests them all to be printed
SMALLEST = *string*	Whether to print the smallest roots instead of the largest (yes, no); default no
METHOD = *string*	Whether to use sums of squares, correlations or variances and covariances (ssp, correlation, variancecovariance); default ssp

Parameters

DATA = *pointers* or *matrices* or *SSPMs*	Pointer of variates forming the data matrix, or matrix storing the variate values by columns, or SSPM giving their sums of squares and products (or correlations) etc
LRV = *LRVs*	To store the principal component loadings, roots, and trace from each analysis
SSPM = *SSPMs*	To store the computed sum-of-squares-and-products or correlation matrix
SCORES = *matrices*	To store the principal component scores
RESIDUALS = *matrices* or *variates*	To store residuals from the dimensions fitted in the analysis (i.e. number of columns of the SCORES matrix, or as defined by the NROOTS option)

You supply the input for PCP using the first parameter; this list may have more than one entry, in which case Genstat repeats the analysis for each of the input structures. Instead of supplying an SSPM, you can supply a pointer containing the set of variates, or a matrix storing the variate values by columns. Genstat will then calculate the sums of squares and products, or correlations, or variances and covariances for the analysis (see option METHOD below).

For example, these two forms of input are equivalent:

```
SSPM [TERMS=Height,Length,Width,Weight] S
FSSPM S
PCP [PRINT=roots] S
```

and

```
PCP [PRINT=roots] !P(Height,Length,Width,Weight)
```

But the first form does mean that you have the sums of squares and products available for later use, in the SSPM S. Here the pointer is unnamed (1.6.3). But you may wish to use a named pointer. For example:

```
POINTER [VALUES=Height,Length,Width,Weight] Dmat
PCP [PRINT=roots] Dmat
```

By default the PCP directive does not print any results: you use the PRINT option to specify what output you require. The printed output is in five sections, each with a corresponding setting, as illustrated in the examples below.

The columns of the matrices of principal component loadings and scores correspond to the latent roots. Each latent root corresponds to a single dimension, and gives the variability of the scores in that dimension. The loadings give the linear coefficients of the variables that are used to construct the scores in each dimension. Example 11.2.1a shows a principal components analysis of four variates of length 12.

Example 11.2.1a

```
  1   UNITS [NVALUES=12]
  2   POINTER [VALUES=Height,Length,Width,Weight] Dmat
  3   READ [PRINT=data,errors,summary] Dmat[]

  4   4.1 5.2 1.2 3.1 4.2 1.5 3.2 5.6 2.3 0.2 0.1 0.2
  5   6.2 4.1 4.1 4.1 2.3 6.2 6.3 5.1 0.2 0.9 4.9 7.3
  6   10.1 5.6 3.2 9.4 1.2 9.8 1.0 1.0 6.1 9.7 1.0 3.7
  7   6.1 9.6 9.7 5.5 2.3 5.0 9.4 8.1 4.5 4.9 0.3 1.8 :

    Identifier    Minimum      Mean    Maximum     Values    Missing
        Height      0.200     4.133     10.100         12          0
        Length      0.200     5.225      9.800         12          0
         Width      0.100     3.700      9.700         12          0
        Weight      0.200     4.575      9.400         12          0
  8   PCP [PRINT=roots,scores,loadings,tests] Dmat

8.....................................................................................................

*****   Principal components analysis   *****

***   Latent Roots   ***

            1           2           3           4
        181.8       130.2        82.5        18.5

***   Percentage variation   ***

            1           2           3           4
        44.01       31.52       19.98        4.49
```

*** Trace ***

   413.2

*** Latent Vectors (Loadings) ***

	1	2	3	4
Height	-0.21529	0.37981	0.78747	-0.43506
Length	-0.25623	0.86524	-0.34389	0.25970
Width	-0.74104	-0.21726	-0.37937	-0.50964
Weight	-0.58211	-0.24474	0.34308	0.69537

*** Significance tests for equality of final K roots ***

No. (K) Roots	Chi squared	df
2	4.61	2
3	7.49	5
4	10.30	9

*** Principal Component Scores ***

	1	2	3	4
1	2.725	0.870	0.425	0.256
2	0.714	-3.340	1.875	-0.029
3	6.897	-3.191	0.149	-1.715
4	-0.177	-0.159	1.700	-1.725
5	-2.087	-0.546	-2.585	0.091
6	-0.521	-6.164	-1.130	1.871
7	-3.819	1.518	6.415	1.112
8	3.541	4.306	-4.085	1.354
9	0.940	5.420	0.734	1.074
10	-6.529	3.002	-1.915	-2.134
11	-5.824	-2.992	-2.319	0.285
12	4.139	1.276	0.738	-0.441

The significance tests are for equality of the k smallest roots: l_i ($i = 1, 2, \dots k$). The test statistic is

$$n - \frac{(2p+11)}{6} \left[\log\left(1/k \sum_{i>k} l_i \right) - 1/k \sum_{i>k} \log l_i \right]$$

where n is the number of units and p is the number of variables. Asymptotically, the statistics have a chi-squared distribution with $(k+2)(k-1)/2$ degrees of freedom. If any latent roots are zero, Genstat excludes them from the calculation of the test statistic; the effective value of p is reduced accordingly.

   If you omit the NROOTS option, Genstat prints by default the results corresponding to all the latent roots. The number of latent roots is the number of variates involved in the input to PCP. The NROOTS option allows you to print only part of the results, corresponding to the first or last r latent roots. You may then want to print the residuals. Example 11.2.1b prints the results corresponding to the first two latent roots; the residuals are formed from the remaining two columns of scores.

Example 11.2.1b

```
  9  PCP [PRINT=scores,residuals; NROOTS=2] Dmat

9...........................................................................

*****  Principal components analysis  *****

***  Principal Component Scores  ***

                            1            2
            1           2.725        0.870
            2           0.714       -3.340
            3           6.897       -3.191
            4          -0.177       -0.159
            5          -2.087       -0.546
            6          -0.521       -6.164
            7          -3.819        1.518
            8           3.541        4.306
            9           0.940        5.420
           10          -6.529        3.002
           11          -5.824       -2.992
           12           4.139        1.276

***  Residuals  ***

            1           0.496
            2           1.875
            3           1.721
            4           2.422
            5           2.587
            6           2.186
            7           6.510
            8           4.304
            9           1.301
           10           2.867
           11           2.337
           12           0.860
```

To print results corresponding to the r smallest latent roots, you must set option NROOTS to r and option SMALLEST to yes. Now if residuals are printed they will be formed from the scores corresponding to the largest roots. The NROOTS and SMALLEST options apply to the latent roots and vectors, the principal component scores and the residuals. So you cannot print directly, for example, the first two columns of scores and the last three columns of loadings. This is rarely required but, if necessary, it can be done by saving the relevant results and printing them separately.

In Example 11.2.1c the three smallest roots are printed, together with the residuals. These correspond to the first column of scores, and can be compared with the scores printed in Example 11.2.1a. You can see that all the residuals are positive: this is because residuals from multivariate analyses generally occupy several dimensions, so they represent distances in multidimensional space and signs cannot be attached to them.

Example 11.2.1c

```
 10   PCP [PRINT=roots,residuals; NROOTS=3; SMALLEST=yes] Dmat

10..................................................................................

*****  Principal components analysis  *****

***  Latent Roots  ***

              1              2              3
         130.23          82.54          18.55

***  Percentage variation  ***

              1              2              3
          31.52          19.98           4.49

***  Trace  ***

         413.2

***  Residuals  ***

              1         2.725
              2         0.714
              3         6.897
              4         0.177
              5         2.087
              6         0.521
              7         3.819
              8         3.541
              9         0.940
             10         6.529
             11         5.824
             12         4.139
```

By default, the PCP directive operates on the SSPM but you can set the METHOD option to correlations to operate on a derived matrix of correlations, as shown in Example 11.2.1d, or to variancecovariance to use variances and covariances. Note that when correlations are analysed the significance-test statistics no longer have an asymptotic chi-squared distribution.

The LRV parameter allows you to save the principal component loadings, the latent roots, and their sum (the trace) in an LRV structure, while the SCORES parameter saves the principal component scores in a matrix. If you have declared the LRV already, its number of rows must be the same as the number of variates supplied in an input pointer or implied by an input SSPM. The number of rows of the SCORES matrix, if previously declared, must be equal to the number of units.

The number of columns of the LRV and of the SCORES matrix corresponds to the number of dimensions to be saved from the analysis, and this must be the same for both of them. If the structures have been declared already, Genstat will take the larger of the numbers of columns declared for either, and declare (or redeclare) the other one to match. If neither has been declared and option SMALLEST retains the default setting no, Genstat takes the number of columns from the setting of the NROOTS option. Otherwise, Genstat saves results for the full set of dimensions. The trace saved as the third component of the LRV structure, however, will contain the sums of all the latent roots, whether or not they have all been saved. Procedure LRVSCREE can be used to produce a "scree" diagram which can be helpful in deciding how many dimensions to save; this is illustrated in 11.3.1.

The SSPM parameter can save the SSPM structure used for the analysis. A particularly convenient instance is when you have supplied an SSPM structure as input but, for example, have set METHOD=correlation: the SSPM that is saved will then contain correlations instead of sums of squares and products.

The RESIDUALS parameter allows you to save the principal component residuals, in a matrix with number of rows equal to the number of units and one column. If the latent roots and vectors (loadings) are saved from the analysis, the residuals will correspond to the dimensions not saved; the same applies if you save scores. If neither the LRV nor scores are saved, the saved residuals will correspond to the smallest latent roots not printed.

Example 11.2.1d

```
  11   LRV [ROWS=Dmat; COLUMNS=2] Latent
  12   SSPM [TERMS=Dmat[]] Corrmat
  13   MATRIX [ROWS=12; COLUMNS=1] Res
  14   PCP [PRINT=roots,scores,tests; METHOD=correlation] Dmat; \
  15     LRV=Latent; SSPM=Corrmat; RESIDUALS=Res

15..............................................................................

*****  Principal components analysis  *****

***  Latent Roots  ***

                1             2             3             4
            1.748         1.209         0.855         0.188

***  Percentage variation  ***

                1             2             3             4
            43.70         30.23         21.37          4.70

***  Trace  ***

        4.000


***  Significance tests for equality of final K roots  ***

Correlation matrix used -
   test statistics are NOT asymptotically chi-squared

     No. (K)      Chi
     Roots      squared        df

        2         5.78         2
        3         8.55         5
        4        11.87         9

***  Principal Component Scores  ***
```

	1	2	3	4
1	0.2477	0.1025	0.0527	0.0317
2	0.0190	-0.2415	0.2655	-0.0057
3	0.6772	-0.2301	0.1819	-0.1701
4	-0.0473	0.0699	0.1737	-0.1709
5	-0.1305	-0.1479	-0.2612	0.0010
6	-0.0340	-0.6395	0.0172	0.1726
7	-0.5549	0.3197	0.5267	0.1211
8	0.4383	0.2537	-0.4546	0.1423
9	0.0577	0.5110	-0.0489	0.1215
10	-0.5569	0.2186	-0.3213	-0.2194
11	-0.4980	-0.3846	-0.2263	0.0113
12	0.3816	0.1683	0.0946	-0.0352

```
16   PRINT Latent[],Res
```

Latent['Vectors']

	1	2
Dmat		
Height	-0.3476	0.6121
Length	-0.1981	0.6896
Width	-0.6201	-0.3067
Weight	-0.6749	-0.2359

Latent['Roots']

1	1.748
2	1.209

Latent['Trace']   4.000

Res

	1
1	0.0615
2	0.2656
3	0.2490
4	0.2437
5	0.2612
6	0.1734
7	0.5404
8	0.4763
9	0.1310
10	0.3891
11	0.2266
12	0.1009

If the variables used to form the SSPM structure are restricted, then the analysis will be subject to that restriction. Similarly, if a pointer to a set of variates is used as input to PCP, then any restriction on the variates will be taken into account by the analysis. If you want principal component scores or residuals to be printed or saved from the analysis, the original data must be available. The matrices to save such results must have been declared with as many rows as the variates have values, ignoring the restriction. You can calculate the analysis from one subset of units, but calculate the scores and residuals for all the units, by using as input to PCP an SSPM structure formed using a weight variate with zeros for the excluded sampling units and unity for those to be included. For example, to exclude a known set of outliers from an analysis, but to print scores for them, these statements could be used:

```
POINTER [NVALUES=5] V
FACTOR [LABELS=!T(No,Yes)] Outlier
READ [CHANNEL=2] Outlier,V[]
CALCULATE Wt = Outlier .IN. 'No'
SSPM [TERMS=V] S
FSSPM [WEIGHT=Wt] S
PCP [PRINT=scores] S
```

Principal component regression is provided by procedure RIDGE.

11.2.2 Canonical variates analysis (CVA)

The CVA directive, for canonical variates analysis, operates on a within-group SSPM. This structure contains information on the within-group sums of squares and products, pooled over all the groups; it also contains the group means and group sizes, from which Genstat can derive the between-group sums of squares and products. The directive finds linear combinations of the original variables that maximize the ratio of between-group to within-group variation, thereby giving functions of the original variables that can be used to discriminate between the groups. The squared distances between group means are Mahalanobis D^2 statistics when all the dimensions are used; otherwise they are approximations. You can form exact Mahalanobis distances with the PCO directive (11.3.1).

CVA performs canonical variates analysis.

Options

PRINT = *strings*	Printed output required (roots, loadings, means, residuals, distances, tests); default * i.e. no printing
NROOTS = *scalar*	Number of latent roots for printed output; default * requests them all to be printed
SMALLEST = *string*	Whether to print the smallest roots instead of the largest (yes, no); default no

Parameters

WSSPM = *SSPMs*	Within-group sums of squares and products, means etc (input for the analyses)
LRV = *LRVs*	Loadings, roots, and trace from each analysis
SCORES = *matrices*	Canonical variate means
RESIDUALS = *matrices*	Distances of the means from the dimensions fitted in each analysis
DISTANCES = *symmetric matrices*	Inter-group-mean Mahalanobis distances

---

You specify the input for CVA using its first parameter, WSSPM, this may contain a list of structures, in which case Genstat repeats the analysis for each of them. The input must be an

SSPM structure, declared with the GROUPS option of the SSPM directive (2.7.2) set to a factor giving the grouping of the units. If the variates used to form this SSPM structure are restricted, then the SSPM is restricted in the same way, and so the CVA directive takes account of the restriction. The other four parameters can be used to save the results.

The three options of the CVA directive control the printed output. By default there is no printed output, and so you should set the PRINT option to indicate which sections you want.

Doran and Hodson (1975) give some measurements made on 28 brooches found at the archaeological site of the cemetery at Munsingen. Seven of these variables are used in the next example; they have been transformed by taking logarithms. For a specified grouping of the 28 brooches into four groups, canonical variates analysis is used to determine possible differences among the groups, and which variables contribute to such differences. (These seven variables are also used in the first example of the CLUSTER directive (11.6.1), and the grouping used here is that obtained from CLUSTER).

Example 11.2.2a

```
    1   UNITS [NVALUES=28]
    2   POINTER [VALUES=Foot_lth,Bow_ht,Coil_dia,Elem_dia,Bow_wdth, \
    3     Bow_thck,Length] Data
    4   FACTOR [LEVELS=4] Groupno
    5   READ Groupno,Data[]
```

Identifier	Minimum	Mean	Maximum	Values	Missing
Foot_lth	2.398	3.278	4.554	28	0
Bow_ht	2.079	2.842	3.296	28	0
Coil_dia	1.792	2.166	2.833	28	0
Elem_dia	1.099	2.026	2.708	28	0
Bow_wdth	3.045	4.064	5.176	28	0
Bow_thck	2.708	3.621	4.357	28	0
Length	3.296	4.003	4.860	28	0

Identifier	Values	Missing	Levels
Groupno	28	0	4

```
   35   SSPM [TERMS=Data[]; GROUPS=Groupno] W
   36   FSSPM W
   37   CVA [PRINT=roots,loadings,means,tests] WSSPM=W
```

```
37...............................................................................
```

```
*****   Canonical variate analysis   *****

***   Latent Roots   ***

            1           2           3
        4.543       3.777       2.537

***   Percentage variation   ***

            1           2           3
       41.85       34.79       23.37

***   Trace   ***

       10.86
```

```
***   Latent Vectors (Loadings)   ***
```

	1	2	3
1	1.130	-2.656	3.397
2	-0.633	1.631	4.799
3	3.501	-1.708	1.450
4	-2.669	-0.623	-2.802
5	-3.468	-0.758	0.757
6	1.859	-2.028	-2.478
7	-1.279	-0.110	-3.598

```
***   Significance tests for dimensionality greater than K   ***
```

K	Chi-squared	df
0	97.60	21
1	60.78	12
2	27.16	5

```
***   Canonical Variate Means   ***
```

	1	2	3
1	2.967	1.998	0.613
2	-0.825	0.122	-1.584
3	1.254	-3.545	0.825
4	-2.835	0.856	2.241

```
***   Adjustment terms   ***
```

	1	2	3
1	-8.40	-19.90	1.94

The pointer Data is declared on lines 2 and 3 to refer to the seven data variables. The factor Groupno specifies the groups (line 4). The matrix of within-group sums of squares and products is declared and formed on lines 35 and 36. The CVA directive (line 37) specifies that the latent roots, the vectors (loadings), and the means of the canonical variate groups are to be printed, together with values for the significance tests for the latent roots that indicate the number of dimensions required.

If there are g groups, at most $g-1$ independent combinations of the variables can be found to discriminate amongst them. However, if there are fewer than $g-1$ variables, v say, then at most v independent combinations can be calculated. Thus there will be at most $\min(g-1, v)$ non-zero latent roots, with associated loadings and canonical variate scores for the group means. In the example above $\min(g-1, v)$ is 3.

The significance tests that are printed are for a significant dimensionality greater than k, that is for the joint significance of the first, second, ..., $(k+1)$th latent roots. This test is printed for $k=0, 1, ... \min(g-1, v)-1$. If the test is "not significant" for $k=r$, then the values of chi-square for $k>r$ should be ignored as the indication is that the remaining dimensions have no interesting structure. The test statistic (Bartlett 1938) is

$$[n - g - \tfrac{1}{2} (v - g)][\sum_{i} \log(l_i + 1) - \sum_{i > k} \log(l_i + 1)]$$

which is asymptotically distributed as chi-squared with $(v-k) \times (g-k-1)$ degrees of freedom. Here n is the number of units, g is the number of groups, v is the number of variables, and l_i is the ith latent root. If the coefficient $[n-g-\frac{1}{2}(v-g)]$ is less than zero, there are too few units for the statistics to be calculated and a message is printed to this effect. In any case, the tests should be treated with caution unless $n-g$ is very much larger than v.

The latent vectors, or loadings, are scaled in such a way that the average within-group variability in each canonical variate dimension is 1: thus the within-group variation is equally represented in each dimension. Since the latent roots are the successive maxima of the ratio of between-group to within-group variation, loadings corresponding to roots less than 1 are for dimensions in the canonical variate space that exhibit more within-group variation than between-group variation. In the example, all three roots are greater than 1, suggesting that differences between the four groups exist in all three dimensions; this is in accordance with the significance tests, which indicate a dimensionality greater than 2. It may not be easy to interpret the latent vectors but, for example, the second latent vector here contrasts the second variable (the height of the bow of the brooch) with the others. This suggests that the second canonical variate distinguishes brooches with a relatively narrow shape. The FACROTATE directive (11.2.3) may help you to interpret the loadings. However, canonical variates analysis and principal components analysis can still be useful, even if the loadings cannot be interpreted.

The scores for the means are arranged so that their centroid, weighted by group size, is at the origin. This is done by subtracting a constant term, for each canonical variate dimension, from the scores initially formed as a linear combination of the group means of the original variables. For example, the constant term of -19.90 occurs in the second score for the third mean, -3.545, formed as:

$$-2.656\bar{v}_{13} +1.631\bar{v}_{23} -1.708\bar{v}_{33} -0.623\bar{v}_{43} -0.758\bar{v}_{53} -2.028\bar{v}_{63} -0.110\bar{v}_{73} -19.90$$

where $\bar{v}_{ij}$ is the mean of the ith variable for the jth group. If you ask for the group mean scores to be printed, then the corresponding constant terms are also printed, as shown in Example 11.2.2a above. You can see from the canonical variate means that the second canonical variate separates the third group from the other three.

Results can be printed for a subset of the latent roots by setting the NROOTS and SMALLEST options of CVA. NROOTS specifies the number of roots for which you want the results to be printed. By default these will be the largest roots, unless you set SMALLEST=yes; then the results will be printed for the smallest non-zero roots. When you print a subset of the results, residuals can be formed and printed from the dimensions that are not displayed.

If you ask for distances, they are formed from the group mean scores for the canonical variate dimensions that are printed. If results are printed for the full dimensionality, the distances will be Mahalanobis distances between the groups.

The LRV parameter allows you to save the loadings, latent roots, and their sum (the trace) in an LRV structure, while the SCORES parameter saves the canonical variate means. If you have declared the LRV already, its number of rows must be the same as the number of variates involved in forming the input SSPM. The number of rows of the SCORES matrix, if previously declared, must be equal to the number of groups.

The number of columns of the LRV and of the SCORES matrix corresponds to the number of dimensions to be saved from the analysis, and this must be the same for both of them. If the structures have been declared already, Genstat will take the larger of the numbers of columns declared for either, and declare (or redeclare) the other one to match. If neither has been declared and option SMALLEST retains the default setting no, Genstat takes the number of columns from the setting of the NROOTS option. Otherwise, Genstat saves results for the full set of dimensions. The trace saved as the third component of the LRV structure, however, will contain the sums of all the latent roots, whether or not they have all been saved. Procedure LRVSCREE can be used to produce a "scree" diagram which can be helpful in deciding how many dimensions to save; this is illustrated in 11.3.1.

The RESIDUALS parameter allows you to save the distances of the means from the dimensions fitted in the analysis in a matrix with number of rows equal to the number of groups and one column. If the latent roots and vectors (loadings) are saved from the analysis, the residuals will correspond to the dimensions not saved; the same applies if you save scores. If neither the LRV nor scores are saved, the saved residuals will correspond to the smallest latent roots not printed.

The DISTANCES parameter allows you to save the inter-group-mean Mahalanobis distances in a symmetric matrix.

In Example 11.2.2b, the canonical variate means are saved for plotting. The canonical variate scores for the units of the data matrix are also plotted; to do this we must first save the loadings (the latent vectors) using the LRV parameter, as shown in Example 11.2.2b.

Example 11.2.2b

```
  38  CVA [PRINT=residuals,distances; NROOTS=2] WSSPM=W; LRV=L; SCORES=Meanscrs

38............................................................................

*****  Canonical variate analysis  *****

***  Residuals  ***

                1          0.613
                2          1.584
                3          0.825
                4          2.241

***  Inter-group distances  ***

    1          0.000
    2          4.231          0.000
    3          5.802          4.215          0.000
    4          5.913          2.140          6.007          0.000
                  1              2              3              4

  39  MATRIX [ROWS=28; COLUMNS=Data] Datamat
  40  CALCULATE Datamat$[*; 1...7] = Data[]
  41  MATRIX [ROWS=28; COLUMNS=2] Unitscrs
  42  CALCULATE Unitscrs = Datamat*+L[1]
  43  VARIATE [NVALUES=28] Uscores[1,2]
  44  & [NVALUES=Groupno] Mscores[1,2]
  45  CALCULATE Uscores[1,2] = Unitscrs$[*; 1,2]
```

```
46   & Mscores[1,2] = Meanscrs$[*; 1,2]
47   & Uscores[1,2] = Uscores[1,2]-MEAN(Uscores[1,2])
48   AXES [EQUAL=scale] 3; YTITLE='canonical variate 2'; \
49     XTITLE='canonical variate 1'; STYLE=box
50   PEN 2,3; COLOUR=1; SYMBOLS=Groupno,1
51   DGRAPH [TITLE='Scores for means (x) and units'; WINDOW=3; KEYWINDOW=0]\
52     Uscores[2],Mscores[2]; Uscores[1],Mscores[1]; PEN=2,3
```

The CVA statement (line 38) prints only the residuals and distances. The NROOTS option specifies that the results to be printed are for the two largest latent roots; the residuals thus correspond to the remaining roots, here only the third. The printed distances are formed from the first two canonical variate means, and thus are the distances between the group means on the graph. The matrix Meanscrs and the LRV structure L are defined to store the canonical variate means and the latent roots and vectors, respectively; because the NROOTS option is set to 2, they both have two columns corresponding to the first two canonical variates.

Lines 39-42 obtain a matrix of canonical variate scores for the units. First, the matrix Datamat is set up to hold the data variates as its columns (lines 39 and 40). The matrix of unit scores is formed by post-multiplying Datamat by the loadings, that is by the first element of the LRV structure; L[1] has as many rows as there are variates (here seven), and as many columns as there are roots to be saved (here two). To plot the scores, you must first copy them from the matrices into variates, as illustrated in lines 43-46. Remember that the canonical variate means are arranged so that their weighted centroid is at the origin. The same adjustment needs to be made to the unit scores; this simply involves subtracting the means from the columns of unit scores (line 47).

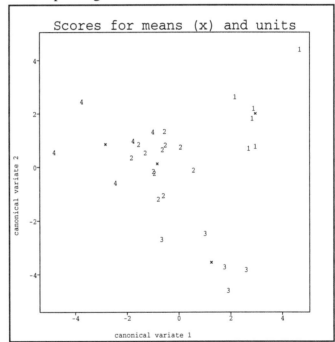

Figure 11.2.2

Figure 11.2.2 shows the resulting unit scores labelled by their group number. The group means are also plotted, as crosses. It can be seen that, in the first two dimensions, groups 1 and 3 are well distinguished from each other, and from groups 2 and 4. There is a slight overlap between groups 3 and 4; however, they are well separated in the third dimension, as shown by the printed mean scores.

11.2.3 Rotation of factor loadings (**FACROTATE**)

Principal components analysis and canonical variates analysis both define a set of dimensions (sometimes called axes) that are linear combinations of the original variables. The individual coefficients of these combinations are called loadings, and can be used to interpret the dimensions. With principal components analysis, the loadings must lie in the range [−1, 1]; this is the situation that we discuss in the initial part of this subsection. The situation with canonical variates analysis is slightly different and is described at the end of this subsection.

When several dimensions are considered it is possible to define an equivalent set of new dimensions, whose loadings are linear combinations of the original loadings. If the absolute values of the loadings for a new dimension are either close to 0 or close to 1, you can interpret the dimension as mainly representing only those original variables with large positive (or negative) loadings. You may sometimes want new dimensions determined by loadings like these, because they are easier to interpret. The methods by which these new dimensions can be obtained are generally known collectively as *factor rotation* because the new dimensions represent a rotation of the axes of the original dimensions. The FACROTATE directive provides two methods of orthogonal factor rotation: varimax rotation and quartimax rotation (Cooley and Lohnes 1971). The default method, varimax rotation, maximizes the variance of the squares of the loadings within each new dimension: the effect of this rotation should be to spread out the squared-loadings to the extremes of their range. Quartimax rotation uses the fourth power of the loadings instead of the second power.

FACROTATE rotates factor loadings from a principal components or canonical variates
analysis according to either the varimax or quartimax criterion.

Options

PRINT = *strings*	Printed output required (communalities, rotation); default * i.e. no printing
METHOD = *string*	Criterion (varimax, quartimax); default vari

Parameters

OLDLOADINGS = *matrices*	Original loadings
NEWLOADINGS = *matrices*	Rotated loadings for each set of OLDLOADINGS

The first parameter, OLDLOADINGS, specifies a list of matrices which provide the input; the columns of each of these matrices should contain the loadings for the original dimensions. The matrices to save the new loadings are listed with NEWLOADINGS parameter; often it will be convenient to use the same structure for output as was used for input.

One way of supplying the loadings for the original variables is by saving the latent roots and vectors from a principal components analysis using the LRV parameter; the first structure of the LRV is then the matrix of loadings. Example 11.2.3a is similar to Example 11.2.1a; however, here the first two latent roots and vectors are saved and used as input to the FACROTATE directive.

Example 11.2.3a

```
    1   UNITS [NVALUES=12]
    2   POINTER [VALUES=Height,Length,Width,Weight] Dmat
    3   READ [PRINT=errors] Dmat[]
    8   LRV [ROWS=Dmat; COLUMNS=2] Latent
    9   PCP [PRINT=loadings] Dmat; LRV=Latent
```

9...

```
*****   Principal components analysis   *****

***   Latent Vectors (Loadings)   ***

                        1           2           3           4
        Height    -0.21529     0.37981     0.78747    -0.43506
        Length    -0.25623     0.86524    -0.34389     0.25970
         Width    -0.74104    -0.21726    -0.37937    -0.50964
        Weight    -0.58211    -0.24474     0.34308     0.69537


   10   FACROTATE [PRINT=rotation,communalities] Latent[1]
```

10..

```
*****   Factor rotation   *****

***   Communalities   ***

                     1
           1      0.1906
           2      0.8143
           3      0.5963
           4      0.3988

***   Rotated factors   ***

                     1           2
           1     -0.0630      0.4320
           2      0.0747      0.8993
           3     -0.7694      0.0660
           4     -0.6312     -0.0172
```

The LRV structure Latent is declared on line 8, and is used on line 9 to save the latent roots and vectors. The full set of latent vectors is printed from the PCP directive to allow you to compare the original loadings with those after rotation. The original loadings seem to tell us that the first new axis is some negative measure of overall size, and that the second is a contrast between the first two variables (Height and Length) and last two (Width and Weight). The new loadings give the first axis as largely consisting of Width and Weight, and the second as largely consisting of Height and Length.

Note that under either method of factor rotation, the total contribution of each of the original variables always remains the same as in the input set of loadings (for mathematical reasons). These contributions are called the *communalities* of the variables, and can be expressed as the sum of the squared loadings: they indicate how much of the variation of each of the original variables is retained in either set of dimensions (whether the original set from the principal component analysis, or the new set from the rotation). For example, the communality for the first variable can be calculated from the set of new dimensions as follows

$$0.1906 = (-0.0630)^2 + (0.4320)^2$$

Equivalently, from the original set, it is

$$0.1906 = (-0.2153)^2 + (0.3798)^2$$

If you keep all the loadings from a principal components analysis, each of the variables will have communality 1. Factor rotation in this case will simply give a set of new loadings, each of which will represent just one of the variables, with loading 1. Thus factor rotation is sensible only if you keep merely the higher-dimensional loadings.

The loadings from canonical variates analysis are not constrained to lie in the range $(-1, +1)$. The factor rotation methods operate in a similar manner as for principal component loadings. Again, the objective is to obtain loading values, such that each is either relatively small or relatively large. Also the communalities of the variables remain the same in the rotated loadings as in the original loadings, and the new loadings are obtained as an orthogonal rotation of the old loadings. However, the complete set of loadings can generally be retained from canonical variate analysis and used for factor rotation, without giving meaningless results. This is because the original dimensions from the canonical variates analysis do not contain all the dimensionality of the original variables, unless the number of variables is less than the number of groups. So a factor rotation of all the dimensions will not merely recover the original variables, as would happen with loadings from principal components analysis. This is illustrated in Example 11.2.3b, where the loadings from Example 11.2.2a are rotated.

Example 11.2.3b

```
   1    UNITS [NVALUES=28]
   2    POINTER [VALUES=Foot_lth,Bow_ht,Coil_dia,Elem_dia,Bow_wdth, \
   3        Bow_thck,Length] Data
   4    FACTOR [LEVELS=4] Groupno
   5    READ [PRINT=errors] Groupno,Data[]
  35    SSPM [TERMS=Data[]; GROUPS=Groupno] W
  36    FSSPM W
  37    LRV [ROWS=Data; COLUMNS=3] L
  38    CVA [PRINT=loadings] WSSPM=W; LRV=L

38..............................................................................

*****  Canonical variate analysis  *****

***  Latent Vectors (Loadings)  ***

                 L['Vectors']
                        1           2           3
            Data
        Foot_lth      1.130      -2.656       3.397
          Bow_ht     -0.633       1.631       4.799
        Coil_dia      3.501      -1.708       1.450
        Elem_dia     -2.669      -0.623      -2.802
        Bow_wdth     -3.468      -0.758       0.757
        Bow_thck      1.859      -2.028      -2.478
```

Length	-1.279	-0.110	-3.598

```
39  FACROTATE OLDLOADINGS=L[1]; NEWLOADINGS=L[1]
40  PRINT L[1]
```

	L['Vectors']		
	1	2	3
Data			
Foot_lth	-0.135	-4.381	0.810
Bow_ht	-0.210	-1.670	4.823
Coil_dia	2.513	-3.254	-0.612
Elem_dia	-2.560	2.192	-2.001
Bow_wdth	-3.530	0.111	0.837
Bow_thck	1.074	-0.471	-3.512
Length	-1.041	2.606	-2.593

Rather than print the rotated loadings directly from the analysis (line 39), the program saves and prints them separately (line 40). This might be appropriate if you intend to calculate canonical variate scores for the units, in the rotated factor space, as was shown in the second example of canonical variates analysis (11.2.2). If you do intend to do this, you will also have to calculate new canonical variate means in the rotated factor space; however, this is easy to do as they are simply the group means of the rotated scores for the units.

11.2.4 Multivariate analysis of variance and covariance

Procedure MANOVA performs multivariate analysis of variance or covariance for the data variates listed by the Y parameter.

The model for the analysis is specified by options of the procedure: TREATMENTSTRUCTURE specifies a model formula to define the treatment terms for the analysis, and BLOCKSTRUCTURE specifies the block model of the design (as with the BLOCKSTRUCTURE directive in an ordinary analysis of variance). For unstratified designs (that is, those with a single error term), BLOCKSTRUCTURE can be omitted. The COVARIATES option specifies a pointer containing any covariates.

The other options control printing and saving of output. PRINT indicates the output required from the multivariate analysis of covariance, with settings ssp to print the sums of squares and products matrices, and tests to print the various test statistics (Wilk's Lambda, with chi-square and F approximations, the Pillai-Bartlett trace, Roy's maximum root test and the Lawley-Hotelling trace). APRINT, UPRINT, and CPRINT control output from the univariate analyses of each of the y-variates, corresponding to options PRINT, UPRINT, and CPRINT, respectively, of the ANOVA directive. The final option, LRV, allows a pointer to be saved containing an LRV structure for each treatment term, and then, if required, one for the covariates. If a term is estimated in more than one stratum, the LRV is taken from the stratum that occurs last in the BLOCKTERMS pointer. The structures in the LRV hold the canonical variate loadings, roots, and trace for the respective treatment term.

Example 11.2.4

```
  1  LIBHELP [PRINT=options,parameters] 'MANOVA'

Procedure        MANOVA

Help['options']
PRINT       = strings    Printed output required from the multivariate
                         analysis of covariance (ssp, tests); default tests
APRINT      = strings    Printed output from the univariate analyses of variance
                         of the y-variates (as for the ANOVA PRINT option);
                         default *
UPRINT      = strings    Printed output from the univariate unadjusted analyses
                         of variance of the y-variates (as for the ANOVA UPRINT
                         option); default *
CPRINT      = strings    Printed output from the univariate analyses of variance
                         of the covariates (as for the ANOVA CPRINT option);
                         default *
TREATMENTSTRUCTURE =    formula  Treatment formula for the analysis
BLOCKSTRUCTURE      =    formula  Block formula for the analysis
COVARIATES          =    pointer  Covariates for the analysis (if any)
LRV         = pointer    Contains elements first for the treatment terms and
                         then the covariate term (if any), allowing the LRV's
                         to be saved from one of the analyses; if a term is
                         estimated in more than one stratum, the LRV is taken
                         from the lowest stratum in which it is estimated

Help['parameters']
Y           = variates  Y-variates for an analysis

  2  " Example of the use of procedure MANOVA: data from
 -3     Chatfield & Collins (1986) Introduction to Multivariate Analysis,
 -4     pages 142, 147, 149, 156."
  5  FACTOR [LEVELS=3; VALUES=3(1...3)] Block
  6  & [VALUES=(1...3)3] Treat,Plot
  7  VARIATE [NVALUES=9] V[1...3]
  8  READ [PRINT=errors] V[]
 12  MANOVA [PRINT=ssp,tests; TREATMENTSTRUCTURE=Treat; \
 13     BLOCKSTRUCTURE=Block/Plot; LRV=!P(TLRV)] V[]

***** Multivariate analysis of variance *****

*** Block stratum   ***

****** Error SSP matrix singular.

*** Residual SSP matrix, with   2   degrees of freedom ***

     1      0.7800
     2      0.0300      0.7800
     3      0.9600      0.6600       1.6800

              1            2             3

*** Block . Plot stratum   ***

*** Treat ***

*** SSP-matrix, with   2   degrees of freedom ***
```

```
1        1.680
2        1.380         1.140
3       -1.260        -1.080         1.260

             1             2             3
```

*** Tests ***

```
        Wilk's  Lambda      0.004313
     Approximate Chi sq        16.34      d.f.    6
     Approximate F test         9.48      on      6    and    4  d.f.
  Pillai-Bartlett trace         1.361
Roy's maximum root test        0.9932
  Lawley-Hotelling trace        146.2
```

*** Residual SSP matrix, with   4   degrees of freedom ***

```
1        0.4600
2        0.0300         0.3000
3       -0.4000        -0.4800         1.0600

             1             2             3
```

```
13   " Print the canonical variates information stored from the MANOVA."
14   PRINT TLRV[]
```

```
             TLRV['Vectors']
                      1             2             3

       V[1]       -10.846         0.576         2.111
       V[2]       -21.135         1.558        -2.955
       V[3]       -13.538         2.857        -0.422


             TLRV['Roots']

          1       145.61
          2         0.58
          3         0.00

TLRV['Trace']        146.2
```

11.2.5 Canonical correlation analysis

Procedure CANCOR provides canonical correlation analysis (see, for example, Mardia, Kent, and Bibby 1979 or Digby and Kempton 1987). The data for the procedure consists of two pointers specified by the PVARIATES and QVARIATES parameters; these contain two sets of variates. The variates may have missing values, or be restricted. Any unit for which any of the variates is missing will be excluded from the analysis, and any restrictions on the variates must be consistent. The other parameters allow results to be saved from the analysis.

Printed output is controlled by the option PRINT with settings: correlations to print the canonical correlations (also expressed as percentages, and cumulative percentages, of their total), pcoeff to print the canonical correlation coefficients for the P-set of variates, qcoeff to print the canonical correlation coefficients for the Q-set of variates, pscores to print the canonical correlation scores for the units calculated from the P-set of variates, and qscores to print the canonical correlation scores for the units calculated from the Q-set of variates.

Example 11.2.5

```
  1   LIBHELP [PRINT=options,parameters] 'CANCOR'

      Procedure        CANCOR

Help['options']
PRINT     = strings    Printed output from the analysis (correlations, pcoeff,
                       qcoeff, pscores, qscores); default * i.e. no output

Help['parameters']
PVARIATES     = pointers   Pointer to P-set of variates to be analysed
QVARIATES     = pointers   Pointer to Q-set of variates to be analysed
CORRELATIONS = diagonal    Stores the canonical correlations from each
                matrices   analysis
PCOEFF        = matrices   Stores the coefficients for the P-set of variates
QCOEFF        = matrices   Stores the coefficients for the Q-set of variates
PSCORES       = matrices   Stores the unit scores from the P-set of variates
QSCORES       = matrices   Store the unit scores from the Q-set of variates

  2   " Example of how to use procedure CANCOR:
 -3     data from Table 3.7 of Digby and Kempton (1987)."
  4   TEXT [VALUES='1d','3a','3d','4a','4d','7a','7d','8a','8d','9a','9d', \
  5     '10a','10d','11/1a','11/1d','11/2a','11/2d','14a','14d','16a','16d',\
  6     '17a','17d','18d'] Plot
  7   POINTER [VALUES=N,Nstar,P,K,Lime] Treats
  8   & [VALUES=Axis_1,Axis_2,Axis_3,Axis_4] Species
  9   VARIATE [NVALUES=Plot] Treats[],Species[]
 10   READ [PRINT=errors] Treats[]
 15   READ [PRINT=errors] Species[]
 24   CALCULATE Species[] = Species[] / 100
 25   CANCOR [PRINT=correlations,pcoeff,qcoeff] Treats; Species

*****   Canonical Correlation analysis   *****

***   Canonical correlations   ***
```

	CA_Corrs	%Corrs	Cum%Corr
1	0.9804	35.99	35.99
2	0.8994	33.02	69.01
3	0.5907	21.69	90.70
4	0.2533	9.30	100.00

```
***   Loadings for the P-set of variates   ***
```

	P_Coeff			
Treats	1	2	3	4
N	-0.1515	-0.0031	-0.0813	-0.0857
Nstar	-0.0264	0.1443	-0.0232	-0.3538
P	-0.0409	0.1077	-0.1249	-0.1487
K	-0.0794	0.1956	0.3124	0.3109
Lime	0.1112	0.2150	-0.2632	0.1681

```
***   Loadings for the Q-set of variates   ***
```

	Q_Coeff			
Species	1	2	3	4

```
Axis_1    0.01003   -0.06995   -0.01411   -0.02015
Axis_2   -0.09108   -0.00145   -0.00622   -0.02793
Axis_3   -0.03738   -0.00317   -0.15526    0.03726
Axis_4    0.03252    0.01647   -0.07699   -0.11913
```

11.3 Ordination from associations

The term *ordination* is used mainly in biometrics, particularly in ecology, where it usually refers to attempts to order a set of objects along some environmental gradient. Archaeologists use the term *seriation* to refer to the same set of techniques, whilst the phrase *multidimensional scaling* is used in some other areas. There is no fixed statistical terminology for these methods; however, they have in common an attempt to "order" a set of objects in one dimension with a generalization to give some useful distribution of the objects in multidimensional space. Several of the well-known ordination methods are available in Genstat, principal components analysis (11.2.1) and correspondence analysis (11.8.1). These methods operate with data in the form of a data matrix or a two-way table. A more general method is principal coordinates analysis, or metric scaling, which operates with data in the form of a symmetric matrix of associations; you can produce the associations using the methods in 11.4. Principal coordinates analysis is provided in Genstat by the PCO directive.

Given a symmetric matrix, A say, with values representing the associations amongst a set of n units, principal coordinates analysis (Gower 1966) attempts to find a set of points for the n units in a multidimensional space so that the squared distance between the ith and jth points is given by:

$$d_{ij} = a_{ii} + a_{jj} - 2a_{ij}$$

If A is a similarity matrix (see 11.4.1) then a_{ii} and a_{jj} are both equal to 1, and so this is equivalent to:

$$d_{ij} = 2 \times (1 - a_{ij})$$

Thus similar units are placed close together and dissimilar units are further apart.

Often the data consist of distances rather than similarities (11.4.3). If B is such a matrix, so that b_{ij} gives the observed distance between the ith and jth units, then the preliminary transformation

$$A = - B \times B / 2$$

will give points with inter-point squared distance

$$d_{ij} = a_{ii} + a_{jj} - 2a_{ij}$$
$$= 0 + 0 - 2 \times (b_{ij} \times b_{ij} / 2)$$
$$= b_{ij}^2$$

Therefore the analysis will give points that generate the supplied distances.

The coordinates of the points are arranged so that their centroid, or mean position, is at the origin. Furthermore they are arranged relative to their principal axes, so that the first dimension of the solution gives the best one-dimensional fit to the full set of points, the first two dimensions give the best two-dimensional fit, and so on. The analysis also gives the distances of the points from their centroid, the origin. Associated with each dimension of the set of coordinates is a latent root which is the sum of squares of the coordinates of all the points in that dimension.

For n units, if there is an exact solution it will be in at most $n-1$ dimensions. However, such a solution may not always be available, because the matrix of distances derived from the associations may not be Euclidean: that is, the distances may not be reproducible by points in a Euclidean space of any number of dimensions. If an incomplete solution results, either because the Euclidean property does not hold or because not all the dimensions are to be used, then a residual can be calculated for each unit; this residual is the difference between (a) the distance from the point for that unit in the incomplete solution to the centroid, and (b) the equivalent distance derived from the original data. When the Euclidean property does not hold, some of the residuals may be complex numbers; Genstat represents these as missing values.

If you regard a set of p variables of length n as giving the coordinates of a set of n points in p dimensions, then you can construct the symmetric matrix with values that give the Euclidean distance between the n points (for example B above). If this matrix is then transformed to an association matrix as

$$A = -B \times B/2$$

the principal coordinates analysis of the association matrix will give identical results to a principal components analysis of the original set of variables.

Another special case of principal coordinates analysis occurs when a within-group SSPM structure is to be analysed. Now you can calculate Mahalanobis squared distances amongst the group means as

$$d_{ij} = (x_i - x_j)' \, W^{-1} \, (x_i - x_j)$$

where x_i is the row vector of means for the ith group, and W is the pooled within-group covariance matrix. These squared distances can be transformed to associations, and used as input to principal coordinates analysis to obtain an ordination of the groups. In general, results from this will be different from those of canonical variates analysis, since the ordination operates on a Mahalanobis distance matrix unweighted by group size, whereas the CVA directive operates on a matrix of between-group sums of squares and products, weighted by group size.

Having obtained an ordination, you may sometimes want to add points to the ordination for additional units. For example, with canonical variates analysis, Genstat gives the scores for the group means; you may want to add points to the group-mean ordination for each of the units. As shown in the example of 11.2.2, it is easy to take the data for the new units, apply the centring of the analysis, and use the loadings matrix to get coordinates for the new units.

When you use principal coordinates analysis to analyse an association matrix, there is no loadings matrix; so the method illustrated in 11.2.2 cannot be used to calculate coordinates. However, if you know the squared distances of the new units from the old, the technique of Gower (1968) can be used to add points to the ordination for the new units. You can do this in Genstat by using the ADDPOINTS directive, together with results saved from the preceding PCO directive.

The assumption that the squared inter-point distance is directly related to the values in the association matrix may be too strict with some types of data, for example in psychology. This has led to a family of methods known as *non-metric scaling* or *multidimensional scaling*, several variants of which are provided by the MDS directive (11.3.3).

11.3.1 Principal coordinates analysis (PCO)

PCO performs principal coordinates analysis, also principal components and canonical
variates analysis (but with different weighting from that used in CVA) as special
cases.

Options

PRINT = *strings*	Printed output required (roots, scores, loadings, residuals, centroid, distances); default * i.e. no printing
NROOTS = *scalar*	Number of latent roots for printed output; default * requests them all to be printed
SMALLEST = *string*	Whether to print the smallest roots instead of the largest (yes, no); default no

Parameters

DATA = *identifiers*	These can be specified either as a symmetric matrix of similarities or transformed distances or, for the canonical variate analysis, as an SSPM containing within-group sums of squares and products etc or, for principal components analysis, either as a pointer containing the variates of the data matrix or as a matrix storing the variates by columns
LRV = *LRVs*	Latent vectors (i.e. coordinates or scores), roots, and trace from each analysis
CENTROID = *diagonal matrices*	Squared distances of the units from their centroid
RESIDUALS = *matrices* or *variates*	Distances of the units from the fitted space
LOADINGS = *matrices*	Principal component loadings, or canonical variate loadings
DISTANCES = *symmetric matrices*	Computed inter-unit distances calculated from the variates of a data matrix, or inter-group Mahalanobis distances calculated from a within-group SSPM

The PCO directive is used for principal coordinates analysis. This method encompasses
principal components analysis and a form of canonical variates analysis as special cases as
explained above.

There are six sections of output from PCO:

roots	prints the latent roots and trace;
scores	prints the principal coordinate scores;
loadings	when the directive is being used for principal components analysis or canonical variates

analysis, this specifies that the loadings from the analysis are to be printed;

residuals   prints the residuals, this is relevant only if results are to be printed corresponding to only some of the latent roots;

centroid   prints the distances (not squared distances) of each unit from their overall centroid;

distances   prints the matrix of inter-unit distances (not squared distances).

The NROOTS and SMALLEST options control the printed output of roots, scores, loadings, and residuals. By default, results are printed for all the roots, but you can set the NROOTS option to specify a lesser number. If option SMALLEST has the default setting no these are taken to be the largest roots, but if you set SMALLEST=yes the results are for the smallest non-zero roots. The inter-unit distances are unaffected by the setting of the NROOTS option.

In its simplest form, the PCO directive needs to be supplied with a symmetric matrix, with values giving the associations amongst a set of objects. This could, for example, be a similarity matrix (11.4.1). The DATA parameter provides the symmetric matrix of associations and the PRINT option specifies what is to be printed.

Nathanson (1971) gives squared distances amongst ten types of galaxy: those of an elliptical shape, eight different types of spiral galaxy, and irregularly-shaped galaxies. The spiral types vary from those that are mainly made up of a central core (coded as types SO and SBO) to those that are extremely tenuous (Sc and SBc). Example 11.3.1a below uses these data to form an ordination of the ten galaxy types. It also illustrates the use of the LRVSCREE procedure to produce a "scree" diagram of the latent roots, to help determine how many roots to consider. The roots are printed together with their values expressed as per-thousands and as cumulative per-thousands. The scree diagram, which is represented by lines of asterisks, may help detect sudden decreases between the roots. LRVSCREE can also print first, second, and third differences between the roots. The final part of the example produces a graph of the ordination, shown in Figure 11.3.1.

Example 11.3.1a

```
 1    TEXT [VALUES=E,SO,SBO,Sa,SBa,Sb,SBb,Sc,SBc,I] Gal
 2    SYMMETRICMATRIX [ROWS=Galaxies] Galaxy
 3    READ [PRINT=data,errors] Galaxy

 4    0
 5    1.87 0
 6    2.24 0.91 0
 7    4.03 2.05 1.51 0
 8    4.09 1.74 1.59 0.68 0
 9    5.38 3.41 3.15 1.86 1.27 0
10    7.03 3.85 3.24 2.25 1.89 2.02 0
11    6.02 4.85 4.11 3.00 2.13 1.71 1.45 0
12    6.88 5.70 5.12 3.72 3.01 2.97 1.75 1.13 0
13    4.12 3.77 3.86 3.93 3.27 3.77 3.52 2.79 3.29 0 :

14    CALCULATE Galaxy = -Galaxy/2
```

```
  15   PCO [PRINT=roots,scores,centroid] Galaxy; LRV=PCOlrv

15...........................................................................
```

***** Principal coordinates analysis *****

*** Latent Roots ***

```
    PCOlrv['Roots']
          1           2           3           4           5           6
        6.662       3.058       1.267       1.171       0.737       0.516
          7           8           9          10
        0.381       0.291       0.109       0.000
```

*** Percentage variation ***

```
    PCOlrv['Roots']
          1           2           3           4           5           6
        46.94       21.55        8.93        8.25        5.19        3.64
          7           8           9          10
         2.69        2.05        0.77        0.00
```

*** Trace ***

```
  PCOlrv['Trace']
      14.19
```

*** Latent vectors (coordinates) ***

```
                    1           2           3           4           5
           1     -1.3965      0.6742     -0.4808     -0.2564     -0.0072
           2     -1.0082     -0.1916      0.2521     -0.0488     -0.2665
           3     -0.8176     -0.3197      0.2581     -0.2306     -0.1209
           4     -0.1744     -0.6571      0.0324      0.0699      0.5732
           5     -0.0114     -0.5111     -0.0315      0.1844      0.2450
           6      0.4237     -0.4417     -0.5654      0.5320     -0.2897
           7      0.8244     -0.3341      0.5082     -0.2136     -0.3104
           8      0.9375      0.2451     -0.3141     -0.0592     -0.1534
           9      1.1167      0.4324     -0.1205     -0.5713      0.2104
          10      0.1057      1.1036      0.4615      0.5937      0.1195
                    6           7           8           9
           1     -0.0422      0.1080      0.1334     -0.1166
           2      0.3960     -0.1314      0.0950      0.1501
           3     -0.3759     -0.0046     -0.3260      0.0324
           4     -0.1177      0.1796      0.1790      0.0944
           5      0.1582     -0.3563     -0.0828     -0.1802
           6      0.0839      0.2260     -0.1229      0.0145
           7     -0.0376      0.1792      0.1915     -0.1381
           8     -0.3087     -0.3187      0.1693      0.0968
           9      0.2703      0.0838     -0.1915      0.0410
          10     -0.0263      0.0344     -0.0450      0.0058
```

* Vectors corresponding to zero or negative roots are not printed *

*** Centroid distances ***

```
                    1           2           3           4           5
                  1.657       1.181       1.074       0.940       0.740
                    6           7           8           9          10
                  1.065       1.132       1.140       1.392       1.346
```

```
16   LRVSCREE PCOlrv

No      Root    %%    Cum    %   Scree Diagram (* represents 2%)

 1     6.662    469   469    47  ***********************
 2     3.058    215   685    22  ***********
 3     1.267     89   774     9  *****
 4     1.171     83   857     8  ****
 5     0.737     52   909     5  ***
 6     0.516     36   945     4  **
 7     0.381     27   972     3  **
 8     0.291     21   992     2  *
 9     0.109      8  1000     1  *
10     0.000      0  1000     0

Scale:  1 asterisk represents 2 units.

17   CALCULATE PCOscore[1,2] = PCOlrv[1]$[*; 1,2]
18   AXES [EQUAL=scale] 3; YORIGIN=0; XORIGIN=0
19   PEN 1; SYMBOLS=Galaxies
20   DGRAPH [TITLE='Principal coordinate analysis'; WINDOW=3; KEYWINDOW=0] \
21       PCOscore[2]; PCOscore[1]
```

Line 2 declares a symmetric matrix to hold the galaxy data; the rows (and columns) are labelled by the codes from Nathanson (1971). Line 14 transforms the data to associations; the data are already in the form of squared distances, so there is no need to square them. Line 15 specifies that the PCO directive is to print the latent roots, the scores for the 10 galaxy types, and their distances from their centroid. The first two latent roots are much larger than the others, and so we can infer that a good ordination of the galaxy types can be found from the first two columns of scores (or dimensions).

Ignoring for the moment the score for the irregular galaxies (0.1057), the first column of scores follows a trend from the elliptical galaxies, through

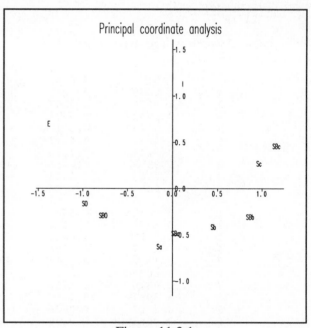

Figure 11.3.1

the densely packed spiral types, to the tenuous spiral types. The irregularly shaped galaxies are placed somewhere near the middle of the others on this first principal axis.

The second axis places the irregular galaxies at the top of the ordination; the other types again roughly follow a trend, but now it is curved. Remember that at most nine dimensions are needed to obtain an exact solution for 10 points; so here the last latent root is zero, and only nine columns of scores are printed.

Instead of a symmetric matrix of associations, the input to PCO can be a pointer whose values are the identifiers of a set of variates, or a matrix storing the variates by columns. Now the PCO directive will construct the matrix of inter-unit squared distances, and will base the analysis on associations derived from this. As described above, this is equivalent to a principal components analysis; however, the results are derived by analysing the distance matrix rather than an SSPM. When there are more units than variates, using PCO for principal components analysis is less efficient than using the PCP directive; however, if there are more variates than units the PCO directive is more efficient.

When PCO is used for principal components analysis, all the variates must be of the same length and none of their values may be missing; any restrictions on the variates are ignored.

Suppose that we have data, as parts per million, for 12 chemical elements measured on eight insects. Analysing the 12 variates with the PCP directive will form the matrix of sums of squares and products for the 12 variates, and use that for the analysis. In Example 11.3.1b the more efficient approach is adopted, analysing the 8-by-8 inter-insect distance matrix instead.

Example 11.3.1b

```
  1   UNITS [NVALUES=8]
  2   POINTER Elements; VALUES=!P(Na,Mg,P,S,Cl,K,Ca,Zn,Fe,Si,Al,Cu)
  3   READ [PRINT=errors] Elements[]
 12   CALCULATE Elements[] = LOG(Elements[]+1)
 13   PCO [PRINT=roots,scores,distances] Elements
```

13...

***** Principal coordinates analysis *****

*** Latent Roots ***

1	2	3	4	5	6	7
25.960	11.437	3.795	1.549	0.790	0.617	0.056

*** Percentage variation ***

1	2	3	4	5	6	7
58.73	25.87	8.58	3.50	1.79	1.39	0.13

*** Trace ***

44.20

*** Latent vectors (coordinates) ***

	1	2	3	4	5
1	-1.0057	1.9782	-0.8397	0.4943	0.0315
2	2.6013	-0.2070	-0.4511	0.4229	-0.3377
3	2.2071	1.7375	0.7367	-0.6271	0.0455
4	-1.7203	-0.6858	-0.8343	-0.7711	-0.3456
5	-1.6349	0.0188	-0.0597	-0.0440	0.6029
6	-1.3564	-0.8063	0.7473	0.3134	-0.1868
7	-1.1926	-0.2210	0.9985	0.1934	-0.1663
8	2.1015	-1.8145	-0.2976	0.0183	0.3565

```
                      6              7
            1      -0.1066         0.0856
            2       0.1168        -0.1242
            3       0.0908         0.0243
            4      -0.0282        -0.0018
            5       0.1580        -0.1229
            6       0.4868         0.0779
            7      -0.5364        -0.0393
            8      -0.1813         0.1003

***   Distance matrix   ***

     1    0.000
     2    4.263   0.000
     3    3.764   2.573   0.000
     4    3.060   4.529   4.894   0.000
     5    2.361   4.388   4.362   1.607   0.000
     6    3.291   4.204   4.502   2.030   1.520   0.000
     7    2.929   4.124   4.072   2.253   1.584   1.228   0.000
     8    4.967   1.909   3.780   4.161   4.196   3.859   3.939   0.000
          1       2       3       4       5       6       7       8
```

The data are defined on lines 1 and 2, and input on line 3. You can see from the report from READ that the amounts of the 12 elements differ considerably from each other. Often with such data, logarithms are taken before any analysis; this has been done on line 12. The PRINT option in the PCO statement (line 13) requests printing of the latent roots, the scores for the eight insects, and the matrix of inter-insect distances. These are shown above. You should note that the printed distances are not squared distances, even though the analysis has been calculated from squared distances.

The third type of input to PCO is an SSPM structure. This must be a within-group SSPM: that is, you must have set the GROUP option of the SSPM directive (2.7.2) when the SSPM was declared. Now the PCO directive will calculate the Mahalanobis distances amongst the group means, and base the analysis on them. As described above, this will give results similar to a canonical variates analysis. The representation of distances in four dimensions will be better than that of CVA, but CVA will be better if you are interested in loadings for discriminatory purposes. In Example 11.3.1c, we analyse the same data as in the examples of CVA (11.2.2). These consist of seven variables measured on 28 brooches; the brooches are classified into four groups.

Example 11.3.1c

```
   1   POINTER [VALUES=Foot_lth,Bow_ht,Coil_dia,Elem_dia,Bow_wdth, \
   2      Bow_thck,Length] Data
   3   FACTOR [LEVELS=4] Groupno
   4   READ [PRINT=errors] Groupno,Data[]
  34   SSPM [TERMS=Data[]; GROUPS=Groupno] W
  35   FSSPM W
  36   PCO [PRINT=roots,scores,distances] W
```

```
36.........................................................................

*****   Principal coordinates analysis   *****
```

```
***   Latent Roots   ***

             1              2              3
         19.91          16.85           6.98

***   Percentage variation   ***

             1              2              3
         45.52          38.52          15.95

***   Trace   ***

        43.73

***   Latent vectors (coordinates)   ***

                         1              2              3
           1         -1.816          2.980          0.631
           2          0.571          0.038         -2.262
           3         -2.162         -2.815          0.560
           4          3.407         -0.204          1.071

***   Distance matrix   ***

    1      0.000
    2      4.767          0.000
    3      5.806          4.855          0.000
    4      6.133          4.383          6.172          0.000
               1              2              3              4
```

The first part Example 11.3.1c, up to line 35, is the same as Example 11.2.2a. The PCO statement (line 36) prints the latent roots, the scores (that is canonical variate means for the four groups), and the matrix of inter-group Mahalanobis distances. The printed distances are Mahalanobis distances, not Mahalanobis squared distances.

The second and subsequent parameters of PCO allow you to save the results. The number of units that determine the sizes of the output structures differs according to the input to PCO. For a matrix or a symmetric matrix the number of units is the number of rows of the matrix, for a pointer it is the number of values in the variates that the pointer contains, while for an SSPM the number of units is the number of groups.

The latent roots, scores, and trace can be saved in an LRV structure using the LRV parameter. If you have declared the LRV already, its number of rows must equal the number of units.

If the input to PCO is a pointer, a matrix, or an SSPM, the principal component or canonical variate loadings can be saved in a matrix using the LOADINGS parameter. The number of rows of the matrix is equal to the number of variates (either those specified by an input pointer or those specified in the SSPM directive for an input SSPM structure), or the number of columns in an input matrix.

The number of columns of the LRV and of the LOADINGS matrix corresponds to the number of dimensions to be saved from the analysis, and this must be the same for both of them. If the structures have been declared already, Genstat will take the larger of the numbers of columns declared for either, and declare (or redeclare) the other one to match. If neither has been declared and option SMALLEST retains the default setting no, Genstat takes the number of columns from the setting of the NROOTS option. Otherwise, Genstat saves results for the full set of dimensions. The trace saved as the third component of the LRV structure, however, will contain the sums of all the latent roots, whether or not they have all been saved.

The distances of the units from their centroid can be saved in a diagonal matrix using the CENTROID parameter. The diagonal matrix has the same number of rows as the number of units, defined above. The RESIDUALS parameter allows you to save residuals, formed from the dimensions that have not been saved, in a matrix with one column and number of rows equal to the number of units. Finally, the inter-unit distances can be saved in a symmetric matrix using the DISTANCES parameter. The number of rows of the symmetric matrix is again the same as the number of units.

11.3.2 Adding points to an ordination (ADDPOINTS)

ADDPOINTS adds points for new objects to a principal coordinates analysis.

Option

PRINT = *strings*	Printed output required (coordinates, residuals); default * i.e. no printing

Parameters

NEWDISTANCES = *matrices*	Squared distances of the new objects from the original points
LRV = *LRVs*	Latent roots and vectors from the PCO analysis
CENTROID = *diagonal matrices*	Centroid distances from the PCO analysis
COORDINATES = *matrices*	Saves the coordinates of the additional points in the space of the original points
RESIDUALS = *matrices* or *variates*	
	Saves the residuals of the new objects from that space

The input to ADDPOINTS is specified by the first three parameters. The NEWDISTANCES parameter specifies an *s×n* matrix containing squared distances of the *s* new units from the *n* old units. The LRV and CENTROID parameters specify structures defining the configuration of old units; these have usually been produced by a PCO statement (11.3.1).

The PRINT option controls the printed output; by default nothing is printed. The option has two settings:

coordinates	prints the coordinates of the new points;
residuals	prints the residual distances of the new units from the coordinates in the space of the old units.

For example, suppose that three original objects are equidistant, with a squared distance of four units amongst them. An ordination of these squared distances will place the points at the corners of an equilateral triangle of side two units. The coordinates of the three points will be (−0.5774, 1.0000), (−0.5774, −1.0000), and (1.1547, 0.0000). Now suppose that a new object is known to be equidistant from the original objects, at some squared distance *d* from them. If *d* is 4/3 the new object can be located precisely at the centroid of the three original points (that is at the origin), and all the distances in the system will be satisfied exactly. However if *d*>4/3, it would be possible to satisfy all the distances in three dimensions by placing the new object at a squared distance of *d*−4/3 above, or below, the plane in which the original points lie. The fitted coordinates in the space of the original objects will be the projection of the new point onto the plane (that is, at the centroid of the original points); the residual for the new object will be the square root of *d*−4/3. If *d*<4/3 the new distances can be satisfied only by introducing an imaginary third dimension in which squared distance is negative: the fitted coordinates will be the same as above, but the residual will be a complex number, which the ADDPOINTS directive will print and store as a missing value.

The other parameters can be used to save the results. The COORDINATES parameter allows you to specify an *s*×*k* matrix to save the coordinates for the new units; the residuals can be saved in an *s*×1 matrix using the RESIDUALS parameter. The value *k* is determined by the dimensionality of the input coordinates from the preceding PCO statement.

In Example 11.3.2, we use the data from Example 11.3.1a on the different galaxy types, and construct an ordination of the eight spiral forms. Then points for the irregular and elliptical types are added to this ordination. First we need to extract from the data the symmetric matrix of distances for the spiral types and also a matrix giving the distances of the two other types from the spiral types (lines 30 and 32). Remember that the input distances were transformed ready for the PCO in 11.3.1; this transformation is also appropriate for the distances amongst the spiral types used as input to PCO in line 36. However, the ADDPOINTS directive requires squared distances so the reverse transformation is required for the distances of the irregular and elliptical galaxy types from the spiral types (line 33).

Example 11.3.2

```
22   TEXT Gname2,Gname8; VALUES=!T(E,I),!T(SO,SBO,Sa,SBa,Sb,SBb,Sc,SBc)
23   SYMMETRICMATRIX [ROWS=Gname8] G8
24   CALCULATE G8 = Galaxy$[!(2...9)]
25   MATRIX [ROWS=Gname2; COLUMNS=Gname8] G2
26   CALCULATE G2 = Galaxy$[!(1,10); !(2...9)]
27   & G2 = -2 * G2
28   LRV [ROWS=Gname8; COLUMNS=2] L8
29   PCO [PRINT=roots] G8; LRV=L8; CENTROID=C8
```

```
29..........................................................................

*****  Principal coordinates analysis   *****

***  Latent Roots   ***

             1            2           3           4           5           6
         5.006        1.359       0.838       0.724       0.508       0.358
             7            8
         0.216        0.000
```

```
***   Percentage variation   ***

          1             2             3             4             5             6
       55.57         15.08          9.30          8.04          5.64          3.98
          7             8
        2.40          0.00

***   Trace   ***

        9.009

   30  ADDPOINTS [PRINT=coordinates,residuals] G2; LRV=L8; CENTROID=C8

30..........................................................................

*****  Adding points to a principal coordinates analysis  *****

***   Coordinates of added points   ***

                     1             2
          1       1.1003       -0.5186
          2      -0.1787       -0.4406

***   Residuals   ***

          1       1.445
          2       1.474
```

11.3.3 Multidimensional scaling (MDS)

The MDS directive carries out iterative scaling, including metric and non-metric scaling. The input data consists of a symmetric matrix whose values may be interpreted, in a general sense, as distances between a set of objects. The matrix is specified by the DATA option; thus only one matrix can be analysed each time the MDS directive is used.

The objective of the MDS directive is to find a set of coordinates whose inter-point distances match, as closely as possible, those of the input data matrix. When plotted, the coordinates provide a display which can be interpreted in the same way as a map: for example, if points in the display are close together, their distance apart in the data matrix was small.

The algorithm invoked by the MDS directive uses the method of steepest descent to guide the algorithm from an initial configuration of points to the final matrix of coordinates that has the minimum stress of all configurations examined.

Printed output is controlled by the PRINT option; by default nothing is printed. There are six possible settings:

coordinates	prints the solution coordinates, rotated to principal coordinates;
roots	prints the latent roots of the solution coordinates;
distances	prints the inter-unit distances, computed from the solution configuration;

fitteddistances	prints the fitted values from the regression of the inter-unit distances on the distances in the data matrix, the regression may be monotonic or linear through the origin, depending on the setting of the METHOD option;
stress	prints the stress of the solution coordinates;
monitoring	prints a summary of the results at each iteration.

The METHOD option determines whether metric or non-metric scaling is given. The algorithm involves regression of the distances, calculated from the solution coordinates, against the dissimilarities in the symmetric matrix specified by the DATA option. With the default setting, METHOD=nonmetric, monotonic regression is used; if METHOD=linear, the algorithm uses linear regression through the origin.

The stress function to be minimized can be selected using the STRESS option. There are three possibilities.

ls (least squares):
$$\sum_i \sum_j (d_{ij} - \hat{d}_{ij})^2 / \sum_i \sum_j d_{ij}^2$$

lss (least-squares-squared):
$$\sum_i \sum_j (d_{ij}^2 - \hat{d}_{ij}^2)^2 / \sum_i \sum_j d_{ij}^4$$

logstress:
$$\sum_i \sum_j (\log d_{ij} - \hat{\log} d_{ij})^2$$

where the d_{ij} are the elements of the input dissimilarity matrix and the $\hat{d}_{ij}$ are the fitted values from the regression by the METHOD option.

The TIES option allows you to vary the way in which tied data values in the input data matrix are to be treated. By default, the treatment of ties is primary, and no restrictions are placed on the distances corresponding to tied dissimilarities in the input data matrix. In the secondary treatment of ties, the distances corresponding to tied dissimilarities are required to be as nearly equal as possible. Kendall (1977) describes a compromise between the primary and secondary approaches to ties: the block of ties corresponding to the smallest dissimilarity are handled by the secondary treatment, the remaining blocks of ties are handled by the primary treatment. This tertiary treatment of ties is useful when the dissimilarities take only a few values. For example, in the reconstruction of maps from abuttal information, the dissimilarity coefficient takes only two values: zero if localities abut, and one if they do not. The block of ties associated with the dissimilarity of zero are handled by the secondary treatment, and the block of ties with dissimilarity one by the primary treatment.

The WEIGHT option can be used to specify a symmetric matrix of weights. Each element of the matrix gives the weight to be attached to the corresponding element of the input data matrix. If the option is not set, the elements of the data matrix are weighted equally. The most important use of the option occurs when the matrix of weights contains only zeros and ones; the zeros then correspond to missing values in the input data matrix, allowing incomplete data matrices to be scaled. Up to about two thirds of the data matrix may be missing before the algorithm breaks down. This enables experimenters to design studies in which only a subset of all the dissimilarities need to be observed. This is particularly useful when there are a large number of units; if the number of units is m, say, a complete $m \times m$ data matrix requires $m(m-1)/2$ dissimilarities to be observed.

Since the algorithm is an iterative one, making use of the method of steepest descent, there is no guarantee that the solution coordinates found from any given starting configuration has the minimum stress of all possible configurations. The algorithm may have found a local, rather than the global, minimum. This problem may be partially overcome by using a series of different starting configurations. If several of the solutions arrive at the same lowest stress solution, then you may be reasonably confident of having found the global minimum. The NSTARTS option determines the number of starting configurations to be used. The starting configuration used on the first start can be specified by the INITIAL option; if this is not set, the default is to take the principal coordinate solution obtained from a PCO analysis of the input dissimilarity matrix. Subsequent starting configurations are found by perturbing each coordinate of the first starting configuration by successively larger amounts. This strategy generally results in at least one starting configuration that does not get entrapped in a local minimum: however there can be no guarantee that the global minimum for the stress function has been found. Experience suggests that, for safety, the NSTARTS option should be set equal to at least 10. By default NSTARTS=1.

The MAXCYCLES option determines the maximum number of iterations of the algorithm. The default of 30 should usually be sufficient. However, it may be necessary to set a larger value for very large data matrices or when using the logstress setting of the SCALING option. The monitoring setting of the PRINT option may be used to see how convergence is progressing.

The NDIMENSIONS parameter must be set to a scalar (or scalars) to indicate the number(s) of dimensions in which the multidimensional scaling is to be performed on the data matrix. An MDS statement with a list of scalars will carry out a series of scaling operations, all based on the same matrix of dissimilarities, but with different numbers of dimensions.

The remaining parameters of the MDS directive allow output to be saved in Genstat data structures. The COORDINATES parameter can list matrices to store the minimum stress coordinates in each of the dimensions given by the NDIMENSIONS parameter, and the STRESS parameter can specify scalars to store the associated minimum stresses. The parameters DISTANCES and FITTEDDISTANCES can specify symmetric matrices to store the distances computed from the coordinates matrix and the fitted distances computed from the monotonic or linear regressions, respectively.

Example 11.3.3 shows the use of non-metric multidimensional scaling with the inter-galaxy distances of Example 11.3.1a, printing the stress, the coordinates, and the roots. The remainder of the example plots the two-dimensional solution obtained as Figure 11.3.3a, and also the "Shepard diagram" in Figure 11.3.3b. This shows the distances that have been computed from the solution obtained – the distances between the points in Figure 11.3.3a – plotted as crosses against the actual distances in the input data, and also the fitted monotonic regression line using circles to show the fitted values. The small distances, typically of the points in Figure 11.3.3a from their immediate neighbours, have been fitted well, as have most of the large distances.

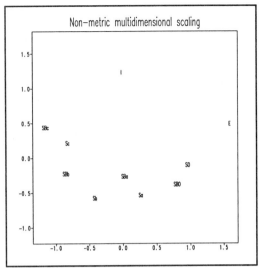

Figure 11.3.3a

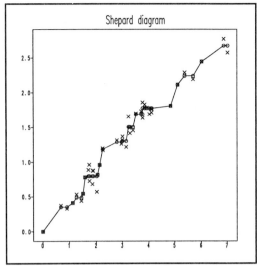

Figure 11.3.3b

Example 11.3.3

```
 1   TEXT [VALUES=E,SO,SBO,Sa,SBa,Sb,SBb,Sc,SBc,I] Galaxies
 2   SYMMETRICMATRIX [ROWS=Galaxies] Galaxy
 3   READ [PRINT=errors] Galaxy
14   MDS [PRINT=roots,coordinates,stress; DATA=Galaxy] NDIMENSIONS=2; \
15     COORDINATES=MDScoord; DISTANCES=MDSdist; FITTED=MDSfit
```

15..

***** Multidimensional scaling *****

*** Least-squares scaling criterion ***
* Distances fitted using monotonic regression (non-metric MDS)   *
* Primary treatment of ties   *

*** Stress   0.0469

*** Coordinates   ***
                MDScoord
                   1   2
           1   1.5703   0.4606
           2   0.9222   -0.1292
           3   0.7476   -0.4075
           4   0.2286   -0.5635
           5   -0.0260   -0.2914
           6   -0.4531   -0.6071
           7   -0.9008   -0.2548
           8   -0.8583   0.1835
           9   -1.2015   0.4082
          10   -0.0290   1.2012
```
```

```
***   Latent Roots   ***
                   1             2
                7.125         2.875
    16   CALCULATE Score[1,2] = MDScoord$[*; 1,2]
    17   VARIATE Actual,OnMDS,FitMDS; Galaxy,MDSdist,MDSfit
    18   PEN 1,2,3; SYMBOLS=Galaxies,1,2; METHOD=(point)2,line; \
    19     LINESTYLE=1; SIZE=2,1,1
    20   AXES [EQUAL=scale] 3; STYLE=box
    21   DGRAPH ['Non-metric multidimensional scaling'; WINDOW=3; \
    22     KEYWINDOW=0] Score[2]; Score[1]
    23   AXES 4; STYLE=box
    24   DGRAPH ['Shepard diagram'; WINDOW=4; KEYWINDOW=0] \
    25     OnMDS,FitMDS; Actual; PEN=2,3
```

11.4   Measures of association

As was explained at the beginning of this chapter, many forms of multivariate analysis operate on symmetric matrices that give similarities between all pairs of samples: these are termed *Q-methods*. The FSIMILARITY directive is concerned with forming similarity matrices, essentially using the method described by Gower (1971). The similarity coefficient that is calculated allows variables to be qualitative, quantitative, or dichotomous, or mixtures of these types; values of some of the variables may be missing for some samples. The values of a similarity coefficient vary between zero and unity, though some authors express them as percentages in the range 0-100%. Two samples have a similarity of unity only when both have identical values for all variables; a value of zero occurs when the values for the two samples differ maximally for all variables. Thus similarity is the complement of dissimilarity, and to convert a similarity s_{ij} into a dissimilarity you can evaluate expressions like $1-s_{ij}$ or $\sqrt{(1-s_{ij})}$. Whether a set of dissimilarities obeys the metric axioms (particularly the triangle inequality), or can be regarded as being generated by distances between pairs of points in a multidimensional Euclidean space, depends on the particular coefficient and on the data themselves. Genstat can evaluate similarities using many of the standard similarity coefficients for qualitative and quantitative variables; Gower and Legendre (1986) discuss some of the properties of these coefficients. In Genstat the resulting similarity matrices are ordinary symmetric matrices, so you can use the standard matrix operations (4.10); their main use in multivariate analysis is for principal coordinates analysis (11.3.1), or other forms of metric scaling or non-metric scaling, or for hierarchical cluster analysis (11.5).

11.4.1   Forming similarity matrices (FSIMILARITY)

FSIMILARITY forms a similarity matrix or a between-group-elements similarity matrix or prints a similarity matrix.

Options

PRINT = *string*   Printed output required (similarity, summary);

	default * i.e. no printing
STYLE = *string*	Print percentage similarities in full or just the 10% digit (full, abbreviated); default full
METHOD = *string*	Form similarity matrix or rectangular between-group-element similarity matrix (similarity, betweengroupsimilarity); default simi
SIMILARITY = *matrix* or *symmetric matrix*	
	Input or output matrix of similarities; default *
GROUPS = *factor*	Grouping of units into two groups for between-group-element similarity matrix; default *
PERMUTATION = *variate*	Permutation of units (possibly from HCLUSTER) for order in which units of the similarity matrix are printed; default *
UNITS = *text* or *variate*	Unit names to label the rows of the similarity matrix; default *

Parameters

DATA = *variates*	The data values
TEST = *strings*	Test type, defining how each variate is treated in the calculation of the similarity between each unit (Jaccard, simplematching, cityblock, Manhattan, ecological, Pythagorean, Euclidean); default * ignores that variate
RANGE = *scalars*	Range of possible values of each variate; if omitted, the observed range is taken

FSIMILARITY allows you to form a symmetric matrix of similarities, or a rectangular matrix of similarities between the units in two groups. You can save either form of similarity matrix, using the SIMILARITY option. FSIMILARITY can also be used to print the symmetric matrix of similarities after it has formed it; alternatively, you can input an existing similarity matrix for printing, using the SIMILARITY option.

The DATA parameter specifies a list of variates, all of which must be of the same length. If any of the variates is restricted, or if the factor in the GROUPS option is restricted, then that restriction is applied to all the variates. Any restriction on any other variate must be to the same set of units. The dimension of the resulting symmetric matrix of similarities is taken from the number of units that contribute to the similarity matrix. If you want to print an existing similarity matrix, the DATA parameter (and the TEST and RANGE parameters) should be omitted, and the SIMILARITY option used to input the matrix concerned.

The TEST parameter specifies a list of strings, one for each variate in the DATA parameter list, that define the "type" of each variate. The type of a variate determines how differences in variate values for each unit contribute to the overall similarity between units: Jaccard is appropriate for dichotomous variables, simplematching for qualitative variables and the other settings give different ways for handling quantitative variables. If you want to exclude a variate from contributing, you should specify an empty string (* or ' '). Otherwise, the form of contribution to the similarity is:

Type	Contribution	Weight
Jaccard	if $x_i = x_j = 1$, then 1	1
	if $x_i = x_j = 0$, then 0	0
	if $x_i \neq x_j$, then 0	1
simplematching	if $x_i = x_j$, then 1	1
	if $x_i \neq x_j$, then 0	1
cityblock	$1 - |x_i - x_j| \, / \,$ range	1
Manhattan	synonymous with cityblock	
ecological	$1 - |x_i - x_j| \, / \,$ range	1
	unless $x_i = x_j = 0$	0
Euclidean	$1 - \{(x_i - x_j) \, / \,$ range$\}^2$	1
Pythagorean	synonymous with Euclidean	

The measure of similarity is formed by multiplying each contribution by the corresponding weight, summing all these values, and then dividing by the sum of the weights.

The RANGE parameter contains a list of scalars, one for each variate in the DATA list. This allows you to check that the values of each variate lie within the given range. If any variate fails the range check, FSIMILARITY gives an error diagnostic and terminates without forming the similarity matrix. The range is also used to standardize quantitative variates; this allows you to impose a standard range, for example when variates are measured on commensurate scales. You can omit the RANGE parameter for all or any of the variates by giving a missing identifier or a scalar with a missing value; Genstat then uses the observed range. If PRINT=summary, Genstat prints the name, the minimum value, and the range for each variate.

The three parameters of the FSIMILARITY directive are also used, for the same purposes, in the directives RELATE (11.4.4), HLIST (11.5.3), and HSUMMARIZE (11.5.4).

The METHOD option controls what type of matrix is produced. METHOD=similarity, the default, gives a symmetric matrix of similarities amongst a single set of units. METHOD=betweengroupsimilarity gives a rectangular matrix of similarities between two sets of units. To form a rectangular matrix of similarities, you must also define the grouping of units by setting the GROUPS option (see below).

The PRINT, STYLE, and PERMUTATION options govern the printing of a symmetric matrix of similarities; you cannot use the FSIMILARITY directive to print a rectangular matrix of similarities between group elements. You can either form the similarity matrix within FSIMILARITY, or input it by the SIMILARITY option. To print the similarity matrix you should set option PRINT=similarity. The STYLE option has two settings, full (the default) or abbreviated. The similarity matrix printed in full style has its values displayed as percentages with one decimal place. If you put STYLE=abbreviated, the values of the similarity matrix are printed as single digits with no spaces, the digit being the 10's value of the similarity as a percentage. In both cases, though, the actual similarities in the range 0-1 are stored in the similarity matrix itself. The PERMUTATION option allows you to specify a variate with values corresponding to the order in which you want the rows of the similarity matrix to be printed. The reordering of the rows is most effective when the permutation arises from a hierarchical clustering and corresponds to the dendrogram order (11.5.1).

Example 11.4.1

```
    1    " Data from  Observers Book of Automobiles  1986
   -2
   -3      16 Italian cars and 12 measurements/characteristics
   -4
   -5       1.  engine capacity        c.c.        Engcc
   -6       2.  number of cylinders                Ncyl
   -7       3.  fuel tank               litres      Tankl
   -8       4.  unladen weight          kg          Weight
   -9       5.  length                  cm          Length
  -10       6.  width                   cm          Width
  -11       7.  height                  cm          Height
  -12       8.  wheelbase               cm          Wbase
  -13       9.  top speed               kph         Tspeed
  -14      10.  time to 100kph          secs        Stst
  -15      11.  carburettor/inj/diesel  1/2/3       Carb
  -16      12.  front/rear wheel drive  1/2         Drive
  -17    "
   18    UNITS [NVALUES=16]
   19    VARIATE Engcc,Ncyl,Tankl,Weight,Length,Width,Height,Wbase,Tspeed,Stst,\
   20       Carb,Drive,Vct[1...3]
   21    POINTER Cd; VALUES=!P(Engcc,Ncyl,Tankl,Weight,Length, \
   22       Width,Height,Wbase,Tspeed,Stst)
   23    READ [PRINT=errors] #Cd,Carb,Drive
   41    TEXT [VALUES=Estate,'Arna1.5','Alfa2.5',Mondialqc,\
   42       Testarossa,Croma,Panda,Regatta,Regattad,Uno,\
   43       X19,Contach,Delta,Thema,Y10,Spider] Carname
   44    FACTOR [Carname; LEVELS=16] Fcar; VALUES=!(1...16)
   45    SYMMETRICMATRIX [ROWS=Carname] Carsim
   46    " Form similarity matrix between cars."
   47    FSIMILARITY [SIMILARITY=Carsim; PRINT=similarities] #Cd,Carb,Drive; \
   48       TEST=4(cityblock),4(Euclidean),2(cityblock),2(simplematch)

   48...........................................................................

   ** Similarity matrix: Carsim **
```

Estate	1	----								
Arna1.5	2	97.6	----							
Alfa2.5	3	81.5	80.0	----						
Mondialqc	4	57.6	54.6	76.2	----					
Testarossa	5	38.9	35.5	56.1	82.7	----				
Croma	6	79.4	77.6	76.2	76.8	56.7	----			
Panda	7	82.0	85.5	61.8	29.6	10.3	54.6	----		
Regatta	8	98.1	96.9	82.3	58.9	39.4	82.0	80.1	----		
Regattad	9	83.9	82.2	67.5	52.5	32.9	75.6	75.6	84.4	----	
Uno	10	88.4	90.9	69.3	40.9	21.1	65.0	96.0	86.6	81.5	----
X19	11	87.0	85.8	82.8	57.8	42.5	60.2	75.8	83.6	70.0	78.7
Contach	12	46.2	43.2	61.8	70.9	88.5	44.0	21.5	45.8	30.9	30.2
Delta	13	95.9	95.1	83.7	58.5	39.3	81.4	81.3	95.9	80.6	87.1
Thema	14	78.5	76.5	75.4	77.4	57.1	98.7	52.9	81.1	74.8	63.7
Y10	15	89.5	92.4	69.2	37.7	19.0	62.1	92.9	87.5	74.0	92.5

Spider	16	77.8	76.1	82.3	74.6	58.3	78.2	62.9	77.2	70.6	67.3
		1	2	3	4	5	6	7	8	9	10

		1	2	3	4	5	6
X19	11	----					
Contach	12	53.1	----				
Delta	13	82.6	47.1	----			
Thema	14	59.1	44.0	80.2	----		
Y10	15	83.0	30.5	88.4	60.5	----	
Spider	16	86.0	50.6	78.8	77.2	70.4	----
		11	12	13	14	15	16

You use the GROUPS option to specify a partition of the units into two groups, by giving a factor with two levels. The units with level 1 of the factor correspond to the rows of the matrix, while the units with level 2 correspond to the columns. As already mentioned, you cannot print this matrix using the FSIMILARITY directive; instead you must save the matrix and then use PRINT (3.2).

The UNITS option allows you to label the rows of the output similarity matrix if the variates of the DATA parameter do not have any unit labels, or if you want to use different labels from those labelling the units of the variates. This labelling also applies to the rows and columns of a matrix of similarities between group elements.

11.4.2 Forming similarities between groups (REDUCE)

Sometimes you may want to regard an n-by-n similarity matrix S as being partitioned into b-by-b rectangular blocks. For example, the cars in 11.4.1 could be classified by their manufacturer. You might then want to form a reduced matrix of similarities, between the different manufacturers instead of between the individual members of the full set of cars. A further example is when there are b soil samples for each of which information is recorded on several soil horizons, possibly different in the different samples. The n sampling units are the full set of horizons that have been observed for the soil samples. The similarity matrix S can be computed for these in the usual way (11.4.1). However you may be more interested in obtaining a reduced similarity matrix between the b soil samples. To do this you have to arrange for each of the b^2 blocks of the full matrix to be replaced by a single value. Each diagonal block must be replaced by unity. Several possibilities exist for replacing the off-diagonal blocks: for example, the maximum, minimum, or mean similarity within the block. Alternatively you can take the view that at least the first horizons of each of two soil samples should agree; you would then replace the block by its first value. Rayner (1966) suggested a more complex method which recognized that certain horizons might be absent from some soil samples; this leads to finding successive optimal matches, conditional on the constraint that one horizon cannot match a horizon that has already been assigned to a higher level; after finding these optima, an average is taken for each horizon. This is termed the *zigzag* method. Again Genstat produces a symmetric similarity matrix, which you can use subsequently for matrix operations or in the appropriate multivariate directives.

REDUCE forms a reduced similarity matrix (referring to the GROUPS instead of the original units).

Options

PRINT = *string*	Printed output required (similarities); default * i.e. no printing
METHOD = *string*	Method used to form the reduced similarity matrix (first, last, mean, minimum, maximum, zigzag); default firs

Parameters

SIMILARITY = *symmetric matrices*	
	Input similarity matrix
REDUCEDSIMILARITY = *symmetric matrices*	
	Output (reduced) similarity matrix
GROUPS = *factors*	Factor defining the groups
PERMUTATION = *variates*	Permutation order of units (for METHOD = firs, last, or zigz)

The SIMILARITY parameter specifies the similarity matrix for the full set of *n* observations; this must be present and have values. The REDUCEDSIMILARITY parameter specifies an identifier for the reduced similarity matrix, of order *b*; this will be declared implicitly if you have not declared it already. The factor that defines the classification of the units into groups must be specified by the GROUPS parameter. The units can be in any order, so that for example the units of the first group need not be all together nor given first. The labels of the factor label the reduced similarity matrix.

The PERMUTATION parameter, if present, must specify a variate. It defines the ordering of samples within each group, and so must be specified for methods first, last, and zigzag. Within each group, the unit with the lowest value of the permutation variate is taken to be the first sample, and so on. Genstat will, if necessary, use a default permutation of one up to the number of rows of the similarity matrix.

If you set option PRINT=similarities, the values of the reduced symmetric matrix are printed as percentages.

The METHOD option specifies how the reduced similarity matrix is to be formed. In Example 11.4.2, the similarity matrix for each car is reduced to a similarity matrix for each manufacturer as represented by the factor Maker. The METHOD option is set to mean. The resulting matrix is printed, and finally stored in the symmetric matrix Makersim.

Example 11.4.2

```
49   " Form reduced similarity matrix for makers."
50   FACTOR [LABELS=!t(Fiat,'Alfa Romeo',Lancia,Ferrari,Lamborghini,\
51     Pinninfarina)] Maker; VALUES=!(2,2,2,4,4,1,1,1,1,1,1,5,3,3,3,6)
```

```
52    SYMMETRICMATRIX [ROWS=Maker] Makersim
53    REDUCE [PRINT=similarities; METHOD=mean] Carsim; \
54      REDUCEDSIMILARITY=Makersim; GROUPS=Maker
```

***** Similarity matrix reduced to groups defined by Maker, using the mean similarity within each group *****

** Reduced similarity matrix: Makersim **

```
Fiat          1     ----
Alfa Romeo    2     82.1   ----
Lancia        3     79.5   84.0   ----
Ferrari       4     43.3   53.1   48.2   ----
Lamborghini   5     37.6   50.4   40.5   79.7   ----
Pinninfarina  6     73.7   78.7   75.5   66.5   50.6   ----

                     1      2      3      4      5      6
```

11.4.3 Forming associations using CALCULATE

An appropriate similarity coefficient can be calculated by FSIMILARITY (11.4.1) for most sets of data. However, many different coefficients of similarity, or distance, have been suggested (see, for example, Gower and Legendre 1986). FSIMILARITY does not cover all of these, but you will generally be able to form the others by using CALCULATE (5.1). Sometimes you may need to convert similarities to dissimilarities (distances), or vice versa. This can be done in many ways; the most common are $D=1-S$ and $D=\sqrt{(1-S)}$, but $D=-\log(S)$ can also be useful. So there are also situations where you may need to transform such matrices using CALCULATE. For example, by putting

```
    FSIMILARITY [SIMILARITY=Smat] V[1...9]; TEST=Euclidean
```

the symmetric matrix Smat will contain similarities constructed from Euclidean squared distances standardized by the ranges of the variates. If you do not want standardization by range, Euclidean distances can be obtained from the PCO directive (11.3.1); but these may then have to be transformed to similarities, for example if you want to use hierarchical cluster analysis (11.5). If Smat has been obtained from the PCO directive, its values should be squared first, to get Euclidean squared distances, and then transformed to similarities:

```
    CALCULATE Smat = Smat*Smat
    & Smat = 1-Smat/MAX(Smat)
```

The FSIMILARITY directive allows variates of different types; for example, dichotomous variates (with values 0 or 1) can have the TEST parameter set to Jaccard or simplematching. Other variates with values on a continuous scale can have the TEST parameter set to cityblock or Euclidean. When both types of variates are present, the resulting similarities will be a weighted average of the component similarities. For example, with five dichotomous variates, Binary[1...5], and three continuous variates, Cont[1...3]

```
    FSIMILARITY [SIMILARITY=Mixed] Binary[1...5],Cont[1...3]; \
      TEST=(Jaccard)5,(cityblock)3
```

will give the similarity matrix Mixed as a weighted average of the Jaccard similarity matrix constructed from Binary[1...5] and the city-block similarity matrix constructed from Cont[1...3]. If, instead of the city-block coefficient, you want to use the unstandardized Euclidean coefficient, you must construct this yourself, as shown above, and then do the averaging:

```
SYMMETRIC [ROWS=N] Jaccard,Euclid,Mixed
FSIMILARITY [SIMILARITY=Jaccard] Binary[1...5]; TEST=jaccard
PCO Cont[]; DISTANCES=Euclid
CALCULATE Euclid = Euclid*Euclid
& Euclid = 1-Euclid/MAX(Euclid)
& Mixed = (5*Jaccard+3*Euclid)/8
```

Gower (1985b) lists 15 different similarity coefficients that have been used for dichotomous variables. Of these, only the simple-matching and Jaccard coefficients can be formed directly with FSIMILARITY; these are the most commonly used. However, a further seven similarity coefficients can be formed using either, or both, of these two. For example, for the five variates Binary[1...5] the Czekanowski coefficient can be calculated from the Jaccard coefficient, using these statements:

```
FSIMILARITY [SIMILARITY=Jaccard] Binary[1...5]; TEST=jaccard
CALCULATE Czekanow = 2 * Jaccard / (1 + Jaccard)
```

Gower (1985b) gives details of the other relationships.

The city-block and Euclidean measures of distance are special cases of the Minkowski distance, which for some positive value of t is:

$$d_{ij} = \left[\sum_k \left(\frac{|x_{ik} - x_{jk}|}{r_k} \right)^t \right]^{1/t}$$

where r_k is usually the range of the kth variable. Although similarities derived from this distance cannot be formed with FSIMILARITY directly, the symmetric matrix Minkwski giving such similarities can be formed from the variates X[1...p] using these statements:

```
CALCULATE Minkwski=0
FOR Thisx=X[1...p]
  FSIMILARITY [SIMILARITY=Temp] Thisx; TEST=cityblock
  CALCULATE Minkwski = Minkwski+Temp**t
ENDFOR
CALCULATE Minkwski = EXP(LOG(Minkwski)/t)
```

11.4.4 Relating associations to data variables (RELATE)

One way of interpreting the principal coordinates obtained from a similarity matrix is by relating them to the original variates of the data matrix. For each coordinate and each data variate, an F-statistic can be computed as if the variate and the coordinate vector were independent. This is not the case but, although the exact distribution of these pseudo F-values is not known, they do serve to rank the variates in order of importance of their contribution to the coordinate vector.

Qualitative variates are treated as grouping factors, and the mean coordinate for each group is calculated. Only 10 groups are catered for; group levels above 10 are combined. The pseudo F-statistic gives the between-group to within-group variance ratio. Missing values are excluded.

Quantitative variates are grouped on a scale of 0-10 (where zero signifies a value up to 0.05 of the range), and mean coordinates for each group are calculated. The printed pseudo F statistic is for a linear regression of the principal coordinate on the ungrouped data variate, after standardizing the data variate to have unit range; the regression coefficient is also printed.

RELATE relates the observed values on a set of variates to the results of a
        principal coordinates analysis.

Options

COORDINATES = *matrices*	Points in reduced space; no default i.e. this option must be specified
NROOTS = *scalar*	Number of latent roots for printed output; default * requests them all to be printed

Parameters

DATA = *variates*	The data values
TEST = *strings*	Test type, defining how each variate is treated in the calculation of the similarity between each unit (Jaccard, simplematching, cityblock, Manhattan, ecological, Pythagorean, Euclidean); default * ignores that variate
RANGE = *scalars*	Range of possible values of each variate; if omitted, the observed range is taken

---

The parameters of the RELATE directive are the same as those of the FSIMILARITY directive, and are described in 11.4.1. However, you do not need to supply the complete list of data variates (with their corresponding types and ranges), only those that you wish to relate to the PCO results. In Example 11.4.3 we examine two of the original variates.

The COORDINATES option must be present and must be a matrix. It represents the units in reduced space. Usually the coordinates will be from a principal coordinates analysis (11.3.1). The number of rows of the matrix must match the number of units present in the variates, taking account of any restriction.

The output from RELATE can be extensive. You may not be interested in relating the variates to the higher dimensions of the principal coordinates analysis even though you may have saved these in the coordinate matrix. The NROOTS option can request that results for only some of the dimensions are printed, for example NROOTS=3 for the first three dimensions as in Example 11.4.3. If NROOTS is not specified, RELATE prints information for all the saved dimensions: that is, for the number of columns of the coordinates matrix.

Example 11.4.3

```
55   " Produce output from ordination of Carsim and RELATE
-56    matrix of coordinates to the original variates "
57   LRV [ROWS=Carname; COLUMNS=6] Carpco; VECTORS=Carvec
58   PCO [PRINT=roots] Carsim; LRV=Carpco
```

58...

***** Principal coordinates analysis *****

*** Latent Roots ***

1	2	3	4	5	6
2.3578	0.8407	0.4220	0.3180	0.2171	0.1795
7	8	9	10	11	12
0.1022	0.0559	0.0504	0.0405	0.0277	0.0207
13	14	15	16		
0.0194	0.0127	0.0121	0.0000		

*** Percentage variation ***

1	2	3	4	5	6
50.42	17.98	9.02	6.80	4.64	3.84
7	8	9	10	11	12
2.19	1.19	1.08	0.87	0.59	0.44
13	14	15	16		
0.41	0.27	0.26	0.00		

*** Trace ***

    4.677

* Some roots are negative - non-Euclidean distance matrix *

```
59   RELATE [COORDINATES=Carvec; NROOTS=3] Weight,Carb; \
60      TEST=cityblock,simplematch
```

60...

**** Relate principal coordinates to original data ****

Variate: Weight
Minimum: 720.0   Range: 786.0   Test type: City block
Data scaled by factor of 0.01272

	F	*	0	1	2	3	4	5
Counts		0	1	2	2	3	2	1
Vector 1	335.8	0.0000	-0.4145	-0.4600	-0.1931	-0.2148	-0.0469	0.1521
Vector 2	0.1	0.0000	-0.1075	-0.1815	-0.1519	0.0847	0.0395	0.4599
Vector 3	0.0	0.0000	-0.1475	-0.0786	0.1350	-0.0672	0.1742	-0.0559

	6	7	8	9	10
Counts	2	0	0	2	1
Vector 1	0.1288	0.0000	0.0000	0.6205	0.8082
Vector 2	0.2076	0.0000	0.0000	-0.1494	-0.1349
Vector 3	0.1034	0.0000	0.0000	-0.0508	-0.1613

** Regression coefficients.   0.0016   0.0001   0.0000

```
    Variate: Carb      Test type: Simple Matching

              F       *       1       2       3
    Counts           0      10       5       1
    Vector 1   4.2  0.0000 -0.1567  0.3558 -0.2118
    Vector 2   4.4  0.0000 -0.1131  0.1900  0.1812
    Vector 3   0.6  0.0000  0.0007  0.0336 -0.1745
```

In Example 11.4.3, the coordinates for the cars in a reduced space of six dimensions are saved in the matrix, Carvec. The first three coordinates account for 71.2% of the trace.

11.5   Hierarchical cluster analysis

One of the main uses of similarity matrices is for hierarchical cluster analysis, which is provided by the HCLUSTER directive. The aim of cluster analysis is to arrange the *n* sampling units into more or less homogeneous groups. HCLUSTER offers several possibilities. The general strategy is best appreciated in geometrical terms, with the *n* sampling units represented by points in a multidimensional space. In *agglomerative* methods, these points initially represent *n* separate clusters, each containing one member. At each of *n*−1 stages, two clusters are fused into one bigger cluster, until at the final stage all units are fused into a single cluster: this process can be represented by a hierarchical tree whose nodes indicate what fusions have occurred. The methods fuse the two closest clusters and vary in how *closest* is defined. In *single-linkage* cluster analysis, *closest* is defined as the smallest distance between any two samples from different clusters; in *centroid* clustering it is the smallest distance between cluster centroids; and so on (see Gordon 1981 for a full discussion).

Genstat will display the tree fitted to a given similarity matrix, and provides a scale to show the level of similarity at which the fusions have occurred; such a scaled tree is termed a *dendrogram*. The endpoints of the dendrogram correspond to the units in some permuted order; you can save this order, for example to use in the FSIMILARITY directive (11.4.1). Of course, a hierarchical tree does not by itself provide a classification. This can be derived by cutting the dendrogram at some arbitrary level of similarity; each cluster then consists of those samples occurring on the same detached branch of the dendrogram. A factor can be formed to indicate cluster membership.

11.5.1   The HCLUSTER directive

HCLUSTER   performs hierarchical cluster analysis.

Options

PRINT = *strings*	Printed output required (dendrogram, amalgamations); default * i.e. no printing
METHOD = *string*	Criterion for forming clusters (singlelink, nearestneighbour, completelink, furthestneighbour, averagelink, mediansort, groupaverage); default sing

| CTHRESHOLD = *scalar* | Clustering threshold at which to print formation of clusters; default * i.e. determined automatically |

Parameters

SIMILARITY = *symmetric matrices*	
	Input similarity matrix for each cluster analysis
GTHRESHOLD = *scalars*	Grouping threshold where groups are formed from the dendrogram
GROUPS = *factors*	Stores the groups formed
PERMUTATION = *variates*	Permutation order of the units on the dendrogram
AMALGAMATIONS = *matrices*	To store linked list of amalgamations

The input for HCLUSTER is provided by the SIMILARITY parameter, as a list of symmetric matrices, one for each analysis. These matrices can be formed by FSIMILARITY (11.4.1), by REDUCE (11.4.2), or by CALCULATE (4.1 and 11.4.3). Missing values are allowed in the similarity matrix only with the single-linkage method.

The GTHRESHOLD and GROUPS parameters must be either both present or both absent. When you are deriving a classification, the level of similarity at which the dendrogram is to be cut is specified by the scalar value in the GTHRESHOLD parameter. The level is given as a percentage similarity. The resulting cluster membership is saved in a factor, whose identifier is specified by the GROUPS parameter. The factor will be declared implicitly, if necessary, and it will have its number of levels set to the number of clusters formed and its number of values taken from the number of rows of the corresponding symmetric matrix.

The PERMUTATION parameter allows you to specify a variate to save the order in which the units appear on the printed dendrogram. Genstat will define it to be a variate automatically, if necessary, with number of values is taken from the number of rows of the corresponding similarity matrix. Conventionally, the first unit on the dendrogram is unit 1 and so the first value of the variate of permutations will be 1.

The AMALGAMATIONS parameter can specify a matrix to store information about the order in which the units form groups, and at what level of similarity. At any stage in the process of agglomeration, each group is represented by the unit with the smallest unit number: for example, a group containing units 2, 5, 17, and 22 is represented by unit 2. This means that the final merge is always between a group indexed by unit 1 and a group indexed by another unit. Since there are $n-1$ stages of agglomeration, the matrix will have a number of rows one less than the number of rows of the input similarity matrix. Each row represents a joining of two groups and consists of three values. The first two values are the numbers indexing the two groups that are joining, and the third value is the level of similarity. So the matrix has three columns. The matrix will be declared implicitly, if necessary.

HCLUSTER can print two pieces of information. The first gives details of each amalgamation, followed by a list of clusters that are formed at decreasing levels of similarity. The second is the dendrogram. The PRINT option allows you to control which of these are printed. If METHOD=singlelink and the PRINT setting includes amalgamations, the minimum spanning tree (11.5.2) will be printed instead of the stages at which the clusters merge. This is because information from forming the minimum spanning tree is used to form the single linkage clustering.

Alternatively, if you save the AMALGAMATIONS matrix, you can use procedure DDENDROGRAM to display the dendrogram using high-resolution graphics.

The METHOD option has seven possible settings; these determine how the similarities amongst clusters are redefined after each merge. The default singlelink, which has synonym nearestneighbour, gives single linkage. The setting completelink (synonym furthestneighbour) defines the distance between two clusters as the maximum distance between any two units in those clusters. The setting averagelink defines the similarity between a cluster and two merged clusters as the average of the similarities of the cluster with each of the two. For groupaverage, an average is taken over all the units in the two merged clusters. Median sorting (Gower 1967) is best thought of in terms of clusters being represented by points in a multidimensional space; when two clusters join, the new cluster is represented by the midpoint of the original cluster points.

The CTHRESHOLD option is a scalar which allows you to define the levels of decreasing similarity at which the lists of clusters are printed with their membership. The decreasing levels of similarity are formed by repeatedly subtracting the CTHRESHOLD value from the maximum similarity of 100%. For example, setting CTHRESHOLD=10 will list the clusters formed at 90% similarity, 80%, and so on. At each level, those units that have not joined any group are also listed. If you do not set this option, the default value will be calculated from the range of similarities at which merges occur, to give between 10 and 20 separate levels.

Example 11.5.1

```
  61   HCLUSTER [PRINT=dendrogram; METHOD=averagelink] Carsim; \
  62     GTHRESHOLD=70; GROUPS=Cargrp; PERMUTATION=Carperm; \
  63     AMALGAMATIONS=Caramalg

**** Average linkage cluster analysis ****

**** Dendrogram ****
  ** Levels    100.0  90.0  80.0  70.0  60.0  50.0

Estate          1   ..
Regatta         8   ..)
Arna1.5         2   ..)
Delta          13   ..).....
Panda           7   ..        )
Uno            10   ..)..     )
Y10            15   .....)..).....
Regattad        9   ..............)..
Alfa2.5         3   ..........        )
X19            11   ........    )     )
Spider         16   ........)..).....)..
Mondialqc       4   ............      )
Croma           6   ..           )    )
Thema          14   ..)..........).....)........
Testarossa      5   ........                )
Contach        12   ........).................)..........

  64   FSIMILARITY [PRINT=similarity; SIMILARITY=Carsim; \
  65     PERMUTATION=Carperm; STYLE=abbreviated]
```

```
65................................................................
```

```
** Abbreviated similarity matrix: Carsim **

Estate          -
Regatta         9-
Arna1.5         99-
Delta           999-
Panda           8888-
Uno             88989-
Y10             889899-
Regattad        8888787-
Alfa2.5         88786666-
X19             888877878-
Spider          7777667788-
Mondialqc       55552435757-
Croma           787856677677-
Thema           7878566775779-
Testarossa      33331213545855-
Contach         444423336557448-
```

11.5.2 Displaying and saving information from a cluster analysis (HDISPLAY)

HDISPLAY displays results ancillary to hierarchical cluster analyses: matrix of mean similarities between and within groups, a set of nearest neighbours for each unit, a minimum spanning tree, and the most typical elements from each group.

Options

PRINT = *strings*	Printed output required (neighbours, tree, typicalelements, gsimilarities); default tree

Parameters

SIMILARITY = *symmetric matrices*	
	Input similarity matrix for each cluster analysis
NNEIGHBOURS = *scalars*	Number of nearest neighbours to be printed
NEIGHBOURS = *matrices*	Matrix to store nearest neighbours of each unit
GROUPS = *factors*	Indicates the groupings of the units (for calculating typical elements and mean similarities between groups)
TREE = *matrices*	To store the minimum spanning tree (as a series of links and corresponding lengths)
GSIMILARITY = *symmetric matrices*	
	To store similarities between groups

You can use the HDISPLAY directive to print ancillary information useful for interpreting cluster analyses, and to save information to use elsewhere in Genstat, for example for plotting.

The SIMILARITIES parameter specifies a list of symmetric similarity matrices. These are operated on, in turn, to produce the output requested by the PRINT option and to save the information specified by other parameters. Since the interpretations of the remaining parameters are closely linked to the different settings of the PRINT option, each setting is discussed below with the relevant parameters.

The NNEIGHBOURS parameter gives a list of scalars indicating how many neighbours will appear in the printed table of nearest neighbours.

The NEIGHBOURS parameter can specify a list of identifiers to store details of nearest neighbours. These will be declared implicitly, if necessary, as matrices. The rows of the matrices correspond to the units; there should be an even number of columns. The values in the odd-numbered columns represent the neighbouring units in order of their similarity, while the values in the even-numbered columns are the corresponding similarities. If you have declared the matrix previously and it does not have enough columns, then NEIGHBOURS stores as many neighbours as possible. If there is an odd number of columns in the matrix, the last column is not filled. If the matrix is declared implicitly, the number of columns will be twice the value of the NNEIGHBOURS scalar.

If the PRINT option includes the setting neighbours, Genstat prints a table of nearest neighbours for every sample, together with their values of similarity. The number of neighbours printed is determined by the value of the NNEIGHBOURS scalar; if NNEIGHBOURS is not set, the table is not printed. This information is also useful for interpreting clusters and ordinations. In Example 11.5.2a, the table is printed for three nearest neighbours, and the matrix Carneig is given values corresponding to the first two nearest neighbours.

Example 11.5.2a

```
  66   MATRIX [ROWS=Carname; COLUMNS=4] Carneig
  67   HDISPLAY [PRINT=neighbours] Carsim; NNEIGHBOURS=3; NEIGHBOURS=Carneig
```

```
  **** Neighbours table derived from Carsim ****
Estate          1         8   98.1        2   97.6       13   95.9
Arna1.5         2         1   97.6        8   96.9       13   95.1
Alfa2.5         3        13   83.7       11   82.8        8   82.3
Mondialqc       4         5   82.7       14   77.4        6   76.8
Testarossa      5        12   88.5        4   82.7       16   58.3
Croma           6        14   98.7        8   82.0       13   81.4
Panda           7        10   96.0       15   92.9        2   85.5
Regatta         8         1   98.1        2   96.9       13   95.9
Regattad        9         8   84.4        1   83.9        2   82.2
Uno            10         7   96.0       15   92.5        2   90.9
X19            11         1   87.0       16   86.0        2   85.8
Contach        12         5   88.5        4   70.9        3   61.8
Delta          13         8   95.9        1   95.9        2   95.1
Thema          14         6   98.7        8   81.1       13   80.2
Y10            15         7   92.9       10   92.5        2   92.4
Spider         16        11   86.0        3   82.3       13   78.8
```

```
  68   PRINT Carneig
```

	Carneig			
	1	2	3	4
Carname				
Estate	8.000	0.981	2.000	0.976
Arna1.5	1.000	0.976	8.000	0.969
Alfa2.5	13.000	0.837	11.000	0.828
Mondialqc	5.000	0.827	14.000	0.774
Testarossa	12.000	0.885	4.000	0.827

Croma	14.000	0.987	8.000	0.820
Panda	10.000	0.960	15.000	0.929
Regatta	1.000	0.981	2.000	0.969
Regattad	8.000	0.844	1.000	0.839
Uno	7.000	0.960	15.000	0.925
X19	1.000	0.870	16.000	0.860
Contach	5.000	0.885	4.000	0.709
Delta	8.000	0.959	1.000	0.959
Thema	6.000	0.987	8.000	0.811
Y10	7.000	0.929	10.000	0.925
Spider	11.000	0.860	3.000	0.823

The GROUPS parameter specifies a factor to divide the units of each similarity matrix into clusters. You may have formed the factor from a previous hierarchical cluster analysis (11.5.1). This parameter must be set if the PRINT option includes the settings typicalelement or gsimilarities.

If the PRINT option includes the setting typicalelement, Genstat prints the average similarity of each group member with the other group members. This is to help you identify typical members of each group: typical members will have relatively large average similarities compared to those of the other members. Within each group, members are printed in decreasing order of average similarity. In Example 11.5.2b, the cars are listed in the order of their mean similarity with the other members of the group to which they belong.

Example 11.5.2b

```
69  HDISPLAY [PRINT=typical] Carsim; GROUPS=Maker

**** Most typical members ****
**** Similarity matrix: Carsim ****

Fiat
Regatta         8      83.3
Uno            10      81.6
Regattad        9      77.4
Panda           7      76.4
X19            11      73.7
Croma           6      67.5

Alfa Romeo
Estate          1      89.5
Arna1.5         2      88.8
Alfa2.5         3      80.7

Lancia
Delta          13      84.3
Y10            15      74.4
Thema          14      70.4

Ferrari
Testarossa      5      82.7
Mondialqc       4      82.7

Lamborghini
Contach        12     100.0
```

```
Pinninfarina
Spider          16    100.0
```

The GSIMILARITY parameter specifies a list of symmetric matrices in which you can save the mean between-group and within-group similarities. Any structure that you have not declared already will be declared implicitly to be a symmetric matrix with number of rows equal to the number of levels of the factor in the GROUPS parameter.

If the PRINT option includes the setting gsimilarities, Genstat prints the mean similarities between-groups and within-groups. Self-similarities are excluded. Example 11.5.2c forms the group similarity matrix based on the groups in the factor Maker, prints the matrix and saves the values in the symmetric matrix Cargsim.

Example 11.5.2c

```
70   HDISPLAY [PRINT=gsimilarity] Carsim; GROUPS=Maker; \
71      GSIMILARITY=Cargsim

**** Mean similarities between and within groups ****
**** Similarity matrix: Carsim ****

** Between and within groups similarity matrix: Cargsim **

Fiat           1     76.6
Alfa Romeo     2     82.1  86.4
Lancia         3     79.5  84.0  76.4
Ferrari        4     43.3  53.1  48.2  82.7
Lamborghini    5     37.6  50.4  40.5  79.7  ----
Pinninfarina   6     73.7  78.7  75.5  66.5  50.6  ----

                      1     2     3     4     5     6

72   PRINT Cargsim

               Cargsim

        Fiat   0.7665
  Alfa Romeo   0.8209      0.8635
      Lancia   0.7952      0.8401     0.7636
     Ferrari   0.4328      0.5313     0.4817   0.8266
 Lamborghini   0.3760      0.5036     0.4054   0.7971   1.0000
Pinninfarina   0.7369      0.7873     0.7547   0.6647   0.5059   1.0000

               Fiat Alfa Romeo     Lancia    Ferrari Lamborghini Pinninfarina
```

The TREE parameter can specify a matrix to save the minimum spanning tree. The matrix is set up with two columns and number of rows equal to the number of units. For each unit, the value in the first column is the unit to which that unit is linked on its left; the second column is the corresponding similarity. The first unit is not linked to any unit on its left, as it is always the first unit on the tree; so the first row of the matrix contains missing values.

Setting the PRINT option to tree prints the minimum spanning tree associated with the similarity matrix specified the SIMILARITY parameter. The minimum spanning tree (MST) is not a Genstat structure, but it can be kept in the form described above: that is, in a matrix with two columns. An MST is a tree connecting the *n* points of a multidimensional representation of the sampling units. In a tree every unit is linked to a connected network and there are no closed loops; the special feature of the MST is that, of all trees with a sampling unit at every node, it is the one whose links have minimum total length. The links include all those that join nearest neighbours; the MST is closely related to single linkage hierarchical trees (11.5.1). Minimum spanning trees are also useful if you superimpose them on ordinations (11.3) to reveal regions in which distance is badly distorted; if neighbouring points, as given by the MST, are distant in the ordination then something is badly wrong (see Gower and Ross 1969). In Example 11.5.2d, the MST is printed and then saved in the structure Cartree which has been declared implicitly as a matrix.

Example 11.5.2d

```
 73  HDISPLAY [PRINT=tree] Carsim; TREE=Cartree

**** Minimum spanning tree ****
**** Similarity matrix: Carsim ****

Estate  Arna1.5      Y10    Panda      Uno
    1......   2...... 15......   7...... 10
    (  97.6      92.4     92.9     96.0
    (
    (  Regatta     Croma    Thema Mondialq Testaros  Contach
    (......   8......   6...... 14......   4......   5...... 12
    (  98.1   (  82.0     98.7     77.4     82.7     88.5
    (         (
    (         ( Regattad
    (         (......   9
    (         (  84.4
    (         (
    (         (  Delta  Alfa2.5
    (         (...... 13......   3
    (            95.9     83.7
    (
    (     X19    Spider
    (...... 11......  16
       87.0      86.0

** Total length     1343.4

 74  PRINT Cartree

                  Cartree
                      1            2
     Carname
      Estate          *            *
     Arna1.5       1.000        0.976
     Alfa2.5      13.000        0.837
    Mondialqc     14.000        0.774
   Testarossa      4.000        0.827
        Croma      8.000        0.820
        Panda     15.000        0.929
      Regatta      1.000        0.981
     Regattad      8.000        0.844
          Uno      7.000        0.960
```

X19	1.000	0.870
Contach	5.000	0.885
Delta	8.000	0.959
Thema	6.000	0.987
Y10	2.000	0.924
Spider	11.000	0.860

11.5.3 The HLIST directive

HLIST lists the values of the data matrix in a condensed form, either in their original order or, more usefully, in the order determined by a cluster analysis (11.5.1). This representation can be very helpful for revealing patterns in the data, associated with clusters, or for an initial scan of the data to pick out interesting features of the variates.

HLIST lists the data matrix in abbreviated form.

Options

GROUPS = *factor*	Defines groupings of the units; used to split the printed table at appropriate places and to label the groups; default *
UNITS = *text* or *variate*	Names for the rows (i.e. units) of the table; default *

Parameters

DATA = *variates*	The data values
TEST = *strings*	Test type, defining how each variate is treated in the calculation of the similarity between each unit (Jaccard, simplematching, cityblock, Manhattan, ecological, Pythagorean, Euclidean); default * ignores that variate
RANGE = *scalars*	Range of possible values of each variate; if omitted, the observed range is taken

---

The parameters of the HLIST directive are the same as those of the FSIMILARITY directive (11.4.1), and are described there. The DATA and RANGE parameters are treated in the same way for HLIST as for FSIMILARITY, but TEST acts slightly differently.

For the TEST parameter, which governs the type of the variate, HLIST distinguishes only between qualitative variates (Jaccard or simplematching) and quantitative variates (other settings). The values of qualitative variates are printed directly. If the range of a quantitative variate is greater than 10, the printed values are scaled to lie in the range 0 to 10. This scaling is done by subtracting the minimum value from the variate, dividing by the range and then multiplying by 10. If the range is less than 10, the values are printed unscaled; so variates with values that are all less than 1 will appear as 0 in the abbreviated table. The values are printed with no decimal places, and in a field-width of 3. In Example 11.5.3a, you can see the effect of scaling the quantitative variates, and not scaling the qualitative variates.

Example 11.5.3a

```
75   HLIST [UNITS=Carname] #Cd,Carb,Drive; \
76      TEST=4(cityblock),4(Euclidean),2(cityblock),2(simplematch)

76.......................................................................
```

***** Key to condensed data matrix *****

	Variate	Minimum	Range	Test type	
1	Engcc	965.0	4202.0	City block	(3)
2	Ncyl	4.000	8.000	City block	(3)
3	Tankl	35.00	85.00	City block	(3)
4	Weight	720.0	786.0	City block	(3)
5	Length	338.0	121.0	Euclidean	(5)
6	Width	149.0	51.0	Euclidean	(5)
7	Height	107.0	39.0	Euclidean	(5)
8	Wbase	216.0	50.0	Euclidean	(5)
9	Tspeed	134.0	157.0	City block	(3)
10	Stst	4.900	14.000	City block	(3)
11	Carb	1.000	2.000	Simple Matching	(2)
12	Drive	1.000	1.000	Simple Matching	(2)

**** Variates listed in condensed form ****

Variate		1	2	3	4	5	6	7	8	9	10	11	12
Test		3	3	3	3	5	5	5	5	3	3	2	2
Range		10	8	10	10	10	10	10	10	10	10	2	1

		1	2	3	4	5	6	7	8	9	10	11	12
Estate	1	1	0	1	3	6	2	6	5	2	4	0	1
Arna1.5	2	1	0	1	1	5	2	8	5	2	3	0	1
Alfa2.5	3	3	2	1	5	7	2	8	7	4	2	0	0
Mondialqc	4	5	4	6	9	9	5	4	9	7	1	1	0
Testarossa	5	9	8	10	10	9	9	1	7	10	0	1	0
Croma	6	2	0	4	5	9	5	9	10	4	2	1	1
Panda	7	0	0	0	0	0	0	10	0	0	8	0	1
Regatta	8	1	0	2	3	7	3	8	5	2	3	0	1
Regattad	9	1	0	2	3	7	3	8	5	1	10	2	1
Uno	10	0	0	0	0	2	1	9	4	0	8	0	1
X19	11	1	0	1	2	4	1	2	0	2	4	0	0
Contach	12	10	8	10	9	6	10	0	5	9	0	0	0
Delta	13	1	0	1	3	4	2	7	6	3	2	0	1
Thema	14	2	0	4	5	10	5	9	10	5	1	1	1
Y10	15	0	0	1	0	0	0	9	0	2	4	0	1
Spider	16	2	0	1	4	6	2	4	2	3	2	1	0

The UNITS option allows you to change the labelling of the units in the table, as shown in Example 11.5.3a. You can specify a text or a pointer or a variate.

You can use the GROUPS option to specify a factor that will split the units into groups. The table from HLIST is then divided into sections corresponding to the groups. If the factor has labels, these are used to annotate the sections; otherwise a group number is used.

Example 11.5.3b

```
77   HLIST [GROUPS=Maker; UNITS=Carname] #Cd,Carb,Drive; \
78      TEST=4(cityblock),4(Euclidean),2(cityblock),2(simplematch)

78...........................................................................

***** Key to condensed data matrix *****

        Variate    Minimum      Range      Test type
   1      Engcc      965.0      4202.0      City block         (3)
   2       Ncyl      4.000       8.000      City block         (3)
   3      Tankl      35.00       85.00      City block         (3)
   4     Weight      720.0       786.0      City block         (3)
   5     Length      338.0       121.0      Euclidean          (5)
   6      Width      149.0        51.0      Euclidean          (5)
   7     Height      107.0        39.0      Euclidean          (5)
   8      Wbase      216.0        50.0      Euclidean          (5)
   9     Tspeed      134.0       157.0      City block         (3)
  10       Stst      4.900      14.000      City block         (3)
  11       Carb      1.000       2.000      Simple Matching    (2)
  12      Drive      1.000       1.000      Simple Matching    (2)
```

```
**** Variates listed in condensed form, grouped by Maker ****

            Variate    1    2    3    4    5    6    7    8    9   10   11   12
            Test       3    3    3    3    5    5    5    5    3    3    2    2
            Range     10    8   10   10   10   10   10   10   10   10    2    1

Fiat
Croma         6    2    0    4    5    9    5    9   10    4    2    1    1
Panda         7    0    0    0    0    0    0   10    0    0    8    0    1
Regatta       8    1    0    2    3    7    3    8    5    2    3    0    1
Regattad      9    1    0    2    3    7    3    8    5    1   10    2    1
Uno          10    0    0    0    0    2    1    9    4    0    8    0    1
X19          11    1    0    1    2    4    1    2    0    2    4    0    0

Alfa Romeo
Estate        1    1    0    1    3    6    2    6    5    2    4    0    1
Arna1.5       2    1    0    1    1    5    2    8    5    2    3    0    1
Alfa2.5       3    3    2    1    5    7    2    8    7    4    2    0    0

Lancia
Delta        13    1    0    1    3    4    2    7    6    3    2    0    1
Thema        14    2    0    4    5   10    5    9   10    5    1    1    1
Y10          15    0    0    1    0    0    0    9    0    2    4    0    1

Ferrari
Mondialqc     4    5    4    6    9    9    5    4    9    7    1    1    0
Testarossa    5    9    8   10   10    9    9    1    7   10    0    1    0

Lamborghini
Contach      12   10    8   10    9    6   10    0    5    9    0    0    0

Pinninfarina
Spider       16    2    0    1    4    6    2    4    2    3    2    1    0
```

11.5.4 Relating the groups to the original data variables (HSUMMARIZE)

The HSUMMARIZE directive helps you to see which clusters, if any, are distinguished by each variate. It requires a factor to define the clusters, as well as the original data variates, together with their types and, optionally, their ranges. From this it prints a frequency table for each variate. Each table is classified by the grouping factor and the different values of the variate.

For qualitative variates (TYPE settings Jaccard or simplematching) the values are integral, and for each group Genstat calculates an interaction statistic labelled chi-squared. This statistic does not have a significance level attached to it, but it does draw attention to groups for which the distribution is markedly different from the overall distribution.

For quantitative variates values are rounded to the nearest point on an 11-point scale (0-10). The interaction statistic is analogous to Student's t, and it draws attention to the groups for which the mean variate value is markedly different from the overall means (again with no significance level attached). Missing values are ignored in the computation of these statistics.

HSUMMARIZE forms and prints a group by levels table for each test together with appropriate summary statistics for each group.

Option

GROUPS = *factor*	Factor defining the groups; no default i.e. this option must be specified

Parameters

DATA = *variates*	The data values
TEST = *strings*	Test type, defining how each variate is treated in the calculation of the similarity between each unit (Jaccard, simplematching, cityblock, Manhattan, ecological, Pythagorean, Euclidean); default * ignores that variate
RANGE = *scalars*	Range of possible values of each variate; if omitted, the observed range is taken

---

The parameters of the HSUMMARIZE directive are the same as those of the FSIMILARITY directive (11.4.1). As with HLIST (11.5.3), the HSUMMARIZE directive distinguishes only between qualitative variates (TYPE settings Jaccard or simplematching) and quantitative variates (other settings of TYPE). The GROUPS option specifies a factor that splits the units into clusters.

As the output from this directive can be very long, only two tables are shown in Example 11.5.4; these illustrate the difference between tables for qualitative and quantitative variates. The grouping factor is taken from the HCLUSTER example in 11.5.1. Each entry in the table gives the number of units from a particular group that have a particular value of the variate.

Example 11.5.4

```
79   HSUMMARIZE [GROUPS=Cargrp] Weight,Carb; \
80      TEST=cityblock,simplematch
```

```
**** Grouped data frequency tables for each variate ****

Variate: Weight
Minimum: 720.0        Range: 786.0      Test type: City block
Data scaled by factor of 0.01272

Cargrp      *   0   1   2   3   4   5   6   7   8   9  10

1           0   3   1   1   4   1   1   0   0   0   0   0
2           0   0   0   0   0   0   2   0   0   0   1   0
3           0   0   0   0   0   0   0   0   0   0   1   1

   Total    0   3   1   1   4   1   3   0   0   0   2   1

Cargrp    Total    Mean        t

1            11    2.18     -1.75
2             3    6.33      1.33
3             2    9.50      2.48

   Total    16    3.88

Variate: Carb    Test type: Simple Matching

Cargrp      *   0   1   2   Total     Chi-sq

1           0   9   1   1      11      2.53
2           0   0   3   0       3      6.60
3           0   1   1   0       2      0.40

   Total    0  10   5   1      16
```

11.6   Non-hierarchical classification

A common statistical problem is to divide the units of a data set into some number of mutually exclusive groups, or classes. Usually you would hope that the groups will be reasonably homogeneous, and distinct from each other. When you do not know the most natural number of classes in advance, you might be interested in several classifications into different numbers of groups: you can then inspect these, and make a decision about the most acceptable number of groups. One way of achieving such groupings is to take the results of a hierarchical classification (11.5), and cut the dendrogram at appropriate levels to obtain groupings into several numbers of classes. However, the statistical properties of the resulting groups are not at all clear, and the hierarchical nature of the groupings into various numbers of classes can impose undue constraints. An alternative approach is to optimize some suitably chosen criterion directly from the data matrix, to obtain one or more non-hierarchical classifications.

Non-hierarchical classification methods differ according to the criterion that they optimize and in the algorithm used to search for an optimum value of the chosen criterion. In Genstat one of four different criteria may be optimized, and the optimization algorithm uses one of two different strategies.

Which criterion to choose depends on the type of data. Suppose first that they can be considered as being a mixture of k multi-Normal distributions, with the same variance-covariance matrix. Then the maximum-likelihood estimate of this matrix is given when the grouping into k classes minimizes the determinant of the within-class variance-covariance matrix, pooled over the k groups (Friedman and Rubin 1967); in other words, the optimization criterion is to minimize this determinant.

When only two groups are to be formed, the criterion above is equivalent to maximizing the Mahalanobis distance between the two classes. However, when the number of groups to be formed is greater than two, maximizing the total Mahalanobis distance between the classes will generally give different results to minimizing the determinant of the pooled within-class dispersion matrix. Maximizing the total Mahalanobis distance is the second available criterion.

The third criterion maximizes the total Euclidean distance between the classes; this is equivalent to minimizing the total within-class sum of squares: that is, the trace of the pooled within-class dispersion matrix. This third criterion can be thought of as a simpler variant of the first, that does not rely on the assumptions of multi-Normality or equal within-class dispersion.

The fourth criterion gives maximal predictive classification (Gower 1974). It is relevant when all the data are binary: that is, when they take only two values, usually designated by zero and one. Within each class, the *class predictor* is defined to be a list with one entry for each variate: the ith entry is whichever value (zero or one) is more frequent in the class for the ith variate. The criterion, W, to be maximized is the sum over the classes of the number of agreements between units of each class and their class predictor. When several different classifications give the same maximum value for W, a subsidiary criterion B is minimized. Whereas W measures within-class homogeneity, B measures between-class heterogeneity: it is the sum of the number of correct predictions for each unit when predicted by any of the class predictors of the classes other than the one to which the unit is assigned.

The algorithm used in Genstat to search for optimal values of the chosen criterion proceeds as follows. Starting from some initial classification of the units into the required number of groups, the algorithm repeatedly transfers units from one group to another so long as such transfers improve the value of the criterion. When no further transfers can be found to improve the criterion, the algorithm switches to a second stage which examines the effect of swopping two units of different classes. The algorithm alternates between the two types of search until neither gives any improvement. Searching for swops is computationally more expensive than searching for transfers, so only one swop is performed each time before the algorithm switches to search for transfers. However, using only swops has the advantage that the group sizes remain constant: if this is what you want, you can direct Genstat to search only for swops.

There is no guarantee that the classification resulting from the above algorithm will be globally optimal: to be sure of that, you would need to try all possible classifications of the units into the required number of groups. All that is known is that no improvement can be made to the criterion by either of the types of transfer strategy. The chance that the algorithm will produce a near-optimal classification can be much improved by providing a good initial classification. You could obtain this from a hierarchical classification method, or by examining a set of principal component scores from the data. The effect of trying different initial classifications can be interesting, and provides some information on the closeness to optimality.

11.6.1 The CLUSTER directive

CLUSTER forms a non-hierarchical classification.

Options

PRINT = *strings*	Printed output required (criterion, optimum, units, typical, initial); default * i.e. no printing
DATA = *matrix* or *pointer*	Data from which the classification is formed, supplied as a units-by-variates matrix or as a pointer containing the variates of the data matrix
CRITERION = *string*	Criterion for clustering (sums, predictive, within, Mahalanobis); default sums
INTERCHANGE = *string*	Permitted moves between groups (transfer, swop); default tran (implies swop also)
START = *factor*	Initial classification; default * i.e. splits the units, in order, into NGROUPS classes of nearly equal size

Parameters

NGROUPS = *scalars*	Numbers of classes into which the units are to be classified: note, the values of the scalars must be in descending order
GROUPS = *factors*	Saves the classification formed for each number of classes

---

By default the CLUSTER directive will not print any results; you must set the PRINT option to indicate which sections of output you want, and whether these should also be printed for the initial classification. The possible settings are as follows.

criterion	prints the optimal criterion value.
optimum	prints the optimal classification.
units	prints the data with the units ordered into the optimal classes.
typical	prints a typical value for each class: for maximal predictive classification this is the class predictor; for the other methods it is the

initial

class mean.

if this is set the requested sections of output are also printed for the initial classification.

The DATA option supplies the data to be classified: the single structure must be either a matrix, with rows corresponding to the units and columns to the variables, or a pointer whose values are the identifiers of the variates in the data matrix. Note that CLUSTER always operates on a matrix, and so will copy the variate values into a matrix if you supply a pointer as input; thus for large data sets it is better to supply a matrix.

The CRITERION option specifies which criterion CLUSTER is to optimize, the default being sums. The four settings are:

sums	maximize the between-group sum of squares;
predictive	maximal predictive classification;
within	minimize the determinant of the pooled within-class dispersion matrix;
mahalanobis	maximize the total Mahalanobis squared distance between the groups.

The INTERCHANGE option specifies which types of interchange (transfers or swops) are to be used. The default is transfer, which is taken to imply that both transfers and swops are used, since a swop is simply two transfers. If you set INTERCHANGE=swop, only swops are used. If INTERCHANGE=* the algorithm does not attempt to improve the classification from the initial classification; you might want this, in conjunction with the PRINT=initial setting, to display the results for an existing classification which you do not wish to improve.

The START option should be used to supply a factor to define the initial classification. If START is not specified, CLUSTER will divide the units, in order, into roughly equal-sized groups. For example, with 97 units to be classified into 10 groups, the first 10 units will be put into the first group, the 11th to 20th into the second group, and so on; the last three groups will contain only nine units each. Procedure CLASSIFY provides another way of forming an initial classification for *k* classes. It finds the *k* units that are furthest apart in the multi-dimensional space defined by the data variates. These are then used as the nuclei for the classes, with each remaining unit being allocated to the class containing the nearest nucleus.

The first parameter, NGROUPS, is used to specify the number of classes to be formed. Any single-valued structure can be supplied here. Often you would want several classifications from a single data set, into different numbers of groups. In this case the NGROUPS parameter should be a list of the numbers of groups in descending order. For the initial classification of the second classification, CLUSTER takes the optimal classification from the first number of groups, and does some reallocation of units to make a smaller number of groups. This is repeated, as often as required, to provide initial classifications for all the later analyses; hence the need to specify the numbers in descending order. The second parameter, GROUPS, is used to specify a list of identifiers of factors to save the optimal classifications.

Doran and Hodson (1975) give some measurements made on 28 brooches found at the archaeological site of the cemetery at Munsingen. Seven of these variables, transformed to logarithms, are used in Example 11.6.1a.

Example 11.6.1a

```
  1   UNITS [NVALUES=28]
  2   POINTER [VALUES=Foot_lth,Bow_ht,Coil_dia,Elem_dia,Bow_wdth, \
  3      Bow_thck,Length] Data
  4   READ Data[]
```

Identifier	Minimum	Mean	Maximum	Values	Missing
Foot_lth	2.398	3.278	4.554	28	0
Bow_ht	2.079	2.842	3.296	28	0
Coil_dia	1.792	2.166	2.833	28	0
Elem_dia	1.099	2.026	2.708	28	0
Bow_wdth	3.045	4.064	5.176	28	0
Bow_thck	2.708	3.621	4.357	28	0
Length	3.296	4.003	4.860	28	0

```
 33   CLUSTER [PRINT=criterion,optimum,initial; DATA=Data] 5,4,3
```

33..

***** Non-hierarchical Clustering *****

*** Sums of Squares criterion ***

*** Initial classification ***

*** Number of classes = 5

*** Class contributions to criterion ***

```
            1           2           3           4           5
          7.623       5.335       1.434       6.251       7.286
```

*** Criterion value = 27.93006

*** Classification of units ***

```
    1   1   1   1   1   1   2   2   2   2   2   2   3
    3   3   3   3   3   4   4   4   4   4   5   5   5
    5   5
```

*** Optimum classification ***

*** Number of classes = 5

*** Class contributions to criterion ***

```
            1           2           3           4           5
          2.205       1.715       1.965       2.361       2.633
```

*** Criterion value = 10.87892

*** Classification of units ***

```
    4   5   3   1   1   5   5   2   1   1   4   2   3
    3   3   3   3   2   3   4   2   2   3   1   1   5
    5   4
```

```
***  Initial classification  ***

***  Number of classes = 4

***  Class contributions to criterion  ***

               1           2           3           4
             2.205       3.839       6.580       2.361

***  Criterion value = 14.98485

***  Classification of units  ***

     4    2    3    1    1    3    3    2    1    1    4    2    3
     3    3    3    3    2    3    4    2    2    3    1    1    3
     3    4


***  Optimum classification  ***

***  Number of classes = 4

***  Class contributions to criterion  ***

               1           2           3           4
             4.394       1.715       3.670       3.119

***  Criterion value = 12.89720

***  Classification of units  ***

     4    3    1    1    1    3    3    2    1    4    4    2    1
     1    1    1    1    2    3    4    2    2    1    1    1    3
     3    4


***  Initial classification  ***

***  Number of classes = 3

***  Class contributions to criterion  ***

               1           2           3
            11.931       4.174       3.670

***  Criterion value = 19.77412

***  Classification of units  ***

     1    3    1    1    1    3    3    2    1    1    1    2    1
     1    1    1    1    2    3    1    2    2    1    1    1    3
     3    2


***  Optimum classification  ***

***  Number of classes = 3

***  Class contributions to criterion  ***

               1           2           3
            15.279       1.714       2.633
```

```
***   Criterion value = 19.62666

***   Classification of units   ***

      1       3       1       1       1       3       3       2       1       1       1       2       1
      1       1       1       1       2       1       1       2       2       1       1       1       3
      3       1
```

The seven variables, represented by the pointer Data, are defined on lines 1 and 2 and are given values in line 3. The PRINT option of the CLUSTER statement (line 33) specifies that the criterion value and optimal classification are to be printed, and that the criterion value and initial classification are to be printed before the transfer and swop algorithm is used. The criterion to be optimized is the default, namely the minimum sum of squares within groups. The DATA option supplies the seven variables, via their pointer. The first parameter specifies that classifications are to be formed into five, then four, then three, groups.

No initial classification has been supplied, so the CLUSTER directive assigns the units to five classes, as described above. Thus the first six units are in class 1, and so on. This classification is printed near the beginning of the output from CLUSTER. It is preceded by the value of the minimum within-class sum of squares criterion for this classification, and a breakdown of this value into the contributions from each class; each such contribution is the sum of squares within a class. At the optimal classification, Genstat prints the criterion value obtained, and its contributions from each class. You can see that the optimal classification obtained is quite different from the initial classification: in fact only 12 of the 28 units are in the same class that they started in.

To obtain an initial classification into four groups the CLUSTER directive reassigns each unit in group 5 to the nearest group: there are five such units, and four of them are closest to group 3. If you examine the initial and optimal classifications into four groups, and the optimal classification into five groups, you will see that many of the units of group 3 have transferred to group 1. This suggests that the optimal fifth group has become the third group; and that the old third and first groups have merged. The initial classification into three groups is similarly formed by reassigning the units in the fourth optimal group: of the five units involved, four are reassigned to group 1. This suggests that group 1 is becoming dominant. In fact little improvement is made to the criterion by forming the optimal classification for three groups; only two units move, both to the first group.

Example 11.6.1b illustrates the maximal predictive criterion. Remember that this method has a subsidiary criterion, B, as well as the main criterion W. The criterion W measures within-class consistency, and has separate contributions from each class; the criterion B measures between-class distinctness and has a contribution from all possible pairs of groups.

Example 11.6.1b

```
1    POINTER [NVALUES=4] Y
2    VARIATE [NVALUES=30] Y[]
3    READ [PRINT=errors; SERIAL=yes] Y[]
8    CLUSTER [PRINT=criterion,optimum,typical; DATA=Y;   \
9      CRITERION=predictive] NGROUPS=5,2; GROUPS=Optimum[5,2]
```

```
9........................................................................

*****  Non-hierarchical Clustering  *****

***  Maximal Predictive criterion  ***

***  Equally optimum classifications  ***

*** Criterion value = 104.00000

*** Criterion B = 49.00000

     3     4     2     1     1     5     3     4     3     5     1     1     2
     4     3     5     5     1     3     4     2     5     2     5     3     5
     3     1     5     1

     3     4     2     1     1     5     3     4     3     5     1     1     2
     3     3     5     5     4     3     4     2     5     2     5     3     5
     3     1     5     1


***  Optimum classification  ***

***  Number of classes = 5

***  Class contributions to criterion  ***

              1          2          3          4          5
           25.00      14.00      28.00      12.00      25.00

***  Criterion value = 104.00000

***  Class contributions to criterion B  ***

              1          2          3          4          5
     1     0.000     10.000      3.000     13.000     18.000
     2     6.000      0.000     10.000     10.000      2.000
     3     4.000     20.000      0.000     14.000     12.000
     4     6.000      9.000      6.000      0.000      3.000
     5    17.000      7.000     15.000     11.000      0.000

***  Criterion B = 49.00000

***  Classification of units  ***

     3     4     2     1     1     5     3     4     3     5     1     1     2
     3     3     5     5     1     3     4     2     5     2     5     3     5
     3     1     5     1

***  Class predictors  ***

              1     2     3     4
     1        0     1     0     0
     2        0     0     1     1
     3        1     0     1     1
     4        0     0     0     1
     5        1     1     0     0


***  Optimum classification  ***

***  Number of classes = 2
```

```
***   Class contributions to criterion   ***

                      1                2
                   43.00            44.00

***   Criterion value = 87.00000

***   Class contributions to criterion B   ***

                      1                2
           1        0.000           18.000
           2       17.000            0.000

***   Criterion B = 35.00000

***   Classification of units   ***

     2      2      2      1      1      1      2      2      2      1      1      1      2
     2      2      1      1      1      2      2      2      1      2      1      2      1
     2      1      1      1

***   Class predictors   ***

                      1      2      3      4
           1          1      1      0      0
           2          1      0      1      1


10   TABULATE [PRINT=counts; CLASSIFICATION=Optimum[5,2]; MARGINS=yes]

                       Count
     Optimum[2]          1                2          Count
     Optimum[5]
           1             7                0            7
           2             0                4            4
           3             0                8            8
           4             0                3            3
           5             8                0            8

        Count           15               15           30
```

Lines 1-3 define and read the data, using the pointer Y to specify four variates each of 30 values. The required non-hierarchical classifications are specified on lines 8 and 9. For each classification the criterion values are printed, together with the optimal classification, and the typical units for each group (that is, the class predictors). The GROUPS parameter has been used to specify factors to hold the optimal classifications.

When the CLUSTER directive has found an optimal classification, it will report all the classifications that it can find with the same optimum (provided that you have asked for the optimal classification to be printed). Several equivalent optimal classifications may often occur with maximal predictive classification, and may occur occasionally with the other criteria. When equally optimal classifications are reported, they are preceded by the criterion value together with the value of the subsidiary criterion (if relevant). If you compare the various optimal classifications printed in Example 11.6.1b, you can see that there is some ambiguity over the allocation of the 14th and 18th units.

After the details of the equally optimal classifications, Genstat prints the breakdown of the

W and *B* criteria for the optimal classification that was found first. The (i,j)th cell of the table of class contributions to criterion *B* shows the number of correct predictions for units in group *i* when predicted by the class predictor of class *j*. For example, amongst the four units in the second group, six dichotomous values (out of 16) are correctly predicted by the first class predictor. You can check this quite easily by comparing the first class predictor (0,1,0,0) with the printed units of group 2.

The results for maximal predictive classification into two groups show a loss of within-class consistency, but improved between-class distinctness. Gower (1974) gives suggestions on how such difficulties may be resolved; for example, maximizing *W−B* would lead to choosing the five-group classification. One preliminary to comparing two classifications is to tabulate them. This has been done on line 10, using as input the factors saved from the CLUSTER statement (for details of the TABULATE directive see 4.11.1). The table printed at the end of the output shows that the first group of the classification into two groups is formed from groups 1 and 5 of the five-group classification; group 2 is formed from groups 2, 3 and 4.

As mentioned already, the results of non-hierarchical classification can vary considerably according to the initial classification. Example 11.6.1c illustrates this, using the same data as Example 11.6.1b.

Example 11.6.1c

```
 11   CLUSTER [PRINT=criterion; DATA=Y; CRITERION=predictive] NGROUPS=6,5

11.............................................................................

*****  Non-hierarchical Clustering   *****

***   Maximal Predictive criterion   ***

***   Optimum classification   ***

***   Number of classes = 6

***   Class contributions to criterion   ***

          1             2             3             4             5             6
       18.00         24.00         19.00         22.00          4.00         22.00

***   Criterion value = 109.00000

***   Class contributions to criterion B   ***

                  1             2             3             4             5
        1       0.000         7.000        12.000         8.000        12.000
        2       7.000         0.000        17.000         9.000         9.000
        3      11.000        14.000         0.000        11.000         9.000
        4      12.000         6.000        10.000         0.000        12.000
        5       2.000         1.000         2.000         2.000         0.000
        6      10.000         8.000         2.000        10.000        14.000
                  6
        1       8.000
        2      11.000
        3       1.000
        4      14.000
```

```
        5         2.000
        6         0.000

***  Criterion B = 50.60000

***  Optimum classification   ***

***  Number of classes = 5

***  Class contributions to criterion   ***

                    1           2           3           4           5
                 18.00       24.00       19.00       22.00       24.00

***  Criterion value = 107.00000

***  Class contributions to criterion B   ***

                    1           2           3           4           5
        1        0.000       7.000      12.000       8.000       8.000
        2        7.000       0.000      17.000       9.000      11.000
        3       11.000      14.000       0.000      11.000       1.000
        4       12.000       6.000      10.000       0.000      14.000
        5       12.000       9.000       4.000      12.000       0.000

***  Criterion B = 48.75000
```

The CLUSTER statement (line 11) specifies that only the criterion value is to be printed, and not the detailed classifications. The number of groups to be formed is first six, then five; thus the initial classification is different from that in Example 11.6.1b. The criterion values are both only slightly better than previously ($W = 107.0$ and $B = 48.75$ compared with $W = 104.0$ and $B = 49.0$); however the contributions from the individual classes are quite different. This example illustrates the difference that the choice of initial classification can make, even with a relatively small number of units. In Example 11.6.1b the initial classification was the default partition into five groups, whereas here it is the classification into six groups, with the sixth group being dispersed.

11.7   Procrustes rotation

Multivariate analyses often give the coordinates of a set of points in some multidimensional space. Typically these are obtained so that certain features of the underlying data are represented by the distances between the points in the multidimensional space. One example is principal components analysis, where the distance amongst the principal component scores represents the Pythagorean distances between the values in the data matrix. Another example is canonical variates analysis, where the distance between the canonical variate scores for the means is the Mahalanobis distance between the groups. The distances amongst a set of points do not change if the origin of the coordinate system is shifted, nor do they change if the axes of the coordinate system are rotated.

Suppose that two sets of points are obtained for the same set of objects but with respect to different coordinate systems. For example, two sets of data concerning the same set of objects may be analysed using principal components analysis to give two sets of principal component scores. Alternatively, one set of data may be analysed using two different methods, again giving two sets of points for the same set of objects. The question that now arises is: can the two sets of points be related to each other without disturbing the relationships contained inside the sets? Since the properties of distance are unchanged by a shift of origin or a rotation of the axes, this question is equivalent to asking whether the coordinate system for one set of points can be shifted and rotated so that they match, as well as possible, the coordinates of the other set of points.

Procrustes rotation, of which there are several variants (Gower 1975b, 1985a), addresses this problem; orthogonal Procrustes rotation is the method most commonly used, and is provided by the ROTATE directive. Suppose that there are two sets of coordinates for n points in r dimensions contained in the $n{\times}r$ matrices X and Y. The X-set is arbitrarily supposed to be a fixed configuration, and the Y-configuration is to be shifted and rotated so that it best matches the X-set. Here *best* means minimizing the sum of the squared distances between the points in the X-set and the matching shifted and rotated points in the Y-set. The best translation (shift of origin) makes the centroids for the two sets of points coincide; this is easily done by translating both sets of points so that their centroids are at the origin. After translation, to find the best rotation involves doing a singular value decomposition (see, for example, Digby and Kempton 1987).

After translation and rotation the goodness of fit can be assessed by the residual sum of squares, which is the sum of squared distances between each X-point and the corresponding Y-point, after translation and rotation. Sometimes the relationships contained inside X and inside Y are similar but are expressed on different scales. You might then want the coordinates in the Y-set to be stretched or contracted by a scaling factor; this can be estimated by least squares. But least-squares scaling should not be used if X and Y are known to be on comparable scales: for example, they may both have come from canonical variates analysis and thus express Mahalanobis distance.

When you cannot say which configuration of points is the fixed set, you might want to know about the results of both Procrustes rotations. The best translation remains the same: both configurations of points are translated so that their centroids coincide, typically at the origin. If the best rotation of Y to X is given by the orthogonal matrix H, then the best rotation of X to Y is the transpose of H. If least-squares scaling is not used, the two residual sums of squares will be the same, unless there is a reflection that has been suppressed. However, if scaling is used, then in general these residuals will differ; you can overcome this by arranging that the two configurations of points, after translation, have the same sum of squares: a convenient value is unity. This initial scaling is particularly desirable when several configurations are to be compared pair by pair.

In general, the best rotation of Y to X may contain a reflection. Usually this is acceptable; however, you may sometimes want to stipulate that the rotation should be a pure rotation and not contain any reflection (Gower 1975a).

Above we have assumed that the two matrices of coordinates have the same number of columns: that is, that the dimensionalities of the two multidimensional spaces are the same. If they differ, Genstat pads out the smaller matrix with columns of zero values, so that it matches the larger.

11.7.1 The **ROTATE** directive

ROTATE does a Procrustes rotation of one configuration of points to fit another.

Options

PRINT = *strings*	Printed output required (rotations, coordinates, residuals, sums); default * i.e. no printing
SCALING = *string*	Whether or not isotropic scaling is allowed (yes, no); default no
STANDARDIZE = *strings*	Whether to centre the configurations (at the origin), and/or to normalize them (to unit sum of squares) prior to rotation (centre, normalize); default cent, norm
SUPPRESS = *string*	Whether to suppress reflection (yes, no); default no

Parameters

XINPUT = *matrices*	Inputs the fixed configuration
YINPUT = *matrices*	Inputs the configuration to be fitted
XOUTPUT = *matrices*	To store the (standardized) fixed configuration
YOUTPUT = *matrices*	To store the fitted configuration
ROTATION = *matrices*	To store the rotation matrix
RESIDUALS = *matrices* or *variates*	
	To store distances between the (standardized) fixed and fitted configurations
RSS = *scalars*	To store the residual sum of squares

The ROTATE directive provides orthogonal Procrustes rotation. You must set the parameters XINPUT and YINPUT, which specify respectively the fixed configuration and the configuration that you want to be translated and rotated; these are called X and Y above. The other parameters are used for saving results from the analysis. For X and Y to refer to the same set of objects they must have the same number of rows, and each object must be represented by the same row in both X and Y. If the XINPUT matrix is $n \times p$ and the YINPUT matrix is $n \times q$, Genstat does the analysis using matrices that are $n \times r$, where r is max(p, q). The smaller matrix is expanded with columns of zeros, as explained above.

The PRINT option specifies which results you want to print; the settings are as follows.

coordinates	specifies that the fixed and fitted configurations are to be printed; note that the fixed configuration is printed after any standardization (see below), and the fitted configuration is printed after standardization and rotation.
residuals	prints the residual distances of the points in the fixed configuration from the fitted points;

rotations   this is after any standardization and rotation.

                                           prints the orthogonal rotation matrix.

sums   prints an analysis of variance giving the sums of squares of each configuration, and the residual sum of squares; if scaling is used, the scaling factor is also printed.

The three other options of the ROTATE directive control the form of analysis. The SCALING option specifies whether you want least-squares scaling to be applied to the standardized YINPUT matrix when finding the best fit to the fixed configuration. You should set SCALING=yes if you want scaling; Genstat will then print the least-squares scaling factor with the analysis of variance. By default there is no scaling.

The STANDARDIZE option specifies what preliminary standardization is to be applied to the XINPUT and YINPUT matrices. It has settings:

centre   centre the matrices to have zero column means;

normalize   normalize the matrices to unit sums of squares.

The default is STANDARDIZE=centre,normalize. The initial centring ensures that the configurations are translated to have a common centroid, and thus automatically provides the best translation of *Y* to match *X*. The normalization arranges that the residual sum of squares from rotating *X* to *Y* is the same as that for rotating *Y* to *X*. Switching off both centring and standardization is rarely advisable, but can be requested by putting STANDARDIZE=*.

With some methods of multivariate analysis, for example the analysis of skew-symmetry (11.8.3), the direction of travel about the origin is important. It is then undesirable to perform a reflection as part of the rotation: the SUPPRESS option can be used to prevent this. The default setting is no, which allows reflection to take place.

As an example, we again consider the galaxies discussed in 11.3. Figures 11.3.1 and 11.3.3a show very similar relationships amongst the galaxy types even though they were produced by different methods, principal coordinates analysis and non-metric multidimensional scaling respectively. Indeed the pictures are almost identical, apart from one being the mirror image of the other. Example 11.7.1 uses Procrustes rotation to assess their similarity. Whereas the scales in Figure 11.3.1 bear a relation to the actual distances input to PCO, those in Figure 11.3.3a need not because in the MDS solution it is only the order of the distance values that is important. So the scaling option of the ROTATE command (lines 10-11) has been set to yes: this also ensures that the sum of squares of the fitted configuration plus that of the residual will equal the sum of squares of the fixed configuration. To assist in the comparison of the two analyses in Example 11.7.1 no normalization is done, and since both input configurations are already centred any standardization has been suppressed. The rotation matrix for a simple reflection would take the form $\begin{pmatrix} -1 & 0 \\ 0 & 1 \end{pmatrix}$ and that from the ROTATE command is very similar to it, although there is also a slight rotation of arccos(0.99888), that is, about 2.7 degrees.

None of the residuals is especially large or small: the second smallest is for the last galaxy type, the Irregulars, which may be because their points are remote from the points for the other galaxy types.

The least-squares scaling factor of 0.9753 is the amount by which the MDS solution has been scaled, after which the sum of squares of its points from the origin is 9.51. The sum of the squared residuals is 0.21, which is also the difference between the sums of squares of the fixed and fitted configurations. Lines 12-16 extract the fixed and fitted coordinates and plot them as Figure 11.7.1, with the larger symbols being used for the fixed points from the PCO analysis. Note that option EQUAL=scale is used in the AXES statement (line 13) – as is appropriate for output from many multivariate analyses.

The second Procrustes rotation in Example 11.7.1 (lines 17-18) is similar to the first, except that reflection has been suppressed. Whilst there is no statistical reason to do this with these configurations of points, it does illustrate what can happen if reflections are suppressed unnecessarily. It is obvious with this example, where only two dimensions are being considered, but with coordinates in more dimensions the effect may be less apparent. The rotation matrix specifies rotation through 180 degrees (apart from 0.5 degree). The sums of squares for the two configurations, and also the scaling factor, are the same as with the first analysis; however the residual is now much larger so that the sums of squares do not add up, as noted below the table.

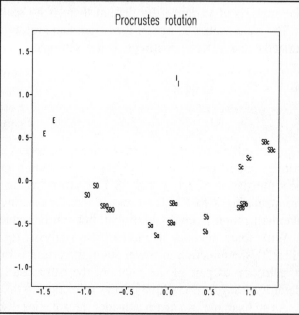

Figure 11.7.1

Example 11.7.1

```
 1   TEXT [VALUES=E,SO,SBO,Sa,SBa,Sb,SBb,Sc,SBc,I] Galaxies
 2   MATRIX [ROWS=Galaxies; COLUMNS=2] Pco,Mds
 3   READ [SERIAL=yes] Pco,Mds
```

Identifier	Minimum	Mean	Maximum	Values	Missing
Pco	−1.396	0.000	1.117	20	0
Mds	−1.202	0.000	1.572	20	0

```
10   ROTATE [PRINT=rotations,residuals,sums; SCALING=yes; STANDARDIZE=*] \
11     XINPUT=Pco; YINPUT=Mds; YOUTPUT=Mdsout
11.................................................................
```

```
*****  Procrustes rotation  *****

***  Orthogonal Rotation  ***

                    1           2
          1    -0.99888     0.04733
          2     0.04733     0.99888

***  Residuals  ***

                    1
          1     0.1914
          2     0.1505
          3     0.0832
          4     0.1389
          5     0.2271
          6     0.1700
          7     0.0600
          8     0.1405
          9     0.1164
         10     0.0696

***  Sums of Squares  ***

Fitted Configuration            9.5124
Residual                        0.2077
--------------------------------------
Fixed Configuration             9.7201

***  Least-squares Scaling factor =       0.9753

   12   CALCULATE Pco1,Pco2,Mdsout1,Mdsout2 = Pco$[*; 1,2],Mdsout$[*; 1,2]
   13   AXES [EQUAL=scale] 3; STYLE=box
   14   PEN 1,2; COLOUR=1; SYMBOLS=Galaxies; SIZE=1.5,0.75
   15   DGRAPH [TITLE='Procrustes rotation'; WINDOW=3; KEYWINDOW=0] \
   16     Pco2,Mdsout2; Pco1,Mdsout1; PEN=1,2
   17   ROTATE [PRINT=rotations,sums; SCALING=yes; STANDARDIZE=*; SUPPRESS=yes]\
   18     XINPUT=Pco; YINPUT=Mds

18.............................................................................

*****  Procrustes rotation  *****

***  Orthogonal Rotation  ***

                    1           2
          1    -0.99996     0.00908
          2    -0.00908    -0.99996

***  Sums of Squares  ***

Fitted Configuration            9.5124
Residual                       11.4846
--------------------------------------
Fixed Configuration             9.7201

***  Least-squares Scaling factor =       0.9753

*  A reflection has been suppressed: sums of squares need not total
```

11.7.2 Generalized Procrustes rotation

Procedure GENPROC can be used when there are several configurations of points to be matched. Generalized Procrustes analysis iteratively matches these to a common centroid configuration using the operations of translation to a common origin, rotation and reflection of axes, and possibly also (if the SCALING option is set to yes) an isotropic scale change. This matching seeks to minimize the sum of the squared distances between the centroid and each individual configuration summed over all points (the Procrustes statistic for each configuration and the centroid, summed over all configurations). The final centroid is referred to principal axes to give a unique consensus configuration.

Generalized Procrustes analysis is widely used in sensory analysis of food, wine, and so on. Further details of the method can be found in the Genstat Procedure Library Manual.

Example 11.7.2

```
   1  LIBHELP [PRINT=options,parameters] 'GENPROC'

      Procedure    GENPROC

Help['options']
PRINT      = strings    Printed output required (analysis, centroid, column,
                        individual, monitoring); default analysis,centroid
SCALING    = string     Whether or not isotropic scaling is allowed (no,yes);
                        default no
METHOD     = string     Method to be used (Gower,TenBerge); default Gower
TOLERANCE  = scalar     The algorithm is assumed to have converged when
                        (last residual sum of squares)
                        - (current residual sum of squares)
                        < TOLERANCE * (number of configurations);
                        default 0.00001
MAXCYCLE   = scalar     Limit on number of iterations; default 50

Help['parameters']
XINPUT     = pointers    Each pointer points to a set of matrices holding
                        the original input configurations
XOUTPUT    = pointers    Each pointer points to a set of matrices to store a
                        set of final (output) configurations
CONSENSUS  = matrices    Stores the final consensus configuration from each
                        analysis
ROTATIONS  = pointers    Each pointer points to a set of matrices to store the
                        rotations required to transform each set of XINPUT
                        configurations to their final (scaled) XOUTPUT
                        configurations
RESIDUALS  = pointers    Each pointer points to a set of matrices to store
                        the distances of a set of scaled XINPUT
                        configurations from its consensus
RSS        = scalars     Stores the residual sum of squares from each analysis
ROOTS      = diagonal    Stores the latent roots from referring the centroid
             matrices    configuration to its principal axis form (consensus)
                        for each analysis
WSS        = scalars     Stores the initial within-configuration sum of squares
                        from each analysis
SCALINGFAC= variates    Stores the isotropic scaling factors for configurations
                        from each analysis

   2  " Data from Gower (1975b).
  -3     Note, however, that in Table 3 the scaling factors printed
```

```
  -4      were SQRT(ro[i]) instead of ro[i], and in Table 4 the
  -5      Between and Within Judges sums of squares were transposed."
   6   MATRIX [ROWS=9; COLUMNS=7] X[1...3]
   7   READ [PRINT=errors; SERIAL=yes] X[]
  38   GENPROC [PRINT=analysis,centroid; SCALING=yes] X
```

Scaling option set

***** Rotation of centroid to principal axes *****

*** Latent roots ***

```
          1    0.609
          2    0.081
          3    0.064
          4    0.027
          5    0.012
          6    0.004
          7    0.002
```

*** Percentage variance ***

```
          1   76.12
          2   10.12
          3    8.05
          4    3.36
          5    1.54
          6    0.56
          7    0.26
```

***** Coordinates of the consensus configuration *****

```
                1         2         3         4         5         6         7

     1   0.0878    0.1798    0.0854    0.0796   -0.0095   -0.0190    0.0074
     2  -0.1453    0.0158    0.0755   -0.0920   -0.0436    0.0188    0.0183
     3   0.1429   -0.0062    0.0618   -0.0536   -0.0233   -0.0188   -0.0333
     4   0.1496   -0.0454    0.1003   -0.0019    0.0822    0.0230   -0.0007
     5   0.1499    0.0935   -0.1329   -0.0559    0.0257   -0.0081    0.0119
     6  -0.3177    0.0650   -0.1190    0.0132    0.0111    0.0175   -0.0177
     7  -0.0944   -0.0572   -0.0044    0.0714   -0.0373    0.0315   -0.0011
     8   0.4636   -0.1240   -0.0753    0.0266   -0.0230   -0.0069    0.0067
     9  -0.4363   -0.1213    0.0086    0.0125    0.0175   -0.0380    0.0086
```

****

***** Analysis of variation for the configurations *****

```
          Scaling  Residual    Total

     1    1.071     0.240      0.931
     2    1.222     0.177      1.033
     3    0.832     0.181      1.036
```

***** Analysis of variation for the entities *****

```
          Consensus Residual    Total

     1    0.162     0.066      0.228
     2    0.114     0.084      0.198
     3    0.087     0.079      0.166
     4    0.125     0.067      0.192
```

5	0.159	0.129	0.287
6	0.361	0.038	0.399
7	0.059	0.026	0.085
8	0.712	0.032	0.744
9	0.622	0.079	0.701

Initial within-configuration s.s.   53254.891

Initial between-configuration s.s.   22114.816

Final residual sum of squares   0.599

Number of steps to convergence 8

An alternative technique, multiple Procrustes analysis, compares the configurations in pairs by ordinary Procrustes rotation to form a symmetric matrix containing the residual sums-of-squares. It then performs a principal coordinate analysis to obtain an ordination representing the individual configurations. An example is given in Digby and Kempton (1987).

11.8   Procedures in the MVA module of the Genstat Procedure Library

Genstat contains many facilities useful for implementing multivariate methods (4.10). These have enabled many further multivariate techniques to be been provided as procedures, which are contained in the MVA module of the Genstat Procedure Library. Below we list the index lines of the relevant procedures in 3[1] of the Library, indicating those that are covered in more detail elsewhere this chapter.

BIPLOT produces a biplot from a set of variates (11.8.2).
CANCOR does canonical correlation analysis (11.2.5).
CLASSIFY obtains a starting classification for non-hierarchical clustering (11.6.1).
CONVEXHULL finds the points of a single or a full peel of convex-hulls.
CORRESP does correspondence analysis, or reciprocal averaging (11.8.1).
DDENDROGRAM draws dendrograms with control over structure and style (11.5.1).
DISCRIMINATE performs discriminant analysis.
DSHADE produces a shaded similarity matrix by high-resolution graphics.
GENPROC performs a generalized Procrustes analysis (11.7.2).
LRVSCREE prints a scree diagram and/or a difference table of latent roots (11.3.1).
MANOVA performs a multivariate analysis of variance or covariance (11.2.4).
MULTMISS estimates missing values for units in a multivariate data set.
RIDGE produces ridge regression and principal component regression analyses (11.2.1).
ROBSSPM forms robust estimates of sum-of-squares-and-products matrices (11.2).
SKEWSYMM provides an analysis of skew-symmetry for an asymmetric matrix (11.8.3).

You can use procedures LIBINFORM or NOTICE to find out about the procedures that are added in later releases of the Library (5.3.1).

11.8.1 Correspondence analysis

Procedure CORRESP can do correspondence analysis, reciprocal averaging or a similar analysis of a biplot-style, as requested by the METHOD option (with setting correspondence, reciprocal, or biplot). Greenacre (1984) describes correspondence analysis; the variations in the method are described by Digby and Kempton (1987).

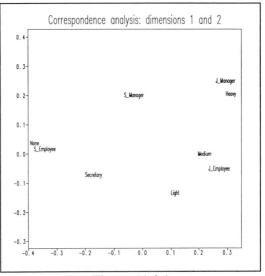

Figure 11.8.1

The data for the procedure is a matrix, specified by the DATA parameter. The matrix must not contain any missing values, and is unchanged on exit from the procedure.

Printed output is controlled by the option PRINT with settings: roots to print the roots, (also the roots expressed as percentages and cumulative percentages), rowscores to print the scores for the rows of the data matrix; rowinertias to print the inertias for the rows of the data matrix, colscores to print the scores for the columns of the data matrix, and colinertias to print the inertias for the columns of the data matrix.

Results from the analysis can be saved using the parameters ROOTS, ROWSCORES, COLSCORES, ROWINERTIAS, and COLINERTIAS. The structures specified for these parameters need not be declared in advance.

The results correspond to p dimensions, where p is one less than the smaller of the numbers of rows and columns of the input data matrix.

Example 11.8.1

```
  1   " Example of how to use procedure CORRESP:
 -2      data from Table 3.1 of Greenacre (1984) "
  3   TEXT [VALUES=S_Manager,J_Manager,S_Employee,J_Employee,Secretary] Staff
  4   & [VALUES=None,Light,Medium,Heavy] Smoke
  5   MATRIX [ROWS=Staff; COLUMNS=Smoke] Smoking; VALUES= \
  6      !( 4, 2, 3, 2,   4, 3, 7, 4,   25,10,12, 4,   18,24,33,13,   10, 6, 7, 2)
  7   PRINT Smoking,'Use CORRESP, printing all results, saving SCORES only';\
  8      FIELDWIDTH=8,*; DECIMALS=0
```

	Smoking			
Smoke	None	Light	Medium	Heavy
Staff				
S_Manager	4	2	3	2
J_Manager	4	3	7	4
S_Employee	25	10	12	4
J_Employee	18	24	33	13
Secretary	10	6	7	2

Use CORRESP, printing all results, saving SCORES only

```
 9  CORRESP [PRINT=roots,rowscores,colscores,rowinertia,colinertia; \
10    METHOD=corresp] Smoking; ROWSCORE=Staff_Sc; COLSCORE=Smoke_Sc
```

***** Correspondence analysis *****

*** Squared singular values ***

	CA_Roots	%Roots	Cum%Root
1	0.07476	87.76	87.76
2	0.01002	11.76	99.51
3	0.00041	0.49	100.00

*** Row Scores and Inertias ***

| | 1 | | 2 | |
Staff	Staff_Sc	R_Inrtia	Staff_Sc	R_Inrtia
S_Manager	-0.0658	0.000247	0.1937	0.002139
J_Manager	0.2590	0.006254	0.2433	0.005521
S_Employee	-0.3806	0.038277	0.0107	0.000030
J_Employee	0.2330	0.024743	-0.0577	0.001520
Secretary	-0.2011	0.005238	-0.0789	0.000807

| | 3 | |
Staff	Staff_Sc	R_Inrtia
S_Manager	-0.0710	0.000287
J_Manager	0.0337	0.000106
S_Employee	0.0052	0.000007
J_Employee	-0.0033	0.000005
Secretary	0.0081	0.000008

*** Column Scores and Inertias ***

| | 1 | | 2 | |
Smoke	Smoke_Sc	C_Inrtia	Smoke_Sc	C_Inrtia
None	-0.3933	0.048892	0.0305	0.000294
Light	0.0995	0.002306	-0.1411	0.004640
Medium	0.1963	0.012381	-0.0074	0.000017
Heavy	0.2938	0.011179	0.1978	0.005066

| | 3 | |
Smoke	Smoke_Sc	C_Inrtia
None	0.0009	0.000000
Light	-0.0220	0.000113
Medium	0.0257	0.000212
Heavy	-0.0262	0.000089

```
11  VARIATE [Staff] Staff_S[1,2]
12  & [Smoke] Smoke_S[1,2]
13  CALCULATE Staff_S[] = Staff_Sc$[*; 1,2]
14  & Smoke_S[] = Smoke_Sc$[*; 1,2]
15  AXES [EQUAL=scale] 3; STYLE=box
16  PEN 2,3; SYMBOLS=Staff,Smoke; SIZE=1.5
17  DGRAPH [TITLE='Correspondence analysis: dimensions 1 and 2'; WINDOW=3;\
18    KEYWINDOW=0] Staff_S[2],Smoke_S[2]; Staff_S[1],Smoke_S[1]; PEN=2,3
```

11.8.2 Biplots

Procedure BIPLOT produces a graphical representation of the relationships between data units and variates, as described by Gabriel (1971).

The data for the procedure consist of a set of variates, contained in a pointer specified by the DATA parameter. The data may be centred at the origin and/or normalized before plotting, by setting the STANDARDIZE option. The variates must not contain any missing values. The values of the variates remain unaltered on exit from the procedure. The setting of the METHOD option indicates form of the biplot: principalcomponent, variate, or diagnostic.

Output is controlled by the PRINT option with settings singular to print the singular values, scores to print the scores, and graph to plot the graph; by default

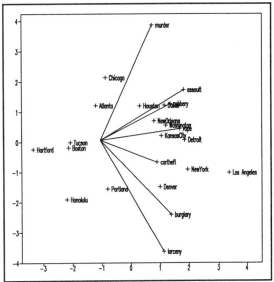

Figure 11.8.2

PRINT=graph. The WINDOW option indicates the window in which to plot a high-resolution graph.

Results from the analysis can be saved using the parameters COORDINATES and VCOORDINATES. The structures specified for these parameters need not be declared in advance.

Example 11.8.2

```
 1   " Data from Hartigan (1975) "
 2   TEXT [VALUES=murder,rape,robbery,assault,burglary,larceny,cartheft] Crime
 3   & [VALUES=Atlanta,Boston,Chicago,Dallas,Denver,Detroit,Hartford,\
 4     Honolulu,Houston,KansasCity,'Los Angeles',NewOrleans,\
 5     NewYork,Portland,Tucson,Washington] City
 6   READ [PRINT=errors] VCrimes[1...7]
24   PEN 1; SYMBOLS=3; LABELS=City
25   PEN 2; SYMBOLS=3; LABELS=Crime
26   AXES 3; XUPPER=4.5
27   BIPLOT [GRAPHICS=high; WINDOW=3] VCrimes
```

11.8.3 Analysis of skew-symmetry

Procedure SKEWSYMM provides the canonical analysis of skew-symmetry described by Gower (1977). The input to the procedure, specified by the parameter DATA, is a (square) asymmetric matrix of associations, *A* say. The rows and columns of *A* usually represent the same set of objects, but in different modes. For example, with migration data, the rows may represent the Countries or States being departed from, and the columns the same locations but being arrived at. The DATA matrix must not contain any missing values.

The results of the analysis are a set of coordinates (scores) for points representing the entities labelling the rows or columns of the DATA matrix. In pairs, these coordinates give positions on a series of planes, also called bimensions. So there is an even number of coordinates for each point; if the DATA matrix has an odd number of rows/columns, there will be one fewer coordinate than the number of rows or columns of the DATA matrix. Also, the "importance" of each plane can be assessed from a set of values (roots) that give the amount of (squared) skew-symmetry explained in each pair of dimensions.

The results are interpreted in terms of the areas of triangles. The skew symmetry between the entities in rows (or columns) p and q is proportional to the area of the triangle OPQ, where O is the origin, and P and Q are the points representing p and q respectively. (For further details see either Gower 1977 or Digby and Kempton 1987.) Within each plane the coordinates are arranged so that their centroid is at $(0,y)$, for $y>=0$, and so that positive row-to-column skew symmetry is represented in a clockwise direction. (Note that in planes other than the first it is residual skew symmetry, after fitting the preceding planes, that is being modelled).

Printed output is controlled by the strings listed for the PRINT option: roots prints the roots (also the roots expressed as percentages and cumulative percentages) and scores prints the scores. Results from the analysis can be saved using the parameters ROOTS and SCORES. The structures specified for these parameters need not be declared in advance. Column labels are provided automatically for the SCORES matrix, but any row labels (useful to identify the entities) are left unchanged.

Example 11.8.3

```
 1    " Example of how to use procedure SKEWSYMM:
-2        data from Table 6.7 of Digby and Kempton (1987) "
 3    TEXT [VALUES=Bare,Lichens,Grasses,Erica,'E/C',Calluna,'C/M', \
 4        Mosses,'C/A','Arctost.'] Vegstate
 5    & [VALUES=B,L,G,E,EC,C,CM,M,CA,A] Labels
 6    MATRIX [ROWS=Vegstate; COLUMNS=Labels] Heath,Coords; VALUES= \
 7      !(15,18,47,15, 5, 1, 1, 1, 5, 3,  0,11,17,27, 0, 8, 1, 6, 3,14,\
 8        0, 0, 5,20, 5, 8, 1, 3, 0, 8,  0, 1, 0,10, 4,21, 3, 7, 0, 0,\
 9        0, 0, 0, 5,10, 5, 4, 0, 5, 0,  4, 1, 0, 7, 2,18,11, 1, 1, 3,\
10        0, 3, 1, 0, 0, 0,101,29,16,3,  0, 0, 0, 0, 0, 3, 7,17, 0, 5,\
11        0, 0, 0, 1, 0, 1, 0, 0, 6, 9,  0, 0, 0,10, 0,21, 0, 2, 5, 7)
12    PRINT Heath; FIELDWIDTH=6; DECIMALS=0
```

	Heath									
Labels	B	L	G	E	EC	C	CM	M	CA	A
Vegstate										
Bare	15	18	47	15	5	1	1	1	5	3
Lichens	0	11	17	27	0	8	1	6	3	14
Grasses	0	0	5	20	5	8	1	3	0	8
Erica	0	1	0	10	4	21	3	7	0	0
E/C	0	0	0	5	10	5	4	0	5	0
Calluna	4	1	0	7	2	18	11	1	1	3
C/M	0	3	1	0	0	0	101	29	16	3
Mosses	0	0	0	0	0	3	7	17	0	5
C/A	0	0	0	1	0	1	0	0	6	9
Arctost.	0	0	0	10	0	21	0	2	5	7

```
13    " Use SKEWSYMM, saving SCORES, printing roots only "
14    SKEWSYMM [PRINT=roots] Heath; SCORES=Coords
```

***** Canonical analysis of Skew-Symmetry *****

*** Squared singular values for each plane ***

	Sk_Roots	%Roots	Cum%Root
1	9605	76.50	76.50
2	2095	16.69	93.19
3	773	6.16	99.35
4	81	0.64	99.99
5	1	0.01	100.00

```
15    CALCULATE Score[1...4] = Coords$[*; 1...4]
16    AXES [EQUAL=scale] 5,6; YLOWER=-3.25,-2.5; YUPPER=8.25,5.5; \
17      XLOWER=-6.25,-2.75; XUPPER=(5.25)2; YORIGIN=0; XORIGIN=0; STYLE=xy
18    PEN 3; SYMBOLS=Vegstate; SIZE=1.5
19    DGRAPH [TITLE='Skew-symmetry analysis: first plane'; WINDOW=5; \
20      KEYWINDOW=0] Score[2]; Score[1]; PEN=3
21    DGRAPH [TITLE='Skew-symmetry analysis: second plane'; WINDOW=6; \
22      KEYWINDOW=0; SCREEN=keep] Score[4]; Score[3]; PEN=3
```

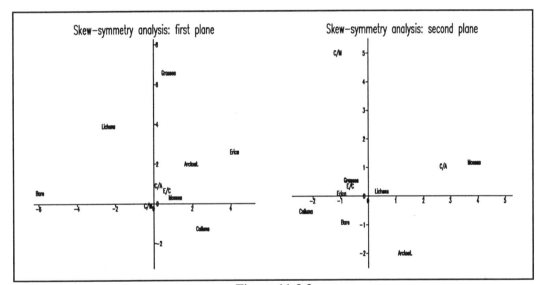

Figure 11.8.3

P.G.N.D.

S.A.H.

12   Analysis of time series

A *time series* in Genstat is a sequence of observations at equally spaced points in time. Each time series is stored in a variate for which the unit number indexes the time points. Genstat cannot deal explicitly with unequal spacing in time. So if you have such a sequence, you will need to do some form of adjustment or interpolation before using the methods described here. Genstat will handle missing values in time series, but these should not represent more than a small fraction of the data.

Usually you will want to describe or model the structure of a series. You can do this without reference to any other variable than the series itself, by examining the relationship between successive measurements. You can also treat a time series as a response variable, which is related to present and past values of explanatory variables that are also time series. *Forecasts* of future values of time series can be derived from these relationships. You can use *filters* to modify time series, for example to smooth them, or to remove trends.

Most of this chapter describes how to analyse time series by the methods advocated by Box and Jenkins (1970). They recommend a modelling procedure involving three stages: model selection (a term used here in preference to that used by Box and Jenkins, which is "identification"), model estimation, and model checking (used here in preference to "verification"). The facilities described in this chapter also provide the basic techniques for spectral analysis, as described by Bloomfield (1976). The Genstat procedure library contains procedures which use the directives described in this chapter, together with graphical presentation of the results, so that standard analyses can be carried out conveniently.

Section 12.1 describes how to derive sample statistics from time series, such as *autocorrelations*: these help you select time-series models. Section 12.2 shows how to calculate the *Fourier transform*, which can be useful for revealing cyclical behaviour; it also describes how to construct the *periodogram*, often called the sample spectrum. Section 12.3 describes *autoregressive integrated moving-average* (ARIMA) models, using the notation of Box and Jenkins. It also describes how these are used as *univariate models*: that is, models to describe the behaviour of a single series. There are directives to let you save the results of estimation, so that you can check models. Once a model has been fitted, you can make forecasts of the future values of the series. Section 12.4 shows how to fit regression models between time series, using an ARIMA model to represent correlated errors. Section 12.5 shows how to extend this to general *transfer-functions* between series: again you can estimate, check, and forecast. Section 12.6 covers the *filtering* of time series by transfer-function models, as used for example in exponential smoothing or seasonal adjustment. Filtering can also be done by ARIMA models, as used in *pre-whitening*. Section 12.7 presents some ways of displaying the properties of the fitted models, such as the theoretical autocorrelations of ARIMA models.

The index for a time-series variate goes from 1 to N, N being the number of observations. However for defining Fourier transformations, the conventional index is $t=0...(N-1)$, and we adhere to this too.

12.1 Correlation

CORRELATE forms correlations between variates, autocorrelations of variates, and lagged cross-correlations between variates.

Options

PRINT = *strings*	What to print (correlations, autocorrelations, partialcorrelations, crosscorrelations); default *
GRAPH = *strings*	What to display with graphs (autocorrelations, partialcorrelations, crosscorrelations); default *
MAXLAG = *scalar*	Maximum lag for results; default * i.e. value inferred from variates to save results
CORRELATIONS = *symmetric matrix*	Stores the correlations between the variates specified by the SERIES parameter

Parameters

SERIES = *variates*	Variates from which to form correlations
LAGGEDSERIES = *variates*	Series to be lagged to form crosscorrelations with first series
AUTOCORRELATIONS = *variates*	To save autocorrelations, or to provide them to form partial autocorrelations if SERIES=*
PARTIALCORRELATIONS = *variates*	To save partial autocorrelations
CROSSCORRELATIONS = *variates*	To save crosscorrelations
TEST = *scalars*	To save test statistics
VARIANCES = *variates*	To save prediction error variances
COEFFICIENTS = *variates* or *matrices*	To save prediction coefficients: in a variate to keep only those for the maximum lag, or in a matrix to keep the coefficients for all lags up to the maximum

The most straightforward use of the CORRELATE directive is to calculate correlation coefficients between a set of variates. In Example 12.1, the PRINT option is set to correlations to display the correlations as a lower-triangular matrix.

Example 12.1

```
  1    "
 -2    EXAMPLE 12.1
 -3    Display the correlations of five time series of United Kingdom Pig
 -4    Production taken from 'Data. A Collection of Problems from Many Fields
```

```
-5     for the Student and Research Worker' by D.F.Andrews and A.M.Herzberg,
-6     Springer-Verlag, 1985.
-7     "
 8  VARIATE [NVALUES=48] Year,Quarter,Gilts,Profit,Slaughter,Cleanpig,Herdsize
 9  OPEN 'ukpig.dat'; CHANNEL=3
10  SKIP [CHANNEL=3] 3
11  READ [CHANNEL=3] Year,Quarter,Gilts,Profit,Slaughter,Cleanpig,Herdsize
```

Identifier	Minimum	Mean	Maximum	Values	Missing
Year	1967	1972	1978	48	0
Quarter	1.000	2.500	4.000	48	0
Gilts	77.0	111.2	140.0	48	0
Profit	5.049	7.064	8.639	48	0
Slaughte	7.87	10.58	14.00	48	0
Cleanpig	2540	3085	3501	48	0
Herdsize	703.0	803.1	922.0	48	0

```
12  CLOSE 'ukpig.dat'; CHANNEL=3
13  CORRELATE [PRINT=correlations] Gilts,Profit,Slaughter,Cleanpig,Herdsize
```

*** Correlation matrix ***

```
    Gilts    1.000
   Profit    0.409    1.000
 Slaughte   -0.522   -0.611    1.000
 Cleanpig   -0.252   -0.396    0.428    1.000
 Herdsize    0.558    0.002   -0.127    0.592    1.000

            Gilts   Profit Slaughte Cleanpig Herdsize
```

Example 12.1 prints the correlations between five time series of quarterly indicators of the pig market. The correlations can be saved in a symmetric matrix using the CORRELATIONS option.

These correlations measure only the simultaneous relationship between the series. More useful are the *autocorrelations* of the series, that is the correlations between values in the series lagged by particular time intervals. The set of autocorrelations for all possible lags is the *autocorrelation function*. You can derive the *partial autocorrelation function* from these. To look at the relationship between two series, you should use the *cross-correlation function* between one series and the other lagged by the various intervals.

The ways of interpreting the correlation functions are described by many standard books about time series. The books by Anderson (1976) and Nelson (1973) are introductory texts, but do not cover the whole range of models covered in this chapter. The book by Box and Jenkins (1970) gives a full description.

12.1.1 Autocorrelation

You can use the CORRELATE directive to display the sample autocorrelation function of a series, either as a table of numbers, or as a graph – called a *correlogram*. In either case, you must specify the maximum lag for which the autocorrelation is to be calculated, *m* say. You can do this either by setting the MAXLAG option to *m*, or by pre-defining the length of a variate to be *m*+1 and including it in the AUTOCORRELATIONS parameter to store the calculated values. If you do not specify the maximum lag, CORRELATE will report a fault diagnostic. Example 12.1.1a plots, saves, and prints the autocorrelations up to lag 30 of the time series of `Gilts` used in Example 12.1.

Example 12.1.1a

```
16   "
17       Show the autocorrelation function of the time series of Gilts from the
18       data in Example 12.1. The values are saved in a variate then printed.
19   "
20
21   CORRELATE [MAXLAG=30; GRAPH=autocorrelations] Gilts; \
22       AUTOCORRELATIONS=Giltsacf
```

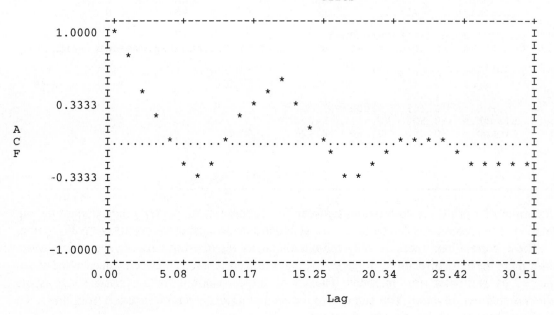

```
23   PRINT [ORIENTATION=across] Giltsacf
```

Giltsacf	1.0000	0.7870	0.4772	0.2384	-0.0118
Giltsacf	-0.2620	-0.3557	-0.2099	-0.0006	0.1668
Giltsacf	0.3155	0.4775	0.5060	0.3458	0.1484
Giltsacf	-0.0234	-0.1529	-0.2819	-0.3262	-0.2356
Giltsacf	-0.1215	-0.0409	-0.0136	0.0397	0.0448
Giltsacf	-0.0599	-0.1757	-0.1977	-0.1886	-0.2104
Giltsacf	-0.1734				

Genstat includes the autocorrelation at lag 0 in the autocorrelation function; this is always unity. The formula used for the sample autocorrelation at lag k is

$$r_k = (1 - k/n) \times C_k / C_0$$

where

$$C_k = \frac{1}{n_k} \sum_{t=1}^{n-k} (y_t - \bar{y})(y_{t+k} - \bar{y})$$

The number n_k is the number of terms included in the sum. The series can contain missing values, but the summation excludes any product that involves any missing values at all. The value $\bar{y}$ is the ordinary sample mean of the whole series, and n is the number of non-missing values in the series. You can restrict a series, but the restricted set must consist of a contiguous set of units. Thus, you can look at the autocorrelation function derived from just the first section of a series, or from just the last section, or from a section in the middle; but you cannot use restriction to exclude a section from the middle of the series, or to exclude just individual observations.

The AUTOCORRELATIONS parameter allows you to save the calculated autocorrelations. If you want to display a correlogram in a different form from the standard one produced by the GRAPH option, you must save the autocorrelations and plot them explicitly using either the GRAPH or DGRAPH directives. You will then need to define the variate of lags from *0* to *m*.

Example 12.1.1b shows the use of procedure BJIDENTIFY to calculate and plot the autocorrelations of the series from the examples above. As well as the autocorrelations, the original series is also plotted, together with the partial autocorrelations and the sample spectrum described in 12.1.2 and 12.2.2. Information about this procedure can be obtained using the command LIBHELP 'BJIDENTIFY'.

Example 12.1.1b

```
-24   "
-25     Use the standard library procedure BJIDENTIFY to display the time series
-26      and its sample autocorrelation function for the time series of Gilts,
-27      together with the sample partial autocorrelations and sample spectrum
-28      or periodogram
-29   "
 30
 31   BJIDENTIFY [GRAPHICS=high; WINDOWS=!(1,2,3,4)] Gilts

Analysis of whole of series Gilts, length 48
showing sample acf and pacf up to lag 24
and sample spectrum with frequency range divided into 80      intervals
```

The graphs produced by BJIDENTIFY are shown in Figure 12.1.1b.

The TEST parameter of CORRELATE allows you to save a statistic that can be used to test the hypothesis that the true autocorrelation is zero for positive lags. It is defined as

$$S = n \sum_{k=1}^{m} r_k^2$$

Provided n (the number of data values) is large and m (the maximum lag) is much smaller than n, then under the null hypothesis, S has a chi-squared distribution with m degrees of freedom. Thus, a large value of S provides evidence of autocorrelation in a time series.

You can calculate autocorrelation functions for several series in one statement by specifying several variates with the SERIES parameter.

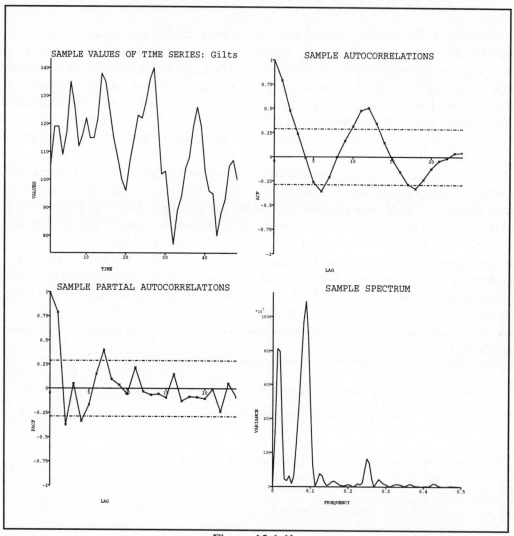

Figure 12.1.1b

12.1.2 Partial Autocorrelation

Genstat forms partial autocorrelations from an autocorrelation function. The value at lag k is defined as

$$\mathrm{corr}(\, y_t, y_{t-k} \mid y_{t-1}, y_{t-2} \cdots y_{t-k+1} \,)$$

representing the excess correlation between values separated by k timepoints that is not accounted for by the intermediate points; it is denoted by $\phi_{k,k}$ because it is also the value of the last in the set of coefficients in the autoregressive prediction equation:

$$y_t = c + \phi_{k,1}y_{t-1} + \ldots + \phi_{k,k}y_{t-k} + e_{k,t}$$

Genstat calculates these coefficients recursively for $k=1\ldots m$ by

$$\phi_{k,k} = (\ r_k - \phi_{k-1,1}r_{k-1} - \ldots - \phi_{k-1,k-1}r_1\ )\ /\ v_{k-1}$$
$$\phi_{k,j} = \phi_{k-1,j} - \phi_{k,k}\phi_{k-1,k-j}\ ,\ j=1\ldots k-1$$
$$v_k = v_{k-1}\ (1 - \phi_{k,k}^2\ )$$

It starts with $v_0=1$, the quantity v_k being the kth order prediction error variance ratio variance$(e_{k,t})$ / variance(y_t).

Partial correlations provide a valuable alternative way of displaying the autocorrelation structure of a series. You can display the partial autocorrelation function either as a table of numbers, or as a graph as shown in Example 12.1.1b. Two methods are available for doing this. You can supply the series using the SERIES parameter, in which case the autocorrelations are formed first, automatically, and the partial autocorrelations are then derived from them. Alternatively, you can set SERIES=*, and provide the autocorrelations using the AUTOCORRELATIONS parameter. You must specify the maximum lag, either by setting the MAXLAG option, or by pre-defining the length of a variate specified for either the AUTOCORRELATIONS or the PARTIALCORRELATIONS parameter.

You can save the partial autocorrelation function using the PARTIALCORRELATIONS parameter. You can set the VARIANCES and COEFFICIENTS parameters to variates to save the *prediction-error variances* $v_0..v_m$, and the *prediction coefficients 1*, $\phi_{m,1}$ ... $\phi_{m,m}$ for the maximum lag m. Genstat sets the first coefficient to 1, and also the first element of the partial autocorrelation sequence to 1: you should find this to be a useful convention for the lag 0 values. Alternatively, if the COEFFICIENTS parameter is set to a matrix structure, the rows of this matrix will be used to save the prediction coefficients for *all* the orders up to the maximum lag. Example 12.1.2 uses some of the previously calculated autocorrelations to produce partial autocorrelations and the matrix of prediction coefficients. Note that the partial autocorrelations also appear down the diagonal of the matrix. The graph in Figure 12.1.1b suggests that an order of 7 would be appropriate for a predictor, the coefficients being in the row labelled 7 of the matrix.

Example 12.1.2

```
 34    "
-35     The first 10 autocorrelations formed in Example 12.1.1b for the time
-36     series of Gilts are used to calculate the prediction coefficients up to
-37     a maximum lag of 10. These are saved in a matrix and printed.
-38    "
 39
 40    VARIATE [NVALUES=11] Shortacf
 41    CALCULATE Shortacf = Giltsacf$[!(1...11)]
 42    TEXT Laglabels
 43    PRINT [SQUASH=yes; IPRINT=*; CHANNEL=Laglabels] !(0...10); \
 44      FIELDWIDTH=2; DECIMALS=0
 45    MATRIX [ROWS=Laglabels; COLUMNS=Laglabels] Predcoef
 46    CORRELATE SERIES=*; AUTOCORRELATIONS=Shortacf; COEFFICIENTS=Predcoef
 47    PRINT [RLWIDTH=10] Predcoef; FIELDWIDTH=6; DECIMALS=2
```

```
              Predcoef
Laglabel       0     1     2     3     4     5     6     7     8     9    10
Laglabel
         0    1.00  0.00  0.00  0.00  0.00  0.00  0.00  0.00  0.00  0.00  0.00
         1    1.00  0.79  0.00  0.00  0.00  0.00  0.00  0.00  0.00  0.00  0.00
         2    1.00  1.08 -0.37  0.00  0.00  0.00  0.00  0.00  0.00  0.00  0.00
         3    1.00  1.10 -0.43  0.05  0.00  0.00  0.00  0.00  0.00  0.00  0.00
         4    1.00  1.12 -0.57  0.42 -0.34  0.00  0.00  0.00  0.00  0.00  0.00
         5    1.00  1.06 -0.50  0.32 -0.15 -0.17  0.00  0.00  0.00  0.00  0.00
         6    1.00  1.09 -0.48  0.28 -0.07 -0.33  0.15  0.00  0.00  0.00  0.00
         7    1.00  1.03 -0.35  0.30 -0.18 -0.14 -0.28  0.40  0.00  0.00  0.00
         8    1.00  0.99 -0.32  0.32 -0.17 -0.16 -0.25  0.30  0.10  0.00  0.00
         9    1.00  0.98 -0.33  0.33 -0.16 -0.16 -0.26  0.31  0.06  0.04  0.00
        10    1.00  0.99 -0.33  0.34 -0.17 -0.17 -0.27  0.33  0.04  0.09 -0.06
```

CORRELATE will print a warning if you include missing values in an autocorrelation function that you have supplied, or if for some other reason the autocorrelations are invalid. In particular, if a partial autocorrelation value is obtained outside the range (−1, 1), Genstat will truncate the sequence at the previous lag.

12.1.3 Cross-correlation

You can calculate cross-correlations between two series by specifying one series with the SERIES parameter and the other with the LAGGEDSERIES parameter. You must define the maximum lag, as for autocorrelations (12.1.1). You can plot or tabulate the resulting function. Example 12.1.3 shows the correlation between one series and the later values of a second series, along with the correlation of the second series with later values of the first. This second set of correlations may be considered as correlations between the first series and the second series at *negative* lags. The two sets of correlations are displayed in the same graph to emphasize this interpretation.

Example 12.1.3

```
 50   "
-51     Obtains and graphs the crosscorrelations between the series Profit and
-52     Gilts from the data in Example 12.1
-53   "
 54
 55   CORRELATE [MAXLAG=20] SERIES=Profit,Gilts; LAGGEDSERIES=Gilts,Profit;\
 56           CROSSCORRELATIONS= P_G_ccf , G_P_ccf
 57   VARIATE [VALUES=0...20] Lag
 58   CALCULATE Neglag=-Lag
 59   FRAME 1; XLOWER=0.05; XUPPER=0.95; YLOWER=0.45; YUPPER=0.95
 60   AXES WINDOW=1; YTITLE='CCF'; XTITLE='LAG'; YLOWER=-1.0; YUPPER=1.0;\
 61          XLOWER=-21; XUPPER=21; STYLE=grid
 62   PEN 1; LINESTYLE=1; METHOD=line; SYMBOL=2
 63   DGRAPH [TITLE='Cross correlations between Profit and Gilts'; \
          WINDOW=1; KEYWINDOW=0] Y=P_G_ccf,G_P_ccf; X=Lag,Neglag; PEN=1
```

The graph produced by Example 12.1.3 is displayed in Figure 12.1.3.

Missing values are allowed, as for autocorrelations. Genstat calculates the sample cross-correlation between the first series x_t and the lagged series y_t at lag k using:

$$r_k = (1 - k/n) \, C_k / (s_x \, s_y)$$

where

$$C_k = \frac{1}{n_k} \sum_{t=1}^{n-k} (x_t - \bar{x})(y_{t+k} - \bar{y})$$

The series x_t and y_t may be of different lengths. The summation includes all possible terms, but excludes any product containing missing values; the number n_k is the number of terms included in the sum. The values $\bar{x}$ and $\bar{y}$ are the sample means, and s_x, s_y are the sample standard deviations. The number n is the minimum of the number of values of x and of y, excluding missing values. You can restrict either series to a set of contiguous units: if both are restricted, their restrictions must match.

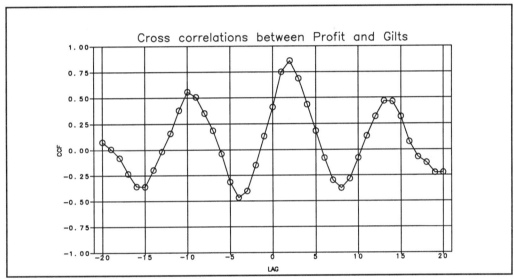

Figure 12.1.3

You can save the cross-correlation function using the CROSSCORRELATIONS parameter. You can also save a test statistic using the TEST parameter; this is used similarly to the statistic described in 12.1.1 to test for lack of lagged cross-correlation in one direction of the relationship between two series. However the test is valid only if each of the series has a zero autocorrelation function. Cross-correlations take precedence in the storage. Thus if you request both autocorrelations and cross-correlations in a single CORRELATE statement, the stored test statistic will relate to the cross-correlations: that for the autocorrelations will not be stored.

12.2 Fourier transformation

This section describes various types of Fourier transformation. These allow you to do most types of spectral analysis with a few Genstat statements. You may want to put these into procedures (5.3) for repeated use. The standard Library contains two procedures that use Fourier transformations. BJIDENTIFY plots the sample spectrum as illustrated in Example 12.1.1b. SMOOTHSPECTRUM, which can be used to calculate and plot smoothed spectrum estimates, is illustrated in Example 12.2.1.

The Fourier or spectral analysis of time series is described comprehensively by Bloomfield (1976) and Jenkins and Watts (1968). The Fourier transformation of a series calculates the coefficients of the sinusoidal components into which the series can be analysed. There are four types of transformation described below, which are appropriate for different types of symmetry in the series. You may often want the length of the variate holding the supplied series to determine implicitly a natural grid of frequencies at which values of the transform are calculated. Genstat will do this if you have not previously declared the identifier supplied for the transform. Alternatively you may want to determine the transform at a finer grid of frequencies, and you can achieve this by declaring a transform variate that is as long as you require. You can do this only for the two types of Fourier transform that apply to real series.

You can also recover the series corresponding to a particular transform; that is, you can invert a transformation.

The conventional index for the series that is being transformed is $0...(N-1)$ in the defining formulae, so that the first element corresponds to the origin for the sinusoidal components in the analysis.

FOURIER calculates cosine or Fourier transforms of real or complex series.

Option

PRINT = *strings*	What to print (transforms); default *

Parameters

SERIES = *variates*	Real part of each input series
ISERIES = *variates*	Imaginary part of each input series
TRANSFORM = *variates*	To save real part of each output series
ITRANSFORM = *variates*	To save imaginary part of each output series
PERIODOGRAM = *variates*	To save periodogram of each transform

---

Series of real numbers are stored in single variates, and series of complex numbers in pairs of variates. You can use the FOURIER directive to calculate the cosine transform of the real series { a_t, $t=0...N-1$ } stored in a variate A by

```
FOURIER [PRINT=transform] A
```

You calculate the Fourier transform of the complex series { a_t+ib_t, $t=0...N-1$ } by storing the values a_t in one variate, A say, the corresponding values b_t in another, B say, and giving the statement:

```
FOURIER [PRINT=transform] A; ISERIES=B
```

You can restrict the series specified by either the SERIES or ISERIES parameter to a contiguous set of units – as for the CORRELATE directive (12.1). Genstat applies the transformation only to the restricted series of values. Similarly, you may supply restricted variates with the TRANSFORM and ITRANSFORM parameters to save the transform: Genstat will then carry out the transformation so as to supply the required number of values (if that is possible according to the rules at the end of 12.2.2). There must be no missing values in the variates in the SERIES or ISERIES parameters, unless you exclude them by a restriction.

Genstat carries out the Fourier transformation using a fast algorithm which relies on the order of the transformation being highly composite (de Boor 1980). In practice, an appropriate order is a round number such as 300 or 6000, consisting of a digit followed by zeroes. If, however, the order has a large prime factor, the transformation may take much longer. For example, a transformation of order 499 is about 25 times slower than one of order 500. In the description below, therefore, we clearly state the order of each form of the transformation, to illustrate a sensible choice of size.

12.2.1 Cosine transformation of a real series

This can be used to calculate the spectrum from a set of autocorrelations. Suppose the variate R contains the values $r_0 \ldots r_n$, and the variate F is to hold the calculated values $f_0 \ldots f_m$ of the spectrum. These values correspond to angular frequencies of $\pi j/m$; that is, periods of $2m/j$, for $j=0\ldots m$. You apply the transformation by putting

```
FOURIER R; TRANSFORM=F
```

If F has not been declared previously, this statement defines it automatically as a variate with $n+1$ values (so $m=n$). If F has been declared to have $m+1$ values, then m must be greater than or equal to n; otherwise Genstat will redeclare F to have $n+1$ values.

The transform is defined when $m>n$ by

$$ f_j \;=\; r_0 + \sum_{k=1}^{n} \left\{ 2\, r_k \cos\!\left(k\, \frac{\pi j}{m} \right) \right\} $$

When $m=n$ the final term in this sum is

$$ r_n \cos(\pi j) = r_n\,(-1)^j $$

and it appears without the multiplier 2. The order of the transformation is $2m$.

If R contains sample autocorrelations, you must multiply it by a variate holding a lag window in order to obtain a smooth spectrum estimate (see Bloomfield 1976, page 166; or Jenkins and Watts 1968, page 243).

Example 12.2.1 uses the SMOOTHSPECTRUM procedure to calculate and plot an estimate of the spectrum of a time series of annual temperature measurements. The graph produced by SMOOTHSPECTRUM is shown below in Figure 12.2.1. The lag window method of smoothing is specified as an option. Error limits for the estimate are included in the graph. The frequency scale is given in cycles per unit time. There is evidence for cycles of periods just over 3 years and 2 years.

Example 12.2.1

```
  1   "
 -2      Smooth spectrum estimation for a series of annual measurements of
 -3      Central England Average Temperature using a Library Procedure.
 -4      Data taken from Manley, G. (1974). Central England temperatures:
  5      monthly means 1659-1973. Quart.J.Met.Soc., 100, 378-405.
 -6   "
  7   VARIATE [NVALUES=315] Cetave
  8   OPEN 'CETAVE.DAT'; CHANNEL=2
  9   READ [CHANNEL=2] Cetave
```

Identifier	Minimum	Mean	Maximum	Values	Missing
Cetave	6.800	9.140	10.600	315	0

```
 10  SMOOTHSPECTRUM [METHOD=lagwindow; BANDWIDTH=0.07; GRAPHICS=high] Cetave
```

Analysis of whole of series Cetave, length 315
Bandwidth used for estimate is   0.07132
Degrees of freedom of estimate is 44
Frequency division of estimates is 70
Probability value used for limits is 0.900
Upper and lower multipliers for limits are 1.477   0.7275
Lag window smoothing used with cut-off lag 26

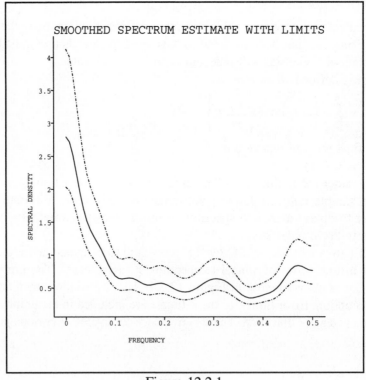

Figure 12.2.1

12.2.2 Fourier transformation of a real series

This can be used to calculate the periodogram of a time series. Suppose the variate X of length N contains the supplied series values $x_0...x_{N-1}$. The result of the transformation is a set of coefficients $a_0...a_m$ of the cosine components and $b_0...b_m$ of the sine components of the series, held in variates A and B, say. Normally the number of such components is related to the length of the series by taking $m=N/2$ if N is even or $m=(N-1)/2$ if N is odd. Then the coefficients correspond to angular frequencies of $2\pi j/N$, which is the same as saying that they correspond to periods N/j for $j=0...m$. Since by definition $b_0=0$, and $b_m=0$ if N is even, there are N "free" coefficients in A and B (which you can think of as the real and imaginary parts of a complex transform with values a_j+ib_j). You can save the periodogram values $p_0...p_m$ in a variate P, say: these are the squared amplitudes of the sinusoidal components, and are calculated by Genstat as $p_j = a_j^2+b_j^2$.

You obtain the transform by putting

```
FOURIER X; TRANSFORM=A; ITRANSFORM=B; PERIODOGRAM=P
```

If you want only the periodogram, you can put

```
FOURIER X; PERIODOGRAM=P
```

If you have not declared A previously Genstat defines it automatically, here as a variate of length $m+1$ where m has the default value defined above. If you have previously declared A, it should have length greater than or equal to $m+1$; otherwise Genstat declares it to have this length. In any case, B and P should have the same length as A, and will be declared (or redeclared) if required.

In the usual case when A, B, or P has the default length $m+1$, the transform is defined by:

$$a_j = \sum_{t=0}^{N-1} \left\{ x_t \cos(t \frac{2\pi j}{N}) \right\} ; \qquad j=0...m$$

$$b_j = \sum_{t=0}^{N-1} \left\{ x_t \sin(t \frac{2\pi j}{N}) \right\} ; \qquad j=0...m$$

In this case, the order of the transformation is N. If A, B, and P have length $m'+1$ with $m'>m$, Genstat computes the results at a finer grid of frequencies $2\pi j/N'$, $j=0...m'$ where $N'=2m'$. These replace $2\pi j/N$ in the above defining sums. The upper limit on the sums remains as $N-1$, although internally Genstat treats it as $N'-1$ with the extra values of $x_N...x_{N'-1}$ being taken as zero. The order of the transformation is then N'. There are various conventions used for scaling the periodogram with factors $2/m$, $1/m$, or $1/\pi m$. You can apply these by using a CALCULATE statement (4.1) after the transformation. You may also want to apply mean correction to the series before calculating the periodogram. Figure 12.1.1b showed the sample spectrum of the time series Gilts. This is just the scaled periodogram calculated using FOURIER as described above. The graph shows a strong peak at frequency 0.08 corresponding to the obvious cycle of period approximately 12 quarters. It also reveals a peak at frequency 0.25 which reflects an annual pattern of period 4 quarters. This is difficult to detect simply by looking at the graph of the series.

12.2.3 Fourier transformation of a complex series

This is the most general form of the Fourier transformation; the other three types are essentially special cases in which some coefficients are zero or have a symmetric structure. Suppose variates X and Y contain values $x_0 \ldots x_{N-1}$ and $y_0 \ldots y_{N-1}$, which may be viewed as the real and imaginary parts of the series $\{ x_t+iy_t, t=0 \ldots N-1 \}$. The results of the transformation are coefficients $a_0 \ldots a_{N-1}$ and $b_0 \ldots b_{N-1}$ which can be held in variates A and B, say: these may similarly be considered as parts of complex coefficients $a_t+ib_t, t=0 \ldots N-1$.

You can do the transformation by putting

```
FOURIER SERIES=X; ISERIES=Y; TRANSFORM=A; ITRANSFORM=B
```

Both X and Y must be variates with the same length N. Similarly A and B must have length N, and if they do not Genstat will declare (or redeclare) them as variates of length N. The order of the transformation is N.

The results are defined by

$$a_j = \sum_{t=0}^{N-1} \left\{ x_t \cos(t\,\frac{2\pi j}{N}) - y_t \sin(t\,\frac{2\pi j}{N}) \right\} ; \qquad j=0\ldots m$$

$$b_j = \sum_{t=0}^{N-1} \left\{ x_t \sin(t\,\frac{2\pi j}{N}) + y_t \cos(t\,\frac{2\pi j}{N}) \right\} ; \qquad j=0\ldots m$$

or equivalently in complex form by

$$(a_j + ib_j) = \sum_{t=0}^{N-1} (x_t+iy_t)\, e^{(it\frac{2\pi j}{N})}$$

The complex transform can be used in cross-spectral analysis.

You can view a Fourier transformation as an orthogonal matrix transformation. Hence its inverse is another Fourier transformation (apart from some simple scaling). You can use this to calculate convolutions. In particular, the correlations of a time series can be obtained by applying the inverse cosine transformation to the periodogram. Example 12.2.3 shows that a repeated Fourier transformation returns the original series – with appropriate scaling.

Example 12.2.3

```
  1   "
 -2      Repeat a Fourier transformation on random numbers.
 -3   "
  4   SCALAR Nvalues; VALUE=25
  5   CALCULATE Rstart,Istart = URAND(6672,0; Nvalues)
  6   FOURIER Rstart; ISERIES=Istart; TRANSFORM=Rmiddle; ITRANSFORM=Imiddle
  7   CALCULATE Rmiddle,Imiddle = Rmiddle,Imiddle * 1,-1 / SQRT(Nvalues)
  8   FOURIER Rmiddle; ISERIES=Imiddle; TRANSFORM=Rfinish; ITRANSFORM=Ifinish
  9   CALCULATE Rfinish,Ifinish = Rfinish,Ifinish * 1,-1 / SQRT(Nvalues)
 10   PRINT Rstart,Istart,Rmiddle,Imiddle,Rfinish,Ifinish; DECIMALS=4
```

Rstart	Istart	Rmiddle	Imiddle	Rfinish	Ifinish
0.4236	0.6865	2.5847	-2.6468	0.4236	0.6865
0.4458	0.7316	0.0363	-0.5219	0.4458	0.7316
0.3443	0.5548	-0.1036	-0.2434	0.3443	0.5548
0.0174	0.7045	0.4952	0.2670	0.0174	0.7045

0.0388	0.7507	-0.0092	-0.3748	0.0388	0.7507
0.7562	0.9707	0.0938	-0.2235	0.7562	0.9707
0.3171	0.7538	-0.0380	0.0790	0.3171	0.7538
0.5931	0.6838	-0.0152	0.4113	0.5931	0.6838
0.9229	0.0015	-0.0863	-0.4419	0.9229	0.0015
0.9485	0.5462	0.1806	-0.2726	0.9485	0.5462
0.3938	0.1294	-0.2906	0.0565	0.3938	0.1294
0.6251	0.4935	0.1896	-0.0188	0.6251	0.4935
0.4973	0.7353	0.1773	0.2825	0.4973	0.7353
0.1379	0.2087	-0.1829	-0.3463	0.1379	0.2087
0.2643	0.6310	-0.4662	-0.2856	0.2643	0.6310
0.9029	0.1571	0.2154	0.3256	0.9029	0.1571
0.3597	0.1690	-0.2685	-0.0011	0.3597	0.1690
0.6736	0.4674	-0.1093	0.3781	0.6736	0.4674
0.7469	0.2263	-0.1546	-0.3584	0.7469	0.2263
0.9657	0.8123	-0.2118	-0.0051	0.9657	0.8123
0.0724	0.4666	-0.0544	0.0780	0.0724	0.4666
0.4650	0.6966	0.1744	-0.0849	0.4650	0.6966
0.5000	0.5380	0.5275	0.4453	0.5000	0.5380
0.6257	0.7017	-0.2402	0.3814	0.6257	0.7017
0.8854	0.4171	-0.3260	-0.3117	0.8854	0.4171

12.2.4 Fourier transformation of a conjugate sequence

It is easiest to think of the Fourier transform of a conjugate sequence as the reverse of the transformation of a real series (12.2.2), with the roles of the series and the transform interchanged. For the true inverse transformation some simple scaling is also required.

Thus if variates A and B of length $m+1$ are supplied containing values $a_0 \ldots a_m$ and $b_0 \ldots b_m$, which may be viewed as parts of complex coefficients a_j+ib_j, the result of the transformation is a single real series $x_0 \ldots x_{N-1}$ held in a variate X of length N.

X can be declared to have length $N=2m$ or $N=2m+1$ (corresponding to the case N even or odd in 12.2.2). The value of b_0 must be zero; also if $N=2m$, the value of b_m must be zero. If either of these conditions is not satisfied, Genstat sets the values of these elements to zero and gives a warning. If X has not been declared previously (or has been declared with a length equal to neither $2m$ nor $2m+1$), then it is declared (or redeclared) with a length governed by whether b_m is 0: $N=2m$ if $b_m=0$, and $N=2m+1$ if $b_m{\neq}0$. The value of b_0 is checked to be zero as before.

You can obtain the transform using the statement

```
FOURIER SERIES=A; ISERIES=B; TRANSFORM=X
```

The definition of the transform is, in the case $N=2m+1$,

$$x_t = a_0 + \sum_{j=1}^{m} 2 \left\{ a_j \cos(t \frac{2\pi j}{N}) + b_j \sin(t \frac{2\pi j}{N}) \right\}$$

In the case $N=2m$, the final term in the sum is simply

$$a_m \cos(t\pi) = a_m (-1)^t$$

and it appears without the multiplier 2. The order of this transformation is N.

12.3   ARIMA modelling

An ARIMA model is an equation relating the present value y_t of an observed time series to past values. The equation includes lagged values not only of the series itself, but also of an unobserved series of *innovations*, a_t ; you can interpret the innovations as the error in predicting y_t from past values y_{t-1}, y_{t-2} .... The usual statistical model assumes that the innovations are a series of independent Normal deviates with mean zero and constant variance. The residuals obtained from fitting the model can be used to estimate the innovations.

A time-series model is specified by three things: the orders, which are the numbers of lagged values that appear in the equation; the parameters, which are the associated coefficients; and, optionally, the actual values of the lags, if these differ from the progression 1...*m*, where *m* is the number of lags. For example, consider the model

$$\nabla y_t - c = \phi_1(\nabla y_{t-1} - c) + a_t - \theta_1 a_{t-1} - \theta_2 a_{t-2}$$

This equation is for the first differences, ∇y_t, of the data, and so has *differencing order d*=1. The *constant term c* represents the mean of ∇y_t. The model has *autoregressive order p*=1 with one parameter ϕ_1, and *moving-average order q*=2 with parameters θ_1 and θ_2.

Example 12.3a fits this model to a series of length 150, and produces forecasts of the next 10 points.

Example 12.3a

```
   3   "
  -4   Fit an ARIMA(1,1,2) model to the series of daylengths, 1821-1970.
  -5   Display the correlations, check the residuals, and forecast till 1980.
  -6   Data from Shi-fang et al. (1977).
  -7   "
   8   OPEN 'daylength.dat'; CHANNEL=2
   9   READ [CHANNEL=2] Daylength

     Identifier   Minimum      Mean   Maximum    Values   Missing
     Daylengt     -347.00     63.88    421.00       150         0

  10   TSM Erp; ORDERS=!(1,1,2)
  11   ESTIMATE Daylength; TSM=Erp

11...........................................................................


***** Time-series analysis *****

  Output series: Daylengt
    Noise model: Erp
                           autoregressive    differencing  moving-average
             Non-seasonal               1               1               2


             d.f.     deviance
  Residual    145       36959.


*** Autoregressive moving-average model ***
```

Innovation variance 251.9

	ref.	estimate	s.e.
Transformation	0	1.00000	FIXED
Constant	1	3.98	4.52

* Non-seasonal; differencing order 1

	lag	ref.	estimate	s.e.
Autoregressive	1	2	0.380	0.104
Moving-average	1	3	-0.5565	0.0897
	2	4	-0.6194	0.0794

```
  12   TKEEP RESIDUALS=Erpres
  13   CORRELATE [MAXLAG=50; GRAPH=autocorrelations] Erpres
```

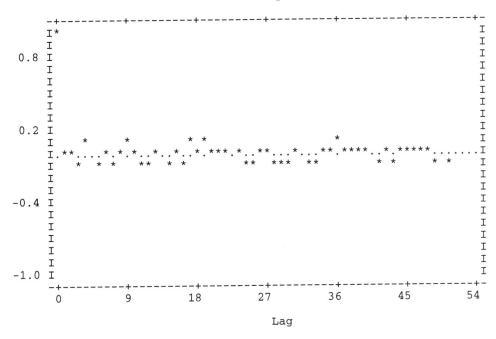

Erpres

Lag

```
  14   FORECAST [MAXLEAD=10]
```

14...

*** Forecasts ***

Maximum lead time: 10

Lead time	forecast	lower limit	upper limit
1	297.0	270.9	323.1
2	305.8	248.9	362.7
3	311.6	216.6	406.5
4	316.2	188.3	444.2
5	320.5	164.4	476.5

6	325.	144.	505.
7	329.	126.	531.
8	333.	111.	555.
9	337.	96.	577.
10	341.	83.	598.

The TSM statement specifies the orders (p,d,q) of the model as $(1,1,2)$, and names the model Erp (for Earth rotation period). The parameters of the model could also have been specified here; but they have been omitted because they have yet to be estimated. The initial values for c, ϕ_1, θ_1, and θ_2 are therefore set by Genstat to zero (the default).

The ESTIMATE statement fits the model to the series by an iterative process, and, in this example, the maximum number of iterations and the convergence criterion are determined by default. The results display the estimated *innovation variance* (or residual variance) and estimates of the other model parameters together with their standard errors. Note that the model also allows for a transformation parameter, which by default is not estimated and has the fixed value of 1.0 indicating no transformation.

The TKEEP statement accesses the variate of residuals a_i; these can also be thought of as the estimated innovations. CORRELATE is used to plot their autocorrelations as a way of checking that the fitted model accounts for all the correlation in the data.

Finally the FORECAST statement prints the forecasts of the next 10 values of the series together with their 90% probability limits.

You can use the RESTRICT directive (4.4.1) to fit models to unbroken sub-series of the data. Genstat automatically estimates missing values in a time series together with the model parameters: all these estimates are allowed for in the number of degrees of freedom.

Further examples of all these directives are shown in 12.3.6. There is a standard Library procedure BJESTIMATE which allows most of the analyses in Example 12.3a to be carried out by issuing a one-line command.

Example 12.3b illustrates this by fitting an ARIMA model to the first 40 points of the series of Gilts from Examples 12.1. If a subset of the series is used in procedure BJESTIMATE, graphs of forecasts are produced for any later timepoints. In this example, the last 8 points are forecast and plotted with the actual values as displayed in Figure 12.3b. A comparison of the forecasts to the actual data provides a simple validation of the fitted model. The residuals are also analysed using procedure BJIDENTIFY within BJESTIMATE, producing the graphs shown in Figure 12.3c. Here only the series and the model orders are specified. The model contains a seasonal part; this is described in 12.3.1.

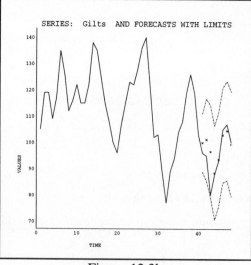

Figure 12.3b

Example 12.3b

```
 65   "
-66    Fit a seasonal ARIMA model to the first 10 years of the quarterly time
-67    series of Gilts used in Example 12.1 using the standard Library
-68    Procedure BJESTIMATE. The final 2 years of data are forecast as a form
-69    of cross-validation of the model, and the residuals are analysed.
-70   "
 71   BJESTIMATE [GRAPHICS=high; WINDOWS=!(1,2,3,4)] SERIES=Gilts; LENGTH=40;\
 72     ORDERS=!(2,0,1,0,1,1,4)
```

Analysis of series x: first 40   values of series Gilts, length 48

72...

*** Convergence monitoring ***

Cycle	Deviance	Current parameters				
1	17055.00	0.	0.	0.	0.	0.
2	3674.779	-0.75527	0.46359	-0.16023	-0.46359	0.42937
3	2272.354	-1.3668	0.95044	-0.16879	-0.25086	0.99756
4	1980.256	-0.70203	1.4846	-0.75820	0.24560	0.90583
5	1776.430	-2.0191	1.6210	-0.85357	0.63853	0.75873
6	1768.455	-1.8406	1.6282	-0.87250	0.64516	0.77478
7	1767.741	-1.9147	1.6198	-0.86499	0.64200	0.77340

Convergence at cycle 7

***** Time-series analysis *****

Output series: x
Noise model: amod

	autoregressive	differencing	moving-average
Non-seasonal	2	0	1
Period 4	0	1	1

	d.f.	deviance
Residual	31	1768.

*** Autoregressive moving-average model ***

Innovation variance 46.07

	ref.	estimate	s.e.
Transformation	0	1.00000	FIXED
Constant	1	-1.893	0.741

* Non-seasonal; no differencing

	lag	ref.	estimate	s.e.
Autoregressive	1	2	1.625	0.104
	2	3	-0.8691	0.0906
Moving-average	1	4	0.649	0.197

* Seasonal; period 4; differencing order 1

```
                 lag ref.     estimate           s.e.
Moving-average     4   5        0.777            0.172


Analysis of residual series follows...
```

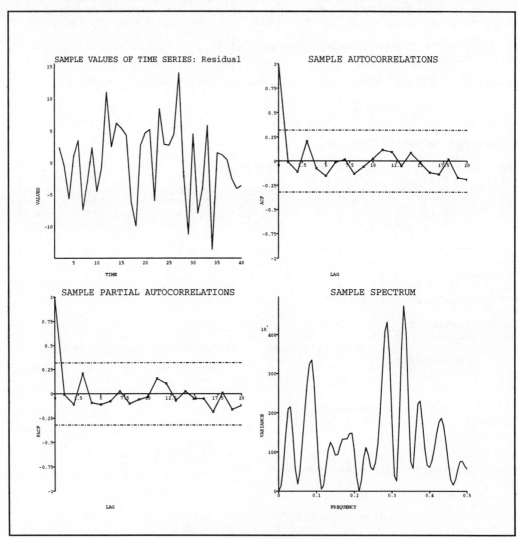

Figure 12.3c

12.3.1 ARIMA models for time series

The TSM directive is described in 2.7.3. Here we describe how to use it for ARIMA models, which correspond to the default setting of its MODEL option (MODEL=arima). The definition of transfer-function models is described in 12.5.1.

In many applications you will need only a simple form of the directive, such as:

```
TSM Erp; ORDERS=!(1,1,2)
```

Notice that TSM simply sets up a named Genstat structure which you can then use in directives such as ESTIMATE. It can also, for example, be saved in a backing-store file (3.5) for further use. In that sense it is analogous to a TERMS statement (8.2.2), which sets up a maximal model for regression analysis, or a TREATMENTSTRUCTURE statement (9.1.1), which sets up a treatment model for analysis of variance.

If a TSM identifier, say Erp, has been declared, you can print the whole model in a descriptive format with the statement:

```
PRINT Erp
```

You can refer to the variates corresponding to the ORDERS, PARAMETERS, and LAGS of the TSM by Erp[1], Erp[2], and Erp[3], or for example by Erp['Orders']. Thus the autoregressive order can be assigned to a scalar P by:

```
CALCULATE P = Erp[1]$[1]
```

since Erp[1] holds the orders of the TSM and its first element is the number of autoregressive parameters.

You can change the values of a TSM at any time, for example by CALCULATE statements. Genstat checks that the TSM values specify a valid model whenever they are used in a time-series directive such as ESTIMATE. However, you must be careful if you change the values of a TSM that you are currently using to fit a model. For example, you could get strange results if you changed the parameter values of the model between the ESTIMATE and FORECAST statements in Example 12.3a.

Using the notation of Box and Jenkins (1970), the simple non-seasonal ARIMA model for the time series y_t is

$$\phi(B) \{\nabla^d y_t^{(\lambda)} - c\} = \theta(B) a_t$$

where B is the backward shift operator $B^p y_t = y_{t-p}$,
∇ is the differencing operator $\nabla y_t = y_t - y_{t-1}$, $\nabla^d y_t = \nabla^{d-1}(y_t - y_{t-1})$, and

$$\phi(B) = 1 - \phi_1 B - \dots - \phi_p B^p$$
$$\theta(B) = 1 - \theta_1 B - \dots - \theta_q B^q$$

The parameter λ specifies a Box-Cox power transformation defined by

$$y_t^{(\lambda)} = (y_t^\lambda - 1) / \lambda, \quad \lambda \neq 0$$
$$y_t^{(0)} = \log(y_t)$$

However, in the default case when λ is fixed and not estimated, the value $\lambda=1$ implies no transformation and then $y_t^{(1)} = y_t$ rather than $y_t - 1$. If $\lambda \neq 1$ or if λ is to be estimated, then Genstat will not let you have values of $y_t \leq 0$. The usual case however is that $\lambda=1$ and is not to be estimated, so that y_t may take any values.

The ORDERS parameter is a list of variates, one for each of the models. For each simple ARIMA model, the variate contains the three values p, d, and q.

The PARAMETERS parameter is a list of variates, one for each of the models. For each simple ARIMA model, the variate contains $(3+p+q)$ values: λ, c, σ_a^2, $\phi_1...\phi_p$, $\theta_1...\theta_q$. You must always include the first three parameters. The parameter σ_a^2 is the innovation variance.

Whenever a TSM is used, Genstat checks its values. The orders must all be non-negative.

The parameters λ and c can take any values, but σ_a^2 must be non-negative. The next $p+q$ values specify the autoregressive and moving-average parameters: they must satisfy the stationarity and invertibility conditions for ARIMA models (see Box and Jenkins 1970). An exception is that before estimation the model parameters may be unset, in which case Genstat sets them to default values. You can omit the PARAMETERS parameter, in which case an unnamed structure is defined to contain the default values. However, you should usually specify the variate of parameters, and if possible assign good preliminary values before estimation (see 12.7.1) as this will speed up the model fitting process.

For convenience when setting the values of parameters, you may wish first to declare scalars or variates containing the separate components:

```
SCALAR Lam,C,Ivar; VALUES=1,4,200
VARIATE [VALUES=0.4] Phi
& [VALUES=-0.5,-0.6] Theta
```

Then to pack these into the parameter variate, you can put

```
VARIATE [VALUES=Lam,C,Ivar,#Phi,#Theta] Erpar
```

Similarly, in order to extract the components after estimation, you can use the EQUATE directive (4.3):

```
EQUATE Erpar; NEWSTRUCTURES=!P(Lam,C,Ivar,Phi,Theta)
```

The LAGS parameter is a list of variates, one for each of the models. For each simple ARIMA model, this variate contains $p+q$ values, one corresponding to each of the autoregressive and moving-average parameters. Genstat then modifies the ARIMA model by defining

$$\phi(B) \;=\; 1 - \phi_1 B^{l_1} - \ldots - \phi_p B^{l_p}$$

$$\theta(B) \;=\; 1 - \theta_1 B^{m_1} - \ldots - \theta_q B^{m_q}$$

The LAGS parameter for this model contains $l_1...l_p$, $m_1...m_q$. The sequences of lags $l_1...l_p$ must be positive integers that are strictly increasing; the default values are $1...p$ if LAGS is not set. The same rule applies to $m_1...m_q$.

The seasonal ARIMA model for the time series y_t is an extension of the simple model, to the form

$$\phi(B)\,\Phi(B^s)\,\{\,\nabla^d \nabla_s^D y_t^{(\lambda)} - c\,\} = \theta(B)\,\Theta(B^s)\,a_t$$

where the extra, seasonal, operators associated with seasonal period s are of three types:

$$\Phi(B^s) \;=\; 1 - \Phi_1 B^s - \ldots - \Phi_P B^{Ps}$$

which is seasonal autoregression of order P;

$$\nabla_s^D$$

which is seasonal differencing of order D; and

$$\Theta(B^s) \;=\; 1 - \Theta_1 B^s - \ldots - \Theta_Q B^{Qs}$$

which is seasonal moving average of order Q.

When seasonal terms are to be included, you must extend the ORDERS parameter so that it contains p, d, q, P, D, Q, and s. Even if the non-seasonal part of the model has $p=d=q=0$, these parameters must still be included at the beginning of the list. The seasonal orders must satisfy $P{\geq}0$, $D{\geq}0$, $Q{\geq}0$ and $s{\geq}1$.

You must also extend the PARAMETERS parameter to contain:
$$\lambda,\ c,\ \sigma_a^{\ 2},\ \phi_1...\phi_p,\ \theta_1...\theta_q,\ \Phi_1...\Phi_P,\ \Theta_1...\Theta_Q$$
You can modify the seasonal model to allow other lags:
$$\Phi(B^s)\ =\ 1\ -\ \Phi_1 B^{L_1}\ -\ ...\ -\ \Phi_P B^{L_P}$$
$$\Theta(B^s)\ =\ 1\ -\ \Theta_1 B^{M_1}\ -\ ...\ -\ \Theta_Q B^{M_Q}$$

The sequence of lags $L_1...L_P$ must be strictly increasing and must be positive-integer multiples of the period s; the default values are s, $2s$ ... Ps. The same rules apply to $M_1...M_Q$. For any seasonal model, you must extend the LAGS parameter, if supplied, so that it contains
$$l_1\ ...\ l_p,\ m_1\ ...\ m_q,\ L_1\ ...\ L_P,\ M_1\ ...\ M_Q.$$
You can use multiple seasonal periods, by extending the variate of ORDERS with further seasonal orders P', D', Q', and s'. You must correspondingly extend the variates of PARAMETERS and LAGS. It is also possible to set the seasonal periods to 1, which means you can estimate non-seasonal models with factored operators.

You can declare an ORDERS variate to have more values than is necessary, provided that the extra values are filled with zeroes, and that the number of values is $3+4k$, k being the number of seasonal periods. The same applies to PARAMETERS and LAGS variates, except that Genstat ignores the extra values whatever they may be. Thus you can extend a simple model to a seasonal model, simply by resetting the extra values.

Finally note that you can use the same ORDERS, PARAMETERS, and LAGS variates in more than one TSM.

12.3.2 The ESTIMATE directive

ESTIMATE estimates parameters in Box-Jenkins models for time series.

Options

PRINT = *strings*	What to print (model, summary, estimates, correlations, monitoring); default mode,summ,esti
LIKELIHOOD = *string*	Method of likelihood calculation (exact, leastsquares, marginal); default exac
CONSTANT = *string*	How to treat the constant (estimate, fix); default esti
RECYCLE = *string*	Whether to continue from previous estimation (yes, no); default no
WEIGHTS = *variate*	Weights; default *
MVREPLACE = *string*	Whether to replace missing values by their estimates

	(yes, no); default no
FIX = *variate*	Defines constraints on parameters (ordered as in each model, tf models first): zeros fix parameters, parameters with equal numbers are constrained to be equal; default *
METHOD = *string*	Whether to carry out full iterative estimation, to carry out just one iterative step, to perform no steps but still give parameter standard deviations, or only to initialize for forecasting by regenerating residuals (full, onestep, zerostep, initialize); default full
MAXCYCLE = *scalar*	Maximum number of iterations; default 15
TOLERANCE = *scalar*	Criterion for convergence; default 0.0004
SAVE = *identifier*	To name save structure, or supply save structure with transfer-functions; default * i.e. transfer-functions taken from the latest model

Parameters

SERIES = *variate*	Time series to be modelled (output series)
TSM = *TSM*	Model for output series
BOXCOXMETHOD = *string*	How to treat transformation parameter in output series (fix, estimate); default fix
RESIDUALS = *variate*	To save residual series

The main use of ESTIMATE is to fit parameters to time-series models, although you can also use it to initialize for the FORECAST directive, even when the model parameters are already known. In many applications of estimating a univariate ARIMA model, you will need only a simple form of the directive, such as:

```
ESTIMATE Daylength; TSM=Erp
```

Examples of ESTIMATE are given at the beginning of 12.3 and in 12.3.6.

The SERIES parameter specifies the variate holding the time series data to which the model is to be fitted.

The TSM parameter specifies the ARIMA model that is to be fitted to the time-series data. This TSM must already have been declared and its ORDERS must have been set. If the LAGS parameter of the TSM has been set, the lags must have been given values. However, if the PARAMETERS of the TSM model have been set, these need not have been declared previously nor given values. When the parameter values are not set, default values are used: these are all zero, except for the transformation parameter, which is set to 1.0 if it is not to be estimated (see BOXCOXMETHOD and FIX below). Any parameter values that you do specify will be used as initial values for the parameters in the model; Genstat replaces any missing values by the default values. If any group of autoregressive or moving-average parameters do not satisfy the required conditions for stationarity or invertibility, all the parameters to be estimated are reset by Genstat to the default values. After ESTIMATE, the parameters of the TSM contain the estimated parameter values.

The BOXCOXMETHOD parameter allows you to estimate the transformation parameter λ.

The RESIDUALS parameter saves the estimated innovations (or residuals). As explained in the description of the LIKELIHOOD option in the next section, the residuals are calculated for $t=t_0...N$, where $t_0=1+p+d-q$ for a simple ARIMA model. If $t_0>1$, missing values will be inserted for $t=1...t_0-1$.

The PRINT option controls printed output. If you specify monitoring, then at each cycle of the iterative process of estimation, Genstat prints the *deviance* (12.3.3) for the current fitted model, together with the current estimates of model parameters. The format is simple with the minimum of description, to let you judge easily how quickly the process is converging; see Example 12.4a. The other settings of PRINT control output at the end of the iterative process. If you specify model, the model is briefly described, giving the identifier of the series and the time-series model, together with the orders of the model. If you specify summary, the deviance of the final model is printed, along with the residual number of degrees of freedom. If you specify estimates, the estimates of the model parameter are printed in a descriptive format, together with their estimated standard errors and reference numbers. If you specify correlations, the correlations between estimates of parameters are printed, with reference numbers to identify the parameters; see Example 12.3.4.

The LIKELIHOOD option specifies the criterion that Genstat minimizes to obtain the estimates of the parameters: this is described in the next section. The default setting exact is recommended for most applications.

You can use the CONSTANT option to specify whether Genstat is to estimate the constant term c in the model. If CONSTANT=fix, the constant is held at the value given in the initial parameter values; this need not be zero.

The RECYCLE option allows a previous ESTIMATE statement to continue; this can save computing time. If RECYCLE=yes, the most recent ESTIMATE statement is continued, unless the SAVE option has been set to the save structure from some other ESTIMATE statement. The SERIES and TSM settings are then taken from this previous ESTIMATE statement: Genstat ignores any specified in the current statement. Most of the settings of other parameters and options are carried over from the previous statement, and new values are ignored. However, there are some exceptions. You can change the RESIDUALS variate, you can reset MAXCYCLE to the number of further iterations you require, and you can change the settings of TOLERANCE and PRINT. You can also change the values of the variate in the WEIGHTS option; you can thus get reweighted estimation. You can change the values of the SERIES itself, although you cannot change missing values; if the MVREPLACE option was previously set to yes, you must put the original missing values back into the SERIES variate before the new ESTIMATE statement.

The WEIGHTS option includes in the likelihood a weighted sum-of-squares term

$$\sum_{t=t_0}^{N} w_t\, a_t^2$$

where w_t, $t=1...N$ are provided by the WEIGHTS variate. The values of w_t must be strictly positive. If $t_0<1$, where $t_0=1+d+p-q$, then w_t is taken as 1 for $t<1$.

The MVREPLACE option allows you to request any missing values in the time-series to be replaced by their estimates after estimation. Genstat will always estimate the missing values, irrespective of the setting of MVREPLACE; so you can also obtain these estimates later from TKEEP (12.3.5).

The FIX option allows you to place simple constraints on parameter values throughout the estimation. The units of the FIX variate correspond to the parameters of the TSM, excluding the innovation variance. The values of the FIX variate are used to define the parameter constraints and must be integers. If an element of the FIX variate is set to 0, the corresponding parameter is constrained to remain at its initial setting. If an element is not 0, and the value is unique in the FIX variate, the parameter is estimated without any special constraint. If two or more values are equal, the corresponding parameters are constrained to be equal throughout the estimation. The number that you give to a parameter by FIX will appear as the reference number of the parameter in the printed model and correlation matrix. This option overrides any setting of CONSTANT and BOXCOXMETHOD. Example 12.3.2a uses the FIX option to constrain some of the parameters in the model fitted in Example 12.3a.

Example 12.3.2a

```
 17   "
-18     Fix parameters in ARIMA(1,1,2) model for daylength:
-19     transformation fixed at 1, Constant unconstrained, AR parameter
-20     fixed at previous estimate, MA parameters constrained to be equal.
-21   "
 22   ESTIMATE [FIX=!(0,1,0,2,2)] Daylength; TSM=Erp

22...............................................................................

***** Time-series analysis *****

  Output series: Daylengt
  Noise model: Erp
                         autoregressive    differencing  moving-average
               Non-seasonal           1               1               2


               d.f.     deviance
  Residual      147       37102.


*** Autoregressive moving-average model ***

Innovation variance 249.5

                     ref.     estimate           s.e.
Transformation         0      1.00000           FIXED
Constant               1         3.97            4.51

* Non-seasonal; differencing order 1

                  lag ref.    estimate           s.e.
Autoregressive      1   0     0.380134          FIXED
Moving-average      1   2      -0.5906          0.0596
                    2   2      -0.5906          0.0596
```

The MAXCYCLE option specifies the maximum number of iterations to be performed.

The TOLERANCE option specifies the convergence criterion. Genstat decides that convergence has occurred if the fractional reduction in the deviance in successive iterations is less than the specified value, provided also that the search is not encountering numerical difficulties that force the step length in the parameter space to be severely limited. You can use monitoring to judge whether, for all practical purposes, the iterations have converged. Genstat gives warnings if the specified number of iterations is completed without convergence, or if the search procedure fails to find a reduced value of the deviance despite a very short step length. Such an outcome may be due to complexities in the likelihood function that make the search difficult, but can be due to your specifying too small a value for TOLERANCE.

The SAVE option allows you to save the *time-series save structure* produced by ESTIMATE. You can use this in further ESTIMATE statements with RECYCLE=yes, or in FORECAST statements. It can also be used by the TDISPLAY and TKEEP directives. Genstat automatically saves the structure from the most recent ESTIMATE statement, but this is over-written when the next ESTIMATE statement is executed, unless you have used SAVE to give it an identifier of its own. You can access the current time-series save structure by the SPECIAL option of the GET directive (13.1.2), and reset it by the TSAVE option of the SET directive (13.1.1).

The METHOD option has four possible settings. The default setting is full which gives the usual estimation to convergence or until the maximum number of iterations has been reached.

With the setting METHOD=initialize, ESTIMATE carries out only the residual regeneration steps (that is, calculation of a_t for $t=t_0...N$) which are needed before FORECAST (12.3.6) can be used. If the model has just been estimated using the default full setting, this is unnecessary. The setting initialize is useful when the time series is supplied with a known model and a minimal amount of calculation is wanted to prepare or initialize for forecasting. None of the model parameters are changed, and no standard errors of parameter estimates are available. Missing values in the series *are* estimated so this setting provides an efficient way of getting their values when the time series model is known; they can then be obtained using TKEEP (12.3.5). The deviance value is also available from TKEEP (12.3.5). This setting is therefore useful for efficient calculation of deviance values when you want to plot the shape of the deviance as a function of parameter values. Example 12.3.2b below illustrates this by producing the contour plot (shown in Figure 12.3.2b) of the log deviance for the daylength model fitted in Example 12.3a. All parameters have their estimated values except the two moving-average parameters. These vary over a grid of 800 points. Values corresponding to non-invertible models are skipped and the contours plotted inside the triangular region of invertible model parameters.

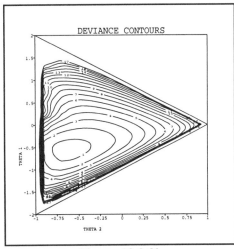

Figure 12.3.2b

Example 12.3.2b

```
 3    "
-4         The deviance function for the model fitted to the series of
-5         daylengths in EXAMPLE 12.3a is plotted as the moving average
-6         parameters are varied.
-7    "
 8    VARIATE [NVALUES=150] Daylength
 9    READ Daylength
```

Identifier	Minimum	Mean	Maximum	Values	Missing
Daylengt	-347.00	63.88	421.00	150	0

```
30    " Set the model parameters to their previously estimated values "
31    VARIATE [VALUES=1,3.98,251.9,0.380,-0.5565,-0.6194] Modpar
32    TSM Moderp; ORDERS=!(1,1,2); PARAMETERS=Modpar
33    SCALAR R,Large,Mdev; VALUE=0.999,12000000,0
34    " Set up a grid of parameter values over which to evaluate
-35        the deviance "
36    CALCULATE Vth1,Vth2 = !(-20...20),!(-10...10)*0.099999
37    " Define the matrix to hold the deviance values "
38    MATRIX [ROWS=41; COLUMNS=21] Devgrid
39    FOR Drow=1...41; Dth1=#Vth1
40        FOR Dcol=1...21; Dth2=#Vth2
41            "Check that the parameters lie within the invertibility region"
42            IF ((ABS(Dth2)<R).AND.((ABS(Dth1)/ABS(1-Dth2))<R))
43                CALCULATE Modpar$[5,6] = Dth1,Dth2
44                ESTIMATE [PRINT=*; METHOD=initialize] Daylength; Moderp
45                TKEEP DEVIANCE=Mdev
46            ELSE
47            "Set the deviance to a large value if the parameters are not
-48            invertible"
49                CALCULATE Mdev = Large
50            ENDIF
51            CALCULATE ELEMENT(Devgrid; Drow; Dcol) = Mdev
52        ENDFOR
53    ENDFOR
54    " Use log deviances so as to reveal the lower contours "
55    CALCULATE Devgrid = LOG(Devgrid)
56    PEN 2,4; LINESTYLE=1; METHOD=closed,line; SYMBOLS=0; COLOUR=2,4
57    FRAME WINDOW=1; YLOWER=0.05; YUPPER=0.95; XLOWER=0.05; XUPPER=0.95
58    AXES WINDOW=1; STYLE=none
59    DCONTOUR [WINDOW=1; INTERVAL=0.2; UPPERCUTOFF=14.0; KEYWINDOW=0] \
60        Devgrid; PEN=2
61    AXES WINDOW=1; YTITLE='THETA 1'; XTITLE='THETA 2';\
62        YLOWER=-2; YUPPER=2; XLOWER=-1; XUPPER=1; STYLE=box
63    DGRAPH [WINDOW=1; KEYWINDOW=0; SCREEN=keep; TITLE='DEVIANCE CONTOURS']\
64        !(2,0),!(-2,0);!(-1,1); PEN=4
```

With the setting METHOD=zerostep the effect is the same as for initialize except that ESTIMATE also calculates the standard errors of the parameters as if they had just been estimated. These can be used together with other quantities available from TKEEP (12.3.5) to construct confidence intervals and carry out tests on the parameter values, which remain unchanged except that the innovation variance in the ARIMA model is replaced by its estimate conditional on all other parameters.

The setting METHOD=onestep gives the same results as specifying the option MAXCYCLE=1 in ESTIMATE. It is convenient for carrying out quick tests of model parameters as illustrated in Example 12.3.2c. The model fitted in Example 12.3a is extended to have three autoregressive parameters, with the new parameters set to zero and the old parameters kept at their estimated values. Then after one step of ESTIMATE the estimates of the new autoregressive coefficients at lags 2 and 3 can be compared with their standard errors to see if there is evidence that they should be retained in the model. In this case the evidence is insufficient. Although iteration to convergence would be very quick for this example, the onestep setting can save time when checking a complicated model for a variety of possible extensions.

Example 12.3.2c

```
 34    "
-35       The model previously fitted to the series of daylengths in Example
-36       12.3a is extended to include two more autoregressive parameters, the
-37       old parameters being kept at their estimated values. The option
-38    METHOD=onestep of ESTIMATE is used to assess whether the new parameters
-39       should be retained in the model.
-40    "
 41
 42    " Save the previous model parameters and redefine the model with
-43       higher autoregressive orders and extended parameter variate. "
 44    CALCULATE Modpar = Erp['Parameters']
 45    & Modparx = !(Modpar$[1,2,3,4],0,0,Modpar$[5,6])
 46
 47    " Save the parameter values "
 48    VARIATE Oldparx; VALUES=Modpar
 49    TSM Erp; ORDERS=!(3,1,2); PARAMETERS=Modparx
 50    ESTIMATE [METHOD=onestep] Daylength; TSM=Erp

 50................................................................................


******** Warning (Code TS 21). Statement 1 on Line 50
Command: ESTIMATE [METHOD=onestep] Daylength;TSM=Erp
The iterative estimation process has not converged
The maximum number of cycles is 1


***** Time-series analysis *****

   Output series: Daylengt
   Noise model: Erp
                         autoregressive   differencing  moving-average
            Non-seasonal            3               1               2


              d.f.      deviance
   Residual    143       36553.


*** Autoregressive moving-average model ***

Innovation variance 252.5
```

	ref.	estimate	s.e.
Transformation	0	1.00000	FIXED
Constant	1	4.05	4.51

* Non-seasonal; differencing order 1

	lag	ref.	estimate	s.e.
Autoregressive	1	2	0.319	0.155
	2	3	0.166	0.161
	3	4	-0.102	0.139
Moving-average	1	5	-0.608	0.136
	2	6	-0.544	0.138

```
 51
 52    " Calculate and print the changes in the parameter values excluding
-53      the transformation and innovation variance parameters. "
 54    CALCULATE Delpar = Modparx-Oldparx
 55    & Del = Delpar$[!(2,4,5...8)]
 56    PRINT Del


      Del
  0.07414
 -0.06133
  0.16587
 -0.10153
 -0.05192
  0.07503
```

12.3.3   Technical information about how Genstat fits ARIMA models

This section describes the estimation of ARIMA models in more detail. You may want to skip this if you are doing fairly routine work.

The first step in deriving the likelihood for a simple model is to calculate

$$w_t = \nabla^d y_t - c , \qquad\qquad\qquad t = 1+d \dots N$$

This has a multivariate Normal distribution with dispersion matrix $V\sigma_a^2$, where V depends only on the autoregressive and moving-average parameters. The likelihood is then proportional to

$$\{ \ \sigma_a^{2m} \, | \, V \, | \ \}^{-\frac{1}{2}} \exp\{ \ -w'V^{-1}w/2\sigma_a^2 \ \}$$

where $m=N-d$. In practice Genstat evaluates this by using the formula

$$w'V^{-1}w \ = \ W + \sum_{t=t_0}^{N} a_t^2 \ = \ S$$

where $t_0=1+d+p-q$. The term W is a quadratic form in the p values $w_{1+d-q} \dots w_{p+d-q}$: it takes account of the starting-value problem for regenerating the innovations a_t, and avoids losing information as would happen if the process used only a conditional sum-of-squares function. If $q>0$, Genstat introduces unobserved values of $w_{1+d-q} \dots w_d$ in order to calculate the sum S. Genstat uses linear least-squares to calculate these q starting values for w, thus minimizing S. We shall call them *back-forecasts*, though if $p>0$ they are actually computationally convenient linear functions of the proper back-forecasts. We shall call S the sum-of-squares function: it is the sum of the quadratic form and the sum-of-squares term, and is identical to the value expressed by Box and Jenkins as

$$\sum_{t=-\infty}^{N} a_t^2$$

using infinite back-forecasting; that is, using:

$$W = \sum_{t=-\infty}^{t_0-1} a_t^2$$

The values a_t for $t=t_0...N$ agree precisely with those of Box and Jenkins.

To clarify all this, consider examples with no differencing; that is, $d=0$. If $p=0$ and $q=1$ then $W=0$ and $t_0=0$, and one back-forecast w_0 is introduced. If $p=1$ and $q=0$ then $W=(1-\phi_1^2)w_1^2$ and $t_0=2$, and no back-forecasts are needed. If $p=q=1$ then $W=(1-\phi_1^2)w_0^2$ and $t_0=1$, and so one back-forecast w_0 is needed. In this case the proper back-forecast is in fact $w_0/(1-\theta_1\phi_1)$.

The value of $|V|$ is a by-product of calculating W and the back-forecast. For example, if $p=0$ and $q=1$, then
$$|V| = (1 + \theta_1^2 + ... + \theta_1^{2N})$$
If $p=1$ and $q=0$,
$$|V| = 1 / (1 - \phi_1^2)$$
and if $p=q=1$,
$$|V| = 1 + (\phi_1 - \theta_1)^2 (1 + \theta_1^2 + ... + \theta_1^{2N-2}) / (1 - \phi_1^2)$$
Concentrating the likelihood over σ_a^2 by setting $\sigma_a^2=S/m$ yields a value proportional to $\{ |V|^{1/m} S \}^{-m/2}$.

The default setting of the LIKELIHOOD option is exact. In this case the concentrated likelihood is maximized, by minimizing the quantity
$$D = |V|^{1/m} S$$
which is called the deviance.

The setting leastsquares specifies that Genstat is to minimize only the sum-of-squares term S. This criterion corresponds to the back-forecasting sum-of-squares used by Box and Jenkins, and will in many cases give estimates close to those of the exact likelihood. However, some discrepancy arises if the series is short or the model is close to the invertibility boundary. This is because of limitations on the back-forecasting procedure, as described in the algorithms of Box and Jenkins. The deviance value D that Genstat prints is, with this setting, simply S.

The setting marginal is described in Section 12.4.

When you use exact likelihood, the factor $|V|^{1/m}$ reduces bias in the estimates of the parameter; you would get bias if you used leastsquares instead. However, $|V|^{1/m}$ is generally close to one, unless the series is short or the model is either seasonal or close to the boundaries of invertibility or stationarity. The leastsquares setting is therefore adequate for most long, non-seasonal sets of data; using it may reduce the computation time by up to 50%. When you specify that Genstat is to estimate the parameter λ of the Box-Cox transformation, Genstat also includes the Jacobian of the transformation in the likelihood function. The result is an extra factor $G^{-2(\lambda-1)}$ in the definition of the deviance, G being the geometric mean of the data,

$$G = \left(\prod_{t=1}^{N} y_t \right)^{\frac{1}{N}}$$

Note that this is not included unless λ is being estimated, even if $\lambda \neq 1$.

You can treat differences in $N\log(D)$ as a chi-squared variable in order to test nested models: this is supported by asymptotic theory, and by experience with models that have moderately large sample sizes. Similarly, you can select between different models by using $N\log(D)+2k$ as an information criterion, k being the number of estimated parameters. But both of these test procedures are questionable if the estimated models are close to the boundaries of invertibility or stationarity. Provided all the models that are being compared have the same orders of differencing, with the differenced series being of length m, it is recommended that $m\log(D)$ be used rather than $N\log(D)$ in these tests since $m\log(D)$ is precisely minus two multiplied by the log-likelihood as defined above.

12.3.4 The **TDISPLAY** directive

TDISPLAY displays further output after an analysis by ESTIMATE.

Options

PRINT = *strings*	What to print (model, summary, estimates, correlations); default mode, summ, esti
CHANNEL = *scalar*	Channel number for output; default * i.e. current output channel
SAVE = *identifier*	Save structure to supply fitted model; default * i.e. that from the last model fitted

No parameters

You can use TDISPLAY to print further output from an ESTIMATE statement. However, if the ESTIMATE statement used the setting METHOD=initialize you will not be able to print the standard errors or correlations between the parameter estimates (see 12.3.2).

The PRINT option has the same interpretation as in ESTIMATE, except that information is not available to monitor convergence. Example 12.3.4 illustrates TDISPLAY in a continuation of Example 12.3.2a.

Example 12.3.4

```
 25   "
-26     Illustrates output from TDISPLAY
-27   "
 28   ESTIMATE Daylength; TSM=Erp
```

28...

```
***** Time-series analysis *****

   Output series: Daylengt
     Noise model: Erp
                          autoregressive   differencing  moving-average
              Non-seasonal              1              1               2


               d.f.     deviance
   Residual     145       36960.


*** Autoregressive moving-average model ***

Innovation variance 251.9

                        ref.      estimate        s.e.
Transformation           0        1.00000        FIXED
Constant                 1           3.98          4.52

* Non-seasonal; differencing order 1

               lag ref.      estimate        s.e.
Autoregressive   1   2          0.380        0.105
Moving-average   1   3         -0.5581       0.0901
                 2   4         -0.6181       0.0797

   29   TDISPLAY [PRINT=correlations]

29...........................................................................


***** Time-series analysis *****

*** Correlations ***

   1   1.000
   2   0.007  1.000
   3   0.004  0.662  1.000
   4  -0.008  0.497  0.559  1.000

             1      2      3      4
```

The CHANNEL option allows you to send the output to another output channel (see 3.3).

You can use the SAVE option to specify the time-series save structure (from ESTIMATE) from which the output is to be taken. By default TDISPLAY uses the structure from the most recent ESTIMATE statement.

12.3.5 The **TKEEP** directive

TKEEP saves results after an analysis by ESTIMATE.

Option

SAVE = *identifier* — Save structure to supply fitted model; default * i.e. that from last model fitted

Parameters

OUTPUTSERIES = *variate*	Output series to which model was fitted
RESIDUALS = *variate*	Residual series
ESTIMATES = *variate*	Estimates of parameters
SE = *variate*	Standard errors of estimates
INVERSE = *symmetric matrix*	Inverse matrix
VCOVARIANCE = *symmetric matrix*	
	Variance-covariance matrix of parameters
DEVIANCE = *scalar*	Residual deviance
DF = *scalar*	Residual degrees of freedom
MVESTIMATES = *variate*	Estimates of missing values in series
SEMV = *variate*	Standard errors of estimates of missing values
COMPONENTS = *pointer*	Variates to save components of output series
SCORES = *variate*	To save scores (derivatives of the log-likelihood with respect to the parameters)

An ESTIMATE statement produces many quantities that you may want to use to assess, interpret, and apply the fitted model. The TKEEP directive allows you to copy these quantities into Genstat data structures. If the METHOD option of the ESTIMATE statement was set to initialize, then the results saved by the options SE, INVERSE, VCOVARIANCE, and SCORE are unavailable. However, you can save the estimates of the missing values and their standard errors. The residual degrees of freedom in this case does not make allowance for the number of parameters in the model, but does allow for the missing values that have been estimated.

The OUTPUTSERIES parameter specifies the variate that was supplied by the SERIES parameter of the ESTIMATE statement; this can be omitted.

You can use the RESIDUALS parameter to save the residuals in a variate, exactly as in the ESTIMATE directive.

The ESTIMATES parameter can supply a variate to store the estimated parameters of the TSM. Each estimated parameter is represented once, but the innovation variance is omitted entirely. Genstat includes only the first of any set of parameters constrained to be equal using the FIX option of ESTIMATE. The order of the parameters otherwise corresponds to their order in the variate of parameters in TSM, and is unaffected by any numbering used in the FIX option.

The SE parameter allows you to specify a variate to save the standard errors of the estimated parameters of the TSM. The values correspond exactly to those in the ESTIMATES variate. Parameters in a time series model may be aliased. This is detected when the equations for the estimates are being solved, and the message ALIASED is printed instead of the standard error when the PRINT option of ESTIMATE or TDISPLAY includes the setting estimates. The corresponding units of the SE variate are set to missing values.

The INVERSE parameter can provide a symmetric matrix to save the product $(X'X)^{-1}$, where X is the most recent design matrix derived from the linearized least-squares regressions that were used to minimize the deviance. The ordering of the rows and columns corresponds exactly to that used for the ESTIMATES variate. The row of this matrix corresponding to any aliased parameter is set to zero except that the diagonal element is set to the missing value.

The VCOVARIANCE parameter allows you to supply a symmetric matrix for the estimated variance-covariance matrix, $\hat{\sigma}_a^2(X'X)^{-1}$, of the TSM parameters. The ordering of the rows and columns and the treatment of aliased parameters corresponds exactly to that used for the ESTIMATES variate.

The DEVIANCE parameter specifies a scalar to hold the final value of the deviance criterion defined by the LIKELIHOOD option of ESTIMATE.

The DF parameter saves the residual number of degrees of freedom, defined for a simple ARIMA model by $N-d-$(number of estimated parameters). If a seasonal model is used, this number is further reduced by Ds.

The MVESTIMATES parameter specifies a variate to hold estimates of the missing values of the series, in the order they appear in the series. You can thereby obtain forecasts of the series, by extending the SERIES in ESTIMATE with a set of missing values. This is less efficient than using the FORECAST directive, but it does have the advantage that the standard errors of the estimates take into account the finite extent of the data, and also the fact that the model parameters are estimated.

The SEMV parameter can supply a variate to hold the estimated standard errors of the missing values of the series, in the order they appear in the series.

The COMPONENTS parameter is used when there are explanatory variables, and is described in 12.5.4.

The SCORE parameter can specify a variate to hold the model scores. The scores are usually defined as the first derivatives of the log likelihood with respect to the model parameters. To get these, the scores supplied by TKEEP should be scaled by dividing by the estimated residual variance and reversing its sign. The elements of the SCORE variate correspond exactly to the parameters as they appear in the ESTIMATES variate. After using ESTIMATE to fit a time series model, the scores should in theory be zero provided the model parameters do not lie on the boundary of their allowed range. The scores are used within ESTIMATE to calculate the parameter changes at each iteration.

Example 12.3.5 is very similar to Example 12.3.2c which printed the parameter changes when using ESTIMATE with METHOD=onestep. Here the setting is METHOD=zerostep. The matrix obtained from INVERSE and the variate from SCORE are multiplied to give values very close to the parameter changes. This is not always the case because ESTIMATE shortens the step if the new parameters would have been outside their allowed range. A test statistic is calculated, as a quadratic form in the scaled score and the matrix obtained from VCOVARIANCE. Under the null hypothesis that the two new parameters have been set to their true values, the distribution of this statistic is chi-squared on two degrees of freedom. The value obtained is consistent with this.

Example 12.3.5

```
  31    "
 -32    The model previously fitted to the series of daylengths in Example
 -33    12.3a is extended to include two more autoregressive parameters,
 -34    the old parameters being kept at their estimated values. The score
 -35    is saved after using ESTIMATE with the option METHOD=zerostep. The
```

```
-36    Inverse matrix is also saved and used to calculate a variate of
-37    parameter corrections. The Variance-Covariance matrix is saved and
-38    used with the scaled score to form a test statistic to assess whether
-39    the new parameters should be retained in the model.
-40    "
 41
 42    " Save the previous model parameters and redefine the model with
-43     higher autoregressive orders and extended parameter variate. "
 44    CALCULATE Modpar = Erp['Parameters']
 45    & Modparx = !(Modpar$[1,2,3,4],0,0,Modpar$[5,6])
 46    TSM Erp; ORDERS=!(3,1,2); PARAMETERS=Modparx
 47    ESTIMATE [METHOD=zerostep] Daylength; TSM=Erp

47.............................................................................
```

***** Time-series analysis *****

```
  Output series: Daylengt
    Noise model: Erp
                          autoregressive    differencing   moving-average
            Non-seasonal              3               1                2


               d.f.      deviance
  Residual      143        36959.
```

*** Autoregressive moving-average model ***

Innovation variance 255.4

	ref.	estimate	s.e.
Transformation	0	1.00000	FIXED
Constant	1	3.98	4.55

* Non-seasonal; differencing order 1

	lag	ref.	estimate	s.e.
Autoregressive	1	2	0.380	0.158
	2	3	0.000	0.180
	3	4	0.000	0.141
Moving-average	1	5	-0.557	0.132
	2	6	-0.619	0.130

```
 48    TKEEP SCORE=Sc; INVERSE=W; VCOVARIANCE=V
 49    PRINT Sc

         Sc
        0.0
       -1.9
     1127.0
     -533.2
       19.3
     -127.1

 50    " Calculate and print the parameter correction variate. "
 51    CALCULATE Del = PRODUCT(W; Sc)
 52    PRINT Del
```

```
              Del
               1

      1      0.07477
      2     -0.06208
      3      0.16752
      4     -0.10249
      5     -0.05255
      6      0.07579
 53   " Form the scaled score and test statistic "
 54   CALCULATE Scsc = Sc/Modpar$[3]
 55   SCALAR Tstat
 56   CALCULATE Tstat = QPRODUCT(T(Scsc); V)
 57   PRINT Tstat

      Tstat
      0.9377
```

As in TDISPLAY, You can use the SAVE option to specify the time-series save structure from which the output is to be taken. By default TKEEP uses the structure from the most recent ESTIMATE statement.

12.3.6 The **FORECAST** directive

FORECAST forecasts future values of a time series.

Options

PRINT = *strings*	What to print (forecasts, limits, setransform, sfe); default fore,limi
CHANNEL = *scalar*	Channel number for output; default * i.e. current output channel
ORIGIN = *scalar*	Number of known values in FORECAST variate; default 0
MAXLEAD = *scalar*	Maximum lead time beyond the origin (i.e number of forecasts); default * i.e. value taken from length of FORECAST variate
PROBABILITY = *scalar*	Probability level for confidence limits; default 0.9
UPDATE = *string*	Whether existing forecasts are to be updated (yes, no); default no
FORECAST = *variate*	To save forecasts of output series; default *
SETRANSFORM = *variate*	To save standard errors of forecasts (on transformed scale, if defined); default *
LOWER = *variate*	To save lower confidence limits; default *
UPPER = *variate*	To save upper confidence limits; default *
SFE = *variate*	To save standardized forecast errors; default *
COMPONENTS = *pointer*	Contains variates to save components of forecast

SAVE = *identifier*   Save structure to supply fitted model; default * i.e. that from last model fitted

Parameters

FUTURE = *variates*   Future values of input series

METHOD = *strings*   How to treat future values of input series (observations, forecasts); default obse

In many applications of forecasting with univariate ARIMA models, you will need only a simple form of the directive. For example

```
FORECAST [MAXLEAD=10]
```

will cause Genstat to print forecasts for 10 lead times, that is, the next 10 time points after the end of your data. However, you must already have used ESTIMATE to specify the time series to be forecast, and the model to be used for forecasting. This information is supplied by the SAVE option; if SAVE is not specified, FORECAST uses the information from the most recent ESTIMATE statement. Once you have used ESTIMATE, you can give successive FORECAST statements to incorporate new observations of the time series, and to produce forecasts from the end of the new data.

If the time series is supplied with a known model (that is, one with all its orders and parameters specified) you can use ESTIMATE with option setting METHOD=initialize before you use FORECAST. This will carry out just sufficient calculations, in particular the regeneration of the model residuals, for FORECAST to be used. The model parameters will not be changed – not even the innovation variance. This setting of METHOD restricts the structures, such as parameter standard errors, that can be accessed using TDISPLAY and TKEEP after ESTIMATE. The SAVE structure created by using ESTIMATE with METHOD=initialize thus requires less space than that produced by the other settings.

The formal parameters of FORECAST are relevant only when the time-series model incorporates explanatory variables, and are described in 12.4.3.

The best way to understand the options of FORECAST is by example. Example 12.3.6a illustrates how to use ESTIMATE to initialize for FORECAST, with a series of 132 points and using a previously estimated model.

Example 12.3.6a

```
   3   "
  -4    Forecast number of airline passengers in 1960, using a seasonal
  -5    ARIMA model whose parameters have already been estimated,
  -6    and based on numbers observed 1949-59.
  -7    Data from Box and Jenkins (1970) page 304.
  -8   "
   9   OPEN 'airline.dat'; CHANNEL=2
  10   UNITS [NVALUES=132]
  11   READ [CHANNEL=2] Apt
```

Identifier	Minimum	Mean	Maximum	Values	Missing
Apt	104.0	262.5	559.0	132	0

```
12   VARIATE [VALUES=0,1,1, 0,1,1,12] Ord
13   & [VALUES=0,0,0.00143, 0.34, 0.54] Par
14   TSM Airpass; ORDERS=Ord; PARAMETERS=Par
15   ESTIMATE [PRINT=model; METHOD=initialize] Apt; TSM=Airpass
```

15...

***** Time-series analysis *****

```
Output series: Apt
  Noise model: Airpass
                      autoregressive    differencing   moving-average
        Non-seasonal              0               1               1
        Period 12                 0               1               1
```

```
16   FORECAST [MAXLEAD=12; FORECAST=Fcst12]
```

16...

*** Forecasts ***

Maximum lead time: 12

Lead time	forecast	lower limit	upper limit
1	419.6	394.3	446.5
2	398.9	370.2	429.7
3	466.7	428.6	508.1
4	454.4	413.5	499.5
5	473.9	427.5	525.3
6	547.6	490.1	611.8
7	623.3	553.8	701.5
8	631.7	557.4	716.0
9	527.2	462.1	601.4
10	462.8	403.1	531.2
11	407.1	352.6	470.2
12	452.7	389.7	525.8

The FORECAST option specifies that the forecast values are to be stored in the variate Fcst12: you could then, for example, display them graphically.

Example 12.3.6b shows the use of the procedure BJFORECAST to construct and plot these same forecasts with their error limits.

Example 12.3.6b

```
33   "
-34   Use the Library Procedure BJFORECAST to calculate and display the
-35   forecasts of the last 12 values based upon the previous 132.
-36   "
37   VARIATE [VALUES=0,1,1, 0,1,1,12] Ord
38   VARIATE [VALUES=0,0.0,0.00143,0.34,0.54] Par
39   TSM Airpass; ORDERS=Ord; PARAMETERS=Par
40   FRAME 1; YLOWER=0.1; YUPPER=0.9; XLOWER=0; XUPPER=1
41   BJFORECAST [GRAPHICS=high] Apt; TSM=Airpass; ORIGIN=132; \
42     TIMERANGE=!(133,144)
```

Forecasts from fixed origin 132   over time range 133   to 144
with probability limits of size 0.900   using   whole   of series

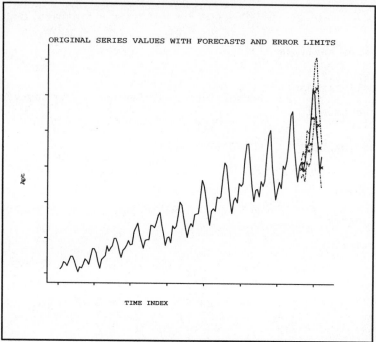

Figure 12.3.6b

Now suppose that a further set of observations of the time series has become available, for example a variate New6 containing the next six values of the series. In order to revise the forecasts, you can incorporate this new information as follows.

Example 12.3.6c

```
  19    "
 -20    Read observed numbers for January to June 1960, and give revised
 -21    forecasts for these months with standardized forecast errors.
 -22    "
  23    READ [PRINT=data] New6

  24    417.0    391.0    419.0    461.0    472.0    535.0:
  25    FORECAST [PRINT=sfe; ORIGIN=6; MAXLEAD=0; FORECAST=New6]
```

25..

*** Forecasts ***

   Forecast origin: 6

Maximum lead time: 0

Lead time	forecast	s.f.e.
-5	417.0	-0.16
-4	391.0	-0.42
-3	419.0	-2.46
-2	461.0	2.39
-1	472.0	0.33
0	535.0	-0.40

The setting PRINT=sfe now causes Genstat to print the standardized errors of the forecast. These are the innovation values that are generated as each successive new observation is incorporated, divided by the square root of the TSM innovation variance. They provide a useful check on the continuing adequacy of the model. For example, excessively large values (compared to the standard Normal distribution) may indicate that you should revise the model. The ORIGIN option specifies the number of new values to be incorporated, the last of these becoming the origin from which any further forecasts might be made. The MAXLEAD option is set to 0, thus preventing new forecasts being produced. The FORECAST option is used to specify the variate containing the new observations of the time series.

Revised forecasts of the next six values of the series can then be produced by a further statement, as shown in Example 12.3.6d.

Example 12.3.6d

```
 28  "
-29    Forecast for July to December 1960.
-30  "
 31  FORECAST [MAXLEAD=6; UPDATE=yes; FORECAST=Fcst6]

31.................................................................................

*** Forecasts ***

Maximum lead time: 6
```

Lead time	forecast	lower limit	upper limit
1	612.1	575.2	651.4
2	620.4	575.8	668.4
3	517.7	475.5	563.7
4	454.5	413.5	499.5
5	399.8	360.7	443.2
6	444.6	397.9	496.7

You use the UPDATE option to incorporate new observations – advancing the origin for future forecasts to the end of whatever new observations have been supplied in previous FORECAST statements since the last ESTIMATE statement. Thus, the FORECAST statement in Example 12.3.6d incorporates the six values supplied in the variate New6 in the previous FORECAST statement. The UPDATE option allows you to alternate between incorporating new observations and producing new forecasts, as observations become available.

By setting the ORIGIN and MAXLEAD options you can both incorporate new observations and produce new forecasts in a single FORECAST statement.

Example 12.3.6e

```
  10   "
 -11    Incorporate new observations and forecast ahead in one statement.
 -12   "
  13   READ [PRINT=data] New6fcst6
  14   417.0   391.0   419.0   461.0   472.0   535.0
  15   *   *   *   *   *   *   :
  16   FORECAST [ORIGIN=6; MAXLEAD=6; FORECAST=New6fcst6]

16.........................................................................................
```

*** Forecasts ***

Forecast origin: 6

Maximum lead time: 6

Lead time	forecast	lower limit	upper limit
-5	417.0	*	*
-4	391.0	*	*
-3	419.0	*	*
-2	461.0	*	*
-1	472.0	*	*
0	535.0	*	*
1	612.1	575.2	651.4
2	620.4	575.8	668.4
3	517.7	475.5	563.7
4	454.5	413.5	499.5
5	399.8	360.7	443.2
6	444.6	397.9	496.7

The ORIGIN option here specifies that the first six values of the FORECAST variate are to be incorporated as new observations. The MAXLEAD option specifies that the next six values are to hold the forecasts produced from the last of these new values.

You can use this form of statement repeatedly, but the ORIGIN value must increase (or stay the same) unless the UPDATE option is used. Without the setting UPDATE=yes, successive statements will incorporate only those new values of the series that occur beyond the previous ORIGIN and up to the new ORIGIN: the forecasts are revised from the new ORIGIN. Setting UPDATE=yes lets you act as though the previous ORIGIN value were zero.

The PROBABILITY option determines the width of the error limits on the forecast. It defines the probability that the actual value will be contained within the limits at any particular lead time. Note that the limits do not apply simultaneously over all lead times.

The SETRANSFORM option specifies a variate to store the standard errors that Genstat used in calculating the error limits of the forecasts, starting at lead time 1. These are the standard errors of the transformed series, according to the value of the Box-Cox transformation parameter; they are functions of the model only, not of the data.

The LOWER option specifies a variate to store the lower limits of the forecasts. This must be the same length as the FORECAST variate. The FORECAST directive puts the values of the lower limit into the variate, matching the forecasts in the FORECAST variate. The UPPER option similarly allows the upper limits to be saved. Note that the limits are constructed as symmetric percentiles, assuming Normality of the transformed time series. Similarly, the forecast is a median value – not necessarily the mode or the mean, unless the transformation parameter is 1.0.

The SFE option specifies a variate to save the standardized errors of the forecasts: see above. The variate must be the same length as the FORECAST variate. The FORECAST directive places values of the errors in the variate, matching the new observations in the FORECAST variate.

The COMPONENTS option is relevant only when the time-series model incorporates explanatory variables, and is described in 12.5.5.

12.4 Regression with autocorrelated (ARIMA) errors

At the beginning of Chapter 8, we noted that regression analysis is not valid if the residuals cannot be assumed to be independent. When modelling observations of a variable that are taken at successive points in time, it is likely that there will be some dependence. A simple check for this is to fit a regression model as in Chapter 8, and then calculate the sample autocorrelation function (12.1.2) of the residuals from the regression. If you think that there might be appreciable autocorrelation, you should try fitting the regression model using an ARIMA model for the errors, as described in this section.

We shall use as an example a time series y_t of daily gas demand (corrected for the effects of days of the week), and a corresponding indicator x_t of the coldness of the days, compiled from temperature, windspeed, and so on. Example 12.4a fits a regression between the variates Demand and Coldness which hold 104 consecutive values of the two series. A first-order autoregressive model, AR(1), is specified for the errors: that is, the model is
$$y_t = c + b\,x_t + e_t$$
$$e_t = \phi_1\,e_{t-1} + a_t$$
where a_t is the series of independent innovations of the errors e_t. We have set PRINT=monitoring in the ESTIMATE statement to show the course of the convergence.

Example 12.4a

```
  2   "
 -3    Regress daily gas demand on coldness, using an AR(1) model for errors.
 -4   "
  5   OPEN 'demand.dat','cold.dat'; CHANNEL=2,3
  6   READ [CHANNEL=2] Demand
```

Identifier	Minimum	Mean	Maximum	Values	Missing
Demand	239.3	348.7	471.8	104	0

```
  7   & [CHANNEL=3] Coldness
```

Identifier	Minimum	Mean	Maximum	Values	Missing
Coldness	-117.30	-49.87	42.60	104	0

```
  8   TSM Erm; ORDERS=!(1,0,0)
  9   TRANSFERFUNCTION Coldness
 10   " Monitor convergence."
```

```
 11  ESTIMATE [PRINT=monitoring,estimates] Demand; TSM=Erm; BOXCOX=estimate
11..................................................................................
```

*** Convergence monitoring ***

Cycle	Deviance	Current parameters			
1	12803380.	0.	1.00000	0.	0.
2	8684909.	1.7447	1.2880	499.04	-0.45365
3	209142.5	3.0618	1.2890	748.64	0.75300
4	38869.89	5.3578	1.3048	1517.3	0.83100
5	28399.29	5.3906	1.3060	1925.6	0.79741
6	27741.35	5.6954	1.3124	1921.7	0.73305
7	27618.37	5.6692	1.3059	1883.2	0.72642
8	27601.49	5.6040	1.3032	1858.0	0.71415
9	27571.62	5.2138	1.2902	1734.7	0.71193
10	27538.84	4.7622	1.2747	1599.7	0.71021
11	27506.65	4.4128	1.2613	1493.9	0.70996
12	27475.54	4.0239	1.2457	1376.4	0.70890
13	27445.39	3.7474	1.2332	1291.5	0.70889
14	27416.51	3.4102	1.2173	1188.4	0.70777
15	27388.23	3.1955	1.2057	1121.4	0.70800

```
******** Warning (Code TS 21). Statement 1 on Line 11
Command: ESTIMATE [PRINT=monitoring,estimates] Demand; TSM=Erm; BOXCOX=estimate
The iterative estimation process has not converged
The maximum number of cycles is 15
```

***** Time-series analysis *****

*** Transfer-function model 1 ***

Delay time 0

	ref.	estimate	s.e.
Transformation	0	1.00000	FIXED
Constant	0	0.	FIXED

* Non-seasonal; no differencing

	lag	ref.	estimate	s.e.
Moving-average	0	1	2.898	0.896

*** Autoregressive moving-average model ***

Innovation variance 2488.

	ref.	estimate	s.e.
Transformation	2	1.1894	0.0514
Constant	3	1029.	270.

* Non-seasonal; no differencing

	lag	ref.	estimate	s.e.
Autoregressive	1	4	0.7067	0.0714

The TSM statement specifies the AR(1) model for the errors. The TRANSFERFUNCTION statement here merely specifies the explanatory variate. You could use this directive to specify a response model that includes lagged effects of the explanatory variate (12.5.2), but in Example 12.4a, the response model is a simple linear regression: this is the default.

The warning shows that the convergence criterion has not been reached within 15 iterations. To satisfy the criterion, we could either increase the limit on the number of iterations by setting the option MAXCYCLE=25, say, or initialize the parameters to rough estimates of the parameters in the model, perhaps using the FTSM directive (12.7.1 and 12.7.2). The statements that follow ESTIMATE in this program use the best parameter values found by ESTIMATE, without further comment.

The ESTIMATE statement simultaneously estimates the regression coefficients c and b and the AR parameter ϕ_1. Also in this case, a Box-Cox transformation is estimated for the response variate, Demand. Note in the printed results that the estimate of b appears under "Transfer-function model 1", as a moving-average parameter at lag 0. By default, Genstat fixes the transformation and constant parameters associated with the explanatory variables to be 1.0 and 0.0. Alternatively, you could estimate these parameters, as described in 12.5.

The constant term c in the regression is included in the results for the autoregressive moving-average model, as is the transformation parameter of the Demand variable, and the estimate of ϕ_1.

You can obtain forecasts of the demand series, by specifying future values of the explanatory variable. In Example 12.4b, the variate Newcold contains the next seven values of coldness.

Example 12.4b

```
 14    "
-15     Forecast gas demand for the next week, given values for coldness.
-16    "
 17    READ [CHANNEL=3] Newcold

       Identifier   Minimum      Mean   Maximum    Values   Missing
         Newcold     -138.3    -102.3     -75.6         7         0
 18    FORECAST [MAXLEAD=7] Newcold

18.......................................................................

*** Forecasts ***

Maximum lead time: 7

       Lead time      forecast  lower limit  upper limit
               1         318.6        290.9        346.0
               2         294.3        259.6        328.1
               3         313.9        277.0        350.0
               4         324.5        286.5        361.7
               5         278.4        238.5        317.2
               6         261.7        221.0        301.2
               7         299.2        259.5        338.0
```

Genstat constructs the forecasts by calculating the predicted linear response at the Newcold values, and adding it to the forecast values of the autocorrelated errors. The forecast limits take this into account.

In practice you would be unlikely to know the future values of explanatory variables. Exceptions are where the variable has a fixed deterministic form such as in a trend, or a cycle, or an intervention variable; or when the variable is under the control of the experimenter, as when sales are related to prices; or when the analysis is retrospective, as in this example. You can predict the explanatory variables in various ways. For example, ordinary weather forecasts are used in practice to forecast gas demand. You cannot usually include the uncertainties in predicting the explanatory variables in the error limits of the forecast. These uncertainties would usually be assessed by trying out different future values of the explanatory variables. Thus the FORECAST statement in the example could be repeated with a variety of future values. But there is one case where you can allow for the uncertainty of predicting the explanatory variables. This is when the future values of the explanatory variables are predictions obtained using univariate ARIMA models. Then you can allow for the errors by setting the ARIMA parameter of the TRANSFERFUNCTION directive, and the METHOD parameter of the FORECAST directive.

12.4.1 The **TRANSFERFUNCTION** directive

TRANSFERFUNCTION specifies input series and transfer function models for subsequent estimation of a model for an output series.

Option

SAVE = *identifier*	To name time-series save structure; default *

Parameters

SERIES = *variates*	Input time series
TRANSFERFUNCTION = *TSMs*	Transfer-function models; if omitted, model with 1 moving-average parameter, lag 0
BOXCOXMETHOD = *strings*	How to treat transformation parameters (fix, estimate); default fix
PRIORMETHOD = *strings*	How to treat prior values (fix, estimate); default fix
ARIMA = *TSMs*	ARIMA models for input series

For regression with autocorrelated errors, you should use TRANSFERFUNCTION to specify the variates that are to be the explanatory variables in a subsequent ESTIMATE statement. Thus in many applications you will need only a simple form of the directive, such as

    TRANSFERFUNCTION Coldness

The first parameter, SERIES, specifies a list of variates holding the time series of explanatory variables.

The BOXCOXMETHOD parameter allows you to estimate separate power transformations for the explanatory variables: the variable x_t is transformed to

$$x_t^{(\lambda)} = (x_t^{\lambda} - 1) / \lambda , \qquad\qquad \lambda \neq 0$$
$$x_t^{(0)} = \log(x_t)$$

The default is no transformation, corresponding to $x_t^{(\lambda)} = x_t$. You can choose whether the transformations are to be fixed or estimated, by specifying one string for each explanatory variable.

The ARIMA parameter allows you to associate with each explanatory variable a univariate ARIMA model for the time-series structure of that variable. If you think such a model is inappropriate, then you should give a missing value in place of the TSM identifier, or leave this parameter unset. You can use these models in any subsequent FORECAST statement to incorporate, into the error limits of the forecasts, an allowance for uncertainties in the predicted explanatory variables; the allowance assumes that the future values of the explanatory variables are forecasts obtained using these ARIMA models (12.4.3).

The TRANSFERFUNCTION and PRIORMETHOD parameters are not relevant in this context, and are described in 12.5.2.

The SAVE option allows you to name the time-series save structure created by TRANSFERFUNCTION. You can use this identifier in a later ESTIMATE statement, and eventually in a FORECAST statement. If you do not name the save structure Genstat will use the most recent save structure, which will be overwritten each time a new TRANSFERFUNCTION statement is given.

12.4.2 Extensions to the ESTIMATE directive for regression with ARIMA errors

The SERIES parameter of ESTIMATE now specifies the response variate, and the TSM parameter specifies the ARIMA model for the errors. Note however, that the transformation parameter of this ARIMA model is used to define a transformation for the response variable, not the errors, and the BOXCOXMETHOD parameter controls its estimation.

The constant term in the ARIMA model corresponds to the usual regression constant term only if there is no differencing specified by the ARIMA model; otherwise it is equivalent to a constant term in a regression between the differenced series.

The PRINT option is the same as described in 12.3.2. But note that the regression estimates for the explanatory variables are printed in a sequence of simple transfer-function models, followed by the ARIMA error model, as shown in Example 12.4a.

The LIKELIHOOD option settings exact and leastsquares are essentially the same as for univariate ARIMA modelling in 12.3. The likelihood for the model is defined as that of the univariate error series e_t which is defined in general by

$$e_t = y_t - b_1 x_{1,t} - \dots - b_m x_{m,t}$$

(the x_i being m explanatory variables). The constant term therefore appears in the model after any differencing of e_t; for example

$$\nabla e_t = c + (1 - \theta_1 B)a_t$$

You can get bias in the estimates of the parameters of an ARIMA model because the regression is estimated at the same time. You can guard against this by specifying LIKELIHOOD=marginal. This can be particularly important if the series are short or if you use many explanatory variables (Tunnicliffe Wilson 1989). The deviance is now defined as

$$D = S \, (\, | \, X'V^{-1}X \, | \quad | \, V \, | \,)^{1/m}$$

where m is reduced by the number of regressors (including the constant term) and the columns of X are the differenced explanatory series: the other terms are as in the exact likelihood described in 12.3.3.

You can use this setting also for univariate ARIMA modelling, when the constant term is the only explanatory term. Furthermore, Genstat deals with missing values in the response variate by doing a regression on indicator variates; these too are included in the X matrix. However, you cannot use marginal likelihood and estimate a transformation parameter in either the transfer-function model or an ARIMA model. Neither can you use it if you set the FIX option in ESTIMATE. In these cases Genstat automatically resets the LIKELIHOOD option to exact.

At every iteration with the setting LIKELIHOOD=marginal, the regression coefficients are the maximum-likelihood estimates conditional upon the estimated values of the parameters of the ARIMA model: these are also the generalized least-squares estimates, conditioned in the same way. This is so even if MAXCYCLE=0; that is, the coefficients of the regression are re-estimated even at iteration 0. Therefore you must not use the marginal setting with the option METHOD=initialize to initialize for FORECAST. You can compare deviance values that were obtained using marginal likelihood only for models with the same explanatory variables and the same differencing structure in the error model.

You can use the setting CONSTANT=fix with marginal likelihood. You can use the FIX option to impose constraints across any or all of the parameters of the regression and the ARIMA model. In order to do this, you may find it easiest to use ESTIMATE without the FIX option first, so that you can ascertain the ordering of the parameters; then give a second statement with the option set. The variate specified in the FIX option must have one element for each parameter that is printed with a reference number. These are, in order, three parameters for each explanatory variate, followed by the ARIMA model parameters. Genstat uses the variate to provide a parameter numbering as described for the FIX option in 12.4.2. Note that this numbering overrides the BOXCOXMETHOD parameter and the CONSTANT option. Thus you can constrain the transformation parameters to be equal for all or some of the variables. You can also estimate a constant term for an input series. For details of this see 12.5.3.

The results of ESTIMATE, accessible by TDISPLAY and TKEEP, are essentially the same as in univariate models. The variate of parameter estimates and associated structures now refers to the whole set of parameters in the order in which they are printed. The variate of missing-value estimates holds first the values from the response variate, and then those from the explanatory variate, in the order in which they were listed in the SERIES parameter of TRANSFERFUNCTION.

12.4.3 Extensions to the FORECAST directive for regression with ARIMA errors

A FORECAST statement for regression with ARIMA errors must be preceded by a TRANSFERFUNCTION statement and an ESTIMATE statement: these initialize the save structure of the time series that is to be used by FORECAST. You use option METHOD=initialize of

ESTIMATE to do this as described in 12.3.6.

You use the FUTURE parameter to specify a list of variates, corresponding to the list of variates specified by the SERIES parameter of TRANSFERFUNCTION. These variates must all have the same length. They hold future values of the explanatory variables to be used either for constructing forecasts of the response variable, or for incorporating new observations in order to revise the forecasts. The use of these future values is similar to the use of the FORECAST variate as described in 12.3.6. For example, let Fcdem be a variate of length seven in Examples 12.4a or 12.4b. The statement

```
FORECAST [MAXLEAD=7; FORECAST=Fcdem] FUTURE=Newcold
```

would cause forecasts of the next week's demand figures to be placed in Fcdem. Suppose that in a week's time, the actual demand had been recorded and was held in the variate Newdem. Then in order to revise the forecasts, you must first incorporate this new information by

```
FORECAST [ORIGIN=7; MAXLEAD=0; FORECAST=Newdem] FUTURE=Newcold
```

Note that if Newcold had previously contained forecasts from an ARIMA model, say, you would have to alter it to contain the recorded values before this statement. You can get revised forecasts of the next week's demand by once more amending Newcold, to hold the values for the coming week, and then using

```
FORECAST [UPDATE=yes; MAXLEAD=7; FORECAST=Fcdem] FUTURE=Newcold
```

An alternative to the previous two statements would be to use variates of length 14, with Newcold holding the seven values just recorded followed by the seven values for the coming week. Similarly Newdem should hold the last seven days' demand, followed by seven missing values. The statement

```
FORECAST [ORIGIN=7; MAXLEAD=7; FORECAST=Newdem] FUTURE=Newcold
```

would then incorporate the first seven values (up to the ORIGIN setting) of each variate, and use the last seven values (specified by MAXLEAD) of Newcold to place revised forecasts into the last seven values of Newdem.

You can use the METHOD parameter when some or all of the future values of the explanatory variables are forecasts obtained using univariate ARIMA models. You can amend the error limits of the forecasts for the response variable to allow for the uncertainty in these future values, but you need to assume that there is no cross-correlation between the errors in these predictions. The list of strings specified by the METHOD parameter indicates for each explanatory variable whether such an allowance should be made. The future values of a series are by default treated as known values if no corresponding ARIMA model is present, or if the transformation parameter of the ARIMA model is not equal to the value used in the regression model for that series. You can change the settings of the METHOD parameter in successive FORECAST statements.

12.5 Multi-input transfer-function models

A transfer-function model allows for lagged effects of an explanatory variable on the response variable, as well as for autocorrelated errors. Using the notation of Box and Jenkins (1970), including a transfer-function model with an ARIMA model for a response variable gives the equation

$$y_t = v(B)x_t + \psi(B)a_t$$

where we shall now call y_t the output series and x_t the input series. You can have several input series, so we shall call the full model for y_t a multi-input model, corresponding to the term "multiple regression" used in Chapter 8. Writing $y_t = z_t + n_t$ where $z_t = v(B)x_t$ and $n_t = \psi(B)a_t$, we shall call z_t the component due to input x_t, and n_t the noise component. An ARIMA TSM is used to represent the structure of n_t, and a transfer-function TSM to represent the structure of z_t as a function of x_t. For example, consider the lagged response, with $|\delta| < 1$:

$$y_t = \omega(x_{t-1} + \delta x_{t-2} + \delta^2 x_{t-3} + \dots) + n_t.$$

Then $v(B) = \omega B / (1 - \delta B)$.

Example 12.5 fits this model to a series of length 40, and produces forecasts of the next eight points.

Example 12.5

```
 -5   "
 -6    One-input transfer-function model relating level of gilts to profits.
 -7   "
  8   VARIATE [VALUES=1...40] Time
  9   UNITS Time
 10   " Read data on gilts and profits from separate files."
 11   OPEN 'GILTS.DAT','PROFITS.DAT'; CHANNEL=2,3
 12   READ [CHANNEL=2] Gilts
```

```
    Identifier    Minimum       Mean    Maximum    Values    Missing
        Gilts     -26.253      1.037     27.971        40          0
```

```
 13   & [CHANNEL=3] Profits
```

```
    Identifier    Minimum       Mean    Maximum    Values    Missing
      Profits    -1.80720    0.02747    1.48670        40          0
```

```
 14   " Set up transfer-function model with delay time 1 and one AR-type
-15     parameter."
 16   TSM [MODEL=transfer] Tf; ORDERS=!(1,1,0,0); PARAMETERS=!(1,0,0,0.1)
 17   TRANSFERFUNCTION Profits; TRANSFER=Tf
 18   " Set up ARIMA model for the noise, with one AR parameter."
 19   TSM Ar; ORDERS=!(1,0,0); PARAMETERS=!(1,0,0,0)
 20   ESTIMATE Gilts; TSM=Ar
```

```
20..........................................................................
```

```
***** Time-series analysis *****
```

```
  Output series: Gilts
    Noise model: Ar
                              autoregressive    differencing   moving-average
          Non-seasonal              1                0               0
  Input series: Profits        ; transfer function: Tf
          Non-seasonal              1                0               0
```

```
                  d.f.      deviance
  Residual          36         900.6
```

```
*** Transfer-function model 1 ***

Delay time 1

                      ref.      estimate         s.e.
Transformation          0       1.00000         FIXED
Constant                0          0.            FIXED

* Non-seasonal; no differencing

                  lag ref.      estimate         s.e.
Autoregressive     1    1        0.6273         0.0805
Moving-average     0    2        8.74           1.16


*** Autoregressive moving-average model ***

Innovation variance 24.52

                      ref.      estimate         s.e.
Transformation          0       1.00000         FIXED
Constant                3         -1.06          2.87

* Non-seasonal; no differencing

                  lag ref.      estimate         s.e.
Autoregressive     1    4        0.740          0.118


   21  " Save the components of the series in variates."
   22  TKEEP COMPONENTS=!P(Fprofits,Noise)
   23  PEN 1,2; COLOUR=1; METHOD=line,point; SYMBOLS=0,1; LINE=1
   24  DGRAPH [TITLE='Fitted series with original data'] Fprofits,Gilts; \
   25    Time; PEN=1,2
   26  " Read future values of profits, and forecast corresponding gilts."
   27  READ [CHANNEL=3] Nprofits

   Identifier   Minimum      Mean   Maximum    Values   Missing
     Nprofits   -1.1645   -0.1374    0.4904         8         0

   28  FORECAST Nprofits

28.....................................................................................


*** Forecasts ***

Maximum lead time: 8

   Lead time      forecast  lower limit  upper limit
           1        -6.50       -14.64         1.65
           2       -10.37       -20.51        -0.24
           3       -17.20       -28.27        -6.12
           4       -16.11       -27.67        -4.55
           5       -12.25       -24.06        -0.44
           6        -4.39       -16.34         7.56
           7         1.10       -10.93        13.13
           8         4.14        -7.93        16.21
```

In this example, the first TSM statement defines the orders of the transfer-function model, the initial values of parameters δ and ω being given as 0.0 and 0.1 respectively. The second TSM statement defines the autoregressive error structure. The TRANSFERFUNCTION statement then specifies the input series to be Profits, and gives the associated transfer-function model. The ESTIMATE statement specifies the output series and the noise model.

After the model has been estimated, the TKEEP statement accesses the two components of Gilts. The first of these, Fprofits, is plotted together with Gilts, to reveal how well the output series has been modelled by the input series.

Finally, new values of the input series are used to construct forecasts of the output series, using the FORECAST directive.

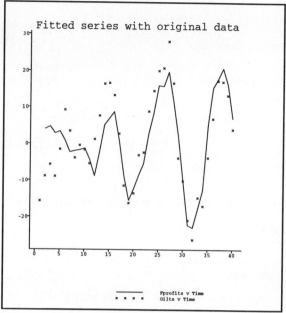

Figure 12.5

12.5.1 Declaring transfer-function models

The basic structure of the TSM directive, and of the models that it defines, is given in 12.3.1. Here we describe the ORDERS, PARAMETERS, and LAGS variates for the option setting MODEL=transferfunction.

The simple non-seasonal transfer-function model relates a component z_t of the output series to the corresponding input series x_t, by the equation

$$\delta(B)\ \nabla^d z_t\ =\ \omega(B)\ B^b\ \{x_t^{(\lambda)} - c\}$$

where

$$\delta(B)\ =\ 1\ -\ \delta_1\ B\ -\ ...\ -\ \delta_p\ B^p$$
$$\omega(B)\ =\ \omega_0\ -\ \omega_1\ B\ -\ ...\ -\ \omega_q\ B^q\ .$$

The integer $b > 0$ defines a pure *delay*, and the integer $d > 0$ defines the order of differencing in the transfer function.

The parameter λ specifies a Box-Cox power transformation for the input series, and the parameter c specifies a reference level for the transformed input. There is no mean correction of the input series when transfer-function models are estimated, and you should use a value of c close to the series mean so as to improve the numerical conditioning of the estimation procedure. However, if the input series x_t is trend-like rather than stationary, you could alternatively use a value for c close to the early series values, because this reduces the transient errors that arise when the transfer function is applied. The PRIORMETHOD parameter of TRANSFERFUNCTION, described below, provides further means of handling these transients.

The parameters λ and c are not estimated unless you specify otherwise by the

BOXCOXMETHOD parameter of TRANSFERFUNCTION or the FIX option of ESTIMATE. Often c in the transfer-function model is aliased with the constant term in the ARIMA errors, and so they should not both be estimated. In some circumstances, however, they both could be estimated, for example in a differenced transfer-function model with stationary noise.

The ORDERS parameter for the simple transfer-function model described above specifies a variate containing the four values b, p, d, and q.

The PARAMETERS parameter specifies a variate containing $3+p+q$ values: λ, c, δ_1, ... δ_p, ω_0, ω_1 ... ω_q. You must always include the parameters λ, c, and ω_0. When you use a transfer-function model, Genstat will check its parameter values. In particular the operator $\delta(B)$ must satisfy the stability or stationarity condition.

The LAGS parameter is optional, and may be used to change the lags associated with the parameters, from the default values of 1 ... p, 1 ... q. The variate of lags contains values corresponding to the parameters δ_1 ... δ_p, ω_1 ... ω_q. They have the same interpretation as the lags in ARIMA models, and must satisfy the same conditions as specified in 12.3.1. Note that there is no lag associated with ω_0, because the delay b provides the necessary flexibility for this.

You can also have seasonal extensions of transfer-function models:

$$\delta(B)\Delta(B^s)\nabla^d\nabla_s^D z_t = \omega(B)\Omega(B^s)B^b\{x_t^{(\lambda)} - c\}$$
$$\Delta(B^s) = 1 - \Delta_1 B^s - ... - \Delta_P B^{Ps}$$
$$\Omega(B^s) = 1 - \Omega_1 B^s - ... - \Omega_Q B^{Qs}$$

Note that there is no Ω_0 coefficient, because ω_0 is always present in the model and provides sufficient flexibility.

The ORDERS parameter here contains b, p, d, q, P, D, Q, and s, and the PARAMETERS parameter contains λ, c, δ_1 ... δ_p, ω_0 ... ω_q, Δ_1 ... Δ_P, Ω_1 ... Ω_Q. You can analogously extend the LAGS parameter. You can have extensions to multiple seasonal periods, as for ARIMA models.

12.5.2 Extensions to the TRANSFERFUNCTION directive for multi-input models

This directive specifies several input series and the associated transfer-function model to be used in a subsequent ESTIMATE statement to fits a multi-input model to an output series.

The SERIES and BOXCOXMETHOD parameters are as described in 12.4.1.

The TRANSFERFUNCTION parameter specifies the transfer-function TSMs that are to be associated with the input series. A missing value in place of a TSM identifier causes Genstat to treat the corresponding input series as a simple explanatory variable, equivalent to a transfer-function model with orders (0,0,0,0).

The PRIORMETHOD parameter specifies, for each input series, how Genstat is to treat the transients associated with the early values of the transfer-function response. In calculating the input component z_t from the input x_t, Genstat has to make assumptions about the unknown values of x_t which came before the observation period. The default is that x_t (or generally $x_t^{(\lambda)}$) is assumed to be equal to the reference constant c of the transfer-function model. The pattern of the transient can be controlled by introducing a number $\max(p+d,b+q)$ of nuisance parameters to represent the combined effects of all earlier input values on the observed output. Setting PRIORMETHOD=estimate specifies that these nuisance parameters are estimated so as to minimize the transients. You should, however, be careful in using this. Often all you will have to do is make a sensible choice of the reference constant c. Estimating the transients is best done as a final stage in refining the model; earlier, this may give poor numerical conditioning.

12.5.3 Extensions to the ESTIMATE directive for multi-input models

ESTIMATE fits a multi-input model to output series that have a specified model for the output noise. The input series and transfer-function models must have been specified in an earlier TRANSFERFUNCTION statement.

The PRINT option is the same as before, but note that the transfer-function models are printed in a descriptive format similar to the ARIMA model, with parameter reference numbers used throughout.

The LIKELIHOOD option settings exact and leastsquares are similar to the settings described in 12.4.2 for regression with ARIMA errors. For example, with a single input, the likelihood is defined as that for the univariate noise series n_t, calculated as $n_t = y_t - z_t$.

The marginal likelihood is permitted only when all the transfer-function models are equivalent to simple regression.

You can use the FIX option as described in 12.3.2 and 12.4.2, to impose constraints among the parameters while the model is being estimated. These constraints operate here across the whole set (in order) of the parameters of the transfer-function models and of the ARIMA model, excluding the innovation variance. You can thus use this option to estimate the constant term in a transfer-function model (but bear in mind the remarks in 12.5.1 about possible aliasing).

12.5.4 Extensions to the TKEEP directive for multi-input models

After a multi-input model has been fitted using ESTIMATE, you can use the COMPONENTS parameter to access the components of the output series that are due to the various input series; you can also access the output noise. In simple regression, the input components are proportional to the input series. But the component resulting from a transfer-function model may be quite different from this. You can examine these components separately, or sum them to show the total fit to the output series that is explained by the input series. Note that the fitted values may appear to be offset from that output series, because the constant term is part of the noise component, and so is not included. Example 12.5 includes a graph of the output component due to the single input. You may want to examine the output noise component. For example, if you thought that the ARIMA model for the output noise was inadequate, you could investigate the noise component with the univariate ARIMA modelling methods described earlier in this chapter.

12.5.5 Extensions to the FORECAST directive for multi-input models

FORECAST for multi-input models is the same as for regression models with ARIMA errors (1.4.3). But it does have one further useful option.

The COMPONENTS option specifies a pointer to variates in which you can save components of future values of the output series. There is a variate for each input component and for the output noise component. These variates correspond exactly to the variates that were specified by the FUTURE parameter for the input series, and by the FORECAST variate for the output series; corresponding lengths must match. The values that the variates hold can therefore be components of the forecasts of the output series, or can be new observations. The can be used to investigate the structure of forecasts.

If the input series ARIMA model and the transfer-function model have differing transformation parameters, then the METHOD option reverts to its default action of treating the values of any future input series as known quantities rather than forecasts.

12.6 Filtering time series

Filtering is a means of processing a time series so as to produce a new series. The purpose is usually to reveal some features and remove other features of the original series. Filters in Genstat are one-sided: that is, each value in the new series depends only on present and past values of the original series. However, you can do two-sided filtering by using the SHIFT and REVERSE functions of CALCULATE (5.2.1).

A filter is defined by a time-series model. For example, consider the exponentially weighted moving average (EWMA) filter

$$y_t = \lambda\, y_{t-1} + (1 - \lambda)\, x_t$$

which smoothes x_t to produce y_t. Then

$$y_t = \{(1 - \lambda) / (1 - \lambda B)\}\, x_t.$$

You can represent this by a transfer function applied to x_t. Example 12.6 applies this filter to smooth a time series of annual temperatures in Central England, taking $\lambda=0.8$: the mean of the series is subtracted from the series before smoothing and restored afterwards. The smoothed series, Smtemp, is shown with the original data in Figure 12.6. This is one way to reduce transient errors at the start of the smoothed series.

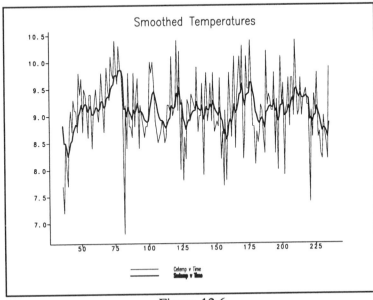

Figure 12.6

Example 12.6

```
   30   "
  -31    Smoothing a series of Central England Temperatures using an
  -32    exponentially weighted filter.
  -33    To illustrate the end-effect problems of filtering a subset of the
  -34    data is used
  -35   "
   36   VARIATE [VALUES=36...235] Time
   37   CALCULATE Cetemp = Cetave$[Time]
   38   & Tmean = MEAN(Cetemp)
   39   & Mcetemp = Cetemp-Tmean
   40   TSM [MODEL=transfer] Ewma; ORDERS=!(0,1,0,0); PARAMETERS=!(1,0,0.8,0.2)
   41   FILTER Mcetemp; NEWSERIES=Smtemp; FILTER=Ewma
   42   CALCULATE Smtemp = Smtemp+Tmean
   43   FRAME 1; YLOWER=0.2; YUPPER=0.9; XLOWER=0; XUPPER=1
   44   PEN 1,2; METHOD=line; LINE=0,1; SYMBOL=0; COLOUR=1; THICKNESS=0.5,2.0
   45   DGRAPH [TITLE='Smoothed Temperatures'] Cetemp,Smtemp; Time; PEN=1,2
```

In this example the filter is defined by a transfer-function model. Alternatively, you can use an ARIMA model to define a filter, in which case the model pre-whitens the series. Suppose, for example, an AR(1) model is specified, with parameter ϕ_1; the result of applying this to a series x_t is to generate a series a_t:

$$a_t = x_t - \phi_1 x_{t-1}$$

Such an operation is usefully applied to whiten a series before calculating its spectrum, or to whiten a pair of series before calculating their cross-correlation.

12.6.1 The **FILTER** directive

FILTER filters time series by time-series models.

Option

PRINT = *strings*	What to print (series); default *

Parameters

OLDSERIES = *variates*	Time series to be filtered
NEWSERIES = *variates*	To save filtered series
FILTER = *TSMs*	Models to filter with respect to
ARIMA = *TSMs*	ARIMA models for time series

The OLDSERIES and NEWSERIES parameters of FILTER specify respectively the time series to be filtered, and the series that result from filtering. A new series must not have the same identifier as the series from which it was calculated. Genstat interprets any missing values in the old series as zero. But if you use the ARIMA parameter (see below), Genstat replaces them by interpolated values when it calculates the filtered series; the missing values remain in the old series.

The FILTER parameter specifies the TSMs to be used for filtering. If the TSM is a transfer-function model (12.5.1), the new series y_t is calculated from the old series x_t by

$$y_t = \{ \, \omega(B)B^b \, / \, \delta(B)\nabla^d \, \} \, x_t.$$

The filter does not use the power transformation nor the reference constant. This lets you apply a single filter conveniently to a set of time series, for which different transformations and different constants might be appropriate. You can always use the CALCULATE directive to apply a transformation to a series before using FILTER.

If the TSM is an ARIMA model (12.3.1), then the new series a_t is calculated from the old series y_t by

$$a_t = \{ \, \phi(B)\nabla^d \, / \, \theta(B)) \} \, y_t.$$

Note that the TSM does not have to be the model appropriate for y_t. Again, Genstat ignores the parameters λ, c, and σ_a^2; you can set them to 1,0,0, for example.

The ARIMA parameter specifies a time-series model for the old series. The purpose is to reduce transient errors that arise in the early part of the new series: these arise because Genstat does not know the values of the old series that came before those that have been supplied. If you do not use this parameter, then Genstat takes these earlier values to be zero. This can cause unacceptable transients which can only be partially removed by procedures such as mean-correcting the old series. If you do use the ARIMA parameter, then Genstat uses the specified model to estimate (or back-forecast) the values of the old series earlier than those that have been supplied.

You do not have to have a good ARIMA model for the old series in order to achieve worthwhile reductions in the transients. Thus a model with orders (0,1,1) and parameters (1,0,0,0.7) would estimate the prior values to be constant, at a level that is a backward EWMA of the early values of the series. Example 12.6.1a is a continuation of Example 12.6, in which the ARIMA parameter is used. The results are shown in Figure 12.6.1a:

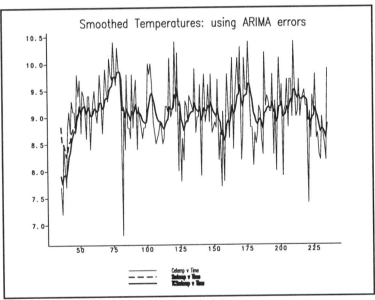

Figure 12.6.1a

the smoothed series, TCSmtemp, fits the series much more closely at the start; the old version of the smoothed series, Smtemp, is also shown on the graph (using a dashed line), to reveal the difference at the start of the series.

Example 12.6.1a

```
50   "
-51   Filter the temperatures using an ARIMA model to reduce the transients.
-52   "
53   TSM Back; ORDERS=!(0,1,1); PARAMETERS=!(1,0,0,0.7)
54   FILTER Cetemp; NEWSERIES=TCSmtemp; FILTER=Ewma; ARIMA=Back
55   PEN 3; METHOD=line; LINE=2; SYMB=0; COLOUR=1; THICKNESS=2
56   DGRAPH [TITLE='Smoothed Temperatures: using ARIMA errors'] \
57      Cetemp,Smtemp,TCSmtemp; Time; PEN=1,3,2
```

For a seasonal monthly time series, an appropriate ARIMA model could have orders (0,1,1,0,1,1,12) and parameters (1,0,0,0.7,0.7). However you must give the supplied model a transformation parameter $\lambda=1$. Any other value for λ breaks the assumption of linearity that underlies the calculations for correcting the transients. The constant term in the ARIMA model can be non-zero, and should be if that is appropriate for the old series. Note that the ARIMA model does not define the filter.

If you specify the ARIMA parameter, Genstat uses this model to interpolate any missing values in the old series before it calculates the new series. Suppose for example that the filter is the identity, defined by a transfer-function model with orders (0,0,0,0) and parameters (1,0,0); then the new series will be the old series with any missing values replaced.

Example 12.6.1b shows how a two-sided filter arises by smoothing the smoothed series a second

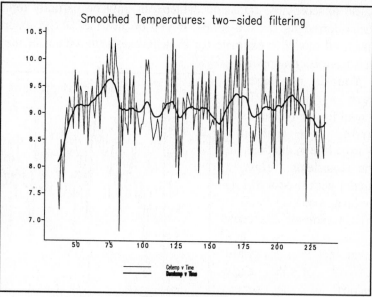

Figure 12.6.1b

time *after* it has been reversed. The ARIMA model has its moving average parameter set to zero because this is appropriate for the series to which the filter is now applied. The result is reversed again and displayed using DGRAPH, see Figure 12.6.1b.

Example 12.6.1b

```
61   "
-62   Carry out two-sided filtering by applying the filter to the
-63   smoothed series in reverse.
```

```
-64   "
 65   CALCULATE Rsmtemp = REVERSE(TCSmtemp)
 66   & Back[2]$[4] = 0
 67   FILTER Rsmtemp; NEWSERIES=Dsmtemp; FILTER=Ewma; ARIMA=Back
 68   CALCULATE Dsmtemp = REVERSE(Dsmtemp)
 69   DGRAPH [TITLE='Smoothed Temperatures: two-sided filtering'] \
 70     Cetemp,Dsmtemp; Time; PEN=1,2
```

12.7 Forming preliminary estimates and displaying models

The ESTIMATE directive (12.3.2) carries out a lot of computation to find the best estimates of the parameters of a time-series model. The amount of computation can be reduced if you provide rough initial values for the parameters, especially when there are many of them. You can get Genstat to do this by using the FTSM directive. FTSM obtains moment estimators of a simple kind, by solving equations between the unknown parameters of the ARIMA or transfer-function model and the autocorrelations or cross-correlations calculated from the observed time series. Sometimes these equations have no solution, or their solution provides values inconsistent with the constraints demanded of the parameters. If so, Genstat sets the corresponding parameters to missing values. The form of the directive is the same for ARIMA and transfer-function models, but the interpretation is slightly different. So we describe the two cases separately.

The TSUMMARIZE directive helps you investigate time-series models by displaying various characteristics. These are the theoretical autocorrelation function of an ARIMA model, and the pi-weights and psi-weights; also the impulse-response function of a transfer-function model. TSUMMARIZE can derive the expanded form of a model, in which all seasonal terms are combined with the non-seasonal term.

12.7.1 Preliminary estimation of ARIMA model parameters

FTSM forms preliminary estimates of parameters in time-series models.

Option

PRINT = *strings*	What to print (models); default *

Parameters

TSM = *TSMs*	Models whose parameters are to be estimated
CORRELATIONS = *variates*	Auto- or cross-correlations on which to base estimates for each model
BOXCOXTRANSFORM = *scalars*	Box-Cox transformation parameter
CONSTANT = *scalars*	Constant term
VARIANCE = *scalars*	Variance of ARIMA model, or ratio of input variance to output variance for transfer model

---

A typical FTSM statement might be

```
FTSM [PRINT=model] Yatsm; CORRELATIONS=Yacf; BOXCOX=Ytran; \
     CONSTANT=Ymean; VARIANCE=Yvar
```

You must previously have declared the time-series model Yatsm to be of type ARIMA with appropriate orders, and lags if you need to specify them. Genstat takes this model to be associated with observations of a time series y_t. The aim of the directive is to set the values of the variate of model parameters equal to preliminary estimates derived from the variate Yacf and scalars Ytran, Ymn, and Yvar.

The variate Yacf should contain sample autocorrelations $r_0 \dots r_m$. You should obtain these from the original time series, stored in variate Y say, by first using the CALCULATE directive to transform Y according to the Box-Cox equations with transformation parameter Ytran (if you do indeed want a transformation). You should then form the differences of the transformed series, according to the degrees of differencing already set in the model; you can use the DIFFERENCE function with the CALCULATE directive for this (4.2.1). Finally, you should use the AUTOCORRELATIONS parameter of the CORRELATE directive (12.1.2) to store the autocorrelations of the resulting series in Yacf. Often you will have done these operations already in order to produce Yacf for selecting a model.

At the same time, you can supply the scalars Ytran, Ymean, and Yvar to set the first three elements of the parameters variate of Yatsm; these cannot be set using Yacf alone. The scalar Ytran should be the parameter used to transform Y, and Genstat will copy it into the first element of the variate of parameters. Genstat will copy the scalar Ymean into the second element, which is the constant term of the model; the recommended value for this is the sample mean of the series from which Yacf is calculated, but you may prefer the value 0. The scalar Yvar is used to set the innovation variance, which is the third element of the variate of parameters. The recommended value is the sample variance of the series from which Yacf is calculated. If you set Yvar to 1.0, then Genstat will set the innovation variance to the variance ratio Variance(e)/Variance(y), as estimated from Yacf according to the model.

If any of the BOXCOX, CONSTANT, or VARIANCE parameters is not set, Genstat will leave unchanged the corresponding value in the variate of parameters of the model. The only exception to this rule is if a parameter is missing. Then Genstat initially sets the transformation parameter to 1.0 (corresponding to no transformation), and the constant to 0.0; the innovation variance is left missing.

12.7.2 Preliminary estimation of transfer-function model parameters

A typical FTSM statement for a transfer-function model might be

```
FTSM [PRINT=model] Xytsm; CORRELATIONS=Xyccf; BOXCOX=Xtran; \
     CONSTANT=Xmean; VARIANCE=Xyvratio
```

You must previously have declared the time-series model Xytsm to be of type transferfunction with appropriate orders, and lags if you need to specify them. Genstat assumes that this model represents the dependence of an output series y_t on an input series x_t in a multi-input model. The directive sets the values of the parameters of the model equal to preliminary estimates derived from Xyccf, Xtran, Xmean, and Xyvratio.

You should put into the variate Xyccf an estimate of the impulse-response function of the model, from which Genstat will derive the parameters. This estimate is usually a sample cross-correlation sequence $r_0 \ldots r_m$ obtained from variates Y and X1 containing observations of y_t and x_t according to one of the following four rules:

(a) In the simple case, the differencing orders of Xytsm are all zero, and you do not want to use any Box-Cox transformation of either y_t or x_t. Then the cross-correlations should be those between variates Alpha and Beta, say, derived from X and Y by filtering (or pre-whitening), as described in 12.6.2. The ARIMA model that you used for the filter should be the same for X and Y, and you should choose it so that the values of Alpha represent white noise.

(b) If the differencing orders of Xytsm are not zero, then before you calculate the cross-correlations you should further difference the series Beta as specified by these orders.

(c) If a Box-Cox transformation is associated with y_t, you should apply it to Y before the filtering. However this transformation parameter must not be associated with Xytsm: you should assign it to the univariate ARIMA model that you have specified for the error term (12.3.2).

(d) If a Box-Cox transformation is associated with x_t, it must be the same as the one you used in the ARIMA model for x_t from which the series Alpha was derived. The scalar Xtran must contain this transformation parameter. Genstat copies it into the first element of the parameter variate of Xytsm. If the Box-Cox parameter is unset, Genstat leaves the transformation parameter of Xytsm unchanged; it is set to 1.0 if it was originally missing.

Genstat copies the scalar Xmean into the second element of the variate of parameters. The recommended value is the sample mean of X after any transformation has been applied. If you do not set the CONSTANT parameter, Genstat leaves the constant parameter of Xytsm unchanged; it is set to 0.0 if it was originally missing.

You use the scalar Xyvratio to obtain the correct scaling of non-seasonal moving-average parameters in Xytsm. All the other autoregressive parameters and moving-average parameters are invariant under scale changes in y_t and x_t. You should set the scalar to the ratio of the sample variances of the variates from which the cross-correlations were calculated; that is, Variance(Beta)/Variance(Alpha). If you do not set this, Genstat uses the value 1.0.

You can use FTSM to go backwards from autocorrelations to the original time-series model. If you apply it to the autocorrelations that were constructed from a time-series model by means of TSUMMARIZE (12.7.3), it will recover the parameters of the model exactly, provided the model is non-seasonal. If the model contains seasonal parameters, with seasonal period s, the parameters will not be recovered exactly, except in one special circumstance: that is, when the non-seasonal part of the model, considered in isolation from the seasonal part, has a theoretical autocorrelation function that is zero beyond lag $s/2$. Otherwise, the non-seasonal and seasonal parts of the model interact, and so Genstat loses accuracy in the recovered parameters. When you use sample autocorrelations, this loss of accuracy tends to be small in comparison with the sampling fluctuations of the estimates. But if s is small, say $s=4$ for quarterly data, the loss could be serious. Exactly the same considerations apply to transfer-function models.

12.7.3 The **TSUMMARIZE** directive

TSUMMARIZE displays characteristics of time series models.

Options

PRINT = *strings*	What to print (autocorrelations, expansion, impulse, piweight, psiweight); default *
GRAPH = *strings*	What to display with graphs (autocorrelations, impulse, piweight, psiweight); default *
MAXLAG = *scalar*	Maximum lag for results; default 0

Parameters

TSM = *TSMs*	Models to be displayed
AUTOCORRELATIONS = *variates*	To save theoretical autocorrelations
IMPULSERESPONSE = *variates*	To save impulse-response function
STEPFUNCTION = *variates*	To save step function from impulse
PIWEIGHTS = *variates*	To save pi-weights
PSIWEIGHTS = *variates*	To save psi-weights
EXPANSION = *TSMs*	To save expanded models
VARIANCE = *scalars*	To save variance of each TSM

For an ARIMA model in the TSM parameter, you can set only the AUTOCORRELATIONS, PSIWEIGHTS, and PIWEIGHTS parameters. Also, you can set the IMPULSERESPONSE parameter only for a transfer-function model. You can set the EXPAND parameter for either type of model. The TSMs in any TSUMMARIZE statement must be completely defined; that is, you must have set the orders and parameters, and the lags if you are using them. The only exceptions are that Genstat takes the transformation parameter to be 1.0 if it is missing, and that the innovation variance of an ARIMA model need not be set.

The MAXLAG option specifies the maximum lag to which Genstat is to do calculations: this applies to autocorrelations, psi-weights, pi-weights, and impulse responses. If MAXLAG is unset, the maximum lag is defined implicitly as the length of the first variate in the parameters. However, if the length of this variate is also undefined, the maximum lag cannot be defined and Genstat reports a fault.

You can set the PRINT and GRAPH options independently of the parameters: these store results, and display the various characteristics of models.

The AUTOCORRELATIONS parameter allows you to store the theoretical autocorrelation function of an ARIMA model. Such a model uniquely defines an autocorrelation function whose values $r_0 \dots r_m$ are assigned by Genstat to the variate R, where m is the maximum lag. If the model has differencing parameters $d=D=0$, then the autocorrelation function is that of a series y_t that follows this model.

If either $d>0$ or $D>0$, then the theoretical autocorrelations are calculated as if $d=D=0$, and so they correspond to those of the differenced y_t series. This is because the autocorrelations of y_t are undefined for non-stationary models.

Example 12.7.3

```
  3  "
 -4   Display the autocorrelations of an AR[2] model.
 -5  "
  6  TSM AR[2]; ORDERS=!(2,0,0); PARAMETERS=!(1,15,2.5,0.5,-0.5)
  7  TSUMMARIZE [MAXLAG=12; PRINT=autocorrelations] AR[2]
```

7 . . . .

```
*** Summary of model AR[2]

        Lag            ACF
         0           1.000
         1           0.333
         2          -0.333
         3          -0.333
         4           0.000
         5           0.167
         6           0.083
         7          -0.042
         8          -0.063
         9          -0.010
        10           0.026
        11           0.018
        12          -0.004
```

The PSIWEIGHTS parameter allows you to store the theoretical psi-weights $\psi_0 \ldots \psi_m$ of an ARIMA model. These are used internally by Genstat when error limits are calculated for forecasts obtained using the model. You will need them for example if you want to calculate the variance of the total of the forecast values up to some specified maximum lead time. They are defined for a non-seasonal model by

$$1 + \psi_1 B + \psi_2 B^2 + \ldots = \theta(B) / \{ \phi(B)\nabla^d \}$$

The PIWEIGHTS parameter allows you to store the theoretical pi-weights $\pi_0 \ldots \pi_m$ of an ARIMA model: these show explicitly how past values contribute to a forecast. The weights are defined by:

$$1 - \pi_1 B - \pi_2 B^2 - \ldots = \{ \phi(B)\nabla^d \} / \theta(B)$$

The IMPULSERESPONSE parameter allows you to store the theoretical impulse-response function, $v_0 \ldots v_m$, of a transfer-function model. This function can help you interpret the model. The sequence is defined for a non-seasonal transfer-function model by:

$$v_0 + v_1 B + v_2 B^2 + \ldots = \omega(B)B^b / \{ \delta(B)\nabla^d \}$$

12.7.4 Deriving the generalized form of a time-series model

For an ARIMA model you can combine into one generalized autoregressive operator all the differencing operators, the non-seasonal autoregressive operators, and the seasonal autoregressive operators. The non-seasonal and seasonal moving-average operators may similarly be combined.

Normally you would want this expanded model to help you understand a series. But you might also want to re-estimate the parameters in the expanded model, to test whether the differencing operators or seasonal factors unnecessarily constrain the structure of the original model.

Example 12.7.4

```
11   "
-12    Expand the seasonal ARIMA model used for modelling the number of
-13    airline passengers in Section 12.3.6.
-14   "
 15   VARIATE [VALUES=0,1,1, 0,1,1,12] Ord
 16   & [VALUES=0,0,0.00143, 0.34, 0.54] Par
 17   TSM Airpass; ORDERS=Ord; PARAMETERS=Par
 18   PRINT Airpass
```

Airpass

Innovation variance 0.001430

```
                     parameter
Transformation          0.
Constant                0.
```

* Non-seasonal; differencing order 1

```
                lag    parameter
Moving-average   1     0.340000
```

* Seasonal; period 12; differencing order 1

```
                lag    parameter
Moving-average  12     0.540000
```

```
 19   TSUMMARIZE [PRINT=expansion] Airpass
```

19...

*** Expansion of model Airpass

*** Autoregressive moving-average model ***

Innovation variance 0.001430

```
                     parameter
Transformation          0.
Constant                0.
```

* Non-seasonal; no differencing

```
                lag    parameter
Autoregressive   1      1.00000
                12      1.00000
                13     -1.00000
Moving-average   1      0.340000
                12      0.540000
                13     -0.183600
```

If you have not previously defined one of the identifiers supplied by the EXPANSION parameter, Genstat will automatically define it to be a TSM, and its component variates will be set up to have the length defined by the corresponding model in the TSM parameter.

The expansion does not change the transformation parameter of the model, nor the constant term, nor the innovation variance. If the model that you have supplied contains non-zero differencing orders, then the generalized model does not satisfy the stationarity constraint on the parameters; neither does the constant term have the same interpretation as it had in the supplied model.

The expansion of transfer-function models exactly parallels that of ARIMA models.

G.T.W.
S.J.W.

13   Customizing and extending Genstat

Genstat is designed so that you can give simple commands to perform standard tasks. The defaults of options and parameters are set to correspond to what are believed to be the commonest needs – as judged by the authors of Genstat. But Genstat can be used in many different applications, and so it is inevitable that our choice of defaults will not correspond with the preferences of every individual user. Section 13.1 describes ways in which you can customize the Genstat command language, making it more convenient for the way in which you work. Section 13.2 describes how to customize the Menu System.

   Similarly, Genstat has been designed to make available many statistical techniques, and to handle data in many different ways. But no computer program can do everything that its users may require, and there will always be techniques that are not provided, if only because new techniques are continually being invented. Consequently, several methods are available to allow you to extend Genstat.

   Firstly, Subsection 13.3.1 describes how to communicate with the operating system of the computer. The SUSPEND directive allows you to issue commands to the operating system while running Genstat. This allows simple tasks such as listing the files in a directory to check where you have stored a set of data, or more complicated tasks like running other programs that may communicate with Genstat through data files. An alternative way of setting up files that can be accessed both by Genstat and by other programs is outlined in 13.3.2.

   The command language described in this Manual allows you to write programs to do many tasks not provided as standard directives. Chapter 5 describes how to extend Genstat in this way, and how you can set up libraries so that these programs can be used automatically – just as if they were part of the standard Genstat language. For complicated tasks, it will almost always take less of your time to write a program in the Genstat language than in a general-purpose language like Fortran, though the Genstat program may use more computing time. However, if a task is to be repeated many times, or if it uses a lot of computer resources, it may be more economical to program at least some of the task in Fortran. Sections 13.3 and 13.4 describe three ways of adding your own source code into Genstat. The best method for a particular application will depend on the complexity of the work done by your programs, how convenient you want to make them to use, and how Genstat has been implemented on your computer.

   The PASS directive, described in 13.3.3, provides a means of communicating between Genstat and external programs which may be written in any computing language. You can thus for example make use of algorithms of your own, or from commercial libraries, to carry out specialized tasks not provided in Genstat, while still benefiting from the data-handling and graphical facilities of Genstat. You do not need to modify Genstat itself in any way in order to use PASS; but PASS may not have been implemented on your computer.

   The OWN directive (13.4.1) is available in all versions of Genstat. A skeleton Fortran subprogram is supplied with Genstat; this is called whenever you give an OWN statement. OWN has a fixed and limited syntax, and consequently you need make only simple modifications to the subprogram – so that it calls your source code and passes the information that it uses, to and from Genstat. To use OWN, you need to relink the Genstat system to include the resulting subprogram and your own code; this is an easy task on most computers, and is documented in the *Installers' Note* distributed with Genstat.

The third method is to define new directives (13.4.4). This method provides the full Genstat syntax for implementing your task, but you need to have more knowledge of the internal workings of Genstat. You will also need to relink Genstat, and have access to the Language-Definition File.

Many of the facilities in this chapter require access to computer files that are distributed with Genstat. Details of these files should be available in the documentation of your implementation of Genstat: an *Installers' Note*, designed for whoever mounts Genstat on the computer, and a *Users' Note* designed for anyone who uses it. Copies of these guides are distributed with Genstat.

When you make your own versions of some of these files, such as the Start-up file or one of the Menu-System files, you will need to be able to tell Genstat to use your version rather than the standard version. This may be done by modifying an environmental or logical variable, depending on your operating system. For convenience, we use the term *environmental variable* in this chapter, and give the names relevant for Release 3.1. The names change in a simple way from release to release to make it easy to set up more than one release of Genstat on a computer. Again, the details are in the guides.

13.1   The Genstat environment

The output from the examples in this manual so far was produced in the standard environment. For example, the Genstat statements were not echoed with line numbers when a program was run interactively, but they were when it was run in batch; newlines in the programs were taken as terminators of statements unless a continuation symbol was given; upper-case and lower-case letters were treated as distinct in identifiers. You can change these and other details of the environment of a job by the SET directive (13.1.1). It is also possible to find out the current environment, using the GET directive (13.1.2). This is of most use inside procedures that are designed to work in a general way.

All the definitions of Genstat directives in this manual include details of the default settings of options and parameters. Similarly, the Procedure Library Manual includes the same information for standard procedures. However, you can redefine these defaults at any time with the SETOPTION and SETPARAMETER directives (13.1.3).

It is unlikely that you would want to work with different Genstat environments on different occasions: it could be very confusing. The ideal way to modify the environment is in a start-up file, which is automatically executed whenever you start using Genstat (13.1.4).

13.1.1 The **SET** directive

SET sets details of the "environment" of a Genstat job.

Options

INPRINT = *strings*
Printing of input as in PRINT option of INPUT (statements, macros, procedures, unchanged); default unch

OUTPRINT = *strings*
Additions to output as in PRINT option of OUTPUT (dots, page, unchanged); default unch

DIAGNOSTIC = *string*
Defines the least serious class of Genstat diagnostic which should still be generated (messages, warnings, faults, extra, unchanged); default unch

ERRORS = *scalar*
Number of errors that a job may contain before it is abandoned (0 implies no limit); default is to leave unchanged

FAULT = *text*
Sets the Genstat fault indicator (for example, FAULT=* clears the last fault); default is to leave the indicator unchanged

PAUSE = *scalar*
Number of lines to output before pausing (interactive use only; 0 implies no pausing); default is no change

PROMPT = *text*
Characters to be printed for the input prompt; default is to leave unchanged

NEWLINE = *string*
How to treat newline (significant, ignored); default is no change

CASE = *string*
Whether lower- and upper-case (small and capital) letters are to be regarded as identical in identifiers (significant, ignored); default is no change

RUN = *string*
Whether or not the run is interactive (interactive, batch); by default the current setting is left unchanged

UNITS = *identifier*
To (re)set the current units structure; default is to leave unchanged

BLOCKSTRUCTURE = *identifier*
To (re)set the internal record of the most recent BLOCKSTRUCTURE statement; default is to leave unchanged

TREATMENTSTRUCTURE = *identifier*
To (re)set the internal record of the most recent TREATMENTSTRUCTURE statement; default is to leave unchanged

COVARIATE = *identifier*
To (re)set the internal record of the most recent COVARIATE statement; default is to leave unchanged

ASAVE = *identifier*
To (re)set the current ANOVA save structure; default is to

	leave unchanged
DSAVE = *identifier*	To (re)set the current save structure for the high-resolution graphics environment; default is to leave unchanged
RSAVE = *identifier*	To (re)set the current regression save structure; default is to leave unchanged
TSAVE = *identifier*	To (re)set the current time-series save structure; default is to leave unchanged
VSAVE = *identifier*	To (re)set the current REML save structure; default is to leave unchanged

No parameters

The default of SET is to do nothing: that is, each option by default leaves the corresponding attribute of the environment unchanged. Of course you have to start somewhere, so an initial environment is defined at the start of any Genstat program; the corresponding initial settings of the options of SET, known as the *initial defaults*, are described below.

The INPRINT option controls what parts of a Genstat job supplied in the current input channel are recorded in the current output file; the input channel can be either an input file or the keyboard. Three parts are distinguished: explicit statements; statements, or parts of statements, that you have supplied in macros using either the ## notation (1.9.2) or the EXECUTE directive (1.9.2); and statements that you have supplied in procedures. The initial default is to record nothing if the output is to the screen, otherwise to record the statements. This aspect of the environment can be modified also by the PRINT option of the INPUT directive (3.4.1) and by the INPRINT option of JOB (5.1.1).

The OUTPRINT option controls how the output from many Genstat directives starts: the output can be preceded by a move to the top of a new page, or by a line of dots beginning with the line number of the statement producing the analysis, or by both. If output is directly to the screen, no new pages are given. The initial default is to give neither if output is to the screen, otherwise to give a new page and a line of dots. Alternatively, this aspect can be modified by the PRINT option of the OUTPUT directive (3.4.3) or by the OUTPRINT option of JOB. The lines of dots are produced by the directives for regression analysis, analysis of designed experiments, REML analysis, multivariate analysis, and time series; also from the FLRV, FSSPM, and SVD directives (4.10.1). If you give an analysis statement within a FOR loop (5.2.1), the line number preceding the line of dots is that of the ENDFOR statement rather than of the analysis statement. New pages are produced with any of the above, and with the GRAPH, HISTOGRAM, and CONTOUR directives (6.1).

The DIAGNOSTIC option lets you control the level of diagnostic reporting. You might want to do this within a procedure, to prevent faults being reported to a user who does not need to know in detail what is going on inside the procedure. By initial default, all diagnostics – messages, warnings, and faults – are printed. You can switch off messages by setting DIAGNOSTIC=warning, or switch off both messages and warnings by setting DIAGNOSTIC=fault. If you set DIAGNOSTIC=*, then no diagnostics will appear. The extra setting gives you extra information, in the form of a dump of the current state of the job; but this is likely to be useful only for developers of Genstat. Printing of diagnostics can also be controlled by the DIAGNOSTIC option of JOB.

The ERRORS option controls what Genstat does when many faults happen within a single job while in batch mode. By initial default, up to five errors per job are reported, and successive faults will not generate diagnostic messages. This ensures, for example, that input intended to be read by a READ statement will not generate many lines of diagnostics if execution halts because of a fault before the READ statement. Note, however, that this option does not affect the detailed error messages printed by the READ directive itself: these are controlled separately by the corresponding ERRORS option of READ. In interactive mode, the count of errors is restarted after each successful statement is issued, though the option is unlikely to be useful in this mode.

The FAULT option is provided primarily to allow procedure writers to modify the internal record that is kept of the most recent fault indicator. Setting FAULT=* clears the record; you can then use the GET directive to ascertain whether a fault has occurred since the record was cleared. You can also set the fault indicator to a particular diagnostic, for example

        SET [FAULT='VA4']

A subsequent DISPLAY statement (5.3.2) will then report the chosen fault in the standard way. The fault indicator is automatically cleared at the start of each job.

The PAUSE option lets you specify how many lines of output are produced at a time; you might, for example, want to read the output on a terminal screen before more output replaces it. Obviously this is relevant only in interactive mode, and may not be needed in the implementations of Genstat that provide a scrollable output window. By initial default, all output is sent to the current output channel as soon as it is available. Some computers can store the output, irrespective of whether Genstat itself has a scrollable window, and let you scroll forward and back to read it at leisure: others just provide keys to freeze the output while you are reading a section, and then to continue to the next segment of output. If you set PAUSE=n, then after every n lines of output Genstat gives a prompt:

        Press RETURN to continue

After you have read the displayed section of output, you can press the <RETURN> key to get the next n lines. The counting of lines is restarted each time you give a statement from the keyboard: it is not restarted between separate statements in a macro, procedure, or auxiliary input channel. If you have specified that Genstat should echo input lines, these are included among the n. Once all the output has been displayed, Genstat prompts for further statements.

The PROMPT option specifies the characters used to prompt for interactive input. The initial default is the greater-than character followed by a space "> ". The prompt can also be modified by the PROMPT option of JOB. Other prompts are used by READ, EDIT, HELP, and QUESTION, and these cannot be altered.

The NEWLINE option allows you to cancel the initial default whereby a newline (<RETURN>) is a terminator both for strings within a string list (1.6.2) and for a statement (1.8). Thus, for example, if you specify

```
SET [NEWLINE=ignored]
```

you need no longer use a backslash (\) to continue a statement onto a new line, since <RETURN> is no longer interpreted as the end of a statement. But you will then have to terminate each statement explicitly with a colon.

The CASE option specifies whether upper-case and lower-case letters are to be treated as the same in identifiers. The initial default is that upper and lower case are not the same; thus, an identifier X is distinct from an identifier x. If CASE is set to ignored, then in later statements, both x and X are treated as the same identifier, X. Thus the structure with identifier x cannot be referenced, unless CASE is later reset to significant.

The RUN option controls whether Genstat interprets the program as being in batch or in interactive mode; this assumed mode is independent of whether the program really is being run in batch or interactively. Initially, a program is taken to be in interactive mode only if the first input channel and the first output channel are both connected to a terminal. The setting of the assumed mode has two effects – on recovery from faults, and on how HELP (1.3.1) and EDIT (4.7.2) operate.

The UNITS option provides another way of setting the *units structure* in addition to the UNITS directive described in 2.3.4. The setting can be the identifier of a variate or text structure; this will become the default labelling structure of other variates, texts, or factors with the same length, in those directives that use such labels. The setting can also be a scalar to specify the default number of units. For further details, see 2.3.4. The setting of the UNITS option is lost at the end of each job within a program.

The last eight options of the SET directive specify special *save structures* for the graphical and analysis directives described in Chapters 6, 8, 9, 10, and 12. You can set the options only to an identifier that you have previously established by the SPECIAL option of the GET directive (13.1.2) or by the SAVE options in the various analysis directives themselves. For example, if two sets of regression analyses are in progress in one job, the SET directive can be used to switch between them:

```
MODEL [SAVE=S1] Y1
FIT X1
MODEL [SAVE=S2] Y2
FIT X1
SET [RSAVE=S1]
FIT X1,X2
```

This program fits the regression of Y1 on X1, using save structure S1, then the regression of Y2 on X1 with save structure S2. Finally, the regression of Y1 on X1 and X2 is fitted, because the current regression save structure is changed to S1 before the last FIT statement.

The settings of these last eight options are lost at the end of a job.

13.1.2 The GET directive

The GET directive allows you to access the current settings of the environment. This can be particularly useful in procedures, when details of the environment may need to change and be reset later to their original state. Sometimes it may be sufficient just to use the PRESERVE option of the PROCEDURE directive (5.3.2) for this purpose, but this causes them to be reset only at the end of a procedure. Note also that information about the environment can be displayed, though not saved, with the HELP directive (1.3.1).

GET accesses details of the "environment" of a Genstat job.

Options

ENVIRONMENT = *pointer*	Pointer given unit labels `'inprint'`, `'outprint'`, `'diagnostic'`, `'errors'`, `'pause'`, `'prompt'`, `'newline'`, `'case'`, and `'run'` used to save the current settings of those options of SET; default *
SPECIAL = *pointer*	Pointer given unit labels `'units'`, `'blockstructure'`, `'treatmentstructure'`, `'covariate'`, `'asave'`, `'dsave'`, `'rsave'`, `'tsave'`, and `'vsave'`, used to save the current settings of those options of SET; default *
LAST = *text*	To save the last input statement; default *
FAULT = *text*	To save the last fault code; default *
EPS = *scalar*	To obtain the value of the smallest x (on this computer) such that $1+x > 1$; default *
NJOB = *scalar*	Number of the current job within the program; default *

No parameters

The ENVIRONMENT and SPECIAL options of GET are used to access and save the current settings of the options of the SET directive (13.1.1). The options of SET are divided into two groups. Those that apply to the general environment can be saved using the ENVIRONMENT option: these are INPRINT, OUTPRINT, DIAGNOSTIC, ERRORS, PAUSE, PROMPT, NEWLINE, CASE, and RUN. Those that apply only to the save structures associated with particular directives are saved by using the SPECIAL option: these are UNITS, BLOCKSTRUCTURE, TREATMENTSTRUCTURE, COVARIATE, ASAVE, DSAVE, RSAVE, TSAVE, and VSAVE.

When you use the ENVIRONMENT option, Genstat sets up a pointer (2.6) with units identified by the labels of the corresponding options of SET: these labels are `'inprint'`, `'outprint'`, and so on. Note that these labels must all be specified in lower case. Each unit of this pointer contains one or more strings, or a scalar, to represent the current setting. Thus, the statement

        GET [ENVIRONMENT=Env]

would set up a pointer called Env with elements Env[`'inprint'`], Env[`'outprint'`], and so on. Each element can also be referred to by its position in the pointer; for example, Env[`'inprint'`] is the same as Env[1]. Example 13.1.2 shows what Env would contain in a batch run where the options of SET had not been changed from their default values.

Example 13.1.2

```
  1   GET [ENVIRONMENT=Env]
  2   PRINT [RLWIDTH=18; SQUASH=yes] Env[]; FIELD=18
```

Env['inprint']	statements		
Env['outprint']	dots		
Env['diagnostic']	messages	page	
Env['errors']	5	warnings	faults
Env['pause']	0		
Env['prompt']	>		
Env['newline']	significant		
Env['case']	significant		
Env['run']	batch		

Thus you do not have to know how the environment has been set in order to change it and then restore it; you can use GET to find out about it, and SET to change it back. For example, suppose that you wanted to stop temporarily the echoing of statements to the output file in a batch program. In the following program the first SET statement cancels the echoing, if indeed any echoing is in progress, and the second restores echoing to what it was before the first SET.

```
GET  [ENVIRONMENT=Env]
SET  [INPRINT=*]
```

(more statements)

```
SET  [INPRINT=#Env['inprint']]
```

The SPECIAL option similarly sets up a pointer to save its information. The unit labels of the pointer are 'units', 'blockstructure', and so on. The first element of the pointer is the units structure, or, failing that, the number of units if you have defined it for the current job. Printing the contents of the other elements is not usually informative, as the information is stored in coded form. The last eight elements of the pointer allow you to access the special save structures in the graphical and analysis directives of Chapters 6, 8, 9, 10, and 12. They are most useful for recovering information about an analysis when you have not already specified an explicit save structure. (Otherwise you would have to do the analysis all over again.) For example, in the statements at the end of 13.1.1, if you had not set the SAVE option in the first MODEL statement, you could instead put

```
MODEL Y1
FIT X1
GET  [SPECIAL=S1]
MODEL [SAVE=S2] Y2
FIT X1
SET  [RSAVE=S1['rsave']]
FIT X1, X2
```

Moreover, the SPECIAL option of GET provides the only way of accessing the save structures associated with the analysis-of-variance directives BLOCKSTRUCTURE (9.2.1), COVARIATE (9.3.1), and TREATMENTSTRUCTURE (9.1.1). In other words, there is no SAVE alternative here.

The LAST option is used to save the latest statement that you have input. You can then give the statement again later in the job without having to retype it, though some implementations of Genstat provide a simpler recall facility using the cursor keys. The option has the same effect as setting up a macro (1.9.2) containing a single statement, and is accessed in the same way. For example, the statements

```
PRINT [SERIAL=yes; IPRINT=*; SQUASH=yes] !t('New Data'),Y
GET [LAST=Prdat]
```

(statements)

```
READ Y
```

(data)

```
##Prdat
```

would print the data, Y, under the title New Data and save the PRINT statement in a text called Prdat. After the next data set is read, the heading New Data and the new data set are printed in the same format as the previous data set. (The options of PRINT are described in 3.2.)

The FAULT option is used to save the last fault code as a single string in a text structure. (A complete list of fault code definitions is available from the HELP environment facility.) This option is particularly useful in procedures, in combination with the DIAGNOSTIC and FAULT options of SET, to control the printing of diagnostics.

The EPS option is used to obtain the smallest number, ε, such that $1.0+\varepsilon$ is recognized by your computer to be greater than 1.0; this is an indication of the precision of the computer, which can affect the behaviour of some of the algorithms used by Genstat. EPS can be used, for example, when testing for convergence of iterative algorithms.

The NJOB option provides the current job number within the Genstat program. It is used in the start-up file (13.1.4) to distinguish between statements to be executed just at the start of the program, and those to be executed at the start of each job.

13.1.3 Changing the defaults of options and parameters

SETOPTION sets or modifies defaults of options of Genstat directives or procedures.

Option

DIRECTIVE = *string*   Directive (or procedure) to be modified

Parameters

NAME = *strings*   Option names
DEFAULT = *identifiers*   New default values

SETPARAMETER sets or modifies defaults of parameters of Genstat directives or procedures.

Option

DIRECTIVE = *string*   Directive (or procedure) to be modified

Parameters

NAME = *strings*   Parameter names

DEFAULT = *identifiers*   New default values
_____

These directives change the defaults settings for the specified directive or procedure for the remainder of the current job. If you use one of these directives in your start-up file (13.1.4) you can make the changed default apply in all your use of Genstat.

To achieve any effect, the option and both parameters of either of these directives must be set. The DIRECTIVE option specifies the name of the directive or procedure that is affected, and the NAME parameter indicates the option or parameter whose default is to be changed. The settings are strings, so need not be quoted because all directive and procedure names are valid as unquoted strings. The DEFAULT parameter is then set to a data structure to provide the new default that you want to be assumed. For example, the following statement modifies the PRINT option of the FIT directive which carries out regression analysis (8.1.2).

        SETOPTION [DIRECTIVE=FIT] PRINT; DEFAULT='deviance'

The usual default of the PRINT option in FIT is to print a statement of the model, a summary of the analysis, and the parameter estimates: this corresponds to the setting PRINT=model,summary,estimates. This SETOPTION statement therefore redefines the default so that any subsequent FIT statement in the job will report only the residual deviance unless you explicitly set the PRINT option.

The defined mode of the PRINT option of FIT is "strings" (8.1.2). However, the DEFAULT parameter of SETOPTION expects a data structure (to allow for all the other modes that might occur), and so it must be set to a text structure containing the string (or strings) that you want to be the default. Similarly, if the defined mode of the option or parameter is "numbers", "expression", or "formula", you must supply a variate, an expression structure, or a formula structure containing the new default. If the defined mode is "identifier", the setting of DEFAULT is simply an identifier, which must be of the required type if this is specified in the definition of the directive or procedure.

To reset the PRINT option of FIT back to its usual default, you would need to give the statement

        SETOPTION [DIRECTIVE=FIT] PRINT; DEFAULT=!t(model,summary,\
          estimates)

The SETOPTION and SETPARAMETER directives can also be used to change defaults of any procedure: this may be a procedure in the standard Procedure Library, the Site Library, or a personal library that you have already opened in the current program, or it may be a procedure that you have defined explicitly in the job.

For example, we shall modify the default action of the DESCRIBE procedure in the standard Library. This procedure prints a summary of values in a variate, and by default does not store the summary. It has a parameter called SUMMARIES which you can set if you want the summaries to be stored. The statement

        SETPARAMETER [DIRECTIVE=DESCRIBE] SUMMARIES; DEFAULT=Sum

would change the default action of this procedure, so that after using it without setting the SUMMARIES parameter the summaries would be available in variate Sum.

13.1.4 Start-up files

A start-up file contains Genstat statements that are to be executed at the beginning of every job. Thus in an interactive run they are executed before Genstat prompts you for commands, and in batch before Genstat executes the statements that you have prepared. The standard start-up file, distributed with Genstat, performs three tasks: it defines filenames and channel numbers to control the Menu System (1.1.3); it prints a banner describing the version of Genstat that is being used, and listing the basic commands like HELP and MENU; and it opens a file to keep a record of interactive sessions. You can set up your own start-up file and arrange for it to be executed instead, to define your preferred Genstat environment automatically at the start of each job.

The standard start-up file contains job-control structures to allow separate sets of statements to be executed in batch and interactive modes, and at the start of the first job and subsequent jobs in the program. In fact, little is done in batch mode because none of the three tasks listed above are relevant: the Menu System cannot be used, the banner is designed to help interactive users, and there is no need to keep a record of statements. When running interactively, the record file is opened at the start of the first job; once open, there is no need to open it at the start of subsequent jobs. Similarly, the banner is printed just at the start of the first job. By contrast, the menu definitions are needed in each job, since the data structures that store them are lost when a new job is started.

You can take a copy of the standard file and edit it – perhaps just to remove the banner, or perhaps to insert a SET statement to change the environment. If you prefer an alternative default for an option or parameter of a directive that you use frequently, you might want to insert a SETOPTION or SETPARAMETER statement; if so, you must put it into the first section of the file, which is executed in both modes and at the start of all jobs. You might also include an OPEN statement to provide automatic access to a personal procedure library or backing-store file; or you could define macros to carry out operations you require frequently.

Having created your own start-up file, you can arrange for it to be used in place of the standard one by resetting Genstat's environmental variable G31START as described in the *Users' Note*.

13.2 Extending the Menu System

The standard Menu System, illustrated in 1.3.1, allows people to use many standard techniques in Genstat without having to learn the command language. Furthermore, the Menu System is designed to be extendable, so that knowledgeable users can modify the standard system, or construct completely new systems, for the convenience of themselves or their colleagues.

Section 13.2.1 describes the QUESTION directive, which is the basic tool used in any menu system to prompt a user for information, possibly by choosing from a menu. The standard system consists of several files of Genstat statements, including many QUESTION statements that ask the user for information and perform tasks from a specific repertoire. The method for setting up such systems is described in 13.2.2. The files that make up the system can be copied and modified as required, or translated into other languages, or just referred to as examples when constructing more specialized or more complicated systems.

13.2.1 Defining menus

QUESTION obtains a response using a Genstat menu.

Options

PREAMBLE = *text*	Text posing a question; (no default)
PROMPT = *text*	Text to be used as final prompt; the default prompt specifies the mode of response and lists the default values (if any), in brackets, followed by ">"
RESPONSE = *identifier*	Structure to store response; default * allows a menu to be saved without being executed
MODE = *string*	Mode of response (e, f, p, t, v); default p
DEFAULT = *identifier*	Response to be assumed if just <RETURN> is given; default is to repeat the prompt until a response is obtained
LIST = *string*	Whether a list of responses, rather than a single response, is valid (yes, no); default no
DECLARED = *string*	Whether identifiers must already be declared (yes, no); default no
TYPE = *strings*	Allowed types for identifiers (datamatrix i.e. pointer to variates of equal lengths as required in multivariate analysis, diagonalmatrix, dummy, expression, factor, formula, LRV, matrix, pointer, RSAVE, scalar, SSPM, symmetricmatrix, table, text, TSAVE, TSM, variate); default * meaning no limitation
PRESENT = *string*	Whether the identifier must have values (yes, no); default no
LOWER = *scalar*	Lower limit for numbers; default *, meaning no check
UPPER = *scalar*	Upper limit for numbers; default *, meaning no check
HELP = *text*	Text to be used in response to a general query for the question; default *
SAVE = *pointer*	Saves or reinputs the specification of the menu (which is then used for any options or parameters not redefined)

Parameters

VALUES = *scalars* or *texts*	Possible codes for MODE t; (no default for MODE t; not relevant for others)
CHOICE = *texts*	Text giving explanation of each letter code; (no default for MODE t; not relevant for others)
HELP = *texts*	Text to be used in response to a specific query for a code; default *

The QUESTION directive displays a Genstat menu and obtains a response when in interactive mode. In batch, the directive does nothing. Here is a simple example that asks the user to provide the identifier of a variate structure; this statement is actually in the part of the standard system that provides analysis of variance.

```
QUESTION [PREAMBLE=!t('Y-VARIATE Menu (from ANOVA Menu)',*, \
    'What is the variate to be analysed ?'; RESPONSE=_yvar; \
    DECLARED=yes; TYPE=variate; PRESENT=yes]
```

This statement displays the following Genstat menu:

Example 13.2.1a

```
Y-VARIATE Menu (from ANOVA Menu)

What is the variate to be analysed ?

Identifier >
```

The PREAMBLE option specifies a text structure, whose contents are printed at the beginning of the menu. Following this is the prompt: by default, this consists of a reminder of what type of answer is expected, followed by the greater-than symbol (>). However, there is a PROMPT option that allows any text to be printed instead, before the greater-than symbol.

The RESPONSE option specifies a dummy identifier that will point to the answer given by the user. Note that the identifiers used in the standard menu system all begin with the underline character (_) to reduce the chance of a clash with your own identifiers. Menus can request information in one of five modes; the default is Mode p (pointer), as here, and expects a response to consist of an identifier; but the MODE option can also be set to v (variate), t (text), e (expression), or f (formula). When a correct answer has been received, an unnamed structure of the relevant type (pointer, variate, or whatever, but see later for text mode) is set up, and the dummy in the RESPONSE option is set to point at this unnamed structure.

Thus, if you give the identifier Y in response to the question above, the dummy _yvar will store the identifier of a pointer containing the single identifier Y after the QUESTION statement above has been executed. So in the standard menu system, the statement following this QUESTION statement is

```
ANOVA #_yvar
```

the hash (#) being needed to substitute the values of the unnamed pointer that is stored in the dummy structure _yvar.

By default, a question will expect to receive a single item of the specified mode: identifier, number, string, expression, or formula. However, if the option LIST is set to yes for modes p, v, or t, then a list of items is expected. The unnamed structure set up to store the answer will then contain as many values as there are items in the list.

The other three options in the example above specify restrictions on the answer that will be accepted. The DECLARED option specifies that the identifier must be of a structure that has already been declared. If a previously unused identifier is given, the QUESTION statement will print a warning, and issue the prompt again. Similarly, the TYPE option specifies what type of structure is acceptable; the setting may be a list of types if relevant. The PRESENT option specifies that the structure must already have values. Two further options, LOWER and UPPER, can be used to specify limits for numbers given in response to questions of mode v.

Most menus in the standard system are of mode t, and resemble more closely what most people think of as a menu than does the simple display above. Such menus require extra information to be specified using parameters of the QUESTION directive. The VALUES parameter should be set to a list of text structures, each of which stores a single string that is to be accepted as an answer to the question. The CHOICE parameter should be set to another list of text structures, each storing a single string that is to be displayed by the side of the corresponding code in the menu to explain it. Example 13.2.1b, shows a further question from the standard system, generated by the statement:

```
QUESTION [PREAMBLE=!t('INPUT menu',*, \
  'Where are the data values ?'); RESPONSE=_cdsourc; \
  MODE=t; DEFAULT='b'] VALUES='b','s','t'; CHOICE= \
  'in a binary file previously set up using Genstat', \
  'in a character-type file, with values separated by spaces', \
  'to be typed at the terminal'
```

Example 13.2.1b

```
INPUT menu

Where are the data values ?

b          in a binary file previously set up using Genstat
s          in a character-type file, with values separated by spaces
t          to be typed at the terminal

Code (b,s,t; Default:  b) >
```

The codes must obey the rules for unquoted strings (1.5.2): that is, they must start with a letter and consist only of letters and digits. Only the first eight characters will be displayed, and only the first eight characters of the answer will be checked – all eight must match. Usually, of course, it is convenient to use single-letter codes.

Note that mode t cannot be used to ask the user for an arbitrary string, for example to provide a label for output. To request such information, you must use mode p, and set TYPE=text; the user must then supply the string in quotes, or supply the identifier of a text structure that already stores the string.

The response to a question of mode t is stored not as a text, but as a variate each value of which is the number of the corresponding code as listed in the VALUES parameter. Usually, of course, a menu of mode t will be set with LIST=no, the default, and so the variate will contain only a single number. This can be used to control subsequent action in the menu system, most conveniently with a CASE statement. For example, the statements in the standard system following the above QUESTION statement could look like the following.

```
CASE _cdsourc
```

" Statements to deal with code b "

```
OR
```

" Statements to deal with code s "

```
OR
```

" Statements to deal with code t "

```
ENDCASE
```

The DEFAULT option is used here to specify a default answer if the user just types RETURN; it can be set for any mode of question. The HELP option and parameter of the QUESTION directive allow you to provide help text to guide the person answering the question. The SAVE option allows you to declare a menu without executing it, and also to execute a menu that has already been stored (see the standard Base Menu File, G3MNBASE, for an example).

13.2.2 Extending the system

The standard Genstat Menu System Release 3[1], consists of eleven files of Genstat statements, together with some definitions in the standard Start-up File, including the procedure MENU that enters the system. The system will expand in later releases, so only an outline is given here. When the statement

```
MENU
```

is given, the procedure changes input channel to the file G3MNBASE. This contains the QUESTION statement that displays the Base Menu. There is then a CASE statement which results in a further change of input channel to one of the files G3MNINPU (input), G3MNCALC (calculations), G3MNTABU (tabulation) or G3MNPICT (pictures), or poses a further question to discover what type of statistical analysis is required. The latter question can change input channel to one of the files G3MNREGR (regression), G3MNANOV (analysis of designed experiments), G3MNMULS (multivariate analysis based on SSPs), G3MNMULD (multivariate analysis base on distances), G3MNTIME (time series), or G3MNTEST (simple statistical tests).

The definitions in the Start-up File are designed to allow modifications – both for extension of the system, and for changes necessary for some implementations of Genstat. For example, the filenames used in the Genstat OPEN statements that pass control to the various menu files above are intended for operating systems like VMS that have the concept of an environmental variable (see the introduction to this chapter). On systems without this concept, the Start-up File can be amended to supply specific names for the files, to be used throughout the system.

The provision of the Genstat Menu System involves no additional overheads to the command-mode operation of Genstat beyond the extra Fortran source code required to interpret the QUESTION directive. Extension of the system, or addition of alternative systems, will have no effect on the operation of Genstat in command mode. The only overhead is the time spent in designing the extensions or additions, and the storage-space required for the resulting files of Genstat statements.

To extend the standard system, someone familiar with the command language should have no difficulty in modifying the existing QUESTION statements to provide extra branches. The best way to find out how to do this is to look at the standard files used by the Menu System. All you need to do is to take a copy of the file controlling the part of the system that you want to modify, edit, and add to the Genstat statements in that file, and then set the relevant environmental variable so that your version of the file rather than the standard version will be used by Genstat.

To add an additional menu system requires the provision of files of commands constructed in the same way as the Standard Menu System. There should be a base file, that will be entered from the MENU procedure, and this should contain commands to switch input to other files as necessary. The construction of the standard Base Menu File should be closely followed. For a simple addition, the alternative base file may be enough. For a complicated system, it may be preferable to have several levels of files in the hierarchy – though the number of levels is limited by the number of input files that can be open simultaneously (usually five).

The MENU procedure is defined in the Procedure Library in such a way as to allow additional systems to be added without the need for modifications to this file. For example, assume that the file G3MNBAS1 is a new base menu file. The user then needs to type

        MENU ['G3MNBAS1']

to enter the new system instead. The statement

        MENU

will enter the standard system as usual. For reference, the definitions of file names and channel numbers are given below, as in the Start-up File.

```
" Set up texts containing logical names of standard menu and record
  files. "
TEXT _flmnbas,_flmnsub,_flcomnd,_flreslt,_flstore; \
  VALUES='G3MNBASE',!t(G3MNINPU,G3MNCALC,G3MNTABU,G3MNPICT, \
  G3MNREGR,G3MNANOV,G3MNMULS,G3MNMULD,G3MNTIME,G3MNTEST), \
  'G31COMND','G31RESLT','G31STORE'
" Set up scalars with channel numbers for input, output and
  backing-store. "
SCALAR _chmnbas,_chmnsub,_chcomnd,_chreslt,_chstore; VALUE=4,3,5,4,5
```

13.3   Communicating with other programs

Genstat is designed as a general statistical package, and so contains facilities for most of the statistical methods that you may need; but there are many other possible requirements. Some of these will use information that you can produce with Genstat; others will generate information that you can analyse with Genstat. Therefore you need to be able to connect Genstat to other programs.

One example involves data bases. These are collections of information on a computer, structured to allow convenient access using a tailored command language. If you want to use Genstat to analyse data from a data base, you will need to extract the information using the data-base language, analyse it with Genstat, and then perhaps add some of the results to the data base.

The simplest method of communication between programs is via files. One program might extract data and store it in a file – either in character form or in binary (3.6.3, 13.3.2); this program might be written in the data-base language. A second program, written in the Genstat language, might then read that file, process the data, and perhaps form another file; and so on.

If you are content to work step-by-step, first running one program then another, you can simply use the facilities described in Chapter 3 for storing and accessing information in files. However, you may prefer to run programs concurrently. The SUSPEND directive, described in 13.3.1, allows you temporarily to halt Genstat and to do other work on the computer before continuing to run Genstat. You can arrange to communicate information between Genstat and other programs that are run while Genstat is halted, simply by reading and writing files as outlined in Section 13.3.2.

The PASS directive, described in 13.3.3, works like SUSPEND, but is designed to deal automatically with the transfer of information between Genstat and separate programs. You can set up the programs using Fortran, incorporating a Fortran subprogram that is distributed with Genstat to deal with communication, or use some other computing language able to read the data file used to transfer the information.

13.3.1 Giving commands to the operating system

SUSPEND suspends execution of Genstat to carry out commands in the operating system. This directive may not be available on some computers.

Options

SYSTEM = *text*	Commands for the operating system; default: prompt for commands (interactive mode only)
CONTINUE = *string*	Whether to continue execution of Genstat without waiting for commands to complete (yes, no); default no

No parameters

The SUSPEND directive may not be implemented on all the types of computers for which Genstat is available. To find out, you can either read the *Users' Note* (see the introduction to this chapter), or type

    SUSPEND

This will produce a message saying either that Genstat has been suspended, or that SUSPEND is not implemented. In the latter case, it is not possible to communicate between Genstat and other programs.

If SUSPEND is implemented, after the message you will get a prompt 'SUSPEND>' for an operating-system command. You will also be told what command in the operating system will return you to Genstat. For example, with the VMS system on VAX computers you type

```
LOGOUT
```

Example 13.3.1a uses SUSPEND with the VAX/VMS operating system. Genstat puts some data in a file, then another program called PROC processes the data and forms another file, and finally Genstat accesses the new data.

Example 13.3.1a

```
> OPEN 'Result1'; CHANNEL=2; FILETYPE=output
> PRINT [CHANNEL=2] Data[]
> CLOSE 2; FILETYPE=output
> SUSPEND


You have suspended Genstat and you can now give commands to
the operating system after the prompt SUSPEND>
Type LOGOUT to return to Genstat.
SUSPEND> PROC Result1 Result2
SUSPEND> LOGOUT
  Process GENSTATUSER logged out at 16:05 on 04-MAR-1993

> OPEN 'Result2'; CHANNEL=2; FILETYPE=input
> READ [CHANNEL=2; END=*] Data[]
```

An alternative, perhaps faster, way of creating files for communication is outlined in 13.3.2.

The SYSTEM option of SUSPEND allows you to give operating-system commands without explicitly returning to the operating system; Example 13.3.1b performs the same task as Example 13.3.1a.

Example 13.3.1b

```
> PRINT [CHANNEL=2] Data[]
> CLOSE 2; FILETYPE=output
> SUSPEND [SYSTEM='PROC Result1 Result2']

> OPEN 'Result2'; CHANNEL=2; FILETYPE=input
> READ [CHANNEL=2; END=*] Data[]
```

If you do not want to wait for an operating-system command to be executed before returning to Genstat, you can set the CONTINUE option together with the SYSTEM option. You can do this only if a single operating-system command is given; that is, if the text set in the SYSTEM option contains only one string. This would not be a good idea in the above example, because the result of the processing program is required by the Genstat statement following the SUSPEND statement, but here is an example where an output file is entered in a line-printer queue during a Genstat program:

```
OPEN 'Result1'; CHANNEL=2; FILETYPE=output
PRINT [CHANNEL=2] Data[]
CLOSE 2; FILETYPE=output
```

```
SUSPEND [SYSTEM='PRINT/QUEUE=LP2 Result1'; CONTINUE=yes]
CALCULATE Newdata[1] = Data[1]*Data[2]
```

The CALCULATE statement may start executing before the file Result1 has been queued for printing; whether it does or not depends on how much work the computer is doing. If the operating-system command also produces messages to the terminal, they may get interleaved with any output from Genstat that is coming to the terminal; it might then be best to avoid using the CONTINUE option.

13.3.2 Writing files to communicate with other programs

Your computer may impose constraints on the extent to which you can read data written by other programs (perhaps in languages other than Fortran). Firstly, the program that created the data file must be compatible with the one that you are using. Secondly, the style of reading and writing of the data used by the programs must be compatible. Usually there is no problem with data stored in character files, like Result1 and Result2 in Example 13.3.1a. However, you may want to take advantage of a faster style of storage, creating *unformatted* files.

The PRINT and READ directives in Genstat have an UNFORMATTED option that allows you write to or read from unformatted files rather than the usual character files (3.1 and 3.2). To communicate between Genstat and another program with unformatted files, the other program must deal with the files in the same way as Genstat: to understand this, you need some knowledge of the Fortran which underlies Genstat.

Genstat puts data in unformatted files by a simple Fortran WRITE statement of the form

WRITE (UNIT) (ARRAY(K),K=1,N)

ARRAY here is either REAL or INTEGER. The value of N will not exceed a certain limit which on most computers is 1024. Structures with a larger number of values than N are split up into several lines, all but (perhaps) the last of which is of size N.

When Genstat reads unformatted files, an analogous Fortran READ statement is used, with the same assumptions on record length. If you wish to read a file that has already been created, you must know how it is arranged. If it does not conform to these Genstat conventions, you will have to use a Fortran program to re-arrange it suitably.

13.3.3 Executing external programs

On some computers, you can arrange that one program, such as Genstat, calls for another to be executed, passing information directly between the two. You can then cause Genstat to execute your own subprograms without having to modify Genstat in any way. This is done by the PASS directive.

To find out if the PASS directive has been implemented in your version, you can either look at the *Users' Note* (see the introduction to this chapter) or type

PASS

in any Genstat program. You will either get a message saying that the PASS directive has not been implemented, or you will get a Genstat diagnostic telling you that Genstat has failed to initiate a sub-process: this means that PASS has been implemented. If PASS has not been implemented, you could use the OWN directive (13.4.1). Alternatively, you may be able to use the SUSPEND directive (13.3.1). First, you should open a file (3.3.1) and PRINT the values of the structures that you want to send to your own program; you can use a character file, or, for faster communication, an unformatted file (13.3.2). Next CLOSE the file (3.3.2), and give a SUSPEND statement to return to the operating system. You can then run your own program, accessing data from the file, and perhaps putting values back into it or another file. Finally return to Genstat, and use READ to access the results of your program. An example of this method is shown in 13.3.1.

To use the PASS directive when it is available, you must first get access to the GNPASS program which is distributed with Genstat. You then form an executable program consisting of GNPASS, slightly modified as detailed below, and your own subprograms. GNPASS, like Genstat, is written in Fortran 77; however, on many computers, it is possible to use equivalent programs in other computing languages. The GNPASS program deals with communication with Genstat, and passes information to and from your subprograms.

PASS does work specified in subprograms supplied by the user, but not linked into Genstat. This directive may not be available on some computers.

Option

NAME = *text*   Filename of external executable program; default 'GNPASS'

Parameter

pointers   Structures whose values are to be passed to the external program, and returned

---

You can pass the values of any data structures except texts. All the structures needed by your subprograms must be combined in a pointer structure, unless only one structure is needed and it is not a pointer (the same rule applies to OWN; see 13.4.1). The structures must have values before you include them in a PASS statement; if you want to use some of the structures to store results from your subprograms, you must initialize them to some arbitrary values, such as zero or missing. If you specify several pointers in a PASS statement, your subprograms will be invoked several times, to deal in turn with each set of structures stored by the pointers. However, the values of the structures in all the pointers are copied before any work is done by your subprograms. Thus, if you want to operate with PASS on the results of a previous operation by PASS, you must give two PASS statements with one pointer each rather than one statement with two pointers.

As an example, consider using PASS to carry out a simple transformation of a variate, as would be done by the statement

    CALCULATE W = M*(V+S)**2

where V and W are variates, and M and S are scalars. You would need a Fortran subprogram to calculate the values of W from supplied values of M, V, and S. The distributed version of the GNPASS program is accompanied with just such a subprogram, called SQUARE, for the purpose of illustrating how to use PASS. So all you need to do is to compile and link the program and subprogram into an executable program, called GNPASS for convenience. Then you can run Genstat and give the following statements:

```
SCALAR S,M; VALUE=2,10
VARIATE [VALUES=1...10] V
& [VALUES=10(*)] W
PASS !p(V,S,M,W)
```

The PASS statement will cause the GNPASS program to run, and assign the calculated values to the variate W.

Numbers can be used in place of scalars, as usual in Genstat statements:

```
PASS !p(V,2,10,W)
```

To transform the values in both V, as above, and another variate X, with values 10...50 say, you could give the statements:

```
VARIATE [VALUES=41(*)] Y
PASS !p(V,2,10,W),!p(X,2,10,Y)
```

The NAME option is used to specify the filename of the executable program formed from the GNPASS program and your subprograms. By default, the name GNPASS is assumed.

The distributed form of the GNPASS program, if available on your computer, consists of Fortran statements that receive information from Genstat as supplied by a PASS statement, call the SQUARE subprogram, and then send back the information as modified by SQUARE. To make it do the task that you require, you need to edit the program to call your subprograms instead of SQUARE. The documentation for GNPASS is provided as comments within the GNPASS program, so the details are not included here as well. After preparing the Fortran, you need to form it into an executable program. This will require a Fortran compiler, and to be certain of communicating successfully with Genstat, the compiler should be the same as that used in preparing Genstat – this information is given in the *Installers' Note* that accompanies Genstat. It may also be possible to use other source languages, provided the input and output formats of their compilers are compatible with that used by Genstat.

13.4 Adding Fortran subprograms

Two ways are provided for you to incorporate new facilities by adding Fortran subprograms directly into Genstat. The first uses a predefined directive called OWN, described in 13.4.1. It has a limited syntax, but this does not need to be modified to allow the new facilities to work. However, you will need to modify the Fortran subprogram supplied with Genstat, which is designed to call your subprograms and pass information between them and the rest of Genstat.

You can also use the Fortran source provided for the OWN directive in conjunction with the FITNONLINEAR directive (13.4.2). This provides the ability to specify computer-intensive calculations for nonlinear models in Fortran rather than using Genstat expressions.

The TABULATE directive also allows extra Fortran source to be included, to control input of data from files recorded in a complex form. This is described in 13.4.3.

If you want to define your own directives instead of using OWN, rather more changes are required. Again, a template Fortran subprogram is supplied, but it will need more modification than the template for OWN, and you will also need to modify Genstat's Language-Definition File to include the details of your new directives. This is outlined in 13.4.4.

13.4.1 The OWN directive

To implement the OWN directive, you must get access to some of the Genstat source code. The relevant section of the code is named Module X, and is distributed with Genstat to all sites, probably in a file called X.FOR. The module consists of several Fortran 77 subprograms mentioned in this chapter: to implement the OWN directive you need to modify only the subprogram called G5XZXO. In its distributed form, G5XZXO is defined to call the G5XZSQ suprogram, which carries out the simple transformation described in 13.3.3. G5XZXO contains extensive comments that describe the way it works, and the straightforward changes that you would need to make in order to call your own subprograms. These comments are designed to be the complete documentation, and so the details are not repeated in this Manual.

OWN does work specified in Fortran subprograms linked into Genstat by the user.

Option

SELECT = *scalar*	Sets a switch, designed to allow OWN to be used for many applications; standard set-up assumes a scalar in the range 0-9; default 0

Parameters

IN = *identifiers*	Supplies input structures, which must have values, needed by the auxiliary subprograms
OUT = *identifers*	Supplies output structures whose values or attributes are to be defined by the auxiliary subprograms

---

The IN parameter allows you to pass values of data structures into your subprograms. Genstat will check these input structures before calling your subprograms, to ensure that they are of the right type and length for your program, and that they have been assigned values. The OUT parameter copies values calculated by your subprograms into Genstat data structures. You can arrange to define the type and length of these output structures either before or after calling your subprograms.

If the setting of the IN parameter is a list of identifiers, the OWN directive will call your subprograms more than once. Each time it will make available to your subprograms the values of one structure in the IN list, and will take information from the subprograms and put them into the corresponding structure in the OUT list. Therefore, to pass several structures at a time to your subprograms, you must put the structures into pointers. For example,

```
OWN IN=!p(A1,A2,A3),!p(B1,B2,B3); OUT=X,Y
```

will call your subprograms twice, passing information about A1, A2, A3, and X the first time, and about B1, B2, B3, and Y the second time. It does this because !p(A1,A2,A3), for example, is a single structure.

If you want to pass just one pointer to your subprograms, you must ensure that OWN does not treat the pointer as a set of structures each of which is to be passed. You can do this by constructing another pointer to hold just the identifier of the pointer that you want to pass; for example:

```
POINTER [VALUES=A,B,C] S1
OWN IN=!p(S1)
```

The SELECT option allows you to call any number of subprograms independently. Thus, you can set up OWN so that the statements

```
OWN [SELECT=1]
```

and

```
OWN [SELECT=2]
```

do totally unrelated tasks. The standard version of G5XZXO deals only with the default value, 0, of SELECT, and would need to be extended if you wanted to cater for alternative values. However, you should be able to use much of the Fortran that deals with the default setting.

The distributed version of Genstat contains a version of the G5XZXO subprogram that carries out a simple calculation, purely for illustration of how the subroutine works, just as for the PASS directive (13.3.3). The result of

```
OWN IN=!p(V,S,M); OUT=W
```

is to shift, square, and scale the values of V; that is, it does almost the same as

```
PASS !p(V,S,M,W)
```

except that W does not have to be defined in advance. Thus, it is even more like the statement

```
CALCULATE W = M*(V+S)**2
```

The subprogram checks that precisely three structures are given in the pointer specified by the IN parameter, and that they are a variate and two scalars with values already present. It also checks that there is precisely one output structure, a variate; this is implicitly declared by OWN if necessary, based on the length of the input variate. Missing values in the input structures are also checked for and dealt with appropriately. The subprogram calls another one called G5XZSQ actually to carry out the transformation: G5XZSQ is very similar to the SQUARE subprogram provided to illustrate PASS. To modify G5XZXO, you need to alter the details of the checks on the structures and substitute the call for one to your own subprogram.

The standard version of the G5XZXO subprogram will produce Genstat diagnostics if the checks on the input or output structures fail, or if there is not enough workspace. These diagnostics are the standard ones with codes VA, SX, and SP, and are dealt with by a section at the end of the G5XZXO subprogram. You can define your own diagnostics, using the code ZZ. You are not allowed to edit the standard file of error messages that stores the one-line definitions of each diagnostic code. However, you can edit the G5XZPF subprogram which is in module X. This prints extra messages after a ZZ diagnostic; instructions for editing the subprogram are contained as comments in it.

Output from your subprograms is most easily arranged by storing the information that you want in data structures, and printing these with a PRINT statement after the OWN statement. Alternatively, you can give Fortran WRITE statements; there are standard routines in Genstat for outputting numbers and strings, but they are not described here. You should use the correct Fortran unit numbers for output, and this varies between implementations of Genstat. You can find out the unit numbers by giving the command

```
HELP environment
```

in any Genstat program. Remember that a Fortran unit number is not the same as a Genstat channel number (Chapter 3).

13.4.2 Supplying calculations in Fortran for nonlinear models

The FITNONLINEAR directive (8.7.2) has options OWN, INOWN, and OUTOWN which allow you to use Fortran rather than Genstat to program the calculation of models. If the OWN option is set, the G5XZXO subprogram will be called at each step of the iterative process to form values for the explanatory variates, the variate of fitted values, or the scalar function value – depending on the type of model specified in FITNONLINEAR. You should set the INOWN and OUTOWN options to pass data structures to the G5XZXO subprogram, such as the parameters of the model and the variates or scalars to store the calculated values.

The FITNONLINEAR directive and the OWN directive give you very similar types of access to the G5XZXO subprogram. For example, if you have set up the G5XZXO subprogram to calculate values of an explanatory variate from values of two scalar parameters, you could put this statement in a program:

```
OWN IN=!p(A,B); OUT=X
```

The same version of the G5XZXO subprogram could be used also in fitting a model for which the values of the parameters A and B are optimized:

```
FITNONLINEAR [OWN=0; INOWN=A,B; OUTOWN=X] X
```

However, as pointed out in 8.7.2, you cannot arrange for your subprograms to be called several times at each step of the search process: Genstat treats the whole setting of the INOWN option as a single pointer structure, and similarly the setting of OUTOWN.

For the sake of efficiency, you should avoid all the checking within the G5XZXO subprogram after the first step of the search in FITNONLINEAR. The G5XZXO subprogram documents how this can be done, making use of a dummy argument of the subprogram, which passes the number of previous calls to G5XZXO with the current statement.

13.4.3 Specifying data input commands in Fortran for TABULATE

The TABULATE directive (4.11.1) includes a facility to tabulate data that might otherwise be difficult to handle by Genstat, for example hierarchical data or data requiring different operations on different types of units before tabulation. This facility is described here because it requires a knowledge of Fortran and the ability to relink Genstat, just like the OWN directive. However, unlike the extension to FITNONLINEAR described in 13.4.2, TABULATE does not share the code supplied in the G5XZXO subprogram; it uses the G5XZIT subprogram, also distributed in Module X.

G5XZIT is a Fortran subprogram, to be modified by you, which is called from within TABULATE for each unit to be tabulated. It contains switches to tell TABULATE when a data error occurs or when all the data have been read. To use it you have to link your own version of Genstat, as when using the OWN directive. Then your version of G5XZIT will be used instead of the standard version supplied as part of Genstat.

The subprogram can be as simple or as complicated as you like (or need), provided it obeys a few simple rules. A very simple version, reading two variates and two factors, is supplied with Genstat (see below). This should provide sufficient information for you to write your own version, and link it into your own private version of Genstat.

Five options of TABULATE control the execution of G5XZIT: these are OWN, OWNFACTORS, OWNVARIATES, INCHANNEL, and INFILETYPE.

The OWN option should be set to a variate allowing you to communicate between your Genstat code and your G5XZIT subprogram. The OWNFACTORS option provides the list of factors to be read by G5XZIT. It must include the classifying factors needed in the current TABULATE instruction, but it may contain others as well. The OWNVARIATES option should provide a similar list of variates. The INCHANNEL option should be set to the Genstat channel number of the data file, as specified in a previous OPEN statement or in the Genstat command line. The INFILETYPE option specifies whether the data file is character (input) or binary (unformatted).

The documentation of G5XZIT is included with the Fortran and so is not repeated here. The following example shows how TABULATE can be used with the standard version of G5XZIT, which is set up simply to read two variates and two factors from a sequential character file. The two variates are read with Fortran format F4.2, which means that their values must be in a field of four characters and will be scaled by 100; the two factors are read with format I4, so the factor values must be integer levels in a field of four characters. Here is a short data file with values in this format.

```
1100 200   1   2
1200 100   1   2
1300 100   2   3
1400 200   2   3
1500 200   1   1
1600 300   2   1
```

Example 13.4.3 shows how TABULATE can read these values from a file called OWN.DAT and form a tabular summary of the first variate.

Example 13.4.3

```
1   " Declare factors F1 and F2 "
2   FACTOR [LEVELS=2] F1
3   & [LEVELS=3] F2
4   " Open data file containing values for V1, V2, F1 and F2 "
5   OPEN 'OWN.DAT'; CHANNEL=3
6   " Print table of means of variate V1, classified by F1 and F2 "
7   TABULATE [PRINT=means; CLASSIFICATION=F1,F2; OWN=0; \
8     OWNFACTORS=F1,F2; OWNVARIATES=V1,V2; INCHANNEL=3] V1
```

	Mean		
F2	1	2	3
F1			
1	15.00	11.50	*
2	16.00	*	13.50

TABULATE allows only one classification set to be used at a time. If the data set is complicated enough to require G5XZIT, then several tabulations with different classifying sets are likely to be needed. Rather than have a separate branch in G5XZIT for each tabulation, you can put all the factors and all the variates that you will need into the settings of the OWNFACTORS and OWNVARIATES options, and leave TABULATE to extract the ones it needs each time. If you have several TABULATE statements as suggested, you will have to close the data file and re-open it between them.

13.4.4 Defining new directives

This section deals with the definition of new directives. These are not procedures, programmed in the Genstat language as described in Chapter 5, but extensions to that language programmed in Fortran, or in any other language that you can link together with Genstat's Fortran on your computer.

The following steps are needed to implement a new directive. First, you need to make a copy of the Genstat *Language-Definition File*, which is distributed with Genstat. Into this file you should add the definition of the syntax of your directive, using the special directive DEFINE together with OPTION and PARAMETER statements similar to those used for procedures (5.3). You should then modify the G5XZXU subprogram, distributed with Genstat in Module X (13.4.1), and relink Genstat to include your subprograms in the same way as for the OWN directive. You then need to arrange for your new version of Genstat to process your language-definition file, as explained in the *Installers' Note*.

DEFINE create syntax definition for a new directive. Available only within the language-definition file.

Options

CODE = *number*	Internal code number that identifies the Fortran interpreter to be called on execution
SCOPE = *string*	Scope of directive (boot, general, graphics); default general
DUMMY = *string*	Whether to suppress dummy substitution (no, yes); default no

Parameter

string	The directive name

The name that you assign to a new directive must be in upper case. It must not clash with an existing name; that is, the first four letters of the new name must not be the same as the first four letters of a standard directive name. Thus you cannot define a directive called CONTRAST because there is already one called CONTOUR. But if a directive name − new or old − has fewer than four letters, Genstat extends it with spaces. So you could define a directive called TAB, even though there are already directives called TABLE and TABULATE; likewise you could define a directive called OWNER and it would not be mistaken for the standard OWN directive. You must also be aware of possible conflicts with procedure names in your job, and in procedure libraries (1.9.1 and 5.3).

You must set the CODE option to a unique code number for each new directive. The codes for new directives are integers 1501 and upwards, to avoid the standard codes; the Genstat interpreter subtracts 1500 from the code before passing it to the G5XZXU subprogram.

DEFINE has two other options, but these are not usually required when extending Genstat. The DUMMY option allows you to suppress the automatic substitution of dummies. By default, when DUMMY=no, dummies are replaced by the structures to which they point. Some standard directives, however, need to operate with the dummies themselves; for example, the CALCULATE directive is defined with DUMMY=yes so that it can evaluate the UNSET function (4.2.6). The SCOPE option should never need to be set for user-defined directives.

After giving the name of the new directive with a DEFINE statement, you should define its options and parameters by an OPTION and a PARAMETER statement. The rules for these are described in 5.3.2. The only addition is that the OPTION and PARAMETER directives in a language-definition file are allowed some extra modes.

You can set the MODE parameter to w (word) when the option or parameter is to allow the choice of string settings, as for example in a PRINT option. Unnamed structures of mode w, for the VALUES and DEFAULTS parameters, are specified using the syntax !w( ). Mode t should be used only for general string settings, like the VALUES option of the TEXT directive.

The other extra mode is i, which indicates an integer value; this is just like mode v except that the setting is turned into integers by rounding before being passed to G5XZXU.

Here is an example defining a new directive called SQUARE which forms a new variate by squaring the values of an old one, as in the illustrative examples of PASS (13.3.3) and OWN (13.4.1); it can also add a shift before squaring, and can change the sign of the result:

```
DEFINE [CODE=1501] 'SQUARE'
PARAMETER 'OLD','NEW'; MODE=p
OPTION 'SHIFT','CHANGESIGN'; MODE=v,w; VALUES=*,!w(no,yes); \
   DEFAULT=0,!w(no)
```

The G5XZXU subprogram is distributed with Genstat, and is already able to interpret any directive that you can define with the DEFINE directive. Thus, it accesses the individual data structures that have been set in the options and parameters of a statement, and stores their attributes in Fortran arrays.

The documentation of G5XZXU is included as comments in the Fortran, so it is not repeated here. The distributed version of the subprogram is set up to interpret the SQUARE directive as described here, to illustrate how a directive is interpreted. The standard Language-Definition File contains the definition of this directive within a comment, so you would need to modify this file if you actually wanted to use the directive.

S.A.H.
P.W.L.

Appendix 1 List of Genstat directives

ADD adds extra terms to a linear, generalized linear, generalized additive, or nonlinear model 379.

ADDPOINTS adds points for new objects to a principal coordinates analysis 616.

ADISPLAY displays further output from analyses produced by ANOVA 473.

AKEEP copies information from an ANOVA analysis into Genstat data structures 505.

ANOVA analyses y-variates by analysis of variance according to the model defined by earlier BLOCKSTRUCTURE, COVARIATE, and TREATMENTSTRUCTURE statements 467.

ASSIGN sets elements of pointers and dummies 196.

AXES defines the axes in each window for high-resolution graphics 304.

BLOCKSTRUCTURE defines blocking structure of the design and hence strata and error terms 483.

BREAK suspends execution of the statements in the current channel or control structure and takes subsequent statements from the channel specified 253.

CALCULATE calculates numerical values for data structures 128.

CASE introduces a "multiple-selection" control structure 234.

CATALOGUE displays the contents of a backing-store file 118.

CLOSE closes files 104.

CLUSTER forms a non-hierarchical classification 646.

COLOUR defines the red, green and blue intensities to be used for the Genstat colours 317.

COMBINE combines or omits "slices" of a multi-way data structure (table, matrix, or variate) 222.

CONCATENATE concatenates and truncates lines of text structures; can change case of letters 188.

CONTOUR produces contour maps of two-way arrays of numbers (on the terminal/printer) 267.

COPY forms a transcript of a job 105.

CORRELATE forms correlations between variates, autocorrelations of variates, and lagged cross-correlations between variates 670.

COVARIATE specifies covariates for use in subsequent ANOVA statements 493.

CVA performs canonical variates analysis 594.

DCONTOUR draws contour plots on a plotter or graphics monitor 286.

DDISPLAY redraws the current graphical display 302.

DEBUG puts an implicit BREAK statement after the current statement and subsequent statements, until an ENDDEBUG is reached 255.

DEFINE create syntax definition for a new directive 760.

DELETE deletes the attributes and values of structures 62.

DEVICE switches between (high-resolution) graphics devices 300.

DGRAPH draws graphs on a plotter or graphics monitor 276.

DHISTOGRAM draws histograms on a plotter or graphics monitor 280.

DIAGONALMATRIX declares one or more diagonal matrix data structures 48.

DISPLAY prints, or reprints, diagnostic messages 250.

DISTRIBUTION estimates the parameters of continuous and discrete distributions 330.

DKEEP saves information from the last plot on a particular device 321.

DPIE draws a pie chart on a plotter or graphics monitor 283.

DREAD reads the locations of points from an interactive graphical device 294.

DROP drops terms from a linear, generalized linear, generalized additive, or nonlinear model 379.

DSURFACE produces perspective views of two-way arrays of numbers 290.

DUMMY declares one or more dummy data structures 40.

DUMP prints information about data structures, and internal system information 63.

DUPLICATE forms new data structures with attributes taken from an existing structure 67.

D3HISTOGRAM produces three-dimensional histograms 292.

EDIT edits text vectors 189.

ELSE introduces the default set of statements in block-if or in multiple-selection control structures 232, 234.

ELSIF introduces a set of alternative statements in a block-if control structure 232.

ENDBREAK returns to the original channel or control structure and continues execution 253.

ENDCASE indicates the end of a "multiple-selection" control structure 234.

ENDDEBUG cancels a DEBUG statement 255.

ENDFOR indicates the end of the contents of a loop 230.

ENDIF indicates the end of a block-if control structure 232.

ENDJOB ends a Genstat job 229.

ENDPROCEDURE indicates the end of the contents of a Genstat procedure 247.

ENQUIRE provides details about files opened by Genstat 106.

EQUATE transfers data between structures of different sizes or types (but the same modes, numerical or text) or where transfer is not from single structure to single structure 167.

ESTIMATE estimates parameters in Box-Jenkins models for time series 691.

EXECUTE executes the statements contained within a text 252.

EXIT exits from a control structure 236.

EXPRESSION declares one or more expression data structures 41.

FACROTATE rotates factor loadings from a principal components or canonical variates analysis according to either the varimax or quartimax criterion 600.

FACTOR declares one or more factor data structures 44.

FCLASSIFICATION forms a classification set for each term in a formula, breaks a formula up into separate formulae (one for each term), and applies a limit to the number of factors and variates in the terms of a formula 194.

FILTER filters time series by time-series models 724.

FIT fits a linear, generalized linear, or generalized additive model 365.

FITCURVE fits a standard nonlinear regression model 435.

FITNONLINEAR fits a nonlinear regression model or optimizes a scalar function 453.

FLRV forms the values of LRV structures 204.

FOR introduces a loop; subsequent statements define the contents of the loop, which is terminated by the directive ENDFOR 230.

FORECAST forecasts future values of a time series 705.

FORMULA declares one or more formula data structures 41.

FOURIER calculates cosine or Fourier transforms of real or complex series 678.

FRAME defines the positions of windows within the frame of a high-resolution graph 302.

FSIMILARITY forms a similarity matrix or a between-group-elements similarity matrix or prints a similarity matrix 622.

FSSPM forms the values of SSPM structures 209.

FTSM forms preliminary estimates of parameters in time-series models 727.

GENERATE generates factor values, in standard order or using the design-key method, or generates pseudo-factors to describe confounding in partially balanced designs 525.

GET accesses details of the "environment" of a Genstat job 741.

GETATTRIBUTE accesses attributes of structures 66.

GRAPH produces scatter and line graphs on the terminal or line printer 258.

GROUPS forms a factor (or grouping variable) from a variate or text, together with the set of distinct values that occur 184.

HCLUSTER performs hierarchical cluster analysis 632.

HDISPLAY displays results ancillary to hierarchical cluster analyses 635.

HELP prints details about the Genstat language and environment 12.

HISTOGRAM produces histograms of data on the terminal or line printer 263.

HLIST lists the data matrix in abbreviated form 640.

HSUMMARIZE forms and prints a group by levels table for each test together with appropriate summary statistics for each group 643.

IF introduces a block-if control structure 232.

INPUT specifies the input file from which to take further statements 108.

INTERPOLATE interpolates values at intermediate points 179.

JOB starts a Genstat job 227.

LRV declares one or more LRV data structures 57.

MARGIN forms and calculates marginal values for tables 218.

MATRIX declares one or more matrix data structures 47.

MDS performs non-metric multidimensional scaling .

MERGE copies subfiles from backing-store files into a single file 122.

MODEL defines the response variate(s) and the type of model to be fitted for linear, generalized linear, generalized additive, and nonlinear models 362.

MONOTONIC fits an increasing monotonic regression of y on x 182.

OPEN opens files 101.

OPTION defines the options of a Genstat procedure with information to allow them to be checked when the procedure is executed 243.

OR introduces a set of alternative statements in a "multiple-selection" control structure 234.

OUTPUT defines where output is to be stored or displayed 110.

OWN does work specified in Fortran subprograms linked into Genstat by the user 756.

PAGE moves to the top of the next page of an output file 105.

PARAMETER defines the parameters of a Genstat procedure with information to allow them to be checked when the procedure is executed 244.

PASS does work specified in subprograms supplied by the user, but not linked into Genstat 754.

PCO performs principal coordinates analysis, also principal components and canonical variates analysis (but with different weighting from that used in CVA) as special cases 609.

PCP performs principal components analysis 587.

PEN defines the properties of "pens" for high-resolution graphics 309.

POINTER declares one or more pointer data structures 54.

PREDICT forms predictions from a linear or generalized linear model 395.

PRINT prints data in tabular format in an output file, unformatted file, or text 91.

PROCEDURE introduces a Genstat procedure 242.

QUESTION obtains a response using a Genstat menu 746.

RANDOMIZE randomizes the units of a designed experiment or the elements of a factor or variate 534.

RCYCLE controls iterative fitting of generalized linear, generalized additive, and nonlinear models, and specifies parameters, bounds etc for nonlinear models 426.

RDISPLAY displays fit of a linear, generalized linear, generalized additive, or nonlinear model 370.

READ reads data from an input file, an unformatted file, or a text 74.

RECORD dumps a job so that it can later be restarted by a RESUME statement 123.

REDUCE forms a reduced similarity matrix (referring to the groups instead of the original units) 627.

RELATE relates observed values on a set of variates to results of a principal coordinates analysis 630.

REML fits a variance-components model by residual (or restricted) maximum likelihood 522.

RESTRICT defines a restricted set of units of vectors for subsequent statements 172.

RESUME restarts a recorded job 124.

RETRIEVE retrieves structures from a subfile 116.

RETURN returns to a previous input stream (text vector or input channel) 109.

RFUNCTION estimates functions of parameters of a nonlinear model 446.

RKEEP stores results from a linear, generalized linear, generalized additive, or nonlinear model 371.

ROTATE does a Procrustes rotation of one configuration of points to fit another 656.

SCALAR declares one or more scalar data structures 39.

SET sets details of the "environment" of a Genstat job 737.

SETOPTION sets or modifies defaults of options of Genstat directives or procedures 743.

SETPARAMETER sets or modifies defaults of parameters of Genstat directives or procedures 743.

SKIP skips lines in input or output files 106.

SORT sorts units of vectors according to an index vector 176.

SPREADSHEET provides spreadsheet-style input and editing of variates, factors, and texts 9.

SSPM declares one or more SSPM data structures 59.

STEP selects a term to include in or exclude from a linear, generalized linear, or generalized additive model according to the ratio of residual mean squares 385.

STOP ends a Genstat program 229.

STORE to store structures in a subfile of a backing-store file 113.

SUSPEND suspends execution of Genstat to carry out commands in the operating system 751.

SVD calculates singular value decompositions of matrices 201.

SWITCH adds terms to, or drops them from a linear, generalized linear, generalized additive, or nonlinear model 379.

SYMMETRICMATRIX declares one or more symmetric matrix data structures 50.

TABLE declares one or more table data structures 51.

TABULATE forms summary tables of variate values 212.

TDISPLAY displays further output after an analysis by ESTIMATE 700.

TERMS specifies a maximal model, containing all terms to be used in subsequent linear, generalized linear, generalized additive, and nonlinear models 377.

TEXT declares one or more text data structures 43.

TKEEP saves results after an analysis by ESTIMATE 701.

TRANSFERFUNCTION specifies input series and transfer function models for subsequent estimation of a model for an output series 714.

TREATMENTSTRUCTURE specifies treatment terms to be fitted by subsequent ANOVA statements 466.

TRY displays results of single-term changes to a linear, generalized linear, or generalized additive model 383.

TSM declares one or more TSM data structures 60.

TSUMMARIZE displays characteristics of time series models 730.

UNITS defines an auxiliary vector of labels and/or the length of any vector whose length is not defined when a statement needing it is executed 45.

VARIATE declares one or more variate data structures 42.

VCOMPONENTS defines the variance-components model for REML 544.

VDISPLAY display further output from a variance-components analysis 574.

VKEEP copies information from a variance-components analysis into Genstat data structures 577.

Appendix 2 Release 3[1] of the Procedure Library

ABIVARIATE produces graphs and statistics for bivariate analysis of variance (R.F.A. Poultney) 462.

AFALPHA generates alpha designs (R.W. Payne) 463, 525, 529-532.

AFCYCLIC generates block and treatment factors for cyclic designs (R.W. Payne) 463, 525, 529, 532, 533.

AFUNITS forms a factor to index the units of the final stratum of a design (R.W. Payne) 463, 490, 529.

AKAIKEHISTOGRAM prints histograms with improved definition of groups (A. Keen) 165.

AKEY generates values for treatment factors using the design key method (R.W. Payne) 463, 525, 528-530.

ALIAS finds out information about aliased model terms in analysis of variance (R.W. Payne) 512.

ANTORDER assesses order of ante-dependence for repeated measures data (M.S. Ridout and R.W. Payne) 462.

ANTTEST calculates overall tests based on a specified order of ante-dependence (R.W. Payne and M.S. Ridout) 462.

AONEWAY provides one-way analysis of variance for inexperienced users (R.W. Payne) 349, 350.

APLOT plots residuals from an ANOVA analysis (A.D. Todd) 462, 481.

ASWEEP performs sweeps for model terms in an analysis of variance (R.W. Payne) 524.

AUDISPLAY produces further output for an unbalanced design after AUNBALANCED (R.W. Payne) 463.

AUNBALANCED performs analysis of variance for unbalanced designs (R.W. Payne) 463.

BARCHART plots a bar chart using line-printer or high-resolution graphics (S.A. Harding and K.E. Bicknell).

BIPLOT produces a biplot from a set of variates (S.A. Harding) 275, 662, 665.

BJESTIMATE fits an ARIMA model, with forecast and residual checks (G. Tunnicliffe Wilson and S.J. Welham) 686-688.

BJFORECAST plots forecasts of a time series using a previously fitted ARIMA model (G. Tunnicliffe Wilson and S.J. Welham) 707, 708.

BJIDENTIFY displays time series statistics useful for ARIMA model selection (G. Tunnicliffe Wilson and S.J. Welham) 673, 678, 686.

BOXPLOT draws Box plots and schematic diagrams (P.W. Lane and S.D. Langton) 275, 325.

BSUPDATE creates or updates a backing-store subfile (P.W. Lane) 112.

CANCOR does canonical correlation analysis (P.G.N. Digby) 505-507, 662.

CENSOR pre-processes censored data before analysis by ANOVA (P.W. Lane) 462.

CHECKARGUMENT checks the arguments of a procedure (R.W. Payne) 249.

CHISQUARE calculates chi-square statistics for one and two-way tables (P.K. Leech and A.D. Todd) 355.

CLASSIFY obtains a starting classification for non-hierarchical clustering (S.A. Harding) 647, 662.

CONCORD calculates Kendall's Coefficient of Concordance for a set of variates (N.M. Maclaren, H.R. Simpson, and S.J. Welham) 354.

CONVEXHULL finds the points of a single or a full peel of convex-hulls (P.G.N. Digby) 662.

CORRESP does correspondence analysis, or reciprocal averaging (P.G.N. Digby) 662-664.

CUMDISTRIBUTION fits frequency distributions to accumulated counts (R.C. Butler and P. Brain).

DAPLOT plots residuals from ANOVA with interactive identification of outliers (R.J. Reader) 462, 481, 482.

DAYCOUNT converts a date to a daycount, or vice versa (T.J. Cole) 179.

DAYLENGTH calculates daylengths at a given period of the year (R.J. Reader and K. Phelps) 179.

DBARCHART produces barcharts for one or two-way tables (R.C. Butler) 275.

DDENDROGRAM draws dendrograms with control over structure and style (P.G.N. Digby) 275, 634, 662.

DECIMALS sets the number of decimals for a structure, using its round-off (A. Keen) 38.

DESCRIBE saves and/or prints summary statistics for variates (R.C. Butler) 152, 326, 327.

DESIGN helps to select and generate effective experimental designs (A.E. Ainsley, M.F. Franklin, and R.W. Payne) 525.

DHELP displays information about high-resolution graphics (S.A. Harding) 319.

DILUTION calculates Most Probable Numbers from dilution series data (M.S. Ridout and S.J. Welham) 421.

DISCRIMINATE performs discriminant analysis (P.G.N. Digby) 662.

DMST provides high-resolution plot of an ordination with minumum spanning tree (A.W.A. Murray).

DREPMEAS plots profiles and differences of profiles for repeated measures data (J.T.N.M. Thissen).

DSCATTER produces a scatter-plot matrix (S.A. Harding).

DSHADE produces a shaded similarity matrix by high-resolution graphics (S.A. Harding) 662.

DSYMBOL provides interactive definition of graphics symbols (S.A. Harding).

DZOOM interactive zooming of a graphical display (S.A. Harding).

EXTRABINOMIAL fits the models of Williams (1982) to overdispersed proportions (M.S. Ridout and P.W. Goedhart).

FACAMEND permutes the labels and corresponding levels of a factor (J.T.N.M. Thissen).

FACPRODUCT forms a factor with a level for every combination of other factors (R.W. Payne) 184, 529, 530.

FEXACT2X2 does Fisher's exact test for 2×2 tables (M.S. Ridout and M.W. Patefield) 356, 357.

FIELLER calculates effective doses or relative potencies (P.W. Lane) 416.

FILEREAD reads data from a file, assumed to be in a rectangular array (P.W. Lane) 74, 79.

FITPARALLEL carries out analysis of parallelism for non-linear functions (R.C. Butler) 454.

FITSCHNUTE fits a general 4 parameter growth model to a non-decreasing Y-variate (A. Keen).

FTEXT forms a text structure from a variate (A. Keen) 187.

GENPROC performs a generalized Procrustes analysis (G.M. Arnold and R.W. Payne) 660-662.

GINVERSE calculates the generalized inverse of a matrix (S.K. Haywood) 154, 200, 203.

GLM analyses non-standard generalized linear models (P.W. Lane).

GLMM fits a generalized linear mixed model to binomial data (S.J. Welham) 349, 463.

GRANDOM generates random numbers from continuous and discrete distributions (D.M. Roberts and P.W. Lane) 162, 323, 329.

HANOVA does hierarchical analysis of variance/covariance for unbalanced data (P.W. Lane).

HEATUNITS calculates accumulated heat units of a temperature dependent process (R.J. Reader, R.A. Sutherland, and K. Phelps) 179.

INSIDE determines the points inside a specified polygon – for use with DREAD (S.A. Harding).

KOLMOG2 performs a Kolmogorov-Smirnoff two-sample test (N.M. Maclaren, H.R. Simpson, and S.J. Welham) 344, 348, 349.

KRUSKAL carries out a Kruskal-Wallis one-way analysis of variance (N.M. Maclaren, H.R. Simpson, and S.J. Welham) 344, 350, 351.

LATTICE analyses square and rectangular lattice designs (K. Ryder, E.R. Williams, and D. Ratcliff).

LIBEXAMPLE accesses examples and source code of Library procedures (R.W. Payne) 12, 112, 175, 239-241, 275.

LIBFILENAME supplies the name of the Procedure Library help information file (R.W. Payne).

LIBHELP provides help information about Library procedures (R.W. Payne) 12, 175, 221, 239-241,

249, 529-532, 604, 606, 660.

LIBINFORM prints information about the contents of the Procedure Library (R.W. Payne) 11, 239, 240, 529.

LIBMANUAL prints a "Manual" containing information about Library procedures (R.W. Payne) 239, 241.

LINDEPENDENCE finds the linear relations associated with matrix singularities (J.H. Maindonald) 200.

LRVSCREE prints a scree diagram and/or a difference table of latent roots (P.G.N. Digby) 592, 598, 601, 612, 662.

LVARMODEL analyses a field trial using the Linear Variance Neighbour model (D.B. Baird) 462.

MANNWHITNEY performs a Mann-Whitney U test (N.M. Maclaren, H.R. Simpson, and S.J. Welham) 344, 347.

MANOVA performs multivariate analysis of variance and covariance (R.W. Payne and G.M. Arnold) 462, 603-605, 662.

MENU initiates a menu system (P.W. Lane) 749, 750.

MPOWER forms integer powers of a square matrix (P.W. Lane) 200.

MULTMISS estimates missing values for units in a multivariate data set (H.R. Simpson and R.P. White) 662.

NLCONTRASTS fits non-linear contrasts to quantitative factors in ANOVA (R.C. Butler) 504.

NORMTEST performs tests of univariate and/or multivariate normality (M.S. Ridout).

NOTICE gives access to the Genstat Notice Board: news, errors etc. (R.W. Payne) 11, 241.

ORTHPOL calculates orthogonal polynomials (P.W. Lane) 179, 407, 408.

PAIRTEST performs t-tests for pairwise differences (P.W. Goedhart).

PCOPROC performs multiple Procrustes analysis using PCO (P.G.N. Digby).

PDESIGN prints or stores treatment combinations tabulated by the block factors (R.W. Payne) 463, 525, 528-530.

PERCENT expresses the body of a table as percentages of one of its margins (R.W. Payne) 221, 222, 329.

PPAIR displays the results of t-tests for pairwise differences in compact diagrams (P.W. Goedhart).

PREWHITEN pre-whiten a time/space series (A.W.A. Murray).

PROBITANALYSIS fits probit models allowing for natural mortality and immunity (R.W. Payne) 428.

PTDESCRIBE gives summary and second order statistics for a point process (R.P. Littlejohn and R.C. Butler).

QUANTILE calculates quantiles of the values in a variate (P.W. Lane) 179.

RANK produces ranks, from the values in a variate, allowing for ties (J.B. van Biezen and C.J.F. ter Braak) 179.

RCHECK checks the fit of a linear or generalized linear regression (P.W. Lane, R. Cunningham, and C. Donnelly) 366, 369, 372, 373, 446, 451.

REPMEAS checks if a set of repeated measures can be analysed as a split plot (R.W. Payne) 462.

RGRAPH draws a graph to display the fit of a regression model (P.W. Lane) 373, 410, 416, 434, 446, 451.

RIDGE produces ridge regression and principal component regression analyses (A.J. Rook and M.S. Dhanoa) 594, 662.

RJOINT modified joint regression analysis for incomplete variety × environment data (P.W.Lane and K.Ryder).

ROBSSPM forms robust estimates of sum-of-squares-and-products matrices (P.G.N. Digby) 200, 586, 662.

RPAIR gives t-tests for all pairwise differences of means from regression or GLMs (J.T.N.M. Thissen and P.W. Goedhart).

RUNTEST performs a test of randomness of a sequence of observations (P.W. Goedhart).

SAMPLE samples from a set of units, possibly stratified by factors (P.W. Lane) 329, 330.

SIGNTEST performs a one or two sample sign test (E. Stephens and P.W. Goedhart) 346.

SKEWSYMM provides an analysis of skew-symmetry for an asymmetric matrix (P.G.N. Digby) 662, 665-667.

SMOOTHSPECTRUM forms smoothed spectrum estimates for univariate time series (G. Tunnicliffe Wilson and S.J. Welham) 678-680.

SPEARMAN calculates Spearman's Rank Correlation Coefficient (N.M. Maclaren, H.R. Simpson, and S.J. Welham) 353, 354.

SUBSET forms vectors containing subsets of the values in other vectors (R.W. Payne) 172, 175, 176.

TTEST performs a one- or two-sample t-test (S.J. Welham) 344-348.

VCONTRAST estimates linear functions of parameters from a REML analysis (S.J. Welham) 561.

VEQUATE equates across numerical structures (P.W. Goedhart) 171.

VFUNCTION calculates functions of variance components from a REML analysis (S.J. Welham) 569.

VHOMOGENEITY tests homogeneity of variances (R.W. Payne) 462.

VINTERPOLATE performs linear and inverse linear interpolation between variates (R.J. Reader) 179.

VORTHPOL calculates orthogonal polynomial contrasts across variates (J.T.N.M. Thissen).

VPLOT plots residuals from a REML analysis (S.J. Welham) 555.

VREGRESS performs regression across variates (M.W. Patefield and D. Tandy).

VTABLE forms a variate and set of classifying factors from a table (P.W. Goedhart) 179, 184.

VWALD prints Wald tests for fixed terms in a REML analysis (W. Buist and B. Engel).

WADLEY fits models for Wadley's problem, allowing alternative links and errors (D.M. Smith).

WILCOXON performs a Wilcoxon Matched-Pairs (Signed-Rank) test (N.M. Maclaren, H.R. Simpson, and S.J. Welham) 344-346, 348.

XOCATEGORIES sets up factors and variates, and analyzes categorical data from crossover trials (D. Smith and M.G. Kenward).

References

Preface

Digby, P.G.N., Galwey, N.W., and Lane, P.W. (1989). *Genstat 5: a second course*. Oxford University Press.

Lane, P.W., Galwey, N.W., and Alvey, N.G. (1987). *Genstat 5: an introduction*. Oxford University Press.

Payne, R.W., Arnold, G.M., and Morgan, G.W. (editors) (1993). *Genstat 5 Procedure Library Manual Release 3[1]*. Numerical Algorithms Group, Oxford.

Chapter 4

Bowdler, H., Martin, R.S., Reinsch, C., and Wilkinson, J.H. (1968). The QR and QL algorithms for symmetric matrices. *Numerische Mathematik* **11**, 293-306.

Digby, P.G.N. and Kempton, R.A. (1987). *Multivariate analysis of ecological communities*. Chapman and Hall, London.

Eckart, C. and Young, G. (1936). The approximation of one matrix by another of lower rank. *Psychometrika* **1**, 211-218.

Golub, G.H. and Reinsch, C. (1971). Singular value decomposition and least squares solutions. *Numerische Mathematik* **14**, 403-420.

Herraman, C. (1968). Algorithm AS12: Sums of squares and products matrix. *Applied Statistics* **17**, 289-292.

Martin, R.S., Reinsch, C., and Wilkinson, J.H. (1968). Householders tridiagonalisation of a symmetric matrix. *Numerische Mathematik* **11**, 181-195.

Rao, C.R. (1973). *Linear statistical inference and its applications*. Wiley, New York.

Wichmann, B.A. and Hill, I.D. (1982). An efficient and portable pseudo-random number generator. *Applied Statistics* **31**, 188-190.

Chapter 6

Butland, J. (1980). A method of interpolating reasonably-shaped curves through any data. *Proceedings of Computer Graphics* **80**, 409-422.

Cleveland, W.S. and McGill, R. (1987). Graphical perception: the visual decoding of quantitative information on graphical displays of data. *Journal of the Royal Statistical Society, Series B* **150**, 192-229.

McConalogue, D.J. (1970). A quasi-intrinsic scheme for passing a smooth curve through a discrete set of points. *Computer Journal* **13**, 392-396.

Yule, G.U. (1927). On the method of investigating periodicity in disturbed series, with special reference to Wolfer's sunspot numbers. *Philosophical Transactions of the Royal Society of London, Series A* **226**, 267-298.

Chapter 7

Bliss, C.I. (1953). Fitting the negative binomial distribution to biological data. *Biometrics* **9**, 176-196.

Conover, W.J. (1971). *Practical nonparametric statistics*. Wiley, New York.

Eid, M.T., Black, C.A., Kempthorne, O., and Zoellner, J.A. (1954). Significance of soil organic phosphorus to plant growth. *Iowa Agricultural Experiment Station Research Bulletin* **406.**

Hoffman, M.S. (editor) (1992). *World almanac and book of facts*. Pharos Books, New York.

Johnson, N.L. and Kotz, S. (1969). *Discrete distributions*. Houghton Mifflin, Boston, Massachussets.

Johnson, N.L. and Kotz, S. (1970a). *Continuous univariate distributions – 1*. Houghton Mifflin, Boston, Massachussets.

Johnson, N.L. and Kotz, S. (1970b). *Continuous univariate distributions – 2*. Houghton Mifflin, Boston, Massachussets.

McCullagh, P. and Nelder, J.A. (1989). *Generalized linear models* (second edition). Chapman and Hall, London.

Ross, G.J.S. (1987). *Maximum likelihood program*. Numerical Analysis Group, Oxford.

Ross, G.J.S. (1990). *Nonlinear estimation*. Springer-Verlag, New York.

Siegel, S. (1956). *Nonparametric statistics for the behavioral sciences*. McGraw-Hill, New York.

Smith, F.B. and Brown, P.E. (1933). The diffusion of carbon dioxide through soils. *Soil Science* **35**, 413-421.

Snedecor, G.W. and Cochran, W.G. (1989). *Statistical methods* (eighth edition). Iowa State University Press, Ames, Iowa.

Tufte, E.R. (1983). *The visual display of quantitative information*. Graphics Press, Cheshire, Connecticut.

Tukey, J.W. (1977). *Exploratory data analysis*. Addison-Wesley, Reading, Massachussets.

Youden, W.J. and Beale, H.P. (1934). A statistical study of the local lesion method for estimating tobacco mosaic virus. *Contributions from Boyce Thompson Institute* **6**, 437-454.

Chapter 8

Bouvier, A., Gelis, F., Huet, S., Messean, A., and Neveu, P. (1985). *CS-NL*. Laboratoire de Biometrie, INRA-CNRZ, Jouy-en-Josas, France.

Carr, N.L. (1960). Kinetics of catalytic isomerization of *n*-Pentane. *Industrial and Engineering Chemistry* **52**, 391-396.

Cook, R.D. and Weisberg, S. (1982). *Residuals and influence in regression*. Chapman and Hall, New York.

Dobson, A.J. (1990). *An introduction to generalized linear models*. Chapman and Hall, London.

Draper, N.R. and Smith, H. (1981). *Applied regression analysis* (second edition). Wiley, New York.

Finney, D.J. (1971). *Statistical method in biological assay* (third edition). Griffin, London.

Forbes, J.D. (1857). Further experiments and remarks on the measurement of heights by the boiling point of water. *Transactions of the Royal Society of Edinburgh* **21**, 235-243.

Goorin, A.M., Perez-Atayde, A., Gebhardt, M., Andersen, J.W., Wilkinson, R.H., Delorey, M.J., Watts, H., Link, M., Jaffe, N., Frei, E. III, and Abelson, H.T. (1987). Weekly high-dose Methotrexate and Doxorubicin for Osteosarcoma: the Dana-Farber Cancer Institute/The Children's Hospital – Study III. *Journal of Clinical Oncology* **5**, 1178-1184.

Grewal, R.S. (1952). A method for testing analgesics in mice. *British Journal of Pharmacology and Chemotherapy* **7**, 433-437.

Hastie, T.J. and Tibshirani, R.J. (1990). *Generalized additive models*. Chapman and Hall, London.

Lane, P.W. and Hastie, T.J. (1992). Providing for the analysis of generalized additive models within a system already capable of generalized linear and nonlinear regression. *Computational Statistics: Proceedings of the 10th Symposium on Computational Statistics* **1**, 391-396. Physica-Verlag, Heidelberg.

Lane, P.W. and Nelder, J.A. (1982). Analysis of covariance and standardization as instances of prediction. *Biometrics* **38**, 613-621.

Martin, J.T. (1940). The problem of the evaluation of Rotenone-containing plants. V. The relative toxicities of different species of derris. *Annals of Applied Biology* **27**, 274-294.

McCullagh, P. and Nelder, J.A. (1989). *Generalized linear models* (second edition). Chapman and Hall, London.

Ratkowsky, D.A. (1983). *Nonlinear regression analysis*. Dekker, New York.

Ratkowsky, D.A. (1990). *Handbook of nonlinear regression models*. Dekker, New York.

Ross, G.J.S. (1987). *Maximum likelihood program*. Numerical Analysis Group, Oxford.

Ross, G.J.S. (1990). *Nonlinear estimation*. Springer-Verlag, New York.

Seber, G.A.F. (1977). *Linear regression analysis*. Wiley, New York.

Seber, G.A.F. and Wild, C.J. (1989). *Nonlinear regression*. Wiley, New York.

Snedecor, G.W. and Cochran, W.G. (1989). *Statistical methods* (eighth edition). Iowa State University Press, Ames.

Sochett, E.B., Daneman, D., Clarson, C., and Ehrlich, R.M. (1987). Factors affecting and patterns of residual insulin secretion during the first year of Type 1 (insulin-dependent) diabetes mellitus in children. *Diabetologia* **30**, 453-459.

Sprent, P. (1969). *Models in regression and related topics*. Methuen, London.

Weisberg, S. (1985). *Applied linear regression*. Wiley, New York.

Woodley, W.L., Simpson, J., Biondini, R., and Berkeley, J. (1977). Rainfall results, 1970-1975: Florida Area Cumulus Experiment. *Science* **195**, 735-742.

Woods, H., Steinour, H.H., and Starke, H.R. (1932). Effect of composition of Portland cement on heat evolved during hardening. *Industrial and Engineering Chemistry* **24**, 1207-1214.

Chapter 9

Armitage, P. (1974). *Statistical methods in medical research*. Blackwell, Oxford.

Bartlett, M.S. (1937). Some examples of statistical methods of research in agriculture and applied biology (with discussion). *Journal of the Royal Statistical Society, Supplement 4*, 137-183.

Cochran, W.G. and Cox, G.M. (1957). *Experimental designs* (second edition). Wiley, New York.

Cox, D.R. (1958). Planning of experiments. Wiley, New York.

Healy, M.J.R. and Westmacott, M.H. (1956). Missing values in experiments analysed on automatic computers. *Applied Statistics* **5**, 203-206.

John, J.A. and Quenouille, M.H. (1977). *Experiments: design and analysis*. Griffin, London.

John, P.W.M. (1971). *Statistical design and analysis of experiments*. Macmillan, New York.

James, A.T. and Wilkinson, G.N. (1971). Factorisation of the residual operator and canonical decomposition of non-orthogonal factors in analysis of variance. *Biometrika* **58**, 279-294.

John, J.A., Wolock, F.W., and David, H.A. (1972). *Cyclic designs*. National Bureau of Standards, Applied Mathematics Series 62.

John, J.A. (1981). Efficient cyclic designs. *Journal of the Royal Statistical Society, Series B* **43**, 76-80.

John, J.A. (1987). *Cyclic Designs*. Chapman and Hall, London.

Kempthorne, O. (1952). *The design and analysis of experiments*. Wiley, New York.

Lamacraft, R.R. and Hall, W.B. (1982). Tables of incomplete cyclic block designs: $r=k$. *Australian Journal of Statistics* **24**, 350-360.

Nelder, J.A. (1965a). The analysis of randomized experiments with orthogonal block structure. I Block structure and the null analysis of variance. *Proceedings of the Royal Society, Series A* **283**, 147-162.

Nelder, J.A. (1965b). The analysis of randomized experiments with orthogonal block structure. II Treatment structure and the general analysis of variance. *Proceedings of the Royal Society, Series A* **283**, 163-178.

Patterson, H.D. (1976). Generation of factorial designs. *Journal of the Royal Statistical Society, Series B* **38**, 175-179.

Patterson, H.D. and Williams E.R. (1976). A new class of resolvable incomplete block designs. *Biometrika* **63**, 83-92.

Patterson, H.D. and Bailey, R.A. (1978). Design keys for factorial experiments. *Applied Statistics* **27**, 335-343.

Patterson, H.D., Williams E.R., and Hunter, E.A. (1978). Block designs for variety trials. *Journal of Agricultural Science* **90**, 395-400.

Payne, R.W. and Wilkinson, G.N. (1977). A general algorithm for analysis of variance. *Applied Statistics* **26**, 251-260.

Payne, R.W. (1990). Remark AS R82 A remark on AS65: Interpreting structure formulae. *Applied Statistics* **39**, 167-175.

Payne, R.W. and Welham, S.J. (1990). A comparison of algorithms for combination of information in generally balanced designs. *COMPSTAT 90 Proceedings in Computational Statistics*, 297-302. Physica-Verlag, Heidelberg.

Payne, R.W. and Tobias, R.D. (1992). General balance, combination of information and the analysis of covariance. *Scandinavian Journal of Statistics* **19**, 3-23.

Preece, D.A. (1971). Iterative procedures for missing values in experiments. *Technometrics* **13**, 743-753.

Rogers, C.E. (1973). Algorithm AS 65: Interpreting structure formulae. *Applied Statistics* **22**, 414-424.

Snedecor, G.W. and Cochran, W.G. (1980). *Statistical methods* (seventh edition). Iowa State University Press, Ames.

Wilkinson, G.N. (1957). The analysis of covariance with incomplete data. *Biometrics* **13**, 363-372.

Wilkinson, G.N. (1970). A general recursive algorithm for analysis of variance. *Biometrika* **57**, 19-46.

Wilkinson, G.N. and Rogers, C.E. (1973). Symbolic description of factorial models for analysis of variance. *Applied Statistics* **22**, 392-399.

Williams, E.R. (1975). *A new class of resolvable block designs*. Ph.D. Thesis, University of Edinburgh.

Yates, F. (1936). Incomplete randomized blocks. *Annals of Eugenics* **7**, 121-140.

Yates, F. (1937). *The design and analysis of factorial experiments*. Technical Communication No. 35 of the Commonwealth Bureau of Soils. Commonwealth Agricultural Bureaux, Farnham Royal.

Chapter 10

Cox, D.R. and Snell, E.J. (1981). *Applied statistics: principles and examples*. Chapman and Hall, London.

Dempster, A.P., Selwyn, M.R., Patel, C.M., and Roth, A.J. (1984). Statistical and computational aspects of mixed model analysis. *Applied Statistics* **33**, 203-214.

Patterson, H.D. and Thompson, R. (1971). Recovery of inter-block information when block sizes are unequal. *Biometrika* **58**, 545-554.

Robinson, D.L., Thompson, R., and Digby, P.G.N. (1982). REML – a program for the analysis of non-orthogonal data by restricted maximum likelihood. *Compstat 1982 Proceedings in Computational Statistics, Part II (supplement)*, 231-232. Physica-Verlag, Vienna.

Robinson, D.L. (1987). Estimation and use of variance components. *The Statistician* **36**, 3-14.

Searle, S.R. (1971). *Linear models*. Wiley, New York.

Snedecor, G.W. and Cochran, W.G. (1989). *Statistical methods* (eighth edition). Iowa State University Press, Ames.

Snell, E.J. and Simpson. H.R. (1991). *Applied statistics: a handbook of Genstat analyses*. Chapman and Hall, London.

Thompson, R. (1977). The Estimation of Heritability with Unbalanced Data I. Observations Available on Parents and Offspring. *Biometrics* **33**, 485-495.

Welham, S.J. and Thompson, R. (1992). REML likelihood ratio tests for fixed model terms. In: *Royal Statistical Society Conference, University of Sheffield, 9-11 September 1992, Abstracts of Papers*.

Chapter 11

Bartlett, M.S. (1938). Further aspects of the theory of multiple regression. *Proceedings of the Cambridge Philosophical Society* **34**, 33-40.

Bladon, S. (1986). *The new observer's book of automobiles*. Frederick Warne, Harmondsworth.

Campbell, N.A. (1980). Robust procedures in multivariate analysis. I: Robust covariance estimation. *Applied Statistics* **29**, 231-237.

Chatfield, C. and Collins, A.J. (1986). *Introduction to multivariate analysis* (revised edition). Chapman and Hall, London.

Cooley, W.W. and Lohnes, P.R. (1971). *Multivariate data analysis*. Wiley, New York.

Digby, P.G.N. and Kempton, R.A. (1987). *Multivariate analysis of ecological communities*. Chapman and Hall, London.

Doran, J.E. and Hodson, F.R. (1975). *Mathematics and computers in archaeology*. Edinburgh University Press.

Eckart, C. and Young, G. (1936). The approximation of one matrix by another of lower rank. *Psychometrika* **1**, 211-218.

Friedman, H.P. and Rubin, J. (1967). On some invariant criteria for grouping data. *Journal of the American Statistical Association* **62**, 1159-1186.

Gabriel, K.R. (1971). The biplot – graphic display of matrices with application to principal component analysis. *Biometrika* **58**, 453-467.

Gordon, A.D. (1981). *Classification: methods for the exploratory analysis of multivariate data*. Chapman and Hall, London.

Gower, J.C. (1966). Some distance properties of latent root and vector methods used in multivariate analysis. *Biometrika* **53**, 325-338.

Gower, J.C. (1967). Multivariate analysis and multidimensional geometry. *The Statistician* **17**, 13-25.

Gower, J.C. (1968). Adding a point to vector diagrams in multivariate analysis. *Biometrika* **55**, 582-585.

Gower, J.C. (1971). A general coefficient of similarity and some of its properties. *Biometrics* **27**, 857-871.

Gower, J.C. (1974). Maximal predictive classification. *Biometrics* **30**, 643-654.

Gower, J.C. (1975a). Algorithm AS 82: The determinant of an orthogonal matrix. *Applied Statistics* **24**, 150-153.

Gower, J.C. (1975b). Generalized Procrustes analysis. *Psychometrika* **40**, 33-51.

Gower, J.C. (1977). The analysis of asymmetry and orthogonality. In: *Recent developments in Statistics* (editors Barra et al.) 109-123. North Holland, Amsterdam.

Gower, J.C. (1985a). Multivariate analysis: ordination, multidimensional scaling and allied topics. In: *Handbook of applicable mathematics* (editor W. Ledermann), Statistics Vol. VIB (editor E. Lloyd). Wiley, Chichester.

Gower, J.C. (1985b). Measures of similarity, dissimilarity, and distance. In: *Encyclopaedia of statistical sciences, Volume V* (editors S. Kotz, N.L. Johnson, and C.B. Read), 397-405. Wiley, New York.

Gower, J.C. and Legendre, P. (1986). Metric and Euclidean properties of dissimilarity coefficients. *Journal of Classification* **3**, 5-48.

Gower, J.C. and Ross, G.J.S. (1969). Minimum spanning trees and single linkage cluster analysis.

Applied Statistics **18**, 54-64.

Greenacre, M. (1984). *Theory and applications of correspondence analysis*. Academic Press, London.

Hartigan, J.A. (1975). *Clustering algorithms*. Wiley, New York.

Kendall, D.G. (1977). On the tertiary treatment of ties. *Proceedings of the Royal Society of London, Series A* **354**, 407-423.

Krzanowski, W.J. (1988). *Principles of multivariate analysis: a user's perspective*. Oxford University Press.

Manly, B.F.J. (1986). *Multivariate statistical methods: a primer*. Chapman and Hall, London.

Mardia, K.V., Kent, J.T., and Bibby, J.N. (1979). *Multivariate analysis*. Academic Press, London.

Nathanson, J.A. (1971). An application of multivariate analysis in astronomy. *Applied Statistics* **20**, 239-249.

Rayner, J.H. (1966). Classification of soils by numerical methods. *Journal of Soil Science* **17**, 79-92.

Chapter 12

Anderson, O.D. (1976). *Time series analysis and forecasting*. Butterworths, London.

Bloomfield, P. (1976). *Fourier analysis of time series: an introduction*. Wiley, New York.

Box, G.E.P. and Jenkins, G.M. (1970). *Time series analysis, forecasting and control*. Holden-Day, San Francisco.

de Boor, C. (1980). FFT as nested multiplication, with a twist. *SIAM Journal of Scientific and Statistical Computing* **1**, 173-178.

Jenkins, G.M. and Watts, D.G. (1968). *Spectral analysis and its applications*. Holden-Day, San Francisco.

Nelson, C.R. (1973). *Applied time series analysis for managerial forecasting*. Holden-Day, San Francisco.

Shi-fang, L., Shi-guang, L., Shu-hua, Y., Shao-zhong, Y., and Yuan-xi, L. (1977). Analysis of periodicity in the irregular rotation of the Earth. *Chinese Astronomy* **1**, 221-227.

Tunnicliffe Wilson, G. (1989). On the use of marginal likelihood in time-series model estimation. *Journal of the Royal Statistical Society, Series B* **51**, 15-27.

Index

See Appendix 1 for page numbers where each Genstat directive is defined, and Appendix 2 for procedures in Release 3[1] of the Procedure Library.